OCEANOGRAPHY
An Invitation to Marine Science

Books in the Wadsworth Biology Series

Oceanography: An Invitation to Marine Science, 2nd, GARRISON

Essentials of Oceanography, GARRISON

Oceanography: An Introduction, 5th, INGMANSON AND WALLACE

Marine Life and the Sea, MILNE

Biology: Concepts and Applications, 2nd, STARR

Biology: The Unity and Diversity of Life, 7th, STARR AND TAGGART

Laboratory Manual for Biology, PERRY AND MORTON

Human Biology, STARR AND McMILLAN

Introduction to Microbiology, INGRAHAM AND INGRAHAM

Cell Ultrastructure, WOLFE

Molecular and Cellular Biology, WOLFE

Introduction to Cell and Molecular Biology, WOLFE

Living in the Environment, 9th, MILLER

Environmental Science, 5th, MILLER

Sustaining the Earth, 2nd, MILLER

Environment: Problems and Solutions, MILLER

Resource Conservation and Management, MILLER

Plant Physiology, 4th, SALISBURY AND ROSS

Plant Physiology Laboratory Manual, ROSS

Plant Physiology, 4th, DEVLIN AND WITHAM

Exercises in Plant Physiology, WITHAM ET AL.

Plants: An Evolutionary Survey, 2nd, SCAGEL ET AL.

Psychobiology: The Neuron and Behavior, HOYENGA AND HOYENGA

Sex, Evolution, and Behavior, 2nd, DALY AND WILSON

Dimensions of Cancer, KUPCHELLA

Evolution: Process and Product, 3rd, DODSON AND DODSON

OCEANOGRAPHY
An Invitation to Marine Science
Second Edition

TOM GARRISON
Orange Coast College

WADSWORTH PUBLISHING COMPANY

I(T)P® An International Thomson Publishing Company

Belmont • Albany • Bonn • Boston • Cincinnati • Detroit • London • Madrid • Melbourne
Mexico City • New York • Paris • San Francisco • Singapore • Tokyo • Toronto • Washington

To Marsha and Jeanne with love and thanks

Science Publisher: Jack Carey
Development Editor: Mary Arbogast
Editorial Assistant: Kerri Abdinoor
Production Services Coordinator:
   Gary Mcdonald
Production: Rogue Valley Publications,
   Mary Douglas
Designer: Kaelin Chappell
Print Buyer: Karen Hunt
Art Editor: Myrna Engler-Forkner
Permissions Editor: Jeanne Bosschart
Copy Editor: George Dyke
Photo Researcher: Tom Garrison

Technical Illustrators: Electronic Color conversion
   and new illustrations by Precision Graphics.
   Other illustrations by Diphrent Strokes; Parrot
   Graphics; Carlyn Iverson; Tasa Graphic Arts,
   Inc.; Kathryn W. Werhane; Jill Turney; Carole
   Lawson.
Cover Designer: Stephen Rapley
Cover Photograph: © Tony Stone Images/H. Richard
   Johnston
Signing Representative: Joann Ludovici
Compositor: The Clarinda Company
Color Separator: The Clarinda Company
Printer: Von Hoffmann Press

Printed in the United States of America
   2 3 4 5 6 7 8 9 10—01 00 99 98 97 96

For more information, contact Wadsworth Publishing Company:

Wadsworth Publishing Company
10 Davis Drive
Belmont, California 94002, USA

International Thomson Publishing Europe
Berkshire House 168-173
High Holborn
London, WC1V7AA, England

Thomas Nelson Australia
102 Dodds Street
South Melbourne 3205
Victoria, Australia

Nelson Canada
1120 Birchmount Road
Scarborough, Ontario
Canada M1K 5G4

International Thomson Editores
Campos Eliseos 385, Piso 7
Col. Polanco
11560 México D.F. México

International Thomson Publishing GmbH
Königswinterer Strasse 418
53227 Bonn, Germany

International Thomson Publishing Asia
221 Henderson Road
#05-10 Henderson Building
Singapore 0315

International Thomson Publishing Japan
Hirakawacho Kyowa Building, 3F
2-2-1 Hirakawacho
Chiyoda-ku, Tokyo 102, Japan

Library of Congress Cataloging-in-Publication Data

Garrison, Tom, 1942–
    Oceanography : an invitation to marine science / Tom Garrison. —
2nd ed.
        p.   cm.
    Includes bibliographical references and index.
    ISBN 0-534-25728-3
    1. Oceanography.   I. Title.
GC11.2.G37   1995
551.46—dc20   95-35233

BRIEF CONTENTS

DETAILED CONTENTS

PREFACE FOR STUDENTS AND INSTRUCTORS

In recent years, the systematic study of the ocean—marine science—has become increasingly popular with college and university students. They could hardly make a better choice. Inevitably, a general marine science course is interdisciplinary; students are invited to see the *connections* bridging astronomy, physics, chemistry, meteorology, geology, biology, ecology, history, and economics—areas of study they once considered separate. Together, teachers and students learn to see that the rise of a wave is linked to the twinkle of distant starlight, that a soft summer breeze is shaped by the chemistry of a billion microscopic plants, that the slow drift of a continent molds the evolution of countless organisms.

Studying the ocean awakens in us the sense of wonder we all felt as children when we first encountered the natural world. Students in marine science classes often find themselves immersed in wonderful stories and spectacular pictures and surrounded by teachers and technicians whose barely contained enthusiasm for their subject overtakes the clock. There is much to tell. The story of the ocean is a story of change and chance; its history is written in the rocks, the water, and the genes of the millions of organisms that have evolved here.

This comprehensive book is designed to tell that story in a clear and interesting way.

The Second Edition

My aim in writing this book was to produce a compelling, visually exciting text that would enhance students' natural enthusiasm for the ocean. My students have been involved in this book from the very beginning—indeed, it was their request for a readable, interesting, and thorough text that initiated the project. Through the fifteen years I have been writing textbooks, my enthusiasm for oceanic knowledge has increased (if that is possible), forcing my patient reviewers and editors to weed out an excessive number of exclamation points. But enthusiasm does shine through. One student reading the final manuscript of the first edition commented, "At last, a textbook that does not read like stereo instructions." Good!

This second edition builds on the first. As before, a great many students have read and commented on the manuscript. In response to their recommendations, as well as those of professors who have adopted the book and the many reviewers who contributed suggestions for strengthening the first edition, we have produced the book in full color, improved and expanded the illustration program, and incorporated many recent advances in oceanography. Major changes in the text include a greater dependence on satellite data and images (especially those from *TOPEX/Poseidon*), an overview of the objectives and funding levels of large new marine science programs, changes in the organization and presentation of the material on continental margins and ocean basins, combining nearly all El Niño information in one chapter (Ocean Circulation), reporting the effects of recent storms, tsunami, and earthquakes, noting the recent collapse of some fisheries, and extensively rewriting the chapter on coasts and most of the material on tides.

Oceanography is surely the most visual of the natural sciences, and we have incorporated some truly wonderful images. Among the many new and noteworthy illustrations, you will find a remarkable photo of a mid-ocean rift above sea level (Figure 3.20), a U.S. Navy nuclear submarine being used as an oceanographic research vessel (Box 4.1e and Figure 18.27), the first comprehensive image of the world ocean floor derived from gravimetric and sound sampling data (Figure 4.21), Calvin and Hobbes pondering the Coriolis effect (Figure 8.6), a particularly attractive sea monster (Box 13.1), yellowfin tuna swimming at high speed (Chapter 16's opener), a feeding blue whale (Figure 16.26), and the newest double-hulled tanker (Figure 19.21).

The organization of the book is straightforward: Because all matter on Earth except hydrogen and some helium was generated in stars, our story of the ocean starts with stars. We continue with a brief look at the history of marine science (with additional historical information sprinkled through later chapters). The theories of Earth structure and plate tectonics are presented next, as a base on which to build the explanation of bottom features that follows. A survey of ocean chemistry and physics prepares us for discussions of atmospheric circulation, classical physical oceanography, and coastal processes. Our look at marine biology begins with an overview of the problems and benefits of living in seawater, continues with a discussion of the production and consumption of food, and ends with taxonomic *and* ecological surveys of marine organisms. A special chapter comparing and contrasting the polar and tropical regions precedes the last chapters on marine resources and environmental concerns.

Connections between disciplines are emphasized throughout. Marine science draws on several fields of study, integrating the work of specialists into a unified whole. For example, a geologist studying the composition of marine sediments on the deep seabed must be aware of the biology and life histories of the organisms in the water above, the chemistry that affects the shells and skeletons of the creatures as they fall to the ocean floor, the physics of particle settling and water density and ocean currents, and the age and underlying geology of the study area. An oceanographer needs to be acquainted with a broad and beautiful array of information. This book is organized to make those connections from the first.

How This Book Is Organized

A broad view of marine science is presented in 20 chapters, each free-standing (or nearly so) to allow an instructor to assign chapters in any order he or she finds appropriate. Each chapter is preceded by an **overview,** a quick survey of the chapter's content. The overview shows you the topics to be covered and previews the most important points. The chapter itself begins with an attractive **vignette,** a short illustrated tale, observation, or sea story to whet your appetite for the material to come. Some vignettes spotlight scientists at work; others describe the experiences of people or animals in the sea.

The chapters are written in an **engaging style.** Terms are defined and principles developed in a straightforward manner. Some of the more complex ideas are initially outlined in broad brush strokes, then the same concepts are discussed again in greater depth after you have a clear view of the overall situation. When appropriate to their meanings, the derivations of words are shown. **Measurements** are given in both metric and English systems. At the request of a great many students, the units are written out (that is, we write *kilometer* rather than *km*) to avoid ambiguity and for ease of reading.

The photos, charts, graphs, and paintings in the extensive **illustration program** have been chosen for their utility, clarity, and beauty. **Boxes** in each chapter present commentaries of special interest on unique topics or controversies.

Concluding each chapter is a **Q-and-A section.** The Q-and-A items are interesting questions, with their answers, that students have asked me over the years. This material is an important extension of the chapters and occasionally contains key words and illustrations. Each chapter ends with a list of important **terms and con-**cepts, which are also defined in an extensive **glossary** in the back of the book. **Study questions** are also included in each chapter. Writing the answers to these questions will cement your understanding of the concepts presented. The **annotated bibliography** at the end of each chapter will be helpful when you wish to know more about a particular topic.

Appendixes will help you master measurements and conversions, geological time, latitude and longitude, chart projections, and the taxonomy of marine organisms. In case you'd like to join us in our life's work, the last appendix discusses **jobs in marine science.**

The book has been thoroughly **student tested.** You need not feel intimidated by the concepts—this material has been mastered by students just like you. Read slowly and go step-by-step through any parts that give you trouble. Your predecessors have found the ideas presented here to be useful, inspiring, and applicable to their lives. Best of all, they have found the subject to be *interesting!*

Acknowledgments

Many people have helped with this book. Sally Beaty at the Southern California Consortium for Community College Television began the whole thing in 1979 when she asked Ruth Lebow and me to write the study guide for the *Oceanus* telecourse. That led to a great many magazine and newspaper articles, student guides, supplements, and, ultimately, the first edition of this textbook. Jack Carey at Wadsworth, with whom I have worked for 25 years, encouraged me to continue writing through thick and thin, and Cecie Starr at Wadsworth, the best textbook author I know, sent suggestions and encouragement at appropriate intervals. The 74 reviewers have contributed to my continuously growing understanding of marine science, but Donald Lovejoy, Stanley Ulanski, Richard Yuretich, Ronald Johnson, and John Mylroie deserve special recognition for their expert direction. My long-suffering departmental colleagues Dennis Kelly, Jay Yett, Robert Profeta, and Joyce Kai-Mott again should be awarded medals for putting up with me, answering hundreds of my questions, and being so forbearing through the book's lengthy gestation period. Thanks again to our dean, Stanley Johnson, and our college president, David Grant, for supporting this project and encouraging our faculty to teach, conduct research, and be involved in community service. Yet another round of gold medals should go to my family for being patient (well, relatively patient) during those years of days and nights when dad was holed up in his dark reference-littered cave working on The Book. Thanks to all.

The people who provided pictures and drawings have worked miracles to obtain the remarkable images in these pages. To mention just a few: Gerald Kuhn sent classics taken by his friend Francis Shepard, Vincent Courtillot of the University of Paris contributed the remarkable photo of the Aden Rift, Catherine Devine at Cornell provided time-lapse graphics of tsunami propagation, Gunðmundur Sigvaldason of the University of Reykjavík sent a photo of Iceland's Thingvellir Graben, Robert Headland of the Scott Polar Research Institute at Cambridge University sent prints of polar subjects, Charles Hollister at Woods Hole kindly provided seafloor photos from his important books, Andreas Rechnitzer and Don Walsh recalled their exciting days with *Trieste,* and Bruce Hall, Pat Mason, Ron Romanoski, Ted Delaca, William Cochlan, Christopher Ralling, John Shelton, Alistair Black, Howard Spero, Eric Bender, Ken-ichi Inoue, and Norman Cole contributed beautiful slides. Herbert Kauainui Kane again allowed us to reprint his magnificent paintings of Hawaiian subjects. Karen Riedel helped with DSDP core images. James Ingle offered me a desk at Stanford whenever I needed it. NOAA, NASA, USGS, the Smithsonian Institution, the Royal Geographical Society, the U.S. Navy, and the U.S. Coast Guard came through time and again, as did private organizations like Alcoa Aluminum, Shell Oil, Cunard, Grumman Aviation, and the *Los Angeles Times.* The Woods Hole team was also generous: Philip Richardson, David Ross, James Broda, Albert Bradley, and Shelly Lauzon all provided photographs and diagrams. Individuals with special expertise have also been willing to share: Hank Brandli processed satellite digital images of storms, Peter Sloss at the National Geophysical Data Center helped me sort through computer-generated seabed images, Michael Gentry mined the archives of the Johnson Space Center for Earth images, John Maxtone-Graham of New York's Ocean Liner Museum found me a rogue wave picture, Ed Ricketts, Jr., contributed a portrait of his father, and professor Lyndon Land of Texas sent a rare photo of a turbidity current. Thanks to all.

The Wadsworth production team for this book has been ideal in every respect. Mary Arbogast, the world's best developmental editor, is more co-author and friend than editor. Her clear, linear writing style has been my guide for years. Production editor Mary Douglas, invariably optimistic and always on top of everything, has kept production flowing smoothly. Myrna Engler-Forkner, art editor, has a filing system that would be the envy of a medium-sized governmental agency. Jeanne Bosschart, permissions editor, has been able to make sense out of my very large permissions and acquisitions file. Kristin Milotich looks after supplements and ancillary details with enthusiasm. Gary Mcdonald, production services coordinator, made sure we were all running in the same direction. And Jack Carey helped me focus on the Big Picture when I was submerged in a thousand details. What pleasure to work with such professionals as these!

A Goal and a Gift

The goal of all this effort: *To allow you to gain an oceanic perspective.* "Perspective" means being able to view things in terms of their relative importance or relationship to one another. An oceanic perspective lets you see this misnamed planet in a new light, and helps you plan for its future. You will see that water, continents, seafloors, sunlight, storms, seaweeds, and society are connected in subtle and beautiful ways.

The ocean's greatest gift to humanity is intellectual—the constant challenge its restless mass presents. Let yourself be swept into this book and the class it accompanies. Take pleasure in the natural world. Ask questions of your instructors, read some of the references, try your hand at the questions at the ends of the chapters. Be optimistic. Please write to me when you find errors or if you have comments. Above all, enjoy yourself!

Tom Garrison
Orange Coast College
Costa Mesa, CA 92628-5005

REVIEWERS

ERNEST ANGINO, *University of Kansas*

STEVEN R. BENHAM, *Pacific Lutheran University*

LATSY BEST, *Palm Beach Community College*

EDWARD BEUTHER, *Franklin and Marshall College*

WILLIAM L. BILODEAU, *California Lutheran University*

JULIE BRIGHAM-GRETTE, *University of Massachusetts at Amherst*

LAURIE BROWN, *University of Massachusetts*

KEITH A. BRUGGER, *University of Minnesota–Morris*

KARL M. CHAUFF, *St. Louis University*

RICHARD DAME, *University of South Carolina, Columbia*

DAVID DARBY, *University of Minnesota at Duluth*

ROBERT J. FELLER, *University of South Carolina–Columbia*

L. KENNETH FINK, JR., *University of Maine*

DIRK FRANKENBURG, *University of North Carolina, Chapel Hill*

ROBERT R. GIVEN, *Marymount College, Rancho Palos Verdes*

KAREN GROVE, *San Francisco State University*

BARRON HALEY, *West Valley College*

JACK C. HALL, *University of North Carolina at Wilmington*

WILLIAM HAMNER, *University of California, Los Angeles*

WILLIAM HARRISON, *Western Michigan University*

TED HERMAN, *West Valley College*

JOSEPH HOLLIDAY, *El Camino College*

ANDREA HUVARD, *California Lutheran University*

RONALD E. JOHNSON, *Old Dominion University*

JOHN A. KLASIK, *California Polytechnic University, Pomona*

ERNEST C. KNOWLES, *North Carolina State University, Raleigh*

EUGENE KOZLOFF, *Friday Harbor, WA*

RICHARD W. LATON, *Western Michigan University*

RUTH LEBOW, *University of California, Los Angeles Extension*

DOUGLAS R. LEVIN, *Bryant College*

LARRY LEYMAN, *Fullerton College*

TIMOTHY LINCOLN, *Albion College*

DONALD L. LOVEJOY, *Palm Beach Atlantic College*

DAVID C. MARTIN, *Centralia College*

CHRIS METZLER, *Mira Costa College*

JOHN E. MYLROIE, *Mississippi State University*

CONRAD NEWMAN, *University of North Carolina, Chapel Hill*

JAMES G. OGG, *Purdue University*

B. L. OOSTDAM, *Millersville University*

JAN PECHENIK, *Tufts University*

BERNARD PIPKIN, *University of Southern California*

MARK PLUNKETT, *Bellevue Community College*

K. M. POHOPIEN, *Covina, CA*

RICHARD G. ROSE, *West Valley College*

WENDY L. RYAN, *Kutztown University*

ROBERT J. SAGER, *Pierce College, Washington*

ROBERT F. SCHMALZ, *Pennsylvania State University*

DON SEAVY, *Olympic College*

SAM SHABB, *Highline Community College*

WILLIAM G. SIESSER, *Vanderbilt University*

RALPH SMITH, *University of California, Berkeley*

SCOTT W. SNYDER, *East Carolina University*

JAMES F. STRATTON, *Eastern Illinois University*

J. COTTER THARIN, *Hope College*

STANLEY ULANSKI, *James Madison University*

J. J. VALENCIC, *Saddleback College*

RAYMOND E. WALDNER, *Palm Beach Atlantic College*

JILL M. WHITMAN, *Pacific Lutheran University*

P. KELLY WILLIAMS, *University of Dayton*

BERT WOODLAND, *Centralia College*

JOHN H. WORMUTH, *Texas A&M University*

RICHARD YURETICH, *University of Massachusetts at Amherst*

MEL ZUCKER, *Skyline College*

CREDITS AND ACKNOWLEDGMENTS

Chapter 1: Chapter opening courtesy of NASA. **Fig. 1.2** is reprinted with permission of Simon & Schuster, Inc., from the Macmillan College text *Exploring the Planets* by W. Kenneth Hamblin and Eric H. Christiansen. Copyright © 1990 by Macmillan College Publishing Company, Inc. **Fig. 1.6** is reprinted with permission of the Lick Observatory. **Figs. 1.7, 1.10** are reprinted by permission of Mike Seeds. **Fig. 1.8** © 1987 by the AAAS. Reprinted by permission of the publisher and Dr. William J. Welch. **Figs. 1.9a & b** are reprinted by permission of the Anglo-Australian Observatory. **Fig. 1.12** is reprinted by permission of William K. Hartmann. **Fig. 1.15b** is reprinted by permission of Woods Hole Oceanographic Institution.

Chapter 2: Figs. 2.1, 2.7 are reprinted with permission from Doubleday, a division of Bantam, Doubleday, Dell Publishing Group, Inc., from *Sailing Ships* by Bjorn Landstrom, 1978. **Figs. 2.6, 2.8, 2.14** paintings © Herb Kawainui Kane. **Fig. 2.10** is reprinted by permission of the James Ford Bell Library, University of Minnesota. **Fig. 2.12** is reprinted by permission of The United States Naval Academy Museum. **Figs. 2.20, 2.21** courtesy of The Library of Congress. **Fig. 2.22** is reprinted by permission of The Royal Geographical Society, London, Archives from the journal of Lt. Pelham Aldrich kept aboard *H.M.S. Challenger*, 1872–1875. **Fig. 2.28** used by permission of Ben Finney, University of Hawaii at Manoa.

Chapter 3: Chapter opening illustration by W. K. Hartmann and R. Miller from *History of the Earth* by W. K. Hartmann and R. Miller, Workman Publishing Company, 1992. Reprinted by permission of the authors. **Fig. 3.6** is reprinted with permission of Simon & Schuster, Inc., from the Macmillan College text *The Earth: An Introduction to PhysicalGeology*, Fourth Edition by Edward J. Tarbuck and Frederick K. Lutgens. Copyright © 1993 Macmillan College Publishing. **Fig. 3.8** is reprinted by permission of the USGS. **Fig. 3.18** is reprinted by permission of Woods Hole Oceanographic Institution. **Fig. 3.20** is reprinted by permission of V. Courtillot. **Fig. 3.24** from *Ocean Science* by Keith Stowe. © 1983 John Wiley & Sons, Inc. **Fig. 3.28** is reprinted by permission of W. C. Pittman III. **Fig. 3.30** is reprinted by permission of the NOAA/National Geophysical Data Center. **Figs. 3.33a & c** from *Earth and Life Through Time*, Second Edition, by Steven M. Stanley. © 1989 by W. H. Freeman and Company. Reprinted by permission. **Fig. 3.34** is reprinted by permission of Dr. Roger Bilham, University of Colorado.

Chapter 4: Chapter opening illustration courtesy of Robert Detrick/Woods Hole Oceanographic Institution. **Box 4.1a** photo used by permission of Bruce Hall. **Box 4.1b** photo used by permission of Scientific Search Project. **Box 4.1c** photo used by permission of Dan Dion. **Box 4.1d** photo used by permission of JAMSTEC. **Box 4.1e** photo used by permission of Ted Delaca, University of Alaska. **Box 4.1f** photo used by permission of

Robert Ballard/Woods Hole Oceanographic Institution. **Box 4.1g** photo used by permission of T. Kleindinst, A. Bradley/Woods Hole Oceanographic Institution. **Fig. 4.11** is reprinted by permission of Taylor & Francis International Scientific and Educational Publishers. **Fig. 4.12** is reprinted by permission of the Society for Sedimentary Geology. **Fig. 4.13** is reprinted by permission of the U.S. Department of the Navy. **Fig. 4.14** is from Alyn C. Duxbury and Alison B. Duxbury, *An Introduction to the World's Oceans*, Fourth Edition. Copyright © 1994 Wm. C. Brown Communications, Inc., Dubuque, IA. All rights reserved. Reprinted by permission. **Box 4.2** photo is used by permission of Nordic Volcanological Institute. **Fig. 4.15** courtesy of Aluminum Company of America. **Fig. 4.17** photo is used by permission of Woods Hole Oceanographic Institution. **Fig. 4.18** photo is used by permission of Deborah Smith/ Woods Hole Oceanographic Institution and Joe Cann, University of Leeds. **Fig. 4.20** is reprinted by permission of Chet Raymo. **Box 4.3b** photo used by permission of Dr. Andreas Rechnitzer. **Fig. 4.21** photo courtesy of Peter Sloss/National Geophysical Data Center, Boulder, Colorado.

Chapter 5: Chapter opening illustration by R. Miller from *History of the Earth* by W. K. Hartmann and R. Miller, Workman Publishing Company, 1992. Reprinted by permission of the authors. **Figs. 5.1, 5.2, 5.3** and material in **Box 5.1** courtesy of Charles D. Hollister. **Fig. 5.4a** photo courtesy of the USGS. **Fig. 5.4b** courtesy of the Department of Geology, University of Delaware. **Fig. 5.6a** courtesy of Dr. Howard Spero. **Fig. 5.6b** courtesy of Jay Yett. **Figs. 5.6c, 5.8a & b** courtesy of Roger Witmer. **Figs. 5.9a & b** courtesy of Unocal Research. **Figs 5.11b–d** are reprinted with permission of Simon & Schuster, Inc., from the Macmillan College text *Exploration of the Oceans* by John G. Weihaupt. Copyright © 1979 by John G. Weihaupt. **Box 5.2a & b** material courtesy of Dr. James Broda/Woods Hole Oceanographic Institution. **Figs. 5.13, 5.14** are reprinted by permission of the Deep Sea Drilling Project, Texas A & M University. **Fig. 5.17** is from A. G. Fischer, *Geologic History of the Western North Pacific*, published June 5, 1970, Vol. 168, p. 1210. © 1970 AAAS. Reprinted by permission.

Chapter 6: Chapter opening painting by Jacques Louis David, 1788. The Metropolitan Museum of Art, Purchase, Mr. and Mrs. Charles Wrightsman Gift, in honor of Everett Fahy, 1977 (1977.10). Copyright © 1989 by The Metropolitan Museum of Art. **Fig. 6.6** is reprinted by permission of Woods Hole Oceanographic Institution. **Fig. 6.7c** photo courtesy of William Cochlan.

Chapter 7: Box 7.1 photo courtesy of William Cochlan. **Fig. 7.9** from G. P. Kuiper, ed., *The Earth as a Planet*, © 1954 The University of Chicago Press. Reprinted by permission of the publisher. **Fig. 7.17** is reprinted by permission of John Wiley & Sons, Inc., from *Oceanography: Perspectives on a Fluid Earth* by Steve Neshyba, © 1987. **Box 7.2** illustration used by permission of Walter Munk,

Arthur Baggeror and the American Institute of Physics. **Fig. 7.22a** photo courtesy of EG&C Ocean Products. **Fig. 7.22b** courtesy of American Underwater Search & Survey, Ltd.

Chapter 8: Fig. 8.6 *Calvin and Hobbes* © 1994 Watterson. Reprinted with permission of Universal Press Syndicate. All rights reserved. **Fig. 8.10** is reprinted with permission of Simon & Schuster, Inc., from the Macmillan College text *Meteorology*, Fourth Edition by Moran and Morgan. Copyright © 1994 Macmillan College Publishing Company, Inc. **Fig. 8.12** is from "Monsoons," by P. J. Webster, *Scientific American*, Vol. 245, No. 2, August 1981. © 1981 by Scientific American, Inc. All rights reserved. Reprinted by permission of the publisher. **Table 8.2** is from *Earth in Crisis: An Introduction to the Earth Sciences*, Second Edition, Thomas L. Burrus, Herbert J. Spiegel, 1980. C. V. Mosby Co. Reprinted by permission of Thomas L. Burrus. **Fig. 8.16** is reprinted by permission of Tom Loebl, Imaging Publications. **Box 8.1** photo courtesy of Alastair Black. **Fig. 8.18** courtesy of NOAA. **Fig. 8.19** is reprinted by permission of The Bettmann Archive. **Fig. 8.20** photo courtesy of NASA.

Chapter 9: Fig. 9.5a is from *Laboratory Exercises in Oceanography*, Second Edition, by Pipkin, Gorsline, Casey and Hammond. © 1987 by W. H. Freeman and Company. Reprinted by permission. **Fig. 9.7b** courtesy of NASA/JPL. **Box 9.1** photo is reprinted by permission of The Bettmann Archive. **Fig. 9.11** is reprinted by permission of O. Brown, R. Evans, and M. Carle, University of Miami Rosenstiel School of Marine and Atmospheric Science. **Table 9.1** is from M. Grant Gross, *Oceanography: A View of the Earth*, Fifth Edition, © 1990, p. 173. Reprinted by permission of Prentice Hall, Englewood Cliffs, N.J. **Figs. 9.12b, 9.13b** courtesy of NOAA/UCAR, Gene Carl Feldman, NASA-GSFC, Otis Brown, University of Miami. **Fig. 9.25c** photo courtesy of Philip Richardson/Woods Hole Oceanographic Institution. **Box 9.2** photo is reprinted by permission of The Kon-Tiki Museet, Oslo, Norway.

Chapter 10: Fig. 10.13 courtesy of U.S. Navy. **Box 10.2** photo by Dr. W. C. Stillwell. Reprinted by permission of Suzanne M. Stillwell. **Figs. 10.16a–c** courtesy of Ivan Narragon. **Fig. 10.17b** used by permission of The University of Hawaii Press. **Box 10.3** photo courtesy of Ron Romanosky.

Chapter 11: Chapter opening photo courtesy of USGS. **Figs. 11.2a & b** photos courtesy of Thames Barrier Visitor's Centre. **Fig. 11.5** research by Philip L. F. Liv, Seung Nam Seo, and Sung Bum Yoon, Civil and Environmental Engineering, Cornell University. Visualization by Catherine Devine, Cornell Theory Center. Used by permission. **Figs. 11.6a & b, 11.8** photos courtesy of USGS. **Fig. 11.7a** is from *Volcanoes in the Sea: The Geology of Hawaii*, Second Edition, 1990. *The*

OCEANOGRAPHY
An Invitation to Marine Science

1 AN OCEAN WORLD

A Marine Point of View

Earth is misnamed. From space the Earth is brilliantly blue, white in places with clouds and ice, sometimes swirling with storms. Dominating its surface is a single great ocean of liquid water. This ocean affects and moderates temperature and dramatically influences weather. Its creatures directly provide at least 2% of humanity's food. From beneath its floor is pumped about one-third of the world's supply of petroleum and natural gas. The ocean borders most of the planet's largest cities. It is a primary shipping and transportation route and a major recreational resource. The dry land on which nearly all of human history has unfolded is hardly visible from space, for nearly three-quarters of the planet is covered by water. *Oceanus* would surely be a better name for our watery home.

The ocean covers and contains some of the same familiar geological shapes we see on dry land. There are hidden mountain ranges, immense hot springs, volcanoes spewing molten rock in unseen eruptions, earthquakes and avalanches, mineral treasures, and cold, calm plains. Indeed, the geology of the submerged edges of the continents and the deeper ocean floor is proving to be as complex and varied as anything above the surface.

The Earth, its ocean, and its organisms have changed together over the ages. Life originated in the ocean, developing and flourishing there for more than three billion years before venturing onto the unwelcoming continents. All our planet's life forms—terrestrial as well as aquatic—must carry an ocean within themselves. Their blood, their eggs, and the fluids that bathe their cells are all saline. We ourselves are made largely of water with about the same relative proportion of salts as the sea. The first nine months of human life are spent in a water world, a warm supportive ocean that cradles from shock and provides a stable and weightless environment for the complex processes of growth and development. After birth, we view the universe through an ocean—the fluid behind the corneas of our

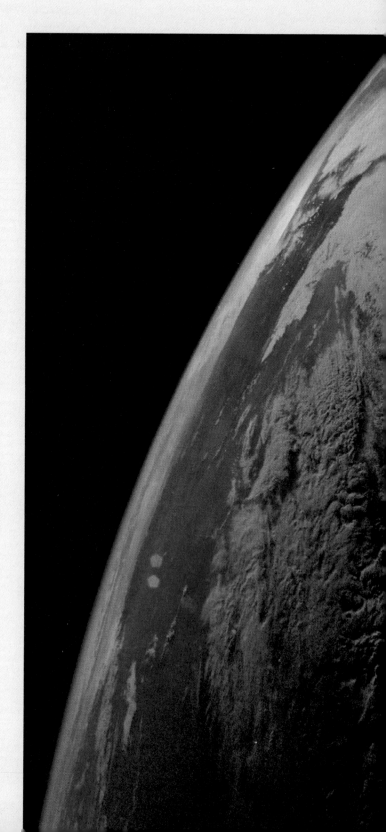

eyes is similar to seawater. In a sense we see everything from a marine point of view.

We have been profoundly influenced by the ocean. It separates enemies but fosters trade; it inspires our spirits of adventure and imagination. Its waters and floor provide important physical and biological resources. Studies of the ocean have extended our scientific understanding of Earth and the universe. An estimated 75% of all Americans now live within 80 kilometers (50 miles)[1] of a coastline (including the Great Lakes). The ocean gives gainful employment to millions of these people, provides transportation for more than 90% of the world's commerce, and silently hides our wastes. Best of all, it gives us hope. We look to the ocean for solitude and inspiration. Mariners, divers, surfers, and anyone else wishing time for quiet reflection know the special peace that the ocean can bring. To fly for hours over the vast Pacific brings a new perspective to anyone content with the poor name Earth.

[1]Throughout this book metric measurements precede English measurements. For a quick review of metric units and their abbreviations, turn to Appendix 1.

CHAPTER OVERVIEW

Earth is a water planet, possibly one of a few in the galaxy. An ocean covering 71% of its surface has greatly influenced its rocky crust and atmosphere. Life on Earth almost certainly evolved in the ocean; the cells of all life-forms are still bathed in salty fluids.

We have learned much about our planet using the scientific method, a systematic way of asking and answering questions about the natural world. Marine science applies the scientific method to the ocean, the Earth of which it is a part, and the living organisms dependent on it.

Most of the atoms that make up the Earth and its inhabitants were formed within stars. Stars form in the dusty spiral arms of galaxies and spend their lives changing hydrogen and helium to heavier elements. As they die, some stars eject these elements into space by cataclysmic explosions. The sun and the planets, including Earth, probably condensed from a cloud of dust and gas enriched by the recycled remnants of exploded stars.

Our ocean is not a remnant of that cloud, however. Most of the ocean formed later, as water vapor trapped in the Earth's outer layers escaped to the surface through volcanic activity during the planet's youth. Life originated in the ocean soon after. Life and the Earth have grown old together.

Earth from space. Our water planet shines blue in the darkness, its surface partly hidden by a turbulent layer of clouds.

ONE WORLD OCEAN

Over 97% of the water on or near Earth's surface is contained in the ocean; less than 3% is held in land ice, groundwater, and all the freshwater lakes and rivers (see **Figure 1.2**). The **ocean**[2] may be defined as the vast body of saline water that occupies the depressions of the Earth's surface. Traditionally, we have divided the ocean into artificial compartments called *oceans* and *seas,* using the boundaries of continents and imaginary lines such as the equator. In fact there are few dependable natural divisions, only one great mass of water. The Pacific and Atlantic oceans and the Mediterranean and Baltic seas, so named for our convenience, are in reality only temporary features of a single **world ocean.** In this book we refer to the ocean *as a single entity,* with subtly different characteristics at different locations but with very few natural partitions. Marine scientists have descended to its greatest depths, have explored a few of its peaks, and have photographed and sampled some of its floor, but we know more about the topography of the far side of the moon than we know about the 70.78% of the Earth covered by water. Despite our years of research, there is much to be learned. The world ocean is a key to the Earth's past and present, and the prime link to its future.

Looking at the ocean as a single unit brings a philosophical advantage. Such a view emphasizes the interdependence of ocean and land, life and water, atmosphere and liquids, and natural and man-made environments.

THE WORLD OCEAN: TWO VIEWS

<u>On a human scale</u> the ocean is impressively large. It covers 361 million square kilometers (139 million square miles) of the Earth's surface (see **Figure 1.3**). The average depth of the ocean is about 3,796 meters (12,451 feet), the volume of seawater is 1.37 billion cubic kilometers (329 million cubic miles), and the average temperature is a cool 3.9°C (39°F). Its mass is a staggering 155 billion billion tons. The deepest spot in the ocean, a gash in the seafloor known as the Mariana Trench, was reached by two researchers in 1960. It is 11,022 meters (36,163 feet) from the surface, 2,192 meters (7,191 feet) deeper than Mount Everest is high. Indeed, landlocked Everest is not the tallest mountain on Earth. That honor belongs to the island of Hawaii, an active volcano whose base rests on seafloor 5,998 meters (19,680 feet) deep and whose snowy crest rises 4,205 meters (13,796 feet) above sea level. The whole immense structure is 10,203 meters (33,476 feet) high!

[2]When an important new term is introduced and defined, it is printed in boldface type. These terms are listed at the end of the chapter and defined in the glossary.

Figure 1.1 A view of the Earth. Most of its surface is covered by a liquid water ocean that averages 3,796 meters (12,451 feet) in depth.

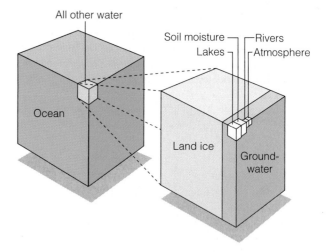

Figure 1.2 The relative amount of water in various locations on or near the Earth's surface. More than 97% of the water lies in the ocean. Ice on land contains about 1.9%, groundwater 0.5%, rivers and lakes 0.02%, and the atmosphere 0.001%.

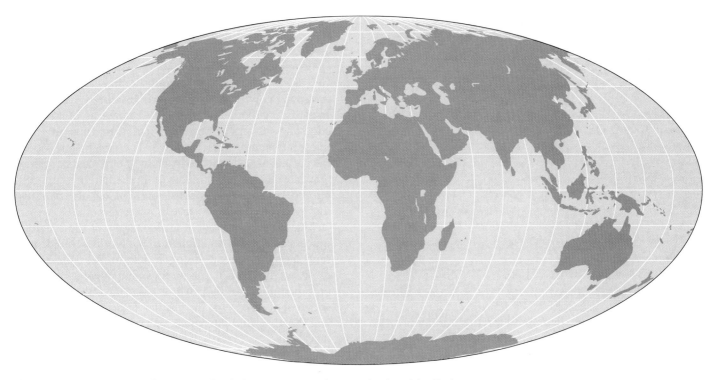

Figure 1.3 The proportion of sea versus land, shown on an equal-area projection of the Earth. An equal-area projection is a map drawn to represent areas in their correct relative proportions.

If the Earth's contours were leveled to a smooth ball, the ocean would cover it to a depth of 2,686 meters (8,810 feet). The volume of the world ocean is currently 11 times the volume of land above sea level; average land elevation is only 840 meters (2,772 feet), but average ocean depth is 4½ times greater! The Pacific Ocean is the Earth's most prominent single feature. **Figure 1.4** shows the distribution of ocean and land in both hemispheres as a function of total area. The Northern Hemisphere is 60.7% sea and 39.3% land; the Southern Hemisphere 80.9% sea and 19.1% land. **Table 1.1** summarizes some basic characteristics of the world ocean.

Virtually every chemical element found on land can be found dissolved in the ocean. The world ocean contains some 5 trillion tons of salts; if dried and spread evenly, that mass would cover the entire planet to a depth of 45 meters (150 feet). Gases also dissolve in seawater; most of the Earth's reservoir of carbon dioxide, a gas critical to plant physiology and climate control, is stored within the ocean.

The ocean makes us human beings appear quite insignificant. On a planetary scale, however, the ocean itself is insignificant. Its average depth is a tiny fraction of the Earth's radius: The blue ink representing the ocean on an 8-inch paper globe is proportionally thicker. The ocean accounts for only slightly more than 0.02% of Earth's mass, or 0.13% of its volume. There is much more

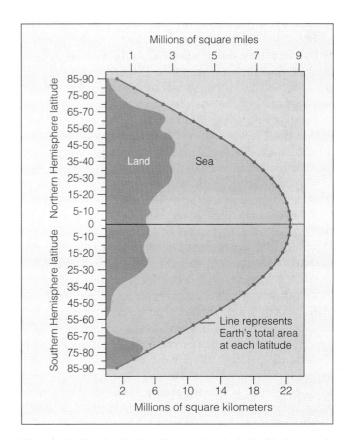

Figure 1.4 The distribution of land and water in the Northern and Southern hemispheres.

Table 1.1	Characteristics of the World Ocean

Area: 361,100,000 square kilometers (139,400,000 square miles)

Volume: 1,370,000,000 cubic kilometers (329,000,000 cubic miles)

Average depth: 3,796 meters (12,451 feet)

Average temperature: 3.9°C (39.0°F)

Average salinity: 34.482 grams per kilogram (0.56 ounce per pound), 3.4%

Most abundant elements (by mass):

Oxygen	(86%)
Hydrogen	(11%)
Chlorine	(1.9%)
Sodium	(1.1%)
Magnesium	(0.1%)

Age: About 4 billion years

Future: Uncertain

water chemically trapped within the Earth's hot interior than there is in its ocean and atmosphere.

THE NATURE OF SCIENCE

Scientists today are asking some critical questions about the origin of the ocean, the age of its basins, and the nature of the life forms it has nurtured. We are fortunate to live at a time when scientific study may be able to answer some of those questions. **Science** is a systematic *process* of asking questions about the observable world and then testing the answers to those questions. Scientists gather and study information (data), but the information itself is not science. Science interprets raw information by constructing a general explanation with which the information is compatible.

Scientists begin with an informed guess called a working **hypothesis,** a speculation about the natural world that can be verified or disproven by observations and experiments. Hypotheses consistently supported by observation or experiment are advanced to the status of **theory,** a statement of relationship accepted by most scientists. The largest constructs, known as **laws,** are principles explaining events in nature that have been observed to occur with unvarying uniformity under the same conditions. Theories and laws in science do not arise fully formed or all at once. Scientific thought progresses as a continuing chain of questioning, testing, and matching theories to observations. A theory is strengthened if new facts support it. If not, the theory is modified or a new explanation is sought. The power of science lies in the ability of the process to operate *in reverse;* that is, in the use of a theory or law to make predictions and anticipate new facts to be observed.

The **scientific method** is the orderly process by which theories are verified or rejected. It is based on the assumption that nature "plays fair"—that is, that the answers to our questions about nature are *ultimately knowable* as our powers of questioning and observing improve. Nothing is ever proven absolutely true by the scientific method. Theories may change as our knowledge and powers of observation change; thus all scientific understanding is tentative. The conclusions about the natural world that we reach by the process of science may not always be popular or immediately embraced, but if those conclusions consistently match observations, they may be considered true.

This book shows some of the results of the scientific process as it has been applied to the world ocean. It presents facts, interpretations of facts, examples, stories, and some of the crucial discoveries that have led to our present understanding of the ocean and the world on which it formed. As the results of science change, so will the ideas and interpretations presented in books like this one.

WHAT IS MARINE SCIENCE?

Whether you live in a coastal city or an inland town, the ocean affects you. The weather you experienced today was shaped by the ocean. Its richness has provided some of your metals, much of the energy needed for electricity and transportation, and about one-half the oxygen you breathe. The ocean's impact on geopolitical and economic matters, and its importance in mythology, literature, and the graphic arts are of critical importance to civilization. And you might have a seafood dinner! Every citizen of Earth should be acquainted with the planet's most obvious feature, and the marine sciences provide the background necessary for logical and ethical use of the ocean and coastal regions.

Marine science (or **oceanography**) is the process of discovering unifying principles in data obtained from the ocean, its associated life-forms, and the bordering lands. It draws on several disciplines, integrating the fields of geology, physics, biology, chemistry, and engineering as they apply to the ocean and its surroundings. *Marine geologists* focus on questions such as the composition of the inner Earth, the mobility of the crust, and the characteristics of seafloor sediments. Some of their work touches on areas of intense scientific and public concern, including earthquake prediction and the distribution of valuable resources. *Physical oceanographers* study and observe wave dynamics, currents, and ocean–atmosphere interaction. Their predictions of long-term climate trends are becoming increasingly important as pollutants change Earth's atmosphere. *Marine biologists* work with

BOX 1.1 ● *Solitude*

The ocean has always challenged the human spirit. Meeting that challenge *alone* is a supreme triumph of humanity over the uncontrollable and unpredictable forces of nature. Whatever the reasons for their voyages, all who travel by sea alone have experienced profound solitude, loneliness, helplessness, and—if things went well—the exultation of success.

The first man to sail alone around the world was Joshua Slocum, a Massachusetts sailor who went to sea when he was 14. He was 51 when he began his solitary voyage. He started from Boston on 24 April 1895 in the *Spray*, an 11-meter (37-foot) sloop he had rebuilt from a derelict hulk. More than three years and 74,000 kilometers (46,000 miles) later, on 3 July 1898, he tied the boat to the cedar spike driven in the bank that held her when she was first launched. "I could bring her no nearer home," he said.

In his book, *Sailing Alone Around the World*, Slocum wrote about the peace and heightened awareness that intimate contact with the ocean can bring.

> The fog lifted just before night, and I was afforded a look at the sun just as it was touching the sea. I watched it go down and out of sight. Then I turned my face eastward, and there, apparently at the very end of the bowsprit, was the smiling full moon rising out of the sea. Neptune himself coming over the bows could not have startled me more. "Good evening, sir," I cried; "I'm glad to see you." Many a long talk since then I have had with the man in the moon; he had my confidence on the voyage.
>
> About midnight the fog shut down again denser than ever before. One could almost "stand on it." It continued so for a number of days, the wind increasing to a gale. The waves rose high, but I had a good ship. Still, in the dismal fog I felt myself drifting into loneliness, an insect on a straw in the midst of the elements. I lashed the helm, and my vessel held her course, and while she sailed I slept.
>
> During these days a feeling of awe crept over me. My memory worked with startling power. The ominous, the insignificant, the great, the small, the wonderful, the commonplace—all appeared before my mental vision in magical succession. Pages of my history were recalled which had been so long forgotten

Joshua Slocum, shown here rounding the tip of Cape Horn in his 37-foot sloop, *Spray*, was the first person to sail around the world alone. After a difficult voyage of more than three years and 46,000 miles, he returned to Newport, Rhode Island, and tied *Spray* to the same cedar spike driven in the bank that held her when she was first launched. "I could bring her no nearer home."

> that they seemed to belong to a previous existence. I heard all the voices of the past laughing, crying, telling what I had heard them tell in many corners of the Earth. The loneliness of my state wore off when the gale was high and I found much work to do. When fine weather returned, then came the sense of solitude, which I could not shake off. . . .

The ocean is as vast, as quiet, as furious, as inspiring as it was in Slocum's day, but our understanding of it has grown immensely. I think he would have enjoyed the insights that nearly a century of progress in marine science has added to the things he saw, heard, and felt.

Source: Joshua Slocum, *Sailing Alone Around the World* (New York: Sheridan House, 1972). Originally published in 1899.

a

b

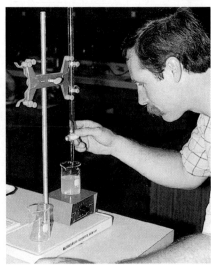

c

Figure 1.5 Marine scientists at work. (a) Researchers steady plankton incubators in heavy seas at the stern of the research vessel *Pt. Sur.* (b) A marine biology student looks for organisms in a bottom grab sample. (c) A marine chemist tests a water sample to find the concentrations of dissolved elements.

the nature and distribution of marine organisms, the impact of oceanic and atmospheric pollutants on the organisms, the isolation of disease-fighting drugs from marine species, and the yields of fisheries. *Chemical oceanographers* study the ocean's dissolved solids and gases, and the relationships of these components to the geology and biology of the ocean as a whole. *Marine engineers* design and build oil platforms, ships, harbors, and other structures that enable us to use the ocean wisely. Other marine specialists study weather forecasting, ways to increase the safety of navigation, methods to generate electricity, and much more.

Virtually all marine scientists specialize in one area of research, but they also must be familiar with related spe-cialties and appreciate the linkages between them. For example, a geologist studying the composition of marine sediments on the deep seabed must be aware of the biology and life histories of the organisms in the water above, the chemistry that affects the shells and skeletons of the creatures as they fall to the ocean floor, the physics of particle settling and ocean currents, and the age and underlying geology of the study area. **Figure 1.5** shows marine scientists in action.

Marine science is a young, vigorous, and very exciting interdisciplinary branch of natural science that often has broad application beyond the ocean. But besides offering solutions to practical problems and contributing to the advancement of knowledge, study of the marine sciences also brings emotional rewards. No one can remain unmoved while reading an eyewitness account of a seismic sea wave striking the Hawaiian islands, or considering the origin of the molecules that make up our bodies, or contemplating the adventures of polar explorers such as Sir Ernest Shackleton, or hearing the musical sounds of communicating whales. The sensory and intellectual richness of the marine sciences makes a deep impression on all of us, and this richness can contribute significantly to our awareness and appreciation of the interconnection of all natural environments.

You don't have to be a marine scientist to appreciate the marine sciences. If, as microbiologist Barry Commoner predicts, we have "perhaps a generation in which to save the environment from the final effects of the violence we have done it," what better use could we make of our time than to learn about the life-giving watery film that covers most of our planet?

THE ORIGIN OF EARTH

We have always wondered about our origins—how the Earth was formed, how the ocean arose, and how life came to be. In the last 50 years researchers using the scientific method have determined a tentative age for the ocean, Earth, and universe. They have developed hypotheses about how matter is assembled, how stars and planets are formed, and even how life may have arisen. Many of the details are still sketchy, of course, but their hypotheses have predicted some important recent discoveries in subatomic physics and molecular biology. Perhaps the most dramatic discoveries in natural science this century have been those dealing with the origin and history of the universe.

The universe apparently had a beginning. The **big bang,** as that event is modestly named, probably occurred about 15 billion years ago. All of the mass and energy of the universe is thought to have been concentrated at a geometric point at the beginning of time, the moment when the expansion of the universe began. We don't know what initiated the expansion, but it continues today and will probably continue for billions of years. If the universe is massive enough, its expansion will eventually stop because of the influence of gravity. The universe will collapse into itself and, perhaps, begin again. If the mass of the universe is too small, as now appears to be the case, it will continue to expand forever.

The very early universe was unimaginably hot, but as it expanded it cooled. About a million years after the big bang, temperatures fell enough to permit the formation of atoms from the energy and particles that had predominated up to that time. Most of these atoms were hydrogen, then as now the most abundant form of matter in the universe. About a billion years after the big bang, this matter began to congeal into the first galaxies and stars.

Figure 1.6 The great galaxy in Andromeda, which is very similar in size and structure to our own Milky Way galaxy. Both are members of a loose aggregation of about 20 galaxies wonderfully called the Local Group.

Galaxies and Stars

A **galaxy** is a huge rotating aggregation of stars, dust, gas, and other debris held together by gravity. There are perhaps 50 billion galaxies in the universe and 50 billion stars in each galaxy. Our galaxy is named the **Milky Way galaxy** (*galaktos* = milk). A galaxy very similar to our own is shown in **Figure 1.6.**

The **stars** that make up a galaxy are massive spheres of incandescent gases. They are usually intermingled with diffuse clouds of gas and debris. In spiral galaxies like the Milky Way, the stars are arrayed in spiral arms radiating from the galactic center. Other galaxies are elliptical or irregular in shape. Our part of the Milky Way is populated with many stars, but distances within a galaxy are so huge that the star nearest the sun is about 42 trillion kilometers (26 trillion miles) away. Astronomers tell us there are as many galaxies in the universe as there are stars in our own galaxy, and more stars in the Milky Way than grains of sand on a beach!

Figure 1.7 A painting of the Milky Way galaxy, as it might look from about 3 million light-years away. (A light-year—the distance light travels in one year—is about 9.6 trillion kilometers or 6 trillion miles.) The cross marks the location of our solar system in the Orion arm. Our sun and its family of planets revolve around the center of the galaxy, completing a revolution every 230 million years.

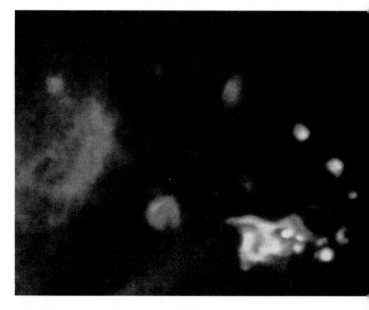

Figure 1.8 A ring of new stars is visible to the right of W49A, the giant collapsing gas cloud that formed them. The cloud is on the far side of our Milky Way galaxy at about the same distance from the galactic center as the sun.

Our sun is a typical star. The sun and its family of planets, called the **solar system,** are located about three-quarters of the way out from the galaxy's center, in a spiral arm called the Orion arm (see **Figure 1.7**). We orbit the galaxy's brilliant core, taking about 230 million years to make one orbit, even though we are moving at about 280 kilometers per second (half a million miles an hour). The Earth has made about 20 circuits of the galaxy since the ocean formed.

The Lives of Stars

Stars form in **nebulae,** cold clouds of dust and gas within galaxies. With the aid of telescopes and infrared sensing satellites, astronomers have observed such clouds in our own and other galaxies. They have seen stars in different stages of development and have inferred a sequence in which these stages occur. The **condensation theory,** a theory based on this inference, explains how stars and planets are believed to form.

The life of a star begins when a diffuse area of a nebula begins to shrink under the influence of its own weak gravity. Gradually, the cloudlike sphere condenses into a knot of gases called a **protostar** (*protos* = first). The original diameter of the protostar may be many times the diameter of our solar system, but gravitational energy causes it to contract. As the protostar shrinks, the compression raises its internal temperature. When the protostar reaches a temperature of about 10 million degrees Celsius (18 million degrees Fahrenheit) nuclear fusion

begins; that is, hydrogen atoms begin to fuse together to form helium, a process that liberates even more energy. This rapid release of energy, which marks the transition from protostar to star, stops the young star's shrinkage. (A group of stars at this stage of development is shown in **Figure 1.8.**) Nuclear fusion of small atoms, not combustion, is what causes a star to shine. Light and heat, by-products of the new star, make the galaxy a little brighter.

After fusion reactions begin, the star becomes stable—neither shrinking nor expanding, and burning its hydrogen fuel at a steady rate. This stable phase does not last forever, however. After a long and productive life, the star has converted a large percentage of its hydrogen to atoms as heavy as carbon or oxygen. When a medium-mass star begins to consume these heavier atoms, its energy output slowly rises and its body swells to a stage aptly named *red giant* by astronomers. The dying giant slowly pulsates, throwing off concentric shells of light gas enriched with these heavy elements. But most of the harvest of carbon and oxygen is forever trapped in the cooling ember at the star's heart.

The life history of a star depends on its initial mass. The scenario we have just outlined describes the life of a star similar in size to our sun. Much more massive stars have shorter but more interesting lives. They, too, fuse hydrogen to form atoms as heavy as carbon and oxygen; but being larger and hotter, their internal nuclear reactions consume hydrogen at a much faster rate. Higher

BOX 1.2 ● *Out in Space, Back in Time*

Time machines are notoriously hard to build. No one has succeeded in constructing one, and the mechanisms proposed by science fiction writers don't seem to work. But you don't need a time machine to see into the past—just go outdoors and look up.

If it's sunny outside, the sunlight striking your head left the sun about 8 minutes ago. You're seeing the sun not as it is *now*, but as it was 8 minutes back in time. If it's a clear night, and you're in the Southern Hemisphere, you may be able to find Alpha Centauri, the star nearest the sun and the brightest star in the constellation of Centaurus. You're not seeing this star as it is now, either; even at a speed of 300,000 kilometers (186,000 miles) per second, its light takes about 52 months to reach your eyes. You're looking 4.3 years back in time.

Brighter objects can be seen at greater distances, and thus farther back in time. The most distant object that can be seen with the unaided eye is the great spiral galaxy in the constellation of Andromeda (shown in Figure 1.6), one of about two dozen nearby galaxies that astronomers call the Local Group. The light recorded in that photograph left the galaxy at about the time our ancestors were first starting to make stone

tools. It had been traveling toward the camera for 2 million years.

The most distant object yet photographed—probably a galaxy in a very early stage of formation—is about 12 billion light-years distant. (That is, its light has been moving toward us for some 12 billion years.) Since the universe is thought to be about 15 billion years old, we're seeing the object as it existed near the start of time.

How far into the past can we go with this time machine? All the way to the beginning, it seems. In the mid-1960s, two Bell Laboratories physicists discovered that a faint echo of the big bang permeates the universe. The energy of this echo "stretched out" as the universe aged and expanded; instead of being light, the energy now exists as radio waves. By chance, some of this radio energy lies in the frequencies we use to broadcast commercial television. About 2% of the "snow" on unused VHF channels results from this energy being picked up by your television antenna. Next time somebody says there's nothing on TV, you might suggest turning on the tube to catch a glimpse of the formation of the universe!

core temperatures permit the formation of atoms up to the mass of iron.

The dying phase of a massive star's life begins when its depleted core collapses in on itself. This rapid compression causes the star's internal temperature to soar. When the infalling material can no longer be compressed, the energy of the inward fall is converted to cataclysmic expansion called a **supernova.** The explosive release of energy in a supernova is so sudden that the star is blown to bits, and its shattered mass is accelerated outward at nearly the speed of light. The explosion lasts only about 10 seconds, but in that short time the nuclear forces holding apart individual atomic nuclei are overcome—and atoms heavier than iron are formed. The gold of your rings, the mercury in a thermometer, and the uranium in nuclear power plants were all created during such a brief and stupendous flash. The atoms produced by a star through millions of years of orderly fusion, *and* the heavy atoms generated in a few moments of unimaginable chaos, are sprayed into space. Every chemical element heavier than hydrogen—most of the

atoms that make up the planets, the oceans, and living creatures—was manufactured by the stars.

Supernova explosions have been witnessed from Earth. Chinese astronomers recorded the position of one such exploding star in the year 1054. Astronomers training telescopes to that point in the sky today see an expanding bubble of stellar debris called the Crab nebula, all that remains of a large star. Another supernova was sighted in Southern Hemisphere skies in February 1987 (see **Figure 1.9**). Data from this explosion support both the model of stellar collapse and the theory of origin of heavy elements just presented.

The Formation of the Solar System

The Earth and its ocean are the indirect result of a supernova explosion. The thin cloud, or **solar nebula,** from which our sun and its planets formed was probably struck by the shock wave and some of the matter of an expanding supernova remnant. Indeed, the turbulence of the encounter may have caused the condensation of our

a   b   c

Figure 1.9 Supernova 1987A was discovered on 23 February 1987. (a) This photograph was taken before the supernova flared. The arrow shows the location of the star that would become the supernova. (b) The supernova at its peak. It was the first supernova visible with the unaided eye since the invention of the telescope. (c) Supernova 1987A as it appeared in 1991 in a photo taken by the Hubble Space Telescope. An expanding yellow ring of glowing gas surrounds the tightly knotted debris from the stellar explosion. The debris contains newly made heavy elements, some of which may eventually condense to form planets.

solar system to begin. The solar nebula was affected in at least two important ways: First, the shock wave caused the condensing mass to spin; and second, the nebula absorbed some of the heavy atoms from the passing supernova remnant. In other words, a massive star had to live its life (constructing elements in the process) and then undergo explosive disintegration in order to seed heavy elements back into the nebular nursery of dust and gas from which our solar system arose. The planets are made mostly of matter assembled in a star (or stars) that disappeared billions of years ago. We are made of that stardust. Our bones and brains are composed of ancient atoms constructed by stellar fusion long before the solar system existed.

By about 5 billion years ago, the solar nebula was a rotating, disk-shaped mass of about 75% hydrogen, 23% helium, and 2% other material (including heavier elements, gases, dust, and ice) (see **Figure 1.10**). Like a spinning skater bringing in her arms, the nebula spun faster as it condensed. Material concentrated near its center became the protosun. Much of the outer material eventually became **planets,** the smaller bodies that orbit a star and do not shine by their own light.

The new planets formed in the disk of dust and debris surrounding the young sun through a process known as **accretion**—the clumping of small particles into large masses. Bigger clumps with stronger gravity pulled in most of the condensing matter. Near the protosun, where temperatures were highest, the first materials to solidify were substances with high boiling points, mainly metals

and certain rocky minerals. The planet Mercury, closest to the sun, is mostly iron because iron is a solid at high temperatures. Somewhat farther out, in the cooler regions, magnesium, silicon, water, and oxygen condensed. Methane and ammonia accumulated in the frigid outer zones. The Earth's array of water, silicon–oxygen compounds, and metals results from its middle position within that accreting cloud. The planets of the outer solar system—Jupiter, Saturn, Uranus, and Neptune—are composed mostly of methane and ammonia ices because those gases can congeal only at cold temperatures.

The period of accretion lasted perhaps 50 to 70 million years. The protosun became a star when its internal temperature became high enough to fuse atoms of hydrogen into helium. The violence of these nuclear reactions sent radiation sweeping past the inner planets, clearing the area of excess particles and ending the period of rapid accretion. Gases like those we now see on the giant outer planets may once have surrounded the inner planets, but this rush of solar energy and particles stripped them away.

This process might not be rare. **Figure 1.11,** an image taken by the Hubble Space Telescope in 1994, shows a dusty oval disk around a star in the constellation of Orion. The condensing disk, glowing by the light of nearby stars, is about 7½ times the diameter of the solar system. Astronomers estimate it contains at least enough material to make three planets the size of the Earth.

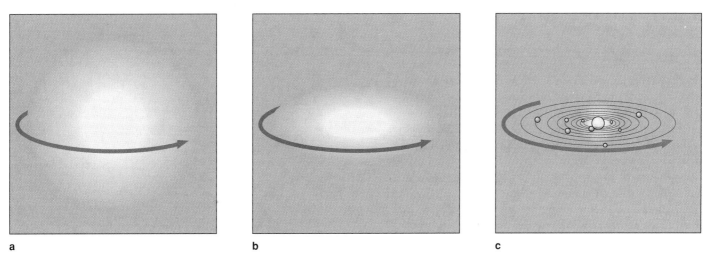

Figure 1.10 The formation of the solar system from the solar nebula. Because the nebula was rotating (a), it contracted into a disk (b). The protosun condensed at the center (c), and when the planets accreted, their orbits all lay in nearly the same plane. In (c), the Earth is shown in blue.

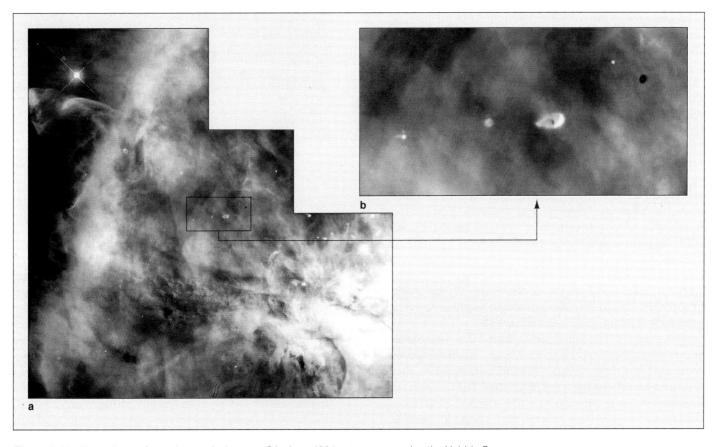

Figure 1.11 Have planets formed around other stars? In June 1994 astronomers using the Hubble Space Telescope announced the discovery of what they believe to be a planetary system forming around a star in the constellation of Orion, some 1,500 light-years from the Earth. A part of the Orion nebula—a "stellar nursery"—is seen in (a). (b) An enlargement of part of the nebula, showing a disk of debris around a young star. Astronomers estimate the dusty oval disk contains at least enough material to make three planets the size of the Earth.

EARTH AND OCEAN

The young Earth, formed by the accretion of cold particles, was probably homogeneous throughout. Then, late in the accretion phase, the Earth's surface was heated by the impact of meteors and other falling debris. This heat, combined with heat from the decay of radioactive elements accumulating within the newly assembled planet, caused the whole planet to melt. Gravity pulled most of the iron inward to form the planet's core. The sinking iron released huge amounts of gravitational energy, which, through friction, heated the Earth even more. At the same time, a slush of lighter minerals—silicon, magnesium, aluminum, and oxygen-bonded compounds—migrated toward the surface, forming the Earth's crust. This important process, called **density stratification,** lasted perhaps 100 million years.

Then the Earth began to cool. Its first surface is thought to have formed about 4.6 billion years ago. Radiation from the energetic young sun stripped away the planet's outermost layer of gases, its first atmosphere; but soon gases that had been trapped inside the forming planet burped to the surface to begin making the present atmosphere. This volcanic venting of volatile substances—including water vapor—is called **outgassing** (**Figure 1.12**). As the hot vapors rose, they condensed into clouds in the cool upper atmosphere. Recent research suggests that millions of tiny, icy comets colliding with the Earth may also have contributed to the accumulating mass of water vapor, this ocean-to-be.

The Earth's surface was so hot that no water could settle there, and no sunlight could penetrate the thick clouds. A visitor approaching from space 4.5 billion years ago would have seen a vapor-shrouded sphere blanketed by lightning-stroked clouds. After millions of years the upper clouds cooled enough for some of the outgassed water to form droplets. Hot rains fell toward the Earth, only to boil back into the clouds again. As the surface became cooler, water collected in basins and began to dissolve minerals from the rocks. Some of the water evaporated, cooled, and fell again. The world ocean was gradually accumulating (see **Figure 1.13**).

These heavy rains may have lasted for 10 million years. Large amounts of water vapor and other gases continued to escape through volcanic vents during that time and for millions of years thereafter. The ocean grew deeper. Evidence suggests that the Earth's crust grew thicker as well, perhaps in part from chemical reaction with oceanic compounds.

The physical expanse and distribution of the early ocean is a matter of some controversy. Most researchers hold that masses of rock have always protruded through the ocean surface to form continents. However, some recent studies suggest that water may have covered the

Figure 1.12 Outgassing. Volcanic gases emitted by fissures add water vapor, carbon dioxide, nitrogen, and other gases to the atmosphere. Volcanism was a major factor in altering the Earth's original atmosphere.

Earth's entire surface for some 200 million years before the continents emerged. Although most of the ocean was in place about 4 billion years ago, ocean formation continues very slowly even today: About 0.1 cubic kilometer (0.025 cubic mile) of new water is added to the ocean each year, mostly as steam flowing from volcanic vents.

The composition of that early atmosphere (often called the primitive atmosphere) was much different from today's, consisting of methane, ammonia, carbon dioxide, water vapor, and other gases. Beginning about 3.5 billion years ago this mixture was gradually altered to the present composition, mostly nitrogen and oxygen, by chemical change and by the oxygen-producing action of photosynthesizing organisms, ancestors of today's green plants.

Eventually sunlight pierced the clouds for the first time. The face of the Earth was exposed in daytime to the radiance of the sun and at night to the twinkle of stars like those that formed its elements.

Figure 1.13 The hot surface of the Earth as it may have looked about 4.2 billion years ago. In this painting, distant rainfall is cooling the surface, and pools of water are collecting in low spots of the solidifying crust. The moon, much closer than at present, looms above the horizon. In fact, there would have been little to see. The early atmosphere was so thick with clouds that no sunlight could have penetrated to the surface; except for lightning strokes and red glowing lava, the embryo ocean and continents would have been in darkness.

THE ORIGIN OF LIFE

Life, at least as we know it, would be inconceivable without large quantities of water. Water has the ability to retain heat, moderate temperature, dissolve many chemicals, and suspend nutrients and wastes. These characteristics make it a mobile stage for the intricate biochemical reactions that allowed life to begin and prosper on Earth.

As early as 1929, biologist J. B. S. Haldane proposed that exposing a primitive atmosphere to ultraviolet radiation or lightning might produce some of the same carbon compounds found in living things. Building on this idea in a classic 1953 experiment, Stanley Miller mixed together water vapor, ammonia, methane, and hydrogen—gases that were thought to be present in the early atmosphere of Earth—and passed an electric spark through them for a week. In that short time, his apparatus produced a number of different amino acids and other organic compounds (**Figure 1.14**). The electric discharge had provided energy for the formation of these simple molecules. Since then other mixtures thought to reflect more accurately the early atmosphere of Earth have been tested, with similar results. When exposed to ultraviolet light, heat, and electrical spark, these mix-

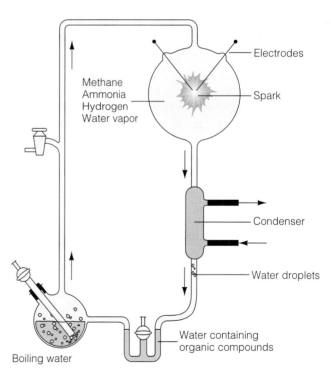

Figure 1.14 Stanley Miller's apparatus for producing organic molecules from a mixture of gases believed to be similar to Earth's primitive atmosphere.

a

Figure 1.15 Environments for biosynthesis? Some scientists believe that life could have originated in coastal tidal pools (a). Weak sunlight and unstable conditions at the Earth's surface, however, may have favored the origin of life near deep-ocean hydrothermal vents (b).

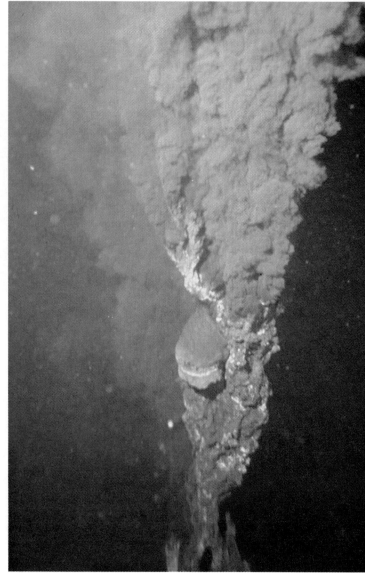

b

tures produce simple sugars and most of the biologically important amino acids. They even produce small proteins and nucleotides (components of the molecules that transmit genetic information between generations).

Did *life* form? No, the compounds that formed are only building blocks of life. But the experiment does tell us something about the commonality and unity of life on Earth. The facts that these crucial compounds can be synthesized so easily and are present in virtually all living forms are probably not coincidental. Those compounds are "permitted" by physical laws and by the chemical composition of this planet. The experiment also underscores the special role of water in life processes. The fact that all life, from a jellyfish to a dusty desert weed, depends on saline water within its cells to dissolve and transport chemicals is certainly significant. It strongly suggests that simple, self-replicating—living—molecules arose somewhere in the early ocean.

The early steps in the evolution of living organisms from simple organic building blocks, a process known as **biosynthesis,** are still speculative (see **Figure 1.15**). One idea, popularized in the 1950s, suggests that life may have originated in shallow tidal pools at the ocean's edge. Evaporation of water from these pools would have

concentrated the amino acid and nucleotide building blocks into a rich organic soup. Grains of sand or tiny bubbles would have provided handy surfaces on which larger chemical combinations could be assembled. Sunlight would have supplied the energy for these reactions. Accumulating in protected pools, reacting aggregates could have become progressively more complex, eventually evolving into biochemical systems capable of reproducing.

Unfortunately for this hypothesis, the sunlight needed to energize the reactions might not have been present; even if it had been, strong sunlight has harmful effects on unprotected large molecules. Planetary scientists now suggest that the sun was faint in its youth. It put out so little heat that the ocean may have been frozen to a depth of around 300 meters (1,000 feet). The ice would have formed a blanket that kept most of the ocean fluid and relatively warm. Periodic fiery impacts by asteroids could have thawed the ice, but between batterings it would have reformed. In 1994, chemists Jeffrey Bada and Stanley Miller suggested that organic material may have formed and then been trapped beneath the ice—protected from the atmosphere, which contained chemical compounds capable of shattering the complex molecules. The first self-sustaining molecules might have arisen deep below the layers of surface ice, on clays or pyrite crystals near warm hydrothermal vents on the ocean floor.

There are other hypotheses. Organic molecules could have ridden to the Earth on icy comets or grown within our planet's solid upper crust. Most researchers agree, however, that whatever the details, some quiet corner of the ocean was the likely place where life began.

A similar biosynthesis cannot occur today. Living things have changed the conditions in the ocean and atmosphere, and those changes are not consistent with any new origin of life. For one thing, green plants have filled the atmosphere with oxygen, a compound that can disrupt any unprotected large molecule. For another, some of this oxygen (as ozone) now blocks much of the ultraviolet radiation from reaching the surface of the ocean. And finally, the many tiny organisms present today would gladly scavenge any large organic molecules as food.

How long ago might life have begun? The oldest fossils yet found, from northwestern Australia, are between 3.4 and 3.5 billion years old (**Figure 1.16**). They are remnants of fairly complex bacteria-like organisms, indicating that life must have originated even earlier, probably only a few hundred million years after a stable ocean formed. Life and the Earth have grown old together; each has greatly influenced the other. The surface of this planet is a nearly continuous living film converting and storing energy in the most subtle and beautiful ways.

AN OCEAN WORLD

Water planets are probably uncommon in the universe, but water itself is not scarce. In our solar system, for example, Mars has polar ice caps, and Jupiter's atmosphere contains hundreds of times more water molecules than are present on the whole Earth. Astronomers have even located water molecules drifting in free space. But *liquid* water is unexpected.

Consider the conditions necessary for a large permanent ocean of liquid water to form on a planet. An ocean world must move in a nearly circular orbit around a stable star. The distance of the planet from the star must be

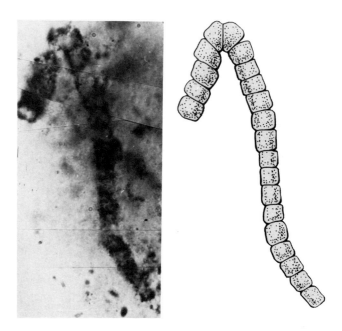

Figure 1.16 Fossil of a bacteria-like organism (with an artist's reconstruction) that photosynthesized and released oxygen into the atmosphere. Among the oldest fossils ever discovered, this microscopic filament from northwestern Australia is about 3.5 billion years old.

just right to provide a temperature environment in which water is liquid. Unlike most stars, a water planet's sun must not be a double or multiple star, or the orbital year would have irregular periods of intense heat and cold. The materials that accreted to form the planet must have included both water and substances capable of forming a solid crust. Volcanoes or steaming vents would be needed to vent water vapor to the surface. Finally, the planet must be large enough that its gravity will keep the atmosphere and ocean from drifting off into space.

Special conditions were also necessary for the formation of life here. Earth's gravity is strong enough to retain an ocean, but not strong enough to crush the life-forms that came from it. The planet has a magnetic field provided by an iron core to deflect radiation that would otherwise harm the genetic instructions of living things. A single moon provides gentle tides to encourage life-forms to leave the ocean and reside on land. The atmosphere is relatively clear—so that sunlight penetrates to the surface—but moist enough to form rains and winds that drive air and ocean currents. Further, the upper air contains ozone, which protects against the most harmful ultraviolet rays. This combination is probably exceedingly rare in the galaxy. We should enjoy and protect the water planet we call home.

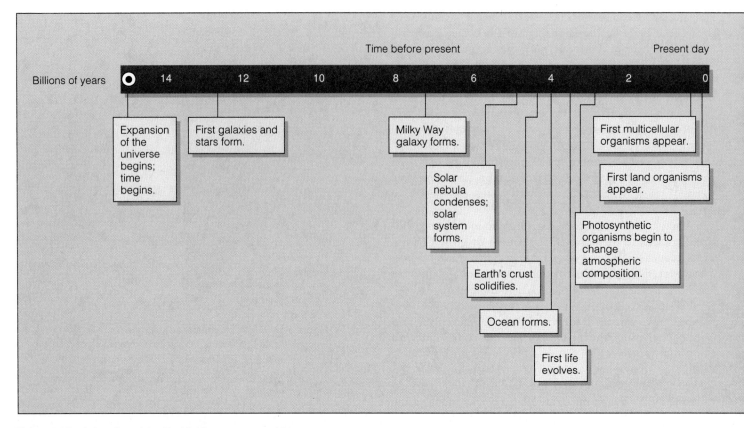

Figure 1.17 A time line of the Earth's history—past and future.

THE DISTANT FUTURE OF EARTH?

Our descendants may enjoy another 5 billion years of Earth as we know it today. But then our sun, like any other star, will begin to die. The sun is not massive enough to become a supernova, but its red giant phase will engulf the inner planets perhaps 6 billion years from now. Its fiery atmosphere may expand to a radius greater than the orbit of the Earth. The ocean and atmosphere, all evidence of life-forms, the crust, and perhaps the whole planet will be recycled into component atoms and hurled by shock waves into space. Our successors, if any, will have perished or fled to safer worlds. Its fuel exhausted and energies spent, the sun will cool to a glowing ember and ultimately to a dark cinder. The history of past and future Earth is shown as a time line in **Figure 1.17.**

Perhaps a new system of star and planets will form from the debris of our remains. But what is the chance that conditions will allow the formation of another ocean world? If there is one, I wonder what its inhabitants will name it.

Q-AND-A[3]

1. Why do we call this planet *Earth?*

   We are terrestrial organisms, so the name is understandable. The modern English word Earth *is derived from the Anglo-Saxon word* eorthe, *which itself comes from the ancient German word for Earth,* Erde. *The earliest German tribes were landlocked people who had no knowledge of the true extent of the world ocean. We seem to be stuck with* Earth.

2. Where did the word *ocean* come from?

   Ocean *derives from the Greek word* okeanos *(oceanus), a word meaning "outer sea" (in contrast to the Greeks' "inner sea," the Mediterranean). The term connotes a large moving river. Okeanos was also a mythical Greek titan who was god of the sea before Poseidon and father of the oceanides, or ocean nymphs. The later Latin name for the ocean was* oceanus. *The Latin word evolved into the Middle English term* ocean, *in which the double-c was pronounced as a* k. *This later became the English word* ocean

---

[3]Each chapter ends with a few questions students have asked me after a lecture or reading assignment. These questions and their answers may be interesting to you, too.

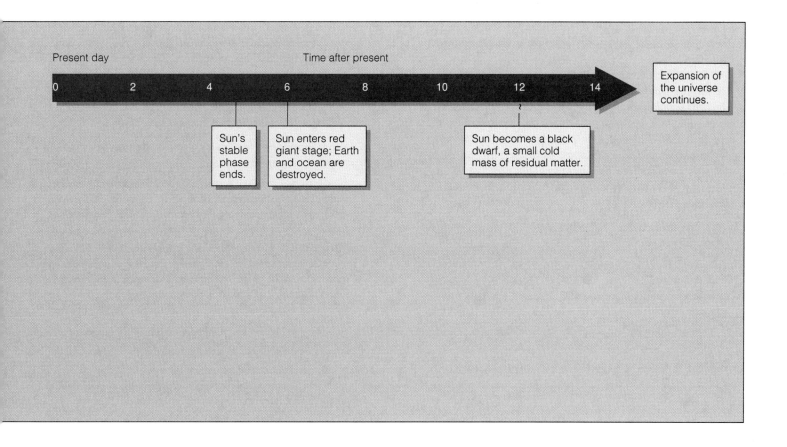

Present day   Time after present

0   2   4   6   8   10   12   14

Expansion of the universe continues.

Sun's stable phase ends.

Sun enters red giant stage; Earth and ocean are destroyed.

Sun becomes a black dwarf, a small cold mass of residual matter.

3. How do scientists know how old the Earth, the solar system, and the galaxy are? How can they calculate the age of the universe?

The age estimates presented in this chapter are derived from interlocking data obtained by many researchers using different sources. One source is meteorites, chunks of rock and metal formed at about the same time as the sun and planets and out of the same cloud. Some meteorites contain gases thought to be remnants of the original solar nebula. Many have fallen to Earth in comparatively recent times. We know from signs of radiation within these objects how long it has been since they were formed. That information, combined with the rate of radioactive decay of unstable atoms in meteorites, moon rocks, and in the oldest rocks on the Earth, allows astronomers to make a reasonably accurate estimate of how long ago these objects formed.

The light from stars can be analyzed to discover their composition. A spectroscope breaks starlight into component colors by passing it through a device that works like a prism. Careful study of the character and distribution of this light

yields information on the star's temperature and composition. The elements present in stars—and the distribution of types of stars in the sky—suggest that most stars go through a predictable life cycle based on their beginning mass.

As for the universe itself, remember that it is expanding. If we assume it has been expanding at a uniform rate, and compensate for slowing because of gravity, we can trace the expansion backward to the point and time of origin, sometime around 15 billion years ago.

4. Life appears to have arisen on Earth soon after the formation of a stable surface. Could life have formed on other planets?

We have no evidence, direct or indirect, of life on other planets around our sun or elsewhere in the universe. Yet it seems provincial to assume that life could have arisen only here. The formation of organic molecules from simple chemicals receiving energy from lightning, heat, ultraviolet light, and other sources may be quite common, and increasing complexity in these compounds may be a universal phenomenon.

5. Would life on other planets resemble life on Earth?

Organisms elsewhere might be very different. Recall that life on this planet probably arose in the ocean, and all life-

forms here carry an ocean of sorts within their bodies. On a planet without water, the organisms would be much different.

For example, on a hypothetical planet with an ammonia ocean, life would not have a structure of cells surrounded by lipid membranes. Lipid membranes are the sheets of fatty molecules that keep the inside of a cell separate from the environment, and ammonia prevents them from forming. Without membranes, cells as we know them are not possible. Notwithstanding this argument, life need not be confined to planets with water. Other life-forms may exist, based on other "brews."

6. You mentioned one hypothesis for the origin of life. Are there others?

Yes, indeed. Divine creation is one, but it is permanently impossible to test that idea using the scientific method. Another is panspermia, which proposes that life formed (or was created) in space and was then distributed through- out the galaxy. Still another involves intelligences "seeding" planets with DNA (or molecules of similar function), later to return and see what useful creatures have resulted. Neither of these ideas can be tested, at least for now. Moreover, both in a sense beg the question; that is, they assume that some extraterrestrial life already existed, but they don't deal with how that life may have arisen.

7. What happened to Joshua Slocum after his single-handed circumnavigation?

He continued to work periodically as a master mariner, but most of his time was spent in writing, lecturing, exhibiting at expositions and fairs, selling curios, farming (a vocation he did not enjoy), and reading. In 1905 he set sail, again in Spray, for the West Indies. After two more trips to the same islands, he departed in 1909, at age 65, in his well-traveled boat to explore South America's Orinoco and Amazon rivers. No trace of him or his boat was ever found.

Terms and Concepts to Remember

accretion	law	science
big bang	marine science	scientific method
biosynthesis	Milky Way galaxy	solar nebula
condensation	nebula	solar system
theory	ocean	star
density	oceanography	supernova
stratification	outgassing	theory
galaxy	planet	world ocean
hypothesis	protostar	

Study Questions

1. Which hemisphere contains the greatest percentage of ocean? Is most of Earth's water in the ocean?

2. Can the scientific method be applied to speculations about the natural world that are not subject to test or observation?

3. Marine biologists sometimes say that all life-forms on Earth, even desert lizards and alpine plants, are marine. Why?

4. How are light elements converted into heavy ones?

5. Will all stars end their lives as supernovas? What happens to the heavy elements made by small stars?

6. Are the ocean and present atmosphere "leftovers" from the original atmosphere of Earth?

7. Where did the Earth's surface water come from?

8. What is biosynthesis? Where do researchers think it might have occurred on our planet? Could it happen again today?

9. Do you think water planets are common in the galaxy?

10. How might Earth be different if an ocean had not formed on its surface?

For Further Study

The World Ocean

Attenborough, D. 1984. *The Living Planet*. Boston: Little, Brown. The companion volume to the BBC/PBS TV series. Chapter 11 on "The Open Ocean" provides a fine introduction to the study of the world ocean.

Benchley, P., and J. Gradwohl, eds. 1995. *Ocean Planet: Writings and Images of the Sea*. New York: Abrams. This companion volume to a major Smithsonian Institution exhibit contains photographs and artwork and writings by Jacques-Yves Cousteau, Rachel Carson, John McPhee, Peter Matthiessen, Benjamin Franklin, and others. Highly recommended.

Carson, R. 1951. *The Sea Around Us*. New York: Houghton Mifflin. The book that introduced many of us to the wonders of the ocean when we were young.

Garrison, T., and R. Lebow. 1993. *Oceanus*. 6th ed. Belmont, CA: Wadsworth. The first lesson overview may be of interest.

Kennish, M. J., ed. 1989. *Practical Handbook of Marine Science*. Boca Raton, FL: CRC Press. An excellent summary of physical and biological oceanographic data, primarily in the form of graphs and tables. A thorough bibliography follows the introduction.

Mangone, G. J. 1986. *Concise Marine Almanac*. Van Nostrand-Reinhold. A wealth of tabular information.

Slocum, J. 1899. *Sailing Alone Around the World.* Reprint. New York: Sheridan House, 1972. Inspiring autobiographical account of the first single-handed circumnavigation.

The Nature of Science

Carey, S. S. 1994. *A Beginner's Guide to Scientific Method.* Belmont, CA: Wadsworth. A brief, nontechnical introduction to the basic methods underlying all good scientific research.

McCain, G., and E. M. Segal. 1988. *The Game of Science.* 5th ed. Pacific Grove, CA: Brooks/Cole. A lighthearted summary that conveys the underlying seriousness of its content.

Morrison, P., and P. Morrison. 1987. *The Ring of Truth: An Inquiry into How We Know What We Know.* New York: Random House. An accomplished physicist discusses how scientists search for answers.

The Origin of the Earth and the Ocean

Cloud, Preston. 1988. *Oasis in Space: Earth's History from the Beginning.* New York: Norton.

Ferris, T. 1988. *Coming of Age in the Milky Way.* New York: Morrow. A beautifully written history of the dawn of our understanding of the universe; very highly recommended.

Frank, L. A., J. B. Sigworth, and J. D. Craven. 1986. "On the Influx of Small Comets into the Earth's Upper Atmosphere: Observations and Interpretations." *Geophysical Research Letters* 13 (no. 4): 303–10. Another source of water for the ocean?

Hamblin, W. K., and E. H. Christiansen. 1990. *Exploring the Planets.* New York: Macmillan. Contains beautiful photographs of the planets taken by robot spacecraft.

Hartmann, W. K., and Chris Impey. 1994. *Astronomy: The Cosmic Journey.* 5th ed. Belmont, CA: Wadsworth. Outstanding general text.

Hawking, S. 1988. *A Brief History of Time.* New York: Bantam Books. A gentle introduction to a challenging topic, with information on the nature of the big bang.

Holland, H. D. 1984. *The Chemical Evolution of the Atmosphere and Oceans.* Princeton: Princeton University Press. Technical.

Jastrow, R. 1978. *God and the Astronomers.* New York: Norton. A look at the overlap of cosmology and religion.

Maran, S. P. 1988. "In Our Backyard, a Star Explodes." *Smithsonian,* April, 46–57. Information on supernova 1987A.

Press, F., and R. Siever. 1986. *Earth.* 4th ed. San Francisco: Freeman. Chapter 1 contains a very clear overview of the formation and early stages of the Earth.

Sagan, C. 1980. *Cosmos.* New York: Random House. The most popular scientific book ever published, and with good reason. A magnificently written and illustrated description of the cosmos. A must for anyone wishing to understand the Earth's place in the universe.

Schopf, J. W., ed. 1983. *Earth's Earliest Biosphere: Its Origin and Evolution.* Princeton: Princeton University Press. A collection of scientific papers on the subject.

Squyres, S. W., and J. F. Kasting. "Early Mars: How Warm and How Wet?" *Science* 265 (no. 5173): 744–48. Early in its history, Mars had a warmer, wetter climate. Evidence suggests that there was never an ocean on Mars, however, and that surface temperatures have rarely risen above 0° Celsius.

Stahler, S. W. 1991. "The Early Life of Stars." *Scientific American,* July, 48–55. Discusses the turbulent youth of stars.

Trefil, J. 1985. *Space, Time, Infinity.* Washington, DC: Smithsonian Books. Great illustrations, clear and well-written text.

Trefil, J. 1995. "Life on Earth—Was It Inevitable?" *Smithsonian* 25 (no. 11): 32–41. In a universe filled with prebiotic compounds, it may be only a small step for some of them to hook up in ways that lead directly to life.

Welsh, W. J. 1987. "Star Formation in W49A: Gravitational Collapse of the Molecular Cloud Core Toward a Ring of Massive Stars." *Science* 238 (no. 4833): 1550–55. Condensation theory in action.

Wetherill, G. W. 1991. "Occurence of Earth-Like Bodies in Planetary Systems." *Science* 253 (no. 5019): 535–38. Mathematical models suggest Earth-like planets may be a common feature of planetary systems.

Woosley, S., and T. Weaver. 1989. "The Great Supernova of 1987." *Scientific American,* August, 32–40. Theories of stellar evolution have been verified by observations, but new questions arise.

The Origin of Life

Bjerklie, D., B. Hillenbrand, and J. O. Jackson. 1993. "Life's Tempestuous Origins." *Time,* 11 October, 68–74. An overview of current thinking on this complex topic. The authors suggest that life evolved faster on Earth than previously thought.

Bracewell, R. 1975. *The Galactic Club.* San Francisco: Freeman. A thoughtful discussion of the possibility of life elsewhere in the universe.

Dyson, F. J. 1985. *Origins of Life.* Cambridge: Cambridge University Press. Edited transcript of the Tarner Lectures given at Cambridge University in January 1985. Dyson suggests some objections to the original Miller experiment.

Flam, F. 1994. "The Chemistry of Life at the Margins." *Science* 265 (22 July): 471–72. Looking at life in the harshest environments of the Earth gives us insight into how life may have arisen here.

Horgan, J. 1991. "In the Beginning . . ." *Scientific American,* February, 116–25. Thoughts on the origin of life by a number of researchers.

Miller, S. L. 1955. "Production of Some Organic Compounds Under Possible Primitive Earth Conditions." *Journal of the American Chemical Society* 77 (no. 2351). A classic paper.

National Academy of Sciences Committee on Science and Creationism. 1984. *Science and Creationism: A View from the National Academy of Sciences.* Washington, DC: National Academy of Sciences. A well-written comparison of biological evolution and "scientific" creationism. (Copies may be obtained from the National Academy of Sciences, 2101 Constitution Avenue NW, Washington, DC 20418.)

Schopf, J. W. 1993. "Microfossils of the Early Archaean Apex Chert: New Evidence of the Antiquity of Life." *Science* 260 (no. 5108): 640–46. Schopf suggests cyanobacterium-like organisms existed at least 3.465 billion years ago.

2 A HISTORY OF MARINE SCIENCE

Making Marine History

History is made a day at a time. Sometimes those who make history realize the significance of a single day's effort, but more often the days blend into weeks and months of hard work punctuated only occasionally by scientific or artistic insights. Still, individual days are important, and none is more important to a discoverer than the day that first piqued his or her involvement in a field that becomes the subject of lifelong study. No matter what follows, that day, spent perhaps with a good teacher and good friends, will remain unique in memory.

Student oceanographers get cold, and wet, and often seasick when they make their first study cruises to sample the ocean they've been learning about in books and classes. They may take samples with sophisticated computerized probes or with buckets and thermometers lowered on a line. They may work from outboard-powered boats or well-equipped research vessels. They struggle with buckets of sediment, cut fragrant piles of frozen fish food for aquarium animals, stand watch in howling storms, steady their microscopes, learn arcane computer commands, memorize the interior of a shark, and are the last ones out of the library at night. They write, they talk, and they party. A few of them experience a flash of commitment and decide to continue. *That* is the moment they will remember most vividly— the day they start to make their own history.

With few exceptions, we have no knowledge of how the travelers and scientists described in this chapter became personally interested in advancing marine science, but we do know that, as before, some of today's student oceanographers will contribute to the oceanography texts of tomorrow. Progress in marine science depends on them, and perhaps on you.

On this page, students and instructors relax after a day of gathering data and specimens aboard a small research vessel. At right, data on temperature, salinity, and dissolved oxygen content are being obtained.

The early history of marine science is closely associated with the history of voyaging. The first marine studies had practical aims: travel, trade, warfare, and facilitating the search for food. Later the search for new knowledge became a goal in itself. The first part of this chapter focuses on marine science for voyaging; it takes a look at some of the voyagers and their voyages, the inventions that made their adventures possible, and some of the discoveries they made. The second part discusses voyaging for science, the rise of oceanographic institutions, and the present thrust of ocean science.

In later chapters, some of the traditional founders of marine science—including Charles Darwin, Fridtjof Nansen, and others—will be discussed, along with their specialties. References to their contributions, along with the major points of this chapter, are shown in a time line at the end of the chapter.

VOYAGING BEGINS

Ours is a restless and inquisitive species. Despite the vast oceans, we have populated nearly every habitable place on Earth. This fact was aptly illustrated when European explorers set out to "discover" the world, only to be met by native peoples at almost every landfall! Clearly, the oceans did not prevent the spread of humanity.

Consider for a moment the difficulty people must have had in migrating from eastern Africa, the probable place of human origin, to the rest of the world. Generations of nomads spread slowly into the Near East, central Europe, and on across the plains of central Asia. They wandered north to the shores of the Bering Sea and, later, over a land or ice bridge to North America, reaching the southern tip of South America about 11,000 years ago. Though most traveled on foot, some people undoubtedly used rafts or boats. Later travelers even colonized remote Pacific islands.

Ocean transportation offers people the benefits of mobility and greater access to food supplies. Any coastal culture skilled at raft building or small-boat navigation would have economic and nutritional advantages over its less adept competitors. From the earliest period of human history, then, understanding and appreciation of the ocean and its life-forms benefited those patient enough to learn.

The first direct evidence we have for **voyaging,** traveling on the ocean for a specific purpose, comes from records of trade in the Mediterranean Sea. The Egyptians organized shipborne commerce on the Nile River, but the first regular ocean traders were probably the Cretans—or the Phoenicians, who inherited maritime supremacy in the Mediterranean after the Cretan civilizations were destroyed by earthquakes and political instability around 1200 B.C. Skilled sailors, Phoenicians carried their wares through the Straits of Gibraltar to markets as distant as Britain and the west coast of Africa. Considering the simple ships they used, this was quite an achievement.

The Greeks began to explore outside the Mediterranean into the Atlantic Ocean around 900–700 B.C. (**Figure 2.1**). Early Greek seafarers noticed a current running from north to south beyond Gibraltar. Believing that only rivers had currents, they decided that this great mass of water, too wide to see across, was part of an immense flowing river. The Greek name for this river was *okeanos.* Our word *ocean* is derived from *oceanus,* a Latin variant of that root. Phoenician sailors were also very much at home in this "river," but like the Greeks they rarely ventured out of sight of land.

As they went about their business, early mariners began to record information to make their voyages easier and safer—the location of rocks in a harbor, landmarks and the sailing times between them, the direction of currents. These first **cartographers** (chart makers) were probably Mediterranean traders who made routine journeys from producing areas to markets. Their first charts (drawn about 800 B.C.) were merely notes to jog their memory for obvious features along the route. Today's **charts** are graphic representations that primarily depict water and water-related information. (*Maps* primarily represent land.)

In this early time other cultures also traveled on the ocean. The Chinese began to engineer an extensive system of inland waterways, some of which connected with the Pacific Ocean to make long-distance transport of goods more convenient. The Polynesian peoples had been moving easily among islands off the coasts of Southeast Asia and Indonesia since 3000 B.C. and were beginning to settle the mid-Pacific islands. Though none of these civilizations had contact with the others, each developed methods of charting and navigation. All these early travelers were skilled at telling direction by the stars and by the position of the rising or setting sun.

Curiosity and commerce encouraged adventurous people to undertake ever more ambitious voyages. But these voyages were possible only with the coordination of astronomical direction-finding (and knowledge of the shape and size of the Earth), advanced shipbuilding technology, accurate graphic charts (not just written descriptions), and, perhaps most important, a growing understanding of the ocean itself. Marine science, the organized study of the ocean, had its origin in the technical studies of voyagers.

SCIENCE FOR VOYAGING

Progress in applied marine science began at the **Library of Alexandria,** in Egypt. Founded in the third century B.C. by Alexander the Great, the library could be considered the first university in the world. Scholars worked and researched there, and students came from around the Mediterranean to study. Written knowledge of all kinds was warehoused around its leafy courtyards. This library constituted history's greatest accumulation of ancient writings. When any ship entered the harbor, the books (actually scrolls) it contained were by law removed and copied; the *copies* were returned to the owner and the originals kept for the library. Caravans arriving overland were also searched. The characteristics of nations, trade, natural wonders, artistic achievements, tourist sights, investment opportunities, and other items of interest to seafarers were catalogued and filed. Manuscripts describing the Mediterranean coast were of great interest. Traders quickly realized the competitive benefit of this information.

Yet marine science was only one of the library's many research areas. For 600 years, it was the greatest repository of wisdom of all kinds and the most influential insti-

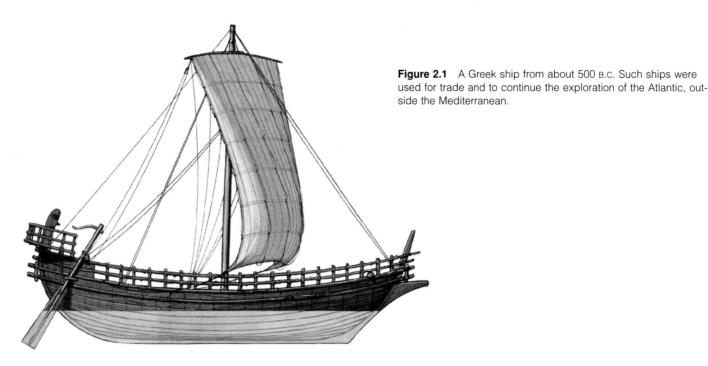

Figure 2.1 A Greek ship from about 500 B.C. Such ships were used for trade and to continue the exploration of the Atlantic, outside the Mediterranean.

tution of higher learning in the ancient world. Here perhaps was the first instance of cooperation between a university and the commercial community, a partnership that has paid dividends for science and business ever since.

Alexandrian Contributions

The second librarian at Alexandria (from 235 B.C. until 192 B.C.) was the Greek astronomer, philosopher, and poet **Eratosthenes of Cyrene.** This remarkable man was the first to calculate the circumference of the Earth. The Greek Pythagoreans had realized Earth was spherical by the sixth century B.C., but Eratosthenes was the first to estimate its true size.

Eratosthenes had heard from travelers returning from Syene (now Aswan, site of the great Nile dam) that at noon on the longest day of the year the sun shone directly onto the waters of a deep vertical well. In Alexandria, he noticed that a vertical pole cast a slight shadow on that day. He measured the shadow angle and found it to be a bit more than 7°, about 1/50 of a circle. He correctly assumed that the sun is a great distance from the Earth; so the sun's rays would approach Syene and Alexandria in essentially parallel lines. If the sun were directly overhead at Syene, but not directly overhead at Alexandria, then the surface of the Earth must be curved. But what was the *circumference* of the Earth?

By studying the reports of camel caravan traders, he estimated the distance from Alexandria to Syene at about 785 kilometers (491 miles). Eratosthenes now had the two pieces of information needed to derive the circumference of the Earth by geometry. **Figure 2.2** shows his method. The precise size of the units of length (stadia) Eratosthenes used is thought to have been 555 meters

(607 yards), and historians estimate that his calculation, made in about 230 B.C., was accurate to within about 8% of the true value. Within a few hundred years most people in the West who had contact with the library or its scholars knew Earth's approximate size.

Even without the contributions of Eratosthenes, the significance of the Library at Alexandria to marine science is immense. In Alexandria traders, explorers, scholars, and students had a place to conduct research and exchange information (and rumors) about the seas. Library researchers invented the astronomical, geometric, and mathematical base for **celestial navigation,** the technique of finding one's position on Earth by reference to the apparent positions of heavenly bodies.

Cartography flourished. The first workable charts representing a spherical surface on a flat sheet were developed by Alexandrian scholars. Latitude and longitude, systems of imaginary lines dividing the surface of the Earth, were invented by Eratosthenes. **Latitude** lines were drawn parallel to the equator, and **longitude** lines ran from pole to pole. He placed the lines through prominent landmarks and important places, creating a convenient though irregular grid (see **Figure 2.3**). Our present regular grid of latitude and longitude was invented by Hipparchus (c.165–c.127 B.C.), a librarian who divided the surface of the Earth into 360 degrees. A later Egyptian–Greek, Claudius Ptolemy (A.D. 90–168), *oriented* charts by placing east to the right and north at the top. Ptolemy's division of degrees into minutes and seconds of arc is still used by navigators.

Ptolemy also attempted to improve on Eratosthenes' surprisingly accurate estimate of the Earth's circumference, but he wrongly calculated a degree as about 50 miles instead of the more correct 70 miles. This error, coupled with his mistake of overestimating the size of Asia,

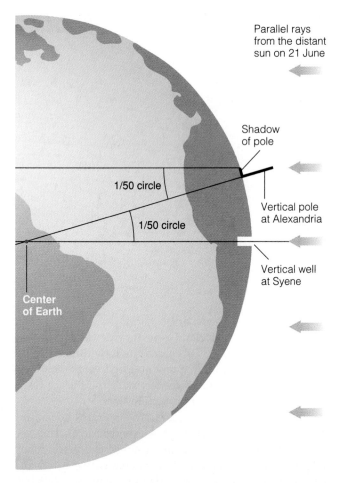

Parallel rays
from the distant
sun on 21 June

Shadow
of pole

1/50 circle

1/50 circle

Vertical pole
at Alexandria

Vertical well
at Syene

Center
of Earth

Figure 2.2 A diagram showing Eratosthenes' method for calculating the circumference of the Earth. As is described in the text, he used simple geometric reasoning based on the assumptions that the Earth is spherical and that the sun is very far away. Using this method, he was able to discover the circumference of the Earth to within about 8% of its true value. (The diagram is not drawn to scale.)

greatly reduced the apparent width of the unknown part of the world between the Orient and Europe. Unfortunately for generations of navigators, Eratosthenes' estimate was forgotten, while Ptolemy's persisted.

Though it weathered the dissolution of Alexander's empire, the Alexandrian Library did not survive the subsequent period of Roman rule. The last librarian was Hypatia, the first notable woman mathematician, philosopher, and scientist. In Alexandria she was a symbol of science and knowledge, concepts the early Christians identified with pagan practices. The mission of the library, as personified by the last librarian, antagonized the governors and citizens of the city of Alexandria. After years of rising tensions, in A.D. 415 a mob brutally murdered her and burned the library with all its contents. Most of the community of scholars dispersed, and Alexandria ceased to be a center of learning in the ancient world. The academic loss was incalculable, and trade suffered because shipowners no longer had a clearinghouse for updating the nautical charts and information they had come to depend on. All that remains of the library today is a remnant of an underground storage room. We shall never know the true extent and influence of its collection of over 700,000 irreplaceable scrolls.

Western intellectual development slackened during the so-called Dark Ages that followed the fall of the Roman Empire in A.D. 476. For almost 1,000 years, until the European Renaissance, much of the progress in medicine, astronomy, philosophy, mathematics, and other vital fields of human endeavor was made by the Arabs or imported by them from Asia. For example, the Arabs used the Chinese-invented compass for navigating caravans over seas of sand. During this time the Vikings raided and explored to the south and west, and the Polynesians continued some of the most extraordinary voyages in history.

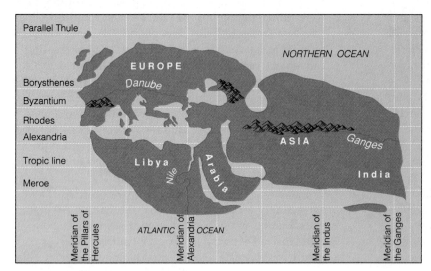

Parallel Thule

NORTHERN OCEAN

EUROPE
Danube

Borysthenes

Byzantium

Rhodes

Alexandria

ASIA

Ganges

Tropic line

Libya

Arabia

Nile

India

Meroe

Meridian of
the Pillars of
Hercules

ATLANTIC OCEAN

Meridian of
Alexandria

Meridian of
the Indus

Meridian of
the Ganges

Figure 2.3 The world, according to a chart from the third century B.C. Eratosthenes drew latitude and longitude lines through important places rather than spacing them at regular intervals as we do today. The Alexandrian perception of the world is reflected in the size of the continents and the central position of Alexandria.

Voyages of the Oceanian Peoples

In the history of human migration, no voyaging saga is more inspiring than that of the **Polynesian** colonizations, the peopling of the central and eastern Pacific islands. A profound knowledge of the sea was required for these voyages, and the story of the Polynesians is a high point in our chronology of marine science applied to travel by sea.

The Polynesians are one of four cultures inhabiting some 10,000 islands scattered across nearly 26 million square kilometers (10 million square miles) of open Pacific Ocean (**Figure 2.4**). The Southeast Asian or

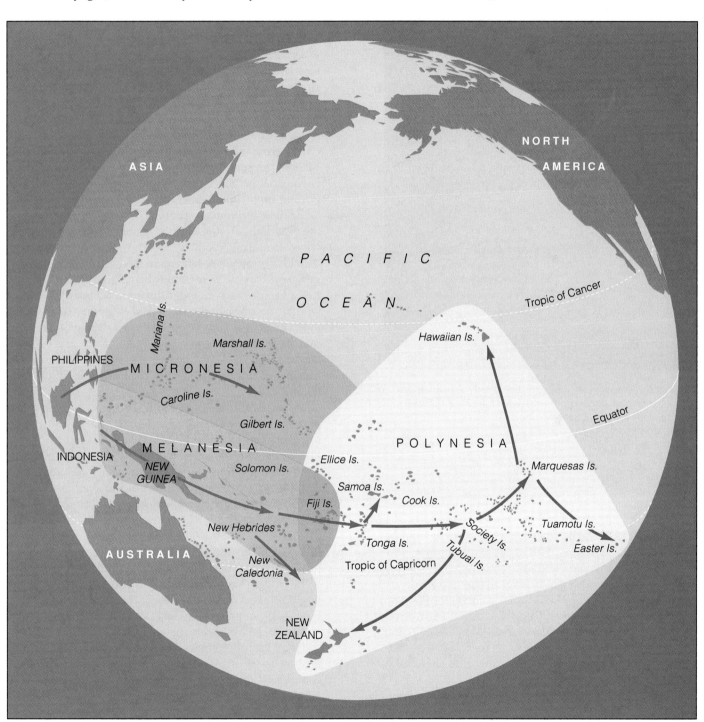

Figure 2.4 The Polynesian Triangle. Ancestors of the Polynesians spread from Southeast Asia or Indonesia to New Guinea and the Philippines by about 20,000 years ago. The mid-Pacific islands have been colonized for about 2,500 years, but the explosive dispersion that led to the settlement of Hawaii occurred about A.D. 450–600. Arrows show a possible direction and order of settlement.

Figure 2.5 A modern Micronesian stick chart. Knots or shells tied at the junctions between bamboo sticks represent islands. Straight strips represent patterns of regular waves; bent strips depict waves curving around islands.

Indonesian ancestors of the Oceanian peoples, as these cultures are collectively called, spread eastward in the distant past. Although experts vary in their estimates, there is some consensus that by 30,000 years ago New Guinea was populated by these wanderers and that by 20,000 years ago the Philippines were occupied. By around 500 B.C. the so-called cradle of Polynesia—Tonga, Samoa, the Marquesas, and the Society Islands—was settled. Oceanian navigators may already have been using shells attached to a bamboo grid to represent the positions of their islands. (A Micronesian stick chart from recent times is shown as **Figure 2.5**.)

For a long and evidently prosperous period the Polynesians spread from island to island until the easily accessible islands had been colonized. Eventually, however, overpopulation and depletion of resources became a problem. Politics, intertribal tensions, and religious

strife shook society. Groups of people scattered in all directions from some of the "cradle" islands during a period of explosive dispersion. Between A.D. 300 and 600 Polynesians successfully colonized nearly every inhabitable island within the vast triangular area shown in Figure 2.4. Easter Island was found against prevailing winds and currents, and the remote islands of Hawaii were discovered and occupied. These were among the last places on Earth to be populated.

How did these risky voyages into unexplored territory come about? Religious warfare may have been the strongest stimulus to colonization. If the losers of a religious war were banished from the home islands under penalty of death, their only hope for survival was to reach a distant and hospitable new land.

Seafaring had been a long tradition in the home islands, but such trips called for radical new technology. Great dual-hulled sailing ships, some capable of transporting up to 100 people, were designed and built. New navigation techniques were perfected that depended on the positions of stars barely visible to the north. New ways of storing food, water, and seeds were devised. Whole populations left their home islands in fleets designed especially for long-distance discovery (**Figure 2.6**). In some cases, fire was nurtured on board in case of landfall on an island that lacked volcanic flame. But a new island was only a possibility, a dream. Their gods may have promised the voyagers safe deliverance to new lands, but how many fleets set out from the troubled homelands only to fall victim to storms, thirst, or other dangers?

Yet in that anxious time the Polynesians honed and perfected their seafaring knowledge. To a skilled navigator, a change in the rhythmic set of waves against the hull could indicate an island out of sight over the horizon. The flight tracks of birds at dusk could suggest the direction of land. The positions of the stars told stories, as did the distant clouds over an unseen island. The smell of the water, or its temperature, or salinity, or color, conveyed information—as did the direction of the wind relative to the sun, and the type of marine life clustering near the boat. The sunrise colors, sunset colors, the hue of the moon—every nuance had meaning, every detail had been passed in ritual from father to son. The greatest Polynesian minds were navigators, and reaching Hawaii was their greatest achievement.

Of all islands colonized by the Polynesians, Hawaii is farthest away, across an ocean whose guide stars were completely unknown to the southern navigators. The Hawaiian Islands are isolated in the northern Pacific. There are no islands of any significance for more than 2,000 miles to the south. Moreover, Hawaii lies beyond the equatorial doldrums, a hot and often windless stretch across which these pioneers must somehow have pad-

Figure 2.6 A Polynesian voyaging canoe. Note the double-hulled design.

dled. And yet some fortunate and knowledgeable people colonized Hawaii sometime between A.D. 450 and 600. Try to imagine their feelings of relief and justification upon reaching a promised paradise under a new night sky. Think of that first approach to the high islands of Hawaii, the first unlimited drink of fresh water, the first solid Earth after months of uncertainty.

Within a hundred years of their first arrival, Hawaiian navigators were routinely piloting vessels on regular return trips to the Marquesas and the Society Islands (Tahiti and others). Some of the trips were undertaken to import needed food species to the newly found islands, but others were made to recruit new citizens and leaders to "green-clad Hawaii."

At a time when seafarers of other civilizations sailed beside the comforting bulk of a charted coast, Polynesians looked to the open sea for sustenance, deliverance, and hope. Their great knowledge of the ocean protected them.

Meanwhile, Back in Europe . . .

The Dark Ages were periodically punctuated by the raids of **Vikings,** bands of Scandinavian adventurers and treasure seekers whose remarkably fast, strong, and stable ships (**Figure 2.7**) enabled them to row up rivers faster than a horse and rider could spread warning. Danish and Norwegian Vikings swept down the coast of Europe; they methodically pillaged Paris, robbed monasteries in Ireland, and looted Britain. The Swedish Vikings foraged as far away as Kiev and Constantinople! In 859 Vikings spent a week or so ashore in Morocco, rounding up prisoners for sale as slaves or to hold for ransom. Sixty-two Viking ships participated, a spectacular display of technology, sea power, seamanship, and navigation.

At first the Europeans were powerless against these marauders, but eventually the need for common defense overcame provincial hostility and xenophobia. One of

Figure 2.7 A Viking ship from around A.D. 900. This painting is a reconstruction drawn from a ship found in 1880 at the bottom of a Norwegian fjord. Sturdy and intended for travel over open water, the ship is 23.3 meters (76 feet) long and 5.25 meters (17 feet) wide. Oarsmen sat on loose benches or chests, and holes for the oars could be covered by small disks, so that water would not flow in when the ship heeled under sail.

the causes of the Renaissance in Europe may have been the experience of banding together for protection against these northern raiders.

As the French, Irish, and British defenses became more effective, the Norwegian Vikings began to look west. Iceland and Greenland had been discovered by ships blown off course during storms. Iceland was colonized by about A.D. 700; Greenland by 995. In an early voyage from Norway to Greenland in 986, a commuter named Bjarni Herjulfsson was blown past his goal by unfavorable winds. For about five days he sailed up and down the coast of a new land (which was, in fact, North America) without landing or making charts. His sketchy reports kindled a real-estate fever; Leif, son of Eric the Red, purchased Bjarni's ship and returned. His party found salmon-filled lakes, vines and grapes, and fodder for cattle in what was probably the northeastern tip of Newfoundland. With a bit of advertising overstatement he called the place *Vinland* ("wine-land").

By A.D. 1000 the Norwegians had colonized Vinland. The settlements were modest, and at first relationships with the native Americans were encouraging. Unlike the Spanish who landed in the New World 500 years later, the Norwegian colonists tried to cooperate with the locals, to learn from them, and to help them in a mutual pact of assistance. Unfortunately misunderstandings

quickly arose, battles ensued, and the colony had to be abandoned in 1020. The Norwegians had neither the numbers, the weapons, nor the trading goods to make the colony a success.

Chinese Contributions

As the Dark Ages continued in Europe, Chinese navigators became more skilled, and their vessels grew larger and more seaworthy. They then set out to explore the other side of the world. Between 1405 and 1433, admiral Zheng He commanded the greatest fleet the world had ever known. At least 317 ships and 37,000 men undertook seven missions to explore the Indian Ocean, Indonesia, and around the tip of Africa into the Atlantic. Their aim: to display the wealth and power of the young Ming dynasty and to "show kindness to people of distant places." The largest ship in the fleet, with nine masts and a length of 134 meters (440 feet) (**Figure 2.8**), was a huge treasure ship carrying objects of the finest materials and craftsmanship. The mission of the fleet was not to accumulate such treasure, but to give it away! Indeed, the primary purpose of these expeditions was to convince all nations with which the fleet had contact that China was the only truly civilized state and beyond any imaginable need for knowledge or assistance.

Figure 2.8 The treasure ship, largest in a vast Chinese fleet whose purpose was to "show kindness to people of distant places." The fleet sailed the Pacific and Indian oceans between 1405 and 1433.

Many technical innovations had been required to make such an ambitious undertaking possible. In addition to the compass mentioned above, the Chinese invented the central rudder, watertight compartments, and sophisticated sails on multiple masts, all of which were critically important for the successful operation of large sailing vessels. Until Europeans adopted the rudder around A.D. 1100, long-distance voyaging in a Western ship large enough to be stable in rough seas was usually difficult. Early Mediterranean traders and, later, the Polynesians and the Vikings had used specialized steering oars held against the right side (steer-board = starboard side) of their boats (as can be seen in Figures 2.1 and 2.6). While this worked well in protected waters, the small area of the steering oar (and the exposed position of the steersman) made it difficult to hold a course on long ocean passages. The centrally mounted submerged rudder solved that problem. Also, dividing the ship into separate compartments below the waterline meant that flooding due to hull damage could be confined to a rela-

tively small area of the ship, and the vessel repaired and saved from sinking. Since sails provided the power to move, advances in sail design could drastically influence the success of any voyage. The Chinese fitted their trapezoidal or triangular sails with battens (pieces of bamboo inserted into stitched seams running the width of the sail) and placed the sails on multiple masts. The sails resembled venetian blinds covered with cloth. It was not necessary for Chinese sailors to climb the masts to unfurl the sails every time the wind changed; everything could be done from the deck with windlasses and lines. The shape of the sails made it easier to sail close to the wind in confined seaways.

Despite having these technical advances, the Chinese abandoned oceanic exploration in 1433. The political winds had changed, and the cost of the "reverse tribute" system was judged too great. In all, until this century, the Chinese made very few contributions to our understanding of the ocean. Still, their voyaging technology filtered into the West and made subsequent discoveries possible.

The Age of Discovery: From Prince Henry to Magellan

Having been jolted by internal awakening and external reality, Renaissance Europeans set out to explore the world by sea. They did not undertake exploration for its own sake, however; any voyage had to have a material goal. Trade between east and west had long been dependent on arduous and insecure desert caravan routes through the central Asian and Arabian deserts. This commerce was cut off in 1453 when the Turks captured Constantinople, and an alternate ocean route was needed.

A European visionary who thought ocean exploration held the key to great wealth and successful trade was **Prince Henry the Navigator,** third son of the royal family of Portugal (**Figure 2.9**). Prince Henry established a center at Sagres for the study of marine science and navigation ". . . through all the watery roads." Although he personally was not well traveled (he went to sea only twice in his life), captains under his patronage explored from 1451 to 1470, compiling detailed charts wherever they went. Henry's explorers pushed south into the unknown and opened the west coast of Africa to commerce. He sent out small, maneuverable ships designed for voyages of discovery and manned by well-trained crews. For navigation, his mariners used the **compass**— an instrument (invented in China in the fourth century B.C.) that points to magnetic north. Although Arab traders had brought the compass from China in the twelfth century, navigators still considered it a magical tool. They concealed the compass in a special box (predecessor to today's binnacle) and consulted it out of plain view. Henry's students knew the Earth was round (but because of the errors of Claudius Ptolemy they were wrong in their estimation of its size).

A master mariner (and skilled salesman), **Christopher Columbus,** "discovered" the New World quite by accident. Native Americans had been living on the continent for about 11,000 years, and the Norwegian Vikings had made about two dozen visits to a functioning colony on the continent 500 years before his noisy arrival; yet Columbus gets the credit. Why? Because his interesting souvenirs, exaggerated stories, inaccurate charts, and promises of vast wealth excited the imagination of royal courts. Columbus made North America a media event without ever sighting it!

Columbus wasn't trying to discover new lands. His intention was to pioneer a sea route to the rich and fabled lands of the East, made famous more than 200 years earlier in the overland travels of Marco Polo. As "Admiral of the Ocean Sea," Columbus was to have a financial interest in the trade routes he blazed. He was familiar with Prince Henry's work and, like all other competent contemporary navigators, knew the Earth

Figure 2.9 Prince Henry of Portugal, the Navigator. In the mid-1400s, Henry established a center at Sagres for the study of marine science and navigation ". . . through all the watery roads."

was spherical. By sailing west he believed he could come close to his eastern destination, the latitude of which he thought he knew. Because he depended on Ptolemy's data, however, Columbus made the *smallest* estimate of the size of the Earth by any navigator in modern history; he assumed the Earth to be only about half its actual size.

Not surprisingly, Columbus mistook the New World for his goal of India or Japan. He thought that the notable absence of wealthy cities and well-dressed inhabitants resulted from striking the coast too far north or south of his desired latitude. He made three more trips to the New World but died still believing that he had found islands off the coast of Asia. He never saw the mainland of North America and never realized the size and configuration of the continents whose future he had so profoundly changed.

Other explorers quickly followed, and Columbus's error was soon rectified. Charts drawn as early as 1507 included the New World (**Figure 2.10**). Such charts perhaps inspired **Ferdinand Magellan (Figure 2.11)**, a Portuguese navigator in the service of Spain, to believe that he could open a westerly trade route to the Orient. Unfortunately, the chart makers estimated that the Americas and the Pacific Ocean were much smaller than they actually are. (Compare Figure 2.10 with Magellan's route, shown in **Figure 2.12**.) In the Philippines Magellan was killed, and his men decided to continue sailing west around the world. Only 34 of the original 260 crew survived, returning to Spain three years after they set out. But they had proved it was possible to circumnavigate the globe.

The Magellan expedition's return to Spain in 1522 marks the end of the Age of European Discovery. An unpleasant era of exploitation of the human and natural resources of the Americas followed. Native empires were

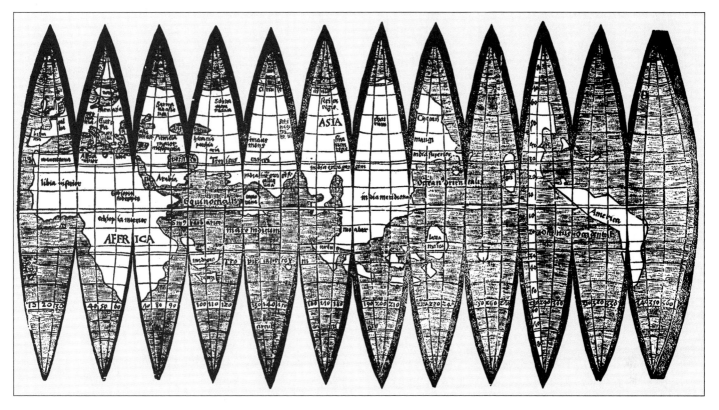

Figure 2.10 The Waldseemüller Map, published in 1507—the first map to name America and to show the New World as separate from Asia. The deep gores are designed to form a globe about 4 inches in diameter.

Figure 2.11 Ferdinand Magellan, a Portuguese navigator in service to Spain, whose expedition was first to circumnavigate the world.

destroyed, and objects of priceless archaeological value were melted into coin to fund European warfare and greed.

VOYAGING FOR SCIENCE

British sea power arose after the Age of Discovery to compete with the colonial aspirations of France and Spain. Sailing ships require dependable supply and repair stations, especially in remote areas. The great powers sent out expeditions to claim appropriate locations, preferably inhabited by friendly natives eager to help provision ships half a globe from home. The French sent Admiral de Bougainville into the South Pacific in the mid-1760s. His 1768 claim for France of what is now called French Polynesia opened the area to the powerful European nations. The British followed immediately.

James Cook

Scientific oceanography begins with the departure from Plymouth Harbor in 1768 of HMS *Endeavour* under command of **James Cook** of the British Royal Navy (**Figure**

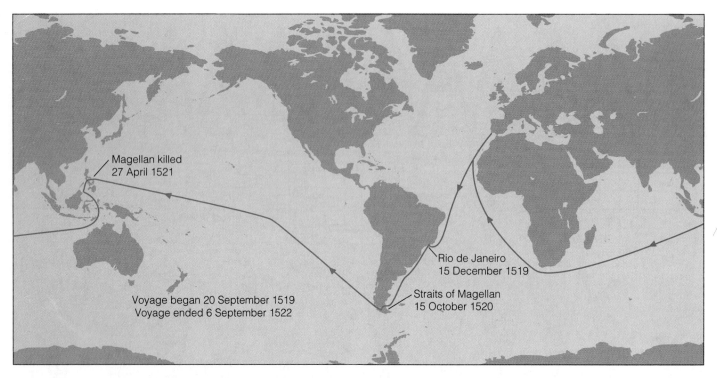

Figure 2.12 Track of the Magellan expedition, the first voyage around the world. Magellan himself did not survive the voyage; only 34 out of 260 sailors managed to return after three years of voyaging.

Figure 2.13 Captain James Cook, Royal Navy.

2.13). An intelligent and patient leader, Cook was also a skillful navigator, cartographer, writer, artist, diplomat, sailor, scientist, and dietician. The primary reason for the voyage was to assert the British presence in the South Seas, but the expedition had numerous scientific goals as well. First, Cook conveyed several members of the Royal Society (a scientific research group) to Tahiti to observe the transit of Venus across the disk of the sun. Their measurements verified calculations of planetary orbits made earlier by Edmund Halley (later of comet fame) and others. Then Cook turned south into unknown territory to search for a hypothetical southern continent, which some philosophers believed had to exist to balance the landmass of the Northern Hemisphere. Cook and his men found and charted New Zealand, mapped Australia's Great Barrier Reef, marked the positions of tens of small islands, made notes on the natural history and human habitation of these distant places, and initiated friendly relations with many chiefs. Cook survived an epidemic of dysentery contracted by ship's company while ashore in Batavia (Djakarta) and sailed home to England around the world in 1771. Because of his insistence on cleanliness and ventilation (and because his provisions included cress, sauerkraut, and citrus extracts), his sailors avoided scurvy—a Vitamin-C deficiency disease that for centuries had decimated crews on long voyages.

The Admiralty was deeply impressed. Cook was promoted to the rank of commander and in 1772 was given

Figure 2.14 First contact: Captain James Cook, commanding HMS *Resolution*, off the Hawaiian island of Kaua´i in 1778. "It required but little address to get them to come along side, but we could not prevail upon any one to come on board; they exchanged a few fish they had in the canoes for anything we offered them, but valued nails, or iron above every other thing; the only weapons they had were a few stones in some of the canoes and these they threw overboard when they found they were not wanted."

command of the ships *Resolution* and *Adventure,* in which he embarked on one of the great voyages in scientific history.[1] On this second voyage he charted Tonga and Easter islands, discovered New Caledonia in the Pacific and South Georgia in the Atlantic. He was first to circumnavigate the world at high latitudes. Though he sailed to 71° south latitude he never sighted Antarctica. He returned home again in 1775.

Promoted to the rank of captain, Cook set off in 1776 on his third, and last, expedition in *Resolution* and *Discovery.* His commission was to find a northwest passage around Canada and Alaska, or a northeast passage above Siberia. He "discovered" the Hawaiian Islands (Hawaiians were there to greet him, of course, as shown in **Figure 2.14**) and charted the west coast of North America. After searching unsuccessfully for a passage across the top of the world, Cook retraced his steps to Hawaii to provision for departure home. On 14 February 1779, after an elaborate farewell dinner with the chief of the island

of Hawaii, Cook and his officers prepared to return to *Resolution* anchored in Kealakekua Bay. The Englishmen somehow angered the Hawaiians and were beset by the crowd. Cook, among others, was killed in the fracas. The tracks of Cook's voyages are shown in **Figure 2.15.**

Cook deserves to be considered a scientist as well as an explorer because of his accuracy, thoroughness, and the completeness in his descriptions. He and the scientists aboard took samples of marine life, land plants and animals, the ocean floor, and geological formations; they also reported their characteristics in their logbooks and journals. His navigation was outstanding, and his charts of the Pacific were accurate enough to be used by the Allies in World War II invasions of the Pacific islands. He drew accurate conclusions, did not exaggerate his findings in his reports, and opened friendly diplomatic relations with many native populations. Cook recorded and successfully interpreted events in natural history, anthropology, and oceanography. Unlike most captains of his day he cared for his men. He was a thoughtful and clear writer. This first marine scientist peacefully changed the map of the world more than any explorer or scientist in history.

---

[1]Sailing master in *Resolution* was the 21-year-old William Bligh, later the object of a famous mutiny.

Figure 2.15 The tracks of Cook's three voyages exploring the Pacific.

Map labels:
Arctic Circle · Siberia · Alaska · ASIA · NORTH AMERICA · 1779 · Tropic of Cancer · Kealakekua (14 February 1779) · Hawaiian Is. · Torres Strait · 1778 · Equator · Batavia · 1771 · 1774 New Hebrides · Tonga · Tahiti · SOUTH AMERICA · 1780 To England · Great Barrier Reef · Coral Sea · New Caledonia · 1773 · Easter Island · Tropic of Capricorn · Queensland AUSTRALIA · 1770 · 1768 From England · 1777 From England · Tasman Sea · NEW ZEALAND · 1774 · 1774 To England · 1773 From England · 1773 · 1769 · Cape Horn · 1774 · Antarctic Circle · ANTARCTICA

First voyage, 1768–1771 → Second voyage, 1772–1775 → Third voyage, 1776–1780

The Longitude Problem

How did Cook (or Columbus, or any ocean explorer) know where he was? Unless explorers could record position accurately on a chart, exploration was essentially useless. They could not find their way home, nor could they or anyone else find the way back to the lands they had discovered.

At night, Columbus and his European predecessors used the stars to find latitude and, as a consequence, knew their position north or south of home. You can do this, too. In the Northern Hemisphere, take a simple protractor and measure the angle between the horizon, your eye, and the north polar star. The protractor reads approximately in degrees of latitude. To find the Indies, for

Figure 2.16 The first chronometers. (a) Harrison's Number One time-keeper. (b) The Number Four timekeeper, which won the £20,000 award offered by the British Board of Longitude. Both are functioning and on display at the National Maritime Museum, Greenwich, England.

example, Columbus dropped south to a line of latitude and followed it west. (This procedure is known to navigators as "running down the latitude.") But to pinpoint a location you need both latitude *and* the east-west position of longitude. (For more about latitude, longitude, and their use in marine navigation, see Appendix III.)

You can find longitude with a clock. First, determine local noon by observing the path of a shadow of a vertical shaft—it is shortest at noon—and set your clock accordingly. After traveling some distance to the west, you will notice that noon according to your *clock* no longer marks the time when the shadow of the *shaft* is shortest at your new location. If "clock" noon occurs three hours before "shaft" noon, you can do some simple math to see how far west of your starting point you have come. The Earth turns toward the east, making one rotation of 360 degrees in 24 hours, so its rotation rate is 15° per hour (360°/24 hours = 15°/hour). The three hours' difference between "clock" noon and "shaft" noon puts you 45° west of your point of origin (3 × 15°). The more accurate the clock (and the measurement of the shaft's shadow), the more accurate your estimate of westward position. In Columbus's time no clocks were accurate enough to make this calculation practical after a few days at sea. How such clocks were invented is a fascinating story.

In 1707 a British battle fleet commanded by Sir Cloudsley Shovel ran aground in the Scilly Islands because they had lost track of their longitude after many weeks at sea; four ships and 2,000 men were lost. This was the most serious in a long string of English maritime disasters involving inaccurate estimation of longitude. In 1714, after lengthy study, the British government offered a £20,000 prize (equivalent to more than $1,800,000 in modern currency) for a method of determining longitude to within one-half degree after 30 days at sea.

The time method described above would work in theory, but existing clocks were too delicate and too inaccurate to meet these standards. The key to the problem was inventing a sturdy clock that ran at a constant rate under any circumstance, even the changeable conditions of a ship at sea.

In 1728, **John Harrison,** a Yorkshire cabinetmaker, began working on such a clock. His radical new timepiece, called a **chronometer,** was governed not by a pendulum (which would be useless in a rolling ship) but by a spring escapement. His first version (**Figure 2.16a**) was tested at sea in 1736, and Harrison was awarded £500 as encouragement to continue his efforts. Over the next 25 years he built three more clocks, culminating in 1760 in his Number Four (**Figure 2.16b**), perhaps the most famous timekeeper in the world.

Figure 2.17 The prime meridian transit circle, Greenwich, England. Zero longitude is marked by the brass rail extending from the instrument in the building toward the foreground.

A sea trial of Number Four was begun in HMS *Deptford* in 1761. Harrison, too old and infirm to accompany the chronometer, sent his son and collaborator to tend the instrument. *Deptford* crossed the Atlantic from England to Jamaica and made a near-perfect landfall. After taking the clock's known error rate into account—its "rate of going" was $2\frac{2}{3}$ seconds a day—the clock was found to be only 5 seconds slow. This would have meant an error in longitude of only 2.3 kilometers (1.4 miles), no small achievement by then-current standards of long-distance navigation.

Technically, Harrison had won the prize—Number Four had more than met the criteria set by the British Board of Longitude—but he was granted only part of the promised reward. Understandably, the officials would not hand over the money until it had been determined that the clock's secrets could be applied to quantity production. Just as understandably, Harrison did not wish his life's work compromised without compensation. He feared (correctly, as it turned out) that once the clockwork was examined by a competent watchmaker, his ideas would be copied. Finally, in 1769, a single copy was made; its success clinched Harrison's achievement. Captain Cook took a chronometer on his last two voyages, but Harrison received the balance of the prize only in 1773 (when he was 80) and then only through the direct intervention of King George III.

All four of Harrison's chronometers are functioning in Britain's National Maritime Museum at Greenwich, in eastern London. Greenwich is an ideal site for the museum; in 1884 the Greenwich meridian, a longitude line at the naval observatory there, became "longitude zero" for the world (**Figure 2.17**). Not since Eratosthenes' selection of Alexandria as the first "longitude zero" had Western nations recognized a common base for positioning.

The Sampling Problem

Marine science advances by the analysis of samples. The chronometer permitted investigators to determine the precise location at which they collected samples of water, bottom sediments, and marine life; but first they had to overcome the difficulty of actually *obtaining* a sample. Sampling of floor sediments or bottom water is not an easy task in the deep ocean. The line used to suspend the sampling device snakes back and forth as currents strike it, and the weight of the line makes it difficult to tell when the sampler has hit bottom. Deploying and recovering the line is laborious and time-consuming, and sometimes the sampling device does not work properly. Early bottom-sampling devices (such as those used by Cook) were simple wax-covered lead weights lowered to shallow bottoms to pick up sediments and test the suitability of anchorages. Later devices took deep-water samples, extracted cores from the sediments, grabbed samples of the bottom, or scooped biological specimens from the ocean floor.

The first researchers to attack the deep-sampling problem successfully were British explorers Sir John Ross and his nephew Sir James Clark Ross (**Figure 2.18**). During an expedition to scout the Northwest Passage in 1818, Sir John Ross obtained a bottom sample from 1,919 meters (3,296 feet) near Greenland by using a clamping sampler to trap the specimen. Sir James Clark Ross, discoverer of the Ross Sea and the area of Antarctica known as Victoria Land, obtained depth **soundings** (depth measurements) of 4,433 meters and 4,893 meters (14,545 feet and 16,054 feet) in the South Atlantic.

Sampling techniques improved through the century. Using a sounding method perfected in the late 1840s by a U.S. Navy midshipman, American commodore Matthew Maury used a long lightweight line and lead weight to discover the Mid-Atlantic Ridge, an important hidden range of mountains. Fridtjof Nansen perfected the deep-water sampling bottle bearing his name near the end of the century. Even today, in spite of modern advances, deep sampling remains difficult. Remotely operated vehicles can work at great depths and return samples and pictures to the surface, but their electronic complexity makes them delicate and expensive to operate.

SCIENTIFIC EXPEDITIONS

Great as his contributions undoubtedly were, Cook's three voyages (and those of the Rosses) were not purely scientific expeditions. These men were British naval officers engaged in Crown business, concerned with charting, foreign relations, and natural phenomena as they applied to Royal Navy matters. The first genuine *only-for-science* expedition may well have been the *Challenger* Expedition of 1872–76, but the United States got into the act first with a hybrid expedition in 1838.

The United States Exploring Expedition

After a 10-year argument over its potential merits, the **United States Exploring Expedition** was launched in 1838. This was primarily a naval expedition, but its captain was somewhat freer in maneuvering orders than Cook had been. The work of the scientists aboard the flagship *Vincennes* and the expedition's five other vessels helped to establish the natural sciences as reputable professions in America. Had it not been for the combative and disagreeable personality of its leader, Lt. Charles Wilkes (**Figure 2.19**), this expedition might have become as famous as those of Cook or the upcoming *Challenger* voyage.

The expedition departed on a four-year circumnavigation. Its goals included showing the flag, whale scouting, mineral gathering, charting, observing, and pure exploration. One unusual goal was to disprove a peculiar theory that the Earth was hollow and could be entered through huge holes at either pole.

Wilkes's team explored and charted a large sector of the east Antarctic coast and made observations that confirmed the landmass as a continent. A map of the Oregon Territory produced in 1841, one of 241 maps and charts drawn by members of the expedition, proved especially valuable when connected to the map of the Rocky Mountains prepared the following year by Captain John C. Fremont. Hawaii was thoroughly explored, and Wilkes led an ascent of Mauna Loa, one of the two highest peaks of Hawaii's largest island. The expedition returned with many scientific specimens and artifacts, which formed the nucleus of the collection of the newly established Smithsonian Institution in Washington, D.C. No evidence of polar holes was found!

Figure 2.18 Sir James Clark Ross, the first researcher to conduct extensive deep soundings.

Figure 2.19 Lt. Charles Wilkes soon after his return from the United States Exploring Expedition.

Figure 2.20 Matthew Fontaine Maury, perhaps the first person for whom oceanography was a full-time occupation. This photograph was probably taken in 1853.

Upon their return, Wilkes and his "scientifics" prepared a final report totaling 19 volumes of maps, text, and illustrations. The report is a landmark in the history of American scientific achievement.

The Work of Matthew Maury

At about the time the Wilkes expedition returned, **Matthew Maury (Figure 2.20)**, a Virginian and fellow U.S. naval officer, became interested in exploiting winds and currents for commercial and naval purposes. After being crippled in a stagecoach accident, in 1842 Maury was given charge of the Navy's Depot of Charts and Instruments. There he studied a huge and neglected treasure trove of ships' logs, with their many regular readings of temperature and wind direction. By 1847 Maury had assembled much of this information into coherent wind and current charts. Maury began to issue these charts free to mariners in exchange for logs of their own new voyages.

Slowly a picture of planetary winds and currents began to emerge. Maury himself was a compiler, not a

scientist, and he was vitally interested in the promotion of maritime commerce. His understanding of currents was built on the work of **Benjamin Franklin.** Nearly 100 years earlier, Franklin had noticed the peculiar fact that the fastest ships were not always the fastest ships; that is, hull speed did not always correlate with out-and-return time on the European run. Franklin's cousin, a Nantucket merchant named Tim Folger, noted Franklin's puzzlement and provided him with a rough chart of the "Gulph Stream" that he (Folger) had worked out. By staying within the stream on the outbound leg and adding its speed to their own, and by avoiding it on their return, captains could traverse the Atlantic much more quickly. It was Franklin who published, in 1769, the first chart of any current (**Figure 2.21**).

But it was Maury who was the first person to sense the worldwide pattern of surface winds and currents. Based on his analysis, he produced a set of directions for sailing great distances more efficiently. Maury's sailing directions quickly attracted worldwide notice: He had shortened the passage for vessels traveling from the American East Coast to Rio de Janeiro by 10 days, and to Australia by 20. His work became famous in 1849 during the California gold rush—his directions made it possible to save 30 days around Cape Horn to California. Applicable U.S. charts still carry the inscription, "Founded on the researches of M. F. M. while serving as a lieutenant in the U.S. Navy." His crowning achievement, *The Physical Geography of the Seas,* a book explaining his discoveries, was published in 1855.

Maury, considered by many to be the father of physical oceanography, was perhaps the first man to undertake the systematic study of the ocean as a full-time occupation.

The *Challenger* Expedition

The first sailing expedition devoted completely to marine science was conceived by Charles Wyville Thomson, a professor of natural history at Scotland's University of Edinburgh, and his Canadian-born student of natural history, John Murray. Stimulated by their own curiosity and by the inspiration of Charles Darwin's 1831–36 voyage in HMS *Beagle,* they convinced the Royal Society and the British government to provide a Royal Navy ship and trained crew for a "prolonged and arduous voyage of exploration across the oceans of the world." Thomson and Murray even coined a word for their enterprise: *oceanography.* Though the term literally implies only marking or charting, it has come to mean the science of the ocean. The government and the Royal Society agreed to the endeavor provided that a proportion of any financial gain from discoveries was handed over to the Crown. This arranged, the scientists made their plans.

HMS *Challenger,* a 2,306-ton steam corvette (**Figure 2.22**), set sail on 21 December 1872 on a four-year voyage that took them around the world and covered 127,600 kilometers (79,300 miles). Although the captain was a Royal Navy officer, the six-man scientific staff directed the course of the voyage. *Challenger's* track is shown as **Figure 2.23.**

One important mission of the *Challenger* **Expedition** was to investigate Edinburgh professor Edward Forbes's contention that life below 300 fathoms (1,800 feet) was impossible because of high pressure and lack of light. The steam winch on board made deep sampling practical, and samples from depths as great as 8,185 meters (26,850 feet) were collected off the Philippines. Through

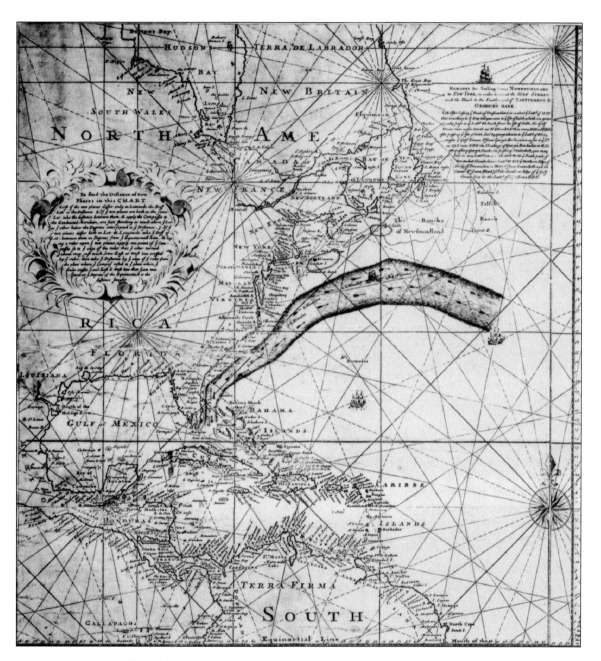

Figure 2.21 Benjamin Franklin's 1769 chart of the Gulf Stream system. His cousin, Timothy Folger, discovered that Yankee whalers had learned to use the Gulf Stream to their advantage. Others, especially English shipowners, were slower to learn. Folger, himself a sea captain, wrote that Nantucket whalers ". . . in crossing it have sometimes met and spoke with those packets who were in the middle of and stemming it. We have informed them that they were stemming a current that was against them to the value of three miles an hour and advised them to cross it, but they were too wise to be counseled by simple American fishermen."

Figure 2.22 Lt. Pelham Aldrich, first lieutenant of HMS *Challenger*, kept a detailed journal of the *Challenger* Expedition. With accuracy and humor he kept this record in good weather and bad, and he had the patience and skill to include watercolors of the most exciting events. This is part of the first page of his journal.

the course of 492 deep soundings with mechanical grabs and nets at 362 stations (including 133 dredgings), Forbes was proven resoundingly wrong. With each hoist, animals new to science were strewn on the deck; in all, staff biologists discovered 4,717 new species! **Figure 2.24** shows one of the large trawl nets used in making some of these discoveries.

The scientists also took salinity, temperature, and water density measurements during these soundings. Each reading contributed to a growing picture of the physical structure of the deep ocean. They completed at least 151 open water trawls and stored 77 samples of seawater for detailed analysis ashore. The expedition collected new information on ocean currents, meteorology, and the distribution of sediments. The locations and profiles of coral reefs were charted. Thousands of pounds of specimens were brought to British museums for study. Manganese nodules, brown lumps of mineral-rich sedi-

ments, were discovered on the seabed, sparking interest in deep-sea mining.

This first pure oceanographic investigation was an unqualified success. The discovery of life in the depths of the oceans stimulated the new science of marine biology. The scope, accuracy, thoroughness, and attractive presentation of the researchers' written reports made this expedition a high point in scientific publication. The *Challenger Report*, the record of the expedition, was published between 1880 and 1895 by Sir John Murray in a well-written and magnificently illustrated 50-volume set; it is still used today. Indeed, it was the 50-volume *Report*, rather than the cruise itself, that provided the foundation for the new science of oceanography. The expedition's many financial spin-offs indicated that pure research was a good investment, and the British government realized quick profits from the exploitation of newly discovered mineral deposits on islands. The *Challenger* Expedition

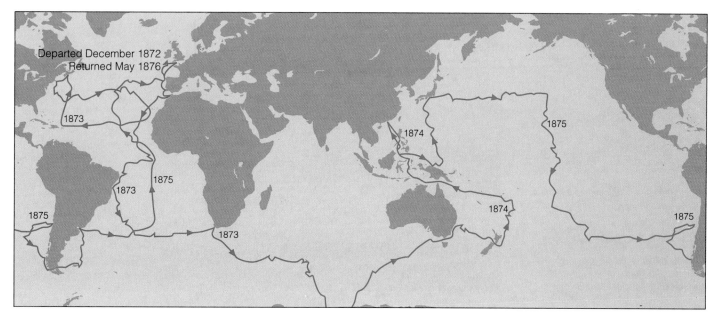

Figure 2.23 HMS *Challenger*'s track, December 1872–May 1876.

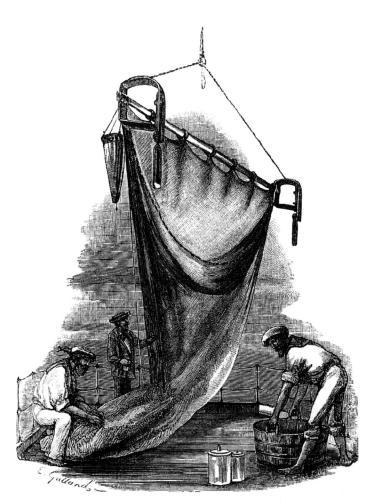

Figure 2.24 Emptying a trawl net, an engraving from the *Challenger Report.*

remains history's longest continuous scientific oceanographic expedition.

With successes like these, the pace of exploration accelerated. American naturalist Alexander Agassiz, sailing in 1877 on the U.S. Coast and Geodetic Survey ship *Blake,* collected data corroborating the *Challenger* material at 355 deep-sea stations. The distribution of manganese nodules was found to be widespread. Further work by Agassiz and his students around the turn of this century on the survey ship *Albatross* helped train a generation of influential American marine biologists. In 1886 the Russians entered the field of marine exploration with the three-year cruise of *Vitiaz* under the leadership of S. O. Makarov; their main contribution was a careful analysis of salinity and temperature of North Pacific water.

TWENTIETH-CENTURY VOYAGING FOR SCIENCE

In this century oceanographic voyages became more technically ambitious and expensive. Scientist-explorers sought out and investigated places that once had been too difficult to reach. New electronic and optical devices aided navigation and sampling. In the last half of the century high-speed shipboard computers made it possible for marine scientists to analyze data while still at sea.

Perhaps polar oceanography made the most dramatic advances early in the century. Newly designed ships and new methods of food storage made polar exploration possible in the last years of the nineteenth century. In

Figure 2.25 The bathyscaphe *Trieste* seen on the surface. *Trieste* reached the ocean's deepest spot in 1960.

1893 Fridtjof Nansen began studying the north polar ocean in *Fram*, a ship designed specifically to withstand the crushing pressure of sea ice. In the next 20 years Nansen and others probed the polar ocean depths. Researchers confirmed the feeding relationships between whales and plankton and collected much data about the whale population of the southern ocean—not out of scientific curiosity, but as a source of oil and baleen (whalebone).

In 1925 the German ***Meteor*** **Expedition**, which crisscrossed the South Atlantic for two years, introduced modern optical and electronic equipment to oceanographic investigation. Its most important innovation was to use an **echo sounder,** a device that bounces sound waves off the ocean bottom, to study the depth and contour of the seafloor (see Figure 3.11). The echo sounder revealed to *Meteor* scientists a varied and often extremely rugged bottom profile rather than the flat floor they had anticipated.

Atlantis, launched in 1931, was the first U.S. research ship built specifically for ocean studies. Investigations by her scientists confirmed Matthew Maury's findings of a mid-Atlantic ridge and helped to discover its extent. The 32-meter (104-foot) schooner *E. W. Scripps,* under the direction of Harald Sverdrup, began a wide-ranging program of chemical, biological, and geophysical exploration off the coast of southern California in 1937. These voyages led to publication in 1942 of *The Oceans,* the first modern reference work on all phases of marine science.

In October 1951 a new HMS *Challenger* began a two-year voyage that would make precise depth measurements in the Atlantic, Pacific, and Indian oceans and in the Mediterranean Sea. With echo sounders, measurements that would have taken the crew of the first *Challenger* nearly 4 hours to complete could be made in seconds. *Challenger II*'s scientists discovered the deepest part of the ocean's deepest trench, naming it Challenger Deep in honor of their famous predecessor. In 1960 U.S. Navy lieutenant Don Walsh and Jacques Piccard descended into the Challenger Deep in *Trieste,* a Swiss-designed, blimplike bathyscaphe (see **Figure 2.25**).

But in many ways the last voyage to be discussed here is the most portentous of all. In 1968 the drilling ship *Glomar Challenger* (see box) set out to test a controversial hypothesis about the history of the ocean floor. It was capable of drilling into the ocean bottom beneath more than 6,000 meters (20,000 feet) of water and recovering samples of seafloor sediments. These long and revealing plugs of seabed rock provided confirming evidence for seafloor spreading and plate tectonics. (The details will be found in Chapter 3.) In 1985, deep-sea drilling duties were taken over by the much larger and more technologically advanced ship *JOIDES Resolution* (see Figure 5.14).[2] She contains equipment capable of drilling in water 8,100 meters (27,000 feet) deep and houses the most completely equipped geological laboratories ever put to sea.

THE RISE OF OCEANOGRAPHIC INSTITUTIONS

The demands of scientific oceanography have become greater than the capability of any single voyage. Oceanographic institutions, agencies, and consortia evolved in

[2]JOIDES stands for Joint Oceanographic Institutions for Deep Earth Sampling

BOX 2.1 ● **Glomar Challenger**

The history of marine science took a long leap forward in 1968 when the drilling ship *Glomar Challenger* returned the first complete cores of deep-sea sediments. A few of these cores—from several sites in the South Atlantic—yielded samples of sediments down to the solid oceanic crust. The oldest sediments, and thus the oceanic crust immediately below, were shown to be surprisingly young—only about 180 million years old. Yet, the oldest continental crust had been dated at more than 3.8 billion years. What could explain this dramatic discrepancy?

Glomar Challenger had been conceived and built in the early 1960s by an international consortium of oceanographic institutions and the U.S. National Science Foundation as part of the Deep Sea Drilling Project. Its goal: to test the then-radical hypothesis that continents are moved across the Earth's surface by seafloor spreading. The relative youth of ocean beds was a central prediction of that theory. Researchers using *Glomar Challenger*'s sophisticated deep-water drilling technology labored for the next 15 years collecting core samples, some of them more than 1,700 meters (5,600 feet) long. The age of these samples confirmed the fact of seafloor spreading, a central principle of the theory of plate tectonics.

In 1983, *Glomar Challenger* was retired, her drilling program successfully completed. She had traveled more than 600,000 kilometers (375,000 miles), had drilled 1,092 holes at 624 sites, and had recovered a total of 96 kilometers (57.6 miles) of deep-sea cores for study. Thanks in large part to her efforts, by the time of her decommissioning the theory of plate tectonics was firmly established in the mainstream of geological understanding.

Glomar Challenger, operated by the Deep Sea Drilling Project from 1968 to 1983. The 122-meter (400-foot) ship used computers to maintain her position with the precision needed to complete the drilling of cores up to a mile long.

a

Figure 2.26 (a) The Woods Hole Oceanographic Institution, Woods Hole, Massachusetts. Marine science has been an important part of this small Cape Cod fishing community since Spencer Fullerton Baird, then assistant secretary of the Smithsonian Institution, established the U.S. Commission of Fish and Fisheries there in 1871. The Marine Biological Laboratory was founded in 1888; the Oceanographic Institution in 1930. (b) The Scripps Institution of Oceanography, La Jolla, California. Begun in 1892 as a portable laboratory-in-a-tent, Scripps was founded by William Ritter, a biologist at the University of California. Its first permanent buildings were erected in 1905 on a site purchased with funds donated by philanthropic newspaper owner E. W. Scripps and his daughter, Ellen.

b

part to ensure continuity of effort. The first of these coordinating bodies was founded by Prince Albert I of Monaco, who endowed his country's oceanographic laboratory and museum in 1906. The most famous alumnus of Albert's Musée Océanographique is Jacques Cousteau, co-inventor in 1943 of the scuba underwater breathing system. Monaco also became the site of the International Hydrographic Bureau, founded in 1921 as an association of maritime nations. This bureau published one of the first general charts of the ocean showing bottom contours.

In the United States, the three preeminent oceanographic institutions are the Woods Hole Oceanographic Institution on Cape Cod, founded in 1930 (and associated with the Massachusetts Institute of Technology and the neighboring Marine Biological Laboratory, founded in 1888); the Scripps Institution of Oceanography, founded in La Jolla, California, and affiliated with the University of California in 1912 (see **Figure 2.26**), and the Lamont–Doherty Earth Observatory of Columbia University, founded in 1949.[3]

[3]There are, of course, other prominent institutions involved in studying the oceans. A list of some of them, and some thoughts on how a student might enter the field, appear in Appendix VI, "Working in Marine Science."

The U.S. government has been active in oceanographic research. Within the Department of the Navy are the Office of Naval Research, the Office of the Oceanographer of the Navy, the Naval Oceanic and Atmospheric Research Laboratory, and the Naval Ocean Systems Command. These agencies are responsible for oceanographic research related to national defense. The National Oceanic and Atmospheric Administration (NOAA), founded within the Department of Commerce in 1970, seeks to facilitate commercial uses of the ocean. NOAA includes the National Ocean Service, the National Weather Service, the National Marine Fisheries Service, and the Office of Sea Grant. The National Aeronautics and Space Administration (NASA), organized in 1958, has also become an important institutional contributor to marine science. For four months in 1978, NASA's *Seasat,* the first oceanographic satellite, beamed oceanographic data to Earth.

More recent contributions have been made by satellites beaming radar signals off the sea-surface to determine wave height, variations in sea-surface contour and temperature, and other information of interest to marine scientists. The first of a new generation of oceanographic satellites was launched in 1992 as a joint effort of NASA and the Centre National d'Études Spatiales (the French space agency). The centerpeice of TOPEX/Poseidon, as the project is known, is a satellite orbiting 1,336 kilometers (835 miles) above the Earth in an orbit that allows coverage of 95% of the ice-free ocean every 10 days. The *TOPEX/Poseidon* satellite (**Figure 2.27**) is supplied with a positioning device that allows researchers to determine its position to within 10 centimeters (4 inches) of the Earth's center. The radars aboard can then determine the height of the sea surface with unprecedented accuracy. Other experiments in this five-year program include sensing water vapor over the ocean, determining the precise location of ocean currents, and determining wind speed and direction. Satellite oceanography is an important frontier, and many discoveries made by satellites are discussed in later chapters.

ALPHABET OCEANOGRAPHY

Oceanography has become big science. Governments and institutions now cooperate in funding research to describe large-scale marine processes, not just describe the nature of a small aspect of the ocean. The largest programs are known by their acronyms—words formed from the first letters of their titles. Here are a few of the most important:

- ODP. The Ocean Drilling Program involves about 20 countries and expends more than $50 million

Figure 2.27 The joint U.S.-French *TOPEX/Poseidon* satellite orbits 1,336 kilometers (835 miles) above the Earth in an orbit that allows coverage of 95% of the ice-free ocean every 10 days. The satellite was launched in 1992 and is supplied with a positioning device that allows researchers to determine its position to within 10 centimeters (4 inches) of the Earth's center. Such accuracy makes possible very accurate determination of sea-surface height by radar transmitters on board.

each year. The successor to the DSDP (Deep Sea Drilling Project) described in Box 2.1, ODP cores the seabed using the drillship *JOIDES Resolution.* ODP's goal is to discover the geological histories of the ocean basins and their margins. ODP's discoveries figure prominently in the history of the ocean floor discussed in Chapter 3.

- WOCE. The aim of the World Ocean Circulation Experiment, part of the world climate research program, is to provide improved models of global ocean circulation (see Chapter 9). It is the largest scientific program ever attempted by physical oceanographers, and it will pool the efforts of scientists from 40 nations. A long-term objective is to predict long-term climate variation. The United States' contribution to WOCE was $19 million in 1992.

- ATOC. Acoustic Thermography of Ocean Climate is planned as a network of acoustic sending and receiving stations around the Pacific Ocean. It was tested in 1991 by a consortium of nine nations; a ship near Heard Island in the southern Indian Ocean produced sounds that were detected off the west coast of the United States, 17,960 kilometers (11,160 miles) away (see Box 7.2). The speed of sound through seawater varies with the temperature of the water, and as its name implies the ATOC project proposes to use sound waves to detect subtle changes in ocean temperature. If it is built, the ATOC network will characterize heat distribution throughout the Pacific and may provide evidence of global warming. There is some concern among marine scientists that the sounds produced by the transmitter will disturb marine animals near the source, and some parts of the project have been redesigned. Tests resumed early in 1995.

- IUSS. In the 1950s the U.S. Navy developed a secret Integrated Undersea Surveillance System that used passive underwater acoustics and sophisticated computers to detect and track submarines over oceanic expanses. Now partially declassified, IUSS (and its SOSUS component—SOund SUrveillance System) consists of a series of ocean-bottom-mounted and towed hydrophones. Acoustic signals they receive are transmitted back to land-based processing centers through underwater communication cables. Information from the hydrophones has been used in the study of underwater earthquakes and the vocalizations of whales. (More on IUSS and SOSUS can be found in Chapter 7.)

- JGOFS. Scientists in the Joint Global Ocean Flux Study have concentrated on the ocean's chemical, physical, and biological processes to increase understanding of the ocean's carbon cycle. Their goals are: to determine what processes control the movement of carbon between the ocean and the atmosphere; and to improve our ability to make global-scale predictions of the likely response of the ocean and atmosphere to human activities.

- TOGA. The Tropical Ocean and Global Atmosphere program is designed to discover variations in the temperature and circulation of the tropical ocean and their relation to atmospheric fluctuations. Study of the causes of the atmospheric and oceanic anomaly known as El Niño (see Chapter 9) is a prime consideration. Together, the U.S. invest-

ment in TOGA and JGOFS is about $30 million annually.

- RIDGE. The Ridge Interdisciplinary Global Experiments (and its international component Inter-RIDGE) are directed at the dynamics of mid-ocean spreading centers. As you will learn in the next chapter, ridges are the boundaries along which new oceanic crust is created. Remotely piloted research vehicles are playing an important role in these studies. The U.S. contribution is about $6 million per year.

- GLOBEC. Global Ocean Ecosystem Dynamics will increase our understanding of the causes of variations in the populations of marine organisms resulting from global climate change. It will also address issues of biological diversity. The U.S. spent about $5 million in 1994.

HISTORY IN THE MAKING

As we have seen, hundreds of marine scientists and their students are using an impressive array of equipment to probe the world ocean. Marine science is by necessity a field science: Ships and distant research stations are essential to its progress. Operating these ships and staffing the research stations is a costly, sometimes dangerous business, yet these researchers are willing to meet the daily challenge. They feel a sense of continuity within their separate specialties because oceanographers must be familiar with the scientific literature, the written history of their fields. History is not an abstract area of interest; rather it is a part of their daily lives. Some of the important milestones in the history of marine science are shown in **Table 2.1.**

One cannot help but feel that the Alexandrian scholars would have appreciated the two-thousand-year effort to understand the oceans. Application of the scientific method to oceanographic studies has yielded many benefits in that time, not the least of which is the satisfaction of knowing a *small* fraction of the story of the Earth and its ocean. The lure of voyaging has been behind many of these discoveries. Scientists in the oceanographic research ships, laboratories, and libraries of the world go on collecting knowledge today. With luck and support, their efforts will continue into the distant future.

Q-AND-A

1. How do we know when people reached certain locations? How do researchers know New Guinea was

Table 2.1   Time Line for the History of Marine Science

Date	Event	Date	Event
4000 B.C.	Egyptian trade on Nile.	1891	Sir John Murray and Alphonse Renard classify marine sediments (see **Chapter 5**).
3800 B.C.	First maps showing water (river charts).	1893	Fridtjof Nansen in Arctic in *Fram* (see **Chapter 18**).
1200 B.C.	Phoenicians trade from Mediterranean to Britain and West Africa.	1900	Richard D. Oldham identifies P and S waves on seismograph (see **Chapter 3**).
1000 B.C.	Polynesians first inhabit Tonga, Samoa.	1906	Prince Albert I of Monaco establishes the Musée Océanographique.
900 B.C.	Greeks first use the term *okeanos,* root of our word *ocean.*	1907	Bertram Boltwood calculates age of Earth by radioactive decay (see **Chapters 1** and **3**).
800 B.C.	First graphic aids to marine navigation.	1911	Roald Amundsen first at South Pole (see **Chapter 18**).
600 B.C.	Greek Pythagoreans assume a spherical Earth.	1912	Alfred Wegener's Frankfurt lectures on continental drift (see **Chapter 3**).
325 B.C.	Pytheas voyages to Britain, links tides to movement of the moon (see **Chapter 11**); Chinese invent the compass.	1912	Scripps Institution allied with the University of California.
300 B.C.	Library founded at Alexandria.	1918	Vilhelm Bjerknes formulates theory of atmospheric fronts (see **Chapter 8**).
230 B.C.	Eratosthenes calculates circumference of the Earth, invents latitude and longitude.	1921	International Hydrographic Bureau founded.
127 B.C.	Hipparchus arranges latitude and longitude in regular grid by degrees.	1925	Departure of *Meteor* Expedition; first echo sounder in operation (see **Chapters 2** and **3**).
A.D. 150	Claudius Ptolemy errs in estimating the Earth's circumference.	1930	Woods Hole Oceanographic Institution founded.
A.D. 415	Alexandrian library destroyed.	1931	*Atlantis* launched.
A.D. 500	Hawaii colonized by Polynesians.	1937	*E. W. Scripps* launched.
A.D. 780	Viking raids begin.	1942	*The Oceans,* first modern reference text, published.
1000	Norwegian colonies in North America.	1943	Jacques Cousteau and Emile Gagnan invent the scuba regulator and tank combination, the "aqualung."
1460	Prince Henry the Navigator dies.		
1492	Columbus's first voyage.	1949	Maurice Ewing forms the Lamont–Doherty Geological Observatory (see **Chapters 2** and **3**).
1522	Magellan's crew completes first circumnavigation.		
1609	Hugo Grotius publishes *Mare Liberum,* the foundation for all modern law of the sea (see **Chapter 19**).	1958	U.S. nuclear submarine *Nautilus* makes first submerged transit of the Arctic ice pack, passes through North Pole (see **Chapter 18**).
1687	Isaac Newton's publication of *Principia Mathematica,* which includes an explanation of the operation of gravity (see **Chapter 11**).	1960	Bathyscaphe *Trieste* carrying Jacques Piccard and Don Walsh reaches bottom of deepest trench at 10,915 meters (35,801 feet) (see **Chapter 4**).
1742	Anders Celsius invents the centigrade temperature scale (see **Chapter 7**).	1962	Rachel Carson's book *Silent Spring* initiates the U.S. environmental movement.
1758	Carolus Linnaeus publishes tenth edition of *Systema Naturae,* in which biological nomenclature is formalized (see **Chapter 15**).	1968	*Glomar Challenger* returns first cores, indicating the age of the Earth's crust. The cores support theories of plate tectonics (see **Chapters 3** and **5**).
1760	John Harrison's Number Four chronometer (see **Chapter 2**).	1969	Santa Barbara, California, oil well blowout captures national attention (see **Chapter 19**).
1768	James Cook's first voyage of discovery (see **Chapter 2**).	1970	National Oceanic and Atmospheric Administration (NOAA) established.
1769	Benjamin Franklin publishes first chart showing an ocean current (see **Chapter 2**).	1970	John Tuzo Wilson writes brief history of the tectonic revolution in geology in *Scientific American* (see **Chapter 3**).
1779	Cook dies in Hawaii.		
1818	John Ross takes first deep-water and sediment samples.	1974	Project FAMOUS (French-American Mid-Ocean Undersea Study) maps and samples the Mid-Atlantic Ridge, a zone of seafloor spreading (see **Chapter 3**).
1831	Charles Darwin departs on five-year voyage aboard HMS *Beagle* (see **Chapter 13**).		
1835	Gaspard Coriolis publishes first papers on an object's horizontal motion across the Earth's surface (see **Chapter 8**).	1977	*Alvin* finds hydrothermal vents in the Galápagos rift (see **Chapters 3, 4,** and **17**).
1836	William Harvey devises a taxonomy of seaweeds (see **Chapter 14**).	1978	*Seasat,* the first satellite dedicated to ocean studies, is launched.
1838	Departure of the United States Exploring Expedition.	1985	*JOIDES Resolution* replaces *Glomar Challenger* in Deep Sea Drilling Project (see **Chapters 2** and **3**).
1847	Hans Christian Oersted observes plankton (see **Chapter 14**).	1985	R. D. Ballard locates wreck of *Titanic.*
1855	Matthew Maury publishes *Physical Geography of the Seas* (see **Chapter 2**).	1987	Observations of supernova 1987A confirm theories of the origin of elements (see **Chapter 1**).
1859	Darwin's *Origin of Species* published (see **Chapter 13**).	1991	JOI researchers bore to a depth of 2 kilometers (1.24 miles) beneath the seafloor near the Galápagos Islands (see **Chapter 3**).
1872	Departure of *Challenger* Expedition.		
1877	Alexander Agassiz begins research in *Blake.*	1992	U.S.-French *TOPEX/Poseidon* satellite launched.
1880	William Dittmar determines major salts in seawater (see **Chapter 6**).	1995	*Keiko,* a small remotely controlled Japanese submersible, sets a new depth record: 10,978 meters (36,008 feet) in the Challenger Deep.
1888	Marine Biological Laboratory founded at Woods Hole, Massachusetts.		
1890	Alfred Thayer Mahan completes *The Influence of Sea Power upon History.*		

inhabited by 30,000 years ago, the tip of South America by 11,400 years ago, and Hawaii by A.D. 450–600?

*Researchers use various methods to date the human artifacts they find. The **radiometric dating** technique, for example, depends on the slow and predictable decrease in radioactivity of naturally radioactive materials. One commonly used form of radiometric dating measures the amount of radioactive carbon in once-living material such as wood or bone. A small amount of the carbon dioxide in the atmosphere contains carbon atoms that have been made radioactive by natural processes. This carbon dioxide is made into glucose and then into cellulose (and wood) by plants. The analysis assumes a fixed ratio of radioactive to non-radioactive carbon dioxide in the air through time. By knowing how the present proportion of radioactive carbon in wood differs from the presumed initial proportion, scientists know how long ago the wood ceased taking up CO_2, and therefore when it stopped living. Animal remains may also be dated this way because animals derive nutrients from photosynthesizing plants. Organic material up to 40,000 years old—including human bones, tools with wood or bone pieces, and food remains—can all be dated using this method.*

2. It would be difficult for humans to walk from Siberia to Alaska today. How was it possible in the past?

The great migrations to the Americas took place at the end of the last ice age, about 13,000 years ago. At that time the large amount of water trapped as ice on the continents caused sea level to fall 100–125 meters (300–400 feet) lower than it is today. The lower sea level exposed land in the Bering Sea and the Aleutian arc between Siberia and Alaska. Lower sea level, combined with the jam of pack ice against the islands themselves, made a passage for migrating game. People followed the animals for food. Both ended up in the "New World" in the process.

3. What's this about a chronometer not having to keep perfect time? I thought you had to know exactly what time it is to be able to calculate your longitude.

Yes, you need accurate time. But a chronometer is valuable not because it necessarily keeps perfect time, but because it loses or gains time at a constant, known rate. Each day the navigator multiplies the number of seconds the clock is known to gain (or lose) by the number of days since the clock was last set—and then adds the total to the time shown on the chronometer's face to obtain the real time. The value of a chronometer lies entirely in its consistency.

4. How do modern navigators find their position at sea?

*Very dull story. They push a few buttons on a box and read their latitude and longitude directly on a screen. This is accomplished by observation of the accelerations experi-*enced by a "black box" since leaving a starting point (inertial navigation) or by analysis of radio transmissions from land stations or satellites. The satellite system is rapidly becoming affordable: For about $350, you can now buy a small, hand-held portable receiver capable of receiving Global Positioning System satellite signals. The GPS system is accurate to about 20 meters! None of these methods is nearly as much fun as the old-fashioned sextant-and-chronometer method, but I suspect that Captain Cook would be **very** impressed by our new tools.*

5. Was Polynesia settled by planned voyages of exploration or by "accident"?

Early twentieth-century interpreters of the migration epics believed that Polynesian navigators "explored the Pacific as a European would a lake." But in the mid-1950s Andrew Sharp, a New Zealand civil servant, challenged that view. His studies concluded that the many islands of Polynesia had been peopled through a long series of accidents by islanders who had become lost, been blown off course, or been driven from their homes by war or famine. Sharp dismissed the reports of intentional exploration and great navigational skills as exaggerations.

*The controversy raged through the 1960s and 1970s, with scientists on both sides of the issue. In 1973 a team of researchers and native sailors decided to design, build, and sail a replica of a voyaging canoe from Hawaii to Tahiti to test the hypothesis that intentional travel had been possible. The double-hulled canoe—unlike any seen on the ocean for hundreds of years—was as close a reproduction as possible given the current state of research into Polynesian craft (see **Figure 2.28**). It would contain no modern navigational tools and would be piloted by Polynesian navigators depending entirely on traditional skills. The canoe was named Hokule´a, "Star of Gladness," after the Hawaiian name for Arcturus, the bright guide star that passes above the Hawaiian chain.*

*Loaded with a full crew, a collection of Polynesian plants, a dog, a pig, two chickens, and enough native food for 30 days at sea, Hokule´a set out for Tahiti on 1 May 1976. For safety reasons a modern yacht shadowed the canoe but shared no navigational data with her crew. On 5 May the Polynesian navigators found that winds and currents were causing the canoe to slip to the west of the intended track. Trimming sails and moving supplies within the hulls changed the canoe's sailing characteristics, and the westward drift was halted. After 30 days at sea they calculated their position within one nautical mile of where the trailing boat's electronically equipped navigator placed them, and three days later on 3 June they sighted Tahiti. Their navigation had been almost perfect (see **Figure 2.29**).*

The success of Hokule´a does not prove that long-distance voyaging was a routine matter to the ancient Poly-

Figure 2.28 The voyage of *Hokule´a. Hokule´a*'s double-hulled design matches as closely as possible that of a Polynesian voyaging canoe from the period of exploration (see Figure 2.6).

nesians; no single experimental voyage can prove a migration theory. Its journey did demonstrate that a Polynesian double-hulled canoe could make long passages against the prevailing winds and currents and that traditional noninstrument navigation could guide a vessel over thousands of miles of open ocean. Yet despite the great skills of their navigators, one cannot help but wonder how many early expeditions failed and how many thousands of voyagers were lost forever in the vastness of the Pacific.

6. Did Columbus discover North America?

No. He never saw North America.

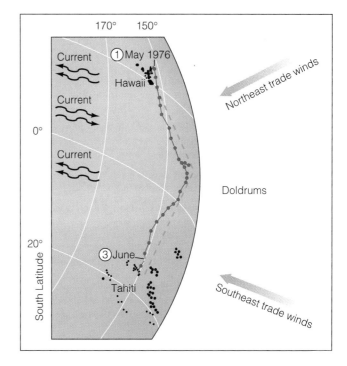

Figure 2.29 *Hokule´a*'s proposed course is marked by a dashed line on the chart and its actual track by a solid line.

Terms and Concepts to Remember

cartographer
celestial navigation
Challenger
   Expedition
chart
chronometer
Columbus,
   Christopher
compass
Cook, James
echo sounder
Eratosthenes of
   Cyrene

Franklin, Benjamin
Harrison, John
Henry the
   Navigator
latitude
Library of
   Alexandria
longitude
Magellan,
   Ferdinand
Maury, Matthew
Meteor Expedition

oceanography
oceanus
Polynesia
radiometric dating
Seasat
sounding
United States
   Exploring
   Expedition
Vikings
voyaging

Study Questions

1. How could you convince a 10-year-old that the Earth is round? What evidence would a child offer that it's flat? How can you counter those objections?

2. What features would be most useful to include in a nautical chart? Why?

3. How did Eratosthenes calculate the approximate size of the Earth? Which of his assumptions was the "shakiest"?

4. How did the Library at Alexandria contribute to the development of marine science? What happened to most of the information accumulated there? Why do you suppose the residents of Alexandria became hostile to the librarians and the many achievements of the library?

5. What were the stimuli to Polynesian colonization? How were the long voyages accomplished? How were Polynesian voyages different from (and similar to) those of the Vikings?

6. What were the main stimuli to European voyages of exploration during the Age of Discovery? Why did it end?

7. If Columbus did not discover North America, then who did?

8. How can you find your approximate latitude and longitude at sea?

9. Imagine that you set your watch at local noon in Kansas City on Monday and then fly to the Coast on Tuesday. You stick a pole into the ground on a sunny day at the beach, wait until its shadow is shortest, and look at your watch. The watch says 10:00 A.M. Are you on the East Coast or the West Coast?

10. What were the goals and results of the United States Exploring Expedition? What U.S. institution greatly benefited from its efforts?

11. What was the first purely scientific oceanographic expedition, and what were some of its accomplishments? What contributions did the earlier, hybrid expeditions make?

12. Sketch briefly the major developments in marine science since 1900. Do individuals, separate voyages, or institutions figure most prominently in this history?

For Further Study

Beaglehole, J. C. 1974. *The Life of Captain James Cook.* Palo Alto, CA: Stanford University Press. The definitive biography of Captain James Cook.

Bellwood, P. S. 1980. "The Peopling of the Pacific." *Scientific American,* November, 174–85.

Bellwood, P. S. 1991. "The Austronesian Dispersal and the Origin of Languages." *Scientific American,* July, 88–93. The voyages of the Polynesians are reflected in their languages.

Boardman, J., et al. 1986. *The Oxford History of the Classical World.* Oxford: Oxford University Press.

Boorstin, D. 1983. *The Discoverers.* New York: Random House. In this elegantly written book, an ex-Librarian of Congress describes our quest to know our world and ourselves. Very informative sections on oceanic exploration and the instruments invented to make discoveries possible.

Ferris, T. 1988. *Coming of Age in the Milky Way.* New York: Morrow. History from a grand perspective. Especially interesting section on the importance of the chronometer.

Finney, B. R. 1979. *Hokule'a: The Way to Tahiti.* New York: Dodd, Mead. The story of the 1970s voyages of a replica Polynesian voyaging canoe.

Garrett, W. E., ed. 1988. "Where Did We Come From?" *National Geographic,* October. The introduction to an issue on "The Peopling of the World," dedicated to human migrations.

Gordon, B. L., ed. 1972. *Man and the Sea: Classic Accounts of Marine Exploration.* New York: Doubleday, American Museum Science Books. Compendium of classic papers from Aristotle, through Franklin and Wegener, to the dives of *Trieste.*

Gould, R. T. 1987. *John Harrison and His Timekeepers.* Greenwich, U.K.: National Maritime Museum. Pamphlet describing how the author restored Harrison's chronometers to working condition and how he researched their history.

Hale, J. R. 1966. *Age of Exploration.* Great Ages of Man. Chicago: Time–Life Books. This clearly written and well-illustrated book is one the best integrated short histories available.

Jennings, J. D., ed. 1979. *The Prehistory of Polynesia.* Cambridge: Harvard University Press.

Kane, H. 1976. *Voyage: The Discovery of Hawaii.* Honolulu: Island Heritage Books. Deftly written and stirringly illustrated book on the Hawaiian colonization. Perhaps hard to find, but not to be missed!

Kane, H. 1991. *Voyagers.* Bellevue, WA: Whalesong. A striking collection of words and paintings by Herbert Kauainui Kane, a leading proponent of Polynesian exploration and interpretation.

LaFay, H. 1972. *The Vikings.* Washington, DC: National Geographic Society. A pictorial history.

Landstrøm, B. 1978. *Sailing Ships.* New York: Doubleday. A master marine architect and artist details the breakthrough ships used in the voyages of discovery. A beautiful book.

Lewis, D. 1972. *We the Navigators.* Honolulu: University of Hawaii Press. A look at Polynesian navigation by a scientist who studied the ancient methods with those few native specialists still alive—and who has tried these methods on the open ocean.

Linklater, E. 1972. *Voyage of the* Challenger. Garden City, NY: Doubleday. Nicely written account, beautifully illustrated with contemporary photographs and excerpts from the logs.

Menard, H. W. 1969. *Anatomy of an Expedition.* New York: McGraw-Hill. The story of the 1965–67 *Nova* expedition. A fine glimpse of a modern oceanographic voyage.

Menard, H. W. 1986. *Islands.* New York: Freeman. The birth, evolution, and death of oceanic islands. Includes information about their colonization.

Morison, S. E. 1942. *Admiral of the Ocean Sea.* New York: Oxford University Press. A positive biography of Columbus.

Morison, S. E. 1978. *The Great Explorers.* New York: Oxford University Press. One of the best discussions of the European discovery of America in print. Morison was himself an American admiral, navigator, and voyager.

Rehbock, P. F. 1992. *At Sea with the Scientifics: The* Challenger *Letters of Joseph Matkin.* Honolulu: University of Hawaii Press. Virtually all of the *Challenger* material was reported by officers and scientists. Here is a below-decks view—the observations and thoughts of an enlisted man. The letters include some honest evaluations of the project and participants, and even a raucous song or two.

Sagan, C. 1980. *Cosmos.* New York: Random House. Shows a reconstruction of the Alexandrian Library and describes its destruction.

Sarton, G. 1959. *A History of Science: Hellenistic Science and Culture in the Last Three Centuries* B.C. Cambridge: Harvard University Press. Heavy going, but a good treatise on Alexandria as a center of learning.

Soule, G. 1970. *The Greatest Depths.* Philadelphia: Macrae Smith. A history of the discovery of, and voyages to, the deepest spots in the ocean.

Temple, R. 1986. *The Genius of China.* New York: Simon & Schuster. A description of 3,000 years of science, discovery, and invention in China—much of which was essential to the explorations undertaken by Western sailors.

Viola, H. J., and C. Margolis, eds. 1987. *Magnificent Voyagers: The United States Exploring Expedition 1838–1842.* Washington, DC: Smithsonian Institution Press.

Wilford, J. N. 1981. *The Mapmakers.* New York: Knopf. An entertaining history of cartography from Eratosthenes to satellite mapping.

3

EARTH STRUCTURE AND PLATE TECTONICS

A Dangerous Breakthrough

Powerful forces inside our planet are continually forming the major features of its surface. These forces determine the outlines and locations of the continents and ocean floors. They build and destroy mountains, raise islands, power volcanoes, form deep trenches, and (through earthquakes) influence the lives of millions of people. Few discoveries in marine science are as exciting to researchers as the recent breakthroughs in our understanding of how these forces work. But intellectual breakthroughs are one thing, physical breakthroughs quite another.

In October 1987 the scientists and crew of the Scripps Institution research vessel *Melville* were present for a breakthrough of sorts—the eruption of an undersea volcano directly beneath their ship! *Melville* was on an expedition to collect rock and water samples from the MacDonald Seamount, a submerged volcano in French Polynesia, located 1,100 kilometers (700 miles) west of Pitcairn Island. When the research team arrived on station they noticed large patches of greenish brown water containing fine particles of volcanic ash, which suggested recent volcanic activity. The crew was lowering

their gear to the top of the seamount, which rises to within 40 meters (130 feet) of the surface, when huge bubbles of gas and steam suddenly engulfed the ship, making "horrendous clangs and clamors" as they burst against *Melville*'s hull. Chocolate-colored water containing steaming lava balls too hot to hold in bare hands streamed to the surface. The ship's depth recorder and hull-mounted water temperature sensor broke down, but no one was injured. The eruption lasted about 5 minutes.

The MacDonald Seamount is the last—and youngest—in a chain of submerged and emergent volcanoes stretching 2,000 kilometers (1,200 miles) across the floor of the South Pacific. Here magma (molten rock) rises from deep inside the Earth. The composition of gases within the young rock would tell *Melville* researchers about the forces and conditions that cause island chains to form, and something about the history of the Pacific floor itself. Samples they obtained that morning helped to confirm recent theories about oceanic *hot spots*, themselves a verification of the theory of plate tectonics. Seldom do researchers have the good fortune to be in exactly the right place at exactly the right time; scientific breakthroughs are generally less threatening to life and property!

In this chapter we investigate the chemical and physical properties of the Earth, the operation of the great forces such as volcanism that have shaped its surface, and some of the smaller forces that continue to surprise scientists. In the next chapter we describe the geologic features generated by these forces.

CHAPTER OVERVIEW

The Earth is composed of concentric spherical layers, with the least dense layer on the outside and the most dense at the core. The layers may be classified by chemical composition into crust, mantle, and core; or by physical properties into lithosphere, asthenosphere, mesosphere, and core. The lithosphere, the outermost solid shell, consists of granitic and basaltic crust bonded to a denser solid region immediately below. It floats on the hot, plastic (deformable) asthenosphere. Geologists have confirmed the existence and basic properties of the layers by analysis of seismic waves that are generated by the forces that cause large earthquakes.

The theory of plate tectonics explains the distribution of earthquake location, the curious jigsaw-puzzle fit of the continents, and the patterns of magnetism in surface rocks. Plate tectonics theory suggests that the Earth's surface is not a static arrangement of continents and ocean, but a dynamic mosaic of jostling lithospheric plates. The plates converge, diverge, and slip past one another, driven by slow, heat-generated currents flowing in the asthenosphere. Most major continental and seafloor features are shaped by plate movement. Plate tectonics explains why our ancient planet has surprisingly young seafloors, the oldest of which is only as old as the dinosaurs (that is, about 1/23 the age of the Earth).

Birth of an island. (Painting by W. K. Hartmann and Ron Miller.)

A LAYERED EARTH

It might seem easy to satisfy our curiosity about the nature of the inner Earth by digging or drilling for samples. The deepest hole drilled so far is being bored by researchers on the Kola Peninsula in Russia. They have reached a depth of 12,063 meters (7.54 miles), where temperatures are 245°C (475°F) and pressure often causes the hole casing to collapse; but the drilling slowly continues. Drilling has also been conducted at sea. The oceanic drilling record is held by scientists of the Joint Oceanographic Institutions' Ocean Drilling Program. In September of 1991, after 12 years of intermittent effort, a drill aboard the ship *JOIDES Resolution* penetrated 2 kilometers (1.25 miles) of seafloor beneath 2.5 kilometers (1.6 miles) of seawater.

No matter where investigators drill, the samples they recover are remarkably similar at all depths: a relatively lightweight solid rock. But the deepest probes have penetrated less than 1/500 of the radius of Earth. Is it reasonable to assume that the Earth consists of lightweight rock all the way through?

Thanks to studies of Earth's orbit begun in the late 1700s, we know the Earth's total mass. This mass is much greater than would have been predicted from even the deepest rocks ever collected. The Earth's interior must therefore contain heavier substances than the rocks from the deep drill holes. We might expect materials from the interior to get denser gradually with increasing depth, but geologists have shown that the density of the materials increases abruptly at specific depths. Thus, they are convinced that the Earth has distinct interior layers, somewhat resembling the inside of an onion.

Density is an expression of the relative heaviness of a substance; it is defined as the mass per unit volume, usually expressed in grams per cubic centimeter (g/cm^3). The density of pure water is 1 g/cm^3. Granite rock is about 2.7 times denser, at 2.7 g/cm^3. The Earth *as a whole* has a density of about 5.5 g/cm^3. As we saw in Chapter 1, it was formed by accretion from a cloud of dust, gas, and stellar debris. Gravity later sorted the components by density, stratifying the Earth into layers. Because each deeper layer is denser than the layer above, we say the Earth is **density stratified.**

Although researchers have never collected samples from below the outermost layer of Earth, they have much indirect evidence about the chemical composition, temperature, and thickness of each layer. They have also gained an understanding of the processes that form the features of Earth's surface by studying the characteristics of the various layers.

Classifying the Layers by Chemical Composition

Early in this century researchers pieced together a view of Earth's interior based on measurements of heat leaking from the interior, the chemical composition of volcanic gases, study of shocks from distant earthquakes, local variations in the pull of gravity, and even the analysis of meteorites thought to have coalesced from the same material as the Earth. They categorized the interior layers of Earth by chemical composition and named the layers *crust, mantle,* and *core* (as **Figure 3.1** shows).

The **crust** is the thin, relatively lightweight outermost layer. It accounts for only 0.4% of the Earth's total mass and less than 1% of its volume. The crust beneath the ocean differs in thickness, composition, and age from the crust of the continents. The thin **oceanic crust** is primarily **basalt**—a heavy, dark-colored rock composed mostly of oxygen, silicon, magnesium, and iron. Its density is about 2.9 g/cm^3. By contrast, the most common material in the thicker **continental crust** is **granite,** a familiar speckled rock composed mainly of oxygen, silicon, and aluminum. Its density is about 2.7 g/cm^3.

The **mantle,** the layer beneath the crust, comprises 68.1% of Earth's mass and 83% of its volume. Mantle materials are thought to contain mainly oxygen, magnesium, and silicon in a 4:2:1 ratio. Its average density is about 4.5 g/cm^3. The mantle is about 2,900 kilometers (1,800 miles) thick.

The **core,** the innermost layer, consists mainly of iron (90%) and nickel, along with silicon, sulfur, and heavy elements. Its average density is about 13 g/cm^3, and its radius is approximately 3,470 kilometers (2,160 miles). The core accounts for about 31.5% of Earth's mass and about 16% of its volume. **Table 3.1** summarizes the characteristics of the layers.

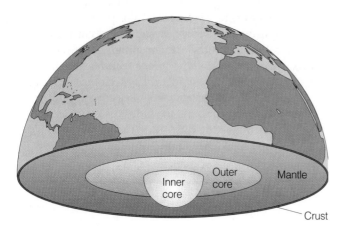

Figure 3.1 The Earth in cross section, showing the major internal layers.

Table 3.1 Characteristics of the Earth's Layers

Layer	Depth from Surface	Primary Composition	Average Temperature	Average Density	Proportion of the Earth's Total Mass	Proportion of the Earth's Total Volume
Crust			~500°C (900°F)		0.4%	<1%
Continental crust	From surface to 35 km (22 mi)	Granite (aluminum silicates)		~2.7 g/cm³		
Oceanic crust	From surface to 11 km (7 mi)	Basalt (magnesium silicates)		~2.9 g/cm³		
Mantle	From 11 km to 2,900 km (from 7 mi to 1,800 mi)	Iron and magnesium silicates	2,500°C (4,500°F)	4.5 g/cm³	68.1%	83%
Core					31.5%	16%
Outer core	From 2,900 km to 5,100 km (from 1,800 mi to 3,200 mi)	Liquid iron and nickel	5,000°C (9,000°F)	11.8 g/cm³		
Inner core	From 5,100 km to 6,370 km (from 3,200 mi to 3,960 mi)	Solid iron and nickel	6,600°C (12,000°F)	16.0 g/cm³		
Whole earth	6,370 km (3,960 mi)			5.5 g/cm³		

Sources: Kennett, 1982; Bott, 1982.

Classifying the Layers by Physical Properties

Subdividing the Earth on the basis of chemical composition does not necessarily reflect the *physical* properties and behavior of rock materials in these layers, however. Different conditions of temperature and pressure prevail at different depths, and these conditions influence the physical properties of the materials subjected to them. The behavior of a rock is determined by three factors: temperature, pressure, and the rate at which a deforming force (stress) is applied. Depending on these three factors a rock may behave in a brittle manner (by breaking), it may deform plastically (flow without fracturing), or it may deform elastically (bend or shrink but return to its original shape once the stress is released).[1]

Recent research has shown that slabs of the Earth's relatively cool and solid crust and upper mantle float—and move—independently of one another over the hotter, partially molten mantle layer directly below. Physical properties are more important than chemical ones in determining this movement; so geologists have devised a classification based on physical properties:

The **lithosphere** (*lithos* = rock)—the Earth's cool, rigid outer layer—may be up to about 100 kilometers (60 miles) thick. It is comprised of the brittle continental and oceanic crusts *and* the uppermost cool and rigid portion of the mantle.

The **asthenosphere** (*asthenos* = soft) is the thin, hot, slowly flowing layer of upper mantle below the lithosphere. Extending to a depth of up to 700 kilometers (430 miles), the asthenosphere is characterized by its ability to deform plastically under stress. Its thick fluidity has been compared to cold taffy.

The **mesosphere** (*mesos* = middle) is the rigid middle and lower mantle extending to the core. Though it is hotter than the asthenosphere, the greater pressure at this depth probably prevents it from flowing. The mesosphere and asthenosphere have a similar chemical composition but very different physical properties.

The **core** is divided into two parts: the outer core is a viscous liquid with a density of about 11.8 g/cm³; the inner core is a solid with a maximum density of 16 g/cm³. Both parts are extremely hot, with an average temperature of about 5,500°C (9,900°F). Recent evidence indicates that the core may be as hot as 6,600°C (12,000°F) at its center, hotter than the surface of the sun! The tremendous pressure at this depth keeps the white-hot inner core solid.

Figure 3.2 compares the chemical and physical classifications of the layers, and **Figure 3.3** shows the litho-

[1]This may help: A candy cane is brittle, Silly Putty is plastic, and a rubber ball is elastic.

Earth's Layers Classified by
Chemical Composition

Earth's Layers Classified by
Physical Properties

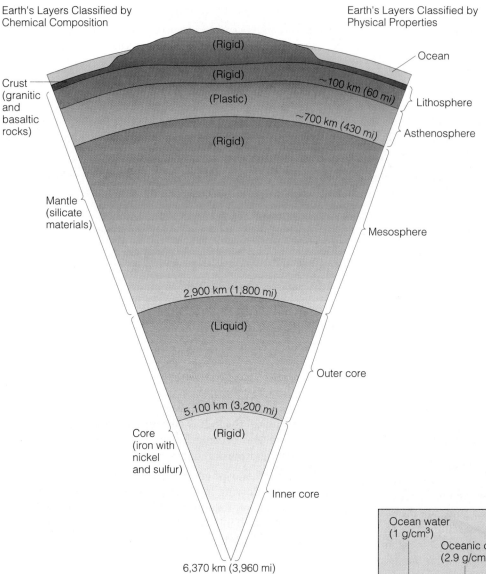

Crust
(granitic
and
basaltic
rocks)

(Rigid)

(Rigid)

(Plastic)

(Rigid)

Ocean

~100 km (60 mi)

Lithosphere

~700 km (430 mi)

Asthenosphere

Mantle
(silicate
materials)

Mesosphere

2,900 km (1,800 mi)

(Liquid)

Outer core

5,100 km (3,200 mi)

(Rigid)

Core
(iron with
nickel
and sulfur)

Inner core

6,370 km (3,960 mi)

Figure 3.2 The chemical and physical organization of the Earth's layers compared. Note that this representation is not to scale. The lithosphere accounts for only about 1.5% of Earth's radius.

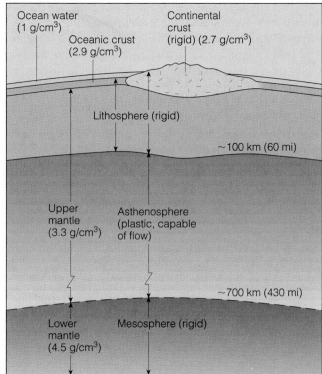

Ocean water
(1 g/cm^3)

Continental
crust
(rigid) (2.7 g/cm^3)

Oceanic crust
(2.9 g/cm^3)

Lithosphere (rigid)

~100 km (60 mi)

Upper
mantle
(3.3 g/cm^3)

Asthenosphere
(plastic, capable
of flow)

~700 km (430 mi)

Lower
mantle
(4.5 g/cm^3)

Mesosphere (rigid)

Figure 3.3 The layering of the lithosphere and asthenosphere in more detail. Densities are shown in grams per cubic centimeter (g/cm^3).

sphere and asthenosphere in more detail. Note in Figure 3.3 that the rigid sandwich of crust and upper mantle—the lithosphere—floats on (and is supported by) the denser plastic asthenosphere. Note also that the structure of oceanic lithosphere differs from that of continental lithosphere. Below the ocean, the crust averages only about 7 kilometers (4 miles) thick, but beneath the continents the crust thickness averages 35 kilometers (22 miles) and increases to 70 kilometers (44 miles) at the highest mountain ranges. The thick granitic continental crust is light enough to project above sea level, but the thin, heavy basaltic oceanic crust is almost always submerged.

Isostatic Equilibrium

Why do large regions of continental crust stand high above sea level? If the asthenosphere is nonrigid and deformable, why don't mountains sink because of their weight and disappear? Another look at Figure 3.3 will help to explain the situation. The mountainous parts of continents have "roots" extending into the asthenosphere. The continental crust and the rest of the lithosphere "float" on the denser asthenosphere. The situation is analogous to buoyancy, the principle that explains why ships float.

Buoyancy is the ability of an object to float in a fluid by displacing a volume of that fluid equal in mass to the floating object's own mass. A steel ship floats because its shape displaces a volume of water equal in weight to its own weight plus the weight of its cargo. Thus, an empty containership displaces a smaller volume of water than the same ship when fully loaded (**Figure 3.4**). The water supporting the ship is not *strong* in the mechanical sense; water does not support a ship the same way a steel bridge supports the weight of a car. Buoyancy, rather than mechanical strength, supports the ship and her cargo.

Any region of a continent that projects above sea level is supported in the same way. As an extreme example, consider the continent containing Mount Everest, highest of Earth's mountains at 8.84 kilometers (29,007 feet) above sea level. Mount Everest and its neighboring peaks are not supported by the *mechanical* strength of the materials within the Earth; nothing on (or in) our planet is that strong. The mountainous upper surface of the continent floats high above sea level because the lithosphere of which it is a part sinks into the plastic asthenosphere until it has displaced a volume of asthenosphere equal in mass to its own mass. The continent's mountains rest at great height, in balance with their subterranean underpinnings but susceptible to rising or falling as erosion or crustal stresses dictate. Lower regions are supported by shallower roots. Like a ship floating in water, the entire continent stands in **isostatic equilibrium** (*isos* = equal, *stasis* = standing).

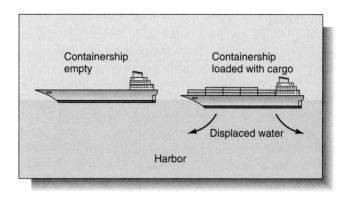

Figure 3.4 The principle of buoyancy. A ship sinks until it displaces a volume of water equal in weight to the weight of the ship and its cargo.

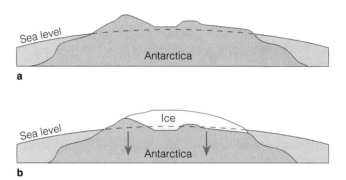

Figure 3.5 The principle of isostatic equilibrium applied to Antarctica. (a) Without an ice cover, the continent would rise higher above sea level, in a situation analogous to the empty containership in Figure 3.4. (b) The great weight of ice blanketing the continent forces its roots deeper into the asthenosphere, so that more of its mass is below sea level. The ice is analogous to the loaded cargo in Figure 3.4.

Antarctica provides an excellent example of how continents maintain isostatic equilibrium in the face of changing conditions. Most of the southern continent is covered by a layer of ice as much as 4,000 meters (13,000 feet) thick. The great weight of the ice has pushed much of the continental crust below sea level (**Figure 3.5**). If this ice were to melt, Antarctica would slowly rise from the water in much the same way a ship rises while being unloaded.

During the most recent ice age, the weight of ice pressed Canada, Scandinavia, and Siberia deeper into the mantle. In the past 18,000 years these areas have gradually rebounded about 500–600 meters (1,600–2,000 feet) as the ice has melted, but they must rise another 200 meters (650 feet) to achieve complete isostatic equilibrium.

Unlike the asthenosphere on which lithosphere floats, crustal rock does not flow at normal surface temperatures. A ship reacts to any small change in weight with a correspondingly small change in vertical position in the water, but an area of continent or ocean floor cannot react to every small weight change because its edges are mechanically bound to adjacent crustal masses. When the force of uplift or downbending exceeds the mechanical strength of the adjacent rock, the rock will fracture along a plane of weakness—a **fault.** The adjacent crustal fragments will move vertically in relation to each other. This sudden adjustment of the crust to isostatic forces by fracturing, or faulting, is one cause of earthquakes. The lithosphere does not always behave as a rigid, brittle solid, however. Where forces are applied slowly enough, some of this material may deform plastically.

Sources of Internal Heat

Heat from within Earth keeps the asthenosphere pliable and the lithosphere in motion. Studies of the sources and effects of Earth's internal heat have contributed to our understanding of the planet's internal construction. In the late 1800s the British mathematician and physicist William Thompson (Lord Kelvin) calculated that the Earth was about 80 million years old, an estimate based on the rate at which the planet would have cooled from an original molten mass. Geologists tried to explain mountain building based on Lord Kelvin's assumption of progressive cooling. In their drying-fruit model, the Earth was considered to have shrunk as it cooled. Mountains were thought to be shrinkage wrinkles—like those seen as a grape transforms into a raisin—and earthquakes were thought to be side effects of this wrinkling. The Earth's true age was not then known, and the "wrinkle" theory depended on rapid cooling of a relatively young Earth. When further research showed that the Earth is about 4.6 billion years old (see Chapter 1), calculations of heat flow indicated that our planet should have cooled almost completely by now. Earthquakes, active volcanoes, and hot springs clearly indicated it had not! There must be other sources of heat energy besides the trapped ancient heat of formation. Better explanations for mountain building and earthquakes were needed.

An important source of heat that was not recognized in Lord Kelvin's time is **radioactive decay.** This process generates heat when unstable forms of elements are transformed into new elements. As we saw in Chapter 1, radioactive decay within the newly formed Earth released heat that contributed to the melting of the original mass. Most of the melted iron sank toward the core, releasing huge amounts of energy. By now almost all of the heat generated by the formation of the core has dissipated, but radioactive elements within the Earth still continue to decay and produce new heat. Today most of the radioactive heating takes place in the crust and upper mantle rather than in the deeper layers.

Some of Earth's internal heat journeys toward the surface by **conduction,** a process analogous to the slow migration of heat along a skillet's handle. Some heat also rises by **convection** in the asthenosphere. Convection occurs when a fluid is heated, expands and becomes less dense, and rises. (Convection causes air to rise over a warm radiator.)

Thus, even after 4.6 billion years, heat continues to flow out from within the Earth. As we shall see, this heat, not raisinlike global shrinkage, builds mountains and volcanoes, causes earthquakes, moves continents, and shapes ocean basins.

THE EVIDENCE FOR LAYERING

It has been known since the mid-1800s that low-frequency waves can travel through the interior of the Earth. The forces that cause **earthquakes** generate low-frequency waves called **seismic waves** (*seismos* = earthquake). Some of these waves radiate through the Earth, reflecting or bending as they travel, and eventually re-appear at the surface. Careful study of the time and location of their arrival at the surface, along with changes in the frequency and strength of the waves themselves, has revealed the information about the chemical and physical nature of Earth's interior discussed earlier in this chapter. (We use the same kind of analysis to select a ripe watermelon. If we tap the outside and hear a *tick*, we suspect the melon isn't ripe. A *thunk* indicates a winner.)

Primary and Secondary Seismic Waves

Two kinds of seismic waves are particularly useful for analyzing the Earth's interior structure. One kind of wave, the **P wave** (or primary wave), is a compressional wave similar in behavior to a sound wave. Rapidly pushing and pulling a very flexible spring (like a Slinky) generates P waves. The **S wave** (or secondary wave) is a transverse wave like that seen in a rope shaken side to side. Both kinds of seismic waves are shown in **Figure 3.6.**

P waves and S waves are generated simultaneously at the source of an earthquake. P waves travel through the Earth nearly twice as fast as S waves; so they arrive first at the **seismometer** (*seismos* = earthquake, *mete* = to measure), an instrument that senses and records earthquakes. Liquids are unable to transmit the side-to-side force of S waves but do propagate compressional P waves. Solid rock transmits both kinds of waves. Thus, analysis of the characteristics of seismic waves returning to the Earth's surface after passage through the interior suggests which parts of the interior are solid, liquid, or plastic. (Note that P waves and S waves don't cause damage. Surface waves of a different kind knock down buildings and endanger people.)

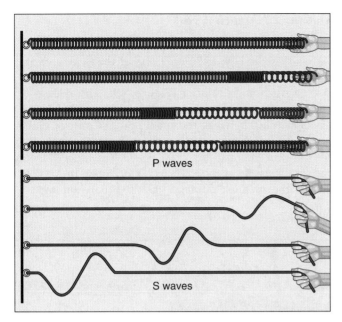

Figure 3.6 P waves (primary waves) are compressional waves like those seen in a Slinky that is alternately stretched and compressed. S waves (secondary waves) are side-to-side waves like those seen in a shaken rope. Both kinds of waves are associated with earthquakes.

Shadow Zones

In 1900 the English geologist Richard Oldham first identified P and S waves on a seismograph. If Earth were perfectly homogeneous, seismic waves would travel at constant speeds from an earthquake, and their paths through the interior would be straight lines (**Figure 3.7a**). Oldham's investigations, however, showed that seismic waves were arriving *earlier* than expected at seismographs far from the quake. This meant that the waves must have traveled *faster* as they went down into the Earth. They must also have been refracted, bent back toward the surface. Oldham reasoned that the waves were being influenced by passage through areas of Earth with different elastic properties than those seen at the surface. Thus it was found that the Earth is not homogeneous and that its elastic properties vary with depth.

In 1906 Oldham made another discovery. He found that P waves arrived at a seismograph farthest away from an earthquake (that is, on the opposite side of the globe) much more slowly than expected. He also found that no S waves survived deep passage through the Earth. Oldham deduced that a dense fluid structure, or core, must exist within the Earth to absorb the S waves and slow down the P waves. He further predicted that a **shadow zone (Figure 3.7b)**, a wide band from which seismic waves were nearly absent, would encircle the side of the Earth opposite the location of the earthquake. The shadow zone would be formed by refraction of the P

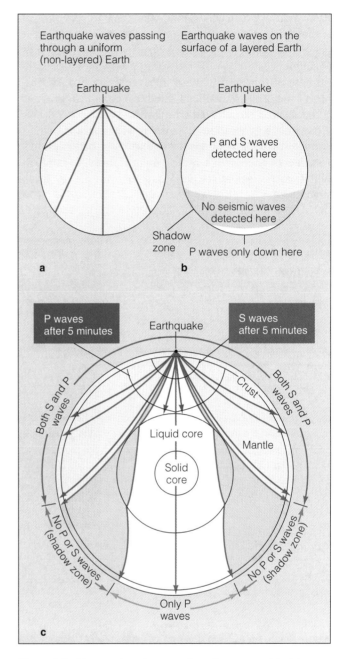

Figure 3.7 How earthquakes contributed to our model of the layered Earth. (a) If the Earth were homogeneous throughout, seismic waves would travel in straight-line paths at constant speed. (b) Actually, the Earth has a dense core, producing a shadow zone in which no seismic waves are detected. (c) The patterns of reflected and refracted waves helped geophysicists deduce the structure of the Earth's layers.

waves by the outer liquid core, as shown in **Figure 3.7c**. The existence of the shadow zones and liquid core was verified by seismographic analysis in 1914.

More sensitive seismographs were developed in the 1930s. In 1935 Dutch seismologist Inge Lehmann suggested that the very faint, very low-frequency P waves discovered opposite the earthquake site had speeded up as they passed through an inner core, indicating that it was a solid (see again Figure 3.7c). Measurements of subtle differences in the pull of gravity, plus a more accurate estimate of the Earth's mass (derived from precise timings of the orbits of artificial satellites) gave further clues to the layered structure of the Earth. By the early 1960s another new generation of sensitive seismographs stood ready to provide geologists with an even better understanding of the Earth's inner configuration. To confirm their theories, scientists needed data from a very large earthquake.

Alaska, 27 March 1964

They did not have to wait long. On the last Friday of March 1964, at 1736 (5:36 P.M.) one of the largest earthquakes ever recorded struck 144 kilometers (90 miles) east of Anchorage, Alaska (**Figure 3.8**). The release of energy ruptured the surface of the Earth for 800 kilometers (500 miles) between the small port of Cordova in the east and Kodiak Island in the west. In some places the vertical movement of the crust was 3.7 meters (12 feet); one small island was lifted 38 feet. Horizontal movement caused the greatest damage: 65,000 square kilometers

Figure 3.8 Alaska earthquake damage, 27 March 1964. (a) The collapse of Fourth Avenue in Anchorage. The earthquake caused the ground to liquefy and slide, undermining buildings and causing the pavement to sink more than 3 meters (10 feet) in places. (b) Wreckage of Government Hill School. The fissures in the playground are about 3.7 meters (12 feet) deep. The water tower in the background somehow withstood the violent forces of the earthquake.

(25,000 square miles) of land abruptly moved west. In 4½ minutes of violent shaking, Anchorage had moved sideways 2 meters (6.6 feet), and the town of Seward had moved 14 meters (46 feet)! A seismic (earthquake-generated) sea wave destroyed two harbors. Over 75% of the state's commerce was disrupted, and thousands of people were made homeless. Damage exceeded $750 million, and, considering the violence of the earthquake, it is a wonder that only 115 lives were lost.

Seismic stations over much of the world saw the extraordinarily large P waves arrive, and many of the 800 seismographs on-line worldwide were physically damaged. The arrival times of the P and S waves at each station were carefully noted. When correlated with the frequency, intensity, and phase characteristics of the waves, these time data helped to confirm the models of Earth layering. The shaken citizens of Anchorage probably didn't derive much comfort from the knowledge gained from their earthquake.

TOWARD A NEW UNDERSTANDING OF EARTH

The Earth's layered internal structure—its brittle lithosphere floating on the hot and viscous asthenosphere—ensures that its surface will be geologically active. Earthquakes and volcanoes attest to the commotion within. But how do internal layering and heat contribute to mountain building, the arrangement of continents, the nature of the seafloors, and the wealth of geological features found everywhere? Are there patterns and order in this apparent chaos? Let us see how we achieved our current understanding of the answers to these questions.

The Age Debate

Geologists try to read the Earth's history in its surface rocks and features. Even today this difficult task is often complicated by seemingly contradictory data. In the past the effort to understand the Earth was especially hindered by an incomplete understanding of its age. Advances in geology had to wait, quite literally, for the time to be right.

At the end of the eighteenth century most European natural scientists believed in a young Earth, one that had formed only about 6,000 years ago. This age had been determined not by an analysis of rocks, but through the genealogy of the Bible's Old Testament. Careful reading of the book of Genesis in 1654 had convinced Irish bishop James Ussher that the Creation had taken place on 26 October 4004 B.C.

James Hutton—a Scottish physician with an interest in geology—decided that the biblical account of the Creation was incorrect because it implied that the landscape was mostly stable and unchanging. Hutton, however, had measured geological changes: the rate at which stream beds eroded, the distribution of sediments by rivers, and the patterns of rocks in the Scottish countryside. From his observations, Hutton concluded that the rate of geological change today is not greatly different from the rate of change in the past. His principle of **uniformitarianism,** formalized in 1788, suggested that all of Earth's geological features and history could be explained by processes identical to ones acting today, and that these processes must have been at work for a very long time.

A few scientists agreed with Hutton. His detractors, however, asked some painful questions: If Earth is very old, and if erosive forces have continued uniformly through time, why isn't Earth's surface eroded flat? Why isn't the ocean brimming with sediment? What post-Creation forces could build mountains?

Believers in another school of thought, **catastrophism,** were able to answer these objections by interpreting the biblical account of the Creation literally. The catastrophists were convinced that Earth was very young and that the biblical flood was responsible for the misleading appearance of Earth's great age. The flood, they maintained, had folded and exposed strata, toppled mountains, filled shallow ocean basins with sediments, and caused many plants and animals to become extinct. This theory had the additional benefit of explaining how seashell fossils could be present on mountaintops.

A further complication was introduced in 1859 with Charles Darwin's publication of *On the Origin of Species.* This work proposed a rational mechanism, natural selection, by which new kinds of living things might come about. Since natural selection required long periods of time to generate the overwhelming variety of life-forms on Earth, biological evidence also suggested an ancient Earth. The arguments intensified.

Scientists attempting to prove or disprove uniformitarianism, catastrophism, and evolution made a wealth of new discoveries during the last half of the nineteenth century. They improved the seismograph, discovered long-distance earthquake (seismic) waves, probed the ocean floors, drew more accurate charts using improved navigational techniques, collected mineral samples from great heights and depths, measured the flow of heat from within the Earth, and identified patterns in the worldwide distribution of fossils. All the theories were reassessed. The evidence from these explorations convinced most researchers that Earth was indeed of great age. The stage was now set for a revolution in geology, the development of what we know today as the theory of plate tectonics. The first steps toward the theory were tentative, however, and some of its proponents were dismissed as lunatics.

A Puzzling Fit

As Leonardo da Vinci noticed on early charts, in some regions the continents looked as if they would fit together like jigsaw-puzzle pieces if the intervening ocean were removed. In 1620 Francis Bacon also wrote of a "certain correspondence" between shorelines on either side of the South Atlantic. In 1885 Edward Suess, a respected German scientist, suggested that the Southern Hemisphere's continents might once have been a single large land mass. He based his belief in part on the similarities of fossils found on these continents, especially fossils of the fern *Glossopteris*. Suess was not taken seriously by his colleagues because he could not explain how the continents had moved. Still, the correspondence of South America and Africa had a certain graphic appeal (**Figure 3.9**).

As they probed the submerged edges of the continents, geologists found that the ocean bottom nearly always sloped gradually out to sea for some distance and then dropped steeply to the deep-ocean floor. They realized that these shelflike continental edges were extensions of the continents themselves. In the few locations where they had measurements, researchers found that the fit between South America and Africa, impressive at the shoreline, was even better along the submerged edges of the continents.

Though so accurate a fit almost certainly could not have occurred by chance, no one had yet proposed a mechanism that could separate whole continents into moving pieces. If such a mechanism did exist, its gradual operation would surely require a great deal of time.

Continental Drift

Into the fray stepped **Alfred Wegener,** a busy German meteorologist, polar explorer, astronomer, and geologist. In a lecture in 1912 he proposed a startling and original theory, **continental drift.** Wegener suggested that all the Earth's land had once been joined into a single supercontinent surrounded by an ocean. He called the land mass **Pangaea** (*pan* = all, *gaea* = Earth) and the surrounding ocean **Panthalassa** (*pan* = all, *thalassa* = ocean). Wegener thought Pangaea had broken into pieces about 200 million years ago. Since then, the pieces had moved to their present positions and were still moving.

Wegener's evidence included the apparent shoreline fit of continents across the North and South Atlantic and new information on offshore contours obtained by contemporary oceanographic expeditions. He pointed to

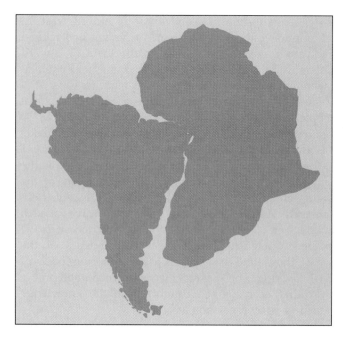

Figure 3.9 Soon after charts began including the New World, some scientists noticed that the coastlines of South America and Africa seemed to fit together in jigsaw-puzzle fashion.

Suess's *Glossopteris* fossils; to areas of erosion apparently caused by the same glacier in tropical areas now widely separated (South Africa, India, and Australia); and to Ernest Shackleton's 1908 discovery of coal, the fossilized remains of tropical plants, in frigid Antarctica. Wegener even suggested that volcanic activity was powered by the friction of continental movement.

Unlike anyone before him, Wegener also proposed a mechanism to account for the hypothetical drift. He believed that the heavy continents were slung toward the equator on the spinning Earth by a centrifugal effect. This force, coupled with the tidal drag on the continents from the combined effects of sun and moon, would account for the phenomenon of drifting continents, he thought.

Wegener was dismissed as a crank. His detractors claimed, with some justification, that he had carefully selected only those data supporting his hypothesis, ignoring contrary evidence. Where, for instance, were the wakes or tracks through old seabed that the migrating continents would leave? But a few geologists sided with Wegener. These "drifters" were hesitant to embrace the centrifugal force theory, yet they were unable to propose an alternate power source that could move the massive granitic continents.

The greatest block to the acceptance of continental drift lay in geologists' view of the Earth's mantle. The available evidence seemed to suggest that a deep, solid

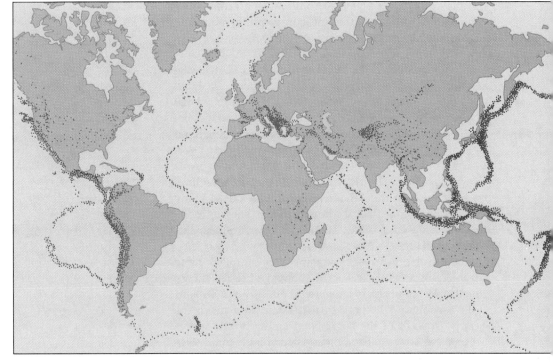

Figure 3.10 Seismic events worldwide, January 1977 through December 1986. The locations of about 10,000 earthquakes are colored red, green, and blue to represent event depths of 0–70 kilometers, 70–300 kilometers, and below 300 kilometers, respectively.

mantle supported the crust and mountains mechanically (not isostatically) from below. Drift would be impossible with this kind of subterranean construction. But then a few perceptive seismic researchers noticed that the upper mantle reacted to earthquake waves as if it were a plastic mass, not a rigid solid. Perhaps such a layer would resemble a slug of iron heated in a blacksmith's forge; it would deform with pressure and even flow slowly. Established geologists, however, dismissed this interpretation, saying that the mountains would simply fall over or sink without rigid underpinnings. By 1926 the "drifters" were in full retreat. When Wegener died on an expedition across Greenland in 1930, his theory was already in eclipse.

The Idea Transformed

The concept of continental drift refused to die, however; those neatly fitted continents provided a haunting reminder of Wegener to anyone looking at an Atlantic chart. In 1935 a Japanese scientist, Kiyoo Wadati, speculated that earthquakes and volcanoes near Japan might be associated with continental drift. In 1940, seismologist Hugo Benioff plotted the locations of deep earthquakes at the edges of the Pacific. His charts revealed the true extent of the **Pacific Ring of Fire,** a circle of violent geologic activity surrounding much of the Pacific Ocean. Seismographs were now beginning to reveal a worldwide pattern of earthquakes and volcanoes. Deep earthquakes did not occur randomly over the Earth's surface but were concentrated in zones that extended in lines along the Earth's crust.

Benioff, Wadati, and others wondered what could cause such an orderly pattern of deep earthquakes. Many of the lines corresponded with a worldwide system of oceanic ridges, the first of which was plotted in 1925 by *Meteor* oceanographers working in the middle of the North Atlantic. **Figure 3.10** is a plot of about 10,000 earthquakes. Notice the odd pattern they form—almost as if the Earth's lithosphere is divided into sections! Benioff's sensitive seismographs also began to gather strong evidence for a partially molten, nonrigid layer in the upper mantle. Could the continents somehow be sliding on that?

Other seemingly unrelated bits of information were accumulating. **Radiometric dating** of sediments and rocks was perfected after World War II. This technique is based on the discovery that unstable, naturally radioactive elements lose particles from their nuclei and change into new stable elements. The radioactive decay occurs at a constant rate, and measuring the ratio of radioactive to stable atoms in a sample provides its age. To the surprise of many geologists, the maximum age of the ocean floor and its overlying sediments was radiometrically dated to less than 200 million years, only about 4% of the age of Earth. The centers of the continents are much older; some parts of the continental crust are more than 3.9 billion years old, about 85% of the age of Earth. *Why was oceanic crust so young?*

Attention quickly turned to the deep-ocean floors, the

complex profiles of which were now being revealed by **echo sounders,** devices that measure depth by bouncing high-frequency sound waves off the bottom (see **Figure 3.11**). In particular, scientists aboard the Lamont–Doherty Geological Observatory deep-sea research vessel *Vema* (a converted three-masted schooner) invented deep survey techniques as they went. They probed the bottom with powerful echo sounders and looked beneath sediments with reflected pressure waves generated by surplus Navy depth charges dropped gingerly overboard. The overall shape of the Mid-Atlantic Ridge was slowly revealed. The ridge's conformance to shorelines on either side of the Atlantic (**Figure 3.12**) raised many eyebrows. Ocean floor sediments were shown to be thickest at the edge of the Atlantic and thinnest near this mid-ocean ridge.

Vema and other research ships also compiled more complete, accurate charts of the submerged edges of continents. At Cambridge University, Sir Edward Bullard used a computer to process these data to achieve the best possible fit of the continental jigsaw-puzzle pieces around the Atlantic. The fit was astonishingly good (**Figure 3.13**).

Mantle studies were keeping pace. The first links in

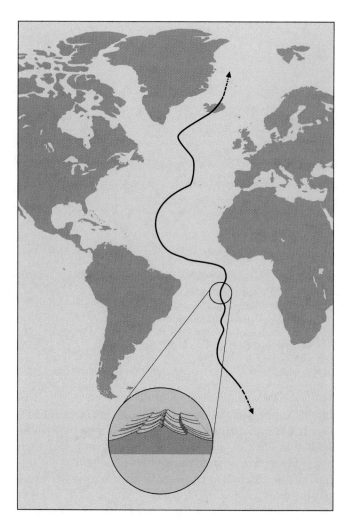

Figure 3.12 The Mid-Atlantic Ridge, showing its conformance to the coastlines of the adjacent continents.

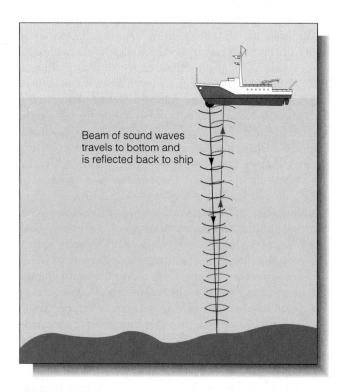

Figure 3.11 Echo sounders sense the contour of the seafloor by beaming sound waves to the bottom and measuring the time required for the sound waves to bounce back to the ship. If the round-trip travel time and wave velocity are known, distance to the bottom can be calculated.

Beam of sound waves travels to bottom and is reflected back to ship

the Worldwide Standardized Seismograph Network, begun during the International Geophysical Year in 1957, were beginning to report data from seismic waves reflected and refracted through the planet's inner layers. This information verified the existence of a layer in the upper mantle that caused a decrease in the velocity of seismic waves. This finding strongly suggested that the layer was not solid. Perhaps the lithosphere was isostatically balanced in this plastic layer, and perhaps continents could move around in it *if* a suitable power source existed.

The Breakthrough: From Seafloor Spreading to Plate Tectonics

In 1960 Professor Harry Hess of Princeton University proposed a radical idea to explain the features of the ocean floor and the fit of the continents. He suggested

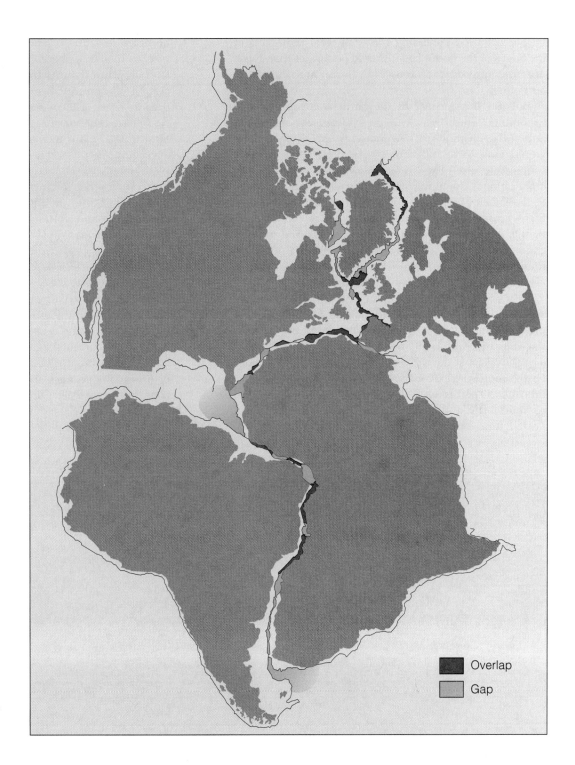

Figure 3.13 The fit of all the continents around the Atlantic, as calculated by Sir Edward Bullard.

that new seafloor develops at the Mid-Atlantic Ridge (and the other newly discovered ocean ridges) and then spreads outward from this line of origin. Continents would be pushed aside by the same forces that cause the ocean to grow. This motion could be powered by **convection currents,** slow-flowing circuits of material within the asthenosphere.

Seafloor spreading, as the new theory was called, pulled many loose ends together. If the mid-ocean ridges were **spreading centers** and sources of new ocean floor rising from the asthenosphere, they should be hot. They were. If the new oceanic crust cooled as it moved from the spreading center, it should shrink in volume and become more dense, and the ocean should be deeper farther from the spreading center. It was. Sediments at the edges of the ocean basin should be thicker than those near the spreading centers. They were, and they were also older.

But did this mean that the Earth was continuously expanding? Since there was no evidence for a growing Earth, the creation of new crust at spreading centers would have to be balanced by the destruction of crust somewhere else. Then researchers discovered that the crust plunges down into the mantle along the periphery of the Pacific. The process is known as **subduction,** and these areas are called **subduction zones** (or Wadati–Benioff zones in honor of their discoverers). The zones of concentrated earthquakes (see again Figure 3.10) were found in regions of crustal formation (spreading centers) and crustal destruction (subduction zones).

In 1965 the ideas of continental drift and seafloor spreading were integrated into the overriding concept of **plate tectonics** (*tekton* = builder; the word *architect* has the same root), primarily by the work of **J. Tuzo Wilson,** a geophysicist at the University of Toronto. In this theory Earth's outer layer consists of about a dozen separate lithospheric **plates,** each about 70 to 100 kilometers (43 to 65 miles) thick, floating on the asthenosphere. When heated from below, the plastic asthenosphere expands,

becomes less dense, and rises. It turns aside when it reaches the lithosphere, however, and it drags the plates laterally until it turns under again to complete the circuit. The large plates include both continental and oceanic crust. The plates, which jostle about like huge flats of ice on a warming lake, are shown and named in **Figure 3.14.** Plate movement is slow in human terms, averaging about 5 centimeters (2 inches) a year. The plates interact at converging, diverging, or slipping junctions, sometimes forcing one another below the surface or wrinkling into mountains. Most of the million or so earthquakes and volcanic events each year occur along plate boundaries.

Through the great expanse of geologic time, this slow movement remakes the surface of the Earth, expands and splits continents, and forms and destroys ocean basins. The light, ancient granitic continents ride high in the lithospheric plates, rafting on the moving asthenosphere below. Plate movement may be caused by friction of asthenosphere convection currents against the plate. In subduction zones, heavy basaltic ocean floor (and its overlying layer of sediment) plunges into the mantle at a subduction zone to be partially remelted. The leading edge of the subducting plate is denser than the upper asthenosphere, and so it is pulled downward into the mantle by

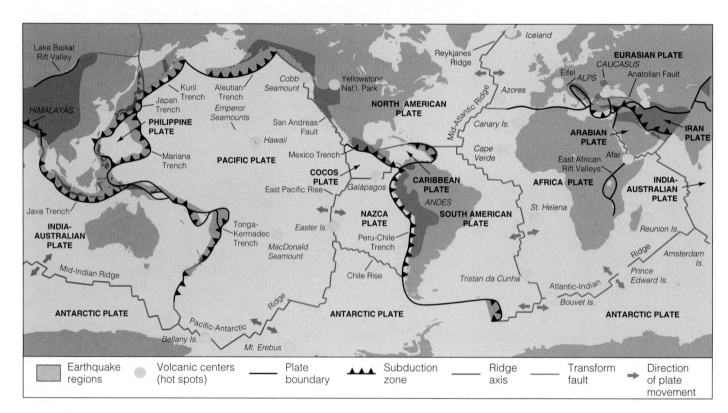

Figure 3.14 The major lithospheric plates, showing their directions of relative movement and the location of the principal hot spots. Note the correspondence of plate boundaries and earthquake locations.

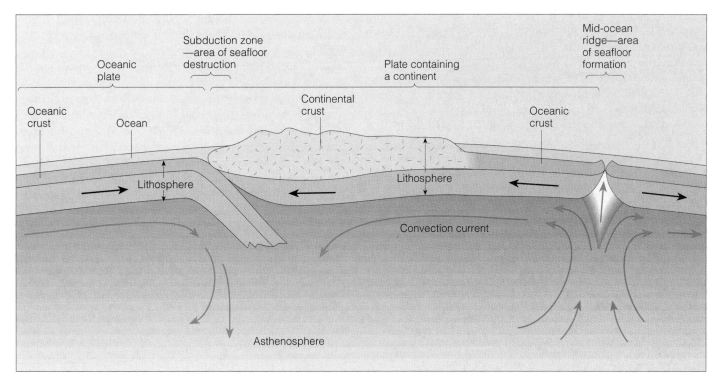

Figure 3.15 An overview of the plate tectonics process.

gravity. Melted material from the subducted plate may separate into lighter components and rise, forming a volcano. No marine sediments are of great age because the ocean floor itself acts as a vast conveyor belt, transporting accumulated sediment to subduction zones where the seafloor sinks into the asthenosphere. Literally and figuratively, it all fits; a cooling, shrinking, raisinlike wrinkling is no longer needed to explain Earth's surface features. **Figure 3.15** presents an overview of the process. This new understanding of the ever-changing nature of the Earth has given fresh meaning to historian Will Durant's warning: "Civilization exists by geological consent, subject to change without notice."

After a series of raucous scientific meetings in 1966 and 1967, the revolution in geology entered a period of rapid consolidation. In 1968 *Glomar Challenger* drilled her first deep-ocean crustal cores and provided the confirmation of plate tectonics. Researchers found supporting data from many sources that tended to confirm Wilson's surprising synthesis. Every scientist had to reexamine his or her specialty in light of this new information. Zoologists found new explanations for the unusual animals of Australia. Biologists discovered a new cause of the isolation required for the formation of new species by natural selection. Paleontologists found an explanation for similar fossils on different continents. Resource specialists could at last explain why coal deposits were buried in

Antarctica. Some geologists were pleased, some were skeptical. All were eager to explore further to prove or disprove the tenets of this new theory.

This historical overview is useful in transmitting the sense of discovery and excitement surrounding the twentieth-century revolution in geology. We now investigate the workings of plate tectonics in more detail.

PLATE TECTONICS: A CLOSER LOOK

Plate tectonics shares the concept of Pangaea with the older theory of continental drift. **Figure 3.16** depicts Pangaea as it may have looked about 200 million years ago, when it was beginning to break up. The present continents of North America and Eurasia, and Africa and South America, are shown as part of the old supercontinent. Plate tectonics can be illustrated by what geologists think happened next.

Because it is much thicker, continental crust is only about half as efficient as oceanic crust in conducting heat from Earth's interior to the surface. Heat from the lower crust and mantle accumulates beneath the continents. For reasons not yet fully understood, excess heat collected along lines beneath Pangaea. This heat, probably generated from the decay of radioactive elements, caused the asthenosphere near the line to expand and

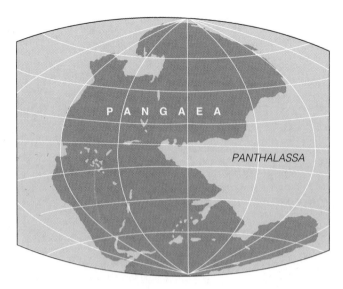

Figure 3.16 Reconstruction of Pangaea as it is thought to have appeared 200 million years ago.

rise, thus lifting and fracturing the lighter, solid lithosphere above. The plate and its embedded supercontinent split in two, and a new plate boundary was formed between the pieces (**Figure 3.17** shows the separation of South America and Africa).

The convection currents in the asthenosphere turned aside when they reached the brittle lower lithosphere. The continental fragments (North America and Eurasia, and South America and Africa) moved apart, and a new ocean basin began to form between the diverging plates. As the broken plate separated at this new spreading center, molten rock called **magma** rose into the crustal fractures. (Magma is called *lava* when found aboveground.) Some of the magma solidified in the fractures; some erupted from volcanoes on the seafloor. Together these processes produced new oceanic crust. As may be seen in **Figure 3.18,** the same process continues today. About 20 cubic kilometers (4.8 cubic miles) of new ocean crust forms each year.

Figure 3.17 The formation of a new plate boundary: the breakup of Pangaea. (a) As the lithosphere began to crack, a rift formed beneath the continent, and molten basalt from the asthenosphere began to rise. (b) As the rift continued to open, the two new continents were separated by a growing ocean basin. Volcanoes and earthquakes occur along the active rift area, which is the mid-ocean ridge. The East African Rift Valley currently resembles this stage. (c) A new ocean basin (shown in green) forms beneath a new ocean. The Red Sea currently resembles this stage. Note also the similarity of (c) with Figure 3.20.

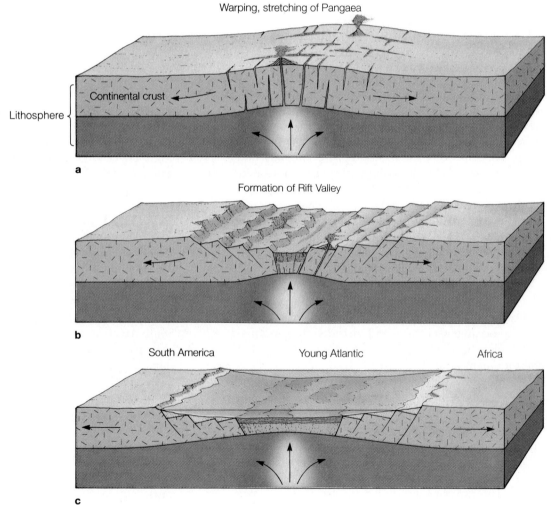

As it moved away from the active spreading center, the oceanic lithosphere cooled and shrank slightly. This shrinkage caused the ocean to become deeper farther from the growing Mid-Atlantic Ridge.

The Atlantic's rate of spreading was—and continues to be—a jerky 5 centimeters (2 inches) a year. This young ocean, which began less than 200 million years ago, therefore was about 25 meters (85 feet) narrower when Columbus sailed than it is today.

Divergent Plate Boundaries

The spreading center at the Mid-Atlantic Ridge is a **divergent plate boundary,** a line along which two plates are moving apart. Oceanic crust forms along divergent plate boundaries. For example, in the South Atlantic a large new ocean basin has formed between the diverging plates (as shown in **Figure 3.19**). A long mid-ocean ridge divided by a central rift valley traverses the ocean floor roughly equidistant from the shorelines in both the North and South Atlantic, terminating beneath the ice cap north of Iceland. The lithospheric plates west of the ridge extend all the way to the eastern edge of the Pacific Ocean and include the North and South American continents. The two plates east of the ridge include Africa and all of Eurasia. The Eurasian Plate wraps around the Earth to contact the western edge of the Pacific Plate. (Look again at Figure 3.14.)

Figure 3.18 The process of seafloor spreading was investigated firsthand in 1979 in an exploration of the Galápagos rift. The heat that the researchers measured, the lack of sediment covering at the ridge, and characteristic pillow-shaped lava formations found on the ridge by the deep submersible *Alvin* were consistent with the theory of plate tectonics.

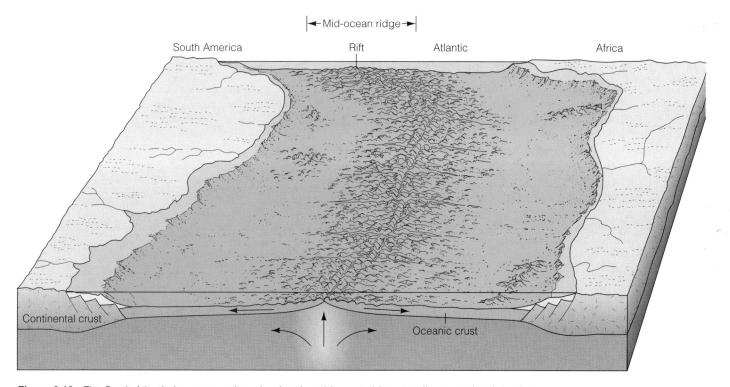

Figure 3.19 The South Atlantic in cross section, showing the mid-ocean ridge at a divergent plate boundary.

Plate divergence is not confined to the Atlantic, nor has it been limited to the last 200 million years. As may also be seen in Figure 3.14, the Mid-Atlantic Ridge has counterparts in the Pacific and Indian oceans. The Pacific floor, for example, diverges along the East Pacific Rise and the Pacific Antarctic Ridge, spreading centers that form the eastern and southern boundaries of the great Pacific Plate. In East Africa, rift valleys have formed relatively recently as plate divergence begins to separate another continent. As has happened in the Red Sea, the ocean will invade when the rift becomes deep enough. **Figure 3.20** shows the characteristic rift shape.

Convergent Plate Boundaries

Crust is destroyed at **convergent plate boundaries,** regions of violent geologic activity where plates are pushing together. South America, embedded in the westward-moving South American Plate, encounters the Pacific's eastward-moving Nazca Plate. The relatively thick and light continental lithosphere of South America rides up and over the heavy oceanic lithosphere of the Nazca Plate, which is subducted into the deep trench that parallels the west coast of South America. **Figure 3.21** is a cross section through these plates.

Figure 3.20 A view down the axis of the divergent boundary immediately south of the Red Sea. The rift is flanked by parallel-running faults, producing a set of escarpments remarkably like those observed at mid-ocean ridges. Compare to Figure 3.17b.

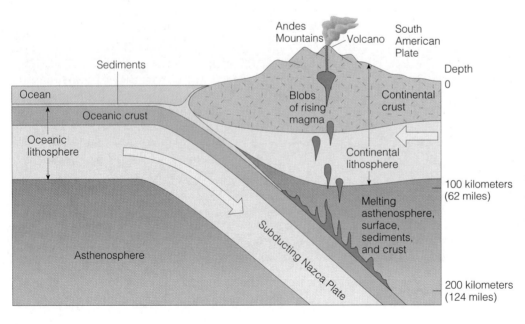

Figure 3.21 A cross section through the west coast of South America, showing the convergence of a continental plate and an oceanic plate. The subducting oceanic plate becomes more dense as it descends, its downward slide propelled by gravity. The oceanic plate melts as it subducts. Blobs of magma rising from the melting zone power Andean volcanoes.

Figure 3.22 The process shown in Figure 3.21 also occurs in the northwestern U.S. Rising magma powered the violent 1980 eruption of Mount St. Helens. The Cascade volcanoes—such as Mount St. Helens, Mount Rainier, Mount Shasta, and Crater Lake—owe their existences to the melting of a subducting plate.

Figure 3.23 The formation of an island arc along a trench as two oceanic plates converge. The arc shape results from the geometric constraints of forces applied to the surface of a sphere. The volcanic islands form as blobs of magma reach the seafloor.

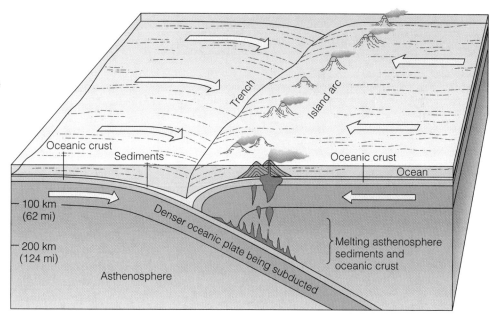

The subducting plate becomes denser as it sinks because pressure at these depths transforms minerals into denser forms and because low-density substances are lost to the surrounding asthenosphere and mantle. Earthquakes are caused when the plate lurches downward or undergoes sudden structural transformation. Some of the oceanic crust and its sediments will melt as the plate plunges downward, forming a magma rich in water and carbon dioxide. In places this magma rises through overlying layers to the surface and causes volcanic eruptions. The active volcanoes of Central America and South America's Andes Mountains are a product of this activity, as are the area's numerous earthquakes. The North American Cascade volcanoes, including Mount St. Helens (**Figure 3.22**), result from similar processes. Most of the subducted crust mixes with the mantle, some of it eventually reaching into the mantle to depths of perhaps 700 kilometers (430 miles)!

In the previous example continental crust met oceanic crust. What happens when two *oceanic* plates converge? One of the colliding plates will usually be older, and therefore cooler and denser, than the other. Pulled by the force of gravity, this heavier plate will slip below the lighter one and sink into the asthenosphere. The ocean bottom is distorted in these areas to form deep trenches, the ocean's greatest depths. Water and carbon dioxide trapped with the melting rock of the subducting plate help to form a relatively light magma and vigorous volcanoes, but the volcanoes emerge from the seafloor rather than from a continent. These volcanoes appear in patterns of curves on the overriding oceanic crust; when they emerge above sea level they form curving chains of islands (see **Figure 3.23**). The arc shape results from the geometric constraints of forces applied on the surface of a sphere. Some geophysicists believe all of Earth's continental crust may have originated from granitic rock

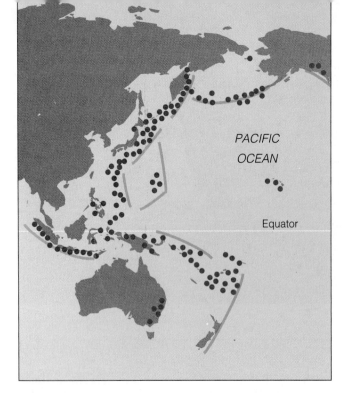

Figure 3.24 The distribution of island arcs, trenches (lines), and volcanoes (dots) in the western Pacific, where two plates converge. The position of volcanoes shown here is approximate.

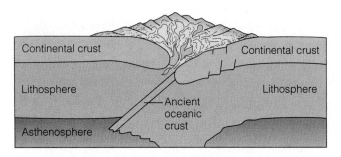

Figure 3.25 A cross section through southern China, showing the convergence of two continental plates. Neither plate is dense enough to subduct; instead, their compression and folding uplift the plate edges to form the Himalayas.

produced in this way. The island arcs may have coalesced to form larger and larger continental masses.

The many island arcs and peripheral trenches of the western and northern Pacific shown in **Figure 3.24** result from the convergence of two oceanic plates. Note the correlation of these areas with Figure 3.10's plot of earthquakes. Subduction at converging oceanic plates was responsible for the great Alaska earthquake of 1964. Plate convergence (and divergence) is faster in the Pacific than in the Atlantic, in a few places reaching a rate of 18 centimeters (7 inches) a year. You can now clearly see the source of the Pacific Ring of Fire.

Two plates bearing continental crust can also converge. The most spectacular example of such a collision,

between the India–Australian and Eurasian plates some 45 million years ago, formed the Himalayas. Neither plate edge is being completely subducted; instead, both are compressed, folded, and uplifted, as **Figure 3.25** shows. Notice the massive supporting "root" beneath the emergent mountain needed for isostatic equilibrium. Here is an explanation for the mountaintop marine fossils catastrophists mistook to be evidence of the biblical flood: The mountains and fossil shells formed on shallow submerged seabeds that were uplifted by plate convergence. The top of Mount Everest is made of rock formed from sediments deposited long ago in a shallow sea!

Transform Plate Boundaries

In some places, crustal plates shear laterally past one another. These areas are called **transform plate boundaries.** Crust is neither produced nor destroyed at this type of junction, but the potential for earthquakes can be great as the plate edges slip past one another. The eastern boundary of the Pacific Plate is a long transform fault system. California's San Andreas Fault (**Figure 3.26**) is merely the most famous of the many faults marking the junction between the Pacific and North American plates. The Pacific Plate moves steadily, but its movement is stored elastically at the North American Plate boundary until friction is overcome. Then the Pacific Plate lurches in abrupt jerks to the northwest along much of its shared border with the North American Plate, an area that includes the major population centers of California. These jerks cause California's famous earthquakes. Because of this movement, coastal southwestern California is gradually sliding north along the rest of North America; some 50 million years from now, it will encounter the Aleutian Trench.

There are, then, two kinds of plate divergences: divergent oceanic crust (such as that in the mid-Atlantic) and divergent continental crust (as in the Rift Valley of East Africa). And there are three kinds of plate convergences: oceanic crust toward continental crust (west coast of South America), oceanic crust toward oceanic crust (northwestern Pacific), and continental crust toward continental crust (Himalaya Mountains). Transform boundaries mark locations at which crustal plates move past one another (San Andreas Fault). Each movement produces a distinct topography, and each zone contains potential dangers for its human inhabitants. The characteristics of plate boundaries are summarized in **Table 3.2.**

**THE CONFIRMATION
OF PLATE TECTONICS**

The theory of plate tectonics has had the same effect on geology that the theory of evolution has had on biology. In each case a catalog of seemingly unrelated facts was

Table 3.2 Characteristics of Plate Boundaries

Plate Boundary		Plate Movement	Seafloor	Events Observed	Example Locations
Divergent plate boundaries	Ocean-ocean	Apart	Forms by seafloor spreading	Ridge forms at spreading center. Ocean basin expands, plate area increases. Many small volcanoes and/or shallow earthquakes.	Mid-Atlantic Ridge, East Pacific Rise
	Continent-continent		New ocean basin may form as continent splits	Continent spreads, central rift collapses, ocean fills basin.	East African Rift, Red Sea
Convergent plate boundaries	Ocean-continent	Together	Destroyed at subduction zones	Dense oceanic lithosphere plunges beneath less dense continental. Earthquakes trace path of downmoving plate as it descends into asthenosphere. A trench forms. Subducted plate partially melts. Magma rises to form continental volcanoes.	Western South America, Cascade Mountains in western U.S.
	Ocean-ocean			Older, cooler, denser crust slips beneath less dense crust. Strong quakes. Deep trench forms in arc shape. Subducted plate heats in upper mantle, magma rises to form curving chains of volcanic islands.	Aleutians, Marianas
	Continent-continent		(N/A)	Collision between masses of granitic continental lithosphere. Neither mass is subducted. Plate edges are compressed, folded, uplifted; one may move beneath the other.	Himalayas, Alps
Transform plate boundaries		Past each other	Neither created nor destroyed	A line (fault) along which lithospheric plates move past each other. Strong earthquakes along fault.	San Andreas Fault; South Island, New Zealand

unified by a powerful central idea. As we will see, many discoveries contributed to our present understanding of plate tectonics, but the most compelling evidence is locked within the floors of the young ocean basins themselves.

Paleomagnetism

A compass needle points toward the magnetic north pole because it aligns with the Earth's magnetic field. Tiny particles of iron-bearing magnetic minerals are found in basaltic magma. These minerals act like miniature compass needles; as they cool to form new seafloor, their magnetic fields align with the Earth's magnetic field. Thus the orientation of the Earth's magnetic field at that particular time becomes frozen in the rock as it solidifies. Any later change in the strength or direction of the Earth's magnetic field will not significantly change the

characteristics of the field trapped within the solid rocks. The "fossil," or remanent, magnetic field of a rock is known as **paleomagnetism** (*palaios* = ancient).

A **magnetometer** measures the amount and direction of residual magnetism in a rock sample. In the late 1950s geophysicists towed sensitive magnetometers just above the ocean floor to detect the weak magnetism frozen in the rocks. When plotted on charts, the data revealed an odd pattern of symmetrical magnetic stripes or bands on both sides of a spreading center (**Figure 3.27a**). The tiny compass needles contained in the rocks of some bands join *with* the Earth's present magnetic orientation to enhance the strength of the local magnetic field, but the needles in rocks in adjacent bands weaken it. What could cause such a pattern?

In 1963 geologists Drummond Matthews and Frederick Vine proposed a clever interpretation. They knew that the Earth's magnetic field reverses at irregular intervals. In a time of reversal a compass needle would point south instead of north, and any particles of magnetic material in fresh seafloor basalt at a spreading center would be imprinted with the reversed field. The alternating magnetic stripes represent rocks with alternating magnetic polarity—one band having normal polarity (magnetized in the same direction as today's magnetic field direction), the next band having reversed polarity (opposite from today's direction). These researchers realized that the pattern of alternating weak and strong magnetic fields was symmetrical because freshly magnetized rocks born at the ridge are spread apart and carried away from the ridge by plate movement (**Figure 3.27b**).

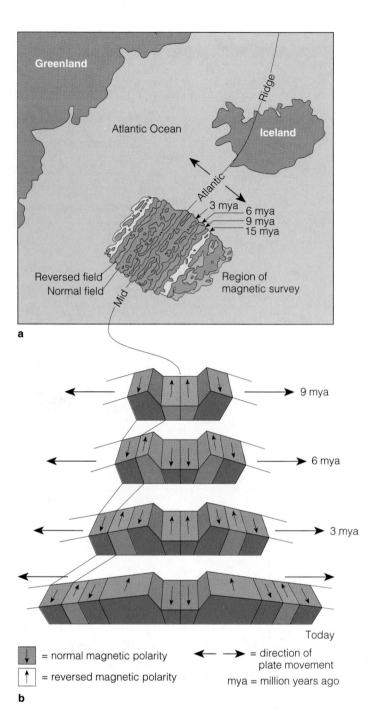

a

b

= normal magnetic polarity

= reversed magnetic polarity

= direction of plate movement

mya = million years ago

Figure 3.27 Patterns of paleomagnetism and their explanation by plate tectonic theory. (a) When scientists conducted a magnetic survey of a spreading center, the Mid-Atlantic Ridge, they found bands of weaker and stronger magnetic fields frozen in the rocks. (b) The molten rocks forming at the spreading center take on the polarity of the planet when they are cooling and then move slowly in both directions from the center. When the Earth's magnetic field reverses, the polarity of new-formed rocks changes, creating symmetrical bands of opposite polarity.

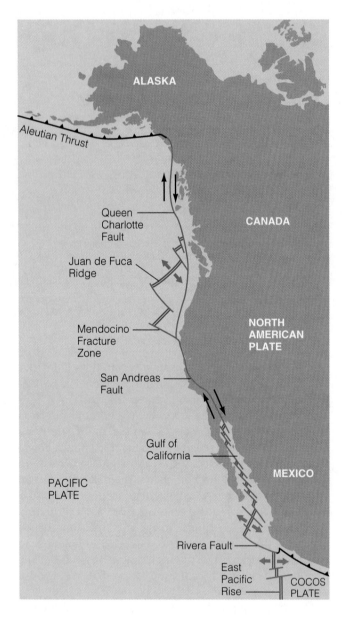

Figure 3.26 A long transform plate boundary, which includes California's San Andreas Fault.

The Earth's magnetic field reverses periodically but at irregular intervals once every 300,000 to 500,000 years. At least nine "flips" have been documented over the last 4.5 million years. No one yet knows what causes these reversals, or what the effects are on the Earth during the change. Similar magnetic patterns have been found on land and independently dated by other means. By 1974, scientists had compiled charts showing the paleomagnetic orientation—and the age—of the seafloors of the eastern Pacific and the Atlantic for about the last 200 million years (see **Figure 3.28**). Plate tectonics beautifully explains these patterns, and

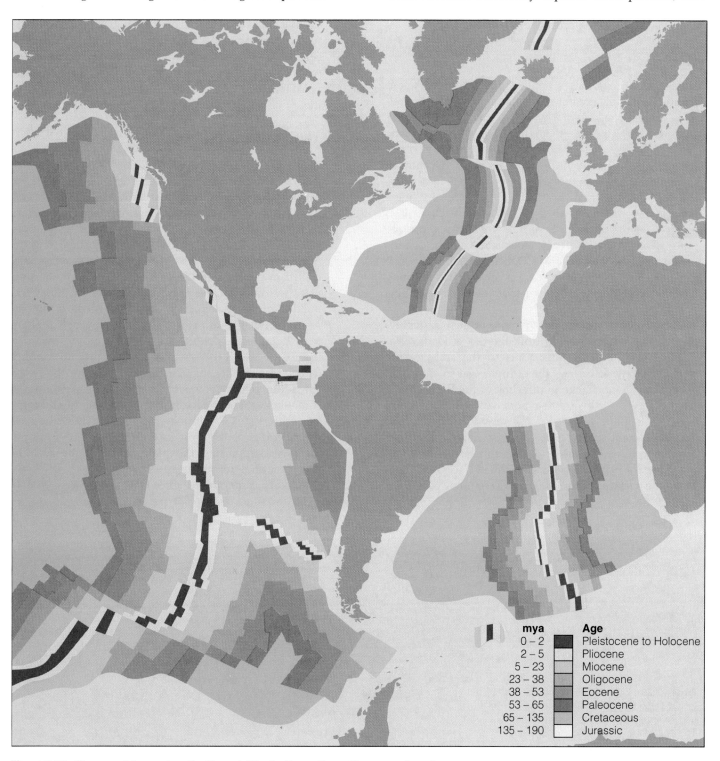

mya	Age
0 – 2	Pleistocene to Holocene
2 – 5	Pliocene
5 – 23	Miocene
23 – 38	Oligocene
38 – 53	Eocene
53 – 65	Paleocene
65 – 135	Cretaceous
135 – 190	Jurassic

Figure 3.28 The age of the eastern Pacific and Atlantic Ocean floors. The magnetic patterns are an expression of seafloor spreading over the last 200 million years. (mya = million years ago.)

the patterns themselves are among the most compelling of all arguments for the theory.

Paleomagnetic data have recently been used to measure spreading rates, to calibrate the geologic time scale, and to reconstruct continents. Paleomagnetism has been among the most productive specialties in geology for the past three decades, but other lines of investigation have also shed light on the process of plate tectonics.

Hot Spots

Hot spots are stationary sources of heat in the upper mantle. Hot spots are not always located at plate boundaries, and no one knows why their source of heat is localized or what anchors them in place. As lithospheric plates slide over these fixed locations, they are weakened from below by rising heat and magma. A volcano can form over the hot spot; but because the plate is moving, the volcano is carried away from its source of magma after a few million years and becomes inactive. It is replaced at the hot spot by a new volcano a short distance away. A chain of volcanoes and volcanic islands results (**Figure 3.29**). The MacDonald Seamount, which gave researchers on *Melville* such an exciting morning (see the beginning of this chapter), is being formed by a hot spot.

Figure 3.30 shows the most famous of these assembly-line chains, which extends from the old eroded volcanoes of the Emperor Seamounts to the still-growing island of Hawaii. In fact, the abrupt bend in the chain was caused by a change in direction of the Pacific Plate, from a largely northward to a more westward movement about 40 million years ago. The next Hawaiian island that will come into being—already named Loihi—is building on the ocean floor to the southeast. Now about 1,000 meters (3,200 feet) beneath the surface, Loihi will break the surface about 30,000 years from now.

There are other hot spots in the Pacific. The island chains formed by their activity also jog in the Hawaiian pattern, indicating that they are positioned on the same lithospheric plate. Chains of undersea volcanoes in the Atlantic, centered on the Mid-Atlantic Ridge, suggest a similar process is at work there. Hot spots can exist beneath continental crust as well; Yellowstone National Park is believed to be over a hot spot beneath the westward-moving North American Plate. Look again at Figure 3.14 to see the locations of hot spots around the world. The configuration and length of all these chains of volcanoes and geothermal sites are consistent with the theory of plate tectonics.

Atolls and Guyots

Atolls are ring-shaped islands of coral reefs and reef-derived sediment centered over submerged inactive volcanoes. The coral animals that build atolls can live only in the upper sunlit layer of seawater. How, then, did their skeletal remains end up at a depth of 1,280 meters (4,222 feet) within Eniwetok Atoll? This surprising discovery was made in 1954 when bore holes were being drilled in preparation for the first hydrogen bomb tests. Plate tectonics suggests an answer. Coral animals can build atop the skeletons of their dead predecessors at a rate of about 1 centimeter (½ inch) each year. Coral animals living on a volcano's flanks can grow upward as the crust beneath the volcano slowly cools and contracts (sinks) during its movement away from the warm spreading center where it formed. A deep column of coral skeletons can accumulate if the rate of sinking is less than about 1 centimeter per year. Thus, the coral record traces plate subsidence for millions of years into the past, supporting the plate tectonics theory. (This process is shown in Figure 18.22.)

Guyots (pronounced ghee-OH) were discovered by

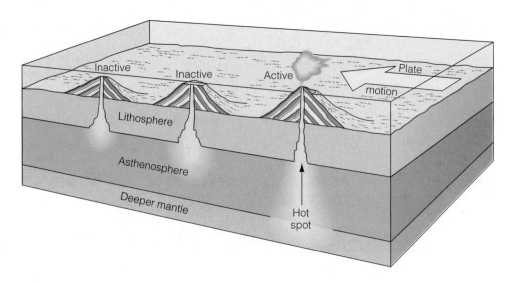

Figure 3.29 Formation of a volcanic island chain by the movement of an oceanic plate over a stationary hot spot. The age of the islands increases toward the left. New islands will continue to form over the hot spot.

Harry Hess during his service as commander of a U.S. Navy transport in the Pacific during the Second World War. With the ship's echo sounder he found chains of odd, flat-topped, submerged volcanic mountains; he named them after Princeton's first professor of geology, Arnold Guyot. More than 500 guyots have been located, and plate tectonics neatly explains the formation of most of them. Like atolls, they were once volcanic peaks (oceanic islands) standing above sea level. As the plate on which they were riding cooled, contracted, and was carried away from the spreading center, they became inactive and stopped growing. They were "shaved" flat by wave action as they sank beneath the ocean surface. Evidence of ancient beaches along their rims suggests that this hypothesis is correct. **Figure 3.31** shows a sinking progression of guyots. They never formed atolls,

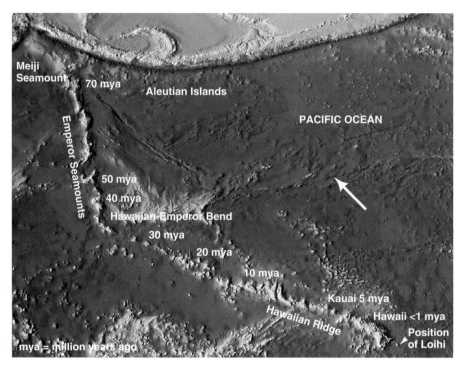

Figure 3.30 The Hawaiian chain, islands formed one by one as the Pacific Plate slid over a hot spot. The oldest known member of the chain, the Meiji Seamount, formed about 70 million years ago (70 mya), and the "bend" in the chain shows that the plate changed direction about 40 million years ago. The island of Hawaii still has active volcanism, but the next island in the chain, Loihi, has begun building on the ocean floor. The arrow shows the current direction of plate motion.

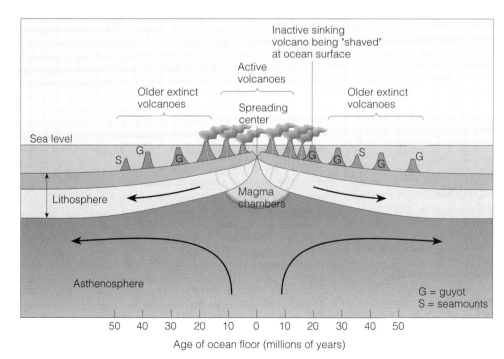

Figure 3.31 The process by which guyots (G) and seamounts (S) form. Guyots have flat tops because they have grown tall enough to be "shaved" by waves at the ocean's surface. Seamounts have a similar origin but retain their more pointed volcano shape because they never reached the surface.

BOX 3.1 ● *Project FAMOUS*

By 1970, the beauty of the plate tectonics theory had won over most of the scientific community, but controversy over the details continued. Chief among these was whether seafloor spreading is an active or passive process. Is active volcanism *pushing* the plates away from the spreading center, or is some other force *pulling* them, allowing new material to rise into the rift? The only way to answer the question was to transport geologists to an active spreading zone for a look.

In 1972 and 1973, 23 French, American, British, and Canadian research ships participated in Project FAMOUS (the French-American Mid-Ocean Undersea Study). Using depth sounders, suspended cameras, and bottom samplers, they scouted a small segment of the Mid-Atlantic Ridge 320 kilometers (200 miles) southwest of the Azores Archipelago. In the summer of 1973, the French submersible *Archimède* made the first on-site photographic survey of the central rift valley. For most of the summer of 1974, *Archimède* was joined at the dive site by the deep-diving *Alvin* and the maneuverable little *Cyana* (designed by Jacques Cousteau). They made more than 50 dives to the floor of the rift valley, a depth

of about 2,700 meters (9,000 feet). The drilling ship *Glomar Challenger* (see Box 2.1 in Chapter 2) also contributed to the project by boring four holes starting 25 kilometers (16 miles) west of the dive site.

Project FAMOUS extended classical field geology to the deep-sea floor for the first time, using research submersibles precisely navigated by a team of scientists. This team sampled water and rock, made hundreds of physical measurements, and took more than 51,000 photographs. They saw that recently solidified lava was restricted to a narrow band, about 1 kilometer (0.6 mile) wide, along the center of the rift valley. Fresh lava hills rose only about 200 to 300 meters (650 to 1,000 feet) above the rugged terrain, leading observers to speculate that pressure from within would not be great enough to push the seafloor aside. These low hills within a central rift, the types and thicknesses of rocks and sediments found on the ridge, the bunlike shape of the pillow lava in the rift (see Figure 3.18), and the patterns of cracks and fissures all supported the pulling conveyor-belt model of seafloor spreading.

---

either, because they began in water too cold for coral or because they moved and sank too rapidly for coral growth to keep up. Indeed, an atoll can become a guyot if its rate of subsidence increases, coral growth slows, or plate motion takes it into colder water.

The Age and Distribution of Sediments

If the ocean basins are genuinely ancient, and if the processes that produce sediments have been operating for most or all of that time, both the thickness and age of sediments on the ocean floor should be great. They are not. The young spreading ridges are almost free of sediment, and the oldest edges of the basins support layers of sediment 15 to 20 times thinner than the age of the ocean itself would suggest.

Powerful low-frequency echo sounders have probed the depth and structure of these sediments in many locations. Core sampling has shown that sediments resting directly on the basalt floor are almost always youngest near the spreading centers and oldest near subduction zones. The oldest sediments of the ocean basins are rarely more than 160 million years old. These data are

consistent with the plate tectonics idea that ocean basins are continuously created and destroyed.

The Oceanic Ridges

The location and configuration of the oceanic ridges are clear evidence of past events. The volcanic nature of ridge islands like Iceland, the shape of the longitudinal rifts splitting the ridge tops, and the sinking of the seabed as new oceanic crust cools and travels outward are all consistent with the theory of plate tectonics. The distribution of transform faults and fracture zones along the oceanic ridges (features described in Chapter 4) also supports plate tectonics theory, as does on-the-spot geological observations such as those made by Project FAMOUS (see **Box 3.1**).

Heat Flow

Heat flows rapidly from the ocean bottom at mid-ocean ridges and slowly in the vicinity of trenches. This suggests the presence of hot new crust at the ridge and cool older crust at the trenches, just as plate tectonics predicts.

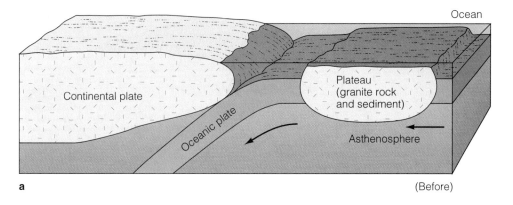

a   (Before)

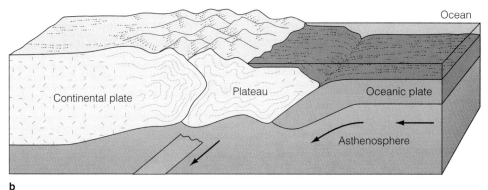

b

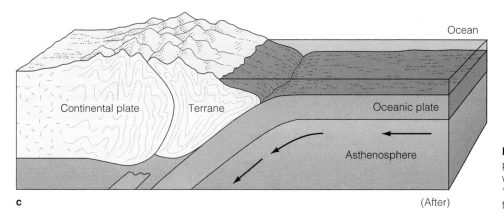

c   (After)

Figure 3.32 Terrane formation. Oceanic plateaus are not subducted into the trench with the oceanic plate. Instead, they are "scraped off," causing uplifting and mountain building as they strike a continent.

Terranes

Buoyant continental and oceanic plateaus, island arcs, and fragments of granitic rock and sediments can be rafted along with a plate and scraped off onto a continent when the plate is subducted. This process is similar to what happens when a sharp knife is scraped across a table top to remove pieces of cool candle wax. The wax accumulates and wrinkles on the knife blade in the same way land masses and ocean sediments accumulate against the face of a continent as the lithosphere in which they are embedded reaches a plate boundary. Plateaus, isolated segments of seafloor, ocean ridges, ancient island arcs, and parts of continental crust that collect on the face of a continent are called **terranes.** The thickness and low density of terranes prevents their subduction. A simplified account of terrane accumulation is diagrammed in **Figure 3.32.**

Terranes are surprisingly common. New England, much of North America west of the Rocky Mountains, and all of Alaska appear to be composed of this sort of crazy-quilt assemblage of material, some of which has evidently arrived from thousands of miles away in the Southern Hemisphere!

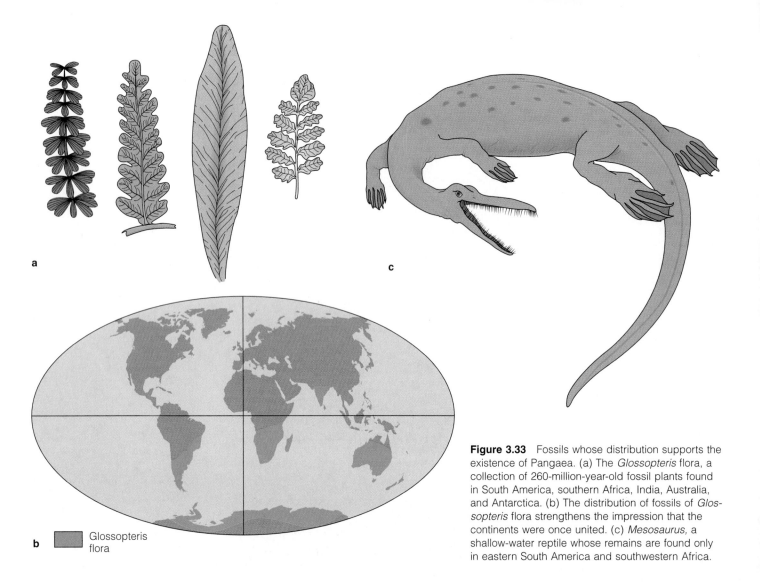

Figure 3.33 Fossils whose distribution supports the existence of Pangaea. (a) The *Glossopteris* flora, a collection of 260-million-year-old fossil plants found in South America, southern Africa, India, Australia, and Antarctica. (b) The distribution of fossils of *Glossopteris* flora strengthens the impression that the continents were once united. (c) *Mesosaurus*, a shallow-water reptile whose remains are found only in eastern South America and southwestern Africa.

Glossopteris flora

Fossils

Suess's *Glossopteris* fossils were part of a complex collection of plants that flourished in low, swampy areas near the margins of glaciers (see **Figure 3.33**). The *Glossopteris* flora, as the fossil plants are called, are common to southern South America, southern Africa, India, Australia, and Antarctica. It is extremely unlikely that identical, intricate, interdependent communities of plants could have arisen simultaneously in all these places. A more likely explanation is a site of common origin in Pangaea. The breakup of Pangaea would account for the present distribution of the *Glossopteris* flora.

Animal fossils also support the idea of an ancient supercontinent. Fossils of *Mesosaurus*, a half-meter- (2-foot-) long aquatic reptile, are found only in eastern South America and southwestern Africa. Again, it is extremely unlikely that this animal could have evolved simultaneously in two such widely separated locations. It is equally unlikely that this small shallow-water reptile could have swum across 5,500 kilometers (2,500 miles) of open ocean to establish itself on both sides of the Atlantic.

Sea Turtles

Sea turtle migration might seem an unlikely confirmation of plate tectonics, but work done by Archie Carr and P. J. Coleman suggests otherwise. A population of green turtles (*Chelonia mydas*) lives off the coast of Brazil but regularly breeds 2,000 kilometers (1,235 miles) away on tiny Ascension Island, a projection of the Mid-Atlantic Ridge. How could these animals make such a long journey to such a small target?

Perhaps the green turtles' distant ancestors had a much easier task when the Atlantic was very small. As seafloor spreading widened the ocean, their original breeding islands sank below sea level, but island after volcanic island erupted in the turtles' path to take their places. Successive generations would need only to extend their travel path directly into the rising sun to accommodate the growing ocean. As the distances grew, turtles adept at homing would have been favorably selected by the environment and would have reproduced most successfully. No other theory explains so well how the turtles' navigational accuracy evolved.

PROBLEMS AND IMPLICATIONS

The theory of plate tectonics reveals much about the nature of the Earth's surface. In case you feel geophysicists have all the answers, however, consider just a few of the theory's unsolved problems:

- Why should long *lines* of asthenosphere be any warmer than adjacent areas?

- If plate movement depends on drag within asthenosphere convection cells, why should the plastic material flow parallel to the plate bottoms for long distances instead of cooling and sinking near the spreading center?

- Is the increasing density of the leading edge of a subducting plate as important as convective friction in making the plate move?

- Are plate movements due entirely to motion of the asthenosphere? Is deeper mantle convection also involved? Are the plates themselves the uppermost part of the convective cell?

- Has seafloor spreading always been a feature of the Earth's surface? Has a previously thin crust become thicker with time, permitting plates to function in the ways described here?

- There is evidence of tectonic movement prior to the breakup of Pangaea. Will the process continue indefinitely, or are there cycles within cycles?

Though there is clearly much to learn, plate tectonics is already an especially powerful predictive theory. Discoveries and insights made by the researchers mentioned in this chapter, and hundreds of others, have borne out the intuition of Alfred Wegener. Our understanding of the process will evolve as more data become available, but there seems very little chance that geologists will ever return to the dominant pre-1960 view of a stable and motionless crust. Plate tectonic theory shows us the picture of an actively cycling Earth and an ever-changing surface, with *a single world ocean* changing shape and shifting positions as the plates slowly move.

The configuration of the ocean basins—discussed in the next chapter—is the result of plate tectonic activity. The variety of these features will make more sense now that you are armed with an understanding of the theory.

Q-AND-A

1. How far beneath the Earth's surface have humans actually ventured?

   Gold miners in South Africa have excavated ore 3,841 meters (12,600 feet, 2.39 miles) beneath the surface, where rock temperature is 55°C (131°F). The miners work in pairs, one chipping the rock and the other aiming cold air at his partner. After a time they trade places.

2. What is the difference between crust and lithosphere?

   Lithosphere includes crust (oceanic and continental) and rigid upper mantle down to the asthenosphere. The velocity of seismic waves in the crust is much different from that in the mantle. This suggests differences in chemical composition or crystal structure, or both. The lithosphere and asthenosphere have different physical characteristics: The lithosphere is generally rigid, but the asthenosphere is capable of slow plastic movement. Asthenosphere and lithosphere also transmit seismic waves at different speeds.

3. What are the most abundant elements in the Earth?

   You may be surprised to learn that oxygen accounts for about 46% of the mass of the Earth's crust. On an atom-for-atom basis the proportion is even more impressive: Of every 100 atoms of Earth's crust, 62 are oxygen. Most of this oxygen is not present as the gaseous element but is combined with other atoms into oxides and other compounds. Most of the familiar crustal rocks and minerals are oxides of aluminum, silicon, and iron (rust, for example, is iron oxide).

   In the Earth as a whole iron is the most abundant element, making up 35% of the mass of the planet, while oxygen accounts for 30% overall. Remember, most of the mass of the universe is hydrogen gas. Oxygen and iron are abundant on Earth only because fusion reactions in stars can transform light elements like hydrogen into heavy ones.

4. Pangaea began to break up about 200 million years ago, a very small fraction of the planet's 4.6-billion-year history. We know that the Earth has had a solid crust for a very long time. Is there evidence of plate tectonic effects *prior to* the breakup of Pangaea?

   Yes. Some evidence remains of the assembly of Pangaea. The Appalachian Mountains of the eastern United States are more ancient than the supercontinent, and they were probably formed by two plates sliding slowly together. Paleomagnetic evidence becomes progressively less certain with age, however, so geologists are uncertain about its interpretation. Yet the likeliest hypothesis is that tectonic forces have been active ever since the Earth's crust formed. Indeed, the radioactivity that heats the interior and helps power asthenosphere convection would have been even stronger earlier in the Earth's history.

5. How do geologists determine the location and magnitude of an earthquake?

   They use the time difference in the arrival of the P and S waves at their instruments to determine the location of an earthquake. At least three seismographs in widely separated locations are needed to get a fix on the location.

The strength of the waves, adjusted for the distance, is used to calculate the earthquake's magnitude. Earthquake magnitude is often expressed on the **Richter scale.** Each full step on the Richter scale represents a 10-fold change in surface wave amplitude and a 32-fold change in energy release. Thus an earthquake with a Richter magnitude of 6.5 releases about 32 times more energy than an earthquake with a magnitude of 5.5, and about 1,000 times that of a 4.5-magnitude quake. People rarely notice an earthquake unless the Richter magnitude is 3.2 or higher, but the energy associated with a magnitude-6 quake may cause significant destruction.

The energy released by the 1964 Alaska earthquake was over a billion times greater than that released by the smallest earthquakes felt by humans, an energy release equal to about twice the energy content of world coal and oil production for an entire year. Very low or very high Richter magnitudes are not easy to measure accurately. The Alaska earthquake's magnitude was initially calculated between 8.3 and 8.6 on the Richter scale, but recent reassessment has yielded an extraordinary magnitude of 9.2. The Earth rang like a great silent bell for 10 days after the earthquake.

6. What's the potential for serious loss of life and property damage due to tectonic plate movement?

Relatively great. About 40% of the world's largest cities lie within 160 kilometers (100 miles) of a plate boundary. By the year 2000, about 290 million people will be living in high-risk areas. About 80% of those at risk live in developing nations, where seismic safety is not a high priority in building design. Another 210 million people are also threatened by active volcanoes; most live along the subduction zones of the Pacific Ring of Fire, where 75% of Earth's 850 active volcanoes are located. Since 1600, there have been approximately 262,000 deaths from volcanic eruptions, about 76,000 in this century. See **Figure 3.34** for a graphic assessment of the risk.

7. Did the most violent earthquakes ever recorded in the continental United States occur in a subduction zone?

No, but they do appear to have been related to plate tectonics. The first occurred on 16 December 1811 at 0200 (2:00 A.M.), near the town of New Madrid on the Missouri frontier; the second, even more powerful, struck the same area on 7 February 1812. According to Walker (1981), the shocks were so powerful that church bells rang as far away as Richmond, Virginia; columns of sand and coal dust shot from the Earth; and cabins and forests near the epicenter were flattened. John James Audubon reported that

". . . the ground rose and fell in successive furrows like the ruffled waters of a lake. The Earth waved like a field of corn before a breeze." The Mississippi River changed course because of the new ground levels in the area.

The quake was centered very deep in the continental crust. The weight of sediments carried for millions of years to the center of the continent from the highlands to the east and west caused the crust to fracture along a deep, ancient fault stretching from Arkansas to Illinois. The fault appears to be a remnant of the ancient assembly of Pangaea. In 1886 Charleston, South Carolina, suffered an earthquake nearly as large as the first New Madrid shock. Most Americans think of California as the earthquake capital of the U.S., but the southeastern part of the country has experienced much larger quakes in historic times.

The Richter scale (and the seismometers needed to gauge it) had not been invented at the time of the New Madrid earthquakes. The magnitude of these earthquakes has been estimated from eyewitness descriptions of the events, by changes to local landscapes, and by the costs of reconstruction. The greatest earthquake ever measured occurred in 1960 in the Pacific subduction zone west of Chile. It measured an astonishing 9.4 on the Richter scale! Earthquakes of this magnitude are, fortunately, very rare.

8. How common are large earthquakes?

About every two or three days, somewhere in the world, there's an earthquake of from 6 to 6.9 on the Richter scale—roughly equivalent to the quake that shook Northridge and the rest of southern California in January 1994; or Kobe, Japan, in January 1995. Once or twice a month, on average, there's a 7 to 7.9 quake somewhere. And there is about one 8 to 8.9 earthquake—similar in magnitude to the 1964 earthquake in Alaska or the 1812 New Madrid quake—each year.

Northridge- and Kobe-sized quakes are small and occur fairly frequently. Large losses of life and property can occur when these earthquakes occur in populated areas. Damage estimates from the Northridge earthquake are approaching $20 billion. In Kobe, more than 5,000 people died, and more than 26,000 were injured. Some 56,000 buildings were destroyed; estimates of the cost of reconstruction are approaching $100 billion.

9. Do other Earth-like planets or moons have hot internal layers and slowly drifting lithospheric plates?

The two bodies studied, the moon and Mars, appear to have much thicker crusts than the Earth. The crusts of the moon and Mars make up about 10% of their planetary masses, while the crust of the Earth constitutes only about 0.4% of its mass. The crust of the moon appears to be contiguous; one researcher has called it a "single-plate" planet. Although plate tectonics, at least in the terrestrial

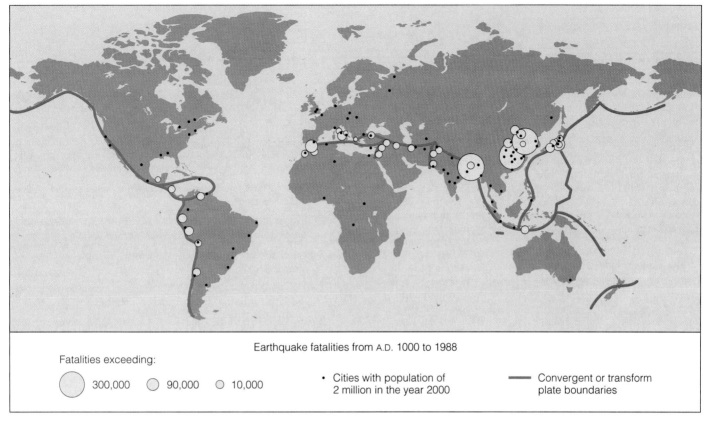

Fatalities exceeding:

○ 300,000   ○ 90,000   ○ 10,000   • Cities with population of 2 million in the year 2000   ── Convergent or transform plate boundaries

Earthquake fatalities from A.D. 1000 to 1988

Figure 3.34 Population centers and earthquake areas. Large cities near plate boundaries in developing countries are likely to have the highest fatalities in a massive earthquake.

sense, almost certainly does not occur on the moon, there is some controversial evidence that the surface of Mars may at one time have consisted of two plates. Mars is pocked by giant volcanoes and marked by gigantic lava plains; it probably had an active interior at some time in the not-too-distant past. The size of martian volcanoes is so huge that some researchers believe hot spots may exist beneath a stationary lithosphere. Could it be that martian plates do not currently move? Mars's current status will have to wait until geologists can send probes or go there themselves.

10. Why does the Earth have a magnetic field?

The Earth's magnetic field is thought to arise from fluid motions and electric currents in the outer core. The likely energy source is heat from the solid inner core causing convection currents in the outer core. These flowing currents of liquid iron, coupled with the rotation of the Earth, form the magnetic field in a process analogous to the way a power station generator produces electricity. The axis of the Earth's magnetic field is tilted about 11° from the axis of the geographic North and South poles. Compass needles point to magnetic north, not to geographic (true) north.

Terms and Concepts to Remember

asthenosphere
atoll
basalt
buoyancy
catastrophism
conduction
continental crust
continental drift
convection
convection current
convergent plate
  boundary
core
crust
density
density
  stratification
divergent plate
  boundary
earthquake

echo sounder
fault
granite
guyot
hot spot
isostatic equilibrium
lithosphere
magma
magnetometer
mantle
mesosphere
oceanic crust
P wave (primary
  wave)
Pacific Ring of Fire
paleomagnetism
Pangaea
Panthalassa
plate
plate tectonics

radioactive decay
radiometric dating
Richter scale
S wave (secondary
  wave)
seafloor spreading
seismic waves
seismometer
shadow zone
spreading center
subduction
subduction zone
terrane
transform plate
  boundary
uniformitarianism
Wegener, Alfred
Wilson, John Tuzo

Study Questions

1. How are Earth's internal layers classified? Which method is more useful in explaining plate tectonics?

2. How is crust different from lithosphere?

3. A ship moving from the Atlantic into the Great Lakes goes from seawater to fresh water. Will the ship sink farther into the water during the passage, stay at the same level, or rise slightly?

4. Some earthquakes are linked to adjustments of isostatic equilibrium. How can this occur? Where would you be likely to experience such an earthquake?

5. Where are the youngest rocks in the seabed? The oldest? Why?

6. What biological evidence supports plate tectonics theory?

7. Would the most violent earthquakes be associated with spreading centers or with subduction zones? Why?

8. Describe the mechanism that powers the movement of the lithospheric plates.

9. What evidence can you cite to support the theory of plate tectonics? What questions remain unanswered? Which side would you take in a debate?

10. Why are the continents about 20 times older than the oldest ocean basins?

For Further Study

Badash, L. 1989. "The Age-of-the-Earth Debate." *Scientific American,* August, 90–96. An excellent summary of the history of the question.

Ballard, R. D., et al. 1975. "Manned Submersible Observations in the FAMOUS Area: Mid-Atlantic Ridge." *Science* 190 (no. 4210): 103–8. An early report of the results of Project FAMOUS.

Bolt, B. A. 1982. *Inside the Earth: Evidence from Earthquakes.* New York: Freeman.

Bott, M. H. P. 1982. *The Interior of the Earth: Its Structure, Constitution, and Evolution.* New York: Elsevier. A prime technical reference.

Burchfiel, B. 1983. "The Continental Crust." *Scientific American,* September, 130–42.

Carr, A., and P. J. Coleman. 1974. "Seafloor Spreading Theory and the Odyssey of the Green Turtle." *Nature* 249: 128–30.

Courtillot, V., and J. Besee. 1987. "Magnetic Field Reversals, Polar Wander, and Core–Mantle Coupling." *Science* 237 (no. 4819): 1140–47. The jerky movement of the magnetic poles may be correlated to episodes of faster or slower continental drift. Includes new information on the transfer of heat from the core to the mantle and on the formation of convection cells.

Dalziel, I. W. D. 1995. "Earth before Pangaea." *Scientific American,* January, 58–63.

Dietz, R. S. 1961. "Continent and Ocean Basin Evolution by Spreading of the Sea Floor." *Nature* 190: 854–57. A historic paper by one of the originators of the theory of seafloor spreading.

Francheteau, J. 1983. "The Oceanic Crust." *Scientific American,* September , 114–29. Companion piece to the Burchfiel article, above.

Frohlich, C. 1989. "Deep Earthquakes." *Scientific American,* January, 48–55. How can rock fail at temperatures and pressures that prevail hundreds of kilometers down? Summarizes Wadati's contributions. (*Note:* See paper by W. H. Green, below.)

Fryer, P. 1992. "Mud Volcanoes of the Marianas." *Scientific American,* February, 46–52. Mantle rock, transformed into mud by water distilled from the subducting Pacific Plate, oozes up along faults near the Mariana Trench to form mountains on the seafloor.

Glen, W. 1982. *The Road to Jaramillo: Critical Years in the Revolution in Earth Science.* Palo Alto, CA: Stanford University Press. An entertaining history of the theory of plate tectonics.

Green, W. H., II. 1994. "Solving the Paradox of Deep Earthquakes." *Scientific American,* September, 64–71. Because the inside of the Earth is plastic and not brittle, geophysicists have known that earthquakes should not happen deep in the Earth. The physics of subducting plates near the bottom of the asthenosphere appears to cause quakes there.

Hamblin, K. W. 1992. *The Earth's Dynamic Systems.* 6th ed. Minneapolis: Burgess Publishing. A systems approach to geological processes; very well illustrated.

Kennett, J. P. 1982. *Marine Geology.* Englewood Cliffs, NJ: Prentice-Hall. A standard text.

Kerr, R. 1984. "The Deepest Hole in the World." *Science* 224 (no. 4656): 1420. Account of a recent attempt to drill into the mantle of the Earth.

Kerr, R. 1992. "Having It Both Ways in the Mantle." *Science* 258 (4 December): 1576–78. New evidence suggests that the mantle might be layered and still mix from top to bottom.

MacDonald, G. A., A. T. Abbott, and F. L. Peterson. 1983. *Volcanoes in the Sea: The Geology of Hawaii.* 2d ed. Honolulu: University of Hawaii Press. Information on Loihi and a new chapter on plate tectonics as it relates to the origin of the islands.

Macdonald, K. C., D. S. Scheirer, and S. M. Carbotte. 1991. "Mid-Ocean Ridges: Discontinuities, Segments and Giant Cracks." *Science* 253 (no. 5023): 986–94. Why are mid-ocean ridges long and continuous rather than divided into shorter segments?

McPhee, J. A. 1993. *Assembling California.* New York: Farrar, Straus & Giroux. A master writer turns his attention to the chaotic terranes of the West.

Menard, H. W. 1986. *The Ocean of Truth: A Personal History of Global Tectonics.* Princeton: Princeton University Press. History of the geological revolution by one of its major scientists. Menard's last book; thoughtful and philosophically written.

Nance, R. D., T. R. Worsley, and J. B. Moody. 1988. "The Supercontinent Cycle." *Scientific American,* July, 72–79. The process of continents breaking and moving appears to be cyclical; great continents break and "heal" at 450-million-year intervals.

Powell, C. S. 1991. "Peering Inward." *Scientific American,* June, 100–111. Excellent review article outlining controversial new views of the Earth's interior.

Press, F., and R. Siever. 1994. *Understanding Earth.* New York: Freeman. An excellent general text with specific emphasis on plate tectonics.

Raymo, C. 1983. *The Crust of Our Earth.* Englewood Cliffs, NJ: Prentice-Hall. Clearly written, well-illustrated nontechnical account.

Schwarzbach, M. 1986. *Alfred Wegener, the Father of Continental Drift.* Madison, WI: Science Tech. A biography (translated from German) that offers many insights into Wegener's life.

Stanley, S. M. 1986. *Earth and Life Through Time.* New York: Freeman. Discusses the biological as well as geological ramifications of plate tectonics.

Walker, B. 1981. *Earthquake.* Chicago: Time–Life Books. Excellent accounts of the Alaska and New Madrid earthquakes.

Wegener, A. 1929. *Die Entstehung der Kontinente und Ozeane* (The Origins of the Continents and Oceans). 4th ed. Reprint. New York: Dover, 1966. Often quite strange, though prophetically insightful. Wegener selected his data carefully to support his hypothesis.

Wenkham, R. 1987. *The Edge of Fire: Volcano and Earthquake Country in Western North America and Hawaii.* San Francisco: Sierra Club Books. An intimate account of life on the edge (and center) of a lithospheric plate, with spectacular photographs and clear writing.

Wesson, P. S. 1971. "Objections to Continental Drift and Plate Tectonics." *Journal of Geology* 80: 185–97. A courageous dissenting view; very instructive, though dated.

White, R. S., and D. P. McKenzie. 1989. "Volcanism at Rifts." *Scientific American,* July, 62–71. A slightly higher temperature of the upwelling mantle may cause spectacular volcanic outbursts.

Note: *Oceanus* 34 (no. 4, Winter 1991–92) is dedicated to the subject of mid-ocean ridges.

CONTINENTAL MARGINS AND OCEAN BASINS

Ticks and Tones

Terrestrial geologists lead a relatively easy life. They may suffer desert heat, tropical rain, or Antarctic cold, but at least they have *direct* access to the rocks and sediments they study. They can look across the land to see how the specimens relate to surrounding rock formations. Marine geologists, on the other hand, almost never have the opportunity to study their targets in context. They rarely get to handle the rocks around the specimens they're after; they rarely even see their specimens in place. The darkness at depth—the blackness beyond the lights—means they never experience spacious vistas of undersea mountain ranges, valleys, or

sediment-covered basins. Often, they must be content to study rocks collected from the ocean bottom by dredges or robots. A few hearty geologists may use scuba equipment to survey a nearshore area and obtain their own study specimens, but they dare not risk exposure in deep water. The frail human body is no match for the cold pressurized depths where important formations lie.

And so marine geologists venture forth in pressure-proof hulls—insulated, temperature controlled, and dry. Once on the bottom, they look through a thick plastic window toward their unreachable goal less than a meter away. Their touch is extended by the arms of

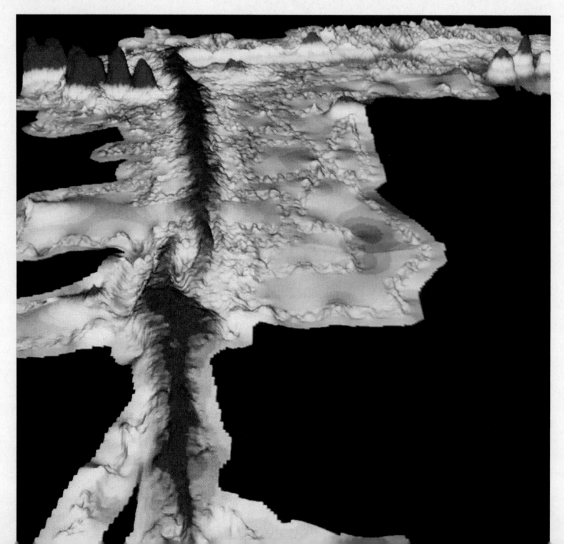

A Sea Beam image of the East Pacific Rise near 9°N, an area where the ocean basin is spreading at the rate of about 8 centimeters (3 inches) a year. Note the unusually offset ridge axis at the center of the image and the volcanoes (here in red false color) at the top.

remote manipulators, their sight by finely detailed television pictures. But the urge to turn over the next stone, to probe under the surface layer, is often impossible to satisfy.

Inevitably, marine geologists have turned to remote sensors to probe places they cannot go. They use robots to return samples, tethered video cameras to record pictures, automated labs to analyze water composition, and echo sounders to coat the ocean floor with ticks and tones.

Remote sensing has a long history. As we have seen, the first systematic deep soundings were made by the Rosses in the early 1800s. Later, after the invention of photography, the first fuzzy images of the ocean floor were taken with leaky cameras attached to very long ropes. The pictures were of poor quality, covered a very small area, and were difficult to place in context. Today the most successful methods for visualizing the seabed use reflected sound instead of light. By the 1950s, precision acoustical sounders were able to measure water depths to an accuracy of about 1.8 meters (6 feet), but the images they produced were little more than wavy lines representing the contour of the bottom directly under the track of the research ship. Despite their limitations, these echo sounders gave oceanographers their first clear overview of basin topography.

In the 1980s sophisticated acoustical equipment, high-speed computers, improved submersibles, and highly accurate satellite navigation systems made it possible to examine the seafloor in unprecedented detail. The Sea Beam multibeam echo sounder system, which generated this relief map of the fast-spreading East Pacific Rise, can cover tens of thousands of square kilometers of the seafloor in a single 30-day cruise. The data collected on the cruise is later processed by land-based computers into this kind of smooth, three-dimensional contour map. In an instant geologists can see patterns that once would have required days or years of concentrated, even dangerous study to discern. As the precision of our equipment improves, the basins begin to yield their deep, dark secrets.

CHAPTER OVERVIEW

The ocean floor can be divided into two regions: continental margins and deep-ocean basins. The continental margin, the relatively shallow ocean floor nearest the shore, consists of the continental shelf and the continental slope. The continental margin shares the structure of the adjacent continents, but the deep-ocean floor away from land has a much different origin and history. Prominent features of the deep-ocean basins include rugged oceanic ridges, flat abyssal plains, occasional deep trenches, and curving chains of volcanic islands. Most of the ocean floor is blanketed with sediment. The processes of plate tectonics, erosion, and sediment deposition have shaped both the continental margins and ocean basins.

THE TOPOGRAPHY OF OCEAN FLOORS

Most people think an ocean basin is shaped like a giant bathtub. They imagine that the continent drops off steeply just beyond the surf and that the ocean is deepest somewhere out in the middle. Until the mid-1800s, natural scientists generally shared this view, but soundings made in the late 1840s revealed shallow areas near the center of the Atlantic, casting doubt on the deepest-at-the-center hypothesis. The *Challenger* Expedition of the late 1870s, which took 492 soundings around the world, found submerged mountains in the mid-Atlantic that were connected into an extensive ridge. In the late 1920s, *Meteor* researchers using echo sounders traced this imposing mid-ocean ridge for much of its length.

Echo sounder studies made before and during World War II discovered other unexpected features. For example, *Challenger* scientists had discovered that shallow extensions of the continents intruded into the Atlantic;

echo soundings showed that this was a worldwide phenomenon. They also found that the deepest parts of the ocean lie near basin *edges* rather than at their centers. Echo sounders powerful enough to probe beneath the layers of sediment on the ocean floor revealed hidden hills, mountains, and steep valleys. These surveys have shown that the seabeds are complex structures indeed; as **Figure 4.1** shows, their contours certainly don't resemble the simple shape of a bathtub.

Surveys with sounding lines and echo sounders can reveal the seabed's shape, but other tools such as the gravimeter must be used to explain the underlying *reasons* for this shape. The **gravimeter,** a sensitive device that measures the pull of gravity at different places on Earth's surface, was developed in the 1930s. The force of gravity depends on the proximity of mass: Nearby heavy objects cause the pull of gravity to be greater, but less massive materials at the same distance cause the pull to be less. If a gravimeter is positioned at a spot over the

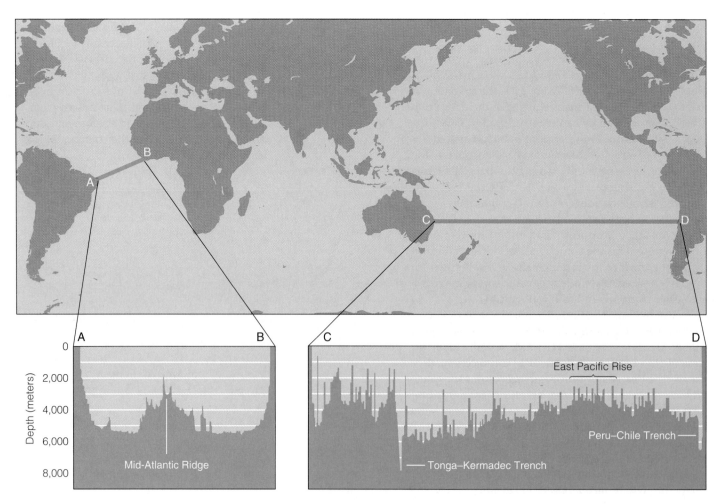

Figure 4.1 Typical cross sections of the Atlantic and Pacific ocean basins. The vertical axis is greatly exaggerated.

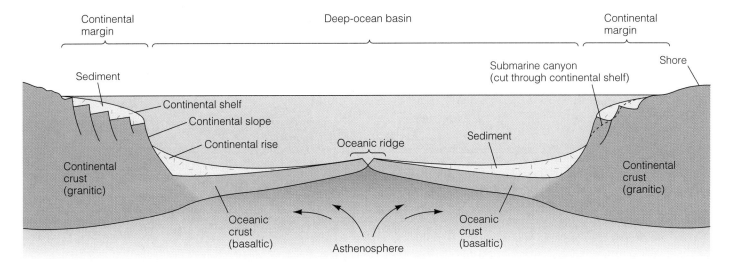

Figure 4.2 Cross section of a typical ocean basin flanked by passive continental margins. (The vertical scale has been exaggerated to emphasize the basin contours.)

ocean floor where the distance to the bottom is known, the gravimeter's reading can be interpreted to indicate the density of the material beneath the bottom. Gravimeter measurements suggested that the undersea edges of continents are made of relatively light granitic rock buried beneath layers of sediment. Readings taken farther from shore indicated that the deep-ocean floor is heavier basalt. The theory of seafloor spreading helped explain why granite and basalt are distributed in that way, and a more detailed picture of the ocean floor began to emerge as plate tectonics theory developed in the early 1960s.

We now have a better understanding of ocean floor shape and structure, thanks in large measure to the wide range of technological tools described in **Box 4.1.** Notice in **Figure 4.2** the transition between the thick granitic rock of the continents and the relatively thin basalt of the deep-sea floor. Near shore the features of the ocean floor are similar to those of the adjacent continents because they share the same granitic basement. The transition to basalt marks the true edge of the continent and divides ocean floors into two major provinces. The submerged outer edge of a continent is called the **continental margin.** The deep-sea floor beyond the continental margin is properly called the **ocean basin.**

Figure 4.3 is a **hypsographic curve** (*hypsos* = height, *graphos* = to write), a plot of the area of Earth's surface above any given elevation or depth above or below sea level. Note that more than half of the Earth's solid surface is at least 3,000 meters (10,000 feet) below sea level. The average depth of the ocean is much greater than the average elevation of the continents: The average depth of the world ocean is 3,790 meters (12,430 feet), but the

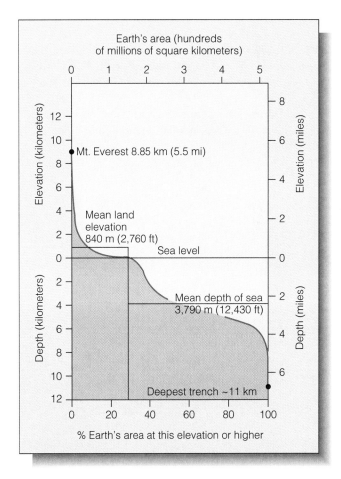

Figure 4.3 The hypsographic curve, a graph showing the distribution of elevations and depths on the Earth.

BOX 4.1 ● *Deep Tools*

Technological advances have permitted researchers to explore the seafloor in person or by using remotely controlled robots. The sophistication and cost of deep-diving devices increases with their depth capability: Scuba outfits cost only a few hundred dollars, but a human-carrying vehicle suitable for a trip to the abyssal plains costs millions.

With the proper mix of gases and adequate training, a scuba diver (**Figure a**) can descend to about 60 meters (200 feet). With special mixtures of breathing gases and dry suits inflated with argon to avoid heat loss, divers have reached a depth of 300 meters (984 feet). Below that depth a small research submarine is better equipped to withstand the cold and pressure. One of the best is the maneuverable, compact, two-person diving saucer used by geologist Francis Shepard and others to do much of the research on canyons reported in this chapter. A California company has recently launched *Deep Flight* (**Figure b**), a more maneuverable successor capable of depths of 1,200 meters (4,000 feet). Protected by a clear dome, the pilot can explore features of the continental shelf and upper continental slope.

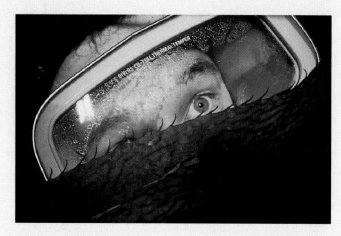

a With the proper mix of gases and adequate training, a scuba diver can descend to about 60 meters (200 feet). A simple scuba setup costs around $500.00. With special mixtures of breathing gases and dry suits inflated with argon to avoid heat loss, divers have reached a depth of 300 meters (984 feet).

b *Deep Flight,* a submersible featuring greatly improved visibility and maneuverability, was launched in 1994. Its depth capability is 1,200 meters (4,000 feet).

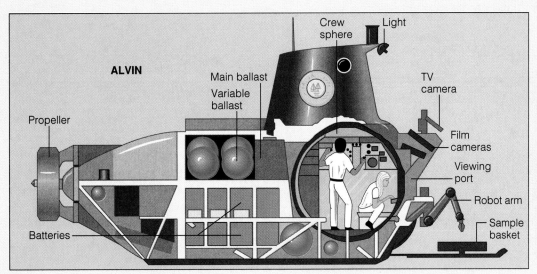

ALVIN

Propeller

Main ballast
Variable ballast

Crew sphere · Light

TV camera

Film cameras

Viewing port

Robot arm

Sample basket

Batteries

c *Alvin*, the best-known and oldest of the six deep-diving manned research submarines now in operation, has made more than 2,500 dives in 30 years of operation. *Alvin* carries three people and is capable of diving to 4,000 meters (13,120 feet).

d *Shinkai 6500*, built in 1981 by Mitsubishi Heavy Industries, is the world's newest and deepest-diving person-carrying submersible. Three people can be accommodated in the vehicle's titanium pressure hull. *Shinkai 6500* reached its design depth of 6,500 meters (21,320 feet) on 11 August 1989.

For descents deeper than mid-slope, however, stronger and more costly submersibles like *Alvin* must be used. *Alvin* (**Figure c**) was commissioned in 1964 and operated jointly by the Woods Hole Oceanographic Institution, the National Science Foundation, the Office of Naval Research, and the National Oceanic and Atmospheric Administration. The most famous of the research submersibles, *Alvin* has explored the Mid-Atlantic Ridge near the Azores at 2,700 meters (9,000 feet), measuring rock temperatures, collecting water samples for chemical analysis, and photographing the rounded basaltic extrusions known as *pillow lava*. Even

Alvin cannot reach all of the abyssal plains of Earth; for really deep voyages a bathyscaphe must be employed. Unfortunately, the only bathyscaphe in existence is *Trieste*, the record-setting vehicle discussed in Box 4.3, and she's in mothballs in Washington, D.C.

The newest deep-diving vehicle capable of carrying a human crew is Mitsubishi Heavy Industries' *Shinkai 6500* (**Figure d**). Now the deepest-diving submersible in operation, *Shinkai 6500* can safely descend to a depth of 6,500 meters (21,300 feet) and thus can reach all but the deepest trench floors, or about 98% of the world ocean bottom.

e Very fast, strong, and silent, a nuclear submarine makes an ideal platform for oceanographic research. The U.S. Navy made one of its nuclear submarines available to marine scientists during the summer of 1993. The research cruise aboard USS *Pargo*, a *Sturgeon*-class submarine strengthened for operation in ice, began in Groton, Connecticut, and ended in Bergen, Norway. More than 9,000 kilometers (5,600 miles) of underway data were collected in the deep Arctic Ocean beneath the pack ice. Water samples were taken for chemical and biological analysis (as seen here), current probes were released and tracked, and buoys were placed to monitor weather and surface water conditions. The data from the cruise will be placed in the public domain and results published in unclassified literature. More cruises are planned.

f *Jason, Jr.,* a small ROV (remotely operated vehicle) that sends information back to operators in the mothership by fiber-optic cable.

g ABE (Autonomous Benthic Explorer), a true robot designed to spend months at a time working alone underwater. ABE will be released to follow a preprogrammed set of research instructions and will resurface up to a year later to share its adventures with its designers.

A modern U.S. Navy nuclear submarine makes a superb manned research vehicle. The USS *Pargo* has recently been made available to marine scientists (**Figure e**). Operating under Arctic ice, *Pargo* made an unprecedented series of measurements in the summer of 1993.

Remotely operated robots are becoming more popular with researchers. Known as ROVs (remotely operated vehicles), they're cheaper to operate and safer than people-carrying vehicles. The first of these, ANGUS (Acoustically Navigated Geophysical Undersea Surveyor), photographed the Galápagos Rift in 1977 (while being towed from a ship) and found hydrothermal vent communities. ANGUS's successor, *Argo,* was developed in the early 1980s; *Argo* sent live television pictures of the deep-sea floor. The small and maneuverable *Jason, Jr.* (**Figure f**) was provided with its own propulsion system and (operated from inside *Alvin*) penetrated a portion of the wreckage of RMS *Titanic,* sending back live color video in 1986.

The state of the art in deep-sea exploration is the true robot. Unlike its predecessors, ABE (Autonomous Benthic Explorer, **Figure g**) will not be tethered to a mothership by a cable. Researchers at Woods Hole Oceanographic Institution launched ABE in the summer of 1995. After a test period, ABE will be set loose to spend months at a time underwater. Instruments guided by an on-board computer will snap photographs, take water temperature, and make other environmental measurements. About the size of a compact car, ABE can move in all directions to look inside crevices, back away from vents gushing superheated water, and navigate narrow canyons. After its programmed run of up to a year, ABE will pop to the surface, broadcast its location, and wait patiently to be picked up by a research vessel. Back in the lab, its computers will communicate with researchers' computers; its adventures will be replayed and its data analyzed.

Perhaps the most imaginative new technology is *telepresence,* the extension of a person's senses by remote manipulators. A scientist might wear a helmet containing small stereo television screens and earphones, and place his hands in special gloves equipped with tactile feedback units. His head and hand movements would be duplicated by a robot on the seafloor, and sensations "felt" by the robot would be relayed back to the scientist through the TVs, earphones, and gloves. He would thus have the sensation of being on the seafloor and could take samples, manipulate tools, or just look around. Other researchers could watch or participate at distant locations via the information superhighway. Oceanographers have set a course for interesting times!

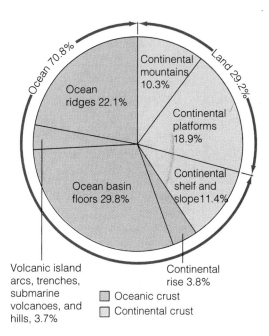

Figure 4.4 Features of Earth's solid surface shown as percentages of the planet's total surface.

average height of the continents is only 840 meters (2,760 feet). This is because continental crust is relatively thicker than oceanic crust, and also because of the difference in density between granitic rock and basalt: The less dense granitic continents "float" in isostatic equilibrium above the level of the dense rock of the ocean basins. The shape of the hypsographic curve results from the smaller volume of continental crust relative to oceanic crust. **Figure 4.4** shows the relative amounts of Earth's surface composed of continental and oceanic structures and the important subdivisions of each.

CONTINENTAL MARGINS

You learned in Chapter 3 that lithospheric plates converge, diverge, or slip past each other. As you might expect, continental margins are greatly influenced by this tectonic activity. Continental margins facing the edges of *diverging* plates are called **passive margins** because relatively little earthquake or volcanic activity is associated with them. They surround the Atlantic; so passive margins are sometimes referred to as *Atlantic-type* margins. Continental margins near the edges of *converging* plates (or near places where plates are slipping past each other) are called **active margins** because of their earthquake and volcanic activity. Because of their prevalence in the Pacific, active margins are sometimes referred to as *Pacific-type* margins.

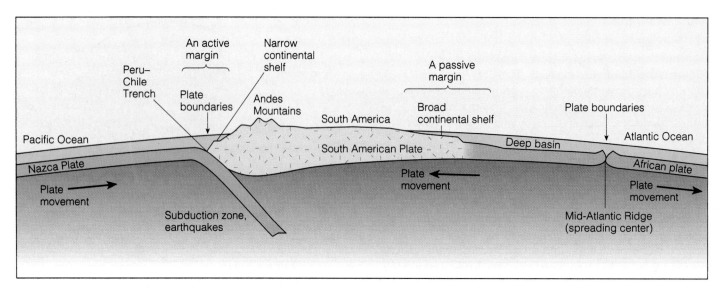

Figure 4.5 Typical continental margins bordering the leading (tectonically active) and trailing (passive) edges of a moving continent. (Exaggerated vertical scale.)

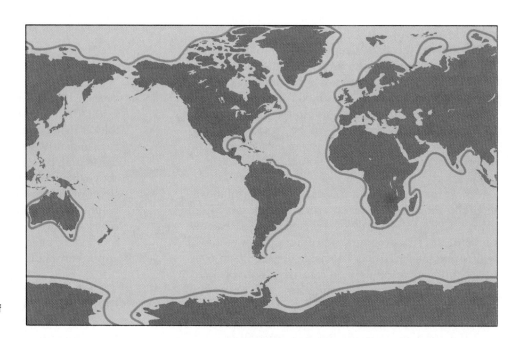

Figure 4.6 The worldwide distribution of passive continental margins.

Active and passive margins west and east of South America are shown in **Figure 4.5.** Note that active margins coincide with plate boundaries but passive margins do not. **Figure 4.6** shows the worldwide distribution of passive margins. Passive margins are also found outside the Atlantic, but active margins are confined mostly to the Pacific.

Continental margins have two main divisions: a shallow, nearly flat continental *shelf* close to shore and a more steeply sloped continental *slope* to seaward.

Continental Shelves

The shallow submerged extension of a continent is called the **continental shelf.** Continental shelves, like the adjacent continents, are underlain by granitic continental crust. They are much more like the continent than like the deep-ocean floor, and they may have hills, depressions, and mineral and oil deposits similar to those on the dry land nearby. Earth's continental shelves are shown in **Figure 4.7.** Taken together, the area of the continental shelves is 7.4% of the Earth's ocean area.

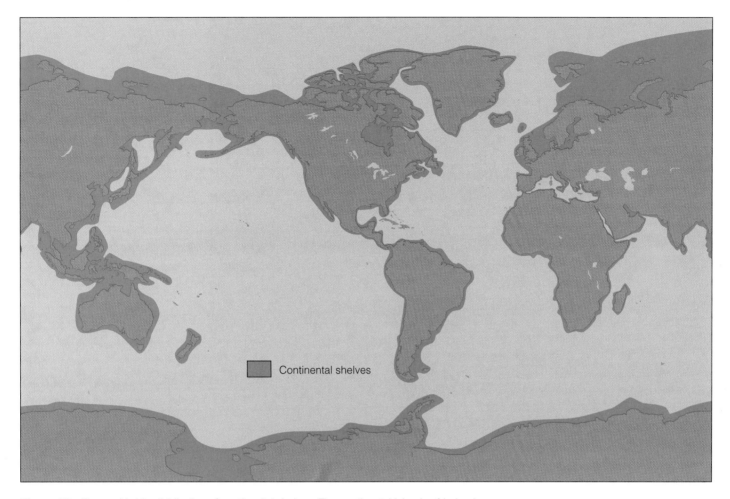

Figure 4.7 The worldwide distribution of continental shelves. The continental islands of Ireland and Great Britain are part of the continent of Europe, and Alaska connects to Siberia.

Figure 4.8 shows a passive-margin continental shelf characteristic of Atlantic Ocean edges. The broad shelf extends far from shore in a gentle incline, typically 1.7 meters per kilometer (about 9 feet per mile), much less than the slope of a well-drained parking lot. Shelves along the margin of the Atlantic Ocean often reach 350 kilometers (220 miles) in width, and end at a depth of about 150 meters (500 feet) where a steeper drop-off begins.

The passive margin shelves of the Atlantic Ocean formed as the fragments of Pangaea were carried away from each other by seafloor spreading. The continental lithosphere, thinned during initial rifting, cooled and contracted as it moved away from the spreading center, submerging the trailing edges of the continents and forming the shelves.

Most of the material comprising a shelf comes from erosion of the adjacent continental mass. Rivers assist in

passive shelf building by transporting huge loads of sediments to the shore from far inland. In some places the sediments accumulate behind natural dams formed by ridges of granitic crust. The weight of the sediment isostatically depresses the continental edges and allows the sediment load to grow even thicker. Sediment at the outer edge of a shelf can be up to 15 kilometers (9 miles) thick and 150 million years old.

The width of a shelf is usually determined by proximity to a plate boundary. You can see in Figure 4.5 that the shelf at the *passive* margin (east of South America) is broad, but the shelf at the *active* margin (west of South America) is very narrow. The widest shelf, 1,280 kilometers (800 miles) across, lies north of Siberia in the tectonically quiet Arctic Sea. Shelf width depends not only on tectonics, but also on marine processes: Fast-moving ocean currents can sometimes prevent sediments from accumulating. For example, the east coast of Florida has

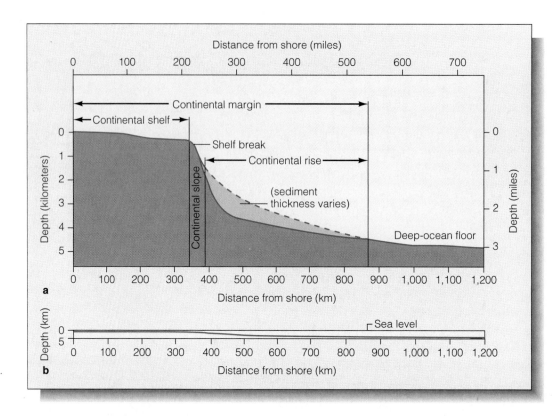

Figure 4.8 The features of a passive continental margin.
(a) Vertical exaggeration 150:1.
(b) No vertical exaggeration.

a very narrow shelf because there is no offshore dam and because the swift current of the nearby Gulf Stream scours surface sediment away.

The shelves of the active Pacific margins are generally not as broad and flat as Atlantic shelves. An example is the abbreviated shelf off the west coast of South America, where the steep western slope of the Andes Mountains continues nearly uninterrupted beneath the sea into the depths of the Peru–Chile Trench (see again Figure 4.5). Active-margin shelves have more varied topography than passive-margin shelves; the character of continental shelves at an active margin is determined more by faulting, volcanism, and tectonic deformation than by sedimentation. Off the coast of southern California, for example, the shelves are broken into a series of deep depressions and high ridges that were formed by faulting and folding. The crests of some of the ridges emerge as islands off the southern California coast. The islands of Catalina, San Clemente, Anacapa, and others are examples of such high ridges.

A few Pacific shelves are broad, however. As in the Atlantic, natural offshore dams trap sediments and form shelves; but in the Pacific the sediment-trapping dams are more commonly offshore chains of volcanoes or lines of coral reefs. Volcanic activity east of China and Southeast Asia has formed a broad basin that is now filling with sediments and is one of the largest shelves in the Pacific (see Figure 4.7).

Because of their gentle slope, continental shelves are greatly influenced by changes in sea level. About 120 million years ago, during the Cretaceous period, high rates of seafloor spreading were associated with the expansion in volume of the mid-ocean ridges. The enlarged ridges displaced seawater and caused sea level to rise. Late Cretaceous sea level may have been about 300 meters (1,000 feet) higher than it is at present, flooding about 35% of the continents and resulting in proportionally greater shelf areas. In contrast, around 18,000 years ago—at the height of the last **ice age** (major period of glaciation)—the ridges had become less active, and massive ice caps covered huge regions of the continent. The water that formed the thick ice sheets was derived from the ocean, and sea level fell about 120 meters (400 feet) below its present position (see **Figure 4.9**). The continental shelves were almost completely exposed, and the surface area of the continents was about 18% greater than it is today. Land mammals inhabited the emerged shelves—more than 50 teeth of mastodons and mammoths have been collected from the continental shelf off the East and Gulf coasts of the United States. Rivers and waves cut into the sediments that had accumulated during periods of higher sea

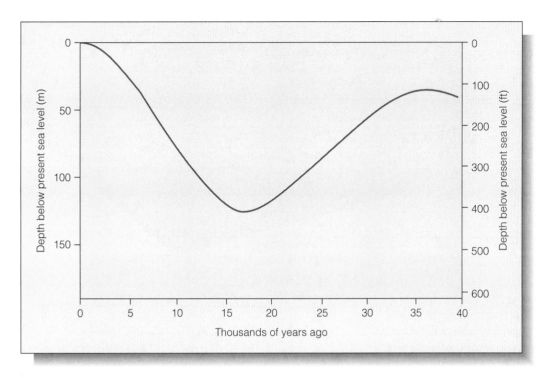

Figure 4.9 Changes in sea level over the last 40,000 years, as traced by radiometric dating of shallow-water marine organisms plotted against the depth at which they were recovered. The dip below present sea level occurred during the last ice age, when much water was trapped in ice sheets, leaving less liquid water for the ocean.

level, and they transported some coarse sediments to their present locations at the shelves' outer edges. Sea level began to rise again when the ice caps melted, and sediments again began to accumulate on the shelves. (More on the history of sea level change will be found in the discussion of coasts in Chapter 12.)

The continental shelves have been the focus of intense exploration for natural resources. Because shelves are the submerged margins of continents, any deposits of oil or minerals along the continental edges are likely to continue offshore. Water depth over shelves averages only about 75 meters (250 feet); so large areas of the shelves are accessible to mining and drilling activities. Many of the techniques used to find and exploit natural resources on land can also be used on the continental shelves. Resource development requires intense scientific investigations, and our understanding of the geology of the shelves has benefited greatly from the search for offshore oil and natural gas.

Continental Slopes

The **continental slope** is the transition between the gently tilting continental shelf and the deep-ocean floor.

Continental slopes are formed of sediments that reach the built-out edge of the shelf and fall over the side. At active margins a slope may also include marine sediments scraped off a descending plate during subduction. The inclination of a typical continental slope is about 4° (or 70 meters per kilometer, 370 feet per mile), equal to the steepest slope allowed on the interstate highway system. As Figure 4.8b implies, even the steepest of these slopes is not precipitous: A 25° slope is the greatest incline yet discovered. In general, continental slopes at active margins are steeper than those at passive margins. Continental slopes average about 20 kilometers (13 miles) wide and end at the continental rise, usually at a depth of about 3,700 meters (12,000 feet). The bottom of the continental slope is the true edge of a continent.

The **shelf break** marks the abrupt transition from continental shelf to continental slope. The depth of water at the shelf break is surprisingly constant—about 150 meters (500 feet) worldwide—but there are exceptions. The great weight of ice on Antarctica, for example, has isostatically depressed that continent, and the depth at the shelf break is 300–400 meters (1,000–1,300 feet). The shelf break in Greenland is similarly depressed.

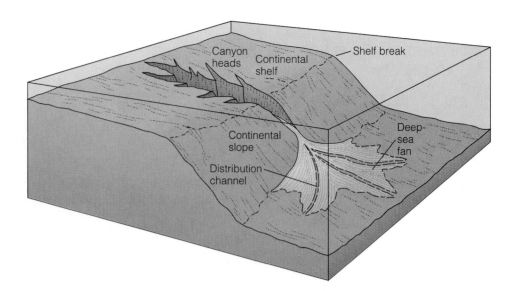

Figure 4.10 A submarine canyon.

Submarine Canyons

Submarine canyons are V-shaped indentations incised into the continental shelf and slope, often terminating on the deep-sea floor in a fan-shaped wedge of sediment (**Figure 4.10**). More than 100 submarine canyons nick the edge of nearly all of Earth's continental shelves. The canyons generally trend at right angles to the shoreline (and shelf edge), sometimes beginning very close to shore. Congo Canyon actually extends into the African continent as a deep estuary at the mouth of the Congo River. These enigmatic features are quite large. In fact, submarine canyons similar in size and profile to Arizona's Grand Canyon have been discovered!

Hudson Canyon, a typical large canyon on a passive margin, is shown in **Figure 4.11**. Like many submarine canyons, Hudson Canyon is located just offshore of the mouth of a river or stream—in this case, New York's Hudson River. Because of their similarity to canyons on land, submarine canyons appear to have been created by erosion; so marine geologists initially thought the canyons were carved into the shelves by stream erosion at times of lower sea level. But most researchers agree that sea level has never fallen more than 200 meters (660 feet) below its present level in the last 600 million years. Stream erosion could account for the shape of the uppermost parts of the canyons. However, since submarine canyons can be traced to depths in excess of 3,000 meters (10,000 feet) below sea level, stream erosion could not have played a direct role in cutting their lower depths. What, then, caused the submarine canyons to form? As is so often the case in the history of marine science, the answer to this riddle was associated with a peculiar event.

A sharp earthquake struck the Grand Banks south of Newfoundland at 1531 (3:31 P.M.) Greenwich time on 18 November 1929. The quake's **epicenter,** the point on the surface of the Earth directly above the focus of an earth-

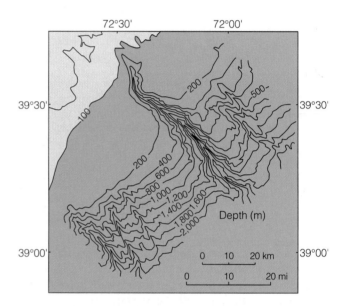

Figure 4.11 A topographic map of Hudson Canyon.

quake, was located on the continental slope at a water depth of about 2,200 meters (7,200 feet). Ordinarily such an earthquake would not attract much attention, but a number of important transatlantic communication cables lay on the ocean floor south of the earthquake area. One by one these cables failed. A segment of the Western Union cable nearest the epicenter broke almost immediately, and a French cable about 160 kilometers (100 miles) south of the Western Union cable parted at 1835 (6:35 P.M.). Another Western Union cable to the south snapped early the next morning, and cables even farther south were damaged later that day. Officials had expected an occasional random failure, but a *sequence* of cable failures, from north to south along 480 kilometers (300 miles) of seafloor in the mid-Atlantic, was baffling.

Local landslides or sediment liquefaction triggered by

the earthquake probably caused the first cable breaks, and although no one thought of it right away, an abrasive underwater "avalanche" almost certainly caused the more distant breaks. These avalanche-like sediment movements, called **turbidity currents,** are caused when turbulence mixes sediments into water above a sloping bottom. The sediment-filled water is denser than the surrounding water; so the thick muddy fluid runs down the slope at speeds up to 27 kilometers (17 miles) per hour. **Figure 4.12** is a rare photograph of a turbidity current.

What is the connection between turbidity currents and submarine canyons? Some geologists have suggested that the canyons have been formed by abrasive turbidity currents plunging down the canyons. Small amounts of debris may cascade continuously down the canyons (**Figure 4.13**), but earthquakes can shake loose huge masses of sediment that rush down the edge of the shelf, scouring the canyon deeper as they go. In this way the canyons can be cut to depths far below the reach of streams even during the low sea levels of the ice ages.

Continental Rises

Along passive margins the oceanic crust at the base of the continental slope is covered by an apron of accumulated sediment called the **continental rise** (see again Figure 4.8). Sediments from the shelf slowly descend to the ocean floor along the whole continental slope, but most of the sediments that form the continental rise are transported to the area by turbidity currents. The width of the rise varies from about 100 to 1,000 kilometers (63 to 630 miles), and its slope is gradual—about one-eighth that of the continental slope. One of the widest and thickest continental rises has formed in the Bay of Bengal at the mouths of the Ganges–Brahmaputra River, the most sediment-laden of the world's great rivers.

DEEP-OCEAN BASINS

Away from the margins of continents the structure of the ocean floor is quite different. Here the seafloor is a blanket of sediments overlying basaltic rocks up to 5 kilometers (3 miles) thick. Deep-ocean basins constitute more than half of the Earth's surface.

The deep-ocean floor consists mainly of oceanic ridge systems and the adjacent sediment-covered plains. Deep basins may be rimmed by trenches or by masses of sediment. Flat expanses are interrupted by islands, hills, active and extinct volcanoes, and active zones of seafloor spreading. The sediments on the deep floor reflect the history of the surrounding continents, the biological productivity of the overlying water, and the ages of the basins themselves.

Figure 4.12 A turbidity current flowing down a submerged slope off the island of Jamaica. The propeller of a submarine caused the turbidity current by disturbing sediment along the slope.

Figure 4.13 A continuous cascade of sediment at the head of San Lucas submarine canyon (off the coast of Baja California, Mexico), which may be eroding the narrow gorge in conjunction with occasional turbidity currents.

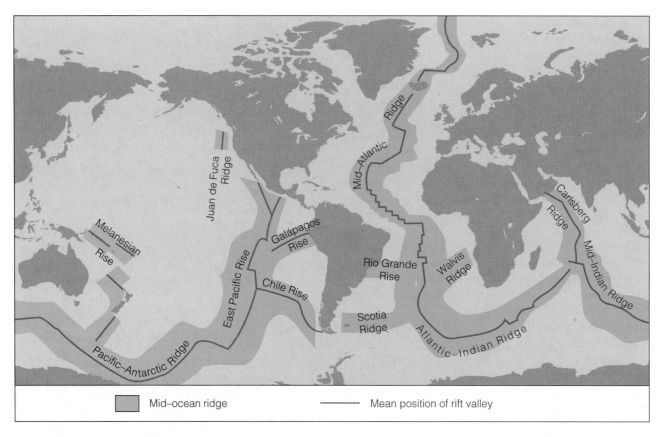

Figure 4.14 The oceanic ridge system.

Oceanic Ridges

If the ocean evaporated, the oceanic ridges would be the Earth's most remarkable and obvious feature. An **oceanic ridge** is a mountainous chain of young basaltic rock at the active spreading center of an ocean. Stretching 65,000 kilometers (41,000 miles), more than 1½ times the Earth's circumference, oceanic ridges girdle the globe like seams surrounding a softball (**Figure 4.14**). The rugged ridges, which often are devoid of sediment, rise about 2 kilometers (1.25 miles) above the seafloor. In places they project above the surface to form islands such as Iceland (see **Box 4.2**), the Azores, and Easter Island. Oceanic ridges and their associated structures account for 23% of the world's solid surface area (all the land above sea level accounts for 29%). Although these features are often called mid-ocean ridges, less than 60% of their length actually exists at mid-ocean.

As we saw in our discussion of plate tectonics, the rift zones associated with oceanic ridges are sources of new ocean floor where lithospheric plates diverge. The oceanic ridges are widest where they are most active. The youngest rock is located at the active ridge center, and rock becomes older with distance from the center. As

crust cools, it shrinks and subsides. Slowly spreading ridges have a steeper profile than rapidly spreading ones because slowly diverging seafloor cools and shrinks closer to the spreading center. **Figure 4.15,** a topographic map of the North Atlantic, clearly shows the great extent of the Mid-Atlantic Ridge, a typical oceanic ridge. Features on the older basalt floor farthest from the spreading center tend to be obscured by a layer of sediment that thickens away from the center.

As can be seen in Figure 4.15, the Mid-Atlantic Ridge is fractured at more or less regular intervals by transform faults. A fault is a break in a rock mass along which movement has occurred, and **transform faults** are planes along which rock masses slide horizontally past one another (**Figure 4.16**). When segments of a ridge system are offset, the fault connecting the axis of the ridge is a transform fault. Shallow earthquakes are common along transform faults. Since the ocean floor cannot expand evenly on the surface of a sphere, plate divergence on the spherical Earth can only be irregular and asymmetrical, and transform faults and fracture zones result. **Fracture zones** are seismically inactive areas that show evidence of past transform fault activity. While segments of a lithospheric plate on either side of a transform fault

BOX 4.2 ● *Life on the Ridge*

The 253,000 inhabitants of the island country of Iceland live on a section of the Mid-Atlantic Ridge that rises above sea level just south of the Arctic Circle. Maintaining civilization on the exposed seafloor at an active spreading center is an interesting experience. There are a few benefits: Irrigation water is piped from hot springs to protect the crops from summer frosts; steam is tapped from underground reservoirs to generate electrical power to all residents in their homes; and nearly every community is supplied with hot-water swimming pools and fresh vegetables, grown year-round in steam-heated greenhouses. But there are two important drawbacks: frequent earthquakes and vigorous volcanoes. It is estimated that the volcanoes of Iceland have produced one-third of the total lava that has flowed onto dry land since the year 1500. With more hot springs and vapor vents than any other country in the world, Iceland is the most geologically active place on Earth.

Iceland is broken by parallel fissures stretching north and south along a central rift zone (**Figure a**); they are similar to fissures on the adjacent submerged ridge. As along the rest of the ridge, tectonic forces are stretching Iceland in the east-west direction, and magma is rising to fill the fissures and become new seafloor. The lava venting from Iceland's volcanoes is chemically identical to lava rising into nearby submerged segments of the Mid-Atlantic Ridge, but the cooling lava doesn't assume the ingotlike pillow shape of submerged lava—for two reasons: First, lava in contact with cold seawater cools and solidifies more rapidly than lava in contact with air; second, the water pressure at the ocean bottom keeps volatile gases in solution during the cooling process, thus maintaining the extruded bunlike shape. Instead, the Icelandic lava blasts from the crevices, explosively releases its gases, and spreads over the desolate countryside.

The central rift zone of Iceland is almost completely covered by lava deposited since the last ice age—that is, within the last 20,000 years. Its forbidding central desert—the Odádahraun, the "Desert of Evil Deeds"—is one of the most lifeless places in the world, so alien-seeming that it was used as a training ground for the American astronauts who walked on the moon. It was here in 1783 that a series of fissures more than 24 kilometers (15 miles) long yawned open and began vomiting lava and corrosive gases. Over the next year 585 square kilometers (226 square miles) of central Iceland was covered by new rock. Sulfurous vapors and acidic

a Thingvellir graben, a rift valley in Iceland. The valley is actually the central part of the Mid-Atlantic Ridge, exposed above sea level.

rains poisoned the crops and killed most of the livestock on the island. One-fifth of the 49,000 inhabitants died of famine that winter.

The section of the Mid-Atlantic Ridge of which Iceland is a part spreads about 15 centimeters (6 inches) per year. Icelanders must contend with considerable geological inconvenience (and occasional bouts of terror), but at least their country is growing bigger by the day.

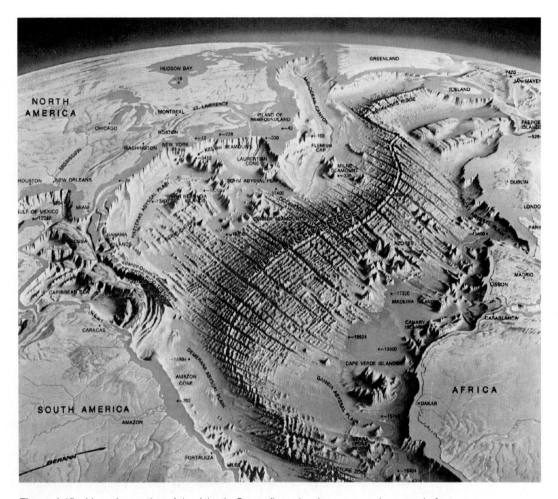

Figure 4.15 Map of a portion of the Atlantic Ocean floor showing some major oceanic features: mid-ocean ridge, transform faults, fracture zones, submarine canyons, seamounts, continental rises, trenches, and abyssal plains. Depths are in feet. The map is vertically exaggerated.

Figure 4.16 Transform faults and fracture zones along an oceanic ridge.

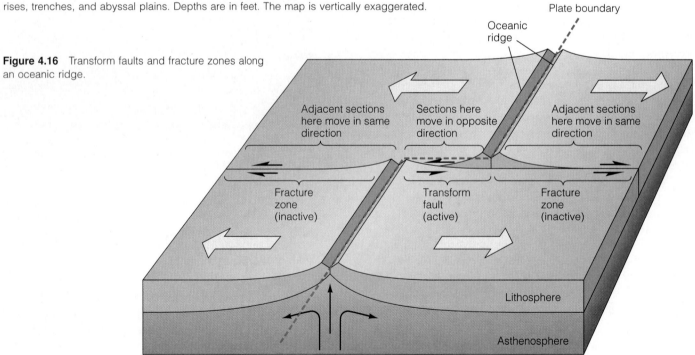

move in *opposite* directions from each other, the plate segments adjacent to a fracture zone move in the *same* direction, as Figure 4.16 shows. Earthquakes rarely occur along fracture zones.

Hydrothermal Vents

Perhaps the strangest features of the ocean basins are the **hydrothermal vents.** Hot springs were discovered on oceanic ridges in 1977 by Robert Ballard and J. F. Grassle of the Woods Hole Oceanographic Institution. Diving in *Alvin* at 3 kilometers (1.8 miles) near the Galápagos Islands along the East Pacific Rise (an oceanic ridge), they came across rocky chimneys up to 20 meters (66 feet) high, from which dark, mineral-laden water was blasting at 350°C (660°F) (**Figure 4.17**). Only the great pressure at this depth prevented the escaping water from flashing to steam. These *black smokers,* as they were quickly nicknamed, fascinate marine geologists. It is believed that water descends through fissures and cracks in the ridge floor until it comes into contact with very hot rocks associated with active seafloor spreading. There the superheated, chemically active water dissolves minerals and gases and escapes upward through the vents by convection.

Since that first discovery, vents have been found on the Mid-Atlantic Ridge east of Florida, in the Sea of Cortez east and south of Baja California, and on the Juan de Fuca Ridge off the coast of Washington and Oregon. Scientists now believe that hydrothermal vents may be very common on oceanic ridges, especially in zones of rapid seafloor spreading. In July 1990 vents were discovered in fresh water, at the bottom of Lake Baikal in southern Siberia. This discovery suggests that the world's oldest and deepest lake may someday become part of the ocean as Asia slowly breaks apart.

Not all vents form chimneys of mineral deposits—some are simply cracks in the seabed, or porous mounds, or broad segments of ocean ridge floor through which warm, mineral-laden water percolates upward. Cooler vents result when hot, rising water mixes with cold bottom water before reaching the surface. Water temperature in the vicinity of most hydrothermal vents averages 8–16°C (46–61°F), much warmer than usual for ocean bottom water, which has an average temperature of 3–4°C (37–39°F). We will study the unusual communities of animals that populate the vents in Chapter 17.

A volume of water equal to the volume of the world ocean is thought to circulate through the hot oceanic crust at spreading centers every 1 to 10 million years! The heat and chemicals issuing from these structures may play important roles in the chemical composition of seawater and the atmosphere, and in the formation of mineral deposits.

Figure 4.17 A black smoker discovered at a depth of about 2,800 meters (9,200 feet) along the East Pacific Rise.

Abyssal Plains and Abyssal Hills

A quarter of Earth's surface consists of abyssal plains and abyssal hills. *Abyssal* is an adjective derived from a Greek word meaning "without bottom." While this is obviously not literally true, you can appreciate how the term came into use following the *Challenger* Expedition's laborious soundings of these extremely deep areas!

Abyssal plains are flat, cold, featureless expanses of sediment-covered ocean floor found on the periphery of all oceans. They are most common in the Atlantic, less so in the Indian Ocean, and relatively rare in the active Pacific, where peripheral trenches trap most of the sediments flowing from the continents. They lie between the continental margins and the oceanic ridges about 3,700 to 5,500 meters (12,000 to 18,000 feet) below the surface (see again Figure 4.15). The Canary Abyssal Plain, a huge plain west of the Canary Islands in the North Atlantic, has an area of about 900,000 square kilometers (350,000 square miles).

Abyssal plains are extraordinarily flat. A 1947 survey by the Woods Hole Oceanographic Institution ship *Atlantis* found that a large Atlantic abyssal plain varies no more than a few meters in depth over its entire area. Such flatness is caused by the smoothing effect of the layers of sediment, which often exceed 1,000 meters (3,300 feet) in thickness. Most of the sediment that forms the abyssal plains appears to be of terrestrial or shallow-water origin, not derived from biological activity in the ocean above. Some of it may have been transported to the plains by turbidity currents. These deep sediment layers mask irregularities in the underlying ocean crust, but echo sounders can "see" through this sediment to reveal the complex topography of the basaltic basin floor below. The broad basaltic shoulders of the Mid-Atlantic Ridge extend beneath this cloak of sediment almost as far as the bordering continental slopes.

Abyssal plain sediments may not be thick enough to cover the underlying basaltic floor near the edges bordering the oceanic ridges. Here the plains are punctuated by **abyssal hills**—small, sediment-covered extinct volcanoes or intrusions of once-molten rock, usually less than 200 meters (650 feet) high. These abundant features are associated with seafloor spreading; they form when newly formed crust moves away from the center of a ridge, stretches, and cracks. Some blocks of the crust drop to form valleys, and others remain higher as hills. Lava erupting from the ridge flows along the fractures, coating the hills. This helps explain why abyssal hills occur in lines parallel to the flanks of the nearby oceanic ridge, and why they occur most abundantly in places where the rate of seafloor spreading is fastest. Abyssal plains and hills account for nearly all of the area of deep-ocean floor that is not part of the oceanic ridge system.

Seamounts and Guyots

The ocean floor is dotted with thousands of "islands" that do not rise above the surface of the sea. These projections are called **seamounts.** Seamounts are circular or elliptical, usually more than a kilometer (0.6 mile) in height, with relatively steep slopes of 20° to 25°. (Abyssal hills, in contrast, are much more abundant, almost always smaller, and not as steep.) Seamounts may be found alone or in groups of from 10 to 100. Though many form at hot spots (see Figure 3.29), most are thought to be submerged inactive volcanoes that formed at spreading centers (see **Figure 4.18** and Figure 3.31). Movement of the lithosphere away from spreading centers has carried them outward and downward to their present positions. As many as 10,000 seamounts are thought to occur in the Pacific, about half the world total.

Guyots are flat-topped seamounts that once were tall enough to approach or penetrate the sea surface. Gener-

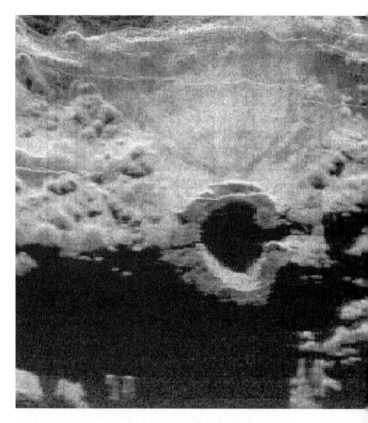

Figure 4.18 A seamount on the Mid-Atlantic Ridge is "photographed" by a high-resolution sonar device towed 3,000 meters (10,000 feet) below the surface

ally they are confined to the west-central Pacific. The flat top suggests that they were eroded by wave action when they were near sea level. Their plateaulike tops eventually sank too deep for wave erosion to continue wearing them down. Like the more abundant seamounts, most guyots were formed near spreading centers and transported outward and downward by seafloor spreading (see again Figure 3.31). As noted in Chapter 3, chains of seamounts and guyots are compelling evidence for the theory of plate tectonics.

Trenches and Island Arcs

A **trench** is an arc-shaped depression in the deep-ocean floor. These flat-bottomed creases in the seafloor occur where a converging oceanic plate is subducted. The water temperature at the bottom of a trench is slightly cooler than the near-freezing temperatures of the adjacent flat ocean floor, reflecting the fact that trenches are underlain by old, relatively cold ocean crust sinking into the upper mantle. Trenches (and their associated island arcs topped by erupting volcanoes) are the most active geological features on Earth. Great earthquakes and tsunamis (huge waves we will discuss in Chapter 11) are

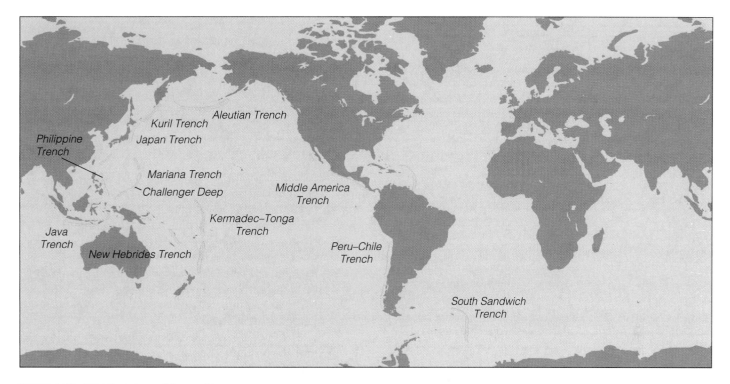

Figure 4.19 Trench regions of the world.

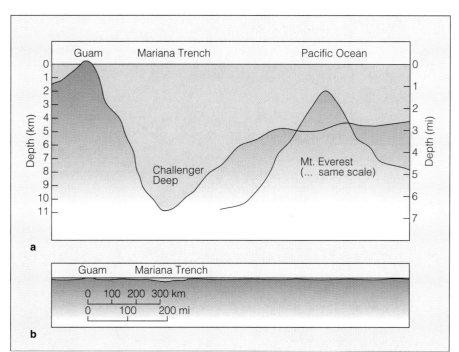

Figure 4.20 The Mariana Trench. (a) Comparing the Challenger Deep and Mount Everest at the same scale shows that the deepest part of the Mariana Trench is about 20% deeper than the mountain is high. (b) The Mariana Trench shown without vertical exaggeration.

often born in them. **Figure 4.19** shows the distribution of the ocean's major trenches. Not surprisingly, most are around the edges of the active Pacific.

Trenches are the deepest places in the Earth's crust, punching 3 to 6 kilometers (1.8–3.7 miles) deeper than the adjacent basin floor. The ocean's greatest depth is the Mariana Trench of the western Pacific, where *Trieste* set the world's deep-diving record (see **Box 4.3**). Near the place where *Trieste* came to rest, the ocean bottom is 11,022 meters (36,163 feet) below sea level, 20% deeper than Mount Everest is high (see **Figure 4.20**). The Mariana Trench is about 70 kilometers (44 miles) wide and 2,550 kilometers (1,600 miles) long, typical dimensions for these structures.

BOX 4.3   ●   Trieste *at the Bottom*

Transporting marine scientists to work often presents daunting obstacles. The most difficult problem is to reach extreme depths, but, amazingly, scientists have visited the bottom of the deepest ocean basin. On 23 January 1960 U.S. Navy lieutenant Don Walsh and Dr. Jacques Piccard descended to a depth of 10,918 meters (6.78 miles) into the Challenger Deep, an area of the Mariana Trench discovered in 1951 by the British oceanographic research vessel *Challenger II*.[1] The vehicle used in the descent was the **bathyscaphe** *Trieste* (**Figures a, b**, and **c**), a deep-diving submersible designed like a blimp with a very strong and thick (and cramped) steel crew sphere suspended below. A blimp uses helium gas for buoyancy, but a gas would be compressed by water pressure; so gasoline, which is relatively incompressible, was used to provide lift.

First, the *Trieste* made a series of increasingly deep practice dives; then the fateful day arrived. The descent out of the ocean's sunlit upper zone was slow at first. The bathyscaphe actually stopped when it encountered the thermocline, a zone of denser water that caused the vessel to become more buoyant. When a little gasoline was valved from the maneuvering tank, the explorers

[1]The position of the Challenger Deep is marked in Figure 4.19, and its depth in relation to Mount Everest is shown in Figure 4.20.

began falling—at a rate of about 4 feet per second—into a dark, calm, cold realm never before seen.

Don Walsh wrote of his experience in 1979:

> At about 600 feet we entered a zone of deepening twilight where colors faded into gray. By 1,000 feet the light had gone completely. We turned out the lights in the sphere to watch for the luminescent creatures that are sometimes visible at this level. We saw very few. Eventually we turned the cabin lights back on and briefly tested the forward lights that throw a beam in front of the observation window. Formless plankton streamed past, giving us a sensation of great speed.
>
> There were minor incidents such as the small leak that always developed in one of the hull connectors—a place where wires from lights and instruments on the outside of the sphere pass through the hull. The leak started at about 10,000 feet. It was an old friend, a tiny drip, drip, drip. I timed the drips and found no change from before, which meant that

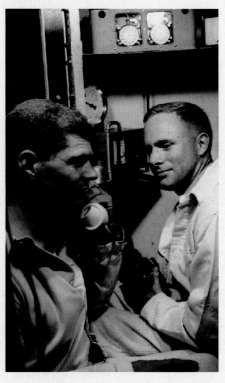

b Andreas Rechnitzer and Don Walsh prepare *Trieste* for a deep dive.

a The bathyscaphe *Trieste*, the pioneering deep submersible, as seen on the surface.

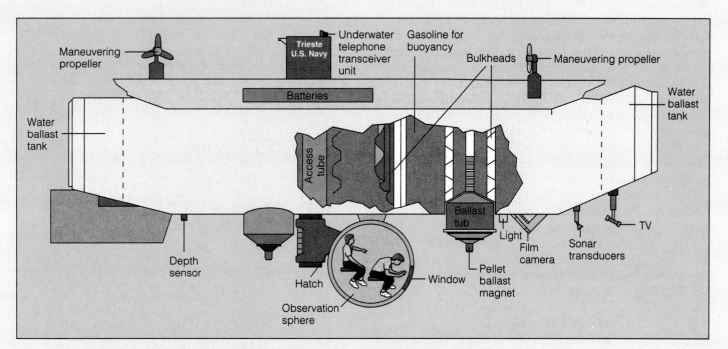

c An engineering cutaway. *Trieste* is 15.7 meters (59.5 feet) long, and the pressure hull protecting her two-person crew is 13 to 18 centimeters (5 to 7 inches) thick. She made 130 dives, including three world-record dives. Her last voyage was in 1963, and *Trieste* is currently on display in the Navy Yard Museum in Washington, D.C.

it had not become more serious. We expected it to disappear at about 15,000 feet, when the water pressure packed the plastic sealer in more tightly—and it did.

At 27,000 feet we checked our rate of descent to 2 feet per second by dumping some shot ballast. We were not too sure of the underwater currents here, and we did not want to go crashing into a wall of the trench by mistake. As we neared 30,000 feet I started thinking about the changes we had planned to make when we got to within 1,000 feet of the bottom—which we now were expecting to find only another 3,500 feet below us. I was running through a mental checklist when we heard and felt a powerful crack. The sphere rocked as though we were on land and going through a mild earthquake.

We waited anxiously for what might happen next. Nothing did. We flipped off the instruments and the underwater telephone so that we could hear better. Still nothing happened. We switched the instruments back on and studied the dials that would tell us if something critical had occurred. No, we had our equilibrium and were descending exactly as before.

We checked our speed to half a foot a second and continued. At that rate, time and distance pass very slowly, and I think for the first time in the dive both of us had the feeling of awe that comes from exploring the totally unknown.

I did not take my eyes off the fathometer, and Jacques never stopped watching out of the tiny porthole with its weak probe of light. No bottom was in sight. . . . Soon Jacques could see a difference in the effect of our light in the water, as the rays reflected off the bottom. As we approached the floor I called the fathometer readings to Jacques in fathoms: "Thirty . . . twenty . . . ten . . ." At eight he called that he could see the bottom.

At 1:10 P.M. we sank gently onto the soft floor. A great cloud of silt rose around us. The fifteen-man, Navy civilian/military team had set a record that could not be broken.

The voyagers took temperature readings and spotted a pair of bottom fish and some crustaceans before ascending. With the possible exception of the moon, no place visited by humans has been more hazardous to explore.

Source: Oceans (no. 6, November–December 1979).

Trenches are curving chains of V-shaped indentations. The trenches are curved because of the geometry of plate interactions on a sphere. The convex sides of these curves face the open ocean (see again Figure 4.19).[1] The trench walls on the island side of the depressions are steeper than those on the seaward side, indicating the direction of plate subduction. The sides of trenches become steeper with depth, normally reaching angles of about 10°–16° before flattening to a floor underlain by thick sediment. (Parts of the concave wall of the Kermadec–Tonga Trench are the world's steepest at 45°.) No continental rise occurs along coasts with trenches because the sediment that would form the rise ends up at the bottom of the trench.

Island arcs, curving chains of volcanic islands and seamounts, are almost always found paralleling the concave edges of trenches. As you may remember from Chapter 3, trenches and island arcs are formed by tectonic and volcanic activity associated with subduction. The descending lithospheric plate contains some materials that melt as the plate sinks into the mantle. These materials rise to the surface as magmas and lavas that form the chain of islands behind the trench. The Aleutian Islands, most Caribbean islands, and the Marianas Islands are island arcs. (See Figure 3.23 for a review of the processes involved in their construction.)

A GRAND TOUR

Researchers at the National Geophysical Data Center have generated a map of the world ocean floor (**Figure 4.21**). The graphic shows all the features discussed in this chapter. These features—and a basic understanding of the geological reasons for their existence—will help you recall the dramatic nature and history of the seafloor that we have discussed in the past two chapters.

[1]To see a result of the deformation of a sphere, press on a Ping-Pong ball with your thumb. The resulting depression will have curved edges; the convex sides of the curves will face your thumb. Note the similarity of that pattern to the position of the trenches in the northern and western Pacific (Figure 4.19).

Figure 4.21 A technological *tour de force* derived from data provided to the National Geophysical Data Center from satellites and shipborne sensors, shows all the features discussed in this chapter. These features—and a basic understanding of the geological reasons for their existence—will help you recall the dramatic nature and history of the seafloor that we have discussed in the past two chapters. Key to features: (3) Aleutian Trench, (6) Clipperton Fracture Zone, (8) East Pacific Rise, (9) Easter Island, (10) Galápagos Rift, (14) Gulf of Aden, (15) Gulf of California, (16) Hawaiian Islands, (17) Juan de Fuca Ridge, (22) Project FAMOUS Area, (25) Mariana Trench, (28) Mid-Atlantic Ridge, (30) Monterey Canyon, (31) Pacific-Antarctic Ridge, (32) Red Sea, (38) Iceland, (39) Azores, (40) Tristan da Cunha.

Q-AND-A

1. I might be interested in exploring some of the continental shelf using scuba. How does scuba work?

    Scuba is an acronym for self-contained underwater breathing apparatus. A scuba diver uses a regulator to valve air from the high-pressure tank on his back into his lungs. At sea level the regulator puts just enough air into the lungs to balance the pressure of air above, so that he can take a breath. At 30 meters (100 feet) depth, the regulator must provide a total of four atmospheres of pressure to allow the same breath—one atmosphere to compensate for the weight of the air, plus three atmospheres to compensate for the weight of the water above. If the diver were suddenly transported to sea level, the imbalance would result in a dangerous—and messy—incident. That's why divers ascend slowly, breathe regularly, and never skimp on regulator quality.

    Scuba divers breathe compressed air, a mixture of nitrogen and oxygen. Pure oxygen is not used because it has dangerous side effects at high pressures. The practical limit for normal scuba operation is about 125 feet. Deeper dives require mixtures of helium and oxygen to prevent the toxic effects of high-pressure nitrogen on the nervous system. If a way could be found to perfuse all of a diver's air-filled spaces with fluid (an oxygen-carrying, low-viscosity silicon compound has been proposed), humans could possibly dive to depths of thousands of feet on helium–oxygen mixtures, requiring only protection from the cold.

2. You mentioned that Walsh and Piccard saw animals at a depth of 10,918 meters (6.78 miles). The pressure there is more than 8 tons per square inch. How could any living thing withstand such crushing pressure?

    It seems impossible until you think about your own situation. Humans live at the bottom of a surprisingly heavy "ocean" of air: A column of air 1 inch square extending from sea level to the top of the atmosphere weighs about 6.3 kilograms (14 pounds). This pressure presents no problem to us because we have the same pressure inside of us pushing out. The animals at great depths also have the same pressure inside as outside; that is, they live in perfect hydrostatic balance. They notice the weight of water above them no more than you notice the weight of air above you.

3. Turbidity currents seem important in forming canyons and distributing deep sediments over abyssal plains. Has anybody ever seen a turbidity current in action?

    Yes, surprisingly. In the late 1940s the Dutch geologist Philip Kuenen produced turbidity currents in his laboratory by pouring muddy water into a trough with a sloping bottom. His observations confirmed nineteenth-century reports that the muddy Rhône River continued to flow in a dense stream along the bottom of Lake Geneva. In the 1960s Robert Dill and Francis Shepard viewed sandfalls in Scripps Canyon from the diving saucer (Figure 4.13), and French researchers have recently photographed these currents in the Mediterranean.

4. Why is Iceland one of only a few places in the world where oceanic crust is found above sea level?

    Oceanic crust is thin and heavy. Because of isostasy, the ocean floor lies at a lower elevation than the thicker, less dense, higher continents. Water filled the lower elevations first, submerging nearly all the basaltic basement we now call ocean floor. In areas of rapid seafloor spreading, the peaks of the mid-ocean ridges are occasionally pushed toward the ocean surface by the large volume of mantle material rising from below. Large quantities of erupted magma (lava) then build the crests above sea level to form islands. The Azores is another place on the Mid-Atlantic Ridge where this is happening.

Terms and Concepts to Remember

abyssal hill	fracture zone	passive margin
abyssal plain	gravimeter	seamount
active margin	guyot	shelf break
bathyscaphe	hydrothermal vent	submarine canyon
continental margin	hypsographic curve	transform fault
continental rise	ice age	trench
continental shelf	island arc	turbidity current
continental slope	ocean basin	
epicenter	oceanic ridge	

Study Questions

1. Why did people think an ocean was deepest at its center? What changed their minds?

2. Why were gravimeter readings important in determining ocean basin structure?

3. What do the facts that (a) granite underlies the edges of continents and (b) basalt underlies deep-ocean basins, suggest? (Hint: Consider thicknesses and densities.)

4. The terms *leading* and *trailing* are also used to describe continental margins. How do you suppose these words relate to *active* and *passive,* or *Atlantic-type* and *Pacific-type* used in the text?

5. What forces control the shape of a continental shelf? A continental slope? A continental rise?

6. What evidence *visible from land* would suggest that a submarine canyon might lie just offshore?

7. Why are abyssal plains relatively rare in the Pacific?

8. Why are trenches and island arcs curved? Is the descent to the bottom steeper on the convex side of the arc or on the concave side? Why the difference? Why do you think most trenches are in the western Pacific? (Hint: Check the position and action of the East Pacific Rise.)

9. Distinguish among abyssal hills, seamounts, guyots, and island arcs.

10. Answer this question if you have already read Chapter 3: Your time machine has been programmed to deliver you to Frankfurt, Germany, on a chilly evening in January 1912, to hear Wegener's lectures on continental drift. What two illustrations from this chapter would you take with you to cheer him up after the lecture? Why did you select those particular illustrations?

11. What is an oceanic ridge? Are they always literally in *mid*-ocean? How are oceanic ridges and trenches related?

For Further Study

Baker, R. 1985. "Submarine Canyons." *California Diver,* July/August, 17–19. Excellent discussion for those qualified divers wishing to investigate these features firsthand.

Burke, K. 1979. "The Edges of the Ocean: An Introduction." *Oceanus* 22 (no. 3, Fall). (Entire issue dedicated to ocean/continent boundaries.)

Cone, J. 1991. *Fire Under the Sea.* New York: Morrow. Popular account of the discovery of volcanic hot springs on the ocean floor and the development of the equipment and techniques necessary for their discovery.

Decker, R., and B. Decker. 1989. *Volcanoes.* New York: Freeman. Information on Icelandic and other volcanoes.

Dill, R. F. 1964. "Contemporary Submarine Erosion in Scripps Submarine Canyon." Ph.D. dissertation, Scripps Institution of Oceanography, University of California, San Diego. Firsthand account of canyon erosion. Available on microfilm or through interlibrary loan.

Edmond, J., and K. Von Damm. 1983. "Hot Springs on the Ocean Floor." *Scientific American,* April, 78–93. Information on the "black smokers" and their associated wildlife.

Emery, K. O. 1969. "The Continental Shelves." *Scientific American,* September, 107–22. Paper by one of the notable researchers in the topic.

Francheteau, J. 1983. "The Oceanic Crust." *Scientific American,* September, 114–29.

Heezen, B., and C. Hollister. 1971. *The Face of the Deep.* New York: Oxford University Press. Magnificent compendium of pictures of the deep-sea bed. Wonderfully written by two of the pioneers in deep-sea research.

Kaharl, V. A. 1990. *Water Baby: The Story of Alvin.* New York: Oxford University Press. Outlines 25 years of remarkable discoveries.

Kennett, J. P. 1982. *Marine Geology.* Englewood Cliffs, NJ: Prentice-Hall. A standard text.

Macdonald, K. C., and P. J. Fox. 1990. "The Mid-Ocean Ridge." *Scientific American,* June, 72–79. The article discusses complex ridge offset patterns and the reason most ridges have an undulating upper surface.

Menard, H. W. 1986. *Islands.* New York: Freeman. A book that combines beautiful illustrations with a distinguished researcher's insight.

Press, F., and R. Siever. 1986. *Earth.* 4th ed. San Francisco: Freeman. Fine general textbook in geology.

Prior, D. B., et al. 1987. "Turbidity Current Activity in a British Columbia Fjord." *Science* 237 (no. 4820): 1330–33. Good recent paper showing how new techniques and equipment are being used in marine research.

Raymo, C. 1983. *The Crust of the Earth: An Armchair Traveler's Guide to the New Geology.* Englewood Cliffs, NJ: Prentice-Hall. Well-illustrated popular account.

Shepard, F. 1969. *The Earth Beneath the Sea.* Rev. ed. New York: Atheneum. Another firsthand account from the grandfather of modern marine geology.

Shepard, F. 1977. *Geological Oceanography.* New York: Crane, Russak. Advanced text by same grandfather. Highly recommended.

Siebold, E., and W. H. Berger. 1993. *The Sea Floor: An Introduction to Marine Geology.* 2d ed. New York: Springer-Verlag. Methods and data are described in this update. Includes new material on physical marine resources.

Walsh, D. 1979. "Twenty Years After the *Trieste* Dive." *Oceans* 12 (no. 6): 37–41. The issue highlights the abyss. Walsh gives a stirring firsthand account of the deepest dive.

Weihaupt, J. G. 1979. *Exploration of the Oceans.* New York: Macmillan. Excellent general reference on floor features.

Willie, P. J. 1976. *The Way the Earth Works.* New York: Wiley. Concise summary of floor anatomy and plate tectonics.

Oceanus 34 (no. 4, Winter 1991–92) is dedicated to the topic of mid-ocean ridges.

5 SEDIMENTS

The Memory of the Ocean

In the early 1960s, researchers aboard the American research ship *Chain* discovered a layer of sediments of unknown composition beneath the Mediterranean Sea. For the next 10 years scientists used continuous seismic profilers (powerful echo sounders) to trace this mysterious layer, which appeared everywhere they probed this inland sea. Unlike most marine sediments, this material reflected back virtually all of the acoustical energy sent toward it. The M-reflector, as it came to be called, covered the basement of the Mediterranean like a thick blanket of snow in a mountain valley.

On 24 August 1970 geologists aboard the deep-sea drilling ship *Glomar Challenger* inspected the first sam-

ples from the layer. They had expected to see finely compacted particles of silts and clays; instead they found pea-sized gypsum (calcium sulfate) gravel, which forms when seawater evaporates. Gypsum occurs in muddy sediments on arid coasts, and on land at the sites of dry lakes and ponds. It had never been reported on the deep-sea floor and does not occur in gravel deposits on the surrounding continents. What could account for its presence?

The mystery deepened as more samples were brought to the surface. One core contained a different form of calcium sulfate, along with other salts commonly found in evaporating seawater. (This core looked

Did huge waterfalls along the Gibraltar precipice refill the Mediterranean Sea? (Painting by Ron Miller.)

like marble and was quickly nicknamed the "Pillar of Atlantis.") Another core showed evidence of algal mats, which form only in water less than 10 meters (33 feet) deep. All these puzzling samples were taken about 200 meters (700 feet) beneath the seabed in 2,000 meters (6,600 feet) of water.

From the evidence contained in these sediment samples, two scientists aboard *Glomar Challenger*, William B. F. Ryan of the Lamont–Doherty Earth Observatory and Kenneth J. Hsü of the Swiss Federal Institute of Technology, pieced together a controversial new history of the Mediterranean Sea. About 6 million years ago, they said, the shallow Strait of Gibraltar—the narrow passage between the Mediterrean and the Atlantic—was blocked by an uplift of the land—a dam, in effect. Water could no longer flow into the Mediterranean; so the sea east of Gibraltar gradually evaporated, leaving behind a desert of gypsum-rich salts—the future M-reflector. Half a million years later, however, either the strait eroded or subsided, or the level of the Atlantic rose — and the ocean flooded back into the basin. The types of fossil microorganisms in the drill cores suggest the rate of refilling was very rapid. In little more than a century, waters equal to 1,000 Niagaras falling across the Gibraltar precipice refilled the Mediterranean Sea!

At least that's how Ryan and Hsü reconstructed the tale of the drill cores. But some scientists who have studied the same evidence doubt that the Mediterranean was ever completely dry. They suggest that most of the floor was studded by alkaline lakes or that perhaps the entire bottom was covered by very shallow, very salty water. The M-reflector could still have formed in this scenario.

Sediments have many stories to tell, some of them still open to interpretation. In a sense sediments are the memory of the ocean because they provide a record—if we can read it—of the recent history of the ocean, and therefore of the planet itself.

CHAPTER OVERVIEW

The ocean floor is covered in most places by layers of sediment. It includes particles from land, from biological activity in the ocean, from chemical processes within water, and even from space. The blanket of marine sediment is thickest at the continental margins and thinnest over the active oceanic ridges. Sediments may be classified by particle size, origin, location, or color. The position and nature of sediments provide important clues to Earth's recent history, and valuable resources can sometimes be recovered from them.

Figure 5.1 Sediment near the crest of the Mid-Atlantic Ridge. The rock outcrops are dusted only lightly with sediment; sponges and gorgonians (relatives of coral) attach to the rocks.

Figure 5.2 Sediment on the tranquil continental rise west of Africa's Sahara Desert. In this geologically quiet area the sediment surface is smooth, and a layer almost 200 meters (660 feet) thick has collected. The small dark object is a sea cucumber.

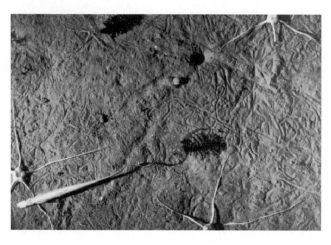

Figure 5.3 Brittle stars and their tracks on the continental slope off New England. The depth is 1,476 meters (4,842 feet).

WHAT SEDIMENTS LOOK LIKE

Sediment is particles of organic or inorganic matter that accumulate in a loose, unconsolidated form. The particles originate from the weathering and erosion of rocks, from the activity of living organisms, from volcanic eruptions, from chemical processes within the water itself, and even from space. Most of the ocean floor is being slowly dusted by a continuing rain of sediments. Accumulation rates vary from a few centimeters per year to the thickness of a dime every thousand years.

Marine sediments occur in a broad range of sizes and types. Beach sand is sediment; so is the mud of a quiet bay and the mix of silt and tiny shells found on the continental margins. Less familiar sediments are the fine clays of the deep-ocean floor, the biologically derived oozes of abyssal plains, and the nodules and coatings that form around hard objects on the seafloor. The origin of these materials—and the distribution and sizes of the particles—depends on a combination of physical and biological processes.

What do sediments look like? That depends on where you look. **Figure 5.1** shows the Mid-Atlantic Ridge, 48°N, at a depth of 2,629 meters (8,626 feet). Sponges and some relatives of coral grow from young rocky outcrops only lightly powdered with sediment. Contrast that rough ridge with the smooth seafloor shown in **Figure 5.2**. The sediment there is about 200 meters (660 feet) thick.

The surface of the sediment is not always smooth. Where bottom currents are swift and persistent, they can cause ripples like those on a streambed. Perhaps the most common irregularities in deep-sea sediments are tracks made by brittle stars (**Figure 5.3**). These widely distributed organisms feed on surface bacteria and fallen particles of organic sediment. Other bottom dwellers, such as sea cucumbers and worms, also leave tracks. Other objects cause irregularities, too. The detritus of human civilization—everything from discarded bottles to barrels of industrial waste—may end up in the sediments of the seabed.

The colors of marine sediments are often quite striking. Sediments of biological origin are white or cream-colored, with deposits high in silica tending toward gray. Deep-sea clays are often a chocolate brown color, from the rusting (oxidation) of iron within the sediments to form iron oxide. Other clays are shades of green or tan. Nodular sediments are a dark sooty brown or black. Some nearshore sediments contain decomposing organic material and smell of hydrogen sulfide, but most are odorless.

A very few areas of the seabed are altogether free of overlying sediments. The water over these areas is not completely sediment-free, but for some reason sediment

Table 5.1 Particle Sizes and Settling Rate in Sediment

Type of Particle	Diameter	Settling Velocity	Time to Settle 4 km (2.5 mi)
Boulder	>256 mm (10 in.)	—	—
Cobble	64–256 mm (>2½ in.)	—	—
Pebble	4–64 mm (⅙–2½ in.)	—	—
Granule	2–4 mm (¹⁄₁₂–⅙ in.)	—	—
Sand	0.062–2 mm	2.5 cm/sec (1 in./sec)	1.8 days
Silt	0.004–0.062 mm	0.025 cm/sec (1/100 in./sec)	6 months
Clay	<0.004 mm	0.00025 cm/sec	50 years[a]

[a]Though the theoretical settling time for individual clay particles is very long, clay particles in the ocean can interact chemically with seawater, clump together, and fall at a faster rate. Small organic particles are often compressed in fecal pellets that can fall more rapidly than would otherwise be possible.

does not collect on the bottom. Strong currents may scour the sediments away; or the seafloor may be too young in these areas for sediments to have had time to accumulate; or hot water percolating upward through a porous seafloor may dissolve the material as fast as it settles.

CLASSIFYING SEDIMENT BY PARTICLE SIZE

Particle size is frequently used to classify sediments. The scheme shown in **Table 5.1** was first devised in 1898 and with refinements has been used by geologists, soil scientists, and oceanographers ever since. In this classification the coarsest particles are boulders, which are more than 256 millimeters (about 10 inches) in diameter. Although boulders, cobbles, and pebbles occur in the ocean, most marine sediments are made of finer particles: **sand, silt, and clay.** The smaller the particle, the more easily it can be transported by streams, waves, and currents. As sediment is transported it tends to be sorted by size; coarser grains, which are moved only by turbulent flow, tend to remain behind finer grains, which are more readily moved. The clays, particles less than 0.004 millimeter in diameter, can remain suspended for very long periods and may be transported great distances by ocean currents before they are deposited.

A layer of sediment can contain particles of similar size, or it can be a mixture of different-sized particles. Sediments composed of particles of one size, say fine sand, are said to be **well-sorted sediments.** Sediments with a mixture of sizes are **poorly sorted sediments.** Sorting is a function of the energy of the environment—the exposure of that area to the action of waves, tides, and currents. Well-sorted sediments occur in an environment where energy fluctuates within narrow limits. Sed-

iments of the calm deep-ocean floor are typically well sorted (see Figure 5.2). Poorly sorted sediments form in environments where energy fluctuates over a wide spectrum. The mix of rubble at the base of a rapidly eroding shore cliff is a good example of poorly sorted sediment. Sediments that have been transported by turbidity currents (which can transport a wide range of grain sizes) tend to be poorly sorted.

CLASSIFYING SEDIMENT BY SOURCE

Another way to classify marine sediments is by their origin. Such a scheme was first proposed in 1891 by Sir John Murray and A. F. Renard after a thorough study of sediments collected during the *Challenger* Expedition. A modern modification of their organization is shown in **Table 5.2.** This scheme separates sediments into four categories by source: terrigenous, biogenous, hydrogenous (or authigenic), and cosmogenous.

Terrigenous Sediments

Terrigenous (*terra* = Earth, *generare* = to produce) sediments are the most abundant. As the name implies, terrigenous sediment originates on the continents or islands near them. Quartz and clay are two common components of terrigenous marine sediments. Quartz, an important mineral in granite, is hard, relatively insoluble, very durable, and can withstand lengthy weathering and transport. Quartz sands washed from the adjacent land are important components of the sediments along continental margins. Feldspar, another important mineral in granite, ultimately decomposes to clay. These tiny particles, which are the chief component of soils, are carried to the ocean by wind, rivers, or streams. Because of their small size, they are easily transported across the

Table 5.2 Classification of Marine Sediments by Source of Particles

Sediment Type	Source	Examples	Distribution	Percent of All Ocean Floor Area Covered
Terrigenous	Erosion of land, volcanic eruptions, blown dust	Quartz sand, clays, estuarine mud	Dominant on continental margins, abyssal plains, polar ocean floors	~45%
Biogenous	Organic; accumulation of hard parts of some marine organisms	Calcareous and siliceous oozes	Dominant on deep-ocean floor (siliceous ooze below about 5 km)	~55%
Hydrogenous (authigenic)	Precipitation of dissolved minerals from water	Manganese nodules, phosphorite deposits	Present with other more dominant sediments	<1%
Cosmogenous	Dust from space, meteorite debris	Tektite spheres, glassy nodules	Mixed in very small proportion with more dominant sediments	0%

Sources: Kennett, 1982; Weihaupt, 1979; Sverdrup, Johnson, and Fleming, 1942.

continental shelf to settle slowly to the deep-ocean floor. Although estimates vary, it appears that about 15 billion metric tons (16.5 billion tons) of terrigenous sediments are transported in rivers to the sea each year, with an additional 100 million metric tons transported annually from land to ocean as fine airborne dust and volcanic ash (see **Figure 5.4**).

Biogenous Sediments

Biogenous (*bio* = life, *generare* = to produce) sediments are the next most abundant marine sediment. The siliceous (silicon-containing) and calcareous (calcium carbonate-containing) compounds that make up these sediments of biological origin were originally brought to the ocean in solution by rivers or dissolved in the ocean at mid-ocean ridges. The siliceous and calcareous materials were then extracted from the seawater by the normal activity of tiny plants and animals to build protective shells and skeletons. Some of this sediment derives from larger mollusk shells or from stationary colonial animals such as corals, but most of the organisms that produce biogenous sediments drift free in the water as plankton. After the death of their owners, the hard structures fall slowly to the bottom and accumulate in layers. Biogenous sediments are most abundant where ample nutrients encourage high biological productivity, usually near continental margins. Over millions of years, organic molecules within these sediments can form oil and natural gas (see Chapter 19 for details).

Hydrogenous Sediments

Hydrogenous (*hydro* = water, *generare* = to produce) sediments are minerals that have precipitated directly from seawater. The sources of the dissolved minerals include submerged rock and sediment, leaching of the fresh crust at oceanic ridges, material issuing from hydrothermal vents, or substances flowing to the ocean in river runoff. As we shall see, the most prominent hydrogenous sediments are manganese nodules, which litter abyssal plains, and phosphorite nodules, seen along some continental margins. Hydrogenous sediments are also called **authigenic** (*authis* = in place, on the spot) because they were formed in the place they now occupy. Though they usually accumulate very slowly, rapid deposition of hydrogenous sediments is possible. The deposition of salts from the evaporating Mediterranean mentioned at the beginning of this chapter occurred relatively quickly.

Cosmogenous Sediments

Cosmogenous (*cosmos* = universe, *generare* = to produce) sediments, which are of extraterrestrial origin, are the least abundant. These particles enter the Earth's high atmosphere as blazing meteors or as quiet motes of dust. Their rate of accumulation is so slow that they never accumulate as distinct layers; they occur as isolated grains in other sediments, rarely constituting more than 1% of any layer. Estimates vary widely, but between 1,000 and 50,000 metric tons (2 million to 110 million

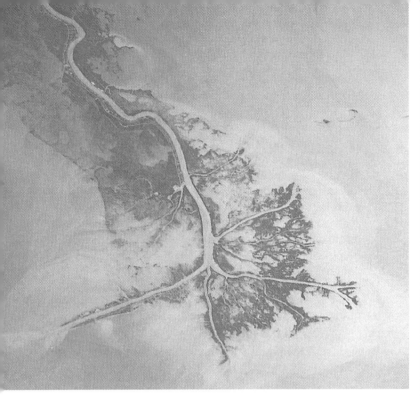

a

Figure 5.4 Sources of terrigenous sediments. (a) Rivers are the main source of terrigenous sediments. This photo, taken from space, shows sediment entering the Gulf of Mexico from the Mississippi River. (b) The wind may transport ash from a volcanic eruption for hundreds of kilometers and deposit it in the ocean. This ash cloud was caused by the summer 1991 eruption of Mount Pinatubo in the Philippines.

b

pounds) of this material fall into the ocean from the high atmosphere every year.

Most extraterrestrial particles that reach the ocean contain iron, and most of these particles dissolve in seawater before reaching the ocean floor. Cosmogenous sediment, however, is composed of iron-rich minerals that have survived the slow descent to the bottom. Occasionally the sediment includes translucent oblong particles of glass known as **tektites** (**Figure 5.5**). Tektites are thought to form from the violent impact of large meteors or asteroids on the crust of Earth or the moon. The impact melts some of the crustal material and splashes it into space; the material melts again as it rushes through the atmosphere, producing the various raindrop shapes shown in the photo. Tektites do not dissolve easily and usually reach the ocean floor. Most are less than 1.5 millimeter ($^1/_{16}$ inch) long.

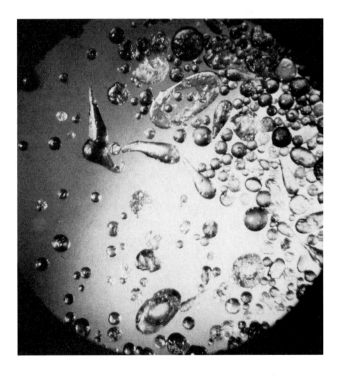

Figure 5.5 Tektites, very rare particles of extraterrestrial origin. These specimens of sculptured glass range from 0.2 to 0.8 millimeter in length. Glassy dust much finer in size, as well as nut-sized chunks, have also fallen on Earth.

Sediment Mixtures

Sediments on the ocean floor only rarely come from a single source; most sediment deposits are a mixture of biogenous and terrigenous particles, with an occasional hydrogenous or cosmogenous supplement. The patterns and composition of sediment layers on the seabed are of great interest to researchers studying conditions in the overlying ocean. Different marine environments have characteristic sediments, and these sediments preserve a record of past and present conditions within those environments.

THE DISTRIBUTION OF MARINE SEDIMENTS

The sediments on the continental margin are generally different in quantity, character, and composition from those on the deeper basin floors. Continental shelf sediments—called **neritic** sediments (*neritos* = of the coast)—consist primarily of terrigenous material. Deep-ocean floors are covered by finer sediments than those of the continental margins, and a greater proportion of deep-sea sediment is of biogenous origin. Sediments of the slope, rise, and deep-ocean floor that originate in the ocean are called **pelagic** sediments (*pelagios* = of the sea). The distribution and average thickness of the marine sediments in each oceanic region are shown in **Table 5.3**. Note that 72% of all marine sediment is associated with continental slopes and rises.

THE SEDIMENTS OF CONTINENTAL MARGINS

The bulk of terrigenous sediment is eroded and carried to streams, where it is transported to the ocean. Currents distribute sand and larger particles along the coast, while wave action carries the silts and clays to deeper water. When the water is too deep to be disturbed by wave action, the sediment comes to rest. Ideally, these processes produce an orderly sorting of particles by size from relatively large grains near the coast to relatively small grains near the shelf break. However, as we saw in the last chapter, shelf deposits are subject to further modification and erosion as sea level fluctuates: Larger particles may be moved toward the shelf edge when sea level is low, as it was in the ice ages. Poorly sorted sediments are also found as glacial deposits. In polar regions glaciers and ice shelves give rise to icebergs. These carry particles of all sizes, and when they melt they distribute their mixtures of rocks, gravel, sand, and silt onto high-latitude continental margins and deep-ocean floor.

Turbidity currents also disrupt the orderly sorting of sediments on the continental margin by transporting coarse-grained particles away from coastal areas and onto the deep-ocean floor.

The rate of sediment deposition on continental shelves is variable, but it is almost always greater than the rate of sediment deposition in the deep ocean. Near the mouths of large rivers, 1 meter (about 3 feet) of sediment may accumulate every 1,000 years. Along the east coast of the United States, however, many large rivers terminate in estuaries, which trap most of the sediment brought to them. The continental shelf of eastern North America is therefore covered mainly by sediments laid down during the last ice age, when sea level was lower.

In addition to terrigenous material, the continental margins almost always contain biogenous sediments. Organic productivity in coastal waters is often quite high, and the skeletal remains of creatures living on the bottom or in the water above mix with the terrigenous sediments and dilute them.

Sediments can build to impressive thickness on continental shelves. In some cases, shelf sediments undergo **lithification:** They are converted into sedimentary rock by pressure or by cementing. If these lithified sediments are thrust above sea level by tectonic forces, they can form mountains or plateaus. The top of Mount Everest, for example, is a shallow-water biogenic marine limestone (a calcareous rock). Much of the Colorado Plateau, with its many stacked layers, was formed by sedimentary deposition and lithification beneath a shallow continental sea beginning about 570 million years ago. The Colorado River has cut and exposed the uplifted beds to form the Grand Canyon. Hikers walking from the Canyon rim down to the river pass through spectacular examples of continental shelf sedimentary deposits. Their journey takes them deep into an old ocean floor!

THE SEDIMENTS OF DEEP-OCEAN BASINS

Sediment thickness is highly variable. When averaged, the Atlantic Ocean bottom is covered by sediments to a thickness of about 1 kilometer (3,300 feet), while the Pacific floor has an average sediment thickness of less than 0.5 kilometer (1,650 feet). There are three reasons for this difference. First, the Atlantic Ocean is smaller in area. Second, the Atlantic is fed by a greater number of rivers laden with sediment. Third, in the Pacific Ocean many oceanic trenches trap sediments moving toward basin centers. Beyond this, the composition and thickness of pelagic sediments also vary with location, being thickest on the **abyssal plains** and thinnest (or absent) on the oceanic ridges.

Table 5.3	The Distribution and Average Thickness of Marine Sediments		
Region	Percent of Ocean Area	Percent of Total Volume of Marine Sediments	Average Thickness
Continental shelves	9%	15%	2.5 km (1.6 mi)
Continental slopes	6%	41%	9 km (5.6 mi)
Continental rises	6%	31%	8 km (5 mi)
Deep-ocean floor	78%	13%	0.6 km (0.4 mi)

Sources: Emery in Kennett, 1982 (Table 11-1); Weihaupt, 1979; Sverdrup, Johnson, and Fleming, 1942.

Turbidites

Mudslides occur in many places along the margin of the continental shelf. Loose sediment periodically rushes down the continental slope in turbidity currents, the erosive force of which is thought to help cut submarine canyons (see Chapter 4). These underwater mudflows can reach the continental rise and often continue moving onto an adjacent abyssal plain before eventually coming to rest. The resulting deposits are called **turbidites;** they are graded layers of coarse-grained terrigenous sand interleaved with the finer sediments typical of the deep-sea floor.

Clays

About 38% of deep-sea sediments are clays and other fine terrigenous particles. As we have seen, the finest terrigenous sediments are easily transported by wind and water currents. Microscopic waterborne particles and tiny bits of windborne dust and volcanic ash settle slowly to the deep-ocean floor, forming fine brown, olive-colored, or reddish clays. As Table 5.1 shows, the velocity of particle settling is directly proportional to particle size, and clay particles usually fall very slowly indeed. Terrigenous sediment accumulation on the deep-ocean floor may be less than a millimeter every thousand years.

Oozes

Seafloor samples taken farther from land usually show a greater proportion of biogenous sediments than those obtained near the continental margins. This is not because biological productivity is higher farther from land (the opposite is usually true), but because there is less terrigenous material far from shore, and thus the deposits contain a greater proportion of biogenous material.

Deep-ocean sediment containing at least 30% biogenous material is called an **ooze** (surely one of the most descriptive terms in the marine sciences). Oozes are named after the dominant remnant organism constituting them. The organisms contributing their remains to deep-sea oozes are small, single-celled, drifting, plantlike organisms and the single-celled animals that feed on them. The hard shells and skeletal remains of these creatures are of relatively dense glasslike silica or calcium carbonate (limey) substances. When these organisms die, their shells settle slowly toward the bottom, mingle with fine-grained terrigenous silts and clays, and accumulate as ooze. The silica-rich residues give rise to **siliceous ooze,** the calcium-containing material to **calcareous ooze.**

Oozes accumulate slowly, at a rate of about 1 to 6 centimeters ($\frac{1}{2}$ to $2\frac{1}{2}$ inches) per 1,000 years. But they collect almost 10 times more quickly than deep-ocean terrigenous clays. The production of any ooze therefore depends on a delicate balance between the abundance of organisms at the surface, the number that settle to the bottom, and the rate of accumulation of terrigenous sediment.

Calcareous ooze forms mainly from shells of the amoeba-like **foraminifera (Figure 5.6a and b),** small drifting mollusks called **pteropods,** and tiny algae known as **coccolithophores (Figure 5.6c).** Although these creatures live in nearly all surface ocean water, calcareous ooze does not accumulate everywhere on the ocean floor because the shells are dissolved by seawater. At great depths seawater contains more CO_2 and becomes slightly acid. This acidity, combined with the increased solubility of calcium carbonate in cold water under pressure, dissolves the shells as they fall. At a certain depth, the **calcium carbonate compensation depth (CCD),** the rate at which calcareous sediments accumulate equals the rate at which those sediments dissolve. Below this depth the tiny skeletons of calcium carbonate

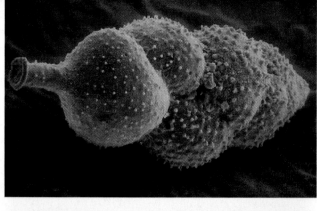

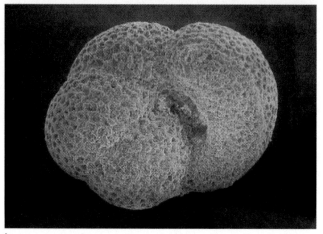

Figure 5.6 Photomicrographs from calcareous ooze. (a) A living foraminiferan, an amoeba-like organism. The shell of this beautiful foram, genus *Hastigerina,* is surrounded by a bubble-like capsule. It is one of the largest of the planktonic species with spines, reaching nearly 5 centimeters (2 inches) in length. (b) The shells of two smaller foraminifera are visible in these scanning electron micrographs: the bottle-shaped, bottom-dwelling *Uvigerina* and the snail-like, planktonic *Globigerina.* (c) Coccoliths, individual plates of coccolithophores, a form of planktonic algae. Because of their tendency to dissolve, calcareous oozes very rarely occur at bottom depths below 4,500 meters (14,800 feet).

Figure 5.7 The dashed line shows the calcium carbonate ($CaCO_3$) compensation depth (CCD). At this depth the rate at which calcareous sediments accumulate equals the rate at which those sediments dissolve. The CCD varies with temperature: The "snow line" is lower in warmer waters and higher in colder waters.

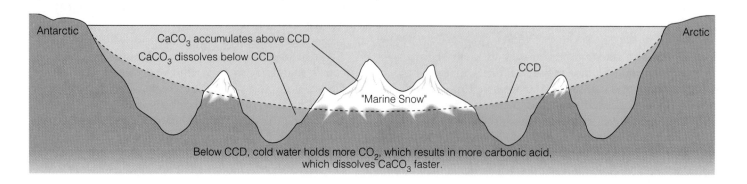

a

b

Figure 5.8 Scanning electron micrographs from siliceous oozes, which are most common at great depths. (a) Shells of radiolarians, an amoeba-like organism. Radiolarian oozes are found primarily in the equatorial regions. (b) Diatoms, single-celled algae. Diatom oozes are most common at high latitudes.

dissolve as they fall (or soon after they reach the bottom); so no calcareous oozes form. Calcareous sediment dominates at bottom depths less than about 4,500 meters (14,800 feet), the usual calcium carbonate compensation depth. Sometimes a "snow line" can be seen on undersea peaks: Above the line the white sprinkling of calcareous ooze is visible; below it, the "snow" is absent (see **Figure 5.7**). About 48% of all deep-ocean sediments are calcareous oozes.

Siliceous (silicon-containing) ooze predominates at greater depths and in colder polar regions. Siliceous ooze is formed from the hard parts of another amoeba-like animal, the beautiful glassy **radiolarian** (**Figure 5.8a**), and from single-celled algae called **diatoms** (**Figure 5.8b**). The silica within these organisms can dissolve in deep water (which is usually deficient in silicon), but silica dissolves *much* more slowly than calcium carbonate does. Slow dissolution, combined with very high diatom productivity in surface waters, leads to the buildup of siliceous ooze. Diatom ooze is most common in the Antarctic because strong ocean currents and seasonal upwelling in this area support large populations of diatoms. Radiolarian oozes occur in equatorial regions, most notably in the zone of equatorial upwelling west of South America (as will be seen in Figure 5.10). About 14% of all deep-ocean sediments are siliceous oozes.

The very small particles that make up most of these deep-ocean sediments would need between 20 and 50 years to sink to the bottom. By that time they would have drifted a great lateral distance from their original surface position. But researchers have noted that the composition of deep bottom sediments is usually similar to particle composition in the water directly above. How could such tiny particles fall quickly enough to avoid great horizontal displacement? The answer appears to involve their compression into fecal pellets. While still quite small, the fecal pellets of small animals are much larger than the tiny individual skeletons of diatoms, foraminifera, and other plantlike organisms that were consumed; so they fall much faster, reaching the deep-ocean floor in about two weeks.

Some deep-sea oozes have been uplifted by geologic processes and are now visible on land. The calcareous chalk White Cliffs of Dover in eastern England are partially lithified deposits composed largely of foraminifera and coccolithophores. Fine-grained siliceous deposits called *diatomaceous earth* are mined from other deposits. This fossil material is a valued component in flat paints, pool and spa filters, and mildly abrasive car and tooth polishes.

Hydrogenous Materials

Hydrogenous sediments also accumulate on deep-sea floors. They are associated with terrigenous or biogenous sediments and very rarely form sediments by themselves. Most hydrogenous sediments are thought to originate from chemical reactions that occur on particles of the dominant sediment.

The most famous hydrogenous sediments are man-

BOX 5.1 ● *Bathybius*

Ooze may be one of the most graphic words in science, but most oozes aren't slimy or sticky or gelatinous, and they don't slither across the seabed. Despite their name, oozes are submerged deposits containing the hard parts of once-living organisms that collectively look and act like fine sand or silt.

But there *was* once an ooze that lived up to the name. This mythical muck generated quite a controversy and changed the history of marine science even though it never really existed!

The story begins in 1857, when officers aboard HMS *Cyclops* ran a line of soundings from Ireland to Newfoundland and collected small samples of bottom sediment at the same time. The trapped sediments were preserved in alcohol and stored in bottles. Eleven years later, Thomas Henry Huxley, the preeminent British biologist, studied the samples. In every sample he analyzed, Huxley found a grayish-white, filmy ooze coating the solid mud. When stirred, the goo broke up into long strings; one scientist remarked that the stuff looked like an egg white stirred in water. The first chemical analysis indicated calcareous granules suspended in a proteinaceous base. After watching bits of the slime beneath his microscope and seeing the granules slowly shift about, Huxley decided "that the granule-heaps and the transparent gelatinous matter in which they are embedded represent masses of protoplasm." He quickly named the ooze *bathybius* (*bathy* = deep, *bios* = life) and leaped to the conclusion that the deep ocean was full of the glop. He first thought it was part of the body of a giant abyssal organism but later decided it must be a remnant of the "primeval living slime" from which all life may have arisen. He studied bathybius, lectured about it, and wrote articles on its properties for scholarly publications.

Many of Huxley's fellow scientists thought he was mistaken about the nature of bathybius. They had been convinced only recently that animal life was possible in the cold, dark depths past about 300 fathoms (1,800 feet, 550 meters), when submarine telegraph cables encrusted with organisms were raised from the ocean bottom. Even so, a whole abyss full of living jelly was hard to accept. Huxley countered with the opinion that the protoplasmic beast could live on organic matter raining down from above. Besides, Huxley was a friend and supporter of Charles Darwin, and this supposed remnant of the primordial ooze seemed to fit neatly into the developing evolutionary scheme.

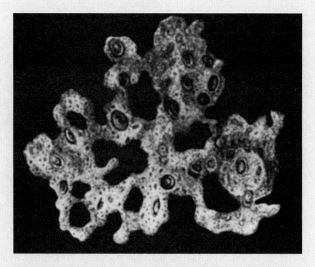

The supposed living slime—primordial ooze—"discovered" by Thomas Henry Huxley in 1868. He believed that this "animal jelly" carpeted the deep floor of the ocean.

Huxley's attentions had stirred up scientific and popular interest in bathybius. This publicity, combined with a growing interest in submerged fossils, chalk deposits, and the newly formulated theory of evolution, was one of the factors that contributed to the rise of oceanography and the commissioning of the *Challenger* Expedition.

Challenger scientists were on the lookout for bathybius, but they didn't find any. What they *did* find was that when too much alcohol was added to some sediment samples, a gelatinous calcium sulfate compound was precipitated from seawater, forming the famous slime. Unpreserved samples showed no evidence of bathybius. Bathybius had never existed—Huxley had been studying an artifact of specimen preservation!

At least Huxley was graceful in his concession. In an open letter written in 1875 to the scientific journal *Nature,* Huxley wrote: "Professor Wyville Thomson . . . informs me that the best efforts of the *Challenger*'s staff have failed to discover bathybius in a fresh state, and that it is seriously suspected that the thing to which I gave that name is little more than sulphate of lime. I am mainly responsible for the mistake. . . ." But the public's appetite for the weird (combined with Huxley's reputation as a first-rate scientist) kept bathybius alive in popular magazines and Sunday supplement articles until the early 1930s.

The story of bathybius demonstrates how important it is for marine scientists to go to sea to obtain and preserve their own samples. There's no substitute for immediate observation and analysis.

ganese **nodules,** which were discovered by the hard-working crew of HMS *Challenger*. The nodules consist primarily of manganese and iron oxides but also contain cobalt, nickel, chromium, copper, molybdenum, and zinc. They form in ways not fully understood by marine chemists, "growing" at an average rate of 1 to 10 millimeters (0.04 to 0.4 inch) per *million* years, one of the slowest chemical reactions in nature. Though most are irregular lumps the size of a potato, some nodules exceed 1 meter (3.3 feet) in diameter. Manganese nodules often form around nuclei such as sharks' teeth, bits of bone, microscopic alga and animal skeletons, or tiny crystals—as the cross section of a manganese nodule in **Figure 5.9a** shows. Bacterial activity may play a role in the development of a nodule. Between 20% and 50% of the Pacific Ocean floor may be strewn with nodules (**Figure 5.9b**). Why don't these heavy lumps disappear beneath the constant rain of accumulating sediment? Possibly the continuous churning of the underlying sediment by creatures living there keeps the dense lumps on the surface, or perhaps slow currents in areas of nodule accumulation waft particulate sediments away.

Challenger scientists also discovered nodules of phosphorite. The first of these irregular brown lumps was taken from the continental rise off South Africa, and phosphorite nodule fields have since been found in shallow water off California, Argentina, and Japan. Phosphorus is an important ingredient in fertilizer, and the nodules may someday be collected as a source of agricultural phosphates. Like manganese nodules, phosphorite nodules are found only in areas with low rates of sediment accumulation.

For now, the market value of the minerals in manganese and phosphorite nodules makes them too expensive to recover. As techniques for deep-sea mining become more advanced and raw material prices grow higher, however, the nodules' concentration of valuable materials will almost certainly be exploited.

Powdery deposits of metal sulfides have been found in the vicinity of hydrothermal vents at oceanic ridges. Hot, metal-rich brines blasting from the vents meet cold water, cool rapidly, and lose the heavy metal sulfides by precipitation. Iron sulfides and manganese precipitates fall in thick blankets around the vents. The cobalt crusts of rift zones also seem to be associated with this phenomenon. These areas may one day be mined for their metal content.

Evaporites are an important group of hydrogenous deposits that include many salts important to humanity. These salts precipitate as water evaporates from isolated arms of the ocean or from landlocked seas or lakes. For thousands of years people have collected sea salts from evaporating pools or deposited beds. Evaporites are forming today in the Gulf of California, the Red Sea, and

the Persian Gulf. The first evaporites to precipitate as water's salinity increases are the carbonates, such as calcium carbonate (from which limestone is formed). Calcium sulfate, which gives rise to gypsum, is next. (The M-reflector at the bottom of the Mediterranean consists of a layer of gypsum-rich evaporites.) Crystals of sodium chloride (table salt) will form if evaporation continues.

Not all hydrogenous calcium carbonate deposits are caused by evaporation, however. A small decrease in the acidity of seawater, or an increase in its temperature, can cause calcium carbonate to precipitate from water of normal salinity. In shallow areas of high biological productivity where sunlight heats the water, microscopic plants use up dissolved carbon dioxide, making seawater slightly less acidic (see Figure 6.13). Molecules of calcium carbonate then may precipitate around shell fragments or other particles. These white, rounded grains are called ooliths (*oon* = egg) because they resemble fish eggs. **Oolite sands**—sands comprised of ooliths—are abundant in many warm, shallow waters such as those of the Bahama Banks.

a

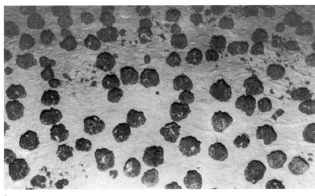

b

Figure 5.9 Manganese nodules. (a) A cross section cut through a manganese nodule, showing the concentric layers of manganese and iron oxides. This nodule is about 11 centimeters (4½ inches) long, a typical size. (b) Cannonball-sized manganese nodules littering the abyssal Pacific.

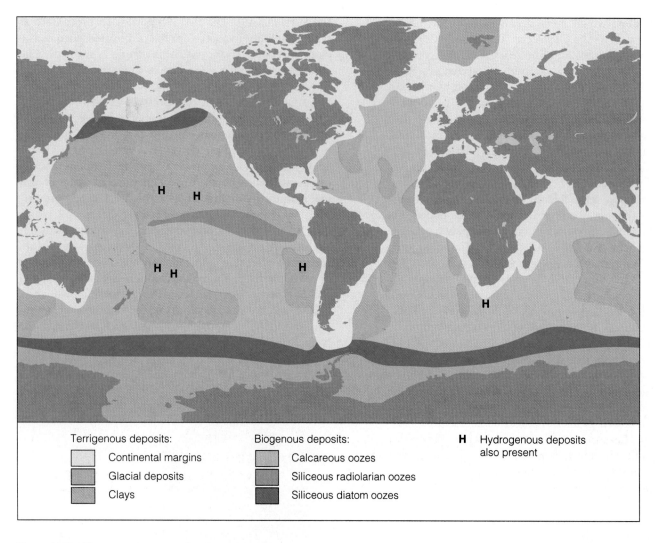

Terrigenous deposits:
- ☐ Continental margins
- ☐ Glacial deposits
- ☐ Clays

Biogenous deposits:
- ☐ Calcareous oozes
- ☐ Siliceous radiolarian oozes
- ☐ Siliceous diatom oozes

H Hydrogenous deposits also present

Figure 5.10 The general pattern of sediments on the ocean floor.

SEDIMENTS: A WORLD OCEAN VIEW

Figure 5.10 is a simplified look at the distribution of marine sediments. Notice especially the lack of radiolarian deposits in much of the deep North Pacific; the strand of siliceous oozes extending west from equatorial South America; and the broad expanses of the Atlantic, South Pacific, and Indian Ocean floors covered by calcareous oozes. As you might expect, the poorly sorted glacial deposits are found only at high latitudes.

This map summarizes more than a century of effort by marine scientists. Studies of sediments will continue because of their importance to natural resource development and because of the details of Earth's history that remain locked beneath their muddy surfaces.

STUDYING SEDIMENTS

Deep-water cameras have allowed researchers to photograph bottom sediments. The first of these cameras was simply lowered on a cable and triggered by a trip wire. Other more elaborate cameras have been taken to the seafloor on towed sleds or deep submersibles.

Actual samples usually provide more information than photographs do. HMS *Challenger* scientists used weighted, wax-tipped poles and other tools attached to long lines to obtain samples, but today's oceanographers have more sophisticated equipment. Shallow samples may be taken using a **clamshell sampler** (named because of its method of operation, not its target; see **Figure 5.11**). Deeper samples are taken by a **piston corer** (**Figure 5.12**),

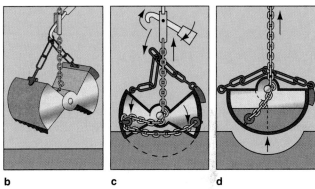

Figure 5.11 (a) A clamshell sampler (b) before sampling, (c) during sampling, and (d) after the sample has been taken. Note that the sample is relatively undisturbed.

a

a

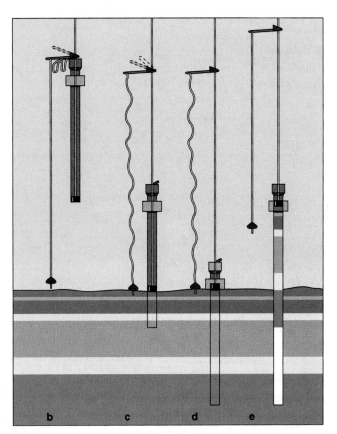

Figure 5.12 (a) A piston corer. (b) The corer is allowed to fall to the bottom. (c) The corer reaches the bottom and continues, forcing a sample partway into the cylinder. (d) Tension on the cable draws a small piston within the corer toward the top of the cylinder, and the pressure of the surrounding water forces the piston deeper into the sediment. (e) The corer and sample being hauled in.

BOX 5.2 ● *Deep-Ocean Cores*

The Ocean Drilling Program (ODP), currently the largest and most successful multinational Earth science research project, is the direct successor to the Deep Sea Drilling Project (DSDP), which began in 1968. The two drilling ships commissioned for the projects pioneered oil drilling technology to retrieve *cores,* long cylinders of sediment and rock extracted from beneath the seafloor. The cores arrive on the deck of *JOIDES Resolution*—the ship currently in active service—in 9.5-meter (30-foot) sections encased in plastic sheaths. Immediately after retrieval, a core is marked to indicate its original location on the seafloor, coded to distinguish top from bottom, measured, and cut into sections for study and storage. Each section is slit lengthwise; one-half is stored for the ODP archives, and the other is taken to the first of many shipboard laboratories for study.

Paleontologists then examine the sediments at the bottom of the core to determine the age of the oldest material. Other researchers measure its density, strength, molecular composition, radioactivity, and ability to conduct heat. Magnetic specialists read paleomagnetic data from the core to determine the ages of the rock fragments and the latitude at which they probably formed. Sensors are also lowered into the hole from which the core was removed, to gather additional information on the physical and chemical properties of the site.

ODP scientists have recently recovered fragments of the oldest remaining seafloor. The sample is about 175 million years old, a relic of the Middle Jurassic period, when the continents were massed in one huge cluster. From this and other cores they have learned about cycles of global climate change, information that will be useful in evaluating the present potential for global warming. They have also discovered how fluids move through the lithosphere, found evidence of an ice-free Antarctica, and noted the influence of plate tectonics on

a Researchers examine a deep core taken off the northwestern coast of Australia.

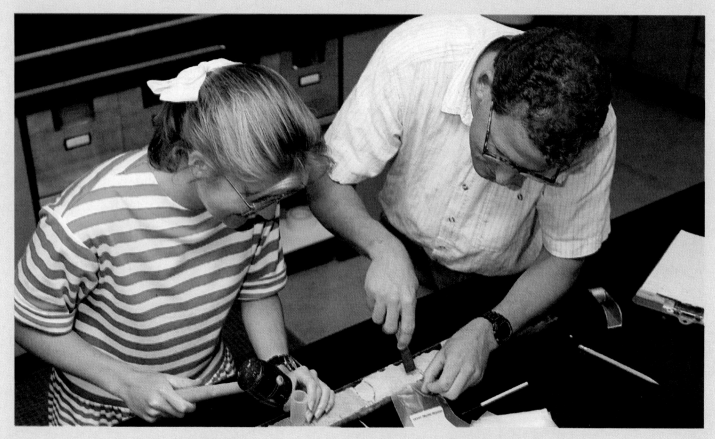

worldwide weather and current patterns. From trapped pollen grains, it is even possible to tell what land plants were thriving on the Earth at the time the core sediments were laid down. As analysis technology evolves, stored cores are restudied and new information obtained.

The size and shape of a typical deep-core section may be seen in **Figure a.** This core was taken 250 meters (820 feet) beneath the seafloor off the northwestern coast of Australia by *JOIDES Resolution* on 22 July 1988. The ocean in the area was around 2,000 meters (6,500 feet) deep.

The material in the core progresses from sandstone (near the bottom of the core, at the right side of the photo) to fine claystone (near the top). Foraminiferans and nanofossils are abundant in the sandstone, and small squidlike fossils, easily visible to the unaided eye, are present in the claystone. The sediments date from the Late Jurassic period, about 140 million years ago. The progression of slowly deposited sediments and fossils suggests that the seabed in this area has slowly subsided, possibly because of tectonic effects.

Details in a core from DSDP Hole 480, leg 64, are shown in **Figure b.** This core was obtained by the now-retired drilling ship *Glomar Challenger* on 1 January 1979 near Guaymas in the Gulf of California. The Gulf is an area of active seafloor spreading (see Figure 3.26), and researchers were interested in sampling the bottom in this place where basaltic magma is intruding into soft, wet, young sediments. The sediments here are very deep and have been accumulating at the extremely rapid rate of about 1,200 meters (4,000 feet) per million years.

Glomar Challenger arrived in the area on 29 December and used a conventional core barrel to penetrate 444 meters (1,456 feet) into diatomaceous ooze and mudstones. Nearly 273 meters (900 feet) of core were recovered from the hole drilled before Hole 480. Because of the disturbance caused by the conventional coring process, researchers were unable to determine the details of fine structure tantalizingly seen in a few lengths of that core.

On 31 December, *Glomar Challenger* moved to a new location 7 kilometers (4.4 miles) northwest. Hole 480, source of the core segment pictured here, was begun later that day. Unlike the previous hole, Hole 480 was drilled using a newly designed hydraulic corer developed by three DSDP engineers. The new device allowed 80% recovery of a core containing essentially undisturbed laminated diatomaceous ooze and muds.

As can be seen in Figure b, the core consists of thin alternating brown and gray bands of sediments. These are believed to be annual couplets, formed about 5 million years ago in response to two seasonal events that still occur in the Gulf: the winter rains that introduce terrigenous clays into the region, and diatom blooms produced by seasonal upwelling and northwest winds. Preservation of these fine details depends upon a low level of free oxygen at the seafloor. Burrowing animals would normally churn these fine layers into mush, but in a low-oxygen environment these animals are absent; so the thin alternating sheets of clay and diatom tests (shells) are preserved.

Note that parts of the core have already been removed for study. Trenches across the width of the core mark places where particular sets of layers have been removed for isotope analysis. A larger sample for paleomagnetic study was taken from the square depression in the sample. Considering the vast expense and skill necessary to retrieve and analyze them, these small, gray, gritty bits of sediment are probably more costly than their weight in fine diamonds!

b This core from the Gulf of California shows thin alternating bands of clay and ooze. Voids in the core show where samples were removed for study.

a device capable of punching through as much as 25 meters of sediment and returning an intact plug of material. Using a rotary drilling technique similar to that used to drill for oil, the drilling ship *Glomar Challenger* (described in Box 2.1) has returned much longer core segments, some more than 1,100 meters (3,608 feet) in length! As we saw in Chapter 2, *Glomar Challenger* drilled more than 1,000 holes for the Deep Sea Drilling Project, a program begun in 1968 with funds from the National Science Foundation. These cores are stored in core libraries, a very valuable scientific resource (**Figure 5.13**). Analysis of sediments and fossils from the Deep Sea Drilling Project cores helped verify the theory of plate tectonics. It has also shed light on the evolution of life-forms and helped researchers to decipher the history of changes in Earth's climate over the last 100,000 years. A newer and larger ship, *JOIDES Resolution* (**Figure 5.14**), is carrying this work forward as part of the Ocean Drilling Program, an international research consortium.

Powerful new continuous seismic profilers have also been used to determine the thickness and structure of layers of sediment on the continental shelf and slope,

Figure 5.13 (*above*) Sediment cores in storage. Cores are sectioned longitudinally, placed in trays, and stored in hermetically sealed cold rooms. The Gulf Coast Repository of the Ocean Drilling Program, located at Texas A&M University (pictured here), stores about 75,000 sections taken from more than 80 kilometers (50 miles) of cores recovered from the Pacific and Indian oceans. Smaller core libraries are maintained at the Scripps Institution in California (Pacific and Indian oceans) and at the Lamont–Doherty Earth Observatory in New York State (Atlantic Ocean).

Figure 5.14 *JOIDES Resolution*, the deep-sea drilling ship operated by the Joint Oceanographic Institutions for Deep Earth Sampling. The vessel is 124 meters (470 feet) long, with a displacement of over 16,000 tons. The rig can drill to a depth of 9,150 meters (30,000 feet) below sea level.

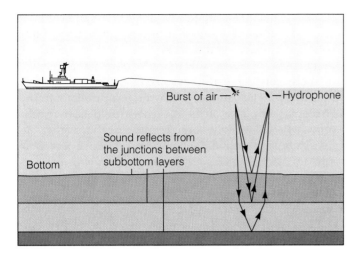

Figure 5.15 A typical method of continuous seismic profiling. A moving ship trails a sound transmitter and receiver at a distance sufficient to minimize interference from the ship's noise. Collapsing bubbles from a burst of compressed air act as a sound source. The sound is reflected from sediment layers beneath the surface and is detected by a sensitive hydrophone for analysis.

and to assist in the search for oil and natural gas (see **Figure 5.15**). Recent improvements in computerized image processing of the echoes returning from the seabed now permit detailed analysis of these deeper layers.

SEDIMENTS AS HISTORICAL RECORDS

In 1899 the British geologist W. J. Sollas theorized that deep-sea deposits could reveal much of the planet's history. In the era before plate tectonics theory this certainly seemed reasonable; the deep-ocean bottom was thought to be a calm, changeless place where an unbroken accumulation of sediment could be probed to discover the entire history of the ocean. Unfortunately for this promising idea, difficulties began to crop up almost immediately. For one thing, the sediments should have been much *thicker* than early probes indicated. If Earth's ocean is truly older than a few hundred thousand years, and if life has existed within it for most of that time, the sediment layer should be thicker than had been observed. Another difficulty lay in the uneven distribution of sediments. Sollas thought that the center of an ocean basin should contain the thickest layers of sediment, yet the raised mid-Atlantic bottom was nearly naked. There didn't seem to be any difference in the nature of the overlying seawater that could account for the variations in thickness and composition of the sediments across the

bottom of the Atlantic. Oozes were especially puzzling: The organisms that form ooze grow well at the surface of the middle Atlantic, yet the mid-Atlantic floor seemed to bear little ooze.

Today we know the tectonic reasons for these discrepancies, but turn-of-the-century geologists were understandably confused. The sedimentologists who followed Sollas learned to read the history of the ocean and atmosphere in the composition, thickness, and other characteristics of each sediment layer. Analysis of the layers by photographs, samples, and sonic profiling is called **stratigraphy** (*stratum* = layer, *graph* = a drawing). The thick, relatively undisturbed sediments of the deep basins, away from zones of seafloor spreading, are the most useful to modern stratigraphers. Light-colored bands of sediment many centimeters thick represent the remains of many once-living organisms, or times when the atmosphere was heavy with airborne ash from volcanic eruptions. Darker layers are rich in clay minerals and may contain hydrocarbons. These and other distinct bands often extend intact for great horizontal distances and can be identified in cores taken at widely separated locations. These bands are very useful in the correlation of units from core to core.

Paleoceanography (*palaios* = ancient), the study of the ocean's past, emerged in the 1930s and 1940s when cores became available from which history could be reconstructed. Paleoceanographers have used the distribution, depth, and composition of sediment layers to tell about cold and warm periods, or about times when ocean life was more or less abundant, or about other conditions in the past. Oozes might have accumulated rapidly when the atmosphere was free of terrestrial dust or volcanic ash. At these times sunlight could easily reach the ocean surface, and plants and animals grew rapidly. Sediments less than about 60,000 years old that contain once-living materials can be dated by radiocarbon analysis. **Figure 5.16** shows a record of recent ice ages reflected in the composition of foraminiferan species and their ages as determined by radiocarbon analysis and other means.

Figure 5.17 shows the age of the Pacific Ocean floor using data obtained largely from analyses of the overlying sediment. Note that sediments get older with increasing distance from the East Pacific Rise spreading center. Geophysicists and paleoceanographers have shown that the reasonable theory of Sollas has fallen short because the ocean basins are continually being created and destroyed by tectonic forces. Therefore, the "memory" of the sediments is not ancient and is in fact continually being erased by ocean floor subduction.

Still, marine sediments in the modern basins can shed light on unexpected details of the last 180 million years of Earth's history. One of the oddest details is the unex-

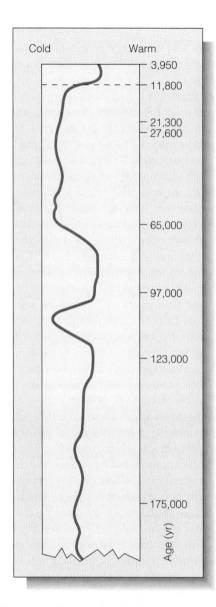

Cold   Warm

Age (yr)

3,950
11,800
21,300
27,600
65,000
97,000
123,000
175,000

Figure 5.16 Interpreting changes in climate from analysis of a deep-sea core. Scientists know that certain species of foraminifera are abundant only when the water temperature is warm, others only when it is cold. By plotting the relative numbers of these species at various depths in a core sample, they can depict major changes in climate, which have been dated at about 11,000 and 65,000 years ago. The dashed line represents the end of the last ice age.

plained extinction of up to 52% of known marine animal species (and the dinosaurs) at the end of the Cretaceous period 65 million years ago. Researchers have proposed hypotheses such as a sudden and violent impact of one or more very large meteors or comets to explain this catastrophe. The clouds of dust and ash thrown into the atmosphere by any of these events would have drastically reduced incident sunlight and greatly affected the lives of organisms and the photosynthetic base of ecosystems. Oceanographers are currently searching for evidence of the cause of the Cretaceous extinctions in layers of deep sediments.

THE ECONOMIC IMPORTANCE OF MARINE SEDIMENTS

Study of sediments has brought practical benefits. In 1993 29% of the world's crude oil and 22% of its natural gas were extracted from the sedimentary deposits of continental shelves and continental rises. Offshore hydrocarbons currently generate annual revenues in excess of $100 billion. Deposits within the sediments of continental margins account for nearly one-third of the world's estimated oil and gas reserves.

In addition to oil and gas, in 1987 sand and gravel valued at more than $350 million were taken from the ocean. This is about 1% of world needs. Commercial mining of manganese nodules has also been considered. In addition to manganese, these nodules contain substantial amounts of iron and other industrially important chemical elements. The high iron content of these nodules has prompted a proposal to rename them *ferromanganese* nodules. We will investigate these resources in more detail in Chapter 19.

Q-AND-A

1. The question of sediment age seems to occupy much of sedimentologists' time. Why?

The dating of sediments has been a central problem in marine science for many years. In 1957, during the International Geophysical Year, sedimentologists designed a coordinated effort to determine sediment age, which included plans for the Glomar Explorer *and* Glomar Challenger *drilling surveys. Their primary interest was to seek confirmation of the hypothesis of seafloor spreading. Cores returned by the Deep Sea Drilling Project in 1968 enabled researchers including J. Tuzo Wilson, Harry Hess, and Maurice Ewing to put the evidence together. Much of the proof for plate tectonics rests on the interpretation of sediment cores.*

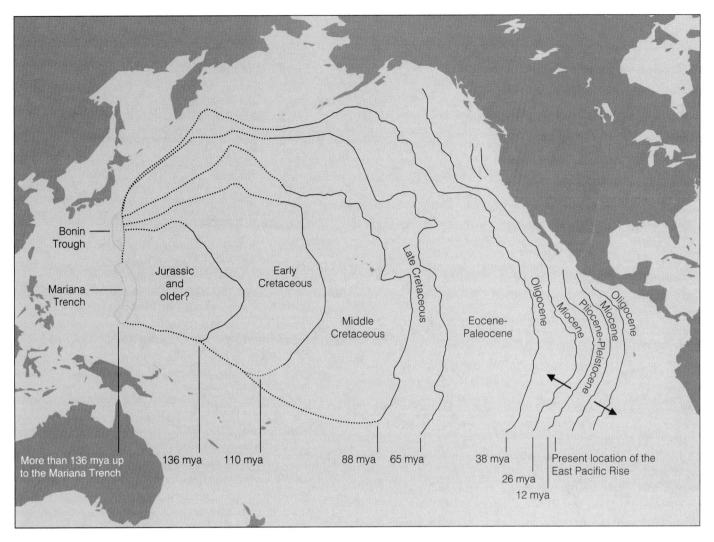

Figure 5.17 The age of portions of the Pacific Ocean floor, based on core samples of sediments just above the basalt seabed, in millions of years ago (mya). The youngest sediments are found near the East Pacific Rise and the oldest close to the eastern side of the troughs and trenches. Contrast this figure with Figure 3.28.

2. Where are sediments thickest?

Sediment is thickest close to eroding land and beneath biologically productive neritic waters, and thinnest over the fast-spreading oceanic ridges of the central South Pacific. The thickest sediments form on continental shelves and are typically more than 1,500 meters (5,000 feet) thick. Remember, much of the rocky material of the Grand Canyon was once marine sediment. The Grand Canyon is more than a mile deep, and the uppermost layer of sedimentary rock has already been eroded completely away!

3. What's the relationship between deep-sea animals and the sediments on which they live?

Life is not abundant on the bottom of the deep ocean. There are no plants at great depths because there is no light, but animals do live there. Some, like the sea cucumber in Figure 5.2 and the brittle stars in Figure 5.3, move slowly along the surface searching for bits of organic matter to eat. Others burrow through the muck in search of food particles. Worms eat quantities of sediment to extract any nutrients that may be present, and then deposit strings of fecal material as they move forward. The deeps are uninviting places, but life is tenacious and survives even in this hostile environment.

Terms and Concepts to Remember

abyssal plain
authigenic
biogenous
calcareous ooze
calcium carbonate
  compensation
  depth (CCD)
clamshell sampler
clay
coccolithophore
cosmogenous
diatom
evaporite
foraminiferan
hydrogenous
lithification
neritic
nodule
oolite sands

ooze
paleoceanography
pelagic
piston corer
poorly sorted
  sediment
pteropod
radiolarian
sand
sediment
siliceous ooze
silt
stratigraphy
tektite
terrigenous
turbidite
well-sorted
  sediment

Study Questions

1. In what ways are sediments classified?

2. List the four types of marine sediments. Explain the origin of each.

3. How are neritic sediments generally different from pelagic ones?

4. Is the thickness of ooze always an accurate indication of the biological productivity of surface water in a given area? (Hint: See next question.)

5. What is the calcium carbonate compensation depth? Is there a compensation depth for the siliceous components of once-living things?

6. What sediments accumulate most rapidly? Least rapidly?

7. Can marine sediments tell us about the history of the ocean from the time of its origin? Why?

8. How do paleoceanographers infer water temperatures, and therefore terrestrial climate, from sediment samples?

9. Where are sediments thickest? Are there any areas of the ocean floor free of sediments?

10. Are sediments commercially important? In what ways?

For Further Study

Arrhenius, G. 1963. "Pelagic Sediments." In *The Sea*, vol. 3, 655–727, edited by M. N. Hill. New York: Interscience. Technical.

Berger, W. H. 1974. "Deep-Sea Sedimentation." In *The Geology of Continental Margins*, edited by C. A. Burk and C. D. Drake. New York: Springer-Verlag. Technical.

Cashman, K. V., and R. S. Fiske. 1991. "Fallout of Pyroclastic Debris from Submarine Volcanic Eruptions." *Science* 253 (no. 5017): 275–80. Debris from frequent undersea volcanic eruptions contributes to the sediments.

Dietz, R. S. 1978. "IFO's (Identified Flying Objects)." *Sea Frontiers* 24 (no. 6): 341–46. Discusses the origin of tektites.

Fischer, A. G., et al. 1970. "Geologic History of the Western North Pacific." *Science* 168: 1210–14. Sediment dates are used to confirm plate movement in the North Pacific.

Heezen, B., and C. Hollister. 1971. *The Face of the Deep*. New York: Oxford University Press. A superb treatise on the seabed, with many wonderful photographs.

Hsü, K. J. 1983. *The Mediterranean Was a Desert: A Voyage of the Glomar Challenger*. Princeton: Princeton University Press. An accessible presentation of evidence for the author's theory that the Mediterranean dried up during the Pliocene epoch.

Kennett, J. P. 1982. *Marine Geology*. Englewood Cliffs, NJ: Prentice-Hall. Excellent and complete general reference.

Linklater, E. 1972. *The Voyage of the Challenger*. New York: Doubleday. An interesting description of the first systematic sediment sampling.

Mangone, G. J. 1986. *Concise Marine Almanac*. New York: Van Nostrand. Oil and gas statistics, and much else of interest.

Prospero, J. 1985. "Records of Past Continental Climates in Deep-Sea Sediments." *Nature* 315 (23 May): 279–80. Subtitle could be "History in the Mucking."

Schlee, S. 1973. *The Edge of an Unfamiliar World: A History of Oceanography*. New York: Dutton. A detailed and readable history, with information on the history of bathybius.

Shackleton, N. J., and C. P. Summerhayes, eds. 1986. *North Atlantic Paleoceanography*. Oxford: Blackwell Scientific. A series of papers exploring recent findings in paleoceanography and the methods used to discover them.

Siebold, E., and W. H. Berger. 1993. *The Sea Floor: An Introduction to Marine Geology*. 2d ed. New York: Springer-Verlag. Methods and data are described in this update. Includes new material on physical marine resources.

Siever, R. 1988. *Sand*. New York: Freeman, Scientific American Library.

Stanley, D. J. 1990. "Med Desert Theory Is Drying Up." *Oceanus* 33 (no. 1, Spring 1990): 14–23. Recent information suggests that Hsü's vision of a deep desert basin in the Mediterranean 5.5 million years ago may not be correct.

Stanley, S. M. 1986. *Earth and Life Through Time*. New York: Freeman. Information on dating of sediments.

Sverdrup, H. U., M. W. Johnson, and R. H. Fleming. 1942. *The Oceans: Their Physics, Chemistry, and General Biology*. New York: Prentice-Hall. Chapter 20 covers marine sedimentation.

Weihaupt, J. G. 1979. *Exploration of the Oceans*. New York: Macmillan. Sediment statistics.

Oceanus 36 (no. 4, Winter 1993–94) was dedicated to 25 years of ocean drilling. Articles include reports by drilling teams of different nations, results from paleoceanographers, findings about plate tectonics and sedimentary processes, and a summary of the latest in deep-sea drilling technology and its spin-offs

6 SEAWATER CHEMISTRY

Cooking Up the Recipe for Water

In the fifth century B.C., the Greek philosopher Empedocles suggested that all matter was composed of four primary substances in various combinations: earth, air, fire, and water. Divide anything into small enough bits, he wrote, and you would end up with one or more of these four elements, none of which could be further divided. A century later this concept was endorsed by Aristotle, a philosopher considered by many to be the father of modern science. Aristotle's stature lent credence to the idea.

By the 1700s, the four-element theory was being seriously challenged. Chemists were making many discoveries that could not adequately be explained using Empedocles' theory. A test that would unequivocally disprove the four-element theory would be to separate one of the four elements into smaller components, and water became the target. On 24 June 1783, the French chemist Antoine-Laurent Lavoisier discovered that water could be subdivided, a landmark event in the history of chemistry.

Lavoisier burned two gases in a special vessel. One gas was produced by dripping acid onto iron or copper. This clear, odorless gas was called "inflammable air" because it burned when ignited. The other gas—also colorless and odorless—was produced by heating a red precipitate of mercury. This second gas was known as the "acid maker" because of its tendency to combine with other substances to make mild

Monsieur and Madame Antoine-Laurent Lavoisier, in a painting by Jacques-Louis David, 1788. M. Lavoisier, a chemist, is credited with the "discovery" of water in 1783. Though a pioneering chemist and one of the founders of the metric system of measurements, Lavoisier could not escape the terrors of the French Revolution. He was sentenced to death by a revolutionary tribunal and guillotined on 8 May 1794.

acids. During the combustion of these gases, droplets of liquid condensed on the cool interior of an attached flask. Tests showed this liquid to be "as pure as distilled water." Lavoisier argued that in chemistry, as in mathematics, the whole is equal to the sum of its parts. "Since only water and nothing else was formed . . . the water was the sum of the [masses] of the two gases from which it was produced."

In short, water was not the irreducible element it had always been thought to be. It was a combination—a compound—of the acid maker (*oxys* = acid, *gen* = making) and the newly named "begetter of water" (*hydro* = water, *gen* = making). Oxygen and hydrogen were found to be simpler substances than water. Elements were clearly more numerous than Aristotle had thought, and water more complex.

We know today that water *is*, in a way, more than the sum of its parts. The parts that Lavoisier described are simple enough: hydrogen, the most abundant atom in the universe, and oxygen, an element that accounts for nearly half the weight of the Earth's crust. But water itself behaves in complex and remarkable ways. The unique properties of water are crucial in determining physical conditions at the Earth's surface.

Because it is familiar and abundant, we don't always appreciate water's unusual characteristics. If liquid methane or ammonia or *any* other flowing substance dominated the surface of the Earth, physical conditions here would be strikingly different. Indeed, life as we know it would be impossible.

As you may recall from Chapter 1, the Earth's surface waters are thought to have escaped from the crust and mantle through the process of outgassing. Outgassing of other substances and the chemical dissolution by water of crustal material have added salts and other solids and gases to the ocean. In this chapter we will first investigate the structure of pure water and then look at the chemistry of seawater, with its many dissolved substances. Chapter 7 discusses some of water's physical properties and the implications of these properties for the ocean and Earth as a whole.

CHAPTER OVERVIEW

Water, a chemical compound composed of two hydrogen atoms and one oxygen atom, is abundant on and within the Earth. The polar nature of the water molecule produces some unexpected chemical properties, one of the most important of which is water's remarkable ability to dissolve more substances than any other natural solvent. Though most solids and gases are soluble in water, the ocean is in chemical equilibrium, and neither the proportion nor amount of most dissolved substances changes significantly through time. Most of the properties of seawater are different from those of pure water because of the substances dissolved in the seawater.

THE WATER MOLECULE

As we have seen, water is a compound. **Compounds** are substances that contain two or more elements in a fixed proportion. The modern definition of an **element** is a substance composed of identical particles, called **atoms,** that cannot be broken into simpler substances by chemical means. Water's familiar chemical formula, H_2O, shows that two atoms of hydrogen (H) are present for each atom of oxygen (O). Some other common compounds are carbon dioxide (CO_2), rust (Fe_2O_3)[1], and simple sugar ($C_6H_{12}O_6$).

A **molecule** is a group of atoms held together by chemical **bonds. Chemical bonds,** the energy relationships between atoms that hold them together, are formed when **electrons**—tiny negatively charged particles found toward the outside of an atom—are shared or transferred between atoms. A water molecule forms when electrons are shared between two hydrogen atoms and one oxygen atom. The bonds formed by shared pairs of electrons are known as **covalent bonds.** Covalent bonds hold together many familiar molecules, including CO_2, CH_4 (methane gas), and O_2 (atmospheric oxygen). Because of the way a water molecule's oxygen electrons are distributed, the overall geometry of the molecule is a bent or angular shape. The angle formed by the two hydrogen atoms and the central oxygen atom is about 105°. The formation of a water molecule is depicted in **Figure 6.1.**

The angular shape of the water molecule makes it electrically asymmetrical, or **polar.** Each water molecule can be thought of as having a positive (+) end and a negative (−) end. This is because **protons**—positively charged particles at the center of the hydrogen atoms—are left partially exposed when the negatively charged electrons bond more closely to oxygen. The polar water molecule acts something like a magnet; its positive end attracts particles having a negative charge, and its negative end attracts particles having a positive charge. When water comes into contact with compounds whose elements are held together by the attraction of opposite electrical charges (most salts, for example), the polar water molecule will separate that compound's component elements from each other. This explains why water can dissolve so many other compounds so easily.

The polar nature of water also permits it to attract other water molecules. When a hydrogen atom (the positive end) in one water molecule is attracted to the oxygen atom (the negative end) of an adjacent water molecule, a **hydrogen bond** forms. The resulting loosely held webwork of water molecules is shown in **Figure 6.2.** Hydrogen bonds greatly influence the properties of

[1]Rust is iron oxide. Fe, the chemical symbol for iron, is derived from *ferrum,* the Latin name for iron.

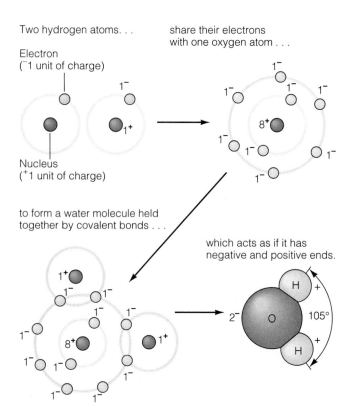

Figure 6.1 The formation of a water molecule.

water by allowing individual water molecules to stick to each other, a property called **cohesion.** Cohesion gives water an unusually high surface tension, which results in a surface "skin" capable of supporting needles, razor blades, and even walking insects. It also causes the capillary action that makes water spread through a towel when one corner is dipped in water. **Adhesion,** the tendency of water to stick to other materials, allows water to adhere to solids, that is, to make them wet. Hydrogen bonds are also what gives pure water and thick ice their pale blue hue.

If hydrogen bonds did not hold water molecules together, water would fly apart to form a gas. Hydrogen sulfide (H_2S) is chemically similar to water but lacks water's ability to form networks of hydrogen bonds. H_2S is therefore a gas rather than a liquid at normal temperatures and pressures. As we shall see in Chapter 7, the polar nature of water and its ability to form hydrogen bonds helps explain the unusual way water responds to heat.

THE DISSOLVING POWER OF WATER

Water is a powerful solvent: It will eventually dissolve nearly any substance. No wonder, then, that seawater and most other liquids in nature are water solutions. A **solution** is made of two components: The **solvent,** usu-

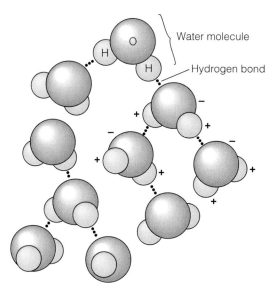

Figure 6.2 Hydrogen bonds in liquid water. The attractions between adjacent polar water molecules form a webwork of hydrogen bonds. These bonds are responsible for cohesion and adhesion, the properties of water that cause surface tension and wetting. Hydrogen bonds among water molecules also make it difficult for individual molecules to escape from the surface.

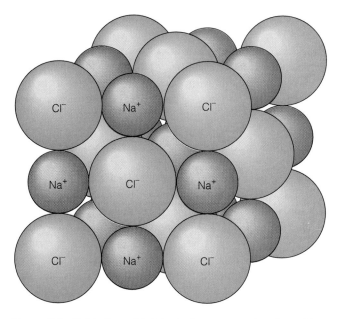

Figure 6.3 Salt in the solid state. In the compound sodium chloride (NaCl), ionic bonds hold the sodium ions (Na^+) and chloride ions (Cl^-) together.

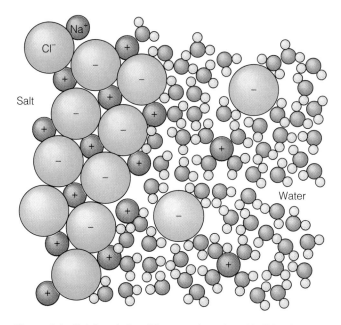

Figure 6.4 Salt in solution. When a salt such as NaCl is placed in water (left side of figure), the positively charged hydrogen end of the polar water molecule is attracted to the negatively charged Cl^- ion, and the negatively charged oxygen end is attracted to the positively charged Na^+ ion. The ions are surrounded by water molecules that are attracted to them, and they become solute ions in the solvent (right side of figure).

ally a liquid, is always the more abundant constituent; the **solute,** often a dissolved solid or gas, is the less abundant. In a true solution (sugar in well-stirred coffee, for example), the molecules of the solute are homogeneously dispersed among the molecules of solvent; that is, the solution has uniform properties throughout. In a **mixture,** different substances are closely intermingled but retain separate identities. The properties of a mixture are heterogeneous; they may vary from place to place within the mixture. Think of noodle soup as a mixture of noodles and liquid.

Water's dissolving power results from the polar nature of the water molecule. Consider how water dissolves sodium chloride (or NaCl), the most common salt.[2] An **ion** is an atom (or small group of atoms) that has an unbalanced electrical charge because it has gained or lost one or more electrons. Unlike the electron *sharing* found in covalently bonded molecules, the sodium atoms in NaCl have *lost* electrons, and chlorine atoms have *gained* them. The resulting ions are linked by the mutual attraction of their opposite electrical charges. The ions of sodium and chloride in NaCl are said to be held together by **ionic bonds (Figure 6.3).** When NaCl dissolves in water (**Figure 6.4**), the polarity of water causes the sodium ion (Na^+) to separate from the chloride ion (Cl^-). The ions move away from the salt crystal, permit-

[2]Na, the chemical symbol for sodium, is derived from *natrium,* the Latin name for sodium.

ting water to attack the next layer of NaCl. *Note that NaCl does not exist as "salt" in seawater;* its components are separated when salt crystals dissolve in water, but they are joined when crystals re-form as water evaporates.

Some other salts form ions in which *groups* of atoms bond together and carry an electrical charge. The sulfate ion (SO_4^{2-}), for example, carries two extra electrons. The charge on the overall ion is therefore −2.

By contrast, oil doesn't dissolve in water even if the two are very thoroughly shaken together. When oil is dispersed in water it forms a mixture because molecules of oil are nonpolar in character. This means that oil has no positive or negative charges to attract the polar water molecule. In a way this is fortunate. Living tissues would readily dissolve in water if the oils within their membranes didn't blunt water's powerful attack.

As we'll see in a moment, seawater is a complex solution of ions and nonionic solutes. These particles can move through still water by **diffusion,** the random movement of particles through a solution. If you were to drop a large crystal of NaCl into a container of pure water, for example, the water would quickly begin to dissolve the salt. At first, the concentration of Na^+ and Cl^- ions would be greater next to the crystal than a short distance away from it; but eventually, the crystal would dissolve completely, and the ions would diffuse evenly through the entire water volume, forming a homogeneous solution. Atmospheric gases dissolve in the ocean surface and diffuse easily through water. Dissolved substances tend to diffuse from regions of high concentration of those substances to regions of low concentration.

When no more of a substance will dissolve in water, the water is said to be *saturated* with that substance. At **saturation** the rate at which molecules of the solute are being dissolved equals the rate at which they are **precipitating** (re-forming into crystals) at another location in the solution.

SEAWATER

About 97.2% of the 1,370 million cubic kilometers (329 million cubic miles) of Earth's surface water is marine. By weight, seawater is about 96.5% water and 3.5% dissolved substances, most of which are salts. The world ocean contains some 5,000 trillion kilograms (5.5 trillion tons) of salt. If the ocean's water evaporated completely, leaving its salts behind, the dried residue could cover the entire planet with an even layer 45 meters (150 feet) thick.

Salinity: Dissolved Solids and Water Together

The total quantity (or concentration) of dissolved inorganic solids in water is its **salinity.** The ocean's salinity varies from about 3.3% to 3.7% by weight, depending on such factors as evaporation, precipitation, and freshwater runoff from the continents, but the average salinity is usually given as 3.5%. Most of the dissolved solids in seawater are salts that have been separated into ions. Sodium and chloride are the most abundant of these.

The many ions present in seawater react with each other (and with water molecules) in complex ways to modify the physical properties of pure water. Consider the following:

1. The heat capacity of water decreases with increasing salinity; that is, less heat is necessary to raise the temperature of seawater by 1° than is required to raise the temperature of fresh water by the same amount.

2. Dissolved salts disrupt the webwork of hydrogen bonding in water. As salinity increases, the freezing point of water becomes lower; the salts act as a sort of antifreeze. Sea ice therefore forms at a lower temperature than ice in freshwater lakes.

3. Because dissolved salts tend to attract water molecules, seawater evaporates more slowly than fresh water. Swimmers usually notice that fresh water evaporates quickly and completely from their skin, but seawater lingers.

4. Osmotic pressure, the pressure exerted on a biological membrane when the salinity of the environment is different from that within the cells, rises with increasing salinity. This is a key factor in transmitting water into and out of cells.

These four properties, which vary with the quantity of solutes dissolved in the water, are called water's **colligative properties** (*colligatus* = to bind together). Because colligative properties are the properties of *solutions,* the more concentrated (saline) water is, the more important these properties become. (Because it is not a solution, pure water has no colligative properties.) As we will see in Chapter 7, seawater's colligative properties are critical in determining many of the ocean's physical properties.

The Components of Salinity

Because about 3.5% of seawater consists of dissolved substances, boiling away 100 kilograms of seawater theoretically produces a residue weighing 3.5 kilograms. Because variations of 0.1% are significant, however, oceanographers prefer to use parts-per-thousand notation (‰) rather than percent (%, parts per hundred) in discussing these materials.[3] The seven ions listed below oxygen and hydrogen in **Table 6.1** make up more than

---

[3]Note that 3.5% = 35‰. If you began with 1,000 kilograms of seawater, you would expect 35 kilograms of residue.

Table 6.1 Major Constituents of Seawater at 35‰ Salinity		
Constituent	Concentration in Parts per Thousand (‰) or Grams per Kilogram (g/kg)	Percent by Weight
Water Itself		
Oxygen	857.8	85.8%
Hydrogen	107.2	10.7%
The Most Abundant Ions		
Chloride (Cl^-)	18.980	1.9%
Sodium (Na^+)	10.556	1.1%
Sulfate (SO_4^{2-})	2.649	0.3%
Magnesium (Mg^{2+})	1.272	0.1%
Calcium (Ca^{2+})	0.400	0.04%
Potassium (K^+)	0.380	0.04%
Bicarbonate (HCO_3^-)	0.140	0.01%
Total	**999.377 g/kg**	**99.9%**

Source: Adapted from Walton-Smith, 1974.

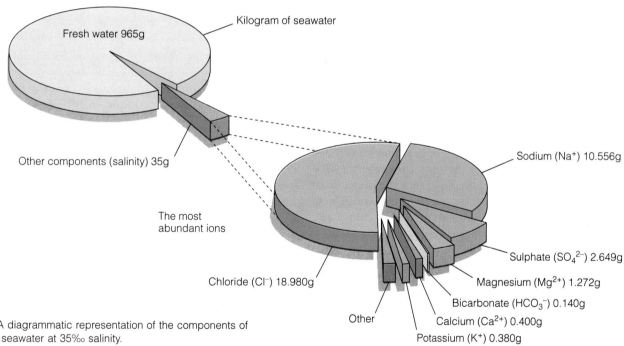

Figure 6.5 A diagrammatic representation of the components of a kilogram of seawater at 35‰ salinity.

99% of this residual material. When seawater evaporates, its ionic components combine in many different ways to form table salt (NaCl), epsom salts ($MgSO_4$), and other mineral salts. **Figure 6.5** shows the proportions of ions in seawater.

Seawater also contains minor constituents. The ocean is sort of an "Earth tea": Nearly every element present in the crust and atmosphere is also present in the oceans,

though sometimes in extremely small amounts. Only 14 elements have concentrations in seawater larger than one part per million. Elements present in amounts less than 0.001‰ (one part per million, or ppm) are known as **trace elements.** These are more easily given in parts per billion (ppb). **Table 6.2** lists some of these minor and trace elements, many of which are crucial to life processes.

Table 6.2   Minor and Trace Elements in Seawater at 35‰ Salinity

Constituent	Concentration in Parts per Thousand (‰), or g/kg		Concentration in Parts per Million, or mg/kg (mg/L)		Concentration in Parts per Billion, or mg/1,000 kg
Minor Elements					
Bromine (Br)	0.065	=	65		
Strontium (Sr)			8		
Boron (B)			4		
Silicon (Si)			3		
Fluorine (F)			1		
Important Trace Elements					
Nitrogen (N)[a]			0.3	=	300
Lithium (Li)					170
Phosphorus (P)					70
Iodine (I)					50
Zinc (Zn)					10
Iron (Fe)					10
Aluminum (Al)					10
Manganese (Mn)					2
Lead (Pb)					0.04
Mercury (Hg)					0.03
Gold (Au)					0.000004

Source: Adapted from Walton-Smith, 1974.

[a]Refers to nitrogen available as a nutrient, not as dissolved gas.

The Source of the Ocean's Salts

Remembering the effectiveness of water as a solvent, you might think that the ocean's saltiness has resulted from the ability of rain, groundwater, or crashing surf to dissolve crustal rock. Much of the sea's dissolved material originated in that way, but is crustal rock the source of all the ocean's solutes? An easy way to find out would be to investigate the composition of salts in river water and compare these figures to those of the ocean as a whole. If crustal rock is the only source, the salts in the ocean should be like those of concentrated river water. But they are not. River water is usually a dilute solution of calcium and bicarbonate ions, while the principal ions in seawater are chloride and sodium. The magnesium content of seawater would also be higher if seawater were simply concentrated river water. The proportions of salts in isolated salty inland lakes, such as Utah's Great Salt Lake or the Dead Sea, are much different from the proportions of salts in the ocean. Thus weathering and erosion of crustal rocks cannot be the sole source of sea salts.

The components of ocean water whose proportions are *not* accounted for by the weathering of surface rocks are called **excess volatiles.** To find the source of these excess volatiles, we must look to the Earth's deeper layers. The upper mantle appears to contain more of the substances found in seawater (including the water itself) than are found in surface rocks, and their proportions are about the same as found in the ocean. As you read in Chapter 3, convection currents slowly churn Earth's mantle, causing the movement of tectonic plates. Because of this activity, some deeply trapped volatile substances escape to the exterior, outgassing through volcanoes and rift vents (see **Figure 6.6**). These excess volatiles include carbon dioxide, chlorine, sulfur, hydrogen, fluorine, nitrogen, and, of course, water vapor. This material, along with residue from surface weathering, accounts for the chemical constituents of today's ocean.

Some of the ocean's solutes are hybrids of the two processes of weathering and outgassing. Table salt, or sodium chloride, is an example. The sodium ions come from the weathering of crustal rocks, while the chlorine ions come from the mantle by way of volcanic vents and outgassing from mid-ocean rifts. As for the lower-than-expected quantity of magnesium ions in the ocean, recent research at a spreading center east of the Galápagos Islands suggests that mid-ocean rifts may play a role in reducing the magnesium content and increasing the calcium content of seawater. The water that circulates through new ocean floor at these sites is apparently stripped of magnesium and a few other elements. The magnesium seems to be incorporated into mineral deposits, but calcium is added as hot water dissolves adjacent rocks.

The Principle of Constant Proportions

In 1865, the chemist Georg Forchhammer noted that although the total *amount* of dissolved solids (salinity) might vary among samples, the *ratio* of major salts in samples of seawater from many locations was constant. In other words, the percentage of various salts in seawater is the same in samples from many places, regardless of how salty the water is. This constant ratio is known as **Forchhammer's principle,** or the **principle of constant proportions.** Forchhammer was also the first to observe that seawater contains fewer silica and calcium ions than concentrated river water—and the first to realize that removal of these compounds by marine animals and plants, to form shells and other hard parts, might account for part of the difference. The English chemist William Dittmar, working with HMS *Challenger* samples 10 years after they had been collected, confirmed Forchhammer's principle of constant proportions. Building on Forchhammer's and Dittmar's work, and taking advantage of improved analytical techniques, researchers have established a reliable way to determine salinity.

Determining Salinity

Water's salinity by weight would seem an easy property to measure. Why not simply evaporate a known weight of seawater and weigh the residue? This simple method yields imprecise results because some salts will not release all the molecules of water associated with them. If these salts are heated to drive off the water, other salts (carbonates, for example) will decompose to form gases and solid compounds not originally present in the water sample.

Modern analysis depends instead on determining the sample's chlorinity. **Chlorinity** is a measure of the total weight of chlorine, bromine, and iodine ions in seawater. Because chlorinity is comparatively easy to measure, and because the proportion of chlorinity to salinity is constant, marine chemists have devised the following formula to determine salinity:

$$\text{Salinity in } \%_o = 1.80655 \times \text{chlorinity in } \%_o$$

Chlorinity is about 19.2‰; so salinity is around 34.7‰.

Seawater samples can be obtained by methods ranging from tossing a clean bucket over the side of the ship to sophisticated tube-and-pump systems. Typically, water samples are collected using a string of sampling bottles similar to those perfected early in this century by the Norwegian scientist and explorer Fridtjof Nansen. A **Nansen bottle** is open at both ends (**Figure 6.7**). A series of bottles is lowered on a wire to known depths. Each bottle is then triggered to close by a brass weight (appropriately called a *messenger*) sent sliding down the line.

Figure 6.6 An active hydrothermal vent spouts superheated seawater. Recent research at a spreading center east of the Galápagos Islands suggests that mid-ocean rifts may play a role in reducing the magnesium content and increasing the calcium content of seawater. The water that circulates through new ocean floor at these sites is apparently stripped of magnesium and a few other elements. The magnesium seems to be incorporated into mineral deposits, but calcium is added as hot water dissolves adjacent rocks. All the water in the ocean is thought to cycle through the seabed at rift zones every 1 to 10 million years.

The bottles are then hauled to the surface and their contents analyzed.

Until recently, marine chemists used a sensitive chemical procedure involving a silver nitrate solution to measure the chlorinity of seawater. Conversion to salinity was then made by a set of mathematical tables. The procedure was calibrated against a standard sample of seawater of precisely known chlorinity. Today's marine scientists use an electronic device called a **salinometer,** which measures the electrical conductivity of seawater (see **Figure 6.8**). Conductivity varies with the concentration and mobility of ions present, and with water temperature. Circuits in the salinometer adjust for water temperature, convert conductivity to salinity, and then display salinity. Salinometers are calibrated against a sample of known conductivity and salinity. The best salinometers can determine salinity to an accuracy of 0.001%. Some salinometers are designed for remote sensing: The electronics stay aboard ship while the sensor coil is lowered over the side. A quick way to measure

a

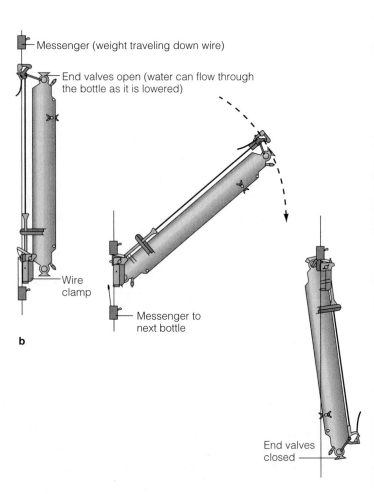

Messenger (weight traveling down wire)

End valves open (water can flow through the bottle as it is lowered)

Wire clamp

Messenger to next bottle

b

End valves closed

c

Figure 6.7 The Nansen bottle. (a) A Nansen bottle being prepared for deployment. (b) When the messenger (a small weight) slides down the line, it triggers the bottle to flip upside down, closing the end valves at the same time. Nansen bottles are being replaced by sampling bottles that are able to hold greater volumes of water and are more resistant to additional triggering, such as the 10-liter Niskin bottles shown in (c). Each Niskin bottle is remotely triggered by a signal from the research ship when the array reaches predetermined depths.

Figure 6.8 This portable salinometer reads temperature, pH, and dissolved oxygen as well as conductivity. Designed to be lowered over the side of a research vessel at the end of a line, the self-powered device contains a small pump that passes water over sensors. This model takes two readings every second and operates down to about 200 meters (660 feet). Data may be retrieved by a cable connected to the research vessel or stored in the salinometer's memory to be retrieved when it is brought back aboard ship. A microprocessor contained within the white tube converts conductivity to salinity, and, when connected to a portable computer, displays the results of all readings as graphs.

salinity in the field is with a **refractometer (Figure 6.9)**, a compact optical device that compares the degree to which light is bent by a sample of seawater to the degree of bending for water of known salinity. Refractometers are not as accurate as good salinometers, however.

Chemical Equilibrium and Residence Times

If outgassing and the chemical weathering of rock are continuing processes, shouldn't the ocean become progressively saltier with age? Landlocked seas and some lakes usually become saltier as they grow older, but the ocean does not. The ocean appears to be in **chemical**

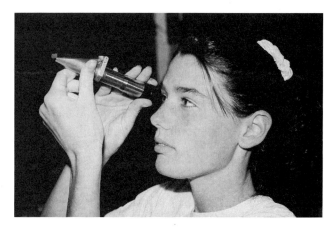

Figure 6.9 A refractometer in operation. The observer places a drop of water on a glass plate on top of the device and then looks into the refractometer to determine the position of a line against a scale. The degree to which the line has been refracted (bent) depends on the salinity of the water sample.

equilibrium; that is, the proportion *and amounts* of dissolved salts per unit volume of ocean are nearly constant. Evidently, whatever goes in must come out somewhere else.

Geologists in the 1950s developed the concept of a *steady state ocean.* The idea suggests that ions are added to the ocean at the same rate as they are being removed. This theory helps explain why the ocean is not growing saltier. The idea was quantified by T. F. W. Barth, who in 1952 devised the concept of **residence time,** the average length of time an element spends in the ocean. Residence time for an element may be calculated by this equation:

$$\text{Residence time} = \frac{\textbf{Amount of element in the ocean}}{\substack{\textbf{Rate at which the} \\ \textbf{element is added to (or} \\ \textbf{removed from) the ocean}}}$$

Additions of salts from the mantle or from the weathering of rock are balanced by subtractions of minerals being bound into sediments. Dissolved salts precipitate out of the water, and the silicon- and calcium carbonate-containing hard parts of living organisms drift slowly down to the seabed. Some of these sediments are removed from the ocean and drawn into the mantle at subduction zones by the cycling of crustal plates. Input (from runoff and outgassing) equals outfall (binding into sediments) for each dissolved component.

The residence time of an element depends on its chemical activity. Atoms (or ions) of some elements, such as aluminum and iron, remain in seawater for a relatively short time before becoming incorporated in sediments; others, such as chloride, sodium, and magnesium, remain in water for millions of years. The

BOX 6.1 ● *Recycling on a Grand Scale*

Ocean water has a residence time. Scientists estimate that about 334,000 cubic kilometers (80,000 cubic miles) of pure water evaporates from the ocean each year. Assuming a total ocean volume of 1,370 million cubic kilometers (329 million cubic miles), we can calculate that about 0.024% of the water in the ocean vaporizes each year. Thus, if it were not replenished by precipitation and runoff from land, the ocean would evaporate completely in about 3,500 years.

For a complete picture, more than oceanic water should be considered. Earth's total surface water volume—including the ocean, lakes, rivers, groundwater, ice caps, and so on—is about 1,399 million cubic kilo-

meters (336 million cubic miles). The total yearly evaporation/precipitation estimate is 396,000 cubic kilometers (95,000 cubic miles) of pure water. From these figures we again calculate a recycling time of about 3,500 years. With the exception of water subducted into the mantle along with sediments at subduction zones, then, water molecules have a residence time in the ocean of about 3,500 years.

Earth's ocean is more than 4 billion years old. Thus, on average, individual water molecules have evaporated from and returned to the ocean more than a million times since the world ocean formed in its basins!

approximate residence times for the major constituents of seawater are shown in **Table 6.3.** Even seawater itself has a residence time (see **Box 6.1**).

Mixing Time

If constituent minerals remain in ocean water longer than the ocean's **mixing time,** they will become evenly distributed through the ocean. Because of the vigorous activity of currents, the mixing time of the ocean is thought to be on the order of 1,000 years. Thus, the ocean has been mixed hundreds of thousands of times during its long history. The relatively long residence times of seawater's major constituents thus assure thorough mixing, the foundation of Forchhammer's principle of constant proportions.

Conservative and Nonconservative Constituents

As we have seen, the major constituent salts maintain constant ratios in seawater, but the quantity and proportions of minor and trace elements can change because of biological activity or local geological events.

Those constituents of seawater that occur in constant proportion or change very slowly through time are **conservative constituents.** Conservative elements have long residence times. Not surprisingly, these are the most abundant dissolved constituents of water—the ones that constitute the bulk of the ocean's dissolved material. The inert gases dissolved in the ocean (and the water of the ocean itself) are also conservative constituents.

Table 6.3	Approximate Residence Times for Constituents of Seawater
Constituent	**Residence Time (years)**
Chloride (Cl^-)	100,000,000
Sodium (Na^+)	68,000,000
Magnesium (Mg^{2+})	13,000,000
Potassium (K^+)	12,000,000
Sulfate (SO_4^{2-})	11,000,000
Calcium (Ca^{2+})	1,000,000
Carbonate (CO_3^{2-})	110,000
Silicon (Si)	20,000
Water (H_2O)	3,500
Manganese (Mn)	1,300
Aluminum (Al)	600
Iron (Fe)	200

Sources: Data from Broecker and Peng, 1982; Bruland, 1983; Riley and Skirrow, 1975.

Nonconservative constituents are those substances dissolved in seawater that are tied to biological or seasonal cycles or to very short geological cycles. They have short residence times. Biologically important nonconservative constituents include dissolved oxygen produced by plants, carbon dioxide produced by animals, silica and calcium compounds needed for plant and animal shells, or the nitrates and phosphates needed for production of protein and other biochemicals. Aluminum, with a residence time of only 600 years, is rapidly removed by adsorption on clay sediment particles; so it is a noncon-

Table 6.4 Major Gases in the Atmosphere and Ocean

Gas	Percent of Gas in Atmosphere, by Volume	Percent of Dissolved Gas in Seawater, by Volume	Concentration in Seawater in Parts per Million, by Weight
Nitrogen (N_2)	78.08%	48%	10–18 ppm
Oxygen (O_2)	20.95%	36%	0–13 ppm
Carbon dioxide (CO_2)[a]	0.035%	15%	64–107 ppm

Sources: Data from Weihaupt, 1979; Hill, 1963.

[a]Also present in seawater as carbonic acid, carbonate ions, and bicarbonate ions.

servative element. Aluminum is very rare in seawater (10 parts per billion). Many trace elements are in great biological demand or tend to form insoluble compounds in water. (Oceanic residence times calculated to be less than 1,000 years—the ocean's mixing time—are probably meaningless. The definition of residence time assumes the existence of steady-state conditions and a well-mixed ocean; so a residence time less than the ocean's mixing time is mainly an indication of an element's reactivity.)

DISSOLVED GASES

Gases in the air readily dissolve in seawater at the ocean's surface. Plants and animals living in the ocean require these dissolved gases to survive. No marine animal has the ability to break down water molecules to obtain oxygen directly, and no marine plant can manufacture enough carbon dioxide to support its own metabolism. In order of their relative abundance, the major gases found in seawater are nitrogen, oxygen, and carbon dioxide (see **Table 6.4**). The proportions of dissolved gases in the ocean are very different from the proportions of the same gases in the atmosphere.

Unlike solids, gases dissolve most readily in *cold* water. A cubic meter of chilly polar water usually contains a greater volume of dissolved gases than a cubic meter of warm tropical water. Dissolved nitrogen gas is a conservative component of seawater, while biologically active oxygen and carbon dioxide are nonconservative components.

Nitrogen

About 48% of the dissolved gas in seawater is nitrogen. (In contrast, the atmosphere is slightly more than 78% nitrogen by volume.) The upper layers of ocean water are usually saturated with nitrogen; that is, additional nitrogen will not dissolve. Living organisms require nitrogen to build proteins and other important biochemicals, but they cannot use the free nitrogen in the atmosphere and ocean directly. It must first be *fixed* into usable

chemical forms by specialized organisms. Though some species of bottom-dwelling bacteria can manufacture usable nitrates from the nitrogen dissolved in seawater, most of the nitrogen compounds needed by living organisms must be recycled among the organisms themselves.

Oxygen

About 36% of the gas dissolved in the ocean is oxygen, but there is about 100 times more gaseous oxygen in the Earth's atmosphere than is dissolved in the whole ocean. An average of 6 milligrams of oxygen is dissolved in each liter of seawater (that is, 6 parts per million of oxygen per liter of seawater, by weight). Yet this small amount of oxygen is a vital resource for animals that extract oxygen with gills. The primary source of the ocean's dissolved oxygen is its photosynthetic plants. Since photosynthesis requires sunlight, most of the available oxygen lies near the ocean's surface. Because marine plants are so abundant and photosynthetically active, much more free oxygen diffuses from ocean to atmosphere than moves from atmosphere to ocean.

Carbon Dioxide

The amount of carbon dioxide (CO_2) in the atmosphere is very small (0.03%) because CO_2 is in great demand by photosynthetic plants as a source of carbon for growth. Carbon dioxide is very soluble in water, though; the proportion of dissolved CO_2 in water is about 15% of all dissolved gases. Because CO_2 combines chemically with water to form a weak acid (carbonic acid, H_2CO_3), water can hold perhaps 1,000 times more carbon dioxide than either nitrogen or oxygen at saturation. Carbon dioxide is quickly used by marine plants, however; so dissolved quantities of CO_2 are almost always much less than this theoretical maximum. Even so, at the present time there is about 60 times as much CO_2 dissolved in the ocean as in the atmosphere. Much more CO_2 moves from atmosphere to ocean than from ocean to atmosphere, in part because some dissolved CO_2 forms carbonate ions, which are locked into sediments, minerals, and the shells and skeletons of living organisms.

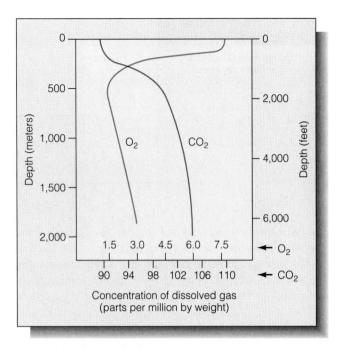

Figure 6.10 How concentrations of oxygen and carbon dioxide vary with depth. Oxygen is abundant near the surface because of the photosynthetic activity of marine plants. Oxygen concentration decreases below the sunlit layer because of the respiration of marine animals and bacteria. In contrast, plants use carbon dioxide during photosynthesis; so surface levels of CO_2 are low. Photosynthesis cannot take place in the dark; so carbon dioxide given off by animals and bacteria tends to build up at depths below the sunlit layer.

Figure 6.10 illustrates how carbon dioxide and oxygen concentrations vary with depth. Carbon dioxide concentrations increase with depth, but oxygen concentrations usually decrease through the mid-depths and then rise again toward the bottom. High concentrations of oxygen at the surface are usually by-products of the photosynthesis of plants in the ocean's brightly lit upper layer. Since plants require carbon dioxide for photosynthesis, surface CO_2 concentrations tend to be low. A decrease in oxygen below the sunlit upper layer usually results from the respiration of bacteria and marine animals, activity that tips the balance in favor of carbon dioxide. Oxygen levels are slightly higher in deeper water because fewer animals are present to take up oxygen reaching these depths and because oxygen-rich polar water is the greatest source of deep water.

ACID–BASE BALANCE

Water can separate to form hydrogen ions (H^+) and hydroxide ions (OH^-). In pure water, these two ions are present in equal concentrations. An imbalance in the ions produces either an acidic or a basic solution. An **acid** is a

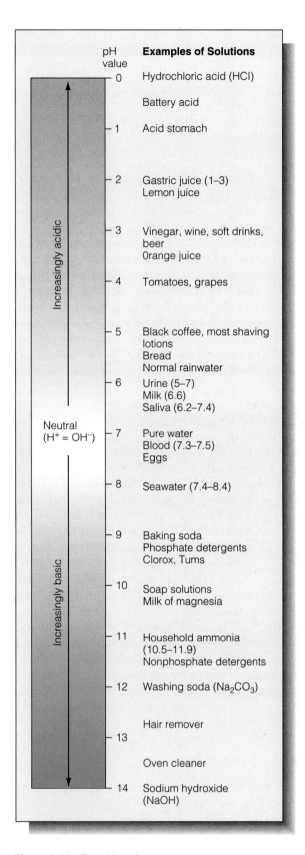

Figure 6.11 The pH scale.

substance that *releases* a hydrogen ion in solution. A **base** is a substance that *combines with* a hydrogen ion in solution. A solution containing a base is also called an **alkaline** solution.

The acidity or alkalinity of a solution is measured in terms of the **pH scale,** which measures the concentration of hydrogen ions in a solution. An excess of hydrogen ions (H^+) in a solution makes that solution acidic. An excess of hydroxide ions (OH^-) makes a solution alkaline. **Figure 6.11** shows a pH scale and the pH of a few familiar solutions. The scale is logarithmic, which means that a change of one pH unit represents a 10-fold change in hydrogen ion concentration. Thus, a modern non-phosphate detergent is 1,000 times more alkaline than seawater, and black coffee is 100 times more acid than pure water. Pure water, which is neutral (neither acidic nor basic) has a pH of 7; lower numbers indicate greater acidity (more H^+), and higher numbers indicate greater alkalinity (fewer H^+).

Seawater is slightly alkaline; its average pH is about 7.8. This seems odd because of the large amount of CO_2 dissolved in the ocean. If dissolved CO_2 combines with water to form carbonic acid, why is the ocean mildly alkaline and not slightly acidic? When dissolved in water, however, CO_2 is actually present in several different forms. Carbonic acid (H_2CO_3) is only one of these. In water solutions, some carbonic acid breaks down to produce the hydrogen ion (H^+), the bicarbonate ion (HCO_3^-), and the carbonate ion (CO_3^{2-}). This sequence is shown in **Figure 6.12.** If the pH of the ocean drops (becomes more acidic), the reaction in Figure 6.12 proceeds to the *left,* raising the pH by removing H^+ ions. If the pH of the ocean rises (becomes more alkaline), the reaction proceeds to the *right,* lowering the pH by releasing H^+ ions. This behavior acts to **buffer** the water, preventing broad swings of pH when acids or bases are introduced. Rivers and lakes lack buffering, making them more susceptible to pH swings when acids or bases are added at waste outfalls.

Though seawater remains slightly alkaline, it is subject to some variation. In areas of rapid plant growth, for example, pH will rise because CO_2 is used by the plants for photosynthesis, and the reactions of Figure 6.12 go to the left, removing free H^+ ions. Because temperatures are generally warmer at the surface, less CO_2 can dissolve in the first place. Thus, surface pH in warm, productive water is usually around 8.5.

At middle depths and in deep water, more CO_2 may be present. Its source is the respiration of animals and bacteria. With cold temperatures, high pressure, and no photosynthetic plants to remove it, this CO_2 will lower the pH of water, making it more acid with depth (see **Figure 6.13**). Thus, deep, cold seawater below 4,500 meters (15,000 feet) has a pH of around 7.5. This lower pH can dissolve calcium-containing marine sediments. (Recall from Chapter 5 that sediments containing calcium carbonate are rarely found in deep water.) A drop to pH 7 can occur at the deep-ocean floor when bottom bacteria consume oxygen and produce hydrogen sulfide.

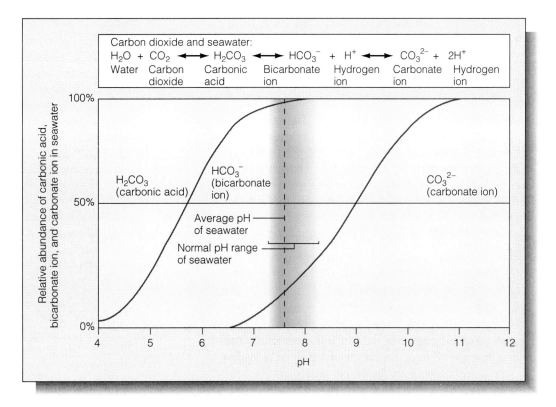

Figure 6.12 Carbon dioxide (CO_2) combines readily with seawater to form carbonic acid (H_2CO_3). Carbonic acid can then lose a H^+ ion to become bicarbonate (HCO_3^-), or two H^+ ions to become carbonate (CO_3^{2-}). As the double-headed arrows indicate, these reactions may move in either direction. The relative abundance of carbonic acid, bicarbonate, and carbonate in seawater is a function of pH. Seawater has an average pH of about 7.8, and the bicarbonate ion is most prevalent. As the graph shows, carbonic acid dominates at low pH and the carbonate ion at higher pH.

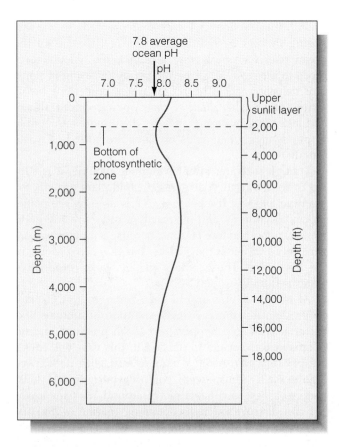

Figure 6.13 The variation in pH with depth.

Q-AND-A

1. If outgassing continues today, why aren't the oceans getting bigger? By now shouldn't they have covered the continents?

   Well, yes, outgassing does continue, but the water is being drawn from the mantle. Some marine geologists believe that the mantle shrinks a little, making the ocean basins get deeper, which accommodates the ocean's greater volume. Also, geological processes have increased the amount of continental crust, and water is drawn back into the mantle at subduction zones. It probably just about balances out.

2. You have noted that there is 100 times more oxygen in the atmosphere than in the ocean. How can that be? Isn't water 86% oxygen by weight?

   Yes, but the oxygen of water (H_2O) is bonded tightly to hydrogen atoms. Unlike atmospheric oxygen, the oxygen in water molecules cannot be used by organisms for respiration. Fish can't extract this oxygen with their gills. Marine animals must depend on dissolved *oxygen, paired molecules of oxygen (O_2) present in the water and free to move through the gill membranes. Compared to the atmosphere, very little of this* free *oxygen is present in the ocean.*

3. You mentioned that hydrogen bonds cause water's blue color. How?

   Pure water is pale blue when a thickness of about 2 meters (7 feet) is viewed against a white background. When water molecules vibrate, adjacent molecules tug and push against their hydrogen-bonded neighbors. This action absorbs a small amount of red light, leaving proportionally more blue light to scatter back to our eyes. The same blue color is also seen in ice formations.

4. There is more than $3.5 million worth of gold (at 1995 prices) in each cubic kilometer of seawater. Why don't we remove some of it and pay off the national debt?

   Easy to say, difficult to do. For one thing, extracting the gold will require sophisticated chemical treatment. But for the sake of argument, let's assume you perfect a magic method of precipitating gold from seawater with a simple wave of your hand. All you have to do is pump water out of the ocean into a large holding tank to do your trick. Unfortunately the energy cost of lifting the water just a few millimeters into the tank would eat up all the profits, and then some. German economists made a study after the First World War to see if this method could be used to retire their war debt. It was shown to be fruitless. So much for get-rich-quick schemes.

5. Some essential resources are relatively abundant in ocean water. Are there any prospects for chemical "mining" of seawater?

   Yes. Commercial extraction of magnesium from seawater began in 1940. Seawater is treated with calcium hydroxide and hydrochloric acid to form magnesium chloride. This substance is dried and electrolytically separated into chlorine gas (which can be used to make more hydrochloric acid) and magnesium metal. Magnesium is an essential component of the aluminum alloys from which aircraft, beverage cans, and automobile parts are manufactured.

   Many nonmetal resources are also obtained from seawater. For example, bromine—an element important in the production of motor fuels, photographic film, dyes, and insecticides—is extracted from concentrated brines or from crystallized sea salts. And, as we'll see in Chapter 19, the salts themselves are among the most important of the ocean's physical resources.

Terms and Concepts to Remember

acid
adhesion
alkaline
atom
base
bond
buffer
chemical bond
chemical
   equilibrium
chlorinity
cohesion
colligative
   properties
compound
conservative
   constituent

covalent bond
diffusion
electron
element
excess volatiles
Forchhammer's
   principle
hydrogen bond
ion
ionic bond
mixing time
mixture
molecule
Nansen bottle
nonconservative
   constituent
pH scale

polar molecule
precipitate
principle of
   constant
   proportions
proton
refractometer
residence time
salinity
salinometer
saturation
solute
solution
solvent
trace element

Study Questions

1. Where did the water of the ocean come from?

2. Why is water a polar molecule? What properties of water derive from its polar nature?

3. How does a solution differ from a mixture?

4. What are some of seawater's colligative properties? Does pure water have colligative properties?

5. Other than hydrogen and oxygen, what are the most abundant elements in seawater?

6. What was the earthly origin of the sodium and chloride ions in common table salt?

7. How is salinity determined? How are modern methods dependent on the principle of constant proportions?

8. How are seawater's conservative constituents different from its nonconservative constituents? Give an example of each.

9. Which dissolved gas is present in the ocean in much greater proportion than in the atmosphere? Why the disparity?

10. What factors affect seawater's pH? How does the pH of seawater change with depth? Why?

For Further Study

Broecker, W. S. 1974. *Chemical Oceanography.* New York: Harcourt Brace Jovanovich. Technical treatment of ocean chemistry.

Broecker, W. S. 1983. "The Ocean." *Scientific American,* September, 146–60. Information on the origin and nature of the ocean's dissolved solids.

Broecker, W. S., and T-H Peng. 1982. *Tracers in the Sea.* Palisades, NY: Columbia University, Lamont–Doherty Geological Observatory. Technical. Source of most of the residence times in Table 6.3.

Bruland, K. W. 1983. "Trace Elements in Seawater." In *Chemical Oceanography,* vol. 8, edited by J. P. Riley and R. Chester. New York: Academic Press. Reactivity vs. residence times for elements with very short calculated residence times (see Table 6.3).

Donovan, A. 1994. *Antoine Lavoisier: Science, Administration, and Revolution.* Blackwell Science Biographies. Cambridge, MA: Blackwell.

Garrison, T., and R. Lebow. 1995. *Oceanus.* 6th ed. Belmont, CA: Wadsworth. Lesson 4 offers a compact summary of the characteristics of water.

Guerlac, H. 1975. *Antoine-Laurent Lavoisier: Chemist and Revolutionary.* New York: Scribner. Information on the "discovery" of water.

Hill, M. N. 1963. *The Sea: Composition of Seawater.* New York: Wiley Interscience. A standard reference on the topic.

Holland, H. D. 1984. *The Chemical Evolution of the Atmosphere and Ocean.* Princeton: Princeton University Press. Useful history lesson.

Horne, R. A. 1969. *Marine Chemistry: The Structure of Water and the Chemistry of the Hydrosphere.* New York: Wiley.

Horne, R. A. 1974. "Water." *Encyclopædia Britannica.* Chicago: Encyclopædia Britannica. A well-written, readable summary.

Kerr, R. A. 1988. "Ocean Crust's Role in Making Seawater." *Science* 239 (no. 4837): 260. The possible role of water circulation through active rift zones in reducing Mg^{2+} and increasing Ca^{2+}.

Libes, S. M. 1992. *An Introduction to Marine Biogeochemistry.* New York: Wiley. A balanced survey of the inorganic and organic aspects of marine chemistry.

MacIntyre, F. 1970. "Why the Sea Is Salt." *Scientific American,* November, 104–15.

McClellan, H. J. 1965. *Elements of Physical Oceanography.* London: Pergamon Press. An advanced text.

Riley, J. P., and G. Skirrow. 1975. *Chemical Oceanography,* vol. 1. 2d ed. New York: Academic Press. Residence time for Cl^- in Table 6.3.

Rubey, W. W. 1951. "Geologic History of Seawater: An Attempt to State the Problem." *Bulletin of the Geological Society of America,* 62: 1111–47. A pivotal paper on the history of seawater and the concept of excess volatiles.

Sverdrup, H. U., et al. 1942. *The Oceans: Their Physics, Chemistry and General Biology.* New York: Prentice-Hall. Excellent, though dated, technical presentation.

Von Arx, W. S. 1962. *An Introduction to Physical Oceanography.* New York: Addison-Wesley.

Wallace, W. J. 1974. *The Development of the Chlorinity/Salinity Concept in Oceanography.* Amsterdam: Elsevier.

Walton-Smith, F. G., ed. 1974. *Handbook of Marine Science,* vol. 1. Cleveland: CRC Press.

Weihaupt, J. G. 1979. *Exploration of the Oceans.* New York: Macmillan. Excellent discussion of dissolved gases in seawater.

Weyl, P. K. 1970. *Oceanography.* New York: Wiley. Clear sections on the comparison of river water with seawater and the geological history of seawater. Source of CO_2/bicarbonates/carbonates data; carbonates and sulfates residence times.

Note: Oceanus 35 (no. 1, Spring 1992) is dedicated to marine chemistry.

OCEAN PHYSICS

Sonar in the Afternoon

Lieutenant Pryor had an interesting problem. Mr. Pryor was a "sound man," an officer assigned to the destroyer USS *Semmes* stationed at the Guantánamo Bay Naval Station in Cuba. During the early 1930s, he and his colleagues had worked to perfect a directional echo-ranging system—which we now call sonar—and now, in 1936, the time had come to test the machinery at sea. The problem: It didn't always work as designed.

Lieutenant Pryor and *Semmes* were given a friendly submarine to chase around the bay. The sonar transmitter aboard the ship projected a beam of sound toward the submarine. The submarine (or any other submerged object) reflected the sound, and the echo was used to determine the object's distance and direction. The new equipment worked well enough during the morning hours, but no matter what the technicians did it seemed unable to function reliably in the afternoon, especially when the sun was hot and the air still.

Pryor theorized that microscopic oxygen bubbles being released by billions of phytoplankton—microscopic plantlike marine organisms—might be interfering with the signals somehow. So he traveled to the Woods

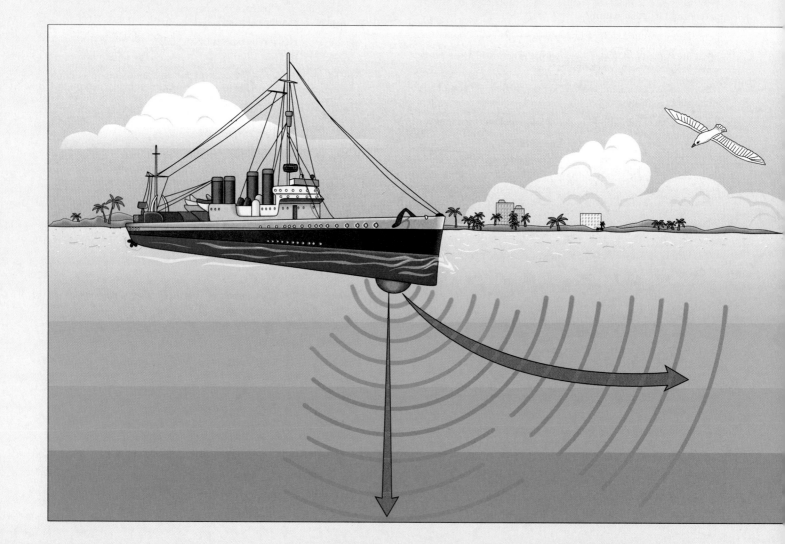

Hole Oceanographic Institution to speak with the institution's director about the so-called "afternoon effect."

The director didn't think bubbling phytoplankton was responsible, but his interest was aroused, and the Woods Hole research ship *Atlantis* joined *Semmes* at Guantánamo Bay the next summer. During an extensive sampling program the two ships found nothing unusual in the water's oxygen content or salinity, but they did notice that sound transmission and reception were most difficult when the topmost layer of the ocean was warm and calm. "I do feel that we are on the track of something," wrote Pryor.

Later that year *Atlantis* and *Semmes* cooperated in sonar studies off Long Island. Woods Hole oceanographers discovered that the afternoon effect occurred because sound waves are bent and distorted by their travel through the well-defined layers of warm and cool water near the ocean surface. Ten years earlier German scientists had shown that the speed of sound in seawater is related to temperature, and tables relating the speed of sound to temperature (and salinity) had been compiled to ensure accurate depth sounding. But echo sounder beams penetrate the ocean's layers vertically, while sonar beams graze their surfaces horizontally. Because of the oblique angle of approach, the sonar beam was bent sharply upward or downward at each water temperature junction. The layers were more homogeneous in the morning before the sun heated the top layer; so sound went out and returned in predictable ways. But in the afternoon, the difference in water temperature of these layers was more pronounced, and sound beams went awry. The implications of this finding for submarine defense were not difficult to grasp. Clearly, naval researchers needed more information about the physical characteristics of the ocean if sonar was to become reliable.

The afternoon effect in action.

The physical characteristics of the world ocean are largely determined by the physical properties of seawater. These properties include water's heat capacity, density, salinity, and ability to transmit light and sound.

The thermal properties of water are responsible for the mild physical conditions at the Earth's surface. Liquid water is remarkably resistant to temperature change; and ice, with its large latent heat of fusion and low density, melts and refreezes over large areas of the ocean, absorbing or releasing heat with no change in temperature. These thermostatic effects, combined with the mass movement of water and water vapor, prevent large swings in Earth's surface temperature.

Changes in temperature and salinity greatly influence water density. Ocean water is usually layered by density, with the densest water on or near the bottom. Density is also important to the ocean's vertical circulation, its acoustical properties, and its buoyant support of organisms.

Sound and light in the sea are affected by the physical properties of water, with refraction and absorption effects playing important roles.

WATER AND HEAT

Perhaps the most important physical properties of water are related to its heat capacity. Its unusual thermal characteristics prevent wide temperature variation from day to night and from winter to summer, permit vast amounts of heat to flow from equatorial to polar regions, and power Earth's great storms, wind waves, and ocean currents.

Four sources of heat have an influence on the ocean: (1) solar energy, (2) heat generated by the radioactive decay of elements within the Earth, (3) heat left over from the Earth's formation, and (4) heat produced artificially by human beings. Of these, solar energy is by far the most important, accounting for more than 3,200 times the contribution of the remaining three. Visible and infrared light absorbed by the ocean, land, or atmosphere turns to heat.

Heat and Temperature

Heat and temperature are not the same thing. **Heat** is energy produced by the random vibration of atoms or molecules. On the average, water molecules in hot water vibrate more rapidly than water molecules in cold water. Heat is a measure of how many molecules are vibrating, and how rapidly they are vibrating. Temperature records only how rapidly the molecules of a substance are vibrating. **Temperature** is an object's response to an input (or removal) of heat. The amount of heat required to bring a substance to a certain temperature varies with the nature of that substance.

An example will help. Which has a higher temperature: a candle flame or a bathtub of hot water? The flame. Which contains more heat? The tub. The molecules in the flame vibrate very rapidly, but there are relatively few of them. The molecules of water in the tub vibrate more slowly, but there are a great many of them; so the total amount of heat energy in the tub is greater.

Temperature is measured in **degrees.** One degree Celsius (°C) = 1.8 degrees Fahrenheit (°F). Though we are more familiar with the older Fahrenheit scale (named after the eighteenth-century German inventor of the mercury thermometer), Celsius degrees are more useful in science because they are based on two of pure water's most significant properties: its freezing point (0°C) and its boiling point (100°C). The Celsius scale—until recently called the centigrade scale because of the 100° interval between base points—was invented by Anders Celsius, a Swedish astronomer, in 1742.

Heat Capacity

Heat capacity is a measure of the heat required to raise the temperature of 1 gram (0.035 ounce) of a substance by 1°C (1.8°F). Different substances have different heat capacities. *Not all substances respond to identical inputs of heat by rising in temperature the same number of degrees.* Heat capacity is measured in calories. A **calorie** is the amount of heat required to raise the temperature of 1 gram (0.035 ounce) of pure water by 1°C.[1]

Because of the unique bonding that occurs among water molecules, the heat capacity of water is among the highest of all known substances. This means that water can absorb (or release) large amounts of heat while changing relatively little in temperature. Anyone who waits by a stove for water to boil knows much about water's heat capacity! By contrast, ethyl alcohol has a much lower heat capacity. If both liquids absorb heat from identical stove burners at the same rate, pure ethyl alcohol (the active ingredient in alcoholic beverages) will rise in temperature about twice as fast as an equal mass of water. Beach sand has an even lower heat capacity: Sand requires as little as 0.2 calorie to rise 1°C (1.8°F). Thus, on sunny days beaches can get too hot to stand on with bare feet, while the water remains pleasantly cool.

Temperature and Density

The uniqueness of pure water becomes even more apparent when we consider the effect of a temperature change on water's density (its mass per unit of volume). Recall from Chapter 3 that the density of pure water is 1 gram per cubic centimeter (1 g/cm^3). Granite rock is heavier, with a density of about 2.7 g/cm^3, and air is lighter, with a density of about 0.0012 g/cm^3. Most substances become denser (weigh more per unit of volume) as they get colder. Cold air sinks and warm air rises because the rapidly vibrating molecules of warm air are farther apart and therefore occupy more space than the same number of cool air molecules. The cold air falls because it is denser. Like air, pure water generally becomes denser as heat is removed and temperature falls; but water's density curve shows a curious anomaly as the temperature approaches the freezing point.

A **density curve** shows the relationship between the temperature or salinity of a substance and its density. Most substances become progressively denser as they cool; their temperature–density relationships are linear (that is, appear as a straight line on graphs). But **Figure 7.1** shows the unusual temperature–density relationship of pure water. Imagine heat being removed from some water placed in a freezer. Initially, the water is at room temperature, point A on the graph. As expected, the density of water increases as its temperature drops along the line from point A toward point B. Approaching point B the density increase slows, reaching a maximum at point B of 1.000 g/cm^3 at 3.98°C (39.2°F). Oddly, water then

---

[1]A nutritional Calorie, the unit we see on cereal boxes, equals 1,000 of these calories. A gram is about 10 drops of seawater.

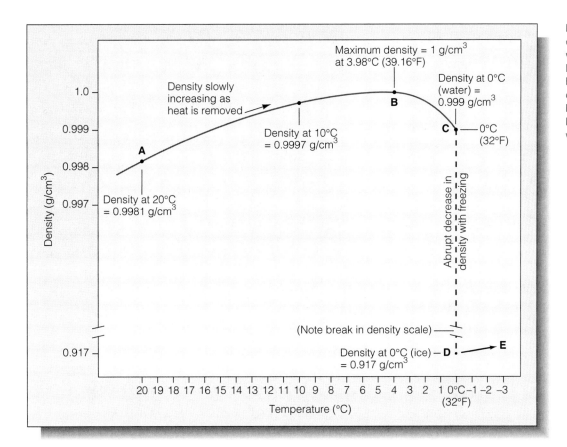

Figure 7.1 The relationship of density to temperature for pure water. Note that points C and D both represent 0°C (32°F) but different densities and thus different states of water. Because the density of ice is less than the density of liquid water, ice floats.

becomes slightly *less* dense as cooling continues, until point C (0°C, 32°F) is reached. At point C the water begins to freeze—to change state by crystallizing into ice.

State is an expression of the internal form of a substance. Water exists in three states: liquid, gas (water vapor), and solid (ice). If the freezer continues to remove heat from the water at point C, the water will change from liquid to solid state. Through this transition from water to ice—from point C to point D—the density of the water *decreases* abruptly. Ice is therefore lighter than an equal volume of water. Ice increases in density as it gets colder than 0°C. No matter how cold it gets, however, ice never reaches the density of liquid water. Being less dense than water, ice "freezes over" as a floating layer instead of "freezing under" like the solid forms of virtually all other liquids.

As we'll see, the implications of water's high heat capacity and the ability of ice to float, are vital in maintaining Earth's moderate surface temperature. First, we look at the transition from point C to point D in Figure 7.1.

Freezing Water

During the transition from liquid to solid state at the **freezing point,** the bond angle between the oxygen and hydrogen atoms in water expands from about 105° to slightly more than 109°. This change allows the hydro-

gen bonds in ice to form a crystal lattice (see **Figure 7.2**). The space taken by 27 water molecules in the liquid state would be occupied by only 24 water molecules in the solid lattice, however; so water has to expand about 9% as the crystal forms. Because the molecules are packed less efficiently, ice is less dense than liquid water—and so it floats. A cubic centimeter of ice at 0°C (32°F) weighs only 0.917 gram, whereas a cubic centimeter of liquid water at 0°C weighs 0.999 gram.

The ice lattice is not a flat sheet but a three-dimensional network. The water molecules in the ice lattice are still linked by hydrogen bonds, but they now form regular hexagons. This hexagonal pattern explains the lovely six-sided symmetry of snowflakes and other ice crystals.

The transition from liquid water to ice crystal (point C to point D in Figure 7.1) requires continued removal of heat energy; the change in state does not occur instantly throughout the mass when the cooling water reaches 0°C (32°F). Again, consider water in a freezer. **Figure 7.3**, a plot of heat removal versus temperature, details the water's progress to ice. As in Figure 7.1, point A represents warm water just placed into the freezer. The removal of heat does not stop when the water reaches point C, but the decline in temperature does. Even though heat continues to be removed, the water will not get colder until all of it has changed state from liquid (water) to solid (ice). Heat may therefore be removed

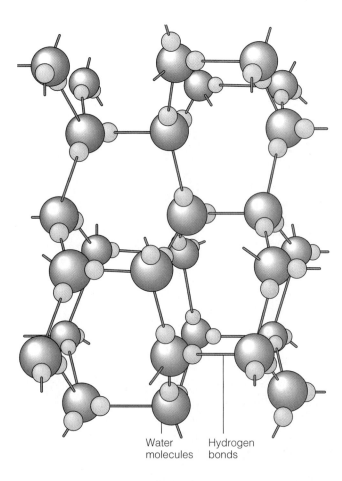

Water   Hydrogen
molecules   bonds

Figure 7.2 The lattice structure of an ice crystal (*left*), showing its hexagonal arrangement at the molecular level. The hexagonal patterns in the snowflakes above are visible reflections of this same structure.

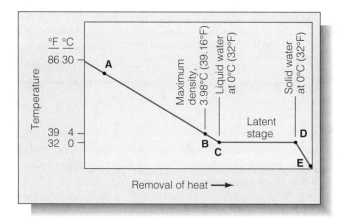

Figure 7.3 A graph of temperature versus heat removal as water freezes (or ice melts). The horizontal line between points C and D represents the latent heat of fusion, when heat is being removed but temperature is not changing. (Note that points A–E on this graph are the same as those in Figure 7.1.)

from water when it is changing state (that is, when it is freezing) without the water dropping in temperature. Indeed, the continued removal of heat is what makes the change in state possible. Heat is released as bonds form to make ice, and that heat must be removed to allow more ice to form.

The removal of heat from point A to point C in Figures 7.1 and 7.3 produces a *measurable* lowering of temperature detectable by a thermometer. Removing just 1 calorie of heat from a gram of liquid water causes its temperature to drop 1°C. This detectable decrease in heat is called **sensible heat** loss. But the loss of heat as water freezes between points C and D is not measurable (that is, not sensible) by a thermometer. Removing a calorie of heat from freezing water at 0°C (32°F) won't change its temperature at all; 80 calories of heat energy must be removed per gram of pure water at 0°C (32°F) to form ice. This heat is called the **latent heat of fusion** (*latere* = to be hidden). The straight line between points C and D in Figure 7.3 represents water's latent heat of fusion. No more ice crystals can form when all the water in the freezer has turned to ice. If the removal of heat continues, the ice will get colder and will soon reach the temperature inside the freezer, point E in Figures 7.1 and 7.3.

Latent heat of fusion is also a factor during thawing. When ice melts, it *absorbs* large quantities of heat (the

same 80 calories per gram) but does not change in temperature until the entire mass has turned to liquid. This is why ice is so effective in cooling drinks. Great amounts of ice form and melt in the polar ocean every year. In addition to providing spectacular scenery (see **Box 7.1**), the process of freezing and thawing moderates global temperature swings.

Evaporating Water

When water evaporates, individual water molecules diffuse into the air (**Figure 7.4**). Since each water molecule is hydrogen-bonded to adjacent molecules, heat energy

BOX 7.1 ● *Icebergs*

Water occurs in three states: liquid, solid, and gas. Oceanographers are most familiar with water's liquid form, but about 6% of the world ocean is covered by ice. A minute fraction of this ice is contained in the fantastic shapes of icebergs. Southern ocean icebergs originate as huge ice sheets attached to the Antarctic continent; lengths of up to 8 kilometers (5 miles) are not unusual, with flat tops rising 45 meters (150 feet) above sea level. Arctic icebergs are typically smaller in area but can be higher, pinnacled (or *castellated*, after their castle turret shapes), and extraordinarily beautiful:

> On the afternoon of the day we knew the storm had passed, I stood on the starboard side of the bridge at a window, with the heavy protective glass lowered. I rested my forearms on the sill, feeling the warmth of the bridge heaters around my legs and a slipstream of cool air past my face. The first icebergs we had seen, just north of the Strait of Belle Isle, listing and guttered by the ocean, seemed immensely sad, exhausted by some unknown calamity. We sailed past them.
>
> I occasionally drew back from the starboard window to make a sketch, or to bring the binoculars up to my eyes. I marveled as much at the behavior of light around the icebergs as I did at their austere, implacable progress through the water. They took their color from the sun, and from the clouds and the water. But they also took their dimensions from the light: the stronger and more direct it was, the greater the contrast upon the surface of the ice, of the ice itself with the sea. And the more finely etched were the dull surfaces of their walls. The bluer the sky, the brighter their outline against it. . . .
>
> Where the walls entered the water, the surf pounded them, creating caverns, grottoes, and ice

> bridges, strengthening an impression of sea cliffs. At the waterline the ice gleamed aquamarine against its own gray-white walls above. Where meltwater had filled cracks or made ponds, the pools and veins were milk-blue, or shaded to brighter marine blues, depending on the thickness of the ice. If the iceberg had recently fractured, its new face glistened greenish blue—the greens in the older, weathered faces were grayer. In twilight the ice took on the colors of the sun: rose, reddish yellows, watered purples, soft pinks. The ice both reflected the light and trapped it within its crystalline corners and edges, where it intensified. . . . How utterly still, unorthodox, and wonderous they seem.

Source: Barry Lopez, *Arctic Dreams* (New York: Scribner's, 1986).

is required to break those bonds and allow the molecule to fly away from the surface. Evaporation cools a moist surface because departing molecules of water vapor carry this energy away with them. (This is how perspiring cools us when we're hot. The heat energy required to evaporate water from our skin is taken away from our bodies, thus cooling us.)

Hydrogen bonds are quite strong, and the amount of energy required to break them—known as the **latent heat of evaporation**—is very high. At 585 calories per gram at 20°C (68°F), water has the highest latent heat of evaporation of any known substance. As before, the term *latent* applies to heat input that does not cause a temperature change but does produce a change of state—in this case from liquid to gas.

About a meter (3.3 feet) of water evaporates each year from the surface of the ocean, a volume of water equivalent to 334,000 cubic kilometers (80,000 cubic miles). The

Figure 7.4 Water vapor is invisible, but as water vapor evaporates from the sea surface and rises into cool air it can condense into the tiny droplets that form clouds and fog. (The tropical Pacific island of Moorea is seen in the distance.)

great quantities of solar energy that cause this evaporation are carried from the ocean by the escaping water vapor. When a gram of water vapor condenses back into liquid water, the same 585 calories is again available to do work. As we shall see in later chapters, winds, storms, ocean currents, and wind waves are all powered by that heat.

Why the big difference between water's latent heat of *fusion* (80 calories per gram) and its latent heat of *evaporation* (585 calories per gram)? Only a small percentage of hydrogen bonds are broken when ice melts, but *all* must be broken during evaporation. Breaking these bonds requires additional energy in proportion to their number.

Seawater and Pure Water

The solids dissolved in seawater change its thermal characteristics, lowering its heat capacity by about 4%. Only 0.96 calorie of heat energy is needed to raise the temperature of seawater by 1°C.

The dissolved solids also interfere with the formation of the ice lattice, acting as "antifreeze" to lower the freezing point. The saltier the water, the lower the freezing point. The temperature of maximum density moves toward the freezing point as salinity increases (see **Figure 7.5**), finally coinciding with the freezing point at a salinity of 24.7‰ (−1.33°C, 29.61°F). Seawater at 35‰, typical ocean salinity, freezes at −1.91°C (28.6°F). Seawater's density simply increases smoothly with decreasing

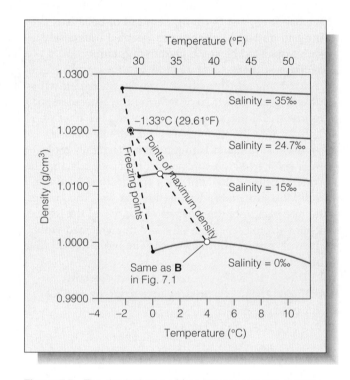

Figure 7.5 The dependence of freezing temperature and temperature of maximum density upon salinity. As we saw in Figure 7.1, pure water is densest at 3.98°C (39.2°F) (point B), and its freezing point is 0°C (32°F). Seawater with 15‰ salinity is densest at 0.73°C (33.31°F), and its freezing point is −0.80°C (30.56°F). The temperature of maximum density and the freezing point coincide at −1.33°C (29.61°F) in seawater with a salinity of 24.7‰. At salinities greater than 24.7‰ the density of water always decreases as temperature increases.

temperature until it freezes. The crystals that form are pure water ice, with the seawater salts excluded. The leftover cold, salty water is very dense.

Seawater evaporates more slowly than fresh water under identical circumstances because the dissolved salts tend to attract and hold water molecules. The latent heat of evaporation, however, is essentially the same for both fresh water and seawater. Salts are left behind as seawater evaporates. The remaining cool, salty water is also very dense.

GLOBAL THERMOSTATIC EFFECTS

The **thermostatic properties** (*therme* = heat, *stasis* = standing still) of water are those properties that act to moderate changes in temperature. Water temperature rises as sunlight is absorbed and changed to heat, but, as we've seen, water has a very high heat capacity; so its temperature will not rise very much even if a large quantity of heat is added. This tendency of a substance to resist change in temperature with the gain or loss of heat energy is called **thermal inertia.** To investigate the impact of water's thermostatic properties on conditions at the Earth's surface, we need to look at the planet's overall heat balance.

Only about 1 part in 2.2 billion of the sun's radiant energy is intercepted by Earth, but that amount averages 7 million calories per square meter per day at the top of the atmosphere—or, for the Earth as a whole, an impressive 17 trillion kilowatts (23 trillion horsepower)! About half of this light reaches the surface. As may be seen in **Figure 7.6,** on a global basis about 26% of the incoming light is reflected back into space from clouds or scattered

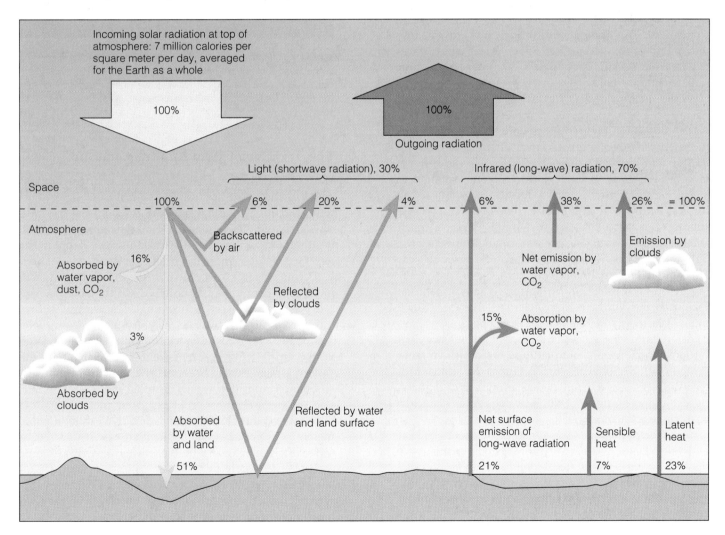

Figure 7.6 The heat budget for the Earth. On an average day, about half of the solar energy arriving at the upper atmosphere is absorbed at the Earth's surface. Light (short-wave) energy absorbed at the surface is converted into heat. Heat leaves the Earth as infrared (long-wave) radiation. Since input equals output over long periods of time, the heat budget is balanced.

by ice, water droplets, or other particles in the atmosphere. Another 19% is absorbed by water vapor, clouds, dust, and CO_2 in the atmosphere. About 4% bounces off the shining sea surface, ice and snow, or rocks and soil. Only 51% is absorbed by the Earth's land and water surface. How much light penetrates the ocean depends greatly on several factors: the angle at which it approaches, the sea state (surface turbulence), the presence of an ice covering or light-colored foam, and others.

The 51% of solar energy striking land and sea is converted to heat and then transferred into the atmosphere by conduction, radiation, and evaporation. The atmosphere, like the land and ocean, eventually radiates this heat back into space in the form of infrared radiation. The heat input and outflow "account" for the Earth can be thought of as a **heat budget.** As in your personal financial budget, income must eventually equal outgo. Over long periods of time the total incoming heat (plus that from earthly sources) equals the total outgoing heat; so the Earth is in **thermal equilibrium:** It is growing neither significantly warmer nor colder.[2] Heat input comes mainly from the sun; heat outflow can occur only as heat radiates into the cold of space.

While the heat budget for the Earth as a whole is in balance, the heat budget for its different latitudes is not. As can be seen in **Figure 7.7,** sunlight striking polar latitudes spreads over a greater area, filters through more atmosphere, and approaches the surface at a low angle, favoring reflection. Polar regions receive no sunlight at all during the depths of local winter. Contrast this to the tropical latitudes. The high solar angle in the tropics distributes the same amount of sunlight over a much smaller area. The more nearly vertical angle at which the light approaches means that it passes through less atmosphere and minimizes reflection. Mid-latitude heating is strongly affected by season: The mid-latitude regions of the Northern Hemisphere receive about 400% more light energy in June than in December. Tropical latitudes thus receive significantly more solar energy than the polar regions, and mid-latitude areas receive more heat in summer than in winter. **Figure 7.8** shows heat received versus heat reradiated as a function of latitude. (It also demonstrates that the heat budget for the Earth as a whole is balanced.)

Since the poles have such a marked deficiency of heat and the equator has such a pronounced surplus, why don't the polar oceans freeze solid and the equatorial ocean boil away? The reason is that currents in the atmosphere and ocean are moving huge amounts of heat between tropics and poles.

[2]Changes in heat balance do occur over short periods of time. Increasing amounts of carbon dioxide and methane in the Earth's atmosphere may be contributing to an increase in surface temperature called the *greenhouse effect.* More on this subject may be found in Chapter 20.

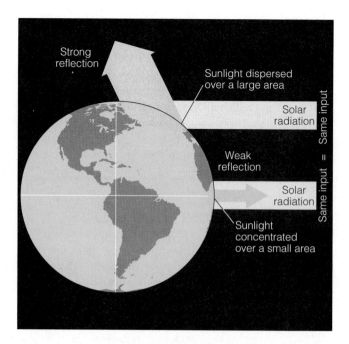

Figure 7.7 How solar energy input varies with latitude. Equal amounts of sunlight are spread over a greater surface area near the poles than in the tropics. Ice near the poles reflects much of the energy that reaches the surface there.

The Transfer of Heat by Water and Air

Water's high heat capacity makes it an ideal fluid to equalize the polar-tropical heat imbalance. Ocean currents and atmospheric weather result from the response of water and air to unequal solar heating. Although weather and currents are discussed in more detail in the next two chapters, here is a brief overview.

Ocean currents carry heat from the tropics (where incoming energy exceeds outgoing) to the poles (where outgoing energy exceeds incoming). The amount of heat transferred in this way is astonishing. For example, "outbound" water in the warm Gulf Stream (a large northward-flowing ocean current just offshore of the eastern United States) is about 10°C (18°F) warmer than "inbound" returning water, meaning that about 10 million calories are transported per cubic meter. Since the flow rate of the Gulf Stream is about 55 million cubic meters per second, some 550 trillion calories are being transported northward in the western North Atlantic each *second!* Nearly half of these calories reach the high latitudes above 40°N. As we will see in Chapter 9, this warmth has a dramatic moderating influence on the winter climate of northwestern Europe.

As impressive as the figures are for ocean currents, the amount of heat transported by water vapor in the atmosphere is even greater. About half of the solar energy entering water results in evaporation. The solar energy required for this evaporation is later surrendered

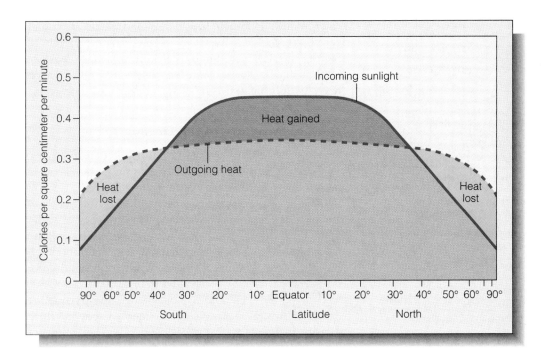

Figure 7.8 Areas of heat gain and loss on Earth's surface. Since the area of heat gained (darker area) equals the area of heat lost (lighter areas), Earth's heat budget is balanced.

during condensation, but usually at a distance from where the initial evaporation occurred. Thus the ocean surface near Cuba may be cooled by evaporation today, the water vapor may then be moved north by winds, and eastern Canada may be warmed by condensation of the same water in a rainstorm later in the week.

Both atmosphere and ocean transfer heat by movement, but water's exceptionally high latent heat of evaporation means that water vapor transfers much more heat (per unit of mass) than liquid water. Weather accounts for about two-thirds of the poleward transfer of heat; ocean currents move the other third.

Thermostatic Effects of Ice

Earth would surely be a much different place if ice were denser than liquid water, if it formed on the ocean bottom, and if it were unavailable for quick interaction with surface heat. Because water expands and floats when it freezes, ice can absorb the morning warmth of the sun, melt, and then refreeze at night—giving back to the atmosphere the heat it stored through the daylight hours. The heat content of the water changes through the day; its temperature does not.

The same principle applies to the seasonal formation and melting of polar ice. More than 18,000 cubic kilometers (4,300 cubic miles) of polar ice, covering some 18 million square kilometers (7 million square miles) of surface, thaws and refreezes each year. Seasonal extremes are moderated by the huge amounts of heat energy that are alternately absorbed and released without a change in temperature. Without these properties of ice, tempera-

tures on the Earth's surface could change dramatically with minor changes in atmospheric transparency or solar output.

TEMPERATURE, SALINITY, AND WATER DENSITY

The density of water is mainly a function of its temperature and salinity. Cold, salty water is denser than warm, less salty water. The density of seawater varies between 1.020 and 1.030 g/cm^3, indicating that a liter of seawater weighs between 2% and 3% more than a liter of pure water (1.000 g/cm^3) at the same temperature. Seawater's density increases with increasing salinity, increasing pressure, and decreasing temperature. **Figure 7.9** shows the relationship between temperature, salinity, and density. Notice that two samples of water can have the *same* density at different combinations of temperature and salinity.

Ocean water tends to form into stable layers, with the heaviest water at the bottom—a form of density stratification. Curiously, even the deepest of these layers originates at the ocean's surface. Very cold and salty water produced during the formation of sea ice at the polar ocean surface is denser than the surrounding water and sinks to the seabed. In a few marginal basins, the most notable being the Mediterranean Sea, evaporation can also produce salty, dense water that sinks toward the bottom until it reaches a layer of water of equal density. In contrast, warm fresh water entering the ocean at a

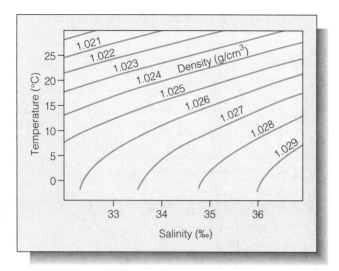

Figure 7.9 The complex relationship among the temperature, salinity, and density of seawater. Note that two samples of water can have the same density at different combinations of temperature and salinity.

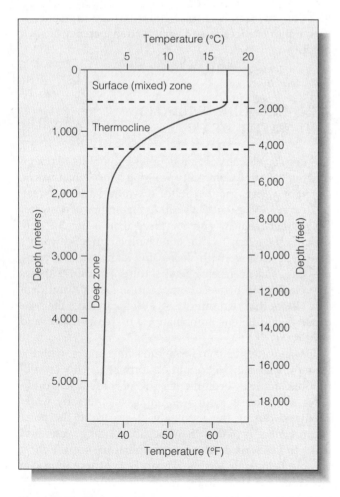

Figure 7.10 Typical temperature variation with depth in the deep ocean at mid-latitudes.

river mouth is much less dense and floats for miles above the cooler salty layers below. Early explorers of South America were amazed to find a surface layer of fresh water far to sea, and they discovered the Amazon River by following this fresh surface layer to its source.

Density Structure of the Ocean

Much of the ocean is divided into three density zones. The **surface zone,** or **mixed layer,** is the upper layer of ocean. Temperature and salinity are relatively constant with depth in the surface zone because of the action of waves and currents. The surface zone consists of water in contact with the atmosphere and exposed to sunlight; it contains the ocean's least dense water, only about 2% of total ocean volume. Depending on local conditions, the surface zone may reach a depth of 1,000 meters (3,300 feet) or be absent entirely. Beneath it is the **pycnocline** (*pyknos* = strong, *clinare* = slope, to lean), a zone in which density increases with increasing depth. This zone isolates surface water from the denser layer below. The pycnocline contains about 18% of all ocean water. The **deep zone** lies below the pycnocline, at depths below about 1,000 meters (3,000 feet) in mid-latitudes (40°S to 40°N). There is little additional change in water density with increasing depth through this zone. This deep zone contains about 80% of all ocean water.

Temperature is the most important factor affecting density at mid-latitudes because the change in temperature with depth is much more pronounced than the change in salinity with depth. **Figure 7.10** shows the general relationship of temperature with depth in the open sea. The surface zone is well mixed, with little decrease in temperature with depth; in the next layer, temperature drops rapidly with depth; beneath it lies the deep zone of cold, stable water. The middle layer, the zone in which temperature changes rapidly with depth, is called the **thermocline** (*therm* = heat). This thermocline is the major cause of the pycnocline mentioned above.

Thermoclines are not identical in form for all areas or latitudes. Temperate and tropical ocean areas cradle a warm surface layer whose bottom boundary is the top of the thermocline. Polar waters, which receive relatively little solar warmth, are not stratified by temperature and generally lack a thermocline. **Figure 7.11** contrasts polar, tropical, and temperate thermal profiles, showing that the thermocline is primarily a mid- and low-latitude phenomenon. Thermocline depth and intensity also vary with season, local conditions (storms, for example), currents, and many other factors. Divers often notice minor thermoclines that form near the surface, but most temperate and tropical oceans also contain a deep main thermocline.

Below the thermocline water is very cold, ranging in temperature from −1°C to 3°C (30.5°–37.5°F). Because this deep and cold layer contains the bulk of ocean water, the average temperature of the world ocean is a chilly 3.5°C (38°F).

Dissolved salts contribute to the ocean's density structure, especially in cool regions where precipitation is high or along coasts exposed to freshwater runoff. The **halocline** (*halos* = salt) is a zone of rapid salinity increase with depth. The halocline often coincides with the thermocline, and the combination produces a pronounced pycnocline. **Figure 7.12** shows how the thermocline and halocline combine to form the pycnocline. **Figure 7.13** records changes in the position of the pycnocline with latitude.

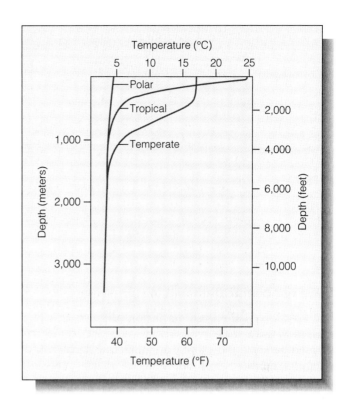

Figure 7.11 Typical temperature profiles at polar, tropical, and middle (temperate) latitudes. Note that polar waters lack a thermocline.

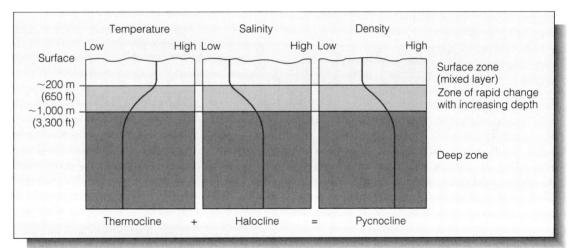

Figure 7.12 Formation of the pycnocline, which depends on the characteristics of the thermocline and halocline.

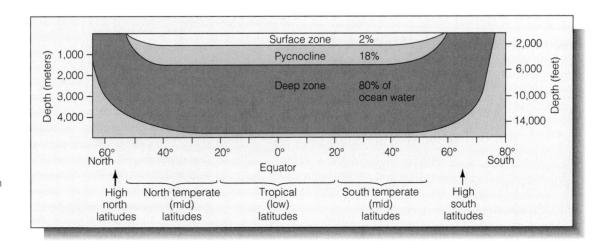

Figure 7.13 Variation in the average depth of density zones with latitude.

Table 7.1	Some Characteristics of the World Ocean Surface, by Latitude		
Characteristic	Tropical Oceanic Waters	Temperate Oceanic Waters	Polar Oceanic Waters
Winter temperature	20–25°C (68–77°F)	5–20°C (41–68°F)	About –2°C (28°F)
Annual variation of temperature	Less than 5°C (9°F)	About 10°C (18°F)	Less than 5°C (9°F)
Average salinity	35‰–37‰	About 35‰	28‰–32‰
Annual variation of air temperature	Less than 5°C (9°F)	About 10°C (18°F)	Up to 40°C (72°F)
Precipitation–evaporation balance	E exceeds P	P exceeds E	P exceeds E

Source: H. Charnock, 1971.

Density Stratification and Water Masses

The layers we have been discussing are distinct water masses. A **water mass** is a body of water with characteristic temperature and salinity and therefore density. Layering by density traps dense water masses at great depths, where they are not exposed to daily heating and cooling, to surface circulation driven by winds and storms, or to light. The pycnocline effectively isolates 80% of the world ocean's water from the 20% involved in surface circulation.

Dense water masses form near polar continental shelves (as cold water freezes and excludes salt) or in enclosed areas such as the Mediterranean Sea (where evaporation exceeds precipitation and river input, raising salinity). These heavy water masses sink, sometimes overlapping one another and often retaining their identity for long periods. Separate water masses below the pycnocline tend not to merge because little energy is available for mixing in these quiet depths. (Water masses are the source of deep ocean currents, as will be discussed in Chapter 9.)

AN OVERVIEW OF OCEAN SURFACE CONDITIONS

Table 7.1 summarizes a few important properties of seawater at polar, temperate, and tropical latitudes. The temperature data show the effects of solar radiation at various latitudes, along with the ratio of evaporation to precipitation. Notice that the temperature of ocean surface water is more variable through the year in the temperate zone than in either the polar or tropical areas, but that temperate zone salinity stays relatively constant. Notice also that evaporation generally exceeds precipitation in the tropics, but that precipitation dominates in temperate and polar zones.

Ocean surface temperature is highest in the Pacific, north and east of Borneo, where westward-driving currents funnel seawater warmed by long exposure to tropical sunlight. The highest open ocean surface temperatures range to 32°C (90°F). Surface salinity is especially high in the North and South Atlantic, where surface water is isolated by the currents flowing around the ocean's periphery. Some of this information is summarized graphically in **Figures 7.14** and **7.15,** figures showing typical ocean surface temperatures and salinities worldwide.

REFRACTION, LIGHT, AND SOUND

The ring of light sometimes seen around the moon and the safe concealment of a submarine may not seem related, but both events depend on **refraction,** the bending of waves. Light and sound are both wave phenomena. When a light wave or sound wave leaves a medium of one density—such as air—and enters a medium of a different density—such as water—at an angle other than 90°, it is bent from its original path. The reason for this bending is that light or sound waves travel at different speeds in the different media.

The situation is analogous to a line of marchers walking along a desert highway with arms intertwined. The marchers can walk faster if they stay on the pavement than if they walk in the sand next to the highway. Their speed on the pavement, then, is greater than their speed in the sand. As long as they stay on the pavement, they won't change direction. But if their marching angle gradually takes them off the edge (into a medium in which their speed is *lower*), the people who reach the sand first will suddenly slow down, and the line will pivot quickly off the highway. They have been refracted. Their progress is depicted in **Figure 7.16a.** Note that the transi-

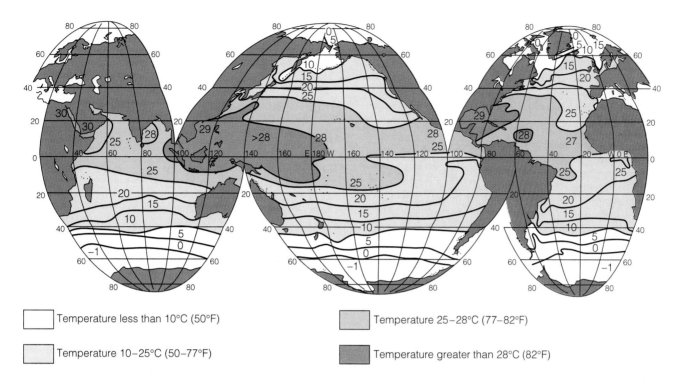

Temperature less than 10°C (50°F)

Temperature 25–28°C (77–82°F)

Temperature 10–25°C (50–77°F)

Temperature greater than 28°C (82°F)

Figure 7.14   Sea surface temperatures in degrees Celsius during Northern Hemisphere summer.

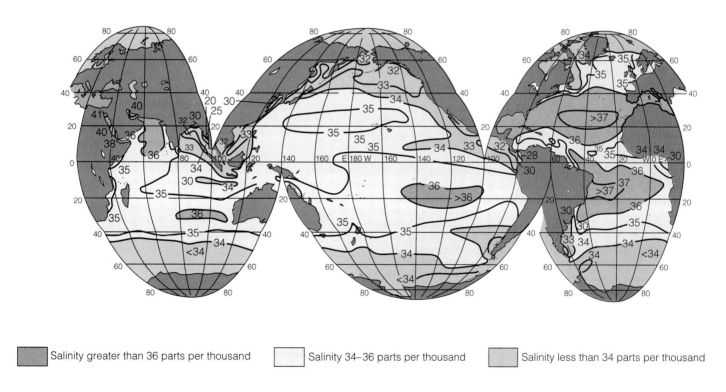

Salinity greater than 36 parts per thousand

Salinity 34–36 parts per thousand

Salinity less than 34 parts per thousand

Figure 7.15   Sea surface salinities in parts per thousand (‰) during Northern Hemisphere summer.

Figure 7.16 An analogy for refraction. The ranks of marchers represent light or sound waves; the pavement and sand represent different media. (a) If the marchers head off the pavement at an angle other than 90°, their path will bend (refract) as they hit the sand because some will be walking more slowly than others. (b) If they march straight off the pavement, the ranks will slow down but not bend as they hit the sand.

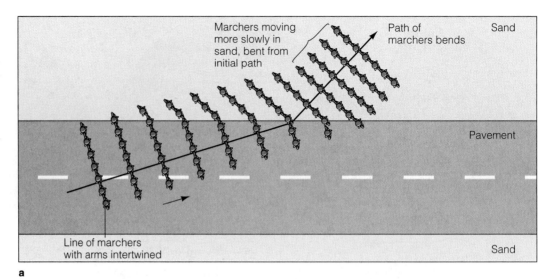

tion from one medium to another must occur at an angle other than 90° for refraction to occur. Our marchers will not change direction if they march straight off the asphalt into the sand. They will still slow down, however (see **Figure 7.16b**).

The refraction of light waves by water happens in much the same way. The speed of light in water is only about three-quarters its speed in air, so water effectively refracts light. (Glass bends light even more.) The degree to which light is refracted from one medium to another is expressed as a ratio called the **refractive index.** The higher the refractive index, the greater the bending of waves between media. The refractive index of water increases with increasing salinity, a fact exploited by the refractometer used to measure salinity shown in Figure 6.9.

Examples of the refraction of light by water are all around you. A pencil sticking out of a glass of water looks bent because of refraction; the submerged steps of a swimming pool ladder look closer than they are because of refraction; and refraction magnifies objects, causing divers to exaggerate the size of the fish that got away.

LIGHT IN THE OCEAN

Light is electromagnetic radiation propagated as small, nearly massless particles. Light has a dual nature: It behaves like both a wave and a stream of particles. The wavelength of light determines its color: Shorter wavelengths are bluer, longer wavelengths are redder. The visible spectrum—the wavelengths that human eyes can detect—is only a small part of the entire electromagnetic spectrum. Except for very long radio waves, water rapidly absorbs nearly all electromagnetic radiation. Only blue and green wavelengths pass through water in any appreciable quantity or distance.

Scattering and Absorption

As we have seen, sunlight has a difficult time reaching and penetrating the ocean; the clouds and the sea surface reflect light, and atmospheric gases and particles scatter and absorb it. Once past the sea surface, light is rapidly weakened by scattering and absorption. **Scattering** occurs as light is bounced between air or water molecules, dust particles, water droplets, or other objects before being absorbed. The greater density of water (along with the greater number of suspended and dissolved particles) makes scattering more prevalent in water than in air. The **absorption** of light is governed by the structure of the water molecules it happens to strike. When light is absorbed, molecules vibrate and the light's electromagnetic energy is converted to heat.

Even perfectly clear seawater is not perfectly transparent. If it were, the sun's rays would illuminate the greatest depths of the ocean, and seaweed forests would fill its warmed basins. The thin film of lighted water at the top of the surface zone is called the **photic zone** (*photo* = light). In clear tropical waters the photic zone may extend to a depth of 200 meters (660 feet), but a more typical value for the open ocean is 100 meters (330 feet). All the production of food by photosynthetic marine plants occurs in this thin, warm surface. Here, water is heated by the sun, heat is transferred from the ocean into the atmosphere and space, and gases are exchanged with the atmosphere. The thermostatic effects we've discussed function largely within this zone. Most

of the ocean's life is found here. The photic zone may be extraordinarily thin, but it is also extraordinarily important.

The ocean below the photic zone lies in blackness. Except for light generated by living organisms, the region is perpetually dark. This dark water beneath the photic zone is called the **aphotic zone** (*a* = without, *photo* = light).

Water Colors

The light energy of some colors is converted into heat nearer to the surface than the light energy of other colors. **Figure 7.17** shows this differential absorption by color. Notice that after 1 meter (3.3 feet) of travel, only 45% of the light energy remains, most of it in the green and blue wavelengths. After 10 meters (33 feet) 85% of the light has been absorbed, and after 100 meters (330 feet) just 1% remains. The dimming light becomes bluer with depth because the red, yellow, and orange wavelengths have already been absorbed. Even in the clearest condi-

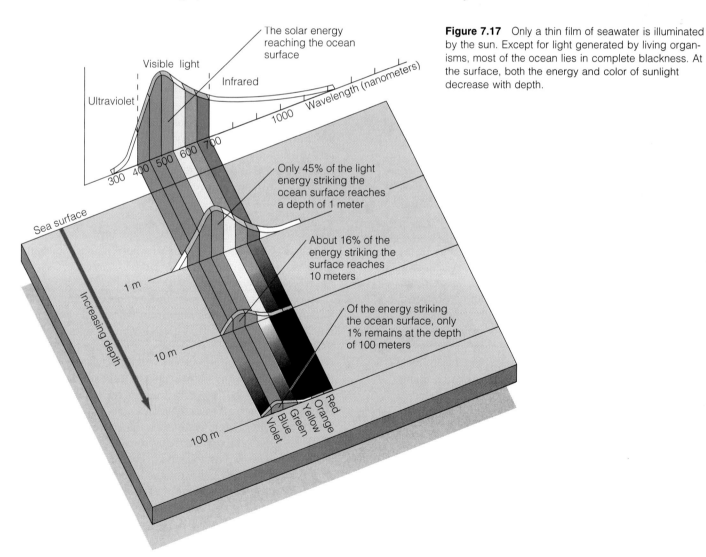

Figure 7.17 Only a thin film of seawater is illuminated by the sun. Except for light generated by living organisms, most of the ocean lies in complete blackness. At the surface, both the energy and color of sunlight decrease with depth.

tions, sunlight rarely penetrates below 250 meters (820 feet).

From above, clear ocean water looks blue because blue light can travel through water far enough to be scattered back through the surface to our eyes. Divers near the surface see an even brighter blue color. Because nearly all red light is converted to heat in the first few meters of ocean water, red objects a short distance beneath the surface look gray. If you were a diver working at a depth of 10 meters (33 feet) and you cut your hand, you would see gray blood rather than red, because there is not enough red light at that depth to reflect from blood's red pigment and stimulate your eye. The underwater pictures of red organisms you've seen are possible only because the diver has brought along a source of white light (which contains all colors).

Suspended particles scatter some colors of light and absorb others. Some sediments reflect yellow light, giving the ocean a yellow cast. The Red Sea gets its name from its abundance of cyanobacteria, small plantlike organisms that contain a reddish pigment. Browns and even purple hues have been attributed to other tiny, single-celled marine life-forms.

SOUND IN THE OCEAN

Sound is a form of energy transmitted by rapid pressure changes in an elastic medium. Sound energy decreases as it travels through seawater because of spreading, scattering, and absorption. Energy loss due to spreading is proportional to the square of the distance from the source. Scattering occurs as sound bounces off bubbles, suspended particles, organisms, the surface, the bottom, or other objects. Eventually sound is absorbed and converted by molecules into a very small amount of heat. Absorption of sound is proportional to the square of the frequency of the sound: Higher frequencies are absorbed sooner. Sound waves can travel for much greater distances through water than light waves can before being absorbed. Because sound travels through water so efficiently, many marine animals use sound rather than light to "see" in the ocean.

The speed of sound in seawater of 35‰ salinity is about 1,500 meters per second (3,345 miles per hour), almost five times the speed of sound in air. The speed of sound in seawater increases as temperature and pressure increase. Sound travels faster at the warm ocean surface than it does in deeper, cooler water. Its speed decreases with depth, eventually reaching a minimum at about 1,000 meters (3,300 feet). Below that depth, however, the effect of increasing pressure offsets the effect of decreasing temperature; so speed increases again. Near the bottom of an ocean basin the speed of sound may actually

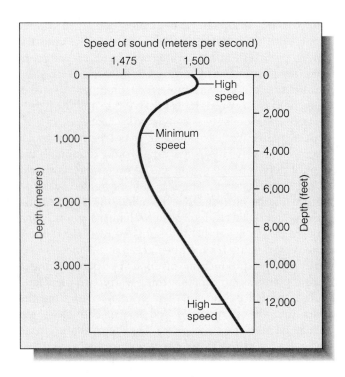

Figure 7.18 The relationship between water depth and the speed of sound.

be higher than at the surface. Though these variations are important in the behavior of oceanic sound, they amount to only 2% or 3% of the average speed of sound in seawater. The relationship between depth and sound speed is shown in **Figure 7.18**.

Sofar Layers and Shadow Zones

The depth at which the speed of sound reaches its minimum varies with conditions, but it is usually located near 1,200 meters (3,900 feet) in the North Atlantic or about 600 meters (2,000 feet) in the North Pacific. Transmission of sound in this minimum-velocity layer is very efficient because refraction tends to cause sound energy to remain within the layer. The outer edges of sound waves escaping from this layer will enter water in which the speed of sound is higher. This will cause the wave to speed up but then pivot back into the minimum-velocity layer, as shown in **Figure 7.19**. Upward-traveling sound waves that are generated within the minimum-velocity layer will tend to be refracted downward, and downward-traveling sound waves will tend to be refracted upward. In short, sound waves bend *toward* layers of lower sound velocity and so tend to stay within the zone. Therefore, loud noises made at this depth can be heard for thousands of kilometers.

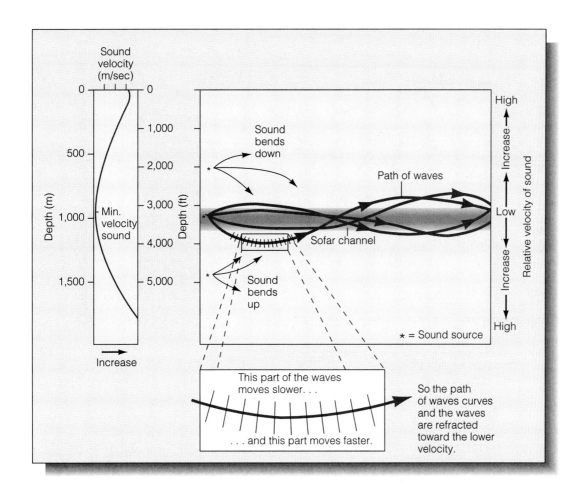

Figure 7.19 The sofar layer, in which sound waves travel at minimum speed. Sound transmission is particularly efficient—that is, sounds can be heard for great distances—because refraction tends to keep sound waves within the channel.

Navy depth charges detonated in the minimum-velocity layer in the Pacific have been heard 3,680 kilometers (2,280 miles) from the explosion. In a recent test, sound generated by a U.S. Navy ship in the Indian Ocean was heard at the Oregon Coast! (See **Box 7.2**.) In the early 1960s, the U.S. Navy experimented with the use of sound transmission in the minimum-velocity layer as a lifesaving tool. Survivors in a life raft would drop a small charge into the water that was set to explode at the proper depth. A number of widely spaced listening stations ashore would compare the differences in the arrival times of the signal and then compute the position of the raft. The project—which has since been abandoned in favor of radio beacons—was called **sofar** (for *so*und *f*ixing *a*nd *r*anging). The minimum-velocity layer has come to be known as the **sofar layer.**

Sound travels slowly in the sofar layer, but it moves rapidly near the bottom of the well-mixed surface layer. Temperature and salinity conditions are homogeneous there; so they do not produce any refraction. Pressure still increases with depth, however, causing a thin high-velocity layer at around 80 meters (260 feet), just above the pycnocline (see again Figure 7.18).

As you read earlier in this chapter, some ships project pulses of sound into the water to search for animals, submarines, or hazards to navigation. Depending on the angle at which the sound waves arrive at the high-velocity layer, they will sometimes split and refract to the surface or bend into the depths. An object beyond the area of divergence may be undetectable; it would be within a **shadow zone,** a region into which very little sound energy penetrates. (The "afternoon effect" discovered by Lieutenant Pryor in the 1930s was caused by a shadow zone.) Shadow zones are of particular interest to submariners—and to the ship captains who use sound to hunt them. As depicted in **Figure 7.20**, a smart submarine captain can use the shadow zone to hide from a pursuer. Based on work done by the crews of *Semmes* and *Atlantis,* naval researchers found the ideal safe depth to be about 125 meters (415 feet).

Sonar

Crews aboard surface ships and submarines employ **active sonar** (*so*und *na*vigation *a*nd *r*anging), the projection and return through water of short pulses (*pings*) of

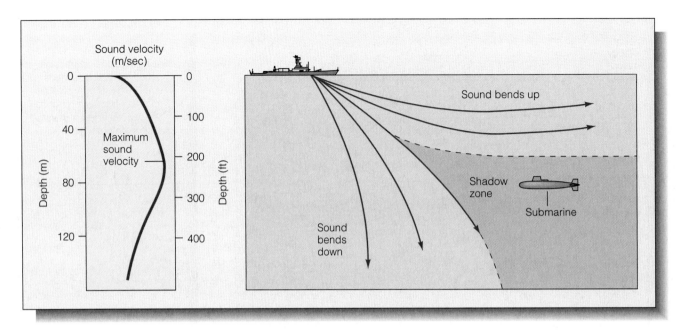

Sound velocity (m/sec)

Maximum sound velocity

Depth (m)

Depth (ft)

Sound bends up

Shadow zone

Submarine

Sound bends down

Figure 7.20 The shadow zone. The thin, high-sound-velocity layer, which forms at a depth of about 80 meters (260 feet), deflects sound. The shadow zone creates a good place for submarines to hide from sonar. (Note the difference in depth scales in Figures 7.18 and 7.19.)

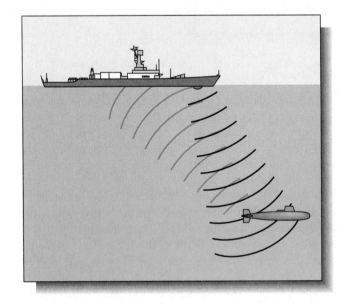

Figure 7.21 The principle of active sonar. Pulses of high-frequency sound are radiated from the sonar array of the sending vessel. Some of the energy of this ping reflects from the submerged submarine and returns to the sending vessel. The echo is analyzed to plot the position of the submarine.

high-frequency sound to search for objects in the ocean (**Figure 7.21**). Much progress has been made since Lieutenant Pryor's work of the 1930s. In a modern system, electrical current is passed through crystals, which respond by producing powerful sound pulses pitched above the limit of human hearing. (Though this high-frequency sound is absorbed rapidly in the ocean, its use greatly improves the resolution of images.) Some of the sound from the transmitter bounces off any object larger than the wavelength of sound employed and returns to a microphone-like sensor. Signal processors then amplify the echo and reduce the frequency of the sound to within the range of human hearing. An experienced sonar operator can tell the direction of the contact, its size and heading, and even something about its composition (whale or submarine or school of fish) by analyzing the characteristics of the returned ping.

Side-scan sonar is a type of active sonar. Operating with as many as 60 transmitter/receivers tuned to high sound frequencies, side-scan systems towed in the quiet water beneath a ship are sometimes capable of near-photographic resolution (see **Figure 7.22**). Side-scan systems are used for geological investigations, archaeological studies, and the location of downed ships and airplanes.

Echo sounders, too, make use of sonar. As you may recall from Chapter 3, an echo sounder transmits a pulse of sound toward the ocean floor, measures the time of its round trip from transducer to seabed and back, computes the depth from the time delay, and then displays the depth on a screen or strip chart (see Figure 3.11). This method can be used to depths of about 5,000 meters (16,500 feet), but attenuation of sound and uncertainty about water conditions (and therefore the speed of sound) generally make it less accurate in deeper water.

For deeper soundings, or to "see" into sediment layers below the surface, geologists use *seismic profilers* employing powerful electrical sparks, explosives, or compressed air to generate a very energetic low-frequency sound pulse. Again, the round-trip travel time of the sound waves is crucial. The low-frequency sound cannot resolve great detail, but the echo can usually provide an image of the sedimentary layers beneath the surface (see Figure 5.15). Low-frequency sound also has the advantage of efficient travel with less absorption.

The first human use of sea sound was passive: Mariners listened through their hulls to the whistles and clicks of whales and other animals. When submarine warfare emerged during World War I, the British invented a simple underwater listening device to detect noises made by enemy submarines. Operators would listen through a sensitive directional microphone for the telltale sounds of a propeller, a torpedo, or even a dropped wrench or slammed hatch cover. The first systems were primitive, but **passive sonar,** as these listening-only devices are now called, is currently undergoing a renaissance. Modern passive devices are much more sophisticated and sensitive than their World War I predecessors. Passive sonar confers the benefit of surprise—unlike active sonar, in which a listener can hear the loud ping long before an operator aboard the sending vessel can hear his faint echo. Usually, it's safer just to listen. The combination of computerized signal processing, microphones towed at a distance from the listener's noisy ship, and better knowledge of ocean physics will improve the usefulness of both kinds of sonar.

Humans are not the only organisms to use sonar. Whales and other marine mammals use clicks and whistles to find food and avoid obstacles. We'll discuss this form of active sonar in Chapter 16.

IUSS

During the cold war, one of the U.S. Navy's highest priorities was the detection and pursuit of submarines. The *I*ntegrated *U*ndersea *S*urveillance *S*ystem (IUSS), a sophisticated and sensitive global network of underwater acoustic sensors, was developed and installed to assist in this task. The network includes thousands of miles of cable, elaborate fixed hydrophone arrays, a small fleet of vessels designed to tow hydrophones, and an extensive shorebased data processing system.

a

b

Figure 7.22 Side-scan sonar. (a) A towed side-scan sonar transceiver. The slit from which sound pulses depart and return is visible near the nose of the device. (b) A side-scan sonar record showing the wreck of the fishing vessel *Margaret Rose* resting on the New England continental shelf. The detailed sonar image shows the raised forecastle, the deck area, and the eastern-rig construction.

A cat-and-mouse game developed from the deployment of IUSS. Submarines became ever quieter, while IUSS became more sensitive. Many targets were detected, but not all were submarines. The fixed component of IUSS, seabed sound sensors known as SOSUS (*So*und *Su*rveillance *S*ystem), was used to study and track whales. In 1993 the Navy declassified and released the first whale data. Among the discoveries:

- Navy analysts tracked a blue whale for 45 days and 1,676 miles (2,683 kilometers) in the mid-Atlantic.

- Estimates were made of the population numbers, seasonal distribution, and migration patterns of five species of marine mammals.

- Acoustic "fingerprints" were used to differentiate among individual whales.

BOX 7.2

Is Climate Change Sofar Away?

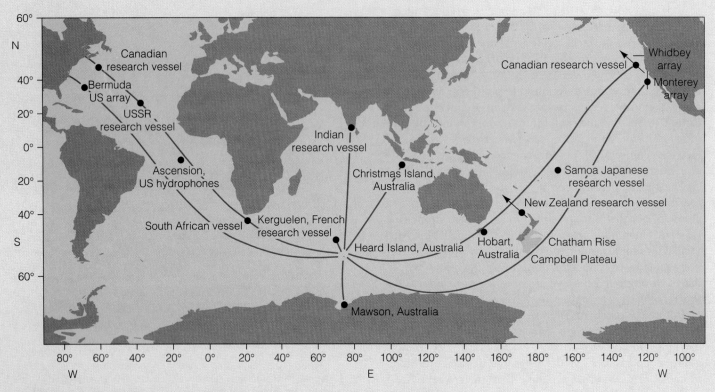

a Paths taken by sound in the Heard Island experiment. The sound sources were suspended from the R/V *Cory Chouest*, located about 50 kilometers (31 miles) southeast of Heard Island. Black circles indicate receiver sites. Ray paths from the source to receivers are along approximate great circle routes.

Early in 1991 underwater sounds generated by a ship near Heard Island in the southern Indian Ocean were detected off the West Coast of the United States, 17,960 kilometers (11,160 miles) away. The foghornlike sounds were produced by powerful underwater loudspeakers suspended 200 meters (650 feet) beneath a research vessel. The ship's five transmitters emitted a booming, low-frequency hum in all directions for an hour and then were silent for an hour. The pattern was repeated through most of one day.

Some of the sound was trapped in the Indian Ocean's minimum-velocity (sofar) layer, in which it traveled both into the Atlantic and around Australia into the

Pacific. The sound was detected by a ship equipped with sensitive hydrophones lowered to the sofar depth off the Oregon coast, near Coos Bay. It took about 3½ hours for the sound to travel that far, nearly halfway around the world. Similarly equipped ships and offshore stations detected the sound at other locations (**Figure a**).

The experiment was part of a plan to measure ocean temperature within a few thousandths of a degree. Since sound travels more rapidly through warmer water, water temperature differences can be calculated by recording the time taken for sound to reach receivers around the world. The temperature of the ocean is an

- The first glimpse of the ocean-basin-wide distribution of marine mammals was obtained.

IUSS is clearly a powerful research tool, and marine scientists hope more data will soon be made available.

important measure of potential global warming, and the sound-transmission experiment collected baseline data for a study designed to help settle the argument over whether carbon dioxide and other *greenhouse gases* are warming the planet artificially. Computer models of Earth's climate suggest that greenhouse gases could cause the ocean to warm by about 0.005°C (0.009°F) per year at a depth of 1,000 meters (3,300 feet). Such a subtle temperature change could not be detected by conventional monitoring, but over a decade this warming would cut a few seconds from the travel time of a sound signal. (More information on the possibilities and consequences of global warming may be found in Chapter 20.)

In mid-1994, nearly 100 researchers at 13 institutions were preparing for a larger two-and-a-half-year, $35 million experiment. Their plans were put on hold, however, by concerns that the low-frequency sound emitted by the transmitters would in some way harm marine organisms, especially whales and other marine mammals that depend on sound for communication and feeding.

Proponents of ATOC (Acoustical Thermography of Ocean Climate), as the experiment is called, point out that the signals will be 100 times quieter than those used in the Heard Island experiment and that the acoustical sources will be placed much deeper. Far less of the signal will reach the shallow waters where most marine mammals live, and the signal will be generated only 2% of the time—20 minutes on and 4 hours off, every fourth day. Additionally, 10% of the budget is earmarked for marine mammal research. Environmental impact statements are being prepared for two possible transmitter sites: off the coast of Kaua'i, Hawaii, and near Point Sur, California.

ATOC researchers feel that the effects on marine life will be minimal and that the potential benefits of the experiment outweigh the small chance of harm. If the experiment is allowed to proceed, some marine mammal specialists want the first months of the test to be devoted to the detection of any changes in marine mammal behavior. The National Marine Fisheries Service is currently collecting information and will make a decision to grant or block the necessary permits in 1995.

Q-AND-A

1. Input of solar radiation in the Northern Hemisphere peaks on 22 June and reaches a minimum on 22 December, because of the Earth's orbital tilt (of which more in Chapter 8). Why, then, do our warmest days occur in August or September and our coldest days in January or February?

   Because of thermal inertia. There is a lag between maximum sunlight and maximum warmth because of water's great heat capacity. The sun must shine on this watery planet for many weeks to raise the summer hemisphere's temperature. Of course water also retains heat well; so the coldest days in the winter hemisphere come well after the darkest ones.

2. Why don't sounds seem to travel easily from air to ocean, or vice versa?

   Sound waves can make the transition from one medium to another with little energy loss only when the speeds of sound waves in the two different media are similar. The speeds of sound in water and air, however, are too different for an efficient transition to be made. Too great a contrast in speed produces reflection of the sound waves at the junction. This is why you can't hear people shouting from the edge of the pool while you're underwater, even though the weak sound of a submerged pebble clicking against the side is very clear and sharp.

   If you place a solid medium in which the speed of sound is intermediate between air and water, the sound can move across one junction and then the other, for a more efficient total transition. Wood works well for this, which explains why ocean noises are easy to hear in wooden boats. Even if the speed of sound in the intermediate medium is higher than in water (as in steel, for instance), some sound will be audible simply because the hard surface provides a good radiating surface for noises coming from the water.

3. I can't tell where sound is coming from when I'm underwater. It seems to be coming from inside my head. Why?

   Our normal sensation of stereo hearing depends in part on the difference in the arrival times of sound from one ear to the other. The speed of sound in water, however, is more than four times greater than its speed in air; so our brains are unable to sense arrival-time differences from sounds originating nearby. You'd need to be more than four times as sensitive—or have a head more than four times as large—to hear stereophonically underwater.

4. Why don't divers see shadows even when the sun is shining directly on the ocean at their location?

   The greater density of water (along with the greater number of suspended and dissolved particles) makes scattering more prevalent in water than in air. A few meters below the surface, light in the ocean seems to come from all directions (rather than primarily from one direction).

5. If water has the highest latent heat of evaporation, why does a drop of alcohol make my hand feel colder when it evaporates than a drop of water does?

   Heat vs. temperature again. The alcohol makes your hand colder, but it evaporates much more quickly. The water stays around longer (takes longer to evaporate), and although it may not cause your skin to become as cold, it will remove more heat.

6. Liquid methane appears to be the only other substance found in quantity in a liquid state in the solar system. If our ocean were of liquid methane rather than liquid water, how would conditions here be different?

   Liquid methane freezes at –183°C (–272°F) and boils at –162°C (–234°F). It has much less heat capacity than liquid water because methane is not dipolar, does not form a great bonded mass as water does, and consequently does not require a large energy input to break the maze of hydrogen bonds to release molecules. Because of methane's low heat capacity, the difference between Earth's polar and tropical daytime temperatures would be drastically greater unless the circulation rate of liquid and vapor currents accelerated to keep pace. Computer modeling yields a very unattractive vision of a planet with a boiling equatorial ocean, crushing atmospheric pressures from the greater amount of vapor in the air, torrential polar methane rainfall, cataclysmic cyclonic storms, and average wind speeds at mid-latitudes of hundreds of kilometers per hour. We miss all this excitement because our ocean has water.

Terms and Concepts to Remember

absorption	latent heat of	sofar layer
active sonar	fusion	sonar
aphotic zone	light	sound
calorie	mixed layer	state
deep zone	passive sonar	surface zone
degree	photic zone	temperature
density curve	pycnocline	thermal
freezing point	refraction	equilibrium
halocline	refractive index	thermal inertia
heat	scattering	thermocline
heat budget	sensible heat	thermostatic
heat capacity	shadow zone	property
latent heat of	side-scan sonar	water mass
evaporation	sofar	

Study Questions

1. How is heat different from temperature?

2. What is the difference between sensible and nonsensible heat? What is meant by latent heat?

3. How does water's high heat capacity influence the ocean? Leaving aside its effect on beach parties, how do you think conditions on Earth would differ if our ocean consisted of ethyl alcohol?

4. Why does ice float? Why is this fact important to thermal conditions on Earth?

5. How is heat transported from tropical regions to polar regions?

6. What factors affect the density of water? Why does cold air or water tend to sink? What role does salinity play?

7. How is the ocean stratified by density? What physical factors are involved? What names are given to the ocean's density zones?

8. How does refraction permit sound to be transmitted in the ocean for thousands of miles?

9. What percent of incoming sunlight reaches the ocean? What happens to that light? Is the heat budget balanced? What would happen if it were not?

10. What factors influence the intensity and color of light in the sea? What factors affect the depth of the photic zone? Could there be a *photocline* in the ocean?

For Further Study

Amato, I. 1993. "A Sub Surveillance Network Becomes a Window on Whales." *Science* 261 (no. 5121): 549–50. Information on IUSS whale data, described as "the acoustic Rosetta stone for whales."

Anikouchine, W. A., and R. W. Sternberg. 1981. *The World Ocean: An Introduction to Oceanography.* 2d ed. Englewood Cliffs, NJ: Prentice-Hall.

Anthes, R. A., et al. 1981. *The Atmosphere.* 3d ed. Columbus, OH: Merrill.

Bowditch, N. 1966. *American Practical Navigator.* Washington, DC: U.S. Navy Hydrographic Office Publication 9. Summaries of the properties of light and sound in the ocean.

Charnock, H. 1971. "Physics of the Ocean." In *Deep Oceans*, 82–120, edited by P. Herring and M. Clarke. London: Praeger. Excellent reference in ocean physics for the interested layman.

Cromie, A. H. 1974. *Physics for the Life Sciences*. New York: McGraw-Hill. Refraction is covered well.

Garrison, T., and R. Lebow. 1995. *Oceanus: The Marine Environment*. 6th ed. Belmont, CA: Wadsworth. Lessons 21 and 22 cover light and sound in the ocean.

Jerlov, N. G., and E. S. Nielsen, eds. 1974. *Optical Aspects of Oceanography*. New York: Academic Press. A standard reference.

King, C. A. M. 1966. *An Introduction to Oceanography*. New York: McGraw-Hill. Water masses, physical properties. Technical.

Long, E. J., ed. 1964. *Ocean Sciences*. Annapolis, MD: Naval Institute Press. Concise general reference to physical properties. Source of history of Project Sofar.

Miller, A. 1976. *Meteorology*. 3d ed. Columbus, OH: Merrill. Excellent brief paperback. Source of heat budget data.

Neshyba, S. 1987. *Oceanography: Reflections on a Fluid Earth*. New York: Wiley. Information on heat transport by ocean currents.

Schlee, S. 1973. *The Edge of an Unfamiliar World: A History of Oceanography*. New York: Dutton. Information on the invention of modern sonar.

Schwoegler, B., and M. McClintock. 1981. *Weather and Energy*. New York: McGraw-Hill. A meteorologist and a physicist present an examination of the weather "machine."

Sverdrup, H. U. 1954. "Oceanography." In *The Earth as a Planet*, 215–57, edited by C. P. Kuiper. Chicago: University of Chicago Press. The thermostatic role of water on Earth; the importance of ocean and atmosphere in transporting heat.

Sverdrup, H., et al. 1942. *The Oceans: Their Physics, Chemistry, and General Biology*. New York: Prentice-Hall. A treasure trove, as always.

Weihaupt, J. C. 1979. *Exploration of the Oceans: An Introduction to Oceanography*. New York: Macmillan. Excellent and complete presentation of physical factors, especially of sound in the sea.

Note: Oceanus 20 (no. 2, Spring 1977) was dedicated to "Sound in the Sea."

ATMOSPHERIC CIRCULATION AND WEATHER

Andrew

The costliest natural disaster in the history of the United States began as a cluster of rolling thunderclouds drifting westward across the coast of central Africa. The Earth's rotation had given the clouds a gentle twist, and warm oceanic air rose into their core. Like water spinning down a bathtub drain, humid air was sucked into a growing vortex of clouds. The moisture condensed to form rain, thus liberating heat energy that caused the storm to grow even larger. On 20 August 1992, near the center of the tropical Atlantic, the storm became Hurricane Andrew.

As the days passed, the hurricane tracked uncertainly toward the west. On two occasions, shearing winds threatened to tear the storm apart; each time it survived and strengthened. Then, on the evening of 23 August, Andrew metamorphosed into a towering black mountain of wind and rain, a rare category-5 storm (**Figure a**). Its spreading violence sucked the sea surface into a 6-meter (19-foot) dome tens of kilometers across. The wind speed near its center rose to 240 kilometers (150 miles) per hour. In the predawn hours of 24 August, the leading edge of the storm touched the

a Hurricane Andrew storms across southern Florida, 24 August 1992.

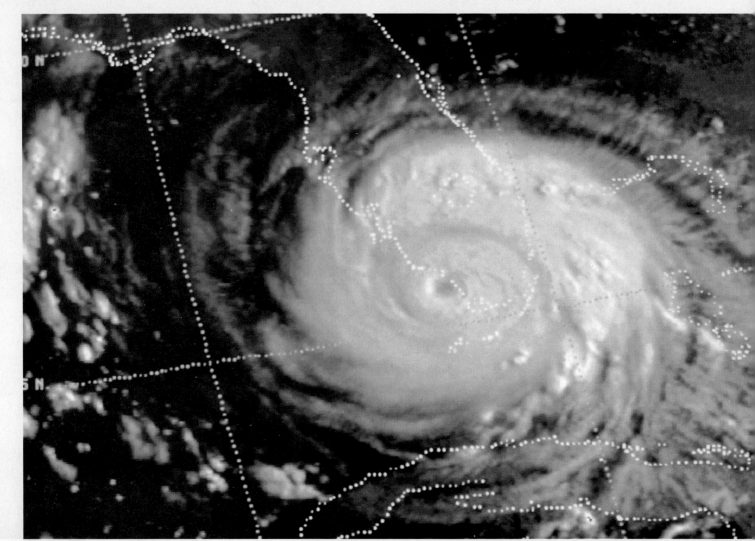

b A neighborhood near Homestead, Florida, bears witness to the shattering force of hurricane winds that reached 350 kilometers (220 miles) per hour.

Florida coast near Biscayne Bay. Eyewitnesses described the rising howl of wind—a tearing, clawing force that entered every window and door, that tore the roofs off homes and children from their parents' grasp, that hurled cars and mobile homes about like toy blocks, and that lasted more than 90 minutes, reaching speeds above 350 kilometers (220 miles) per hour. The division between ocean and land blurred beneath an onrushing wall of windblown seawater. At the height of the storm the hot, wet air glowed yellow-green and was ear-poppingly thin. As can be seen in **Figure b,** devastation was all but absolute.

Before the giant storm died later that week in a series of rattling thunderstorms across the Mississippi Valley, 160,000 people would be homeless, 43 would be dead, and 68,000 businesses would be destroyed; the damage exceeded $30 billion.[1] Scouring currents had rearranged the south Florida coast, sunk ships, shattered harbor installations, uprooted submarine cables, and inundated parts of the Everglades with salt water. Forecasters had not foreseen the ultimate violence of Andrew. Clearly, marine scientists have much to learn about the air–ocean interactions.

[1]As this is written, damage estimates from the January 1994 earthquake in southern California—the second costliest natural disaster in U.S. history—are approaching $20 billion.

The water, gases, and energy at the Earth's surface are shared between the atmosphere and the ocean. The two bodies are in continuous contact, and conditions in one are certain to influence conditions in the other. The interaction of ocean and atmosphere moderates surface temperatures, shapes the Earth's weather and climate, and creates most of the sea's waves and currents.

The atmosphere responds to uneven solar heating by flowing in three great circulating cells over each hemisphere. This circulation of air is responsible for about two-thirds of the heat transfer from tropical to polar regions. The flow of air within these cells is influenced by the rotation of the Earth. To observers on the surface, the Earth's rotation causes moving air (or any moving mass) in the Northern Hemisphere to curve to the right of its initial path, and in the Southern Hemisphere to the left. The apparent curvature of path is known as the Coriolis effect.

Uneven flow of air within cells is one cause of the atmospheric changes we call *weather.* Large storms are spinning areas of unstable air occurring between or within air masses. Extratropical cyclones originate at the boundary between air masses; tropical cyclones, the most powerful of Earth's atmospheric storms, occur within a single humid air mass. The immense power of tropical cyclones is derived from water's latent heat of evaporation.

COMPOSITION AND PROPERTIES OF THE ATMOSPHERE

The lower atmosphere is a nearly homogenous mixture of gases, most plentifully nitrogen (78.1%) and oxygen (20.9%). Other elements and compounds, as listed in **Table 8.1,** make up less than 1% of its composition.

Air is never completely dry; **water vapor,** the gaseous form of water, can occupy as much as 4% of its volume. Sometimes liquid droplets of water are visible as clouds or fog; but more often the water is simply there—invisible, having entered the atmosphere from the ground, plants, and the sea surface (**Figure 8.1**). The residence time of water vapor in the lower atmosphere is about 10 days. Water leaves the atmosphere by condensing into dew, rain, or snow.

Air has mass. A 1-square-centimeter column of air (0.16 square inch), extending from sea level to the top of the atmosphere, weighs about 1.04 kilograms (2.3 pounds). A 1-square-foot column of air the same height weighs more than a ton.

The temperature and water content of air greatly influence its density. Because the molecular movement associated with heat causes a mass of warm air to occupy more space than an equal mass of cold air, warm air is less dense than cold air. But contrary to what we might guess, humid air is *less* dense than dry air at the same temperature—because molecules of water vapor weigh less than the nitrogen and oxygen molecules that the water vapor displaces.

Near the Earth's surface, air is packed densely by its own weight. Air lifted from near sea level to a higher altitude is subjected to less pressure and will expand. As anyone knows who has felt the cool air rushing from an open tire valve, air becomes cooler when it expands. The opposite effect is also familiar: Air compressed in a tire pump becomes warmer. Air descending from high altitude toward sea level warms as it is compressed by the higher atmospheric pressure near the Earth's surface.

Warm air can hold more water vapor than cold air. Water vapor in rising, expanding, cooling air will often condense into clouds (aggregates of tiny droplets) because the cooler air can no longer hold as much water vapor. If rising and cooling continues, the droplets may coalesce into raindrops or snowflakes. The atmosphere will then lose water as **precipitation,** liquid or solid water that falls from the air to the Earth's surface. These rising-expanding-cooling and falling compressing-heating relationships are important in understanding atmospheric circulation, weather, and climate.

WEATHER AND CLIMATE

Weather is the state of the atmosphere at a specific place and time. **Climate** is a long-term average of weather in an area. Both are influenced by the amount of solar radi-

Table 8.1	Average Composition of Dry Air, by Volume
Component	Percentage
Nitrogen	78.084%
Oxygen	20.946%
Argon	0.934%
Carbon dioxide	0.033%
Neon, helium, methane, and other elements and compounds	0.003%
	100.000%

Figure 8.1 Steam fog over the ocean indicates rapid evaporation. Water vapor is invisible, but as water vapor rises into cool air it can condense into visible droplets.

ation the area receives, by local terrain and nearby large bodies of water, by changing geological and biological conditions, and by other factors. Forecasters can often predict weather because of its local and sequential nature, but climate predictions are harder to make because these factors operate over larger areas and longer periods of time.

We know that climate has changed in the relatively recent past (the last ice age peaked only 18,000 years ago), but we are unsure about the causes of the change. We cannot yet predict climate change; data and models on which to base these predictions are still incomplete. Accurate climate predictions will soon be as critical as accurate weather predictions, however. Human-induced changes in the quantities of atmospheric carbon dioxide and ozone will lead to climate changes, but we're not sure what they will be. (More on this important topic may be found in Chapter 20.)

ATMOSPHERIC CIRCULATION

Air does not remain stationary over the Earth. Rather, it flows in large patterns shaped both by variations in solar heating with latitude and season and by the rotation of the Earth. The mass movement of air is known as **wind.**

About half of the energy radiated toward the Earth from the sun is absorbed by the Earth, but this energy is not distributed evenly across the planet's surface. The amount of solar energy reaching Earth's surface per minute, the **insolation rate,** varies with the transparency of the atmosphere, the angle of the sun above the horizon, and the local reflectivity of the surface. The most important factors that affect solar angle are latitude and season.

Uneven Solar Heating and Latitude

More sunlight falls in equatorial regions than strikes the poles. As we saw in Chapter 7, sunlight striking the Earth at high (polar) latitudes spreads its energy over a greater area because its approach angle, its **angle of incidence,** is lower (see Figure 7.7). This lower angle also means that light reaching the surface at the poles has been filtered through a thicker layer of atmosphere. These factors, along with the high reflectivity of ice and snow, combine to produce much less solar heating at the poles. Equatorial regions, on the other hand, are warmed by light approaching at a higher angle of incidence. The sunlight is also subjected to less atmospheric filtration and surface reflection.

Uneven Solar Heating and the Seasons

At mid-latitudes the Northern Hemisphere receives about three times as much solar energy per day in June as it does in December. This difference is due to the $23\frac{1}{2}°$ tilt of the Earth's rotational axis relative to the plane of its orbit around the sun (**Figure 8.2**). The mounting angle you may have noticed in library reference globes indicates this tilt, or **orbital inclination.** The inclination of the axis causes the change of seasons.

The rotating Earth spins around the polar axis. This spin causes the daily rising and setting of the sun. Notice in Figure 8.2 that the axis of the Earth (extension of the line between the North and South poles) points toward the same places in space regardless of the season. Luckily for navigators, the north polar axis currently passes near the bright star Polaris, making Polaris the North Star. As Earth revolves around the sun, the constant tilt of its rotational axis causes the Northern Hemisphere to lean *toward* the sun in June but *away* from it in December.

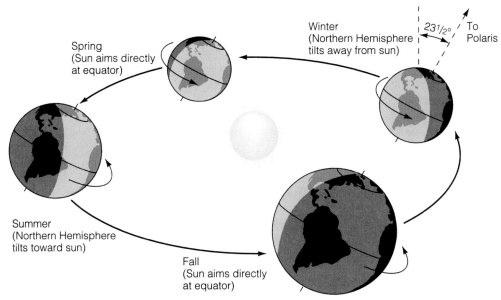

Spring
(Sun aims directly
at equator)

Winter
(Northern Hemisphere
tilts away from sun)

$23\frac{1}{2}°$ To
Polaris

Summer
(Northern Hemisphere
tilts toward sun)

Fall
(Sun aims directly
at equator)

Figure 8.2 The seasons (shown here for the Northern Hemisphere only) are caused by variations in the amount of incoming solar energy as the Earth makes its annual rotation around the sun on an axis tilted by $23\frac{1}{2}°$.

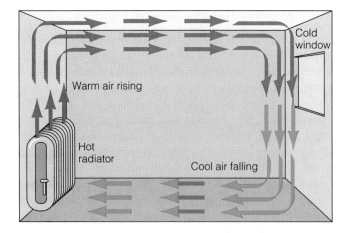

Figure 8.3 A convection current forms in a room when air flows from a hot radiator to a cold window and back.

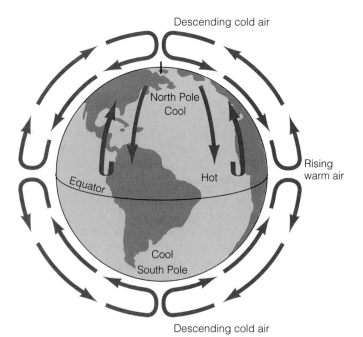

Figure 8.4 A hypothetical model of Earth's air circulation if uneven solar heating were the only factor to be considered. The thickness of the atmosphere is greatly exaggerated in this drawing.

The sun therefore appears higher in the sky in the summer but lower in winter. The inclination of the Earth's axis also causes days to become longer as summer approaches but shorter with the coming of winter. Longer days mean more time for the sun to warm the Earth's surface.

The greatest amount of solar heating is possible where the sun appears directly overhead. This never happens north of 23½° north latitude (23½°N) or south of 23½° south latitude (23½°S). (Can you see why?) Most of the sun's warmth is therefore concentrated between these two latitudes, a zone known as *the tropics.*

Uneven Solar Heating and Atmospheric Circulation

The concentration of solar energy near the equator has important effects on the atmosphere. We know that warm air rises and that cool air sinks. Think of air circulation in a room with a hot radiator opposite a cold window (**Figure 8.3**). Air warms, expands, becomes less dense, and rises over the radiator. Air cools, contracts, becomes more dense, and falls near the cold glass window. The circular current of air in the room, a **convection** current, is caused by the difference in temperature between the ends of the room. A similar process occurs over the surface of the Earth. Surface temperatures are higher at the equator than at the poles, and air can gain heat from warm surroundings. Since air is free to move over the Earth's surface, it would be reasonable to assume that an air circulation pattern like the one shown in **Figure 8.4** would develop over the Earth. In this ideal model, air heated in the tropics would expand and become less dense, rise to high altitude, turn poleward, and "pile up" as it converged near the poles. The air

would then cool and contract by radiating heat into space, sink to the surface, and turn equatorward, flowing along the surface back to the tropics to complete the circuit.

But this is *not* what happens. Global circulation of air is governed by two factors: uneven solar heating *and* the rotation of the Earth. The eastward rotation of the Earth on its axis deflects the moving air or water (or any moving object having mass) away from its initial course. This deflection is called the **Coriolis effect** in honor of **Gaspard Gustave de Coriolis,** the French scientist who worked out its mathematics in 1835. An understanding of the Coriolis effect is important to an understanding of atmospheric and oceanic circulation.

The Coriolis Effect

To an earthbound observer, any object moving freely across the globe appears to curve slightly from its initial path. In the Northern Hemisphere this curve is to the right, or *clockwise,* from the expected path; in the Southern Hemisphere it is to the left, or *counterclockwise.* To earthbound observers the deflection is very real; it isn't caused by some mysterious force, and it isn't an optical illusion or some other trick caused by the shape of the globe itself. *The observed deflection is caused by the observer's moving frame of reference on the spinning Earth.*

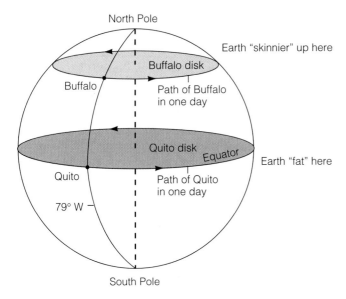

Figure 8.5 Diagram of the thought experiment in the text, showing that Buffalo travels a shorter path on the rotating Earth than Quito does each day.

The influence of this deflection can be illustrated by performing a mental experiment involving concrete objects—in this case, cities and cannonballs—and then by applying the principle to atmospheric circulation. Let's pick as examples for our experiment the equatorial city of Quito, capital of Ecuador, and Buffalo, New York. Both cities are on almost the same line of longitude (79°W); so Buffalo is almost exactly north of Quito (as **Figure 8.5** shows). Like everything else attached to the rotating Earth, both cities make one trip around the world each 24 hours. Through one day, the north-south relationship of the two cities never changes: Quito is *always* due south of Buffalo.

A complete trip around the world is 360°; so each city moves eastward at an angular rate of 15° per hour (360°/24 hours = 15°/hour). Yet, even though their angular rates are the same, the two cities move eastward at different speeds. (**Figure 8.6** puts another spin on this idea.) Quito is on the equator, the "fattest" part of the Earth. Buffalo is farther north at a "skinnier" part. Imagine both cities isolated on flat disks, and imagine the Earth's sphere being made of a great number of these disks strung together on a rod connecting North Pole to South Pole. From Figure 8.5, you can see that Buffalo's disk is smaller and therefore has a smaller circumference than Quito's. Buffalo doesn't have as far to go in one day as Quito because its disk is not as large. That means that Buffalo must move eastward *more slowly* than Quito to maintain its position due north of Quito.

Look at the Earth from above the North Pole in **Figure 8.7.** The Quito disk and the Buffalo disk must turn through 15° each hour (or Earth would rip itself apart), but the city on the equator must move faster to the east to turn its 15° each hour because its "slice of the pie" is larger. Buffalo must move at 1,260 kilometers (787 miles) per hour to go around the world in one day, while Quito must move at 1,658 kilometers (1,036 miles) per hour to do the same.

Now imagine a massive object moving between the two cities. A cannonball shot north from Quito toward Buffalo would carry Quito's eastward component as it goes; that is, regardless of its northward speed, the cannonball is also moving *east* at 1,658 kilometers (1,036

Figure 8.6 Same idea, different approach.

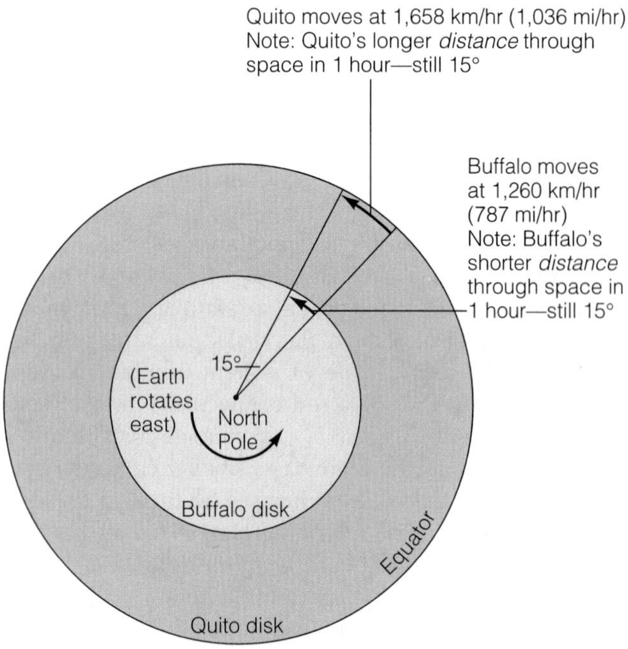

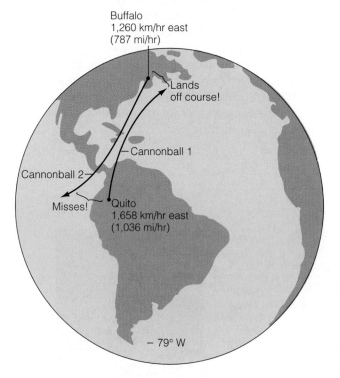

Quito moves at 1,658 km/hr (1,036 mi/hr)
Note: Quito's longer *distance* through space in 1 hour—still 15°

Buffalo moves at 1,260 km/hr (787 mi/hr)
Note: Buffalo's shorter *distance* through space in 1 hour—still 15°

Figure 8.7 A continuation of the thought experiment. A look at the Earth from above the North Pole shows that Buffalo and Quito move at different velocities.

Figure 8.8 The final step in the experiment. As observed from *space*, cannonball #1 (shot northward) and cannonball #2 (shot southward) move as we might expect; that is, they travel straight away from the cannons and fall to Earth. Observed from the *ground*, however, cannonball #1 veers slightly east and cannonball #2 veers slightly west of their intended targets. The effect depends on the observer's frame of reference.

miles) per hour. The fact of being fired northward by the cannon does not change its eastward movement in the least. As the cannonball streaks north, an odd thing happens. The cannonball veers from its northward path, angling slightly to the right (east) (**Figure 8.8**). Actually, this first cannonball is moving just as an observer from space would expect it to, but to those of us on the ground the cannonball "gets ahead of the Earth." As cannonball #1 moves north, the ground beneath it is no longer moving eastward at 1,658 kilometers per hour. During the ball's time of flight, Buffalo (on its smaller disk) *has not moved eastward enough to be where the ball will hit.* If the time of flight for the cannonball is 1 hour, a city 398 kilometers east of Buffalo (1,658 [Quito's speed] – 1,260 [Buffalo's speed] = 398) will have an unexpected surprise. Albany may be in for some excitement!

If a cannonball were fired south from Buffalo toward Quito, the situation would be reversed. This second cannonball has an eastward component of 1,260 kilometers (787 miles) per hour even while it sits in the muzzle. Once fired and moving southward, cannonball #2 travels over portions of the Earth that are moving ever faster in an eastward direction. The ball again appears to veer off course to the right (see Figure 8.8 again), falling into the Pacific to the west of Ecuador. Don't be deceived by the word *appears* in the last sentence. The cannonballs really do veer to the right, or **clockwise.** Only to an observer in space would they appear to go straight, and points on the Earth would appear to move out from underneath them.

The Coriolis effect is a real effect dependent on our rotating frame of reference. Part of that frame of reference involves the direction from which you view the problem. Thus in Figure 8.8, cannonball #2 looks to you as if it is veering left, but to the citizens of Buffalo facing south to watch the cannonball disappear, it moves to the right (west). Coriolis deflection works **counterclockwise** in the Southern Hemisphere because the frame of reference there is reversed. Also, except at the equator (where the Coriolis effect is nonexistent), the Coriolis effect influences the path of objects moving from east to west, or west to east. A quick way to remember the Coriolis effect: In the Northern Hemisphere, objects move clockwise, to the right; in the Southern Hemisphere, objects move counterclockwise, to the left.

Because the Coriolis effect influences any object with mass—*as long as that object is moving*—it plays a large role in the movements of air and water on the Earth. The Coriolis effect is most apparent in mid-latitude situations involving the almost frictionless flow of fluids: between layers of water in the ocean and in the circuits of winds. Does the Coriolis effect influence the directions of cars and airplanes? Yes, but in these cases friction (of tires on pavement, of wings on air) is much greater than the

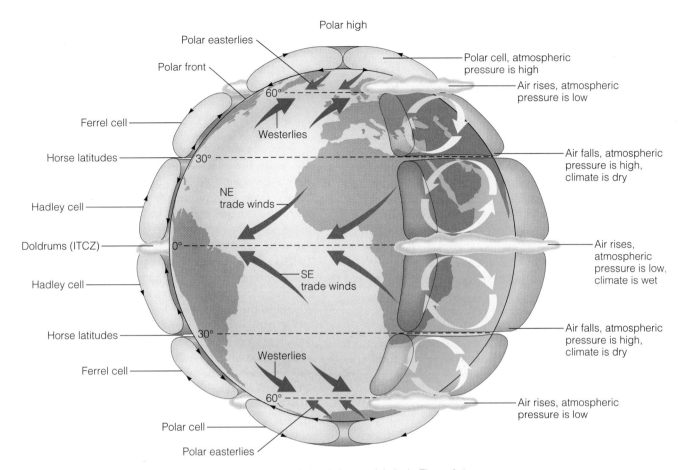

Figure 8.9 Global air circulation as described in the three-cell circulation model. As in Figure 8.4, air rises at the equator and falls at the poles. But instead of one great circuit in each hemisphere from equator to pole, there are three. Note the influence of the Coriolis effect on wind direction.

influence of the Coriolis effect; so the deflection is not observable.

The Coriolis Effect and Atmospheric Circulation Cells

We can now modify our original model of atmospheric circulation (Figure 8.4) into the more correct representation provided in **Figure 8.9.** Yes, air does warm, expand, and rise at the equator; and air does cool, contract, and fall at the poles. But instead of continuing all the way from equator to pole in a continuous loop in each hemisphere, air rising from the equatorial region moves poleward and is gradually deflected eastward; that is, it turns to the right in the Northern Hemisphere and to the left in the Southern Hemisphere. This eastward deflection is caused by the Coriolis effect. (Note that the Coriolis effect does not *cause* the wind; it only *influences* the wind's direction.)

As air rises at the equator it loses moisture by precipitation (rainfall) caused by expansion and cooling. This

drier air now grows denser in the upper atmosphere as it radiates heat to space and cools. When it has traveled about a third of the way from the equator to the pole—that is, to about 30°N and 30°S latitude—the air becomes dense enough to fall back toward the surface. Most of the descending air turns back toward the equator when it reaches the surface. In the Northern Hemisphere the Coriolis effect again deflects this surface air to the right, and the air blows across the ocean or land from the northeast. (This air is represented by the arrows labeled "NE trade winds" in Figure 8.9.) Though it has been heated by compression during its descent, this air is generally still colder than the surface over which it flows. The air soon warms as it moves equatorward, however, evaporating surface water and becoming humid. The warm, moist, less dense air then begins to rise as it approaches the equator and completes the circuit.

Such a large circuit of air is called an **atmospheric circulation cell.** A pair of these tropical cells exists, one on each side of the equator. They are known as **Hadley cells** in honor of George Hadley, the London lawyer and philosopher who worked out an overall scheme of wind circulation in 1735. Look for them in Figure 8.9.

A more complex pair of circulation cells operates at mid-latitudes in each hemisphere. Some of the air descending at 30° latitude turns poleward rather than equatorward. Before this air descends to the surface it is joined at high altitude by air returning from the north. As can be seen in Figure 8.9, a loop of air forms between 30° and about 50°–60° of latitude. As before, the air is driven by uneven heating and influenced by the Coriolis effect. Surface wind in this circuit is again deflected to the right, this time flowing from the west to complete the circuit. (This air is represented by the arrows labeled "Westerlies" in Figure 8.9.) The mid-latitude circulation cells of each hemisphere are named **Ferrel cells** after William Ferrel, the American who discovered their inner workings in the mid-nineteenth century. They, too, can be seen in Figure 8.9.

Meanwhile, air that has grown cold over the poles begins blowing toward the equator at the surface, turning to the west as it does so. At between 50° and 60° latitude in each hemisphere, this air has taken up enough heat and moisture to ascend. However, this polar air is denser than the air in the adjacent Ferrel cell and does not mix easily with it. The unstable zone between these two cells generates most mid-latitude weather. At high altitude the ascending air from 50°–60° latitude turns poleward to complete a third circuit. These are the **polar cells.**

Three large atmospheric circulation cells—a Hadley cell, a Ferrel cell, and a polar cell—therefore exist in each hemisphere. Air circulation within each cell is powered by uneven solar heating and influenced by the Coriolis effect.

WIND PATTERNS

The model of atmospheric circulation described above has many interesting features. At the bands between circulation cells, the air is moving *vertically*, and surface winds are weak and erratic. Such conditions exist at the equator (where air rises and atmospheric pressure is generally low) or at 30° latitude in each hemisphere (where air falls and atmospheric pressure is generally high). Places within these circulation cells where air moves rapidly *horizontally* across the surface from zones of high pressure to zones of low pressure are characterized by strong, dependable winds.

Sailors have a special term for the calm equatorial areas where the surface winds of the two Hadley cells converge: the **doldrums.** The word has come to be associated with a gloomy, listless mood, perhaps reflecting the sultry air and variable breezes found there. Scientists who study the atmosphere call this area the **intertropical convergence zone,** or **ITCZ,** to reflect the influence of wind convergence on conditions near the equator. Strong heating in the ITCZ causes surface air to expand and rise. The humid, rising, expanding air loses moisture as rain, some of which contributes to the success of tropical rain forests.

Sinking air, in contrast, is generally arid. The great deserts of both hemispheres, dry bands centered around 30°, mark the intersection of the Hadley and Ferrel cells. Because evaporation is higher than precipitation in these areas, ocean surface salinity tends to be highest at these latitudes (as can be seen in Figure 18.4). At sea, these areas of high atmospheric pressure and little surface wind are called the **horse latitudes.** Spanish ships laden with supplies for the New World were often becalmed there, sometimes for weeks on end. When the mariners ran out of water and feed for their livestock, they were forced to throw the dead horses over the side.

Of much more interest to sailing masters were the bands of dependable surface winds *between* the zones of ascending and descending air. Most constant of these are the persistent **trade winds,** centered at about 15°N and 15°S latitude. The trade winds are the surface winds of the Hadley cells as they move from the horse latitudes to the doldrums. In the Northern Hemisphere they are the northeast trade winds; the southeast trade winds are the Southern Hemisphere counterpart.[2] The **westerlies,** surface winds of the Ferrel cells centered at about 45°N and 45°S latitude, flow between the horse latitudes and the boundaries of the polar cells in each hemisphere. Thus, the westerlies approach from the southwest in the Northern Hemisphere and from the northwest in the Southern Hemisphere. Sailors outbound from Europe to the New World learned to drop south to catch the trade winds and to return home by a more northerly route to take advantage of the westerlies. Trade winds and westerlies are shown in Figure 8.9.

Cell Circulation: Ideal Versus Actual

The three-cell model of atmospheric circulation discussed above represents an average of air flow through many years over the planet as a whole. Though the model is accurate in a general sense, local details of cell circulation vary because surface conditions are different at different longitudes. The ocean's thermostatic effect is the major factor reducing irregularities in cell circulation over water.

---

[2]Winds are named by the direction from which they blow. A west wind blows from the west toward the east; a northeast wind blows from the northeast toward the southwest.

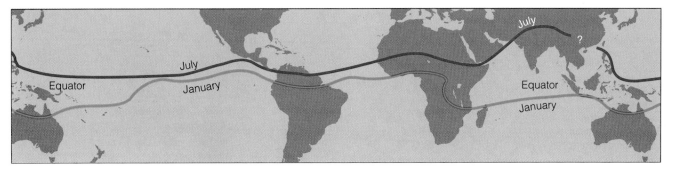

Figure 8.10 Seasonal changes in the position of the ITCZ. The zone reaches its most northerly location in July and its most southerly location in January. Because of the thermostatic effect of water, the seasonal north-south movement is generally less over the ocean than over land.

For example, the equator-to-pole patterns of airflow within circulation cells along the 20°E line of longitude, a meridian crossing Africa and Europe, are much more complex than the patterns of flow along 170°W longitude, which is almost exclusively over the ocean. The ITCZ is also much narrower and more consistent over the ocean than over land. Since the Northern Hemisphere contains much less ocean surface than the Southern Hemisphere, and since land masses have a lower heat capacity than the ocean, seasonal differences in temperature and cell circulation are more extreme in the north. Cell circulation is also much more symmetrical in the Southern Hemisphere.

Another consequence of the markedly different proportions of land to ocean surface in the two hemispheres is the position of the ITCZ. The convergence zone does not coincide with the **geographical equator** (0° latitude). Instead it lies at the **meteorological equator** (or *thermal equator*), an irregular imaginary line of thermal equilibrium between the hemispheres, situated about 5° north of the geographic equator. The positions of the meteorological equator and the ITCZ generally coincide, changing with the seasons, moving slightly farther north in the northern summer and returning toward the equator in the northern winter (**Figure 8.10**). Atmospheric and oceanic circulation in the two hemispheres is approximately symmetrical about the meteorological equator, not the geographical equator. Thus the doldrums, trade winds, horse latitudes, and westerlies shift north in the northern summer, and south in the northern winter.

There are also east-west variations in the expected patterns of the circulation cells. In the northern winter, air above the chilled continental masses of North America and Siberia becomes very cold and dense. This air sinks and forms zones of high atmospheric pressure over the continents. Air over the relatively warmer waters near the Aleutians and Iceland rises and forms zones of low atmospheric pressure. Air flows from the high-pressure zones toward low-pressure zones, modifying the flow of air within the cells. In summer the situation is reversed: Low pressure forms over the heated land masses, and higher pressure forms over the cooler ocean. These effects are most pronounced in the middle latitudes of the Northern Hemisphere, where land and water are present in near-equal amounts.

Figure 8.11 is a general depiction of planetary wind patterns and zones of high and low pressure during the Northern Hemisphere winter. Note that the zones of high and low pressure correspond only roughly to the positions of high- and low-pressure areas modeled in Figure 8.9. Most of the departure from the model is caused by the geographical distribution of land masses and the different responses of land and ocean to solar heating.

The major surface wind and pressure systems of the world, and their prevailing weather conditions, are summarized in **Table 8.2**. These great wind patterns are responsible for about two-thirds of the heat transfer from the tropics to the polar regions. (Ocean currents account for the other third.)

Monsoons

A **monsoon** is a pattern of wind circulation that changes with the season. (The word *monsoon* is derived from *mausim*, the Arabic word for season.) Areas subject to monsoons generally have wet summers and dry winters.

Monsoons are linked to the different heat capacities of land and water, and to the annual north-south movement of the ITCZ. In the spring, land heats more rapidly than the adjacent ocean. Air above the land becomes warmer and so rises. Relatively cool air flows from over the ocean to the land to take its place. Continued heating causes this humid air to rise, condense, and form clouds and rain. In autumn the land cools more rapidly than the adjacent ocean. Air cools and sinks over the land, and

Figure 8.11 The surface wind pattern and mean sea level pressure distribution for the Northern Hemisphere winter. Cold, dense air above the chilled continental masses of Canada and Siberia sinks and forms zones of high atmospheric pressure over the northern continents. Air over the relatively warmer waters near the Aleutians and Iceland rises and forms zones of low atmospheric pressure. Air flows from the high-pressure zones toward low-pressure zones. In summer low pressure forms over the heated land masses, and higher pressure forms over the cooler ocean.

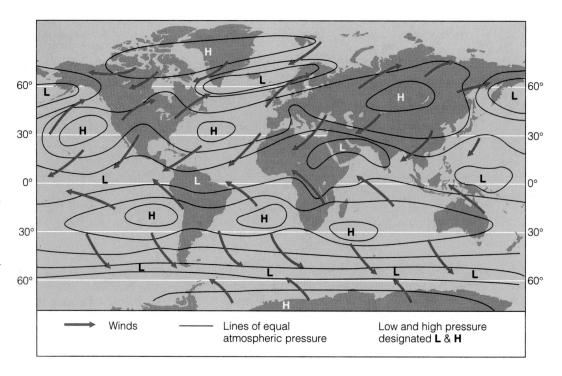

Winds — Lines of equal atmospheric pressure — Low and high pressure designated **L** & **H**

Table 8.2	Major Wind and Pressure Systems and Related Weather				
Region	Name	Pressure	Surface Winds	Weather	
Equator (0°)	Doldrums	Low	Light, variable winds	Cloudiness, abundant precipitation in all seasons, breeding ground for hurricanes. Relatively low sea surface salinity because of rainfall (see Figure 18.4)	
0°–30°N and S	Trade winds	—	Northeast in Northern Hemisphere, southeast in Southern Hemisphere	Summer wet, winter dry; pathway for tropical disturbances	
30°N and S	Horse latitudes	High	Light, variable winds	Little cloudiness; dry in all seasons. Relatively high sea surface salinity because of evaporation (see Figure 18.4)	
30°–60°N and S	Prevailing westerlies	—	Southwest in Northern Hemisphere, northwest in Southern Hemisphere	Winter wet, summer dry; pathway for subtropical high and low pressure	
60°N and S	Polar front	Low	Variable	Stormy, cloudy weather zone; ample precipitation in all seasons	
60°–90°N and S	Polar easterlies	—	Northeast in Northern Hemisphere, southeast in Southern Hemisphere	Cold polar air with very low temperatures	
90°N and S	Poles	High	Southerly in Northern Hemisphere, northerly in Southern Hemisphere	Cold, dry air; sparse precipitation in all seasons	

Source: Burrus & Spiegel, 1980.

Note: Compare to Figure 8.9.

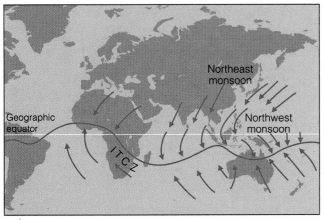

a January

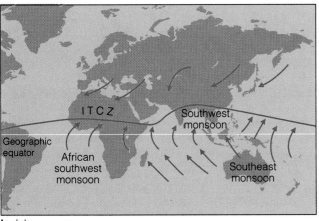

b July

Figure 8.12 Monsoon patterns. During the monsoon circulations of January and July, surface winds are deflected to the right in the Northern Hemisphere and to the left in the Southern Hemisphere.

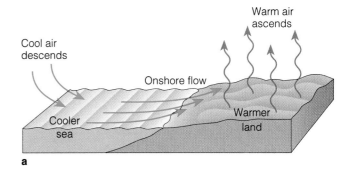

a

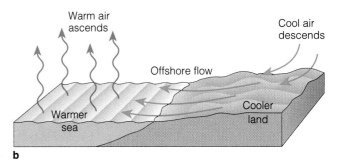

b

Figure 8.13 The flow of air in sea breezes and land breezes. (a) Afternoon onshore sea breeze. (b) Nighttime offshore land breeze.

dry surface winds move seaward. The intensity and location of monsoon activity depends on the position of the ITCZ. Note in **Figure 8.12** that the monsoons follow the ITCZ south in the Northern Hemisphere's winter, and north in its summer.

In Africa and Asia, more than 2 billion people depend on summer monsoon rains for drinking water and agriculture. The most intense summer monsoons occur in Asia. The great landmass of Asia draws vast quantities of warm, moist air from the Indian Ocean (see again Figure 8.12). Southerly winds drive this moisture toward Asia, where it rises and condenses to produce a months-long deluge. Much smaller monsoons occur in North America as warming and rising air over the South and West draws humid air and thunderstorms from the Gulf of Mexico.

Land breezes and sea breezes are small, daily minimonsoons. Morning sunlight falls on land and adjacent

sea, warming both. The temperature of the water doesn't rise as much as the temperature of the land, however. The warmer inland rocks transfer heat to the air, which expands and rises, creating a zone of low atmospheric pressure over the land. Cooler air from over the sea then moves toward land; this is the **sea breeze** (see **Figure 8.13a**). The situation reverses after sunset, with land losing heat to space and falling rapidly in temperature. After a while, the air over the still-warm ocean will be warmer than the air over the cooling land. This air will then rise, and the breeze direction will reverse, becoming a **land breeze** (see **Figure 8.13b**). Land breezes and sea breezes are common and welcome occurrences in coastal areas.

STORMS

Storms are regional atmospheric disturbances characterized by strong winds often accompanied by precipitation. Few natural events underscore human insignificance like a great storm. When powered by stored sunlight, the combination of atmosphere and ocean can do fearful damage.

In Bangladesh, on 13 November 1970, a tropical cyclone (a hurricane) with wind speeds of more than 200 kilometers (125 miles) per hour roared up the mouths of the Ganges River, carrying with it masses of seawater up to 12 meters (40 feet) high. Water and wind clawed at the aggregation of small islands, most just above sea level, that makes up this impoverished country. In only 20 minutes at least 300,000 lives were lost, and estimates ranged to 1 million dead! Property damage was absolute. Photographs taken soon after the storm showed a horizon-to-horizon morass of flooded, deep-gashed ground tortured by furious winds. There was almost no trace of human inhabitants, farms, domestic animals, or villages. Another great storm, which struck in May 1991, killed another 200,000 people. The economy of the shattered country may never fully recover.

A much different type of storm hammered the U.S. East Coast in March of 1993. Mountainous snows from New York to North Carolina (1.3 meters, or 50 inches, at Mount Mitchell), 175-kilometer- (100-mile-) per-hour winds in Florida, and record cold in Alabama (−17°C, or 2°F, in Birmingham) were elements of a four-day storm that spread chaos from Canada to Cuba. At least 238 people died on land; another 48 were lost at sea. At one point more than 100,000 people were trapped in offices, factories, vehicles, and homes; 1.5 million were without electricity. Economic damage exceeded $1 billion.

These two great storms exemplify *tropical cyclones* and *extratropical cyclones* at their worst. As the name implies, tropical cyclones like the hurricane that struck Bangladesh are primarily a tropical phenomenon. Extratropical cyclones—the winter weather disturbances with which residents of the U.S. Eastern Seaboard and other mid-latitude dwellers are most familiar—are found mainly in the Ferrel cells of each hemisphere. (Note that the prefix *extra-* means "outside" or "beyond"; so *extratropical* refers to the location of the storm, not its intensity.)

Both kinds of storms are **cyclones,** huge rotating masses of low-pressure air in which winds converge and ascend. The word *cyclone,* derived from the Greek noun *kyklon* (meaning "an object moving in a circle"), underscores the spinning nature of these disturbances. (Don't confuse a cyclone with a **tornado,** a much smaller funnel of fast-spinning wind associated with severe thunderstorms.)

Air Masses

Cyclonic storms form between or within air masses. An **air mass** is a large body of air with nearly uniform temperature, humidity, and therefore density throughout. Air pausing over water or land will tend to take on the characteristics of the surface below. Cold, dry land causes the mass of air above to become chilly and dry. Air above a warm ocean surface will become hot and humid. Cold, dry air masses are dense and form zones of

Figure 8.14 Vilhelm Bjerknes, the Norwegian meteorologist who, with his son Jacob, formulated the air mass theories on which our understanding of weather are based.

high atmospheric pressure. Warm, humid air masses are less dense and form zones of lower atmospheric pressure.

Air masses can move within or between circulation cells. Density differences, however, will prevent the air masses from mixing when they approach one another. Energy is required to mix air masses. Since that energy is not always available, a dense air mass may slide beneath a lighter air mass, lifting the lighter one and causing its air to expand and cool. Water vapor in the rising air may condense. All of these effects contribute to turbulence at the boundaries of the air masses.

The boundary between air masses of different density is called a **front.** The term was coined by **Vilhelm Bjerknes (Figure 8.14),** a pioneering Norwegian meteorologist who saw a similarity between the zone where air masses meet and the violent battle fronts of the First World War.

Extratropical cyclones form at a front between *two* air masses. Tropical cyclones form from disturbances within *one* warm and humid air mass.

Extratropical Cyclones

Extratropical cyclones form at the boundary between each hemisphere's polar cell and its Ferrel cell—the **polar front.** These great storms occur mainly in the winter hemisphere when temperature and density differences across the polar front are most pronounced. Remember, the cold wind poleward of the front is generally moving from the east; the warmer air equatorward of the front is generally moving from the west (see again Figure 8.9). The smooth flow of winds past each other at the front may be interrupted by zones of alternating high and low atmospheric pressure that bend the front into a series of waves. Because of the difference in wind direction in the air masses north and south of the polar front, the wave shape will enlarge, and a twist will form along

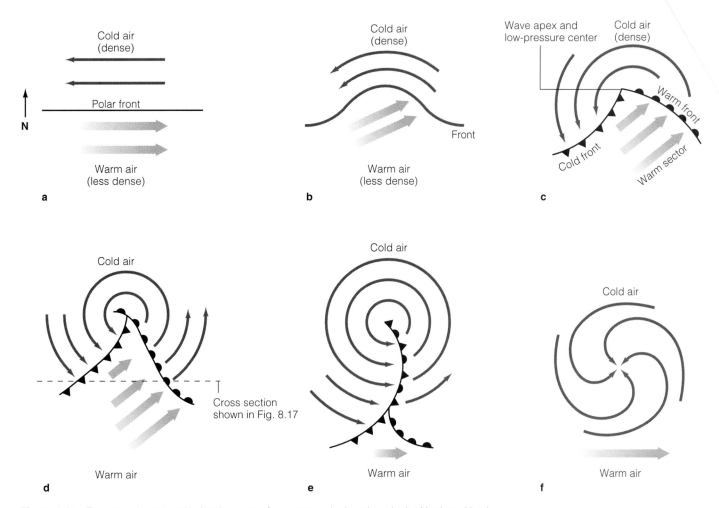

Figure 8.15 The genesis and early development of an extratropical cyclone in the Northern Hemisphere. (The arrows depict air flow.)

the front. The different densities of the air masses prevent easy mixing; so the cold, dense air mass will slide beneath the warmer, lighter one. Formation of this twist in the Northern Hemisphere, as seen from above, is shown in **Figure 8.15.** The twisting mass of air becomes an extratropical cyclone. (The great storm that killed 15 sailors during the 1979 Fastnet race was an extratropical cyclone. This storm and its effects are discussed in **Box 8.1.**)

The twist that generates an extratropical cyclone circulates counterclockwise in the Northern Hemisphere, seemingly in opposition to the Coriolis effect. The reasons for this paradox become clear, however, when we consider the wind directions and the nature of interruption of the airflow between the cells. (In fact, the counterclockwise motion of the cyclone *is* Coriolis-driven because the large-scale airflow pattern at the edges of the cells is generated in part by the Coriolis effect.) Wind speed increases as the storm "wraps up" in much the same way that a spinning skater increases rotation speed

by pulling in his or her arms close to the body. Air rushing toward the center of the spinning storm rises to form a low-pressure zone at the center. Extratropical cyclones are embedded in the westerly winds and thus move eastward. They are typically 1,000 to 2,500 kilometers (625 to 1,600 miles) in diameter and last from two to five days. **Figure 8.16** provides a beautiful example.

Precipitation can begin as the circular flow develops. **Figure 8.17,** a cross section through the colliding air masses of Figure 8.15d, shows why. Precipitation is caused by the lifting and consequent expansion and cooling of the mass of mid-latitude air involved in the twist. As it rises and cools, this air can no longer hold all of its water vapor; so clouds and rain result. When *cold* air advances and does the lifting (as on the left side of Figure 8.17) a *cold front* occurs. A *warm front* happens when *warm* air is blown on top of the retreating edge of cold air (as on the right side of Figure 8.17). The wind and precipitation associated with these fronts are sometimes referred to as **frontal storms.**

Figure 8.16 An extratropical cyclone over Ohio, 25 April 1991. Compare with Figure 8.15d.

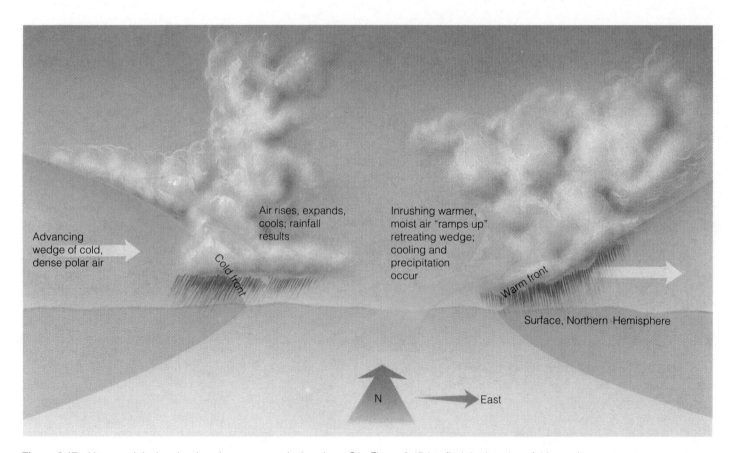

Advancing wedge of cold, dense polar air

Air rises, expands, cools; rainfall results

Inrushing warmer, moist air "ramps up" retreating wedge; cooling and precipitation occur

Cold front

Warm front

Surface, Northern Hemisphere

N

East

Figure 8.17 How precipitation develops in an extratropical cyclone. See Figure 8.15d to find the location of this section.

BOX 8.1

Fastnet: Racing into Disaster

Bad day for a yacht race.

The Earth's atmosphere and its ocean are intimately intertwined, their gases and waters freely exchanged. Gases entering the atmosphere from the ocean have important effects on climate, and gases entering the ocean from the atmosphere can influence sediment deposition, the distribution of life, and some of the physical characteristics of the seawater itself. Water evaporated from the ocean surface and moved by the winds helps minimize worldwide extremes of surface temperature and, through rain, provides moisture for agriculture. The weather that so profoundly affects our daily lives is shaped at the junction of wind and water.

Our understanding of this atmosphere–ocean interaction—perhaps the most complex problem in physical oceanography—is still incomplete. Sometimes this lack of knowledge is merely inconvenient, as when an incorrect forecast results in a rainy picnic or a cloudy day at the beach. But at other times our ignorance can contribute to the loss of hundreds or thousands of human lives and the destruction of property worth billions of dollars.

Consider, for example, the 1979 Fastnet Yacht Race across the shallow waters southwest of the British Isles. The Fastnet race is always a difficult test of seamanship and equipment. The 1979 race began on 11 August in weather so mild that the crews' major concern was sea fog and calm winds. It ended three days later in shrieking winds and towering seas generated by a storm that exploded into the North Atlantic from its birthplace over the American Great Plains. One specialist called the unpredicted storm a "meteorological bomb."

The first morning of the race, a weak storm system began moving from Nova Scotia toward the racecourse. Its actions were not unusual; it behaved as a mild summer storm is expected to behave. The next night, under the cover of darkness, it blended with a jet stream, one of a number of high-altitude bands of strong wind that blow across the Earth's surface. At around noon on 13 August the fast-moving storm suddenly deepened: Its atmospheric pressure plummeted, and its internal winds increased. At this crucial juncture, three merchant ships that were nearest to the low-pressure center somehow reported incorrect pressure and wind speed readings. The conflicting reports confused the forecast, but the race was already under way. That night, after dark, the "bomb" exploded amidst the Fastnet fleet.

What was supposed to be a 960-kilometer (600-mile) sail around a lighthouse off the Irish coast became every sailor's worst nightmare. Twenty-seven hundred men and women in 303 yachts were trapped in a tempest with winds gusting to 147 kilometers (92 miles) per hour. More than one-third of the boats were knocked down by the 15-meter (50-foot) waves, their masts paralleling the ocean surface. Fully one-fourth of them capsized. Sailors who had retreated to the safety of their cabins were battered by tins of food and pieces of equipment as their boats violently rolled and pitched. Twenty-four crews abandoned their boats, five of which sank. Fifteen people died of drowning or hypothermia. It was the greatest disaster in the century-long history of ocean yacht racing.

Were the forecasters responsible for the disaster? A board of inquiry found such "explosively developing meteorological and oceanographic conditions . . . to be essentially unpredictable." Before the race started, no forecaster could have foreseen the violence of this storm. The high winds had changed direction during the tempest and had built waves that moved at an angle through large waves created by the earlier wind direction. Interference between the wave sets created steep and chaotic sea conditions that few yachts could survive. The lesson is clear: Ocean scientists need to learn more about the dynamics of air–ocean interaction. Far more than yacht races are at risk.

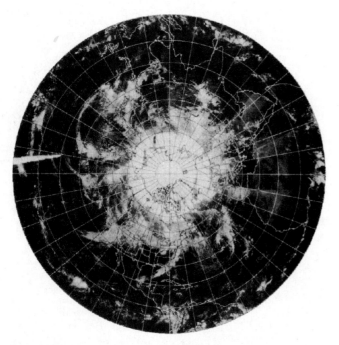

Figure 8.18 A satellite view of the Northern Hemisphere showing the cloud patterns of 4 January 1982. The North Pole is at the center of the picture. The arcs of clouds that pinwheel outward from the center of the image are associated with extratropical cyclones.

Figure 8.19 Storm damage from a particularly severe nor'easter, the Ash Wednesday (7 March) storm of 1962. Massive damage such as this, on Fire Island, New York, is most common for buildings near the shoreline.

Chains of extratropical cyclones swirl over land and ocean in the middle cells of both hemispheres (see **Figure 8.18**). They are the principal cause of weather in the midlatitude regions, where most of the world's people live. When a weather forecaster speaks of a cold front bringing winter snow or rain to your area, the forecaster is describing the approach of a cold, south-sweeping arm of an extratropical cyclone.

North America's most violent extratropical cyclones are the **nor'easters** (northeasters) that sweep the Eastern Seaboard in winter. The name indicates the direction from which the storm's most powerful winds approach. About 30 times a year, nor'easters moving along the mid-Atlantic and New England coasts generate wind and waves with enough force to erode beaches and offshore barrier islands, disrupt communication and shipping schedules, damage shore and harbor installations, and break power lines. About every 100 years a nor'easter devastates coastal settlements. In spite of a long history of destruction, people continue to build on unstable exposed coasts (see **Figure 8.19** and the photograph that opens Chapter 12).

Tropical Cyclones

Tropical cyclones are great masses of warm, humid, rotating air. They occur in all tropical oceans except the equatorial South Atlantic. Large tropical cyclones are called **hurricanes** (*Huracan* = the Taino god of the wind)

in the North Atlantic and eastern Pacific, typhoons (*Taifung* = Chinese for great wind) in the western Pacific, tropical cyclones in the Indian Ocean, and willi-willis in the waters near Australia. To qualify formally as a hurricane or typhoon, the tropical cyclone must have winds of at least 118 kilometers (74 miles) per hour. About 100 tropical cyclones grow to hurricane status each year. A very few of these develop into superstorms, with winds near the core that exceed 250 kilometers (155 miles) per hour! (To imagine what winds in such a storm might feel like, picture yourself clinging to the wing of a twin-engined private airplane in flight!) Tropical cyclones containing winds less than hurricane force are called *tropical storms* and *tropical depressions*.

From above, tropical cyclones appear as circular spirals, as **Figure 8.20** shows. They may be 1,000 kilometers (630 miles) in diameter and 15 kilometers (9.5 miles, 50,000 feet) high. The calm center, or *eye* of the storm—a zone some 13 to 16 kilometers (8 to 10 miles) in diameter—is sometimes surrounded by clouds so high and dense that the daytime sky above looks dark. Farther out, churned by furious winds, the rainband clouds condense huge amounts of water vapor into rain. A tropical cyclone is diagrammed in **Figure 8.21**.

Figure 8.20 Hurricane Elena over the Gulf of Mexico, photographed from the space shuttle *Discovery* on 2 September 1985. The eye of the storm is clearly visible. Note the spiral bands of cloud extending in toward the eye. Compare this oblique view of a tropical cyclone with the vertical view of Hurricane Andrew that opens this chapter.

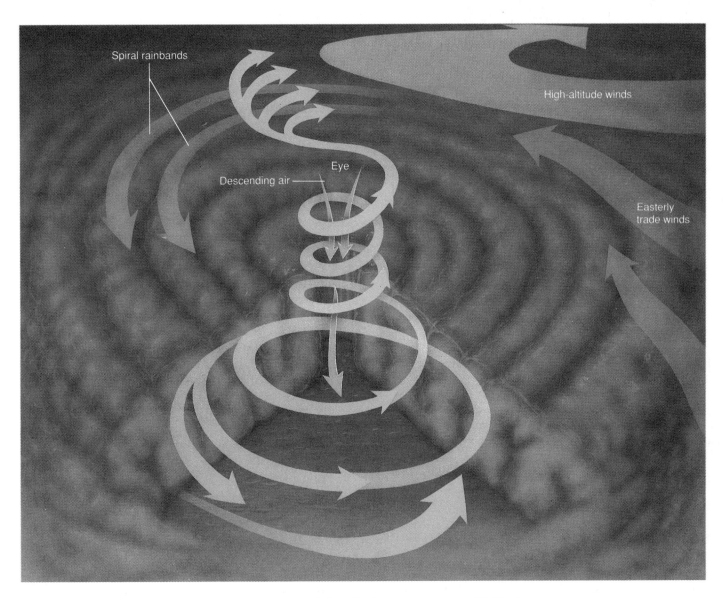

Figure 8.21 The internal structure of a hurricane. The vertical dimension is greatly exaggerated in this drawing.

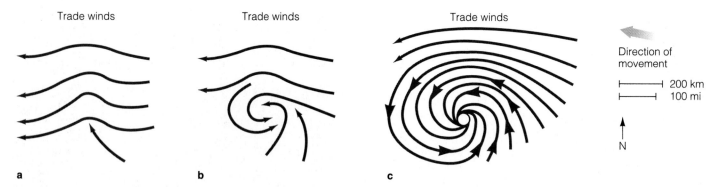

Trade winds   Trade winds   Trade winds

Direction of movement

⊢——————⊣ 200 km
⊢——————⊣ 100 mi

N

a   b   c

Figure 8.22   Steps in the development of a hurricane from an easterly wave.

Unlike extratropical cyclones, these greatest of storms form within *one* warm, humid air mass between 10° and 25° latitude in both hemispheres. (Though air conditions would be favorable, the Coriolis effect closer to the equator is too weak to initiate rotary motion.) The origins of tropical cyclones are not well understood. A tropical cyclone usually develops from a small tropical depression. Tropical depressions form in easterly waves, areas of lower pressure within the easterly trade winds that are thought to originate over a large, warm landmass. When air containing the disturbance is heated over tropical water with a temperature of about 26°C (79°F) or more, circular winds begin to blow in the vicinity of the wave, and some of the warm humid air is forced upward. Condensation begins, and the storm takes shape. Under ideal conditions the embryo storm reaches hurricane status—that is, with wind speeds in excess of 118 kilometers (74 miles) per hour—in two to three days. The first part of the process is shown in **Figure 8.22.**

The centers of most tropical cyclones move westward and poleward at 5 to 40 kilometers (3 to 25 miles) per hour. Typical tracks of these storms are shown in **Figure 8.23.** A trio of Pacific tropical cyclones is shown in **Figure 8.24.** Contrast this photo with Figure 8.18, a satellite photo showing extratropical cyclones in the Northern Hemisphere.

Although its birth process is somewhat mysterious, the source of the storm's power is well understood. Its strength comes from the same seemingly innocuous process that warms a chilled soft-drink can when water from the atmosphere condenses on its surface. As you may recall from Chapter 7, it takes quite a bit of energy to evaporate water into the atmosphere: Water's latent heat of evaporation is very high. That heat energy is released when the water vapor recondenses as liquid. It tends to warm your drink very quickly, and the more humid the air, the faster the condensing and warming.

Think of the situation with a hair dryer. The heat energy generated by the dryer causes water to evaporate rapidly from your hair. When that water recondenses to liquid (on a nearby can of soda, for instance), the original heat used to evaporate the water is released. The cycle of evaporation and condensation has carried heat from the hair dryer to your soda. In tropical cyclones the condensation energy generates air movement (wind), not more heat. Fortunately only 2% to 4% of this energy of condensation is converted into motional energy!

A tropical cyclone is an ideal machine for "cashing in" water vapor's latent heat of evaporation. Warm, humid air forms in great quantity only over a warm ocean. As already noted, tropical cyclones originate in ocean areas having surface temperatures in excess of 26°C (79°F) (see Figure 8.24). As hot, humid tropical air rises and expands, it cools and is unable to contain the moisture it held when warm. Rainfall begins. The rainfall rate in some parts of the storm routinely exceeds 2.5 centimeters (1 inch) per hour, and 20 billion metric tons of water can fall from a large tropical cyclone in a day! Tremendous energy is released as this moisture changes from water vapor to liquid. In one day, a large tropical cyclone generates about 2,400 billion kilowatts of power, equivalent to the electrical energy needs of the entire United States for a year! Thus, solar energy ultimately powers the storm in a cycle of heat absorption, evaporation, condensation, and conversion of heat energy to kinetic energy. This energy is available as long as the storm stays over warm water and has a ready source of hot, humid air. Tropical cyclones may sometimes even leave the tropics, as **Box 8.2** recounts.

Three aspects of a tropical cyclone can cause property damage and loss of life: wind, rain, and storm surge. The destructive force of winds of 250 kilometers (155 miles) per hour—or more—is self-evident. Rapid rainfall can cause severe flooding when the storm moves onto land.

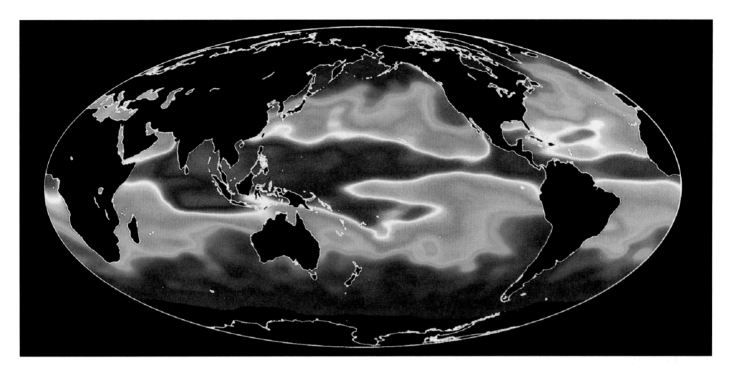

Figure 8.23a Tropical cyclones can develop in the hot, humid air over a sea surface exceeding 26°C (79°F). The base map in this diagram is derived from satellite data showing water vapor in the atmosphere in October 1992. Red zones indicate high humidity and warm temperatures.

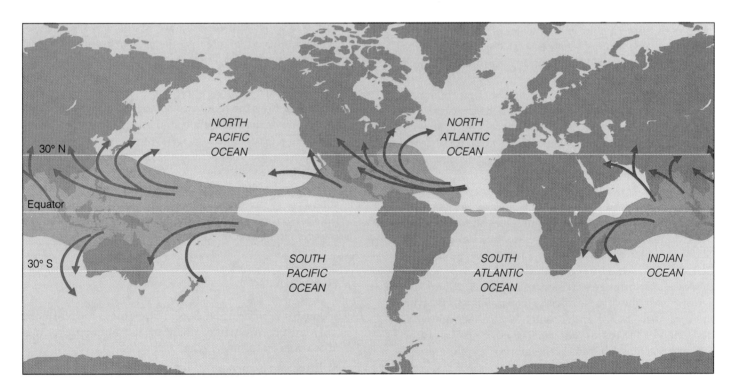

Figure 8.23b The breeding grounds of tropical cyclones are shown as shaded areas. The storms follow curving paths: First they move westward with the trade winds. Then they either die over land or turn eastward until they lose power over the cooler ocean of mid-latitudes. Cyclones are not spawned over the South Atlantic or the southeast Pacific because their waters are too chilly, nor in the still air—the doldrums—within 5° of the equator.

Figure 8.24 Three tropical cyclones lined up between Hawaii and Central America. On the left is Hurricane Kate; in the center is a tropical depression that never reached hurricane force; and on the right is Hurricane Liza, which within hours smashed into La Paz, Mexico, and caused millions of dollars in damage.

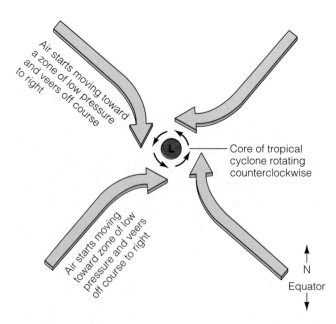

Air starts moving toward a zone of low pressure and veers off course to right

Air starts moving toward zone of low pressure and veers off course to right

Core of tropical cyclone rotating counterclockwise

N

Equator

Figure 8.25 The dynamics of a tropical cyclone, showing the influence of the Coriolis effect. Note that the storm turns the "wrong" way (that is, counterclockwise) in the Northern Hemisphere, but for the "right" reasons.

But the most danger lies in **storm surge,** a mass of water driven by the storm. The low atmospheric pressure at the storm's center produces a dome of seawater that can reach a height of 1 meter (3.3 feet) in the open sea. The water height increases when waves and strong hurricane winds ramp the water mass ashore. If a high tide coincides with the arrival of all this water at a coast or if the coastline converges (as is the case at the mouths of the Ganges in Bangladesh), rapid and catastrophic flooding will occur. Storm surges of up to 12 meters (40 feet) were reported at Bangladesh in 1970. Much of the $3 billion in damage done by Hurricane Hugo to Charleston, South Carolina, in September 1989 was caused by a 5-meter (17-foot) storm surge arriving at high tide. You'll learn more about storm surges in Chapter 11's discussion of large waves.

You may have noticed that tropical cyclones turn counterclockwise in the Northern Hemisphere and clockwise in the Southern Hemisphere. Does this mean that the Coriolis effect does not apply to tropical cyclones? No. This apparent anomaly is caused by the Coriolis deflection of winds approaching the center of a low-pressure area from great distances. **Figure 8.25** provides a view from above a Northern Hemisphere storm, with the core of the storm at the center. Notice the rightward deflection of the *approaching air.* The edge spin given by this approaching air causes the storm to spin counterclockwise in the Northern Hemisphere.

Tropical cyclones last from 3 hours to 3 weeks; most have lives of 5 to 10 days. They eventually run down when they move over land or over water too cool to supply the humid air that sustains them. The friction of a land encounter rapidly drains a tropical cyclone of its

power, and a position above ocean water cooler than 24°C (76°F) is a sure harbinger of the storm's demise. When deprived of energy, the storm "unwinds" and becomes a mass of unstable humid air pouring rain, lightning, and even tornadoes from its clouds. Tropical cyclones can be dangerous to the end: Torrential rain streaming from the remnants of Hurricane Agnes in 1976 caused more than $2 billion in damage, mostly to Pennsylvania. Chesapeake and Delaware bays were flooded with fresh water and sediments, destroying much of the shellfish industry there.

Tropical cyclones are nature's escape valves, flinging solar energy poleward from the tropics. They are beautiful, dangerous examples of the power represented by water's latent heat of fusion.

Q-AND-A

1. Earth's orbit brings it closer to the sun in the Northern Hemisphere's winter than in its summer. Yet it's warmer in summer. Why?

   Earth's orbit around the sun is elliptical, not circular. The whole Earth receives about 7% more solar energy through the half of the orbit during which we are closer to the sun than through the other half. The time of greater energy input comes during our winter, but the entire Northern Hemisphere is tilted toward *the sun during the summer,*

BOX 8.2

BOX 8.2 ● *Tropical Cyclones in the Temperate Zone*

Though they invariably begin over a warm tropical ocean, tropical cyclones can sometimes find their ways to surprisingly high latitudes. In September 1938, a large hurricane reached Montreal. On its way north, its storm surge ravaged 100,000 square kilometers (39,000 square miles) of coastline between Cape Hatteras and New England. Wind and water from the storm killed at least 600 people, destroyed more than 60,000 homes and 26,000 automobiles, and downed 20,000 miles of electric, telephone, and telegraph wires. More than 250 million trees were destroyed in New England alone. The nearshore seabed was radically altered, rendering nautical charts useless. Total damage was estimated at more than a third of a billion dollars, an immense sum for that time. And all this destruction was above 35° north latitude.

Eyewitnesses were astonished to discover a tropical storm in their decidedly *un*tropical environment. One man who watched the 12-meter (40-foot) storm surge approach thought it was a "thick and high bank of fog rolling in fast from the ocean. When it came closer we saw that it wasn't fog, it was water." Earlier that day, another man had taken delivery of an expensive barometer he had ordered from the New York sporting goods

firm of Abercrombie & Fitch. When he took the instrument out of its box, he noticed with dismay that its needle was pegged at "Hurricane." He tapped and shook the barometer to get it working right, but the needle wouldn't budge. He immediately wrote a vigorous letter of complaint to the company and hurried to the post office to mail it. His coastal home and his new barometer were gone when he returned.

Weather watchers cowering inside sturdy brick structures atop Mount Washington, the highest spot in New England, could have confirmed the barometer's message. They watched in amazement as the howling winds reached 300 kilometers (190 miles) per hour, the highest hurricane winds ever measured in North America. Winds of that speed are uncommon even in tornadoes.

The 1938 storm was rare in its great strength and extension to relatively high latitudes. Of course, lower latitudes are at greater risk because storms will be more frequent. A smaller storm was responsible for the worst natural disaster in U.S. history, the Galveston flood of September 1900, in which at least 6,000 people died. Galveston's latitude is 29°N, still 5°30' above the tropic line.

---

which results in much more light reaching it in the summer. Three times more energy enters the Northern Hemisphere each day at midsummer than at midwinter.

2. How did the trade winds get their name? Is their name a reminder of the assistance they provided to shipboard traders anxious to sell their wares in distant corners of the world?

   The trade winds are not named after their contribution to commerce in the days of sail. This use of the word trade *derives from an earlier English meaning equivalent to our adverbs* steadily *or* constantly. *These persistent winds were said to "blow trade."*

3. Does the ocean affect weather at the centers of continents?

   Absolutely. In a sense, all *large-scale weather on Earth is oceanically controlled. The ocean acts as a solar collector and heat sink, storing and releasing heat. Most great*

storms (tropical and extratropical cyclones alike) form over the ocean and then sweep over land.

4. On average, the meteorological equator lies about 5° north of the geographical equator. Why the displacement?

   The meteorological equator lies in the Northern Hemisphere because, overall, that hemisphere is slightly warmer than the Southern Hemisphere. At least three factors are responsible for this. First, the unbroken ice covering of Antarctica reflects more light back into space than the surface of the Arctic Ocean, which contains occasional light-absorbing patches of open water. Second, there is more land surface in the tropical latitudes of the Northern Hemisphere than in the Southern Hemisphere. Since land warms more than water with the same input of solar energy, tropical latitudes are warmer in the Northern Hemisphere. And third, ocean currents transport more warm water toward the north than toward the south.

5. Are any of the results of large-scale atmospheric circulation apparent to the casual observer?

Yes. The view of atmospheric circulation developed in this chapter explains some phenomena you may have experienced. For instance, flying from Los Angeles to New York takes about 40 minutes less than flying from New York to Los Angeles because of westerly headwinds (indicated in Figures 8.9 and 8.11).

Because of these same prevailing westerlies, most storms travel over the United States from west to east. Weather prediction is based on observations and samples taken from the air masses as they move. Forecasting is often easier in the East and Midwest than in the West because more data are available from an air mass when it's over land.

Temperatures are milder (not as hot in summer, not as cold in winter) on the West Coast than on the East Coast,

again because of the prevailing westerlies. The wind blows over water (which moderates temperatures) toward the West Coast, but over land toward the East Coast.

6. Has anything similar to Earth's weather patterns been seen on other planets?

Yes, indeed. Saturn, Jupiter, and Neptune have tremendous cyclonic storms. A few on Jupiter are large enough to be seen from Earth through small telescopes. Tracks of tornadoes have recently been identified on Mars (see Grant and Schultz, 1987). Venus has huge cloud banks suggestive of polar fronts and extratropical cyclones. The Coriolis effect and uneven solar heating are common to all planets with atmospheres in this solar system; so we shouldn't be surprised by similarities.

Terms and Concepts to Remember

air mass	Ferrel cell	nor'easter
angle of incidence	front	(northeaster)
atmospheric	frontal storm	orbital inclination
circulation cell	geographical equator	polar cell
Bjerknes, Vilhelm	Hadley cell	polar front
climate	horse latitudes	precipitation
clockwise	hurricane	sea breeze
convection	insolation rate	storm
Coriolis, Gaspard	intertropical	storm surge
Gustave de	convergence zone	tornado
Coriolis effect	(ITCZ)	trade winds
counterclockwise	land breeze	tropical cyclone
cyclone	meteorological	water vapor
doldrums	equator	weather
extratropical	monsoon	westerlies
cyclone		wind

Study Questions

1. What is the composition of air? Can more water vapor be held in warm air or cool air? What happens when air containing water vapor rises? Which is denser at the same temperature and pressure: humid air or dry air?

2. What causes precipitation (rain and snow)?

3. How does weather differ from climate? Which is easier to predict? Why?

4. What factors contribute to the uneven heating of Earth by the sun?

5. How does the atmosphere respond to uneven solar heating? How does the rotation of the Earth affect the resultant circulation?

6. Describe the atmospheric circulation cells in the Northern Hemisphere. At what latitudes does air move vertically? Horizontally? What are the trade winds? The westerlies? Where are deserts located? Why? What is ocean surface salinity like in these desert bands?

7. How are the geographic equator, meteorological equator, and ITCZ related? What happens at the ITCZ? What is a monsoon? How is monsoon circulation affected by the position of the ITCZ?

8. How do the two kinds of large storms differ? How are they similar? What causes an extratropical cyclone? What happens in one?

9. What triggers a tropical cyclone? From what is its great power derived? What causes the greatest loss of life and property when a tropical cyclone reaches land?

10. If the Coriolis effect causes the clockwise deflection of moving objects in the Northern Hemisphere, why does air rotate counterclockwise around zones of low pressure in that hemisphere?

For Further Study

Anthes, R. A. 1982. *Tropical Cyclones: Evolution, Structure and Effects.* Boston: American Meteorological Society. The definitive technical work on the subject.

Battin, L. J. 1961. *The Nature of Violent Storms.* New York: Doubleday. A classic.

Burrus, T. L., and H. J. Spiegel. 1980. *Earth in Crisis.* St. Louis: Mosby.

Clowes, E. S. 1939. *The Hurricane of 1938 on Eastern Long Island.* Bridgehampton, NY: Hampton Press. Eyewitness accounts of the 1938 hurricane.

Dolan, R., and H. Lins. 1987. "Beaches and Barrier Islands." *Scientific American,* July, 68–77. Discusses the folly of building close to shore.

Emanuel, K. A. 1988. "Toward a General Theory of Hurricanes." *American Scientist* 76 (no. 4): 370–79. The world's fiercest storms are produced by the most benign climates. Fine recent review article with some controversial views on the sources of a tropical cyclone's power.

Fisher, B. 1980. *The Fastnet Disaster and After.* London: Pelham Books. The chilling account of the lethal Fastnet Yacht Race of 1979. An extratropical cyclone at its nastiest.

Friedman, R. M. 1989. *Appropriating the Weather: Vilhelm Bjerknes and the Construction of a Modern Meteorology.* New York: Cornell University Press. An excellent recent history of Bjerknes's contributions.

Gore, R. 1980. "Voyager Views Jupiter." *National Geographic,* January, 2–29. Wonderful photography of extraterrestrial weather and storms. The principles of uneven solar heating and the Coriolis effect certainly apply to Jupiter.

Grant, J. A., and P. H. Schultz. 1987. "Possible Tornado-like Tracks on Mars." *Science* 238 (no. 4817): 883–85. New interpretation of unusual markings on the martian surface.

Ingersoll, A. P. 1983. "The Atmosphere." *Scientific American,* September, 162–75. Energy balance and climate modeling.

Kotsch, W. J., and R. Henderson. 1984. *Heavy Weather Guide.* 2d ed. Annapolis, MD: Naval Institute Press. Updated version of the definitive mariner's guide to heavy weather at sea. Outstanding source of practical information, theory, anecdotal material, and photographs, some of which may cause even the most ardent sailor to invest in desert property.

Lutgens, F. K., and E. J. Tarbuck. 1989. *The Atmosphere: An Introduction to Meteorology.* 4th ed. Englewood Cliffs, NJ: Prentice-Hall. Excellent general text.

McDonald, J. E. 1952. "The Coriolis Effect." *Scientific American,* May, 72–78. Good discussion of theory and practical application.

McWhirter, N. 1978. *The Guinness Book of World Records.* 1978 ed. New York: Bantam Books. Storm statistics.

Moran, J. M., and M. D. Morgan. 1989. *Meteorology: The Atmosphere and the Science of Weather.* New York: Macmillan. Excellent general text.

Rousmaniere, J. 1980. *Fastnet, Force 10.* New York: Norton. The story of the 1979 Fastnet Yacht Race disaster from a participant's point of view.

Schaefer, V. J., and J. A. Day. 1981. *A Field Guide to the Atmosphere.* Peterson Field Guide Series. Boston: Houghton Mifflin. An illustrated guide to atmospheric phenomena.

Stewart, R. W. 1969. "The Atmosphere and the Ocean." *Scientific American,* September, 76–87. Excellent review article.

Webster, P. J. 1981. "Monsoons." *Scientific American,* August, 108–18.

Whipple, A. B. C. 1982. *Storm.* Alexandria, VA: Time–Life Books. Fine general overview of storms. Excellent photos, diagrams, and charts, coupled with lively writing.

Williams, J. 1962. *Oceanography: An Introduction to the Marine Sciences.* New York: Little, Brown. Clear mathematical presentation of the Coriolis effect.

Williams, J. 1992. *The Weather Book.* New York: Vantage Press. Concise overview for the interested layman. Nontechnical.

Williams, J., F. Doehring, and I. Duedall. 1993. "Heavy Weather in Florida." *Oceanus* 36 (no. 1, Spring): 19–26. A discussion of Florida hurricanes and tropical storms over 122 years.

9
OCEAN CIRCULATION

Going with the Flow

The first navigators who sailed past Gibraltar into the Atlantic noticed a persistent southward flow of water, which could drive them off course. Pytheas of Massalia, a Greek ship's captain who explored the northeastern Atlantic in the fourth century B.C., was the first observer to record this slow, continuous movement and to estimate its speed. Though his principal work, *On the Ocean,* is lost, we know of his discoveries from the writings of the historians of his time. Pytheas imagined the

flow was part of an immense river, too wide to sail across. Greek traders plying the eastern Atlantic coast later used the term *okeanos* (oceanus), meaning "great river," to describe it.

Today, we know that the river idea is not far off the mark: The world ocean contains masses of water called **currents,** flowing rivers that move water and heat the way a living organism's circulatory system moves vital substances. The volume of water moving in any one of

Gibraltar from space; the Mediterranean lies beyond.

the ocean's largest surface currents is about 50 times the combined flow of all the Earth's freshwater rivers. The term *current* is usually reserved for water flowing horizontally (that is, parallel to the ocean's surface), but masses of water can also move vertically. Downward movement transports gases to the deep ocean for the maintenance of life there, and upward movement brings fallen nutrients back toward the surface.

The southward-flowing "river" that Pytheas discovered is now called the Canary Current after a cluster of islands lying in its midst. The Canary Current forms as surface water sweeps eastward and southward along the northern boundary of the Atlantic. Though not particularly fast or deep as currents go, the Canary Current moves about 16 million cubic meters (21 million cubic yards) of water per second along the African coast between the Canary and Cape Verde islands. Had Pytheas gone with the flow, his ship would probably have turned westward with the water and headed toward North America. His chroniclers might have had a much different tale to tell had he persisted all the way around the North Atlantic and back to Gibraltar—along a path his *oceanus* would lead.

CHAPTER OVERVIEW

Ocean water circulates in currents. Surface currents affect the uppermost 10% of the world ocean. The movement of surface currents is powered by the warmth of the sun and by winds. Some surface currents are rapid and riverlike, with well-defined boundaries; others are slow and diffuse. Water in surface currents tends to flow horizontally, but it can also flow vertically in response to wind blowing near coasts or along the equator. Surface currents transfer heat from tropical to polar regions, influence weather and climate, distribute nutrients, and scatter organisms. They have contributed to the spread of humanity to remote islands, and they are important factors in maritime commerce.

Circulation of the 90% of ocean water beneath the surface zone is driven by the force of gravity, as dense water sinks and less dense water rises. Since density is largely a function of temperature and salinity, the movement of deep water due to density differences is called *thermohaline circulation*. Currents near the seafloor flow as slow, riverlike masses in a few places, but the greatest volumes of deep water creep through the ocean at an almost imperceptible pace.

The Coriolis effect, gravity, and friction shape the direction and volume of surface currents and thermohaline circulation.

THE FORCES THAT DRIVE CURRENTS

Ocean currents are affected by two kinds of forces: the **primary forces** that start water moving and determine its velocity, and the **secondary forces** and factors that influence the direction and nature of its flow. Primary forces are the stress of wind blowing over the water, thermal expansion and contraction of water, and density differences between water layers. Secondary forces and factors are the Coriolis effect, gravity, friction, and the shape of the ocean basins themselves.

About 10% of the water in the world ocean is involved in **surface currents,** horizontally flowing water at the ocean's surface driven by thermal expansion and wind friction. Most surface currents move water within and above the pycnocline, the zone of rapid density change with depth. Water beneath the pycnocline also circulates, but the power for this slower, deeper circulation comes from the action of gravity on adjacent water masses of different densities. Since density is largely a function of temperature and salinity, circulation due to density differences is called *thermohaline circulation.* We'll investigate thermohaline circulation after a discussion of surface currents.

SURFACE CURRENTS

Solar heating causes water to expand slightly. Because of this, sea level near the equator is about 8 centimeters (3 inches) higher than sea level in temperate ocean areas. Water in the cold regions near the poles cools and contracts by a similar amount. This global difference creates a very slight slope, and warm equatorial water flows "downhill" (poleward) in response to gravity. Poleward-moving surface water tends to lag behind the Earth's eastward rotation, however, and moves toward the ocean's western boundaries. The water's slow travel is also influenced by the Coriolis effect. A sluggish circular flow therefore develops in ocean basins on either side of the equator.

But the difference in temperature between polar and tropical regions is not the major primary force responsible for surface currents. That honor must go to winds. As you read in Chapter 8, surface winds form global patterns within latitude bands (see **Figure 9.1**). Most of Earth's surface wind energy is concentrated in each hemisphere's trade winds and westerlies. Waves on the sea surface transfer some of the energy from the moving air to the water by friction. This tug of wind on the ocean surface begins a more rapid mass flow of water. As a rule of thumb, the friction of wind blowing for at least 10 hours will cause surface water to flow downwind at about 2% of wind speed. The water flowing beneath the wind forms a surface current.

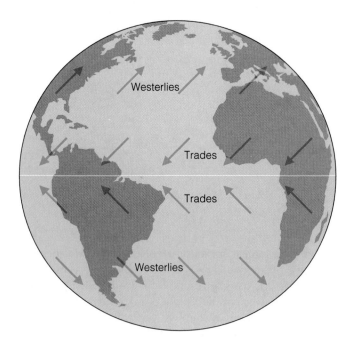

Figure 9.1 Winds, driven by uneven solar heating and Earth's spin, power the movement of the ocean's surface currents.

Because of the Coriolis effect, Northern Hemisphere surface currents flow to the right of the wind direction. Southern Hemisphere currents flow to the left. Intervening continents and basin topography often block continuous flow and help to deflect the moving water into a circular pattern. This flow around the periphery of an ocean basin is called a **gyre** (*gyros* = a circle). Two gyres are shown in **Figure 9.2.**

Flow Within a Gyre

Figure 9.3 shows the North Atlantic gyre in more detail. Though it flows continuously without obvious places where one current ceases and another begins, oceanographers subdivide the North Atlantic gyre into four interconnected currents because each has distinct flow characteristics and temperatures. (Gyres in other ocean basins are similarly divided.) Notice that the east-west currents in the North Atlantic gyre flow to the right of the driving winds; once initiated, water flow in these currents continues in a roughly east-west direction. Where their flow is blocked by continents, the currents turn clockwise to complete the circuit.

Why does water flow around the periphery of the ocean basin instead of spiraling to the center? After all, the Coriolis effect influences any moving mass as long as it moves, so water in a gyre might be expected to curve to the center of the North Atlantic and stop. To understand this aspect of current movement, imagine the forces acting on a surface water particle at 45° north latitude (point A in **Figure 9.4**). Here the westerlies blow

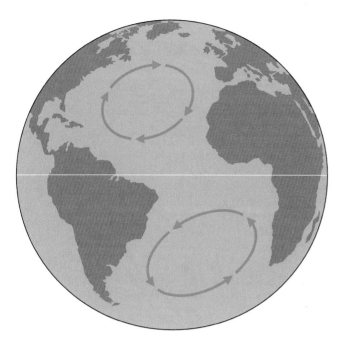

Figure 9.2 A combination of four forces—surface winds, the sun's heat, the Coriolis effect, and gravity—circulates the ocean surface clockwise in the Northern Hemisphere and counterclockwise in the Southern Hemisphere, forming gyres.

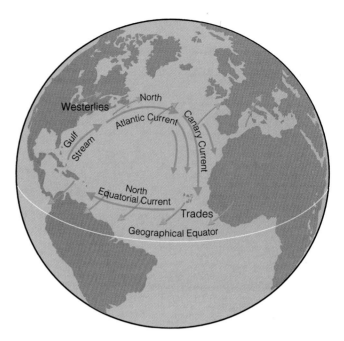

Figure 9.3 The North Atlantic gyre, a series of four interconnecting currents with different flow characteristics and temperatures.

from the southwest, so initially the particle will move toward the northeast. The rightward Coriolis deflection then causes the water to flow almost due east. A particle at 15° north latitude (point B) responds to the push of the trade winds from the northeast, however, and with Coriolis deflection it will flow almost due west.

Research published early in this century indicates that when driven by the wind, the topmost layer of ocean water in the Northern Hemisphere flows at about 45° to the right of the wind direction, a flow consistent with the arrows leading away from points A and B in Figure 9.4. But what about the water in the next layer down? It can't "feel" the wind at the surface; it "feels" only the movement of the water immediately above. This deeper layer of water moves at an angle to the right of the overlying water. The same thing happens in the layer below that, and the next layer, and so on, to a depth of about 100 meters (330 feet) at mid-latitudes. Each layer slides horizontally over the one beneath it like cards in a deck, with each lower card moving at an angle slightly to the right of the one above. Because of frictional losses, each lower layer also moves more slowly than the layer above. The resulting situation, portrayed in **Figure 9.5**, is known as **Ekman spiral** after its Swedish discoverer. The term is somewhat misleading; the water does not spiral downward in a whirlpool-like motion. Rather, the spiral is a way of conceptualizing the horizontal movements in a layered water column, each layer moving in a slightly different horizontal direction. A curious result of Ekman

Figure 9.4 Surface water blown by the winds at point A will veer to the right of its initial path and continue eastward. Water at point B veers right and continues westward.

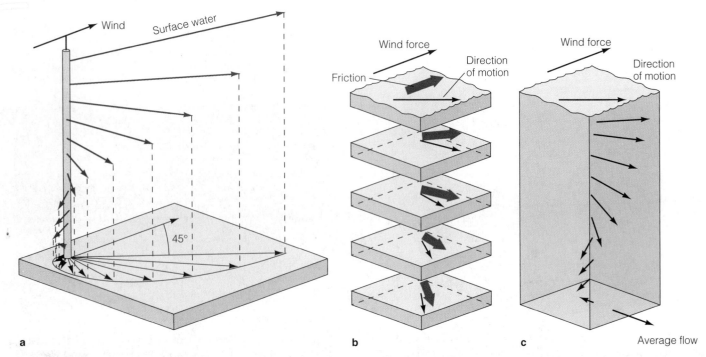

Figure 9.5 Ekman spiral and the mechanism by which it operates. (a) The Ekman spiral model. (b) A body of water can be thought of as a set of layers. The top layer is driven forward by the wind, and each layer below is moved by friction. Each succeeding layer moves with a slower speed and at an angle to the layer immediately above it—to the right in the Northern Hemisphere, to the left in the Southern Hemisphere—until friction becomes negligible. (c) Though the direction of movement varies for each layer in the stack, the theoretical net flow of water in the Northern Hemisphere is 90° to the right of the prevailing wind force. The length of the arrows is proportional to the speed of the currrent in each layer.

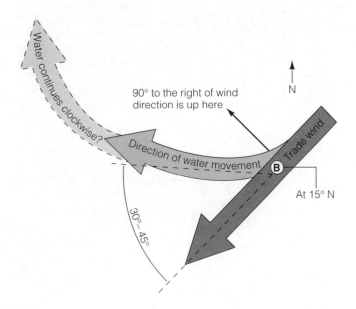

Figure 9.6 The movement of water away from point B in Figure 9.4 is influenced by the Coriolis effect and gravity.

spiral is that water at some depth will be flowing in the opposite direction from the surface current!

The net motion of the water down to about 100 meters, after allowance for the summed effects of Ekman spiral (the sum of all the arrows indicating water direction in the affected layers), is known as **Ekman transport**. In theory, the direction of Ekman transport is 90° to the right of wind direction (in the Northern Hemisphere).

Armed with this information, we can look in more detail at the area around point B in Figure 9.4, which is enlarged in **Figure 9.6.** In nature Ekman transport in gyres is less than 90°; in most cases the deflection barely reaches 45°. This deviation from theory occurs because of an interaction between the Coriolis effect and a secondary force, gravity. Some flowing Atlantic water has turned to the right to form a hill of water: It followed the rightward dotted-line arrow in Figure 9.6. Why does the water now go straight west from point B without turning? Because, as **Figure 9.7a** shows, to turn further right the water would have to move uphill in defiance of gravity, but to turn left in response to gravity would defy the Coriolis effect. So the water continues westward and then clockwise around the whole North Atlantic gyre,

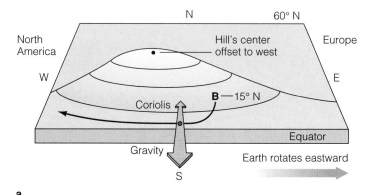

a

Figure 9.7 (a) The surface of the North Atlantic is raised to form a low hill. The eastward rotation of the Earth offsets the center of the hill to the west. Water from point B (see also Figures 9.4 and 9.6) turns westward and flows along the side of this hill. The westward-moving water is balanced between the Coriolis effect (which would turn the water to the right) and gravity (which would turn it to the left). Thus, water in a gyre moves along the outside edge of an ocean basin. (b) The average height of the surface of the North Atlantic is shown in color in this image derived from data taken in 1992 by the *TOPEX/Poseidon* satellite. Red indicates the highest surface, green and blue the lowest. Note that the measured position of the hill is offset to the west as seen in (a). The gradually sloping hill is only 2 meters (6.5 feet) high and would not be apparent to anyone traveling from coast to coast.

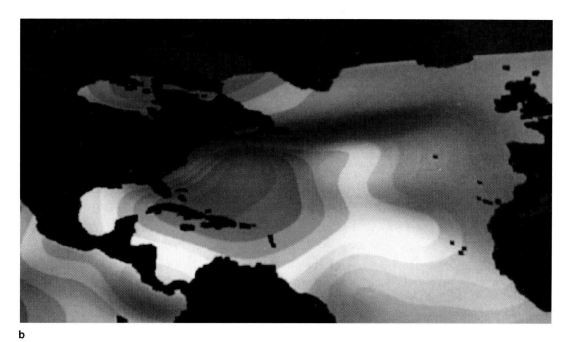

b

dynamically balanced between the force of gravity and Coriolis deflection.

Yes, there really *is* a hill near the middle of the North Atlantic, centered in the area of the Sargasso Sea (see **Figure 9.7b**). This hill is formed of surface water gathered at the ocean's center of circulation. It is not a steep mountain of water—its maximum height is an unspectacular 2 meters (6.5 feet)—but rather a gradual rise and fall from coastline to open ocean and back to opposite coastline. Its slope is so gradual you wouldn't notice it on a transatlantic crossing.

The hill is maintained by wind energy. If the winds did not continuously inject new energy into currents, friction within the fluid mass and with the surrounding ocean basins would slow the flowing water, gradually converting its motion into heat. The balance of wind energy and friction, and of the Coriolis effect and gravity, propels the currents of the gyre and holds them along the outside edges of the ocean basin.

Geostrophic Gyres

Gyres in balance between gravity and the Coriolis effect are called **geostrophic** gyres (*Geos* = Earth, *strophe* = turning), and their currents are geostrophic currents. Because of the patterns of driving winds and the present positions of continents, the geostrophic gyres are largely independent of each other in each hemisphere.

There are six great current circuits in the world ocean, two in the Northern Hemisphere and four in the Southern Hemisphere. They are shown in **Figure 9.8.** Five are geostrophic gyres: the North Atlantic gyre, the South Atlantic gyre, the North Pacific gyre, the South Pacific gyre, and the Indian Ocean gyre. Though it is a closed circuit, the sixth and largest current is technically not a gyre because it does not flow around the periphery of an ocean basin. The **West Wind Drift,** or Antarctic Circumpolar Current, as this exception is called, flows endlessly eastward around Antarctica, driven by powerful, nearly

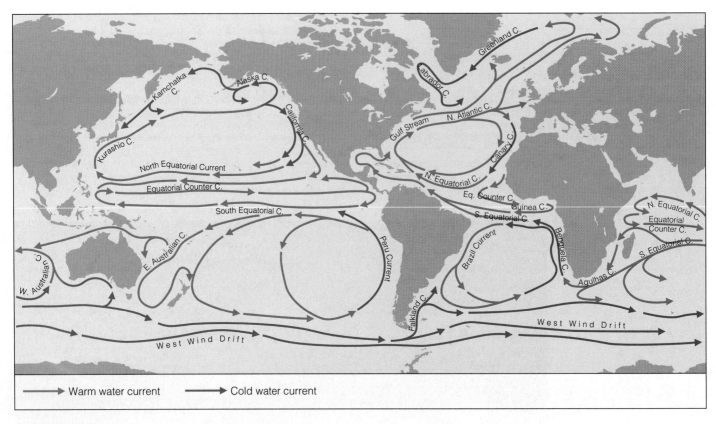

→ Warm water current   → Cold water current

Figure 9.8 The major surface currents of the world ocean.

ceaseless westerly winds. This greatest of all surface ocean currents is never deflected by a continental mass.

We might expect the two gyres in the North and South Pacific (and the two gyres in the North and South Atlantic) to converge exactly at the geographic equator. However, as Figure 9.8 shows, the junction of equatorial currents lies a few degrees north of the geographical equator, at the meteorological equator. As noted in Chapter 8, the meteorological equator and the intertropical convergence zone (the band at which the trade winds converge) are displaced 5°–8° northward primarily because of the heat accumulated in the Northern Hemisphere's greater tropical land surface area. Ocean circulation, like atmospheric circulation, is balanced around the meteorological equator.

Currents Within Gyres

Because of the different factors that drive and shape them, the currents constituting geostrophic gyres have different characteristics. Geostrophic currents may be classified by their position within the gyre as western boundary currents, eastern boundary currents, or transverse currents.

Western Boundary Currents   The fastest and deepest geostrophic currents are found at the western boundaries

of ocean basins (that is, off the east coast of continents). These narrow, fast, deep currents move warm water poleward in each of the gyres. There are five large **western boundary currents:** The Gulf Stream (in the North Atlantic), the Japan or Kuroshio Current (in the North Pacific), the Brazil Current (in the South Atlantic), the Agulhas Current (in the Indian Ocean), and the East Australian Current (in the South Pacific).

The **Gulf Stream** is the largest of the western boundary currents. Studies of the Gulf Stream, such as those undertaken by the drift submarine *Ben Franklin* (see **Box 9.1**), have revealed that off Miami, the Gulf Stream moves at an average speed of 2 meters per second (5 miles per hour) to a depth of over 450 meters (1,500 feet). Water in the Gulf Stream can move more than 160 kilometers (100 miles) in a day. Its average width is about 70 kilometers (43 miles).

The volume of water transported in western boundary currents is extraordinary. The unit used to express volume transport in ocean currents is the **sverdrup (sv),** named in honor of Harald Sverdrup, one of this century's pioneering oceanographers. A sverdrup equals 1 million cubic meters per second.[1] The Gulf Stream flow

---

[1]One million cubic meters is about one-half the volume of the Louisiana Superdome.

BOX 9.1 ● *Streaming Along*

In July 1969 six researchers sealed themselves into a specially designed submarine for a 2,640-kilometer (1,650-mile) drift within the Gulf Stream, the great ocean current that flows northward off the east coast of the United States. Their vehicle, named the *Ben Franklin* in honor of the man who first took a scientific interest in the current, was designed by Dr. Jacques Piccard, the builder of the bathyscaphe *Trieste* (see Box 4.3).

The 15-meter (50-foot) *Ben Franklin* was towed into the Gulf Stream off West Palm Beach, Florida, and positioned at a depth of 150 meters (500 feet). The month-long program of observations was designed to track the Gulf Stream and measure physical conditions within it. The scientists aboard were in regular contact with two surface ships that followed the drifting submarine and monitored its exact position. They kept a nearly continuous record of water temperature, salinity, and density.

The team made surprising discoveries right from the start. First, the submarine drifted more rapidly than expected. The speed of the surface current often exceeds 8 kilometers (about 5 miles) per hour, but oceanographers believed the current would move more slowly at greater depths. It didn't; the surface ships drifted at about the same rate as the submarine. And the drift was not always smooth. Underwater hills looming from the seabed, many of them previously uncharted, disrupted the flow of the submerged current and caused the submarine to rise and fall alarmingly. Gulf Stream eddies

also caused problems. Thirteen days into the mission, a huge eddy spun the submarine 80 kilometers (50 miles) out of the Gulf Stream core, forcing it to surface in order to be towed back into the center of the Stream. These great eddies were found to reach much deeper than previously believed (we now know that they can reach all the way to the ocean bottom).

Another surprise was the general lack of marine life. The crew did sight several species of sharks, a huge jellyfish with tentacles 10 centimeters (4 inches) thick, and an aggressive broadbilled swordfish that jousted with their vessel before swimming away unharmed; but this was less life than they had expected. The biology program did make some strides, however. Tape recordings of underwater sounds of biological origin were studied on board, and some of these recordings were later used in the first detailed analyses of whale vocalizations. Dr. Piccard photographed some unusual bioluminescent organisms, a few of which were new to science. The crew also completed medical and psychological investigations on each other and analyzed the performance of the vehicle's life support systems. These studies were used in later submarine and spacecraft design.

When the *Ben Franklin* finally bobbed to the surface 510 kilometers (310 miles) south of Halifax, Nova Scotia, researchers were confident of their biggest finding: They hadn't known as much about the Gulf Stream as they thought they did.

The submarine *Ben Franklin* enters New York harbor in August 1969 after drifting for 31 days within the Gulf Stream.

is at least 55 sv (55 million meters per second), about 300 times the usual flow of the Amazon, the greatest of rivers. In **Figure 9.9** the surface currents of the North Atlantic gyre are shown with their volume transport (in sverdrups) indicated.

Water in a current, especially a western boundary current, can move for surprisingly long distances within well-defined boundaries. In the Gulf Stream the current-as-river analogy can be startlingly apt: the western edge of the current is often clearly visible. Water within the current is usually warm, clear, and blue, often depleted of nutrients and incapable of supporting much life. By contrast, water over the continental slope adjacent to the current is often cold, green, and teeming with life.

Long, straight edges are the exception rather than the rule in western boundary currents, however. Unlike rivers, ocean currents lack well-defined banks, and friction with adjacent water can cause a current to form waves along its edges. Western boundary currents meander as they flow poleward. The looping meanders sometimes connect to form turbulent rings, or **eddies,** that trap cold or warm water in their centers and then separate from the main flow. For example, *cold-core eddies* form in the Gulf Stream as it meanders southward upon leaving the coast of North America off Cape Hatteras (see **Figure 9.10a**). Warm-core eddies can form north of the Gulf Stream when the warm current loops into the cold water lying to the north (see **Figure 9.10b**). Because of the mechanics of their formation, warm-core eddies rotate clockwise, and cold-core eddies rotate counterclockwise.

The slowly rotating eddies move away from the current and are distributed across the North Atlantic. Some may be 1,000 kilometers (625 miles) in diameter and retain their identity for more than three years. In midlatitudes as much as one-fourth of the surface of the North Atlantic may consist of old cold-core eddy remnants! Both cold and warm eddies are visible in the satellite image of **Figure 9.11**. Recent research suggests that their influence reaches to the seafloor. Warm- and cold-core eddies may be responsible for the slowly moving *abyssal storms* that leave often-observed ripple marks in deep sediments.

Eastern Boundary Currents   There are five **eastern boundary currents** found at the eastern edge of ocean basins (that is, off the west coast of continents): the Canary Current (in the North Atlantic), the Benguela Current (in the South Atlantic), the California Current (in the North Pacific), the West Australian Current (in the Indian Ocean), and the Peru or Humboldt Current (in the South Pacific).

Eastern boundary currents are the opposite of their western boundary counterparts in nearly every way:

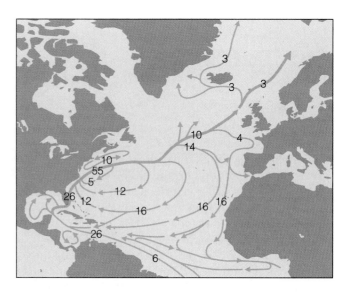

Figure 9.9   The general surface circulation of the North Atlantic. The numbers indicate flow rates in sverdrups (1 sv = 1 million cubic meters of water per second).

They carry cold water equatorward; they are shallow and broad, sometimes more than 1,000 kilometers (625 miles) across; their boundaries are not well defined; and eddies tend not to form. Their total flow is less than that of their western counterparts. The Canary Current in the North Atlantic carries only 16 sv of water at about 2 kilometers (1.2 miles) per hour. The current is so shallow and broad that sailors may not notice it. Contrast the flow rates of the North Atlantic's western and eastern boundary currents in Figure 9.9. **Table 9.1** summarizes the major differences between boundary currents in the Northern Hemisphere.

Westward Intensification   Why should western boundary currents be concentrated and eastern boundary currents be diffuse? One reason is the converging flow of the trade winds on either side of the equator. Water moved by the trades approaches the meteorological equator and is shepherded west, where it piles up at the western edge of the basin before turning swiftly poleward. This concentration of water produces the poleward-moving western boundary currents. In contrast, the westerly winds of each hemisphere do not converge, and water driven by them is not swept along a line of convergence. Coriolis deflection can therefore move some of the eastward-moving water equatorward before the basin's eastern boundary is reached.

A second reason is the rotation of the Earth itself. The hill described in Figure 9.7 is offset to the west because of the Earth's eastward rotation, so water must squeeze closer to the ocean basin's western edge to pass around

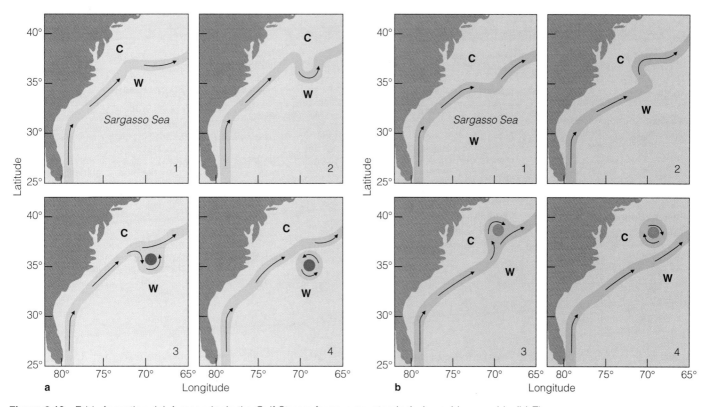

Figure 9.10 Eddy formation. (a) A meander in the Gulf Stream forms a counterclockwise cold-core eddy. (b) The eddy formed here rotates clockwise and has a warm core (C = cold water, W = warm water; blue = cold, red = warm).

Figure 9.11 The Gulf Stream viewed from space. This image is a composite of temperature data returned from NOAA polar orbiting meteorological satellites during the first week of April 1984. The composite image is printed with an artificial color scale: Reds and oranges are a warm 24–28°C (76–84°F); yellows and greens are 17–23°C (63–74°F); blues are 10–16°C (50–61°F); and purples are a cold 2–9°C (36–48°F). The Gulf Stream appears like a red (warm) river as it moves from the southern tip of Florida (1) north along the East Coast. Moving offshore at Cape Hatteras (2), it begins to meander, with some meanders pinching off to form warm-core (3) and cold-core (4) eddies. As it moves northeastward, the water cools dramatically, releasing heat to the atmosphere and mixing with the cooler surrounding waters. By the time it reaches the middle of the North Atlantic, it has cooled so much that its surface temperature can no longer be distinguished from that of the surrounding waters.

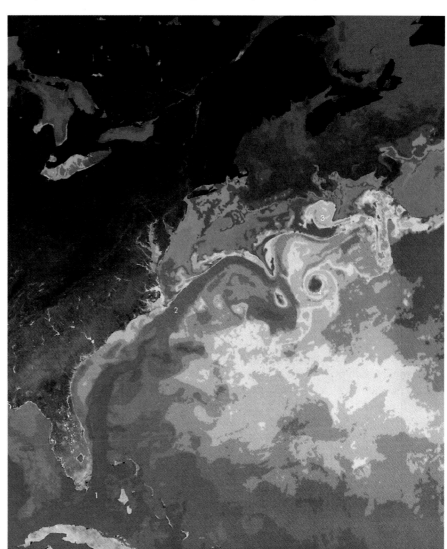

Table 9.1	Boundary Currents in the Northern Hemisphere				
Type of Current (example)	General Features	Speed	Transport (millions of cubic meters per second)	Special Features
Western Boundary Currents Gulf Stream, Japan Current	**Warm** Narrow, <100 km Deep—substantial transport to depths of 2 km	Swift, hundreds of kilometers per day	Large, usually 50 sv or greater	Sharp boundary with coastal circulation system; little or no coastal upwelling; waters tend to be depleted in nutrients, unproductive, waters derived from trade wind belts
Eastern Boundary Currents California Current, Canary Current	**Cold** Broad, ~1,000 km Shallow, <500 m	Slow, tens of kilometers per day	Small, typically 10–15 sv	Diffuse boundaries separating from coastal currents; coastal upwelling common; waters derived from mid-latitudes

the hill at the western boundary. The combined effect on current flow is known as **westward intensification,** a phenomenon clearly visible in Figure 9.9.

Transverse Currents As we have seen, most of the power for ocean currents is derived from the trade winds at the fringes of the tropics and from the mid-latitude westerlies. The stress of winds on the ocean in these bands gives rise to the **transverse currents**—currents that flow from east to west and west to east, thus linking the eastern and western boundary currents.

The trade-wind-driven North and South Equatorial currents in the Atlantic and Pacific are moderately shallow and broad, but each transports about 30 sv westward. Because of the thrust of the trades, Atlantic water at Panama is usually 20 centimeters (8 inches) higher, on average, than water across the isthmus in the Pacific. The Pacific's greater expanse of water at the equator and stronger trade winds develop more powerful westward-flowing equatorial currents, and the height differential between the western and eastern Pacific is thought to approach 1 meter (3.3 feet)!

Westerly winds drive the eastward-flowing transverse currents of the mid-latitudes. Because they are not shepherded by the trade winds, eastward-flowing currents are wider and flow more slowly than their equatorial counterparts. The North Pacific and North Atlantic currents are Northern Hemisphere examples.

As can be seen in Figure 9.8, the westward flow of the transverse currents near the equator proceeds unimpeded for great distances, but the eastward flow of transverse currents at middle and high latitudes in the northern ocean basins is interrupted by continents and island

arcs. In the far south, however, eastward flow is almost completely free. Intense westerly winds over the southern ocean drive the greatest of all ocean currents, the unobstructed West Wind Drift (or Antarctic Circumpolar Current). This current carries more water than any other—at least 100 sv west-to-east in the Drake Passage between the tip of South America and the adjacent Palmer Peninsula of Antarctica.

Countercurrents and Undercurrents

Equatorial currents are typically accompanied by **countercurrents** flowing on the surface in the opposite direction from the main current. As you may remember, at the meteorological equator air is rising and the trade winds do not blow across the ocean. Without persistent winds to drive water to the west, some backward flow of water occurs at the meteorological equator. Some water flows away from the main equatorial currents (which flow westward a bit north and south of the meteorological equator) to return on the surface to the east. Look for this in the Pacific in Figure 9.8.

Countercurrents can also exist *beneath* surface currents. These are sometimes referred to as **undercurrents.** The first undercurrent was discovered in 1951 in the central Pacific by Townsend Cromwell, a researcher employed by the U.S. Fish and Wildlife Service to investigate deep long-line fishing techniques pioneered by the Japanese. The Pacific Equatorial Undercurrent, or Cromwell Current, flows beneath the North Equatorial Current with an average velocity of 5 kilometers (3 miles) per hour at a depth of 100–200 meters (330–660 feet). It is about 300 kilometers (190 miles) wide and carries a volume equivalent to about half the Gulf Stream. It has been traced for over 14,000 kilometers (8,700 miles), from New Guinea to Ecuador. Undercurrents have since been found under most major currents. Their volumes are often a significant percentage of the current above.

Undercurrents can influence conditions at the ocean surface. The ocean around the Galápagos Islands is cold at the surface, even though the islands straddle the equator. For many years the cold was blamed on Antarctic water driven north in the Peru (Humboldt) Current, but recent research has shown that this surface water turns west before reaching the islands. The chill is actually caused by the upwelling of the Pacific Equatorial Undercurrent (the Cromwell Current) near the eastern end of its travel. Once at the surface, the cool water moves toward the west as part of the South Equatorial Current. The Galápagos Islands lie in its path.

Oceanographers are uncertain about what causes undercurrents, but near the equator (where the Coriolis effect is negligible) some water is not deflected north or south when it reaches the west side of the basin. It stays on the equator and underflows the surface current back to the east. This eastward-flowing current is kept on course by the Coriolis effect. If the current veers north from the equatorial region, the Coriolis effect bends it to the right, or south, back to its original course. Should the current wander too far south, the Coriolis effect in the Southern Hemisphere deflects the water to the left (north).

Exceptional Surface Currents

Some currents don't behave in ways you might first expect. Local or seasonal variations in wind and geography sometimes intrude on the flow patterns we have discussed. Three of these exceptions are noted here.

El Niño   Surface winds across most of the tropical Pacific normally move from east to west (see Figures 8.9 and 8.11). The trade winds blow from the normally high-pressure area over the eastern Pacific (near Central and South America) to the normally stable low-pressure area over the western Pacific (north of Australia). However, for reasons that are still unclear, these pressure areas change places at irregular intervals of roughly three to eight years: High pressure builds in the western Pacific, and low pressure dominates the eastern Pacific. Winds across the tropical Pacific then reverse direction and blow from west to east: The trade winds weaken or reverse. This change in atmospheric pressure (and thus in wind direction) is called the **Southern Oscillation.**

The trade winds normally drag huge quantities of water westward along the ocean's surface on each side of the equator, but as the winds weaken these equatorial currents crawl to a stop. Warm water that has accumulated at the western side of the Pacific can then return east along the equator toward the coast of Central and South America. The eastward-flowing warm water usually arrives near the South American coast around Christmas time; hence the current's name, **El Niño,** "the Christ child." The phenomena of the Southern Oscillation and El Niño are coupled; so the terms are often combined to form the acronym **ENSO,** for El Niño/Southern Oscillation. An ENSO event lasts about a year. The effects are felt not only in the Pacific; all ocean areas at trade wind latitudes in both hemispheres can be affected.

Normally, a current of cold water, rich in upwelled nutrients, flows north and west away from the South American continent (see **Figure 9.12**). When the propelling trade winds falter during an ENSO event, warm equatorial water that would normally flow westward in the equatorial Pacific backs up to flow east (see **Figure 9.13**). Much of the flow is in the nature of a greatly strengthened equatorial countercurrent, but oceanographers have recently discovered that the Pacific Equatorial Undercurrent also increases greatly in volume during these times. The normal northward flow of the cold Peru Current is interrupted or overridden by the warm water. The nutrient-laden Peru Current is responsible for the great biological productivity of the ocean off the coasts of Peru and Chile. When this current slows, fish and seabirds dependent on the abundant life it contains die or migrate elsewhere.

During major ENSO events, sea level rises in the eastern Pacific, sometimes by as much as 20 centimeters (8 inches) in the Galápagos. Water temperature also increases by up to 7°C (11°F). The warmer water causes more evaporation, and the area of low atmospheric pressure over the eastern Pacific intensifies. Humid air rising in this zone, centered some 2,000 kilometers (1,300 miles) west of Peru, causes high precipitation in normally dry areas. The increased evaporation intensifies coastal storms, and rainfall inland may be much higher than normal. Marine and terrestrial habitats and organisms can be affected by these changes.

The most severe ENSO event in this century occurred in 1982–83. Effects associated with El Niño were spectacular over much of the Pacific. Australia suffered a year of debilitating drought; agricultural losses were estimated at $2.5 billion. Ecuador and Peru experienced severe flooding, and many rainfall records (and piers) were broken along the coast of California; global economic impact exceeded $8 billion. Sea level at San Diego rose 10 centimeters (4 inches) higher than normal. Two years later, water temperatures off the west coast of the United States were still unusually warm, and many exotic marine species were reported. Some birds that had not been seen in southern California for decades returned.

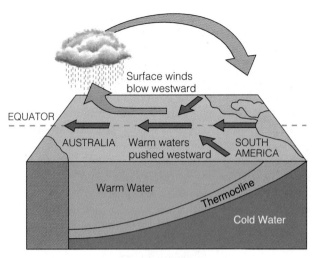

a   **Normal Conditions**

Figure 9.12 A non-El Niño year. (a) Normally the air and surface water flow westward, the thermocline rises, and upwelling of cold water occurs along the west coast of Central and South America. (b) This map from satellite data shows the temperature of the equatorial Pacific on 31 May 1988. The warmest water is indicated by the dark red, and progressively cooler water by yellow and green. Note the coastal upwelling along the coast at the lower right of the map, and the tongue of recently upwelled water extending westward along the equator from the South American coast.

b

Anglers along the west coast were astonished to find fish species typical of Mexican waters dangling from their hooks. The El Niños of 1985–86 and 1992–1994 were not as pronounced.

Studies of the ocean and atmosphere in 1982–83 have given researchers new insight into the behavior and effects of the Southern Oscillation. Some researchers believe that the 1982–83 event was triggered by the violent 1982 eruption of El Chichón, a Mexican volcano, which injected huge quantities of obscuring dust and sulfur-rich gases into the atmosphere. Others have recently suggested that the Southern Oscillation may be triggered by pulses of heat entering the Pacific at crustal spreading zones near the equator. Some of this heat is moved by currents to the ocean surface. The warming of surface water may cause air above the ocean to warm, atmospheric pressure to fall, and the east-to-west trade winds to slacken. Though the exact cause or causes of the Southern Oscillation are not yet understood, subtle changes in the atmosphere permit meteorologists to predict a severe El Niño nearly a year in advance of its most serious effects.

Monsoon Currents Another set of unusual movements occurs in the northern Indian Ocean. From November to March the northeast trade winds propel water in the usual way, and normal surface circulation results. In the summer months, however, the intertropical convergence zone moves north. Winds over the Indian Ocean change direction to blow from the south and southwest (see again Figure 8.12). Water in the relatively small northern Indian Ocean begins to flow clockwise, and the normally westward-flowing North Equatorial Current south of India reverses flow to become the eastward-moving Southwest Monsoon Current. (Due to its temporary nature, this current is not shown in Figure 9.8.)

High-Latitude Currents A third exception to expected surface flow is seen in the many lesser current circuits at high northern latitudes in Figure 9.8. Some of these—such as the Greenland and Labrador currents in the North Atlantic and the Kamchatka and Alaska Currents in the North Pacific—move in apparent contradiction to the influence of the Coriolis effect, gravity, and friction. These currents form when larger currents are split and deflected by collision with a continent. The polar easterly winds at these latitudes keep the water moving westward. The size of these high-latitude currents on the map is exaggerated by the distortion inherent in the Mercator projection (see Appendix IV). The direction and rate of flow of these smaller currents are controlled primarily by the tendency of water to flow

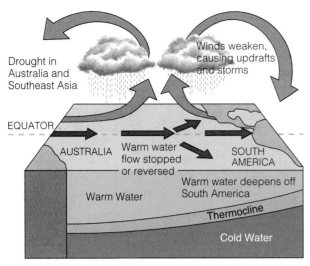

a   **El Niño Conditions**

Figure 9.13 An El Niño year. (a) When the Southern Oscillation develops, the trade winds diminish and then reverse, leading to an eastward movement of warm water along the equator. The surface waters of the central and eastern Pacific become warmer, and storms over land may increase. (b) Sea-surface temperatures on 13 May 1992, a time of El Niño conditions. The thermocline was deeper than normal, and equatorial upwelling was suppressed. Note the absence of coastal upwelling along the coast and the lack of the tongue of recently upwelled water extending westward along the equator.

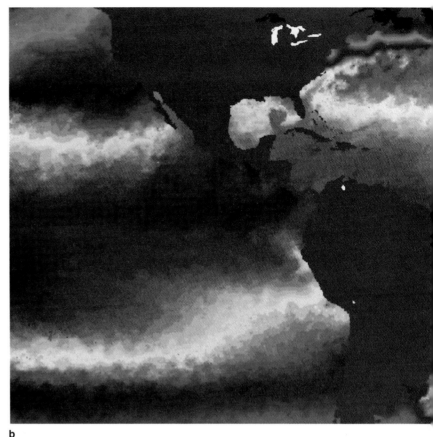

b

around obstacles rather than by the direct action of winds or Coriolis deflection. Technically these are not geostrophic currents, but their importance is often great. The Greenland and Labrador currents, for example, transport icebergs to the Atlantic shipping lanes and deliver nutrients to the rich fishing grounds off New-foundland.

Effects of Surface Currents on Climate

Surface currents distribute tropical heat worldwide. Warm water flows to higher latitudes, transfers heat to the air and cools, moves back to low latitudes, absorbs heat again, and the cycle repeats. The greatest amount of heat transfer occurs at mid-latitudes, where about 10 million billion calories of heat are transferred each second—more than a million times the power consumed by all the world's human population in the same length of time! This combination of water flow and heat transfer from and to water influences climate and weather in several ways.

Commercial airliners flying to Europe from U.S. West Coast cities fly over central Ontario. In winter and spring months, the ground is hidden beneath masses of ice and snow, but passengers can often make out the frozen surface of James Bay. When those airliners land in London or Edinburgh, their passengers find a much milder climate than the barren whiteness of central Canada. Yet

London's latitude of 51°N is the same as that of the southern tip of James Bay. Edinburgh lies at 55°N, the same latitude as Ontario's Polar Bear Provincial Park, a sanctuary for migrating polar bears. Why the difference in climate? The predominant direction of air flow is over the water eastward toward the British Isles. Even in winter Edinburgh and London are bathed in air only recently in contact with the relatively warm North At-lantic Current. Scotland and England therefore have a maritime climate, and their only polar bears are found in zoos. These cities are warmed in part by the energy of tropical sunlight transported to their high latitudes by the Gulf Stream (see again Figure 9.8).

At lower latitudes on an ocean's eastern boundary the situation is often reversed. Mark Twain is supposed to have said that the coldest winter he ever spent was a summer in San Francisco. Summer months in that West Coast city are cool, foggy, and mild, while Washington, D.C., on nearly the same line of latitude (but on the western boundary of an ocean basin), is known for its August heat and humidity. Why the difference? Look at Figure 9.8 and follow the currents responsible. The California Current, carrying cold water from the north, comes close to the coast at San Francisco. As shown in **Figure 9.14,** air normally flows clockwise in summer around an offshore zone of high atmospheric pressure. Wind approaching

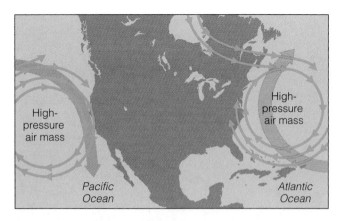

Figure 9.14 General summer air circulation patterns of the east and west coasts of the United States. Warm ocean currents are shown in red, cold currents in blue. Air is chilled as it approaches the west coast and warmed as it approaches the east coast.

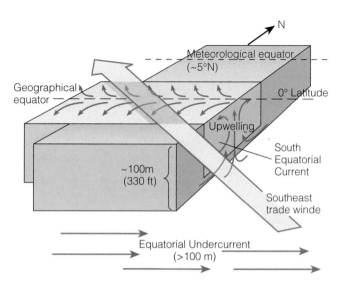

Figure 9.15 Equatorial upwelling. The South Equatorial Current, especially in the Pacific, straddles the geographic equator (see again Figure 9.8). Water north of the equator veers to the right (northward), and water to the south veers to the left (southward). Surface water therefore diverges, causing upwelling. Most of the upwelled water comes from the area above the equatorial undercurrent, at depths of 100 meters or less. As can be seen in Figure 9.12, the phenomenon of equatorial upwelling is pronounced in the Pacific.

the California coast loses heat to the cold sea and comes ashore to chill San Francisco. Summer air often flows around a similar high off the East Coast (the Bermuda High). Winds approaching Washington, D.C., therefore blow from the south and east. Heat and moisture from the Gulf Stream contribute to the capital's oppressive summers. (In winter, on the other hand, Washington D.C. is colder than San Francisco because westerly winds approaching Washington are chilled by the cold continent they cross.)

WIND-INDUCED VERTICAL CIRCULATION

The wind-driven *horizontal* movement of water can sometimes induce *vertical* movement in the surface water. This movement is called **wind-induced vertical circulation.** Upward movement of water is known as **upwelling;** the process brings deep, cold, usually nutrient-laden water toward the surface. Downward movement is called **downwelling.**

Equatorial Upwelling

Because the meteorological equator usually lies about 5° north of the geographic equator, the South Equatorial currents of the Atlantic and Pacific straddle the geographic equator. Though the Coriolis effect is weak near the geographic equator (and absent *at* the geographic equator), water moving in the currents on either side of the geographic equator is deflected slightly poleward and replaced by deeper water (**Figure 9.15**). Thus, **equatorial upwelling** occurs in these westward-flowing equatorial surface currents. Upwelling is an important process because this deep water is often rich in the nutrients needed by marine organisms for growth. The long,

thin band of upwelling and biological productivity extending along the equator westward from South America is clearly visible in Figures 9.12 and 9.20. The layers of radiolarian ooze on the equatorial Pacific seabed (Figure 5.10) are testimony to the biological productivity of surface water there. By contrast, generally poor conditions for growth prevail in most of the open tropical ocean—because strong layering isolates deep, nutrient-rich water from the sunlit ocean surface.

Coastal Upwelling

Wind blowing parallel to shore or offshore can cause what is known as **coastal upwelling.** The friction of wind blowing along the ocean surface causes the water to begin moving, the Coriolis effect deflects it to the right (in the Northern Hemisphere), and the resultant Ekman transport moves it offshore. As shown in **Figure 9.16,** coastal upwelling occurs when this surface water is replaced by deep water rising along the shore. Again, because the new surface water is often rich in nutrients, prolonged wind can result in increased biological productivity. Coastal upwelling along the coast of South America is visible in the lower right corner of Figure 9.12. (Other examples will be seen in Figures 9.20 and 14.3).

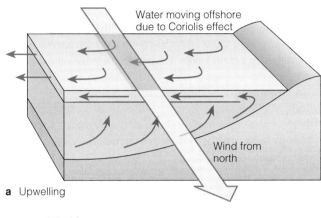

a Upwelling

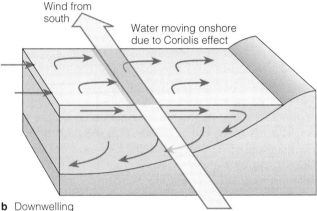

b Downwelling

Figure 9.16 Coastal upwelling and downwelling. (a) In the Northern Hemisphere, coastal upwelling can be caused by winds from the north blowing along the west coast of a continent. Water moving offshore is replaced by cold, deep, nutrient-laden water. (b) A prolonged southerly wind along a Northern Hemisphere west coast can result in downwelling.

Upwelling can also influence weather. Wind blowing from the north along the California coast causes offshore movement of surface water and subsequent coastal upwelling. The overlying air becomes chilled, contributing to San Francisco's famous fog banks and cool summers (see again Figure 9.14). Wind-induced upwelling is also common in the Peru Current, along the west coast of Antarctica's Palmer Peninsula, in parts of the Mediterranean, and near some large Pacific islands.

Downwelling

Water driven toward a coastline will be forced downward, returning seaward along the continental shelf. This downwelling (**Figure 9.16b**) helps supply the deeper ocean with dissolved gases and nutrients, and it assists in the distribution of living organisms. Unlike upwelling, downwelling has no direct effect on the climate or productivity of the adjacent coast.

THERMOHALINE CIRCULATION

The surface currents we have discussed affect the uppermost layer of the world ocean (about 10% of its volume), but horizontal and vertical currents also exist below the pycnocline in the ocean's deeper waters. The slow circulation of water at great depths is driven by density differences rather than by wind energy. Because density is largely a function of water temperature and salinity, the movement of water due to differences in density is called **thermohaline circulation** (*therme* = heat, *halos* = salt). Virtually the entire ocean is involved in slow thermohaline circulation, a process responsible for most of the vertical movement of ocean water.

Water Masses

As you may recall from Chapter 7, the ocean is density stratified, with the most dense water near the seafloor and the least dense near the surface. Each water mass has specific temperature and salinity characteristics. Density stratification is most pronounced at temperate and tropical latitudes because the temperature difference between surface water and deep water is greater there than near the poles.

The water masses possess distinct, identifiable properties. Like air masses, water masses don't mix easily when they meet; instead, they flow above or beneath each other. Water masses can be remarkably persistent and will retain their identity for great distances and long periods of time. Oceanographers name water masses according to their relative position.

In temperate and tropical latitudes, there are five common water masses:

- *Surface water*, to a depth of about 200 meters (650 feet)
- *Central water*, to the bottom of the main thermocline (which varies with latitude)
- *Intermediate water*, to about 1,500 meters (5,000 feet)
- *Deep water*, water below intermediate water but not in contact with the bottom, to a depth of about 4,000 meters (13,000 feet)
- *Bottom water*, water in contact with the seafloor

Surface currents move in the relatively warm upper environment of surface and central water. The boundary between central water and intermediate water is the most abrupt and pronounced.

No matter at what depth they are located, the characteristics of each water mass have been determined by conditions of heating, cooling, evaporation, and dilution that occurred at the ocean *surface* when the mass was formed. The heaviest (and deepest) masses were formed

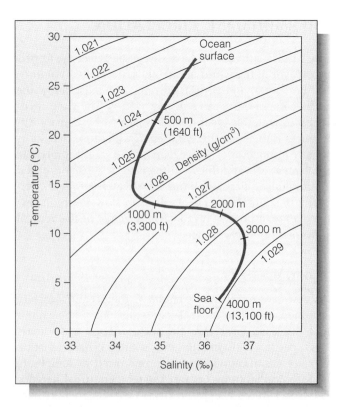

Figure 9.17 A general temperature–salinity (T-S) diagram, with density curves added. The S curve tracks the combination of temperature and salinity (and therefore density) with increasing depth. Note that the top of the S curve represents the temperature and salinity (and therefore density) of water at the ocean surface; the bottom of the curve represents the density of water at the ocean floor. Intermediate depths are shown on the curve.

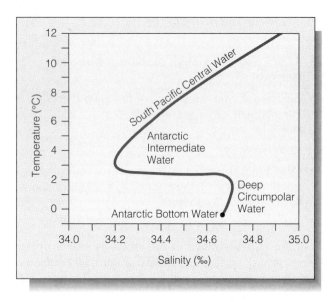

Figure 9.18 A typical T-S diagram for the South Pacific Ocean. As in Figure 9.17, depth increases along the blue line from ocean surface to seafloor.

by surface conditions that caused water to become very cold and salty. Water masses near the surface are warmer and less saline; they may have formed in warm areas where precipitation exceeded evaporation. Water masses at intermediate depths are intermediate in density.

In spite of this differentiation, the relatively cold water masses lying beneath the thermocline exhibit smaller variations in salinity and temperature than the water in the currents that move across the ocean's surface.

The Temperature–Salinity Diagram

Perhaps the best way to visualize ocean layering is with a **temperature–salinity (T-S) diagram** like the one in **Figure 9.17.** The S-shaped curve through the center of the figure shows the temperature and salinity of water at each depth indicated. Note that many combinations of temperature and salinity can yield the same density, and that the density of the water tends to increase with depth. The shape of the curve is governed by the position and nature of water masses. **Figure 9.18** is a typical T-S diagram for the South Pacific Ocean. It shows that salinity increases abruptly at the transition from Antarctic Intermediate Water to Deep Circumpolar Water, and that temperature drops rapidly between Deep Circumpolar Water and Antarctic Bottom Water.

Formation and Downwelling of Deep Water

Antarctic Bottom Water, the most distinctive of all deep-water masses, is characterized by a salinity of 34.65‰, a temperature of –0.5°C (30°F), and a density of 1.0279 grams per cubic centimeter. This water is noted for its extreme density (the densest in the world ocean), for the great amount of it produced near Antarctic coasts, and for its ability to migrate north along the seafloor.

Most Antarctic Bottom Water forms in the Weddell Sea during winter. Sea ice can incorporate only about 15% of seawater's salt, and the salt remaining in the surrounding cold water forms a frigid brine. Between 20 and 50 million cubic meters of this brine forms every second! The water's great density causes it to sink. As it descends it mixes with nearly equal parts of water from the southern West Wind Drift.

The mixture settles along the edge of Antarctica's continental shelf and spreads along the deep-sea bed, creeping north in slow sheets. Antarctic Bottom Water flows many times more slowly than the water in surface currents: In the Pacific it may take 1,000 years to reach the equator. Six hundred years later it may be as far away as the Aleutian Islands at 50°N! Antarctic Bottom Water also flows into the Atlantic Ocean basin, where it flows north at a faster rate than in the Pacific. Antarctic Bottom Water has been identified as high as 40° *north* latitude on the Atlantic floor, a journey that has taken some 750 years.

Some dense bottom water also forms in the northern polar ocean, but the topography of the Arctic Ocean basin prevents most of it from escaping. Thus a much smaller volume of cold, salty water reaches the North Pacific ocean floor. **Figure 9.19** is a typical T-S diagram for the North Pacific. Note that the abrupt transition to a very dense layer of bottom water seen in Figure 9.18 is missing here.

Other distinct deep-water masses exist. Their positions in relation to each other are always determined by their relative densities. North Atlantic Deep Water forms when the relatively warm and salty North Atlantic Ocean cools as cold winds from northern Canada sweep over it. Exposed to the chilled air, water at the latitude of Iceland releases heat, cools from 10°C to 2°C (50°F to 36°F), and sinks. (Transferred to the air, this bonus heat helps to moderate European winters.) As can be seen in **Figure 9.20,** similar water forms at the West Wind Drift in the South Atlantic. Pacific Deep Water also forms in the Pacific, in the West Wind Drift, and along the east coast of the Kamchatka Peninsula. Atlantic or Pacific

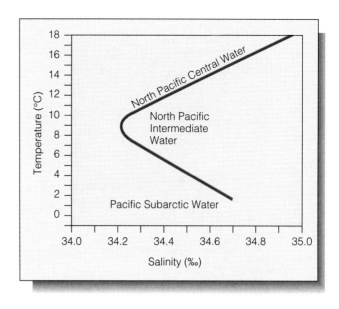

Figure 9.19 A typical T-S diagram for the North Pacific Ocean.

Figure 9.20 Places where surface water sinks and where deep or intermediate water rises.

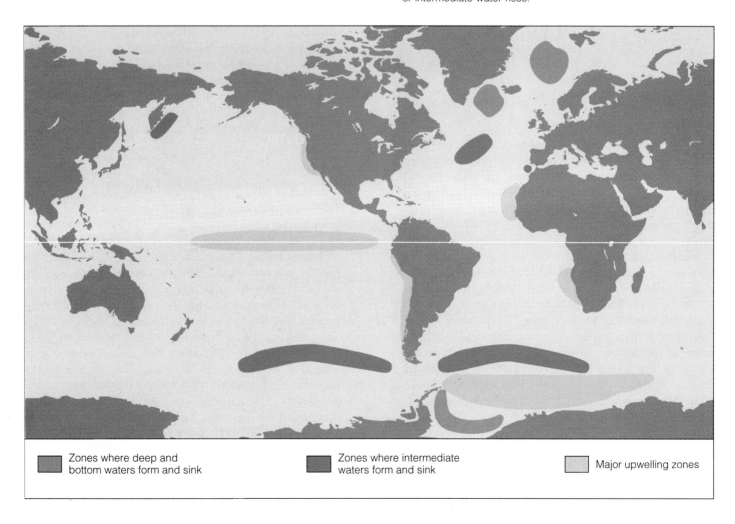

Zones where deep and bottom waters form and sink	Zones where intermediate waters form and sink	Major upwelling zones

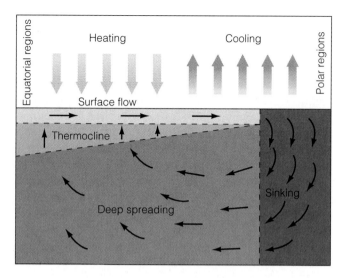

Figure 9.22 The classical model of a pure thermohaline circulation, caused by heating in lower latitudes and cooling in higher latitudes.

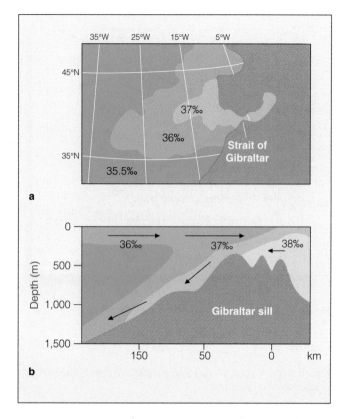

Figure 9.21 Saline flow over the sill of the Mediterranean Sea at Gibraltar. (a) A map showing the Strait of Gibraltar; salinity was measured at 1,000 meters. (b) A vertical section at the strait, showing the movement of water masses of different salinity.

Deep Water is less dense than Antarctic Bottom Water and so floats above it, out of contact with the ocean floor.

A deep-water mass also forms in the enclosed Mediterranean Sea, where surface water is made more saline by the excess of evaporation over freshwater input. About 300,000 cubic kilometers (72,000 cubic miles) more water evaporates annually from the Mediterranean than is replaced by river runoff or precipitation. In the cool winter months, Mediterranean water with a salinity of about 38‰ flows past the lip of Gibraltar and spreads into the Atlantic as Mediterranean Deep Water (**Figure 9.21**). Mediterranean Deep Water underlies much of the central water mass in the Atlantic, and some of this water can be traced as far south as the basins of the Antarctic. Though saltier than Antarctic Bottom Water or Atlantic Deep Water, Mediterranean Deep Water is considerably warmer and therefore not as dense. It will lie atop the layers of denser water at high southern latitudes.

Researchers can determine the age of deep water by analyzing its dissolved oxygen content. Ocean water picks up free oxygen only at or near the surface by con-

tact with the atmosphere or through the action of photosynthetic plants. After leaving the surface, the water gradually loses its oxygen through the respiration of organisms or by chemical reactions with sediments, rocks, or dissolved components. The dissolved oxygen content of a water mass is therefore a rough index of its age—the length of time since the water left the surface. Researchers have also found that water masses slowly mix with the surrounding water as they flow away from their sources, losing their individual identity as they grow older.

Thermohaline Circulation Patterns

The great quantities of dense water sinking at ocean basin edges must be offset by equal quantities of water rising elsewhere. **Figure 9.22** shows an idealized model of thermohaline flow. Note that water sinks relatively rapidly in a small area where the ocean is very cold, but it rises much more gradually across a very large area in the warmer temperate and tropical zones. It then slowly returns poleward near the surface to repeat the cycle. The continual diffuse upwelling of deep water maintains the existence of the permanent thermocline found everywhere at low and mid-latitudes. This slow upward movement is estimated to be about 1 centimeter (½ inch) per day over most of the ocean. If this rise were to stop, downward movement of heat would cause the thermocline to descend and would reduce its steepness. In a sense, the thermocline is "held up" by the continual slow upward movement of water.

Most features of this ideal circulation pattern exist in

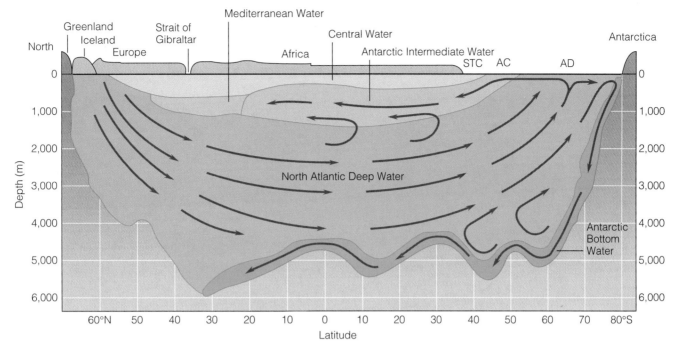

Figure 9.23 The water layers and deep circulation of the Atlantic Ocean. Arrows indicate the direction of water movement. STC indicates the position of the Subtropical Convergence, AC the Antarctic Convergence, and AD the Antarctic Divergence. The surface layer is too thin to show clearly in this scale. The vertical scale is greatly exaggerated. (For a view of the convergence zones from above, please see Figure 18.2.)

nature. **Figure 9.23** shows deep circulation in the Atlantic. The water masses, each of distinct density and sandwiched in layers, are slowly propelled by gravity. Masses butt against one another in **convergence zones,** and the heavier water slides beneath the lighter water.[2] Hundreds of years may pass before water masses complete a circuit or blend to lose their identities. Remember, Antarctic Bottom Water in the Pacific retains its character for up to 1,600 years! The residence time of most deep water is less, however; it takes about 200–300 years to rise to the surface. (By contrast, a bit of surface water in the North Atlantic gyre may take only a little more than a year to complete a circuit.)

Not all thermohaline circulation is so sedate. Ripple marks in sediments, scour lines, and the erosion of rocky outcrops on deep-ocean floors provide evidence that relatively strong, localized bottom currents exist. Some of these currents may move as rapidly as 60 centimeters (24 inches) per second. These relatively fast currents are strongly influenced by bottom topography, and they are sometimes called **contour currents** because their dense

[2]For the location of convergence zones on the ocean surface, please see Figure 18.2.

water flows around (rather than over) seafloor projections. Bottom currents generally move equatorward at or near the western boundaries of ocean basins (below the western boundary surface currents). The subtle density differences between deep-water masses are not capable of moving water at the speed of the wind-driven surface currents. Water in some of these currents may move only 1–2 meters (3–7 feet) per day. Even at that slow speed the Coriolis effect modifies their pattern of flow.

Thermohaline Flow and Surface Flow: The Global Heat Connection

As we have seen, swift and narrow currents along the western margins of ocean basins carry warm tropical surface waters toward the poles. In a few places, as shown in Figure 9.20, the water loses heat to the atmosphere and sinks to become deep water and bottom water. This sinking is most pronounced in the North Atlantic. The cold, dense water moves at great depths toward the Southern Hemisphere, and eventually wells up into the surface layers of the Indian and Pacific oceans. Almost 1,000 years is required for this water to make a complete circuit.

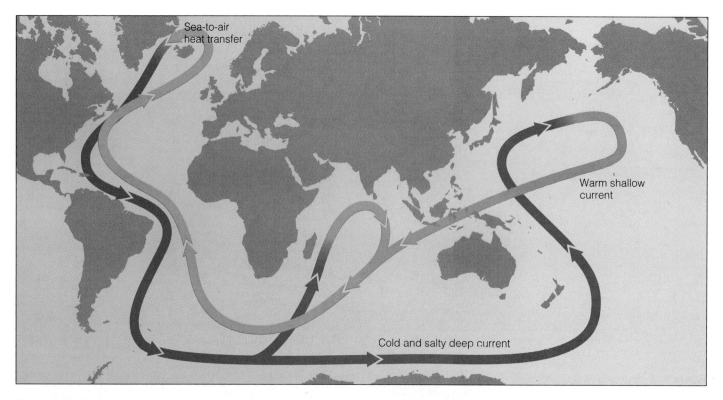

Figure 9.24 A slow global oceanic conveyor belt transports warm surface waters (orange) from the warm areas north of Australia and the Indian Ocean into the North Atlantic, where they cool, sink, and move southward (purple) through the depths of the world ocean. The water gradually upwells to the surface after about 1,000 years.

Marine scientists suspect that the transport of tropical water to the polar regions is part of a global conveyor belt for heat. A simplified outline of the global circuit, the result of two decades of concentrated effort to understand deep circulation, is shown in **Figure 9.24.** This slow circulation straddles the hemispheres and is superimposed upon the more rapid flow of water in surface gyres. Recent analysis of this global circuit suggests that some of the heat warming the European coast enters the ocean in the vicinity of Indonesia and Australia, travels to the Indian Ocean, and enters the Gulf Stream by way of the Agulhas Current rounding the southern tip of Africa. The surface water that leaves the Pacific is driven in part by excess rainfall and river runoff throughout the Pacific basin. The slow, steady, three-dimensional flow of water in the conveyor belt distributes dissolved gases and solids, mixes nutrients, and transports the juvenile stages of organisms between ocean basins.

STUDYING CURRENTS

Traditional methods of measuring currents divide into two broad categories: the float method and the flow method. The **float method** depends on the movement of a drift bottle or other free-floating object. In the **flow method,** the current is measured as it flows past a fixed object.

Surface currents can be traced with drift bottles or drift cards. These tools are especially useful in determining coastal circulation, but they provide no information on the path the drift bottle or card may have taken between its release and collection points. Researchers wishing to know the precise track taken by a drifting object can deploy more elaborate drift devices, such as the drogue arrangement in **Figure 9.25a.** These drogues can be tracked continuously by radio direction finders or radar. Surface currents can also be tracked by noting the difference between the daily expected and observed positions of ships at sea.

Bottle, card, or drogue studies are almost always carefully planned and executed, but not all surface drift releases have been intentional. In May 1990, a violent storm struck the containership *Hansa Carrier* enroute between Korea and Seattle. Twenty-one boxcar-sized cargo containers were lost overboard, among them containers holding 30,910 pairs of Nike athletic shoes. About six months later, shoes from the broken containers began washing up on beaches from British Columbia to Oregon. Because the shoes were not tied together in pairs,

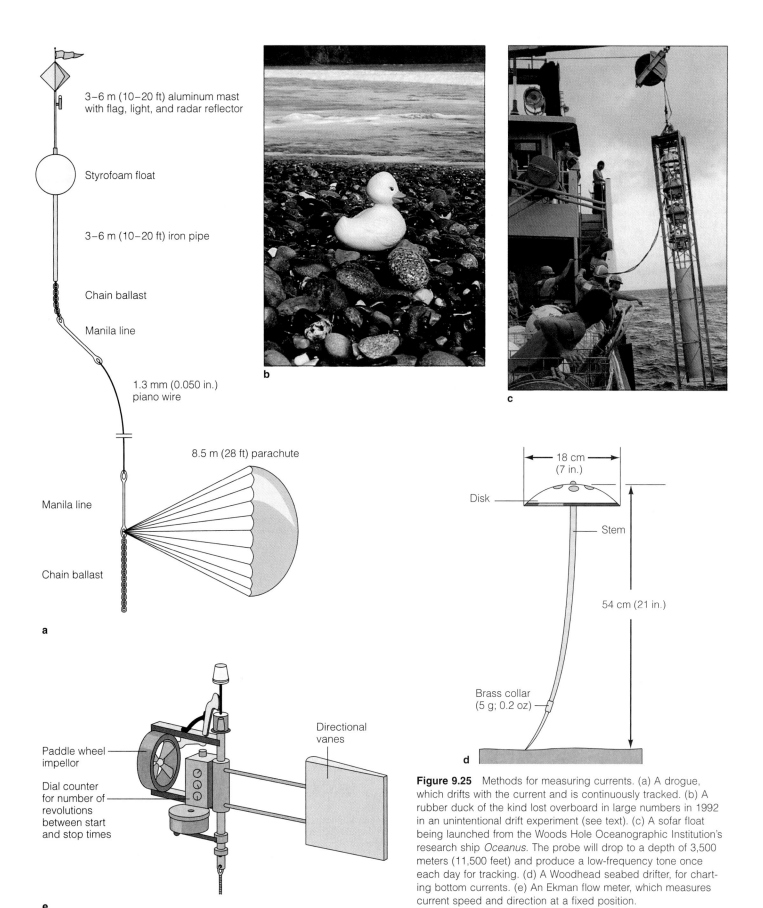

3–6 m (10–20 ft) aluminum mast with flag, light, and radar reflector

Styrofoam float

3–6 m (10–20 ft) iron pipe

Chain ballast

Manila line

1.3 mm (0.050 in.) piano wire

8.5 m (28 ft) parachute

Manila line

Chain ballast

a

b

c

18 cm (7 in.)

Disk

Stem

54 cm (21 in.)

Brass collar (5 g; 0.2 oz)

d

Paddle wheel impellor

Dial counter for number of revolutions between start and stop times

Directional vanes

e

Figure 9.25 Methods for measuring currents. (a) A drogue, which drifts with the current and is continuously tracked. (b) A rubber duck of the kind lost overboard in large numbers in 1992 in an unintentional drift experiment (see text). (c) A sofar float being launched from the Woods Hole Oceanographic Institution's research ship *Oceanus*. The probe will drop to a depth of 3,500 meters (11,500 feet) and produce a low-frequency tone once each day for tracking. (d) A Woodhead seabed drifter, for charting bottom currents. (e) An Ekman flow meter, which measures current speed and direction at a fixed position.

beachcombers placed advertisements in local newspapers and held swap meets to exchange the shoes (which were in excellent condition despite having been exposed to the ocean). Oceanographers noticed the ads and asked the media to request that individuals let them know where and when they found shoes. By knowing the place where the shoes were lost and the places where they were found, researchers have been able to refine their computer models of the North Pacific gyre. Some of the shoes have completed a full circuit of the North Pacific. A similar spill occurred on 10 January 1992, when a storm-beset freighter lost a container filled with 29,000 rubber ducks, turtles, and other bathtub toys in the North Pacific (**Figure 9.25b**). At last report, 400 of the toys have been recovered from 500 miles of Alaskan shoreline.

Deeper currents can also be surveyed by free-floating devices. The Swallow float (named after its developer) is used to detect the drift of intermediate water masses. Adjusted to descend to a specific density (and therefore depth), the Swallow float emits sonar "pings" as it drifts so that a tracking vessel can follow the movement of the water mass in which it is embedded. More sophisticated devices developed in the 1970s and 1980s depend on sofar channels (see Chapter 7) to transmit sound. Low-frequency tones are broadcast from autonomous submerged probes to moored listening stations (see **Figure 9.25c**). These sofar probes have accumulated more than 240 float-years of data in the North Atlantic, at depths from 700 to 2,000 meters (2,300 feet to 6,600 feet). One of these drifting probes has been sending data for nine years!

Bottom currents can also be investigated. The Woodhead seabed drifter (**Figure 9.25d**), a type of drift card for bottom currents, looks like a small weighted parasol. Released in groups, a few of these objects may eventually wash on shore to be returned by interested beachcombers.

Current meters, or flow meters, measure the speed and direction of a current from a fixed position. Most flow meters, such as the Ekman type shown in **Figure 9.25e**, use rotating vanes to measure current speed and a recording compass to measure direction. Bottom-water movements are usually too slow to be measured by flow meters.

Advances in electronics and computer design have made possible several new methods of study. One new device pioneered by the U.S. Office of Naval Research measures current speed by sensing the electromagnetic force generated by seawater as it moves in Earth's magnetic field. Buoys equipped with these sensors can record current speed and direction without dependence on delicate moving parts.

Yet another method, developed for study of thermohaline circulation, senses the presence in seawater of tritium, a radioactive isotope of hydrogen. A small amount of tritium is produced in the upper atmosphere when hydrogen is bombarded by cosmic rays, but most of it was produced by hydrogen bomb testing in the 1960s. Most tritium combines with oxygen to become radioactive water, and some of this water enters the ocean by precipitation or river runoff. At high latitudes this water sinks and, labeled by its tritium content, can be traced by sampling. The speed of deep currents has been measured by analysis of their tritium content.

Satellites can sense sea-surface temperature (see, for example, Figures 9.11, 9.12, 9.13) and topography, ocean color, and even chlorophyll content (an indication of biological productivity and therefore of nutrient content; see Figure 14.15), giving fresh insight into ocean circulation. So far, satellites can observe only the ocean surface, but satellite-borne lasers may someday probe the ocean to greater depths.

Acoustical tomography (*tomos* = to cut or slice, *graph* = to write) uses pulses of low-frequency sound to sense variations in water temperature, salinity, and movement beneath the surface. Experiments begun in 1981 have used multiple transmitters and receivers to investigate the formation and movement of Gulf Stream eddies and deep-ocean circulation at a depth of 760 meters (2,500 feet). Preliminary results suggest that a great deal of the ocean's energy of motion is involved in relatively small fluctuating flows analogous to atmospheric weather systems. Unlike weather systems—which generally extend across 1,000 kilometers (630 miles) and last less than a week—this midscale oceanic turbulence typically extends only 100 kilometers (63 miles) but may persist for more than 100 days.

Researchers are also using acoustic tomography to conduct a long-term study of the formation of North Atlantic Deep Water south of Greenland. The technique may revolutionize our understanding of deep-sea circulation.

Currents are the very heart of physical oceanography. Their global effects, great masses of water, complex flow, and possible influence on human migrations (see **Box 9.2**) make their study of particular importance.

Q-AND-A

1. Water in Northern Hemisphere gyres turns clockwise. But water in my Northern Hemisphere kitchen sink sometimes goes down the drain clockwise and sometimes counterclockwise. Doesn't the Coriolis effect work for sink drains?

BOX 9.2   •   *A Current Controversy*

Anyone looking at an ocean surface circulation chart can imagine that people might hitch rides on the currents. A glance at Figure 9.8 opens intriguing possibilities: Did the Polynesians actually arrive in their islands from South America? Could North African cultures have contributed to the technical achievements of Central and South American civilizations? Instead of walking great distances over land, did native Alaskans arrive by sea from China or Japan?

These ideas have been discussed since the late nineteenth century. In 1947 Thor Heyerdahl, a Norwegian researcher, decided to put one of them to a test. He was convinced that Polynesians were so different from southeast Asians that they could not have been descended from Asian ancestors. He believed that the first Polynesians had drifted—intentionally or accidentally—from the mainland of South America, driven west with the prevailing winds and currents. To demonstrate this theory, he and five companions lashed together locally available balsa logs to form a raft (which they named *Kon-Tiki* in honor of an Inca god) and set out from a Peruvian seaport. After drifting aboard *Kon-Tiki* for about 8,000 kilometers (5,000 miles) and 101 days, Heyerdahl and his friends washed ashore on an atoll in the Tuamotu Islands east of Tahiti. In 1952, Heyerdahl elaborated on his theory, suggesting that the Polynesians were descendants of Caucasoids from Bolivia, who were themselves perhaps of North African origin.

Heyerdahl's theory of the American origin of the Polynesians was popular with the public, but it gathered little support among scholars. He had proven that South Americans *could* have reached Polynesia, but did the Polynesians actually come from South America? As we saw in Chapter 2, modern anthropologists say they

The raft *Kon-Tiki*, on which Thor Heyerdahl and five companions drifted from South America to islands east of Tahiti.

did not. Polynesian languages are similar to the languages of Southeast Asians, not South Americans. Serological studies—investigations of the structure of blood proteins—have also shown parallels betwen Southeast Asian populations and the people of Polynesia. Perhaps most telling, *Kon-Tiki* had to be towed 80 kilometers (50 miles) from shore; though the prevailing winds and currents flow from east to west, they do not commence until at least this distance away from the mainland.

Still, there are tantalizing wisps of connection. The number of similarities in artistic styles and motifs between the Polynesians and mainland Indians stretch the bounds of coincidence. And what of the sweet potato? Its Polynesian name *kumara* is similar to the term *kumar* used by some Quechua tribes of Peru.

Yes, the Coriolis effect applies to the water in your sink. But Coriolis-related accelerations are very, very small compared with forces imparted to the draining water by small imperfections in the shape of the drain (that may favor "wrong way" flow) or to tiny currents already present in the sink water before you pulled the plug.

Remember, the Coriolis effect depends on the difference of eastward speed between the northern lip of the basin and

the southern lip. In an ocean basin the difference can amount to hundreds of miles per hour. The difference in a kitchen-sink basin is minuscule.

2. If the Gulf Stream warms Britain during the winter and keeps Baltic ports free of ice, why doesn't it moderate New England winters? After all, Boston is closer to the warm core of the Gulf Stream than London is.

Yes, but remember the direction of prevailing winds in winter. Winter winds at Boston's latitude are generally from the west, so any warmth is simply blown out to sea. On the other side of the Atlantic the same winds blow toward London. It does get cold in London, but generally winters in London are much milder than those in Boston.

3. Could currents be used as a source of electrical power? With the Gulf Stream so close to Florida, it seems that some way could be devised to take advantage of all that water flow to turn a turbine.

It's been considered. The total energy of the Gulf Stream flowing off Miami has been estimated at 25,000 megawatts! A Woods Hole Oceanographic Institution team has proposed a honeycomblike array of turbines for the layer between 30 and 130 meters (100–430 feet) across 20 kilometers (13 miles) of the current. They estimate a power output of around 1,000 megawatts, equal to the generation potential of two large nuclear power plants. Engineering difficulties would be considerable, however.

4. Are there any *nongeostrophic* currents? Are any currents not noticeably influenced by gravity, the Coriolis effect, uneven solar heating, planetary winds, and so forth?

Yes. Besides the high-latitude currents mentioned in the text, there are small-scale currents that are not noticeably affected. Currents of fresh water from river mouths, rip currents in surf, and tidal currents in small harbors are much more affected by basin and bottom topography than by the Coriolis effect and gravity.

5. Why are western boundary currents strong in *both* hemispheres? I thought things went the other way (counterclockwise) in the Southern Hemisphere. Shouldn't the *eastern* boundary currents be stronger down there?

Western boundary currents are strong in part because the "Coriolis hill" is offset to the west, forcing water to move in a relatively narrow path along the ocean's western boundary. The bulge is offset to the west because the Earth turns toward the east. And the Earth turns eastward in both hemispheres. Western boundary currents are therefore strong in both hemispheres.

6. You mentioned a hill near the middle of the North Atlantic centered in the area of the Sargasso Sea. What is the Sargasso Sea? Does the Sargasso Sea have anything to do with currents?

Between the Mid-Atlantic Ridge and the Bahamas the sea surface is often strewn with seaweeds bearing small grapelike clusters of floats. The seaweed was named sargassum *after the Portuguese word for grapes. The area is known as the Sargasso Sea.*

The Sargasso Sea is usually a calm place. This area forms the top of the "hill," the bulge caused by the Coriolis effect in the North Atlantic. The water trapped here is warm and relatively nonproductive. Early mariners were afraid that their ships would be entangled in the weed and that they would run out of water and food in this becalmed ocean area.

In one way, the presence of the weed and calm conditions changed the history of the world. Columbus's crew were in a mutinous state when they encountered seaweed. Because seaweed usually meant shallow water and nearby land, Columbus was able to convince them to continue for a few more days in hopes of finding the Orient nearby. The rest, as they say, is history.

7. A north wind comes from the north, but a north current is going north. Why the difference?

Traditions die hard, it seems. For thousands of years winds have been named by where they come from. A north wind comes from the north, and a west wind comes from the west. Currents, though, are named by where they are going. A southern current is headed south; a western current is moving west. An exception is the West Wind Drift, which moves eastward. This current, however, is named after the wind that drives it, the powerful polar westerlies.

It may also be a matter of perspective. Ancient peoples took shelter from winds (and referred to the wind's place of origin). But early oceanic travelers were aware of where the currents were carrying them to.

Terms and Concepts to Remember

acoustical tomography	current	ENSO
Antarctic Bottom Water	downwelling	equatorial upwelling
coastal upwelling	eastern boundary current	float method
contour current	eddy	flow method
convergence zone	Ekman spiral	geostrophic
countercurrent	Ekman transport	Gulf Stream
	El Niño	

gyre	temperature–salinity (T-S) diagram	western boundary current
primary forces	thermohaline circulation	westward intensification
secondary forces	transverse current	West Wind Drift
Southern Oscillation	undercurrent	wind-induced vertical circulation
surface current	upwelling	
sverdrup (sv)		

Study Questions

1. What forces are responsible for the *movement* of ocean water in currents?

2. What forces and factors influence the *direction and nature* of ocean currents?

3. What is a gyre? How many large gyres exist in the world ocean? Where are they located?

4. Why does water tend to flow around the periphery of an ocean basin? Why are western boundary currents the fastest ocean currents? How do they differ from eastern boundary currents?

5. What are countercurrents? Undercurrents? How is El Niño thought to be related to these currents?

6. What is the role of ocean currents in the transport of heat? How can ocean currents affect climate? Contrast the climate of a mid-latitude coastal city at a western ocean boundary with a mid-latitude coastal city at an eastern ocean boundary.

7. What are water masses? Where are distinct water masses formed? What determines their relative position in the ocean?

8. What drives the vertical movement of ocean water? What is the general pattern of thermohaline circulation?

9. What methods are used to study ocean currents?

10. How have (or how *might* have) ocean currents influenced history?

For Further Study

Behringer, D., et al. 1982. "A Demonstration of Ocean Acoustic Tomography." *Nature* 299: 121–25.

Bird, J. 1991. "Supercomputer Voyages to the Southern Seas." *Science* 254 (no. 5032): 656–57. A model of ocean circulation in the Southern Hemisphere helps explain the mildness of Britain's winters. The first elucidation of the global heat conveyor.

Broecker, W. S., and G. H. Denton. 1990. "What Drives Glacial Cycles?" *Scientific American,* January, 48–58. Contribution of North Atlantic Deep Water formation to moderate European winters.

Hollister, C. D., and A. Nowell. 1984. "The Dynamic Abyss." *Scientific American,* March, 42–53. Conditions are not as calm as had been previously suspected.

Huyghe, P. 1990. "The Storm Down Below." *Discover,* November, 70–76. Abyssal storms may be caused by the influence of warm- and cold-core eddies.

Kennett, J. 1982. *Marine Geology.* Englewood Cliffs, NJ: Prentice-Hall. Excellent, concise chapter on thermohaline circulation.

Knauss, J. A. 1961. "The Cromwell Current." *Scientific American,* April, 105-16. A look at deep-water currents under the equatorial currents. Useful though dated.

Kunzig, R. 1991. "Can Earth's Internal Heat Drive Ocean Circulation?" *Science* 252 (no. 5013): 1620–21. Heat coming from vents in oceanic ridges may churn abyssal waters.

MacLeish, W. H. 1989. *The Gulf Stream: Encounters with the Blue God.* New York: Houghton Mifflin. The Gulf Stream's story.

MacLeish, W. H. 1989. "Painting a Portrait of the Stream from Miles Above—and Below." *Smithsonian,* March. Satellite imagery, wonderful pictures.

McLellan, H. J. 1965. *Elements of Physical Oceanography.* Oxford: Pergamon Press. Excellent technical overview of thermohaline circulation.

Neelin, J. D., and J. Marotzke. 1994. "Representing Ocean Eddies in Climate Models." *Science* 264 (20 May): 1099–1100. The large surface area of the North Atlantic covered by pinched-off eddies appears to have a greater influence on climate than once thought.

Neshyba, S. 1987. *Oceanography: Perspectives on a Fluid Earth.* New York: Wiley. Insight on the interconnection between surface and deep circulation.

Price, J., R. Weller, and R. Schudlich. 1987. "Wind-Driven Ocean Currents and Ekman Transport." *Science* 238 (no. 4833): 1534–38. Verification of theoretical Ekman transport. Observed transport is consistent with theoretical Ekman transport to within 10%.

Rasmussen, E. M. 1985. "El Niño and Variations in Climate." *American Scientist* 73 (no. 2). Interactions between ocean and atmosphere in the mid-Pacific can affect weather patterns worldwide.

Richardson, P. L. 1991. "SOFAR Floats Give a New View to Ocean Eddies." *Oceanus* 34 (no. 1, Spring): 23–31. Sofar channels permit long-distance tracking of current probes for many years.

Schmitt, R. W., Jr. 1995. "The Ocean's Salt Finger." *Scientific American,* May, 70–75. Small-scale dynamics of water mixing have large-scale consequences for the structure of the ocean.

Spinel, R. C., and P. F. Worcester. 1990. "Ocean Acoustic Tomography." *Scientific American,* October, 94–99. Recent information on the predominance of midscale circulation.

Stewart, R. W. 1969. "The Atmosphere and the Ocean." *Scientific American,* September, 76–87. Excellent review article.

Stommel, H. 1958. "The Circulation of the Abyss." *Scientific American,* July, 85–93. The first coherent theory of thermohaline circulation is delineated here.

Stommel, H. 1965. *The Gulf Stream.* Berkeley: University of California Press. Expert dissection of a western boundary current.

Stommel, H. 1987. *A View of the Sea.* Princeton: Princeton University Press. A famous oceanographer explains ocean currents in an imaginary dialog between himself and an inquisitive chief engineer of an oceanographic research vessel. Currents and the author's view of life are both on display here. A wonderful book.

Sverdrup, H. U., et al. 1942. *The Oceans: Their Physics, Chemistry, and General Biology.* New York: Prentice-Hall. Mathematical analysis of ocean currents.

Warren, B. A. 1971. "Antarctic Deep Water Contribution to the World Ocean." In *Research in the Antarctic,* pub. 93, 630–43, edited by L. O. Quam. Washington, DC: American Association for the Advancement of Science.

Weibe, P. 1982. "Rings of the Gulf Stream." *Scientific American,* March, 60–79. Discussion of the physics and biology of Gulf Stream eddies.

Whitehead, J. A. 1989. "Giant Ocean Cataracts." *Scientific American,* February, 50—61. "Waterfalls" of dense water plunging down continental slopes are likened to cataracts. The role of dense water in maintaining the chemistry and climate of the deep ocean is explored.

Wyrtki, K. 1961. "The Thermohaline Circulation in Relation to the General Circulation in the Oceans." *Deep-Sea Research* 8 (no. 1): 39–64.

Note: Oceanus 37 (no. 1, Spring 1994) is dedicated to Atlantic Ocean circulation.

WAVE DYNAMICS AND WIND WAVES

"These blue-water hills . . ."

In the Spring of 1916 Sir Ernest Shackleton found himself and the members of his failed south-polar expedition marooned on a desolate island at the tip of Antarctica's Palmer Peninsula, far from any hope of rescue. Their expedition ship, aptly named *Endurance*, had been crushed by ice and lay at the bottom of the Weddell Sea. Someone had to go for help.

They had only one chance: to reach a whaling station on South Georgia Island, over 1,300 kilometers (800 miles) away. With the force of the westerlies and the thrust of the West Wind Drift, it would be impossible to return if they were blown past the small island. But Shackleton's largest and strongest boat, the *James Caird*, was only 6.85 meters (22.5 feet) long, and ahead lay a violent stretch of ocean containing the largest waves on Earth (**Figure a**).

Although they had little hope of success, a crew of six set out on the morning of 24 April 1916. In his account of their desperate voyage, the captain of *Endurance*, Frank Worsley, gives us a feeling for the power of the ocean encircling Antarctica:

> In the afternoon [of the fifth day] the swell settled and lengthened out—the typical deep-sea swell of these latitudes. Offspring of the westerly gales, the great unceasing westerly swell of the Southern Ocean rolls almost unchecked around this end of the world in the Roaring Forties and the Stormy Fifties.[1] The highest, broadest, and longest swells in the world, they race on their encircling course until they reach their birthplace again, and so, reinforcing themselves, sweep forward in fierce and haughty majesty, four hundred, a thousand yards, a mile apart in fine weather, silent and stately they pass along. Rising forty or fifty feet and more from crest to hollow, they rage in apparent disorder during heavy gales. . . .
>
> At times, rolling over their allotted ocean bed, in places four miles deep, they meet a shallow of thirty to a hundred fathoms. Their bases retarded by the bank, their crests sweep up in furious anger at this check, until their front forms an almost perpendicular wall of green rushing water that smashes on a ship's deck, flattening steel bulwarks, snapping two-inch steel stanchions, and crushing deckhouses and boats like eggshells. These blue-water hills in a very heavy gale move as fast as thirty-five miles an hour, but striking the banks, the madly leaping crests falling over and onward, probably attain a momentary speed of fifty or sixty miles or more. The impact of hundreds of tons of *solid* water at this speed can only fairly be imagined.
>
> So we held our way; in those valleys and on those ridges alternately. First, half becalmed—a hill of water ahead, another astern—the following hill lifts us, as the boat slides with increasing speed down the ever steepening slope, till with a sudden upward swoop, the sea boiling white around and over us, we are on the summit with a commanding view of a panorama of dark grey and indigo blue rollers, topped and broken with white horses. The crest passing leaves the boat apparently stationary, gravity now dropping us back till the next hollow reaches us, and so on *ad nauseam*.

After an all but unbearable voyage of 16 days they made a perfect landfall, reported the position of the rest of the expedition to the whalers, and assisted with the rescue. Against all odds, not a single life was lost.

[1]*Forties* and *Fifties* here refer to the ocean in the area of 40° and 50° south latitude. Note in Figure 10.10 the extreme height of waves in this area.

Source: F. A. Worsley, *Shackleton's Boat Journey* (New York: Norton, 1977), 116–18.

a Deckhands man the pumps as huge waves pound the British expedition ship *Terra Nova* in the West Wind Drift, December 1910.

Waves transmit energy, not water mass, across the ocean's surface. The speed of ocean waves usually depends on their wavelength, with long waves moving fastest. Arranged from short to long wavelengths (and therefore from slowest to fastest), capillary waves are generated by very small disturbances, wind waves by wind, seiches by the rocking of water in enclosed spaces, tsunami by seismic and volcanic activity or other sudden displacements, and tides by gravitational attraction. The behavior of waves depends largely on the relation between a wave's size and the depth of water through which it is moving. Waves can refract, reflect, break, and interfere with one another.

This chapter introduces the basic properties of all ocean waves and emphasizes the occasionally enormous "small" waves generated by wind. The next chapter extends this information to great waves of longer wavelengths.

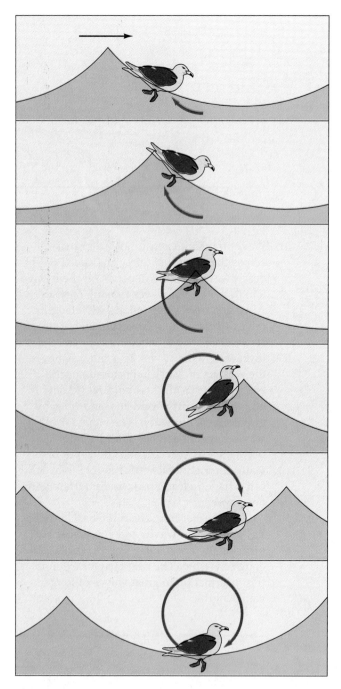

Figure 10.1 A floating sea gull demonstrates that waveforms travel but that the water itself does not. In this sequence, a wave moves from left to right as the gull (and the water in which it is resting) rotates in an imaginary circle: moving slightly to the left up the front of an approaching wave, then to the crest, and finally sliding to the right down the back of the wave.

OCEAN WAVES

To most people an ocean wave in deep water appears to be a massive moving object—a ridge of water traveling across the sea surface. An ocean wave is one of several kinds of **waves,** which are disturbances caused by the movement of energy from a source through some medium (solid, liquid, or gas). As the energy of the disturbance travels, the medium through which it passes moves in specific ways. Sometimes this movement is visible to us as a hump in the medium. This traveling hump, or ridge, produces the appearance of movement we see in a wave. In an ocean wave, *energy* is moving at the speed of the wave, but *water* is not.

Picture a resting sea gull as it bobs on the wavy ocean surface far from shore. The sea gull doesn't move only up and down, or only forward and backward, as waves pass his position. The reason for this is that the wave motion is neither purely transverse (as the movement of the rope in Figure 3.6) nor purely longitudinal (as the movement of the spring in the same figure). The gull moves in *circles:* up-and-forward as the tops of the waves move to his position, down-and-backward as the tops move past. Each circle is equal in diameter to the wave's height. As may be seen in **Figure 10.1,** energy in waves flows past the resting bird, but the gull and his patch of water move only a very short distance forward in each up-and-forward, down-and-back wave cycle. The water on which the bird rests does not move continuously across the sea surface as the wave illusion suggests.

The nearly friction-free transfer of energy from water particle to water particle in these circular paths, or **orbits,** transmits wave energy across the ocean surface and causes the wave form to move. This kind of wave is known as an **orbital wave**—a wave in which particles of the medium (water) move in closed circles as the wave passes. Orbital ocean waves occur at the boundary between two media (between air and water) and between layers of water of different densities. Because the wave form moves forward, these waves are a type of **progressive wave.**

The progressive wave that moved the gull was caused by wind. Other forces can generate much bigger progressive waves, in which water molecules move through much larger circular or elliptical orbits. Some of these waves are so large that they do not appear to us as waves at all, but rather as the slow sloshing of water in a harbor or bay, as dangerous flooding surges of water, or as rhythmic and predictable ocean tides.

Ocean waves have distinct parts. The **wave crest** is the highest part of the wave above average water level; the **wave trough** is the valley between wave crests below average water level. **Wave height** is the vertical distance between a wave crest and the adjacent trough, while

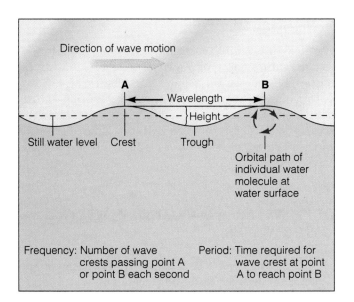

Figure 10.2 The anatomy of a progressive wave.

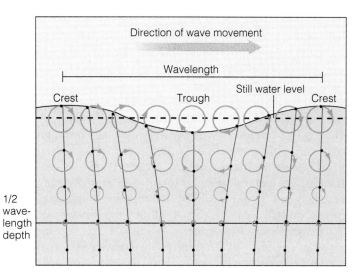

Figure 10.3 The orbital motion of water particles in a wave, which extends to a depth of about one-half of the wavelength.

wavelength is the horizontal distance between two successive crests (or troughs). The relationship among these parts is shown in **Figure 10.2**. The time it takes for two successive wave crests (or troughs) to pass a fixed point, usually measured in seconds, is known as the **wave period. Wave frequency** is the number of waves passing a fixed point per second. Frequency is the inverse of period.

The circular motion of water particles at the surface of a wave continues underwater. As **Figure 10.3** shows, the diameter of the orbits through which water particles move diminishes rapidly with depth. For all practical purposes wave motion in deep-water waves is negligible below a depth of one-half the wavelength. This means that divers in 20 meters of water would not notice the passage of a wind wave of 30-meter wavelength, and they might barely notice the wave if they were at a depth of 15 meters. Since most ocean waves have moderate wavelengths, the circular disturbance of the ocean that propagates these waves affects only the uppermost layer of water. Note that the movement of water in circles doesn't resemble interlocking mechanical gears. Instead, there is a coordinated, uniform circular movement of water molecules in one direction as the waves pass.

CLASSIFYING WAVES

Ocean waves are classified by the *disturbing force* that creates them, the *restoring force* that tries to flatten them, and their *wavelength*. (Wave height is not often used for classification because it varies greatly depending on water depth, interference between waves, and other factors.)

Disturbing Force

Energy that causes ocean waves to form is called a **disturbing force.** Wind blowing across the ocean surface provides the disturbing force for wind waves. Arrival of a storm surge or seismic sea wave in an enclosed harbor or bay, or a sudden change in atmospheric pressure, is the disturbing force for the resonant rocking of water known as a *seiche.* Landslides, volcanic eruptions, and faulting of the seafloor associated with earthquakes are the disturbing forces for seismic sea waves (also known as *tsunami*). The disturbing forces for tides are changes in the direction of gravitational forces among the Earth, moon, and sun, combined with Earth's rotation.

Restoring Force

Restoring force is the dominant force trying to return the water surface to flatness after a wave has formed in it. If the restoring force of a wave were quickly and fully successful, a disturbed sea surface would immediately become smooth, and the energy of the embryo wave would be dissipated as heat. But that isn't what happens. Waves continue after they form because the restoring force overcompensates and causes oscillation. The situation is analogous to a weight bobbing at the bottom of a very flexible spring, constantly moving up and down past its normal resting point.

The restoring force for very small water waves—those with wavelengths less than 1.73 centimeters (0.68 inch)—is cohesion, the property that allows individual water molecules to stick to each other by means of hydrogen bonds (see Figure 6.2). These **capillary waves** are transmitted across a puddle because cohesion, the force that makes the tea creep up on the sides of a teacup, tugs the tiny wave troughs and crests toward flatness. Capillary waves are the first waves to form when the wind blows. These small ripples are important in transferring energy from air to water to drive ocean currents (as we saw in Chapter 9), but they are of little consequence in the overall picture of ocean waves because they are tiny and carry very little energy.

All waves with wavelengths greater than 1.73 centimeters depend on gravity to provide the restoring force. Like the spring weight moving up and down, gravity pulls the crests downward, but the momentum of the water causes the crests to overshoot and become troughs. The repetitive nature of this movement gives rise to the circular orbits of individual water molecules in an ocean wave. These larger waves are called **gravity waves.** Since the circular motion of water molecules in a wave is nearly friction free, gravity waves can travel across thousands of miles of ocean surface without disappearing, eventually to break on a distant shore.

Wavelength

Wavelength is a direct measure of wave size. **Table 10.1** lists the causes and typical wavelengths of wind waves, seiches, seismic sea waves, and tides. **Figure 10.4** shows the relation between disturbing and restoring forces, period, and the relative amount of energy present in the ocean's surface for each wave type. Note that more energy is stored in wind waves than in any of the other wave types.

DEEP-WATER WAVES AND SHALLOW-WATER WAVES

Most of the characteristics of ocean waves depend on the relationship between their wavelength and water depth. Wavelength determines the *size* of the orbits of water molecules within a wave, but water depth determines the *shape* of the orbits. The paths of water molecules in a wind wave are circular only when the wave is traveling

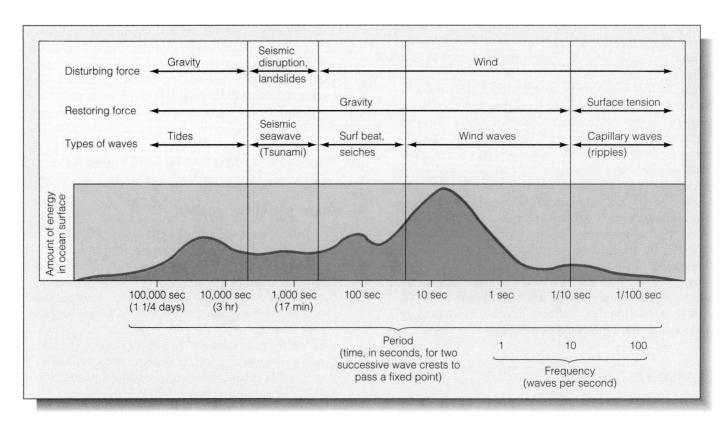

Figure 10.4 Wave energy in the ocean as a function of the wave period. As the graph shows, most wave energy is typically concentrated in wind waves. However, tsunami, rare events in the ocean, can transmit more energy than all wind waves for a brief time.

Table 10.1 Wavelengths and Disturbing Forces of Important Ocean Waves		
Wave Type	Typical Wavelength	Disturbing Force
Wind wave	60–150 m (200–500 ft)	Wind over ocean
Seiche	Large, variable; a fraction of basin size	Changes in atmospheric pressure, storm surge, tsunami
Seismic sea wave (tsunami)	200 km (125 mi)	Faulting of seafloor, volcanic eruption, landslide
Tide	½ circumference of Earth	Gravitational attraction, rotation of Earth

in deep water. A wave cannot "feel" the bottom when it moves through water deeper than one-half its wavelength; too little wave energy is contained in the small circles below that depth. Waves moving through water deeper than one-half their wavelength are known as **deep-water waves.** A wave has no way of knowing how deep the water is, only that it is in water deeper than about one-half its wavelength. For example, a wind wave of 20-meter wavelength will act as a deep-water wave if it is passing through water more than 10 meters deep (see **Figure 10.5a**).

The situation is different for wind-generated waves close to shore. The orbits of water molecules in waves moving through shallow water are flattened by the proximity of the bottom. Water just above the seafloor cannot move in a circular path, only forward and backward. Waves in water shallower than 1/20 their original wavelength are known as **shallow-water waves** (see **Figure 10.5b**). A wave with a 20-meter wavelength will act as a shallow-water wave if the water is less than 1 meter deep.

Transitional waves travel through water deeper than 1/20 their original wavelength but shallower than one-half their original wavelength. In our example, this would be water between 1 meter and 10 meters deep. The orbital motion of water molecules in a transitional wave is shown in **Figure 10.5c.**

Of the four wave types listed in Table 10.1, only wind waves can ever be deep-water waves. To understand why, remember that most of the ocean floor is deeper than 125 meters (400 feet), one-half the wavelength of very large wind waves. The wavelengths of the larger waves are *much* longer; the wavelength of seismic sea waves usually exceeds 100 kilometers (62 miles). No ocean is 50 kilometers (31 miles) deep; so seiches, seismic sea waves, and tides are forever in water that to them is shallow or transitional in depth. Their huge orbit circles flatten against a distant bottom that is always less than half a wavelength away.

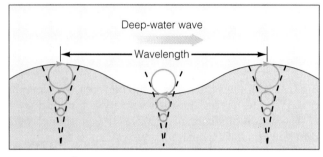

a Depth $\geq \frac{1}{2}$ wavelength

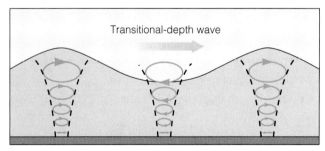

b Depth $\leq \frac{1}{20}$ wavelength

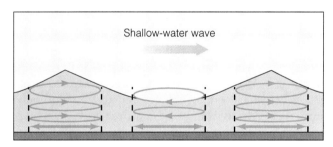

c Depth $\leq \frac{1}{2}$ to $\geq \frac{1}{20}$ wavelength

Figure 10.5 Progressive waves. (a) A deep-water wave; (b) a shallow-water wave; (c) a transitional wave.

In general, the longer the wavelength, the faster the wave energy will move through the water. For deep-water waves this relationship is shown in the formula

$$V = L/T$$

in which V represents velocity, L is wavelength, and T is time, or period (in seconds). Wavelength is difficult to determine at sea, but period is comparatively easy to find: An observer simply times the movement of waves past the bow of a stopped ship, for example. It turns out that if period (T) is known, velocity (V) can be calculated from the relation

$$V \text{ (in meters/sec)} = 1.56T$$

or, if you prefer,

$$V \text{ (in feet/sec)} = 5T$$

Figure 10.6 shows the relationship between wavelengths of deep-water waves and their velocity and period.

The action of shallow-water waves is described by a different equation, which may be written as

$$V = \sqrt{gd} \text{ or } V = 3.1\sqrt{d}$$

where V is velocity (in meters per second), g the acceleration due to gravity (an average of 9.8 meters per second per second), and d the depth of water in meters. The *period* of a wave remains unchanged regardless of the depth of water through which it is moving. As deep-water waves enter the shallows and feel bottom, however, their velocity is reduced and their crests bunch up, so their wavelength shortens.

Listed below are two very different kinds of ocean waves (a comparison of apples and oranges), but notice the general relationship between wavelength and wave velocity: The longer the wavelength, the greater the velocity.

Wind Waves (Deep-Water Waves)
- Period to about 20 *seconds*
- Wavelength to perhaps 600 meters (2,000 feet) in extreme cases
- Speed to perhaps 112 kilometers (70 miles) per hour in extreme cases

Seismic Sea Waves (Shallow-Water Waves)
- Period to perhaps 20 *minutes*
- Wavelength typically 200 kilometers (125 miles)
- Speeds of 760 kilometers per hour (470 miles per hour)

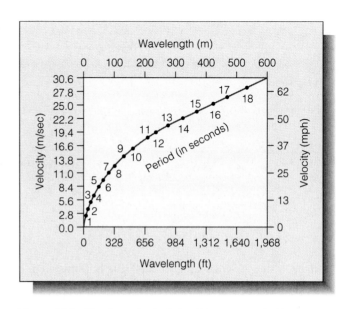

Figure 10.6 The theoretical relationship among velocity, wavelength, and period in deep-water waves. Velocity is equal to wavelength divided by period. If one characteristic of a wave can be measured, the other two can be calculated. The easiest to measure exactly is period.

Remember that it is energy—not the water mass itself—that is moving through the water at the astonishing speed of 760 kilometers (470 miles) per hour in seismic sea waves. (A speed of 760 kilometers per hour is the speed of a commercial jet airliner!)

We will discuss seiches, seismic sea waves, and tides in the next chapter. The rest of this chapter deals primarily with wind waves.

WIND WAVES

Wind waves are gravity waves formed by the transfer of wind energy into water. Most wind waves are less than 3 meters (10 feet) high. Wavelengths from 60 to 150 meters (200 to 500 feet) are most common in the open ocean.

Wind waves grow from capillary waves. Capillary waves form as wind friction stretches the water surface and as surface tension tries to restore it to smoothness. These waves are nearly always present on the ocean. Capillary waves interrupt the smooth sea surface, deflect surface wind upward, slow it, and cause some of the wind's energy to be transferred into the water to drive the capillary wave crest forward. The wind may eddy briefly downwind of the tiny crest, creating a slight partial vacuum there. Atmospheric pressure pushes the trailing crest forward (downwind) toward the trough, adding still more energy to the water surface. The increasing energy in the water surface expands the circular orbits of water particles in the direction of the wind, enlarging the small wave's size. The capillary wave becomes a wind wave when its wavelength exceeds 1.73 centimeters (0.68 inch), the wavelength at which gravity supersedes capillary action as the restoring force.

If the wind wave remains in water deeper than one-half its wavelength, and the wind continues to blow, the wave becomes larger. Its crest is thrust higher into faster wind, extracting even more energy from the moving air. The circular orbits of water particles within the wave grow larger with more energy input; height, wavelength, and period increase proportionally. The irregular peaked waves in the area of wind wave formation are called **sea** (**Figure 10.7**); the chaotic surface is formed by simultaneous wind waves of many wavelengths. When the wind slows or ceases, as it does away from a storm, the wave crests become rounded and regular.

Swell Formation and Dispersion

Because they move faster, waves with longer wavelengths leave the area of wave formation sooner. They outrun their smaller relatives. Mature waves from a storm sort themselves into groups having similar wavelengths and speeds. The process of wave separation, or **dispersion**, produces the familiar smooth undulation of the ocean surface called **swell** (**Figure 10.8**). Swell often move thousands of kilometers from a storm to a shore. Contrary to what you might expect, observers far from the storm would first encounter quick-moving waves of long wavelength, then middle-sized waves, and then slow, small ones. Because the circular movement of

Figure 10.7 Sea, an area of wind wave formation.

Figure 10.8 Swell—mature, regular wind waves sorted by dispersion—off the Oregon coast.

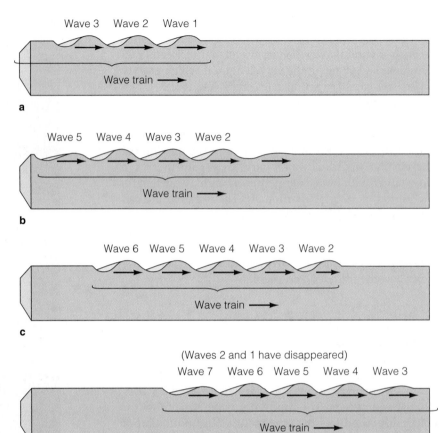

Figure 10.9 The progress of a wave train. (a) The energy in the leading waves (here, waves 1 and 2) is transferred into circular movement in undisturbed water. (b) As waves 1 and 2 are drained of energy, they gradually disappear, but the circular movement forms new waves 4 and 5 at the end of the train.

water particles in deep-water waves is virtually friction free, the waves will continue until they break upon a shore. There they release their absorbed wind energy as random movement, heat, and sound.

Progressing groups of swell with the same origin and wavelength are called **wave trains.** The leading waves in the wave train are drained of energy because they must begin the circular movement of the undisturbed water into which they are intruding. These leading waves gradually disappear, but after the wave train has passed some energy remains behind in the circles to form new waves. New waves thus form behind as the leading waves disappear at the front of the wave train. This process is shown in **Figure 10.9.**

The implications of this detail are surprising. Though each wave moves forward with a velocity proportional to its wavelength, the wave train itself moves forward at only *half* that speed. Groups of waves therefore move ahead at half the speed of individual waves within the group. The half-speed advance of the wave train is called **group velocity.** Note that individual waves do not persist in the ocean. Individual waves last only as long as they take to pass through the group. As deep-water waves move into shallow water, the speed of individual waves within the group slows until wave speed equals group velocity.

Factors Affecting Wind Wave Development

Three factors affect the growth of wind waves. Wind must be moving faster than the wave crests for energy transfer from air to sea to continue; so the mean speed, or **wind strength,** of the wind is clearly important to wind wave development. A second factor is the length of time the wind blows, or **wind duration;** high winds that blow only a short time will not generate large waves. The third factor is the uninterrupted distance over which the wind blows without significant change in direction, the **fetch.**

Table 10.2 shows how wave height, wavelength, period, and wave speed are related to fetch *with wind speed held constant.* Notice that in a very long fetch, wave speed can approach wind speed. Clearly, the waves described in Table 10.2 have grown larger and faster with prolonged input of wind energy.

A strong wind must blow continuously in one direction for nearly three days for the largest waves to

Table 10.2 The Relationship of Fetch to Average Wind Wave Height, Wavelength, Period, and Wave Speed, with Wind Speed Held Constant at 93 km/hr (58 mi/hr)

Fetch	Wave Height	Wavelength	Period	Wave Speed
19 km (12 mi)	2 m (6 ft)	35 m (115 ft)	4 sec	21 km/hr (13 mi/hr)
93 km (58 mi)	3 m (10 ft)	70 m (230 ft)	6 sec	32 km/hr (20 mi/hr)
370 km (230 mi)	5 m (16 ft)	100 m (330 ft)	8 sec	43 km/hr (27 mi/hr)
740 km (460 mi)	7 m (23 ft)	150 m (500 ft)	10 sec	53 km/hr (33 mi/hr)
1,850 km (1,150 mi)	12 m (39 ft)	200 m (660 ft)	11 sec	64 km/hr (40 mi/hr)

Table 10.3 Conditions Necessary for a Fully Developed Sea at Given Wind Speeds, and the Parameters of the Resulting Waves

Wind Conditions			Wave Size		
Wind Speed in One Direction	Fetch	Wind Duration	Average Height	Average Wavelength	Average Period
19 km/hr (12 mi/hr)	9 km (12 mi)	2 hr	0.27 m (0.9 ft)	8.5 m (28 ft)	3.0 sec
37 km/hr (23 mi/hr)	139 km (86 mi)	10 hr	1.5 m (4.9 ft)	33.8 m (111 ft)	5.7 sec
56 km/hr (35 mi/hr)	518 km (322 mi)	23 hr	4.1 m (13.6 ft)	76.5 m (251 ft)	8.6 sec
74 km/hr (46 mi/hr)	1,313 km (816 mi)	42 hr	8.5 m (27.9 ft)	136 m (446 ft)	11.4 sec
92 km/hr (58 mi/hr)	2,627 km (1,633 mi)	69 hr	14.8 m (48.7 ft)	212.2 m (696 ft)	14.3 sec

Source: Data from U.S. Army Corps of Engineers Coastal Research Center, Richmond, Virginia.

develop fully. A **fully developed sea** is the maximum wave size theoretically possible for a wind of a specific strength, duration, and fetch. Longer exposure to wind at that speed will not increase the size of the waves. The first three columns of **Table 10.3** give examples of the combinations of wind speed, fetch, and duration required to form waves of the fully developed sea described in the remaining columns. Note that waves in a fully developed sea are small if wind speed is low. However, if the wind speed is 74 kilometers (46 miles) per hour through 1,313 kilometers (816 miles) for 42 hours—conditions not wholly unrealistic in the Pacific—waves that average 8.5 meters (27.9 feet) high can result. The highest 10% of the waves in this sea will exceed 17.2 meters (56.6 feet) in height! The combination of factors

required to produce truly great seas doesn't occur very often. In the example given at the bottom of Table 10.3, it seems unlikely that a 92-kilometer- (58-mile-) per-hour wind would blow steadily in one direction for 69 hours over 2,627 kilometers (1,633 miles). Of course, as Lieutenant Marggraff's experience in 1933 suggests (see **Box 10.1**), the ocean is full of surprises!

The greatest potential for large waves occurs beneath the strong and nearly continuous winds of the West Wind Drift surrounding Antarctica. The early nineteenth-century French explorer of the South Seas Jules Dumont d'Urville encountered a wave train with heights estimated "in excess" of 30 meters (100 feet) in Antarctic waters, and in 1916 Ernest Shackleton contended with occasional waves of similar size in the West Wind Drift

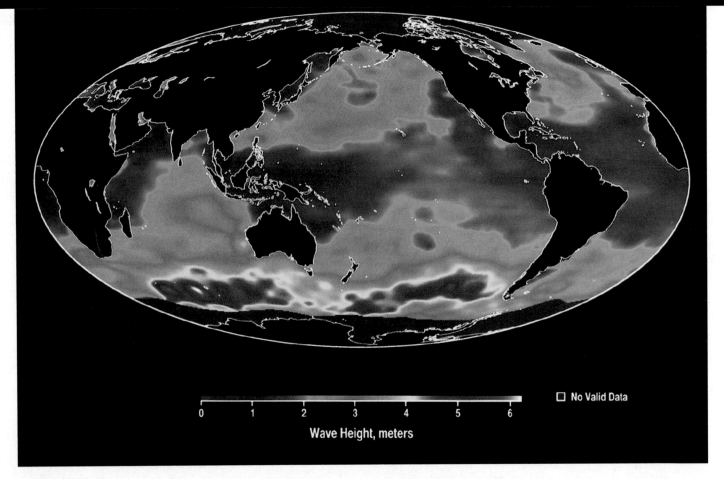

Wave Height, meters

Figure 10.10 Global wave height acquired by a radar altimeter aboard the *TOPEX/Poseidon* satellite in October 1992. In this image, the highest waves occur in the southern ocean, where waves were over 6 meters high (represented in white). The lowest waves (indicated by dark blue) are found in the tropical and subtropical ocean, where wind speed is lowest.

Figure 10.11 An immense wave rushes toward a ship.

during a heroic voyage to remote South Georgia Island in an open boat (see this chapter's vignette and Chapter 18). Satellite observations (like those of **Figure 10.10**) have shown that wave heights to 11 meters (36 feet) are fairly common in the West Wind Drift.

In zones of high winds a less than fully developed sea can also attract attention. Though wind speed within cyclonic storms is often very strong, the circular motion of air doesn't allow long fetches, and fully developed seas rarely occur beneath such storms. Officers standing deck watches during storms rarely quibble with theoretical maximum height vs. observed height, however. Wind waves can be overwhelming even if they are not fully developed (**Figure 10.11**).

BOX 10.1 ● *Theory Meets Experience on the Crest of a Wave*

Think of the ocean and you think of waves. On the night of 7 February 1933, Lieutenant Frederick Marggraff's world consisted of little *but* waves. Mr. Marggraff was standing a predawn watch on the bridge of the USS *Ramapo*, a large U.S. Navy tanker steaming from Manila to San Diego. February 7 was the third full day of a furious storm, and the tanker's crew had become professionally and emotionally interested in the intense wind and mountainous seas sweeping past their ship. A few of the officers may have thought about the so-called 60-foot law, a paragraph in a Navy Hydrographic Office bulletin implying that all sightings of waves over 60 feet high (about 18 meters) were exaggerations. Indeed, despite persistent reports to the contrary, few theorists believed that wind-driven ocean waves exceeded 18 meters in height. Lieutenant Marggraff was about to prove them wrong.

February is the stormiest month in the North Pacific, and *Ramapo* had encountered the largest and most severe storm of the year, a storm made more intense by the coalescence of three low-pressure centers. For days a steady wind had blown at 107 kilometers (67 miles) per hour, and gusts to 126 kilometers (78 miles) per hour often lashed the decks. But the wind blew persistently from one direction, and though the monstrous waves it generated dwarfed the tanker, they were surprisingly orderly in form.

"The conditions for observing the seas from the ship were ideal," wrote *Ramapo*'s executive officer. "We were running directly down the wind with the sea. There were no cross seas and therefore no peaks along wave crests. There was practically no rolling, and the pitching

motion was easy because of the fact that the sides of the waves were much longer than the ship. The moon was out astern and facilitated observations during the night. The sky was partly cloudy." At about three in the morning, Mr. Marggraff observed a train of giant waves looming in the moonlight. As the trough of the first wave approached the ship, he noted that its distant crest was on a level with the crow's nest on the mainmast. At that instant the ship's stern sank into the bottom of the onrushing trough. The next two waves were about the same size. Not surprisingly, such immense waves made an indelible impression on all who witnessed them.

How big were these waves? When the executive officer had some time to spare, he did some calculations. **Figure a** illustrates the attitude of the ship when the largest waves were measured. The height of the waves was determined by using a set of the ship's plans, a calculation of the height of the observer above the sea surface, the draft of the ship, and a sight to the horizon. The largest wave for which a dependable observation had been made was 34 meters (112 feet) high, still a record! After the executive officer published his findings, the paragraph on the 60-foot law was quietly dropped from the next edition of the Hydrographic Office bulletin.

a How the great wave observed from the USS *Ramapo* was measured. An officer on the bridge was looking toward the stern and saw the crow's nest in his line of sight to the crest of the wave, which had just come in line with the horizon. Wave height was calculated based on the geometry of the situation.

Observer

Crow's nest

Line of sight to crest and horizon

11° 50'

34 m
(112 feet)

Wind Wave Height

Wave height is not directly represented in the deep-water wave formula $V = L/T$. During their formation, moderately sized wind waves in the open ocean exhibit a maximum 1:7 ratio of wave height to wavelength (see **Figure 10.12**); this ratio is the **wave steepness.** Waves 7 meters long will not be more than 1 meter high, and waves of 70-meter wavelength will not exceed 10 meters of height. The angle at their crest will not exceed 120°. A peaked appearance usually indicates the continuing injection of wind energy. If a wave gets any higher than the 1:7 ratio for its wavelength it will break, and excess energy from the wind will be dissipated as turbulence; hence the *whitecaps* or *combers* associated with a fully developed sea.

As wind waves mature into swell their wavelengths increase somewhat, and the ratio of wavelength to wave height becomes greater: The waves become less steep. Swell continue in this rounded-top shape until they break on shore.

Wavelength Records

How big can wind waves get? This is an interesting question, and one not dispassionately answered. Calculating the wavelength from wave period (T in $V = L/T$), Bigelow and Edmondson (1953) reported that mariners have sighted waves with wavelengths of 451 meters (1,481 feet) off the west coast of Ireland, 583 meters (1,914 feet) off the Cape of Good Hope, and 829 meters (2,719 feet) in the equatorial Atlantic. (That last monster would have had a period of 25 seconds!) The nineteenth-century French admiral J. Mottez reported a wave with a wavelength of 790 meters (2,600 feet), a period of 23 seconds, and a speed of 123 kilometers (76 miles) per hour in the equatorial Atlantic west of Africa. The wavelength of the *Ramapo* wave discussed in Box 10.1 was calculated at 360 meters (1,180 feet); so these sightings are perhaps exaggerated. I can attest to the fact that smaller waves can also add a great deal of excitement to a deck officer's day (see **Figure 10.13**)!

INTERFERENCE AND ROGUE WAVES

The real situation of wind waves at sea is not as simple as has been suggested above. The ideal vision of one set of waves moving in one direction at one speed across an otherwise smooth surface is almost never observed in the ocean.

Independent wave trains exist simultaneously in the ocean most of the time. Since long waves outrun shorter ones, wind waves from different storm systems can overtake and interfere with one another. One wave doesn't

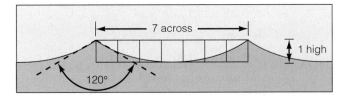

Figure 10.12 A wind wave of moderate size shown during the time of its formation. The ratio of height to wavelength, called *wave steepness*, is 1:7; the crest angle does not exceed 120°.

crawl over the others when they meet. Instead they add to (or subtract from) one another. Such interaction is known as **interference.** In **Figure 10.14a**, a wave of one wavelength is represented as a blue line, a slightly longer wavelength wave as a green line. In the sea surface where these waves coincide (shown in **Figure 10.14b**), you can see the alternation between addition (large crests and troughs) and subtraction (almost no waves at all). The cancellation effect of subtraction is termed **destructive interference**—not because of harm to lives or property, but because wave interference destroys or cancels the waves involved. **Constructive interference** is the additive formation of large crests or deep troughs, the size of which exceeds the size of each of the participating waves.

You have probably noticed that the surf along a coast seems to rise to a few big waves, diminish, and then build again. Surfers wait for the big "sets" to arrive, ride toward the shore, and then use the relatively calm interval to swim out into position for the next big set of waves. Constructive and destructive interferences explain this behavior, called **surf beat.** Constructive interference between waves of different wavelengths creates the sought-after big waves; destructive interference diminishes the waves and makes it easier to swim back out. The characteristics of surf beat explain why in some instances every ninth wave might be quite large. As wavelengths and interference change, though, every seventh wave might be large, or every fifth, or twelfth. Contrary to folklore there is no set ratio.

Interference can have sudden unpleasant consequences on the open sea. In or near a large storm, wind waves at many wavelengths and heights may approach a single spot from different directions. If such a rare confluence of crests occurred at your position, a huge wave crest would suddenly erupt from a moderate sea to threaten your ship. The freak wave—called a **rogue wave**—would be much larger than any noticed before or after, and it would be higher than the theoretical maximum wave capable of being sustained in a fully developed sea. In such conditions one wave in about 1,175 is over three times average height, and one in every 300,000 is over four times average height! **Box 10.2**

Figure 10.13 A U.S. Navy destroyer battles large waves generated by a storm in the Pacific. Despite the battering, the ship is holding a steady enough course to take on fuel and stores from an aircraft carrier.

describes some encounters between ocean liners and rogue waves.

Currents can contribute to the formation of a kind of rogue wave that does not depend on interference. These waves are formed by the interaction of a wind wave and a swift surface current. The southeastern tip of Africa seems to breed a disproportionally large number of these giant waves. The Agulhas current, a strong western boundary current, flows close to shore there, moving in the opposite direction to wind waves generated by high-latitude southwesterly gales. When large waves suddenly hit the 5-kilometer- (3-mile-) per-hour current, they can "trip" on it, rise suddenly to double their original height, and break precipitously. Such waves have broken tankers in half; many lives have been lost.

WIND WAVES APPROACHING SHORE

Most wind waves eventually find their way to a shore and break, dissipating all their order and energy. The process begins with the transition of our now familiar deep-water wave to a transitional wave in water less than half a wavelength deep. **Figure 10.15** outlines the events leading to the break:

1. The wave train moves toward shore. When the water is less than half the wavelength in depth, the wave "feels" bottom.

2. The circular motion of water molecules in the wave is interrupted. Circles near the bottom flatten to ellipses. The wave's energy must now be packed into less water depth; so the wave crests become peaked rather than rounded.

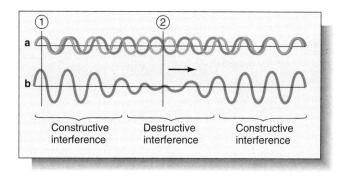

Figure 10.14 Constructive and destructive interference. (a) Two overlapping waves of different wavelength are shown in blue and green. Note the wave shown in blue has a slightly longer wavelength. If both are present in the ocean at the same time, they will interfere with each other to form the composite wave shown in (b). At the position of line 1, the two waves in (a) will constructively interfere to form very large crests and troughs in (b). At the position of line 2, the two waves will destructively interfere, and crests and troughs in (b) will be very small.

3. Friction with the bottom slows the wave. Waves behind it continue toward shore at the original rate. Wavelength therefore decreases, but period remains unchanged.

4. The wave becomes too high for its wavelength, approaching the critical 1:7 ratio.

5. As the water becomes even shallower, the part of the wave below average sea level slows because of friction with the bottom. When the wave was in deep water, molecules at the top of the crest were supported by the molecules ahead (thus transferring energy forward). This is now impossible

BOX 10.2   ● *Rogues at Sea*

In 1942, the Cunard liner RMS *Queen Mary*, then the largest ship in the world at 310 meters (1,020 feet) and 81,000 tons, encountered a rogue wave while transporting 15,000 American soldiers to Glasgow. About 1,100 kilometers (700 miles) off the coast of Scotland, the mountainous wave struck the ship broadside and heeled her past 45°. An eyewitness reported that the *Queen* ". . . listed until her upper decks were awash and those who had sailed in her since she first took to the sea were convinced she would never right herself." She did, but as the captain calmly noted, "She took rather a long time to do so."

Earlier, smaller ocean liners suffered even more severely. An encounter with a rogue wave nearly capsized another Cunard liner, RMS *Carmania*, in 1912. Already reeling from the effects of a huge North Atlantic extratropical cyclone, *Carmania* was struck broadside by a towering rogue wave:

> The ship went through a series of violent 50° rolls that carried kitchen ranges away and sent them grinding back and forth across the tiles. Deck chairs, in the words of one observer, "floated through the air like pieces of paper." Some cabins were made untenable as beds, only recently introduced as a luxury, battered their way through wooden partitions. Passengers thus dispossessed finally lay flat in the ship's square, clutching the base of the pillars.

Modern ships are not immune to the effects of rogue waves. The Italian Lines' beautiful flagship *Michelangelo*

A rare photo of a rogue wave, taken in 1935 from the liner SS *Washington*, eastbound across the North Atlantic. Rogue waves occur when two or more wave crests add together to form a short-lived high wave.

was severely damaged by a rogue wave in a 1966 storm only 1,500 kilometers (900 miles) off New York. Inch-thick glass was shattered on the bridge 24 meters (80 feet) above the waterline, steel railings on the upper decks were ripped away, and the bow was flattened. Many passengers were injured; three died. Even the contemporary Cunard liner *Queen Elizabeth 2* has suffered from rogue wave encounters. After an April 1972 incident in the North Atlantic, the bill for replacement crockery exceeded £2,000!

Source: John Maxtone-Graham, *The Only Way to Cross* (New York: Macmillan, 1972), 297–98.

---

because the *water* is moving faster than the *wave*. As the crest moves ahead of its supporting base, the wave breaks. The break occurs at about a 3:4 ratio of wave height to water depth. (That is, a 3-meter wave will break in 4 meters of water.) The turbulent mass of agitated water rushing shoreward during and after the break is known as **surf.** The **surf zone** is the region between the breaking waves and the shore.

Waves break against the shore in different ways, depending in part on the slope of the bottom. The break can be violent and toppling, leaving an air-filled channel (or *tube*) between the falling crest and the foot of the wave. These **plunging waves (Figure 10.16a)** form when

waves approach a steeply sloping bottom. A more gradually sloping bottom generates a milder **spilling wave** as the crest slides down the face of the wave (**Figure 10.16b**). **Surging waves (Figure 10.16c)** occur when an abrupt beach slope prevents a proper break and water simply rushes ashore. Along with some wind waves, storm surges and seismic sea waves form surging waves.

Slope alone does not determine the position and nature of the breaking wave. The contour and composition of the bottom can also be important. Gradually shoaling bottoms can sap waves of their strength because of prolonged friction against the bottom of the lowest elliptical water orbits. Energy may be lost even more rapidly if the bottom is covered with loose gravel or irregular growths of coral. Masses of moving sea-

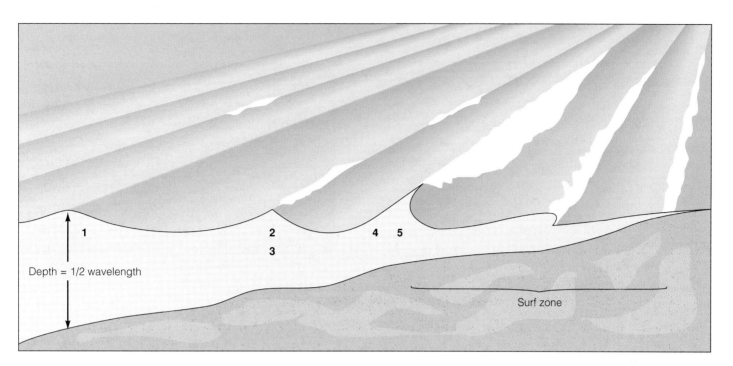

Figure 10.15 How a wave train breaks against the shore. (1) The swell "feels" bottom when the water is shallower than one-half the wavelength. (2) The wave crests become peaked because the wave's energy is packed into less water depth. (3) Friction with the bottom slows the wave, while waves behind it maintain their original speed. Therefore, wavelength shortens, but period remains unchanged. (4) The wave approaches the critical 1:7 ratio of wave height to wavelength. (5) The wave breaks when the ratio of wave height to water depth is about 3:4.

a

b

Figure 10.16 Types of breaking waves. (a) Plunging waves form when the bottom slopes steeply toward shore. (b) Spilling waves form when the bottom slopes gradually. (c) Surging waves form when the beach slopes too abruptly for the waves to break.

c

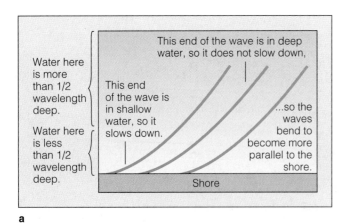

a

b

Figure 10.17 Wave refraction. (a) Diagram showing the elements that produce refraction. (b) Wave refraction around Maili Point, Oahu, Hawaii. Note how the wave crests bend almost 90° as they move around the point.

weeds or jostling chunks of sea ice can also extract energy from a wave. In a few rare cases the shore is configured in such a way that waves don't break at all: The waves have lost virtually all their energy by the time their remnants arrive at the beach.

Wave Refraction

What happens when a wave line approaches the shore at an angle, as it almost always does (**Figure 10.17**)? The line does not break simultaneously because different parts of it are in different depths of water. The part of the wave line in shallow water slows down, but the attached segment still in deeper water continues at original speed; so the wave line bends, or refracts. The bend can be as much as 90° from the original direction of the wave train. This slowing and bending of waves in shallow water is called **wave refraction**.[2] The refracted waves break in a line almost parallel to the shore.

Wave refraction can produce some odd surf patterns. A swell with a wavelength of 600 meters "feels" bottom at 300 meters (1,000 feet), while a swell with a wavelength of 200 meters "feels" bottom at 100 meters (330 feet). You may remember that the average depth at the outer continental shelves is only about 150 meters (500 feet). As these two wave trains approach shore, the longer swell will be slowed by the shelf for perhaps 60 kilometers (38 miles) before the shorter swell is slowed by the bottom. With one wave bending and the other heading in its original direction, the potential for complex interference is great. Thus, waves may break for a time at one spot, cease, and then begin breaking at another spot a few hundred meters down the beach. The

ocean surface might have a checkerboard appearance from the constructive and destructive interference of these crests (see **Figure 10.18**).

Wave Diffraction

Wave diffraction is the propagation of a wave around an obstacle. Unlike wave refraction (which depends on a wave's response to a change in velocity), wave diffraction depends on the interruption of the obstacle to provide a new point of departure for the wave. This is illustrated by the gap in the breakwater in **Figure 10.19**. Wave crests excite the water in the gap. Water moving in the gap generates smaller waves in the harbor, which radiate from the gap and disturb the quiet water. The diffracted waves are much smaller than those in the open ocean, but boats in the harbor might still be jostled.

A more complex case of diffraction is shown in **Figure 10.20,** in which waves are interrupted by a chain of islands. Some of the wave energy "squeezes" through the spaces between the islands. These spaces act as if they were new sources of waves. The waves spread into the space past the island, creating areas of reinforcement and cancellation. Polynesian navigators learned early in their training to sense the disturbances in wave patterns caused by diffraction. The best among them could feel the presence of islands beyond the horizon by subtle irregularities in the rocking of their canoe.

Wave Reflection

The waves discussed so far in this chapter—the familiar ocean waves in which the disturbance travels in one direction along the surface of the transmission medium—are known as *progressive waves*. A vertical barrier such as a seawall, large ship hull, or smooth jetty will

[2]To review the principle of refraction, see Chapter 7, page 164.

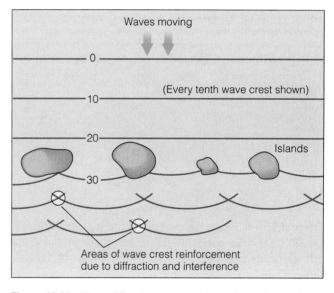

Figure 10.18 A "checkerboard" sea surface off La Jolla, California, the result of wave interference.

Figure 10.19 Diffraction of waves by a breakwater gap at Morro Bay, California.

Figure 10.20 Wave diffraction past an island chain. Polynesian navigators used diffraction patterns to sense the presence of islands out of sight over the horizon (see Figure 2.5).

reflect progressive waves with little loss of energy. If the waves approach the obstruction straight on, the reflected waves will move away from the obstruction in the direction from which they came. This **wave reflection** will cause interference in the form of vertical oscillations called **standing waves.** Because of constructive interference between crests (and troughs), these waves can be dangerous to boats or swimmers near the obstruction. Waves approaching the obstruction at an angle can also reflect; the sea surface near the reflector will not form standing waves, but complicated sea-surface motions

develop. Watch for these effects the next time you visit a solid breakwater or a steep beach.

INTERNAL WAVES

Progressive waves can occur at the junction between air and water, as we have seen, or they can form at the boundary between water layers of different densities. These subsurface waves are called **internal waves.** As is the case with ocean waves at the air–ocean interface,

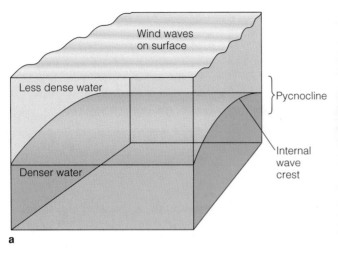

Figure 10.21 Internal waves can form between masses of water with different densities, especially at the base of the pycnocline. (a) The crest of an internal wave. (b) Internal waves diffracting around the Seychelles Islands in the Indian Ocean, photographed by the crew of the space shuttle *Columbia* in January 1990. The waves are visible because their crests have altered the reflectivity of the ocean surface. Patterns of constructive and destructive interference can be seen.

internal waves possess troughs, crests, wavelength, and period. "Desktop-ocean" devices, which are sold at some gift shops, demonstrate internal waves. These sealed bottles contain nonmixing liquids of contrasting colors and slightly different densities. When you tilt the bottle, a very slow internal wave forms at the junction of the liquids, moves slowly to the low end, and breaks.

Normal ocean waves move rapidly because the difference in density between air and ocean is relatively great. Internal waves usually move very slowly because the density difference between the joined media is very small. Internal waves occur in the ocean at the base of the pycnocline, especially at the bottom edge of a steep thermocline (**Figure 10.21a**). The wave height of internal waves may be greater than 30 meters (100 feet), causing the thermocline to undulate slowly through a considerable depth. Their wavelength often exceeds 0.8 kilometer (0.5 mile); periods are typically 5 to 8 minutes. Oceanographers are not certain how these large, slow-motion waves are generated, but wind energy, tidal energy, or ocean currents may be responsible. Surface manifestations of internal waves have been photographed from space (**Figure 10.21b**).

Q-AND-A

1. If so much energy is expended as wind waves break, why doesn't water in the surf zone get hot?

   The amount of energy moving through the ocean in progressive waves is impressive. The energy of a wave is proportional to the square of its height. Inman and Brush (1973) report that each linear meter of a wave 2 meters above average sea level represents an energy flow of about 25 kilowatts (34 horsepower), enough to light 250 100-watt light bulbs; a wave twice as high would contain four times as much energy. A single wave 1.2 meters high striking the west coast of the United States may release as much as 50 million horsepower!

   The energy of wind waves is dissipated mostly as heat in the surf, but since water has such a high latent heat (discussed in Chapter 7), the injection of heat into the surf zone doesn't significantly increase water temperature. The surf zone is also an area of vigorous mixing; so any heat released is quickly distributed through a large volume of water.

   A number of schemes have been proposed to take advantage of this energy before it is dissipated. Among the most

BOX 10.3 ● *Surfing*

More than 2 million Americans surf regularly. The thrill of rushing down the face of a growing, breaking wave is exhilarating! People willingly endure cold, boredom, and some danger for a ride lasting only a few seconds. If you're a good swimmer, give it a try. You will need to paddle your board or swim vigorously to match your speed to that of the advancing wave crest. As the wave rises to break, your forward speed (and sense of timing) will place you on the leading edge of the crest, accelerating downward and forward. The technique takes time to master, but the feeling is worth the effort!

Prejudice compels me to reveal that *real* surfers are body surfers. The buoyant forces on which board surfing depends are diminished in body surfing, but forces within the wave propel the body surfer with greater acceleration. A body surfer's intimate contact with the water and close proximity to the wave surface greatly increase the sensation of movement.

The ultimate trick is to surf the wave completely submerged, as a dolphin does. You push powerfully off the bottom (or swim rapidly toward the surface) as the wave crest approaches, inserting yourself into the freshly breaking wave from below. The rapid flow of water within the toppling wave will ripple your skin, and the sudden acceleration can thrust you from the face of the wave like a wet bar of soap shooting from a fist. Altogether a wonderful oceanic experience!

Body surfing. Many surfers prefer body surfing to board surfing because of the body surfer's proximity to the water and the sensation of great speed and acceleration.

promising is a design proposed by Kvaerner Industrier A/S, Oslo, Norway (see Figure 19.6).

2. What about wind waves and the Coriolis effect? Do wind waves turn right in the Northern Hemisphere as they move across the ocean surface?

   The Coriolis effect has no influence on waves with periods less than about 5 minutes. Large shallow-water waves such as tsunami, seiches, and tides involve the mass movement of water and are influenced by the Earth's rotation. Wind waves, however, with a period rarely exceeding 20 seconds, are not.

3. I love to surf. Where should I plan to live?

   The Pacific has the largest potential fetch distances; so the best chances for large wind waves are there. The biggest wind waves are made in the West Wind Drift, but temperatures there are much too cold for comfortable surfing. Besides, the sea state in areas of continuous high wind is chaotic, and surfing requires orderly waves. It is best to let the wave trains sort out. Hawaii is in the middle of the

Pacific, so wind waves from polar storms in either hemisphere strike its shores only after much sorting by wavelength. Order is assured. The water is warm, too. I vote for Hawaii.

4. Could a method be invented to surf internal waves?

   Think of the possibilities: a wave 30 meters high moving in slow motion with a period of 5 to 8 minutes. It would be a calm and stately ride, and the density differences would require a very delicate balancing act at the interface between the two water densities. Still, with the right equipment to adjust buoyancy and provide air, it might be done.

   Indeed, surfing an internal wave may already have been inadvertently and disastrously accomplished. The slow undulation of an internal wave may have led to the loss in 1963 of USS Thresher, first of a new class of U.S. Navy fast-attack submarines. An equipment failure at high speed coupled with a surprise wave encounter may have driven the new submarine past its designed pressure limit, causing the violent implosion that cost 129 lives.

Terms and Concepts to Remember

capillary waves	plunging wave	$V = \sqrt{gd}$
constructive interference	progressive wave	wave
	restoring force	wave crest
deep-water wave	rogue wave	wave diffraction
destructive interference	sea	wave frequency
	shallow-water wave	wave height
dispersion		wavelength
disturbing force	spilling wave	wave period
fetch	standing wave	wave reflection
fully developed sea	surf	wave refraction
gravity waves	surf beat	wave steepness
group velocity	surf zone	wave train
interference	surging wave	wave trough
internal wave	swell	wind duration
orbit	transitional wave	wind strength
orbital wave	$V = L/T$	wind wave

Study Questions

1. How do particles move in an ocean wave? How is that movement similar to or different from the movement of particles in a wave in a spring or a rope? How does this relate to a *stadium wave*—a waveform made by sports fans in a circular arena?

2. Draw an ocean wave and label its parts. Include a definition of *wave period.*

3. Make a list of ocean waves, arranged by disturbing force and wavelength. What is restoring force? How is a capillary wave different from a gravity wave?

4. What is the general relationship between wavelength and wave speed? How does water movement in a wave change with depth? Can deep-water waves and shallow-water waves exist at the same point offshore (that is, in the same depth of water)?

5. What factors influence the growth of a wind wave? What is a fully developed sea? Where would we regularly expect to find the largest waves? Are waves in fully developed seas always huge?

6. How can a rogue wave be larger than the theoretical maximum height of waves in a fully developed sea?

7. What happens when a wind wave breaks? What factors affect the break? How are plunging waves different from spilling waves?

8. Describe *wave reflection, refraction,* and *diffraction.*

9. How is a progressive wave different from a standing wave? Must standing waves be orbital waves only, or can standing waves also form in shaken ropes or pushed-and-pulled springs?

10. What are internal waves? Are internal waves larger or smaller than most wind waves? Slower or faster?

For Further Study

Bascom, W. 1959. "Ocean Waves." *Scientific American*, August, 89–97. A classic reference on the subject, well known to every oceanography professor and almost every student. Excellent illustrations.

Bascom, W. 1980. *Waves and Beaches.* Rev. ed. Anchor/Doubleday. Perhaps the best work for the lay reader on the subject. Excellent photographs and diagrams, lively and clear writing.

Bigelow, H. B., and W. T. Edmondson. 1947. *Wind Waves at Sea, Breakers and Surf.* Reprinted. Washington, DC: U.S. Navy Hydrographic Office, 1953. A classic now showing its age.

Bowditch, N. 1966. *American Practical Navigator.* Washington, DC: U.S. Navy Hydrographic Office. A one-volume nautical reference of unexcelled brevity and quality. Chapters on waves are clear, direct, and written from a practical mariner's viewpoint.

Chelton, D. B., et al. 1981. "Global Satellite Measurements of Water Vapor, Wind Speed, and Wave Height." *Nature* 294 (no. 5841): 529–32. *Seasat* information on wave height.

Coles, K. A. 1968. *Heavy Weather Sailing.* New York: de Graaf. Outstanding reference, valuable appendixes.

Deacon, G. 1984. *The Antarctic Circumpolar Ocean.* Cambridge: Cambridge University Press. Information on *Seasat* observations of wind waves in the West Wind Drift.

Elachi, C. 1982. "Radar Images of the Earth from Space." *Scientific American*, December, 54–61. Surface waves are clearly visible at radar wavelengths.

Inman, D. L., and B. M. Brush. 1973. "The Coastal Challenge." *Science* 181: 20–31. Power statistics for wind waves.

Kampion, D. 1989. *The Book of Waves.* Santa Barbara, CA: Arpel Graphics. The definitive coffee-table book of waves. Magnificent photography.

Kinsman, B. 1965. *Wind Waves: Their Generation and Propagation on the Ocean Surface.* Englewood Cliffs, NJ: Prentice-Hall. The definitive technical work on the subject.

Maxtone-Graham, J. 1972. *The Only Way to Cross.* New York: Collier/Macmillan. Rogue waves and other stories. One of the best books on the era of the fabled transatlantic liners. Highly recommended for its wonderful writing style.

Noll, G., and A. Gabbard. 1989. *Da Bull: Life over the Edge.* Bozeman, MT: Bangtail Press. Surfer Greg Noll rides Hawaii's largest wave. Historical view of the golden age of board surfing.

Robinson, J. 1976. "Superwaves of Southeast Africa." *Sea Frontiers* 22 (no. 2). Rogue waves at their finest.

Smith, F. G. W. 1970. "The Simple Wave." *Sea Frontiers* 16 (no. 5). Turns out waves are not really very simple.

Stevens, O., et al. 1981. "Sailing Yacht Capsizing." In *Proceedings of Chesapeake Sailing Yacht Symposium, January, 1981.* Society of Naval Architects and Marine Engineers. Terrifying!

Stewart, C. 1968. "Wave Theory and Facts." In *Heavy Weather Sailing,* edited by K. A. Coles. New York: de Graaf. Basic mathematics of waves presented clearly and well.

Stewart, R. H. 1985. *Methods of Satellite Oceanography.* Berkeley: University of California Press. Techniques for imaging waves from space.

Sverdrup, H., et al. 1942. *The Oceans: Their Physics, Chemistry, and General Biology.* Englewood Cliffs, NJ: Prentice-Hall. Thorough mathematical account of waves.

Whitemarsh, R. P. 1934. "Great Sea Waves." *Proceedings of the United States Naval Institute* 60 (no. 8). The commanding officer of *Ramapo*'s account of The Wave.

11 TSUNAMI, SEICHES, AND TIDES

The Gods of Lituya Bay

One of the largest waves ever observed on Earth formed on 9 July 1958. Four of the six eyewitnesses lived to tell about it.

The great wave formed in Lituya Bay, a narrow glacier-fed inlet on the northeast shore of the Gulf of Alaska. The area was explored in 1788 by the French explorer Jean-François La Pérouse, but the only notable oceanic feature mentioned in his account was the strong tidal current encountered at the bay's narrow entrance. Had he inquired, the local inhabitants could have told him of the awesome magic contained within the bay: Waves ". . . as big as gods."

Three pleasure boats were anchored in the bay on the calm evening the great wave appeared. At 10:16 P.M. the crew of the U.S. Geologic Survey ship *Stephen R. Capps,* lying at anchor in another bay about 100 kilometers (63 miles) to the east, felt a strong earthquake along the Fairweather Fault, a fault that angles west along the steep mountains rimming the innermost reaches of Lituya Bay. The earthquake dislodged 61 million cubic meters (80 million cubic yards) of rock, mud, and glacial ice, which plummeted into upper Lituya Bay. Water displaced by this enormous mass of debris generated a huge surge that raced to the Bay's opposite wall, where it climbed 530 meters (1,740 feet) above sea level. (The Sears Tower in Chicago, the world's tallest inhabited structure, is 87 meters—286 feet—shorter than the height of this "splash"!) The displacement also generated a gravity wave that roared out into the bay at a speed between 155 and 210 kilometers (97 and 130 miles) per hour. This wave was at least 49 meters (160 feet) high, roughly the height of a 15-story building.

Mr. Howard Ulrich, aboard the small boat *Edrie,* was probably the first to see the wave in the gathering darkness. Ulrich and his seven-year-old son had been awakened by a deafening roar and had gone on deck to investigate. Ulrich did not comprehend the wave at first; it looked like a low cloud or fogbank extending from shore to shore. The monster began to break as it rounded the north side of an island in the center of the bay, but on the south side, where his boat rested at anchor, the wave had not started to curl. Ulrich was unable to free the anchor, but the line snapped as the boat rose to meet the wave. *Edrie* accelerated upward to the crest, was carried forward over a submerged peninsula, and was deposited into a different section of the now wildly churning bay. Amazingly, the two were able to leave the bay under their boat's own power the next day.

Mr. and Mrs. William Swanson, aboard their fishing trawler *Badger* anchored on the north side of the bay, were not so fortunate. The Swansons were awakened by a violent vibration. At first, William Swanson saw nothing unusual; then a blue wall moved past the northern end of the island toward their boat, traversing the distance in about 90 seconds. The trawler's anchor chain parted, and she ascended the wave bow-first at a terrifying angle. Then the top of the wave broke, engulfing the boat. *Badger* shuddered and then rushed stern-first down the face of the shattered wave, wind howling through her rigging. She was carried at a height of 25 meters (80 feet) over the tops of the highest trees, over the bay mouth bar, and out to sea. The Swansons' boat was fatally broken. A deserted skiff was floating nearby, and they clambered aboard as *Badger* sank. They were rescued by a fishing boat 2 hours later.

The two-person crew of the third boat was unluckiest of all. They perished within the wave. William Swanson last saw the *Sunmore* in the trough of the wave moving toward a rocky cliff. No sign of crew or boat was ever found.

The 49-meter height of the wave was established by survey crews analyzing the line of tree devastation around the periphery of the bay (clearly visible in the photograph). It gives one pause to learn that this wave was comparatively small. Though they have not been directly observed, much larger waves have left trails of smashed timber even farther inland. Evidence of a 61-meter (200-foot) wave was reported in 1899; an 1853 wave reached 120 meters (395 feet) in height; and on 27 October 1936 a landslide-generated wave rose to 150 meters (490 feet) in Lituya Bay!

A panoramic view of Lituya Bay, Alaska, showing destruction caused by a landslide-generated tsunami on 9 July 1958. The surge and giant wave of water generated by a rock slide at the far end of the bay destroyed the forest in the light areas along the shores of the bay.

CHAPTER OVERVIEW

The waves discussed in this chapter are caused by the low atmospheric pressure and wind associated with large cyclonic storms, the rocking of water in confined harbors and bays, the sudden displacement of the water surface by phenomena such as earthquakes and landslides, and gravitational attraction.

Unlike wind waves, these great waves—including storm surges, seiches, tsunami, and tides—have such long wavelengths that they are always in "shallow water" (water that is shallower than half their wavelength). These long waves travel at high speeds. Some of the waves can be destructive, but their ability to cause damage is fortunately not proportional to their wavelengths.

The "gods" of Lituya, known as tsunami, are members of the class of long waves. Although they have a different origin from wind waves, the wavelength, wave height, and wave period of long waves are measured in the same way. And as was the case for wind waves, energy in most large waves passes through the ocean by the orbital movement of water particles.

Listed in order of wavelength, from smallest to largest, the waves described in this chapter are: storm surges, seiches, tsunami, and tides. Even the smallest of these waves is much longer in wavelength than the largest wind waves. Because of their very long wavelengths, each of the waves in this chapter is classified as a shallow-water wave. As we shall see, all of these waves move through the ocean very rapidly.

A WORD ABOUT "TIDAL WAVES"

The term *tidal wave* is often misused to describe some of the waves discussed in this chapter. The popular media and general public tend to label *any* unusually large wave a tidal wave regardless of its origin. Press accounts of storms at sea usually list a rogue wave as a tidal wave, and very large sets of wind waves are called tidal waves by some yacht owners or surfers. The sea waves associated with earthquakes are almost always called tidal waves in media damage reports; certainly the Lituya Bay waves have been misnamed this way. The waves caused by the approach of a tropical cyclone to land may also incorrectly be termed tidal waves. The term *tidal wave* is *not* synonymous with *large wave,* however. As we will see, the only true tidal waves are relatively harmless waves associated with the tides themselves.

STORM SURGES

The abrupt bulge of water driven ashore by a tropical cyclone (hurricane) or frontal storm is called a **storm surge.** The low atmospheric pressure associated with a great storm will draw the ocean surface into a broad dome as much as 1 meter (more than 3 feet) higher than average sea level. This dome of water accompanies the storm to shore, becoming much higher as the water gets shallower at the coast. There the water ramps ashore, driven forward by large, storm-generated wind waves (**Figure 11.1**). The crest of a storm surge can temporarily add up to 7.5 meters (25 feet) to coastal sea level. Water can reach even greater heights when the surge is funneled into a confined bay or estuary.

A storm surge is a short-lived phenomenon. Technically it is not a wave because it is only a crest; so wavelength and period cannot be assigned to it. Water in a storm surge does not come ashore as a single breaking wave but rushes inland in what looks like a sudden, very high, windblown tide. Indeed, storm surges are sometimes called *storm tides.* The wrong combination of low atmospheric pressure, strong onshore winds, high tide, and bottom contour can be very dangerous—especially if estuaries in the area have been swollen by heavy rainfall preceding the storm.

Storm surges have had catastrophic consequences. The frightful tropical storm of November 1970 in Bangladesh (described in Chapter 8) generated a storm surge up to 12 meters (40 feet) high that caused the death of more than 300,000 people. Storm surges associated with extratropical cyclones (frontal storms) can also do tremendous damage. On 1 February 1953, a storm surge and high tide arrived simultaneously against the Dutch coast. Wind waves breached the dikes and flooded the low country, covering more than 3,200 square kilometers (800,000 acres) and drowning 1,783 people. The Dutch

Figure 11.1 A storm surge. The low pressure and high winds generated by a hurricane can produce a storm surge up to 9 meters (30 feet) high. Hurricane Camille, which struck the Mississippi coast in 1969, produced a storm surge 7 meters (23 feet) high. The massive inundation was responsible for a substantial fraction of the $1.4 billion in damage suffered in the area. In August of 1992, Hurricane Andrew struck the coasts of South Florida and Louisiana with sustained winds of 225 kilometers (140 miles) per hour. A 4-meter (13-foot) storm surge was responsible for heavy damage to harbors, coastal towns, and wildlife refuges south of Miami. With losses approaching $30 billion, Hurricane Andrew was the most expensive natural disaster in American history.

Figure 11.2 The Thames tidal barrier in London. The natural funnel shape of the Thames estuary concentrates surges as they near London. The barrier sections can be raised (a) to prevent flooding upstream. Each barrier section is controlled by hydraulic arms within towers. The great size of the $1.3 billion project can be seen in (b).

a

b

anticipate that this coincidence of events will occur only about once in 400 years. The dikes have been rebuilt higher and the land reclaimed from the North Sea at the cost of billions of dollars. On the opposite side of the North Sea, Londoners have spent $1.3 billion on a flood defense system at the mouth of the Thames River. The centerpiece of the project is an immense barrier against storm surges (see **Figure 11.2**). Experts expect the barrier to prevent a devastating flood on an average of once every 50 years.

The U.S. coast is also at risk. In 1900, a storm surge topped the Galveston, Texas, seawall, swept into the city, and killed more than 6,000 area residents. (In contrast, no lives were lost during a similar Texas storm in 1961—because coastal barriers had been constructed and advance warning permitted preparation and evacuation.) In August 1954, the combined effects of a high tide and Hurricane Carol caused water to rise almost 5 meters (16 feet) above normal to flood much of Providence, Rhode Island. Damage exceeded $41 million. City officials have since built a protective surge barrier across upper Narragansett Bay. Anyone living in a low-lying coastal area frequented by violent storms should be aware of the potential danger of storm surge and should heed calls for evacuation if conditions warrant.

SEICHES

When disturbed, water confined to a small space (such as a bucket, a bathtub, or a bay) will slosh back and forth at a specific resonant frequency. The frequency changes with different amounts of water, or with different sizes or shapes of containers. If you carry a shallow tray of water from one place to another, you're careful not to move the tray at its resonant frequency, to avoid a spill. Most of the water's random motion quickly settles down after you place the tray on a table top, but the water in the tray may rock gently for some seconds at this one resonant frequency. That rocking is a **seiche** (pronounced *saysh*).

The seiche phenomenon was first studied in Switzerland's Lake Geneva by eighteenth-century researchers curious about why the water level at each end of the long, narrow lake rises and falls at regular intervals after windstorms. They found that constant breezes tend to push water into the downwind end of the lake. When the wind stops, the water is released to rock slowly back and forth at the lake's resonant frequency, completing a crest-trough-crest cycle in a little more than an hour. At the ends of the lake the water rises and falls a foot or two; at the center it moves back and forth without changing height (**Figure 11.3**). This kind of wave is called a **standing wave** because it oscillates vertically with no forward movement. The point (or line) of no vertical wave action in a standing wave—the place in the lake where the water moves only back and forth—is called a **node**. In Lake Geneva the wavelength of the seiche is twice the length of the lake itself; the node lies at the center. The lake acts like a large version of the water tray in the example above.

Energy can be added rhythmically to a seiche. The process is analogous to adding energy to a child in a swing. Insert a little energy at the right time in each cycle, and the child will swing higher and higher. (As our daughter discovered at an early age, the same thing applies to water in a bathtub. By moving her body in rhythm with the growing wave, she quickly learned how to slop most of the bathwater onto the floor.)

This kind of rhythmic oscillation in a seiche can contribute to lakeside property damage during storms. Lake Michigan began to rock because of successive storms during the first week of February 1987. On 8 February water level at the south end of the lake was swinging a meter (more than 3 feet) above normal. North-northwest winds from a storm driving down the long axis of the lake generated wind waves as high as 3 to 6 meters (10 to 20 feet) above this bulge. The waves caused severe flooding and erosion in Chicago.

This kind of activity can also occur in confined areas of ocean ranging in size from bays and harbors to entire ocean basins. As we have seen, a drop in atmospheric pressure beneath a storm can draw water to a slightly higher level. When the storm moves inland the water is released, and the oscillation begins. The wavelength of the oscillation is a multiple of the length of the ocean basin, often thousands of miles. Tides and surf beat can also cause seiches in bays or harbors, as can internal

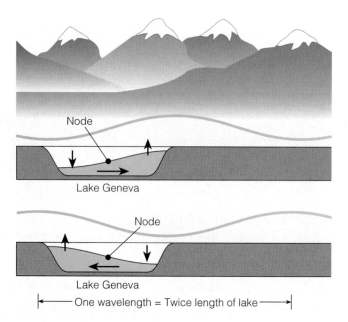

Figure 11.3 Seiche in Lake Geneva. After a windstorm, the lake sloshes back and forth, forming a standing wave (seiche) with a wavelength of 220 kilometers (132 miles)—twice the length of the lake. The period of the seiche is about 72 minutes, and its height is about 0.5 meter (1.5 feet). At the node, water moves sideways and does not rise or fall.

waves moving toward shore. Seiche periods may range from a few minutes to more than a day. In large areas the wavelength of these standing waves may rival that of tsunami.

Damage from seiches along most ocean coasts is rare. The wavelength may be tremendous, but seiche wave height in the open ocean rarely exceeds a few inches. Larger seiches can occur in harbors: Coastal seiches at Nagasaki, on the southern coast of Japan, occasionally reach 3 meters (10 feet). Seiches may disturb shipping schedules by interfering with the predicted arrival times of tides; or they may cause currents in harbors, which could snap mooring lines. In rare instances, seiches contribute to the peril of people on shore, but usually because the seiche is coincidentally accompanied by wind waves of considerable height or by a tsunami.

TSUNAMI AND SEISMIC SEA WAVES

Long-wavelength, shallow-water progressive waves caused by the rapid displacement of ocean water are called **tsunami**, a descriptive Japanese term combining *tsu* (harbor) with *nami* (wave). The word is both singular and plural. Tsunami caused by the sudden vertical movement of the Earth along faults (the same forces that cause earthquakes) are properly called **seismic sea waves**. Tsunami can also be caused by landslides, ice-

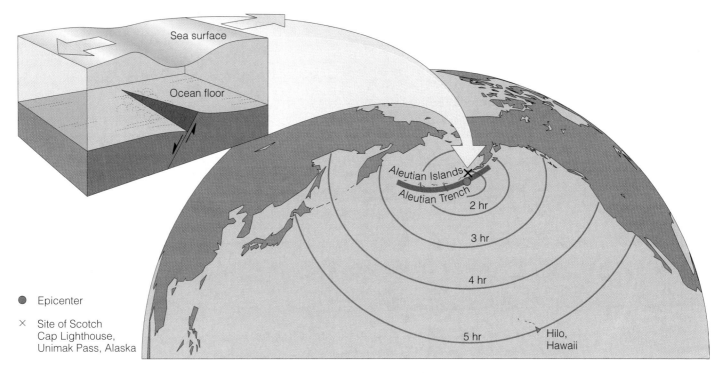

Figure 11.4 The tsunami of 1 April 1946 began when a rupture along a submerged fault lifted the sea surface above. The wave moved outward at a speed of about 212 meters per second (472 miles per hour). At this speed, it took only about 5 hours to travel to the Hawaiian Islands.

bergs falling from glaciers, volcanic eruptions, and other direct displacements of the water surface. Note that all seismic sea waves are tsunami, but not all tsunami are seismic sea waves.

Origins

"Small" tsunami are caused by the displacement of surface water. Although less energy is released by landslides than by most seismic fractures, the resulting sea waves are still very destructive for people or structures near their point of origin. This is especially true if the wave is formed within a confined area, as we saw in the Lituya Bay example at the beginning of this chapter.

Seismic sea waves originate on the ocean floor when Earth movement along faults displaces ocean water. **Figure 11.4** shows the birth of a seismic sea wave in the Aleutians. Rupture along a submerged fault lifts the sea surface above. Gravity pulls the crest downward, but the momentum of the water causes the crest to overshoot and become a trough. The oscillating ocean surface generates progressive waves that radiate from the epicenter in all directions (see **Figure 11.5**). Waves would also form if the fault movement were downward. In that case a depression in the water surface would propagate outward as a trough. The trough would be followed by smaller crests and troughs caused by surface oscillation.

It seems strange to refer to tsunami—waves with wavelengths of up to 200 kilometers (125 miles)—as shallow-water waves. Yet half their wavelength would be 100 kilometers (62 miles), and even the deepest ocean trenches do not exceed 11 kilometers (7 miles) in depth. These immense waves therefore never find themselves in water deeper than half their wavelength. Like any transitional or shallow-water wave, seismic sea waves are affected by the contour of the bottom and are commonly refracted, sometimes in unexpected ways.

Speed

The velocity of a tsunami is given by the formula for the speed of a shallow-water wave:

$$V = \sqrt{gd}$$

Since the acceleration due to gravity (g) is 9.8 meters (32.2 feet) per second per second, and since a typical Pacific abyssal depth (d) is 4,600 meters (15,000 feet), solving for V shows that the wave would move at 212 meters per second (472 mi/hr). At this speed a seismic sea wave will take only about 5 hours to travel from Alaska's seismically active Aleutians to the Hawaiian Islands.

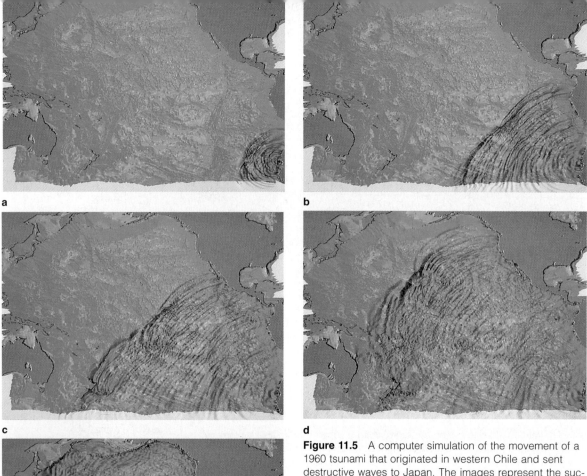

Figure 11.5 A computer simulation of the movement of a 1960 tsunami that originated in western Chile and sent destructive waves to Japan. The images represent the successive positions of the waves (a) 2.5 hours after the earthquake, (b) after 5 hours, (c) after 12.5 hours, (d) after 17.5 hours, and (e) reaching Japan after 22.5 hours of travel.

Encountering Tsunami

We are familiar with the steepness of a wind wave and the short period of a few seconds between its crests. Tsunami are much different. Once a tsunami is generated, its steepness (ratio of height to wavelength) is extremely low. This lack of steepness, combined with the wave's very long period (5 to 20 minutes), enables it to pass unnoticed beneath ships at sea. A ship on the open ocean that encounters a tsunami with a 16-minute period would rise slowly and imperceptibly for about 8 minutes, to a crest only 0.3 to 0.6 meter (1 or 2 feet) above average sea level. It would then ease into the following trough 8 minutes later. With all the wind waves around, such a movement would not be noticed.

As the seismic sea wave crest approaches shore, however, the situation changes rapidly and often dramatically. The period of the wave remains constant, velocity drops, and wave height greatly increases. As the crest arrived at the coast, observers would see water surge ashore in the manner of a very high, very fast tide. In confined coastal waters relatively close to their point of origin, tsunami can reach a height of perhaps 30 meters (100 feet). The wave is a fast, onrushing flood of water, not the huge, plunging breaker of popular folklore.

The wave energy spreads through an enlarging circumference as a tsunami expands from its point of origin. People on shore near the generating shock have reason to be concerned because the energy will not have dissipated very much. On 1 April 1946 a fracture along the Aleutian trench generated a seismic sea wave that quickly engulfed the Scotch Cap lighthouse on Unimak Island in the Aleutians. The lighthouse was completely destroyed, and the five coastguardsmen operating the lighthouse died. (See **Figure 11.6**.)

The same seismic sea wave reached the Hawaiian Islands about 5 hours later. By this time the wave cir-

a   b

Figure 11.6 Tsunami destruction. Scotch Cap lighthouse guards Unimak Pass in Alaska's Aleutian Islands. (a) The lighthouse before 1 April 1946. Its foundation is 15 meters (49 feet) above sea level, the upper plateau 36 meters (118 feet) above sea level. (b) The site after the tsunami. The lighthouse, support buildings, and radio masts are gone, and the slopes heavily washed almost to the plateau. Five coastguardsmen died in the wave.

a

Figure 11.7 Tsunami in progress. (a) Wave front rushes up the Wailuku River in Hilo, Hawaii, during the tsunami of 1 April 1946. Note the steep front, the turbulence of the water behind it, and the placidity of the water in front. The right-hand span of the bridge was destroyed by an earlier wave of the same tsunami. (b) The largest wave of the 1 April 1946 tsunami rushes ashore in Hilo. Terrified residents run for their lives.

b

cumference was enormous and its energy more dispersed. Even so, successive waves surged onto Hawaiian beaches at 15-minute intervals for more than 2 hours. One 9-meter (30-foot) wave struck the town of Hilo, and water rose to a height of 17 meters (56 feet) in an exposed valley near Polulu! At least 150 people were killed in Hawaii that morning, and property damage was in excess of $25 million. Photographs taken in Hilo that morning are shown in **Figure 11.7**.

V-shaped Hilo Bay is an especially dangerous place during a tsunami because its funnel-like outline concentrates the energy of the waves. Fourteen years after the 1946 disaster, a seismic disturbance in the subduction zone off western South America generated a tsunami

BOX 11.1 ● *An Eyewitness Account of a Tsunami*

The 1 April 1946 Hawaii tsunami was observed by marine geologist Francis P. Shepard. An account of his experience was first published in his book *The Earth Beneath the Sea*. An excerpt from this book follows.

The term *tidal wave* has had an ominous sound in Hawaii since April 1, 1946. . . . At that time my wife and I were living in a rented cottage at Kawela Bay on northern Oahu. . . . On the previous day, a Sunday, the beaches and reefs were swarming with people and the cottages alive with activity. Fortunately, almost everybody left to go back to Honolulu that night. Early the next morning we were sleeping peacefully when we were awakened by a loud hissing sound, which sounded for all the world as if dozens of locomotives were blowing off steam directly outside our house. Puzzled, we jumped up and rushed to the front window. Where there had been a beach previously, we saw nothing but boiling water, which was sweeping over the ten-foot top of the beach ridge and coming directly at the house. I rushed and grabbed by camera, forgetting such incidentals as clothes, glasses, watch, and pocketbook. As I opened the door I noticed with some regret that the water was not advancing any farther but, instead, was retreating rapidly down the slope.

By that time I was conscious of the fact that we might be experiencing a tsunami. My suspicions became confirmed as the water moved swiftly seaward, and the sea level dropped a score of feet, leaving the coral reefs in front of the house exposed to view. Fish were flapping and jumping up and down where they had been stranded by the retreating waves. Quickly taking a couple of photographs, in my confusion I accidentally made a double exposure of the bare reef. Trying to show my erudition, I said to my wife, "There will be another wave, but it won't be as exciting as the one that awakened us. Too bad I couldn't get a photograph of the first one."

Was I mistaken? In a few minutes as I stood at the edge of the beach ridge in front of the house, I could see the water beginning to rise and swell up around the outer edges of the exposed reef; it built higher and higher and then came racing forward with amazing velocity. "Now," I said, "here is a good chance for a picture." I took one, but my hand was rather unsteady that time. As the water continued to advance I shot another one, fortunately a little better. . . . As it piled up in front of me, I began to wonder whether this wave was really going to be smaller than the preceding one. I called to my wife to run to the back of the house for protection, but she had already started, and I followed her just in time. As I looked back I saw the water surging over the spot where I had been standing a moment before. Suddenly we heard the terrible smashing of glass at the front of the house. The refrigerator passed us on the left side moving upright out into the cane field. On the right came a wall of water sweeping toward us down the road that was our escape route from the area. We were also startled to see that there was nothing but kindling wood left of what had been the nearby house to the east. Finally, the water stopped coming on and we were left on a small island, protected by the undamaged portion of the house, which, thanks to its good construction and to the protecting ironwood trees, still withstood the blows. The water had rushed on into the cane field and spent its fury.

My confidence about the waves getting smaller was rapidly vanishing. Having noted that there was a fair interval before the second invasion (actually fifteen minutes as we found out later), we started running along the emerging beach ridge in the only direction in which we could get to the slightly elevated main road. As we ran, we found some very wet and frightened Hawaiian women standing wringing their hands and wondering what to do. With difficulty we persuaded them to come with us along the ridge to a place where there was a break in the cane field. As we hurried through this break, another huge wave come rolling in over the reef and broke with shuddering force against the small escarpment at the top of the beach. Then, rising as a monstrous wall of water, it swept on after us, flattening the cane field with a terrifying sound. We reached the comparative safety of the elevated road just ahead of the wave.

There, in a motley array of costumes, various other refugees were gathered. One couple had been cooking their breakfast when all of a sudden the first wave came in, lifted their house right off its foundation, and carried it several hundred feet into the cane field where it set it down so gently that

Two houses washed into a small lake at Kawela Bay, Oahu, Hawaii, during the 1 April 1946 tsunami.

their breakfast just kept right on cooking. Needless to say, they did not stay to enjoy the meal. Another couple had escaped with difficulty from their collapsing house.

We walked along the road until we could see nearby Kawela Bay, and from there we watched several more waves roar on to the shore. They came with a steep front like the tidal bore I had seen move up the Bay of Fundy at Moncton, New Brunswick and up the channels on the tide flat at Mont-Saint-Michel in Normandy. We could see various ruined houses, some of them completely demolished. One house had been thrown into a pond right on top of another. Another was still floating out in the bay.

Finally, after about six waves had moved in, each one apparently getting progressively weaker, I decided I had better go back and see what I could rescue from what was left of the house where we had been living. After all, we were in scanty attire and required clothes. I had just reached the door when I became conscious that a very powerful mass of water was bearing down on the place. This time there simply was no island in back of the house during the height of the wave. I rushed to a nearby tree and climbed it as fast as possible and then hung on for dear life as I swayed back and forth under the impact of the wave. Like the others, this wave soon subsided, and the series of waves that followed were all minor in comparison.

Source: Francis Shepard, *The Earth Beneath the Sea,* revised edition (New York: Atheneum, 1969), 31–35.

that drowned 61 people in Hilo. Wave-cut scars on cliffs north of the town suggest that visits by large tsunami are not rare occurrences.

Note that the destruction in Hawaii was not caused by one wave (as at Scotch Cap), but by a series of waves following one another at regular intervals (see **Box 11.1**). Some energy from the main tsunami wave was distributed into smaller waves ahead of or behind the main wave as it moved. If the epicenter of the displacement responsible for a tsunami is very far away, sea level at shore will rise and fall as these waves arrive. The interval between crests (the wave period) is usually about 15 minutes. Coastal residents far from a tsunami's origin can be lulled into thinking the waves are over; they return to the coastline only to be injured or killed by the next crest. This behavior has contributed to loss of life in Hawaii.

Tsunami in History

Destructive tsunami strike somewhere in the world an average of once each year. An earthquake along the Peru–Chile trench on 22 May 1960 killed more than 4,000 people, and tsunami reaching Japan, 14,500 kilometers (9,000 miles) away, killed 180 people and caused $50 million in structural damage. Los Angeles and San Diego harbors were badly disrupted by seiches excited by the tsunami. In 1992 a tsunami struck the coast of Nicaragua and killed 170 people; 13,000 Nicaraguans were left homeless. In 1993 an earthquake in the Sea of Japan generated a tsunami that washed over areas 29 meters (97 feet) above sea level and killed 120 people (**Figure 11.8**).

Half the time a tsunami's trough will arrive first. This occurred with catastrophic results in the Tagus estuary at Lisbon, Portugal, on 1 November 1755. Many curious people were attracted to the shore when sea level dropped several meters over about 15 minutes. People ventured out to pick up fish, inspect grounded vessels, and look around. Thousands were killed when the first large crest arrived, and would-be rescuers were drowned by later crests, some of which exceeded 10 meters (33 feet) in height! In all, more than 50,000 people died in and around Lisbon.

There have been even greater tsunami disasters. In 1703 at Awa, Japan, 100,000 people died. On 27 August 1883 the explosive volcanic eruption of Krakatoa in Indonesia generated 35-meter (116-foot) waves that destroyed 163 villages and killed more than 36,000 people. Evidence of even larger waves has been discovered. Researchers have found signs of a huge wave, perhaps as high as 91 meters (300 feet), that crashed against the Texas coast 66 million years ago. It may have been caused by a comet or asteroid striking the Gulf of Mexico near Yucatan. The wave scoured the floor of the Gulf;

Figure 11.8 Destruction caused by a tsunami in the Japanese town of Aonae on Okushiri Island in the Sea of Japan in July 1993. Waves washed over areas 29 meters (97 feet) above sea level; 120 people died. This was the largest tsunami to strike Japan in recent times.

picked up sand, gravel, and shark's teeth; and deposited the material in what is now central Texas.

The Tsunami Warning Network

Since 1948, an international tsunami warning network has been in operation around the seismically active Pacific to alert coastal residents of possible danger. Warnings must be issued rapidly because of the speed of these waves through the water. Telephone books in coastal Hawaiian towns contain maps and evacuation instructions for use when the warning siren sounds.

The tsunami warning system was responsible for averting the loss of many lives after the great 27 March 1964 earthquake in Alaska (see Chapter 3). A 3.7-meter (12-foot) wave, probably the fourth crest to reach the coast, swept into Crescent City, California, about 6 hours after the quake. Though more than 300 buildings were destroyed or damaged, five gasoline storage tanks set ablaze, and 27 blocks of the city demolished, there were relatively few casualties.

It has been over 30 years since a tsunami caused substantial damage in the U.S. Some public safety experts suggest we have become complacent about the risks associated with these destructive waves.

TIDES

Tides are periodic, short-term changes in the height of the ocean surface at a particular place, caused by a com-

bination of the gravitational force of the moon and sun and the motion of the Earth. With a wavelength that can equal half the Earth's circumference, tides are the longest of all waves. Unlike the other waves we have met, these huge shallow-water waves are never free of the forces that cause them, and so they act in unusual but generally predictable ways.

Around 300 B.C., the Greek navigator and astronomer Pytheas first wrote of the connection between the position of the moon and the height of a tide, but full understanding of tides had to await Isaac Newton's analysis of gravitation. We now know that the main cause of tides is the gravity of the moon acting on the ocean.

The Moon and Tidal Bulges

The pull of gravity between two bodies is proportional to the masses of the bodies but inversely proportional to the square of the distance between them. This means that gravitational attraction drops off quickly with distance. Gravity tends to pull the Earth and moon together, but inertia—the tendency of moving objects to continue in a straight line—keeps them apart. The Earth and the moon don't smash into each other (or fly apart) because they are in a stable orbit; their mutual gravitational attraction is exactly offset by their inertia. They rotate around a common center of mass that is not in space, but—because Earth's mass is 81 times that of the moon—1,700 kilometers (1,057 miles) *inside* the Earth. This center of mass is shown as point A in **Figure 11.9**.

Though their sums are equal and opposite, the inward

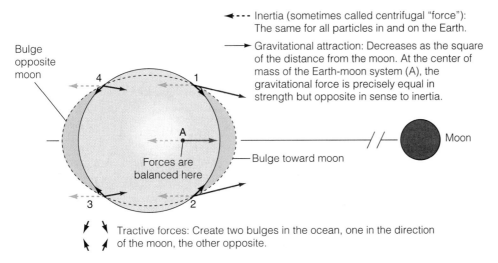

Inertia (sometimes called centrifugal "force"): The same for all particles in and on the Earth.

Gravitational attraction: Decreases as the square of the distance from the moon. At the center of mass of the Earth-moon system (A), the gravitational force is precisely equal in strength but opposite in sense to inertia.

Bulge opposite moon

Forces are balanced here

A

Bulge toward moon

Moon

Tractive forces: Create two bulges in the ocean, one in the direction of the moon, the other opposite.

The two forces that can move the ocean are balanced only at the center of earth; elsewhere the net imbalance is a small force that causes ocean water to converge into two equal "bulges," as shown.

Figure 11.9 The actions of gravity and inertia on particles at five different locations on the Earth. At points 1 and 2, the gravitation attraction of the moon slightly exceeds the outward-moving tendency of inertia, and the imbalance of forces causes water to move along the Earth's surface to converge at a point *toward* the moon. At points 3 and 4, inertia exceeds gravitational force; so water moves along the Earth's surface to converge at a point *opposite* the moon. Forces are balanced only at point A, 1,700 kilometers inside the Earth. (See text for a detailed explanation.)

pull of gravity and the outward-moving tendency of inertia don't always act in exactly the same balanced way on each particle of Earth and moon. In Figure 11.9, the four numbered points represent particles at different places on Earth. Notice that points 1 and 2 are closer to the moon; so gravitational attraction at those points slightly exceeds the outward-moving tendency of inertia. Water there tends to be attracted toward the moon and so is pulled along the ocean surface toward a spot beneath the moon. At points 3 and 4, slightly farther from the moon, inertia exceeds gravitational attraction. Water at those points tends to be flung away from the moon and so moves along the ocean surface toward a spot opposite the moon.

The imbalanced forces doing the pulling and flinging are known as *tractive forces*. Together, the tractive forces cause two bulges in the ocean, one in the direction of the moon, the other in the opposite direction. Only at point A are the inward pull of gravity and the outward-moving tendency of inertia exactly equal and opposite. (Remember that point A is located within the Earth, not at its surface.) The solid Earth cannot move much in response to these forces, but the fluid atmosphere and ocean can. We don't notice the changes in the height of the atmosphere, but changes in water level are visible to coastal observers. In **Figure 11.10**, tractive forces pull water toward a point beneath the moon, and to a point opposite the moon.

Water bulge resulting from inertia (centrifugal "force")

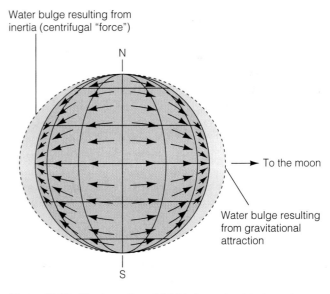

To the moon

Water bulge resulting from gravitational attraction

Figure 11.10 The formation of tidal bulges at points toward and away from the moon.

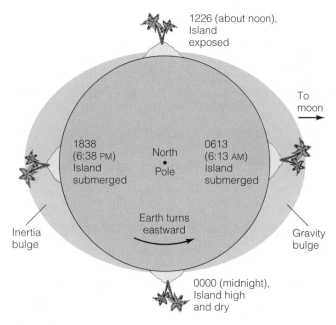

Figure 11.11 How the Earth's rotation beneath the tidal bulges produces high and low tides. Notice that the tidal cycle is 24 hours 50 minutes long because the moon rises 50 minutes later each day.

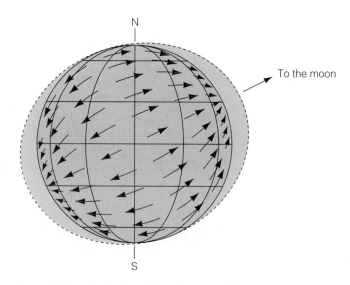

Figure 11.12 The bulges follow the moon. When the moon's position is above the equator, the bulge toward the moon is also located above the equator. The opposite inertia bulge is correspondingly below the equator. (Compare with Figure 11.10.)

How do these bulges cause the rhythmic rise and fall of the tides? The bulges tend to stay aligned with the moon as the Earth spins around its axis. **Figure 11.11** shows the situation in Figure 11.10 as it would look from above the North Pole. As the Earth turns eastward, an island on the equator is seen to move in and out of these bulges through one rotation (one day). The bulges are the crests of the planet-sized waves that cause **high tides. Low tides** correspond to the troughs, the area between bulges. Starting at 0000 (midnight) we see the island in shallow water at low tide. Around six hours later, at 0613 (6:13 A.M.), the island is submerged in the lunar bulge at high tide. At 1226 (about noon) the island is within the tide wave trough at low tide. At 1838 (6:38 P.M.) the island is again submerged, this time in the opposite crest caused by inertia. About an hour after midnight (0050) on the next day the island is back in shallow water, where it began.

The wave crests and troughs that cause high and low tides are actually very small: A 2-meter (7-foot) rise or fall in sea level is insignificant in comparison to the ocean's great size. The bulges (tide wave crests) travel about 1,600 kilometers (1,000 miles) per hour at the equator in an attempt to keep up with the moon. Theoretically, the wavelength of these tide waves is as long as 20,000 kilometers (12,500 miles)! The bulges tend to stay aligned with the moon as the Earth spins around its axis. The key to understanding tides is to see the Earth turning beneath these bulges.

There are complications, of course. For example, the **lunar tides**, tides caused by gravitational and inertial interaction of moon and Earth, complete their cycle in a tidal day (also called a lunar day). A complete tidal day is 24 hours 50 minutes long because the moon, which exerts the greatest tidal influence, rises 50 minutes later each day. Thus, the highest tide also arrives 50 minutes later each day.

Another complication arises from the fact that the moon does not stay right over the equator; each month, it moves from a position as high as 28½° above Earth's equator to 28½° below. If the moon is above the equator, the bulges will be offset accordingly (as in **Figure 11.12**). When the moon is 28½° north of the equator, an island north of the equator will pass through the bulge on one side of the Earth but miss the bulge on the other side. During one day the island passes through a very high tide, a low tide, a lower high tide, and another low tide. This is shown in **Figure 11.13**.

The Sun's Role

The sun's gravity also attracts particles on Earth. Remember, closeness counts for much in determining the strength of gravitational attraction. The sun is about 27 million times more massive than the moon, but the moon is about 387 times closer than the sun; so the sun's influence on the tides is only about half that of the moon's. Smaller solar bulges tend to follow the sun through the

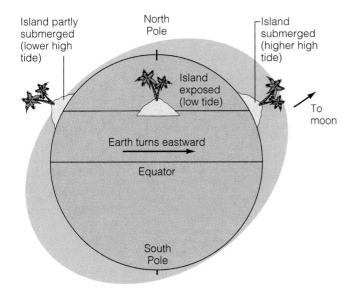

Figure 11.13 How the changing position of the moon relative to the Earth's equator produces higher and lower high tides.

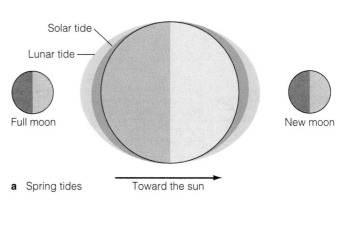

a Spring tides

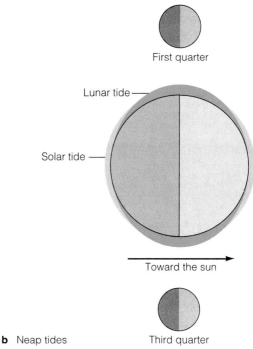

b Neap tides

Figure 11.14 Relative positions of the sun, moon, and Earth during spring and neap tides. (a) At the new and full moons, the solar and lunar tides reinforce each other, making the highest high and lowest low tides. (b) At the first and third quarter moons, the sun, Earth, and moon form a right angle, creating the lowest high and the highest low tides.

day. These are the **solar tides**, caused by the gravitational and inertial interaction of the sun and Earth.

Like the moon, the sun also appears to move above and below the equator (23½° north to 23½° south, as you may recall from Chapter 8); so the position of the solar bulges varies like that of the lunar bulges. The Earth revolves around the sun only once a year, however; so the position of the solar bulges above or below the equator changes much more slowly than the position of the lunar bulges.

Sun and Moon Together

The ocean responds simultaneously to inertia and to the gravitational force of both sun and moon. If Earth, moon, and sun are all in a line (as shown in **Figure 11.14a**), the lunar and solar tides will be additive, resulting in higher high tides and lower low tides. But if the moon, Earth, and sun form a right angle (as shown in **Figure 11.14b**), the solar tide will tend to diminish the lunar tide. Because the moon's contribution is more than twice that of the sun, the solar tide will not completely cancel the lunar tide.

The large tides caused by the linear alignment of sun, Earth, and moon are called **spring tides** (*springen* = to move quickly). During spring tides high tides are very high, and low tides very low. This occurs at two-week intervals corresponding to the new and full moons. (Please note that spring tides don't happen only in the spring of the year.) **Neap tides** (*naepa* = hardly disturbed) occur when the moon, Earth, and sun form a right angle.

During neap tides, high tides are not very high, and low tides not very low. Neap tides alternate at two-week intervals with spring tides. **Figure 11.15** plots tides at two coastal sites through spring and neap cycles.

Because their orbits are not perfect circles, the moon and the sun are closer to the Earth at some times than at others. The difference is significant—some 25,000 kilometers (16,000 miles) in the moon's case. Because the

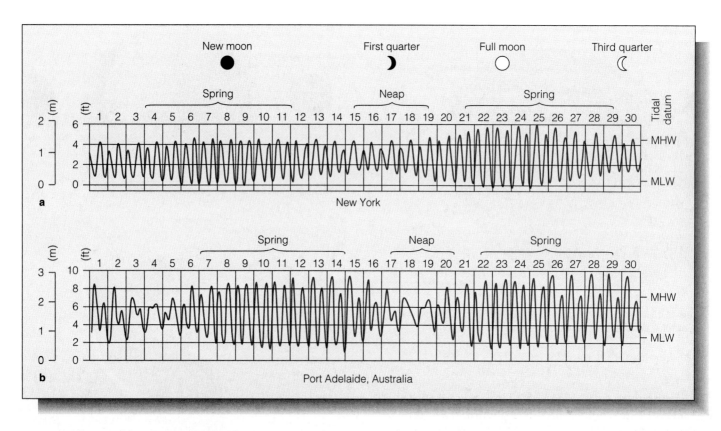

Figure 11.15 Tidal records for a typical month at (a) New York and (b) Port Adelaide, Australia. Note the relationship of spring and neap tides to the phases of the moon. MHW = mean high water; MLW = mean low water.

intensity of gravitational attraction is inversely proportional to the square of the distance between the bodies, the closer moon raises a higher tidal crest. If the moon and sun are over nearly the same latitude, and if the Earth is also close to the sun, extreme spring tides will result.

Tidal Patterns

The wavelength of tides rarely reaches the theoretical maximum of half the circumference of the Earth. As the Earth turns, land masses obstruct the tidal crests, diverting, slowing, and otherwise complicating their movements. This interference produces different patterns in the arrival of tidal crests at different places. In some places tides disappear altogether.

The shape of the basin itself has a strong influence on the patterns and heights of tides. As we have seen, water in large basins can rock rhythmically back and forth in seiches. Tidal crests can stimulate this oscillation, and the configuration of coasts around a basin can alter its rhythm. Because tides are shallow-water waves, frictional forces also play an important role.

For these and other reasons some coastlines experi-

ence **semidiurnal** (twice daily) **tides**: two high tides and two low tides of nearly equal level each lunar day. Others have **diurnal** (daily) **tides**: one high and one low. The tidal pattern is called **mixed** if successive high tides or low tides are of significantly different heights through the cycle. **Figure 11.16** gives an example of each tidal pattern. At Cape Cod two tidal crests arrive per lunar day, a semidiurnal pattern (**Figure 11.16a**). The natural tendency of water in an enclosed ocean basin to rock at a specific frequency modifies the pattern in the Gulf of Mexico; so New Orleans sees one crest per lunar day, a diurnal pattern (**Figure 11.16b**). The west coast of the United States has a mixed tidal pattern: often a *higher* high tide, followed by a *lower* low tide, a *lower* high tide, and a *higher* low tide each lunar day. (That's not as confusing as it first seems—see **Figure 11.16c**.)

The Pacific has a unique pattern of diurnal, semidiurnal, and mixed tides. As **Figure 11.17** shows, it has the most complex of all tidal patterns. The east coast of Australia, all of New Zealand, and much of the west coasts of Central and South America have a semidiurnal tidal pattern. The Aleutians have diurnal tides. The Pacific coasts of North America and some of South America have mixed tides. Why the differences?

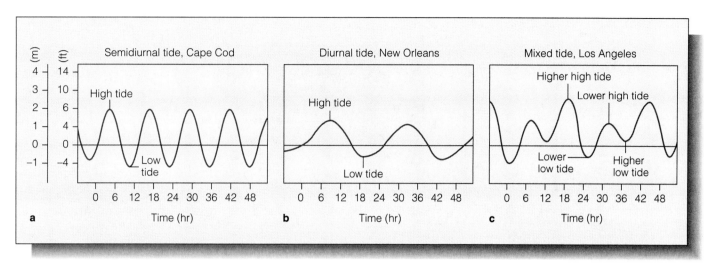

Figure 11.16 Tide curves for the three common types of tides. (a) A semidiurnal tide pattern at Cape Cod, Massachusetts. (b) A diurnal tide pattern at New Orleans, Louisiana. (c) A mixed tide pattern at Los Angeles, California.

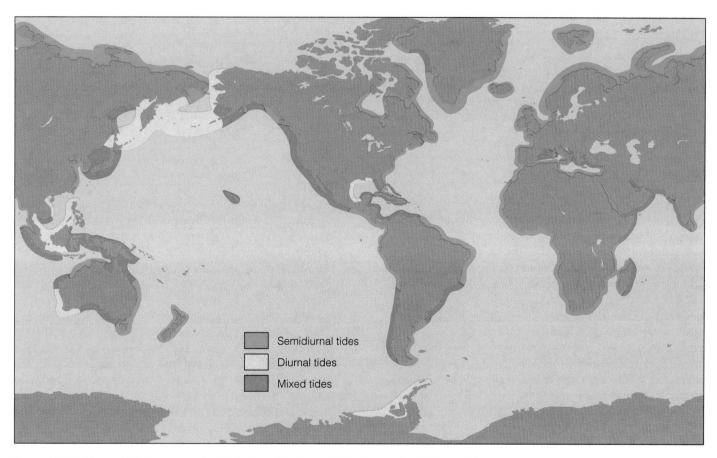

Figure 11.17 The worldwide geographic distribution of the three tidal patterns. Most of the world's ocean coasts have semidiurnal tides.

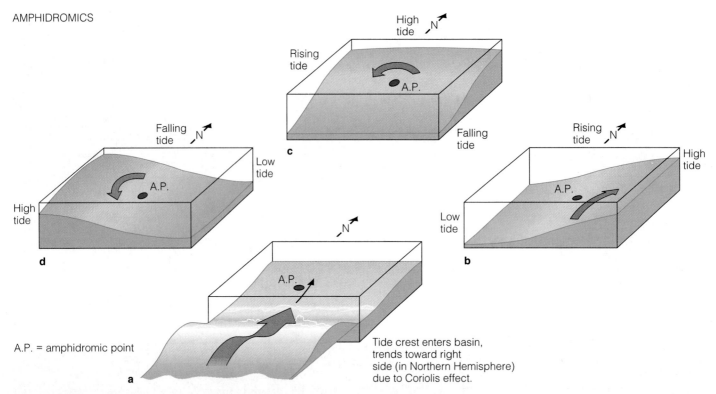

A.P. = amphidromic point

Tide crest enters basin, trends toward right side (in Northern Hemisphere) due to Coriolis effect.

Figure 11.18 The development of amphidromic circulation. (a) A tide wave crest enters an ocean basin in the Northern Hemisphere. The wave trends to the right because of the Coriolis effect (b), causing a high tide on the basin's eastern shore. Unable to continue turning to the right because of the interference of the shore, the crest moves northward, following the shoreline (c) and causing a high tide on the basin's northern shore. The wave continues its progress around the basin in a counterclockwise direction (d), forming a high tide on the western shore and completing the circuit. The point around which the crest moves is an amphidromic point (AP).

Remember the surface of Lake Geneva in our discussion of seiches? The water level at the center of the lake remains at the same height while water at the ends rises and falls (see again Figure 11.3). The long axis of the lake stretches east and west. Because of the Coriolis effect, water moving east at the center of the lake is deflected slightly to the right (to the south). If the lake were larger and the flow of water greater (and thus the Coriolis effect stronger), water would hug the southern shore as it traveled eastward. When the water began to rock the other way—back to the west—the Coriolis effect would move the water to the right toward (and along) the northern shore. Note that the overall movement of water would be counterclockwise.

Water moving in a tide wave tends to stay to the right of an ocean basin for the same reason. As water moves north in a Northern Hemisphere ocean, it moves toward the eastern boundary of the basin; as it moves south, it moves toward the western boundary. A wave crest moving counterclockwise will develop around a node if this motion continues to be stimulated by tidal forces. This rotary motion is shown in **Figure 11.18**.

The node (or nodes) near the center of an ocean basin is called an **amphidromic point** (*amphi* = around + *dromas* = running). An amphidromic point is a no-tide point in the ocean, around which the tidal crest rotates through one tidal cycle. Because of the shape and placement of land masses around ocean basins, the tidal crests and troughs cancel each other at these points. The crests sweep around amphidromic points like wheel spokes from a rotating hub, radiating crests toward distant shores. The tide waves are influenced by the Coriolis effect because a large volume of water moves with the tide wave. The tide waves move counterclockwise around the amphidromic point in the Northern Hemisphere, and clockwise in the Southern Hemisphere. The height of the tides increases with distance from an amphidromic point.

About a dozen amphidromic points exist in the world ocean; **Figure 11.19** shows their location. Notice the complexity of the Pacific, which contains five. It's no wonder that the arrival of tide wave crests at the Pacific's edge produces such a complex mixture of tide patterns, depending on shoreline location.

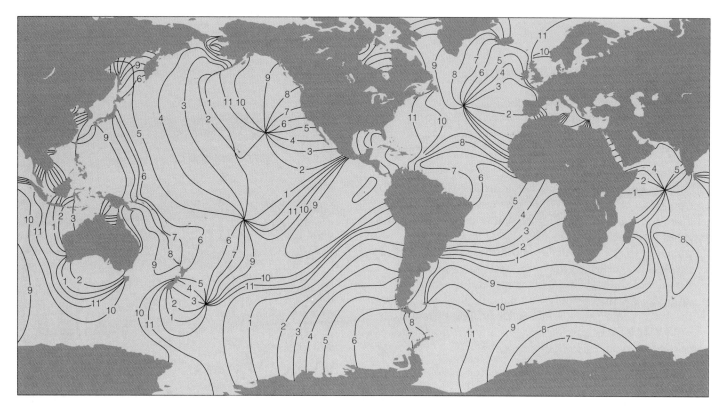

Figure 11.19 Amphidromic points in the world ocean. Tidal ranges generally increase with increasing distance from amphidromic points. Lines radiating from the points indicate tide waves moving around these points, counterclockwise in the Northern Hemisphere and clockwise in the Southern Hemisphere. The numbers represent the positions of a hypothetical tidal crest in hours. (See text for details.)

Tidal Datum and Tidal Range

The reference level to which tidal height is compared is called **tidal datum.** Tidal datum is the zero point (0.0) seen in tide graphs such as Figures 11.15 and 11.16. This reference plane is not always set at **mean sea level**, the height of the ocean surface averaged over a few years' time. On coasts with mixed tides, the zero tide level is the average level of the lower of the two daily low tides (*mean lower low water*, or MLLW). On coasts with diurnal and semidiurnal tides, the zero tide level is the average level of all low tides (*mean low water*, MLW).

The **tidal range** (high-water to low-water height difference) varies with basin configuration. In small areas such as lakes, the tidal range is small. In larger enclosed areas such as the Baltic or Mediterranean seas, tidal range is also moderate. Tidal range is not the same over a whole ocean basin; it varies from the coasts to the centers of oceans. The largest tidal ranges occur at the edges of the largest ocean basins, especially in bays or inlets that concentrate tidal energy because of their shape. Tidal range at the apex of a funnel-shaped sea, gulf, or bay can often be extreme. In the eastern reaches of the Bay of Fundy near Moncton, New Brunswick (Canada), tidal range is especially wide: up to 15 meters (50 feet) from highs to lows (see **Figure 11.20**, page 268). The northern reaches of the Sea of Cortez east of Baja California have a tidal range of about 9 meters (30 feet). Tide waves sweeping toward the narrow southern end of the North Sea can build to great heights along the southeast coast of England and the north coast of France.

A **tidal bore** (*bara* = wave) will form in some inlets (and their associated rivers) exposed to great tidal fluctuation. Here, at last, is a true **tidal wave**—a steep wave moving upstream generated by the action of the tide crest in the enclosed area of a river mouth (see **Figure 11.21**, page 268). Though most are less than 1 meter (3 feet) high, bores may be up to 3 meters (10 feet) high and move from 5 to 7.5 meters per second (11 to 17 miles per hour). Their potential danger is lessened by their predictability. Accurately predicting the arrival of tidal bores is essential to safe navigation. In addition to the Bay of Fundy, tidal bores are common in the Amazon, the Ganges Delta, and England's Severn River. Some

BOX 11.2 ● *Maelstrom!*

According to Homer's account in *The Odyssey*, across Ulysses' homeward path lay a lethal whirlpool named Charybdis:

> We sailed up the narrow strait. On one side was shining Charybdis, who made her terrible ebb and flow of the sea's water. When she vomited it up, like a caldron over a strong fire, the whole sea would boil up in turbulence, and the foam flying spattered the pinnacles of the rocks in either direction; but when in turn again she sucked down the sea's salt water, the turbulence showed all the inner sea, and the rock around it groaned terribly, and the ground showed at the sea's bottom, black with sand. My crew turned sallow with fright, staring into this abyss from which we expected our immediate death.

For millennia, the prospect of navigating past Charybdis terrified sailors in the Strait of Messina, which separates Italy and Sicily. Its activity was greatly reduced in 1908 when the great Calabria–Messina earthquake and tsunami altered the geology of the area.

Edgar Allan Poe wrote about a whirlpool called the maelstrom, which lies among the southern Lofoten Islands off Norway's west coast. The rapid spinning of water in the maelstrom raises its outer edge and depresses the central core. Here is Poe's description, from his 1841 story "A Descent into the Maelstrom":

> The edge of the whirl was represented by a broad belt of gleaming spray; but no particle of this slipped into the mouth of the terrific funnel, whose interior, as far as the eye could fathom it, was a smooth, shining, and jet-black wall of water, inclining to the horizontal at an angle of some forty-five degrees, speeding dizzily round and round with swaying and sweltering motion, and sending forth to the winds an appalling voice, half shriek, half roar, such as not even the mighty cataract of Niagara ever lifts up in its agony to Heaven.

Do these whirlpools genuinely deserve the dread they inspired, or is their reputation largely a figment of artistic imagination (**Figure a**)?

Whirlpools arise in shallow, restricted straits connecting two large bodies of deep water. They are caused by the tides. If the large bodies of water move to different tidal cycles, high tide in one will not occur at the same

a Maelstrom: fiction. A representation by the early twentieth-century English illustrator Arthur Rackham.

time as high tide in the other. Turbulence develops when one mass of water tries to pass the other during changes in tide. Under ideal conditions, bottom contours and the rush of opposing tidal currents can cause seawater in the confined area to spin vigorously. The larger the opposing bodies of water, and the smaller the passage through which the confined water must pass, the greater the vortex caused by the tidal currents.

Maelstrøm is a Norwegian word derived from the Dutch *malen* ("to grind in a circle," as a millstone grinds

b Maelstrom: fact. The maelstrom is located in the strait between the Norwegian islands of Moskenes and Moskenesøya.

grain) and *strøm* ("stream"). Today's maelstrom is a heaving mass of active tidal water connecting Vest Fjord and the Norwegian Sea between the islands of Moskenesøya and Moskenes. Look for the area on a chart at 67°48′N, 12°50′E, about 165 kilometers (100 miles) north of the Arctic Circle. Currents in the passage can reach a speed of 13 kilometers (8 miles) per hour when the tides change, and strong local winds make the passage even more dangerous. The shallow "saddle" separating the large bodies of water to the east and west is only about 20 meters (70 feet) deep, and water rushing over its rocky bottom is driven to spin.

But is the chaos as horrifying as Homer, Poe, Jules Verne, and others have suggested? Well, no. Fishing boats in this rich fishing region carefully avoid the area during times of maximum tidal difference, and no sailor likes to risk his or her life in turbulent confined waters during times of high winds. But as can be seen in **Figure b,** the sucking, roaring, cavernous maw of the maelstrom is more fiction than fact.

Figure 11.20 Tides in the eastern Bay of Fundy on the Atlantic coast of Canada. Tidal range is near 15 meters (50 feet). At the peak of the flood, water rises 1 meter (3.28 feet) in 23 minutes.

Figure 11.21 A tidal bore on the Severn River in southwestern England. A tide crest sweeping up the funnel-shaped Bristol channel from the Atlantic Ocean encounters the increasingly shallow and narrow river passage, and it rushes forward to cause the bore. Like all tidal bores, the Severn bore reaches its greatest height during the days surrounding spring tides. The waves that form the bore in this example are nearly 1 meter (3.3 feet) high and are traveling at about 5 meters per second (11 miles per hour). Note that the smooth waves are followed by turbulent water also moving rapidly upriver. The tidal bore on the Severn River is large enough for surfing, and accomplished board surfers can ride upstream for many miles.

rivers in southern China may have three or four simultaneous bores at different places along the river's length.

Tidal Currents

The rise or fall in sea level as a tide crest approaches and passes will cause a **tidal current** of water to flow into or out of bays and harbors. Water rushing into an enclosed area because of the rise in sea level as a tide crest approaches is called a **flood current**. Water rushing out because of the fall in sea level as the tide trough approaches is called an **ebb current**. (The terms *ebb tide* and *flood tide* have no technical meaning.) **Slack water**, a time of no currents, occurs when the current changes direction.

Anyone who has stood at the narrow mouth of a large bay or harbor cannot help but be impressed with the speed and volume of the tidal current that occurs between tidal extremes. Midway between high and low spring tides, the ebb current rushing from San Francisco Bay strikes the base of the south tower of the Golden Gate Bridge with such force that a bow wave is formed, giving the illusion that the bridge itself is moving rapidly. Tidal currents at the Golden Gate can reach 3 meters per second (about 7 miles per hour) because of

the volume of enclosed water and the narrowness of the channel through which it must escape. Navigators must know the times of tidal currents to safely negotiate any harbor entrance or other narrow strait—in some places, this knowledge may save their lives (see **Box 11.2**).

Tidal currents become more complex in the open sea. One's position relative to an amphidromic point, the shape of the basin, and the magnitude of gravitational forces and inertia must all be considered to calculate the speed and direction of tidal currents over a deep bottom. The velocity of tidal currents is less in the open sea because the water is not confined, as it is in a harbor. Open-sea tidal currents have been measured at a few centimeters per second, and their velocity tends to decrease with depth.

Tidal Friction

The daily rise and fall of the tides consumes a very large amount of energy, energy ultimately dissipated as heat. Most of this energy comes directly from the rotation of the Earth itself, and tidal friction is gradually slowing Earth's rotation by a few hundredths of a second per century. Even such a small change has long-term planetary effects, however. Geologists studying the daily growth rings of fossil corals and clams estimate that the length of the day has grown longer; so the number of days in a year has become less as planetary rotation has slowed. Evidence suggests that 350 million years ago a year contained between 400 and 410 days, with each day being about 22 hours long; and 280 million years ago there were about 390 22½-hour days in a year.

Tidal friction affects other bodies. Tidal forces have locked the rotation of the moon to that of the Earth. As a result, the same side of the moon is always facing the Earth, and a day on the moon is a month long.

Predicting Tides

There are at least 150 tide-generating and tide-altering forces and factors. About seven of the most important of these must be considered if we wish to predict tides mathematically. The interactions of all the forces and factors are so complex that if a previously unknown continent were discovered on Earth, the coastal tide times and ranges on its shores could not be accurately predicted. It is the study of past records that allows tide tables to be projected into the future. This experience permits prediction of tidal height to an accuracy of about 3 centimeters (1.2 inches) for years in advance.

Even so, extraneous factors can affect the estimates. For example, arrival of a storm surge will greatly affect the height or timing of a tide—as will gentle, atmospherically induced seiching of the basin or excitement of large-scale resonances by a tsunami. Even a strong, steady wind on- or offshore will affect tidal height and the arrival time of the crest. Weather-related alterations are sometimes called **meteorological tides** after their origin.

Figure 11.22 Tidal power installation at the Rance estuary in western France.

Power from the Tides

Humans have found ways to use the tides. Ships sail to sea and return to port with the tides. Intentional grounding of a ship with the fall of a tide can provide a convenient, if temporary, drydock. To these traditional uses has been added a potential alternative to our growing dependence on fossil fuels: taking advantage of trapped high tide water to generate electricity.

Tidal power is the only marine energy source that has been successfully exploited on a large scale. The first major tidal power station was opened in 1966 in France on the estuary of the river Rance (see **Figure 11.22**), where tidal range reaches a maximum of 13.4 meters (44 feet). Built at a cost of $75 million, this 850-meter- (2,800-foot-) long dam contains 24 turbo-alternators capable of generating 544 million kilowatt-hours of electricity annually. At high tide, seawater flows from the ocean through the generators into the estuary. At low tide the seawater and river water from the estuary flow out through the same generators. Power is generated in both directions. A smaller but similar installation is generating power on the Annapolis River in Nova Scotia. A much larger generating facility has been proposed for Passamaquoddy Bay, a part of the Bay of Fundy between Maine and New Brunswick, Canada.

Tidal power has many advantages: Operating costs are low, the source of power is free, and no carbon dioxide or other pollutants are added to the atmosphere. But even if tidal power stations were built at every appropriate site worldwide, the power generated would amount to less than 1% of current world needs. And, of course, this method of power generation is not free of trade-offs. The dam and electrical generators can be damaged by storms, and the large, finely made metal valves and vanes at the heart of the plant are easily corroded by seawater. Computer simulations have suggested that

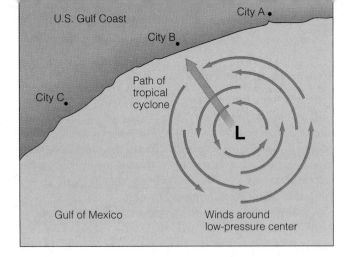

Figure 11.23 A path of a tropical cyclone.

installing a dam would change the resonance modes of a bay or estuary—and therefore the height of the tide wave. Studies also suggest that sensitive planktonic and benthic marine life would be disrupted and even that increased tidal friction would cause a tiny decrease in the rate of Earth's rotation.

Q-AND-A

1. **Figure 11.23** shows a large tropical cyclone approaching the low-lying U.S. Gulf Coast. Now that you're an expert on waves, which of the three cities—A, B, or C—would you suggest be evacuated first? Why?

   City C is the safest because the winds in the counterclockwise turning storm will actually blow waves and water offshore as the storm approaches. City B, directly in the path of the storm, might seem to be in the worst danger, but winds won't drive the storm surge directly on shore at the height of the storm. City A is in the biggest trouble because the storm surge, though not as high as that at City B, will be driven onshore there by intense winds.

2. One of my friends has discovered I'm studying waves. He demands to know about the world's *largest* wave. What's the record, anyway?

   If your friend classifies the largest wave by wavelength, the answer is easy: Tides take the prize, with a wavelength that is ideally half the Earth's circumference. Tides also win the speed race: Their crests move at speeds up to 1,600 kilometers (1,000 miles) per hour.

   If your friend classifies the largest wave by height, things become more complicated. As you read in Chapter 10, the crew of Ramapo *measured a wave 34 meters (112 feet) high in 1933. In general, the highest wind waves probably develop in the West Wind Drift, the nasty area of windy ocean ringing Antarctica. Multiple storms there, some with huge fetch distances, can cause very large waves and opportunity for constructive interference between them. In 1916 during an agonizing small-boat trip from an island on which their expedition was marooned, Frederic*

   Worsley, captain of Ernest Shackleton's ship Endurance, *encountered occasional tremendous waves greater than any he had encountered in a lifetime at sea. He estimated a few were more than 30 meters (100 feet) high!*

   Some tsunami may be higher, however. As we have seen, landslides in Lituya Bay, Alaska, have generated waves in the 50-meter (165-foot) range. And a wave of about 530 meters (1,740 feet) formed briefly as water surged up the opposite side of the bay on 9 July 1958. (It was more a titanic splash than a classic tsunami; one writer has likened the effect to dropping a sack of concrete into a filled bathtub.) The greatest height recently recorded for a tsunami in the ocean was 85 meters (278 feet) in 1971 off Ishigaki, Japan. But don't forget the 91-meter (300-foot) tsunami thought to have occurred 66 million years ago on the coast of what is now Texas.

3. Are there tides in the solid Earth?

   Yes. Even the Earth isn't stiff enough to resist the tidal pulls of the moon and sun. Bulges occur in the solid Earth just as they appear in the ocean or the atmosphere. The crests (bulges) of the Earth are, of course, much smaller; 25 to 30 centimeters (10 to 12 inches) is about average. They pass unnoticed beneath us twice a day. On the other hand, tidal variability in the height of the atmosphere has been measured in miles.

4. In my newspaper one of today's low tides is listed as –1.0, 1445. What does that mean?

   In the United States, it means that the water will be 1.0 feet below tidal datum (that is, below mean lower low water [MLLW]—the long-term average position of the lower of the daily low tides) at 2:45 P.M. local time. This might be a good afternoon to spend at the shore digging for clams because a tide that low would expose intertidal organisms only rarely seen above water. See Figure 17.7 for more information on the relationship between tidal height and exposure.

5. Are there tides in a glass of water?

   Yes. They're too small to detect, but they do exist. Each molecule of water in the glass responds to the same planetary forces that affect molecules in the ocean.

6. Will people in California be in great danger from a seismic sea wave when a major earthquake occurs on the San Andreas Fault?

   Probably not, for two reasons. First, San Andreas quakes are usually lateral-displacement quakes: The ground moves suddenly sideways rather than up or down. Tsunami are usually generated by vertical movement. Second, the motion along the San Andreas Fault will be parallel to the coast, not at right angles to it. If the ground movement were toward the coast (as could happen north of San Francisco), the "shove" might result in a wave. There may be some slopping about at the coastline during a large earthquake, but probably not a classic tsunami.

Terms and Concepts to Remember

amphidromic point	mixed tide	standing wave
diurnal tide	neap tide	storm surge
ebb current	node	tidal bore
flood current	seiche	tidal current
high tide	seismic sea wave	tidal datum
low tide	semidiurnal tide	tidal range
lunar tide	slack water	tidal wave
mean sea level	solar tide	tide
meteorological tide	spring tide	tsunami

Study Questions

1. Though they move through all the ocean, the waves described in this chapter are referred to as shallow-water waves. How can that be?

2. What combination of events makes a storm surge most destructive? Are all storms approaching a coast associated with a storm surge? Is a storm surge a progressive wave?

3. What factors affect the wavelength and period of a seiche? Do seiches themselves often cause damage?

4. What causes tsunami? Are all seismic sea waves tsunami? Are all tsunami seismic sea waves? How fast do tsunami travel? Do they move in the same way or at the same speed in a confining bay as they would in the open ocean?

5. Are tsunami ever dangerous if encountered in the open sea? What happens when they reach shore?

6. What causes the rise and fall of the tides? What celestial bodies are most important in determining tides?

7. What is a high tide and a low tide? A spring tide and a neap tide?

8. What are the most important factors influencing the heights and times of tides? What tidal patterns are observed? Are there tides in the open ocean? If so, how do they behave?

9. How does the latitude of a coastal city affect the tides there—or does it?

10. From what you learned about tides in this chapter, where would you locate a plant that generated electricity from tidal power? What would be some advantages and disadvantages of using tides as an energy source?

For Further Study

Bascom, W. 1980. *Waves and Beaches.* Rev. ed. New York: Anchor/Doubleday. One of the best general references available. Clear and lively writing about the influences of long-wavelength waves on coastal formations.

Bowditch, N. 1970. *American Practical Navigator.* Washington, DC: U.S. Navy Hydrographic Office. A one-volume nautical reference of unexcelled brevity and precision. The chapters on large waves are clear and direct.

Defant, A. 1958. *Ebb and Flow: The Tides of Earth, Air and Water.* Ann Arbor: University of Michigan Press.

Dudley, W. C., and M. Lee. 1988. *Tsunami!* Honolulu: University of Hawaii Press. Combines firsthand accounts of survivors with scientific information on tsunami.

Folger, T. 1994. "Waves of Destruction." *Discover,* May, 66–73. Information on tsunami, including the September 1992 Nicaraguan event.

Giese, G. S., and D. C. Chapman. 1993. "Coastal Seiches." *Oceanus* 36 (no. 1, Spring): 38–46. Internal waves approaching the shore may be responsible for some occasionally destructive seiches.

Greenberg, D. A. 1987. "Modeling Tidal Power." *Scientific American,* November, 128–38. What effect would a dam to extract power from tidal crests have in upper New England?

LeBlanc, J. 1986. "Tales of the Wild Ocean." *Oceans* 19 (no. 4). Stories of the ocean at its most amplitudinally violent.

LeProvost, C., A. F. Bennett, and D. E. Cartwright. 1995. "Ocean Tides for and from TOPEX/Poseidon." *Science* 267 (no. 5198): 639–42. Satellite data on tides are compared with data from 78 ground stations. Data from the satellite are shown to be nearly as accurate as those from the stations, a "ground truth" allowing researchers to trust satellite readings of sea surface elevation in the open sea.

Macdonald, G. A., and A. T. Abbott. 1970. *Volcanoes in the Sea: The Geology of Hawaii.* Honolulu: University of Hawaii Press. Tsunami have shaped the Hawaiian coastal landscape.

McCredie, S. 1994. "When Nightmare Waves Appear Out of Nowhere to Smash the Land." *Smithsonian,* March, 28–39. Disturbing photos of tsunami, including the July 1992 event in Japan.

Miller, D. J. 1960. "The Alaska Earthquake of July 10, 1958: Giant Wave in Lituya Bay." *Bulletin of the Geological Society of America* 50 (April): 253–66.

Pipkin, B., et al. 1987. *Laboratory Exercises in Oceanography.* 2d ed. New York: Freeman. Long-wave calculations and examples.

Shepard, F. P. 1969. *The Earth Beneath the Sea.* New York: Atheneum. A widely admired and wonderfully written book for the general reader. The excerpt discussing the 1 April 1946 tsunami is from this source.

Shepard, F. P. 1977. *Geological Oceanography.* New York: Crane, Russak. Accessible, well written and illustrated. More comprehensive than *The Earth Beneath the Sea.*

Smith, F. G. W. 1970. "The Simple Wave." *Sea Frontiers* 16 (no. 5). Turns out it's not very simple.

United States Geological Survey, Professional Paper 354-C. n.d. *Giant Waves in Lituya Bay, Alaska.* Excerpts from official reports and firsthand accounts.

Van Dorn, W. G. 1974. *Oceanography and Seamanship.* New York: Dodd, Mead. Excellent drawings and diagrams for sailors interested in the technical workings of the ocean.

Von Arx, W. S. 1962. *An Introduction to Physical Oceanography.* New York: Addison-Wesley. Excellent technical discussion of tides, good illustrations.

COASTS

A Fragile Seam Between Sea and Land

Coastal areas join land and sea. The place where ocean meets land is usually called the **shore,** but the term **coast** refers to the larger zone affected by the processes occurring at this boundary. A sandy beach might form the shore in an area, but the coast (or coastal zone) includes the marshes, sand dunes, and cliffs just inland of the beach, as well as the sand bars and troughs immediately offshore. The world ocean is bounded by about 440,000 kilometers (273,000 miles) of shore.

Because of its proximity to both ocean and land, a coast is subject to natural events and processes common to both realms. A coast is an active place. Here is the battleground on which wind waves break and expend their energy. Tides sweep water on and off the rim of land, rivers drop most of their terrigenous sediments at the coasts, and storms batter the continents. The *shape* of a coast is a product of many processes: uplift and subsidence, the wearing-down of land by **erosion,** and the redistribution of material by sediment transport and deposition. The *location* of a coast depends primarily on global tectonic activity.

Coasts are dynamic, changeable places, especially when viewed on a human scale. Henry David Thoreau wrote that to the people of Massachusetts the sea is their garden, and the dog that growls at their door is the Atlantic Ocean. Few stretches of offshore water are

Beach erosion has toppled this home on Nantucket Island off the coast of Massachusetts

more dangerous and turbulent than the winter Atlantic off New England, and this particular dog has an expensive appetite for the same shores favored by thousands of tourists and longtime seasonal residents. The resort town of Scituate, south and east of Boston, illustrates the ephemeral nature of some coasts. In February 1978 a particularly strong nor'easter, combined with exceptionally high tides, wreaked havoc on the town's sandy shore. Waves washed over the coastal zone, heavily built with vacation homes and businesses. The private property loss from the 36-hour storm was great, but in this one storm the cost to public structures like seawalls, groins, and piers amounted to millions of dollars *more* than the value of the private property. In the 350 years since English settlers began recording the history of this stretch of shore, it has been inundated by storms and severely damaged at least 84 times. Since 1938 the Scituate seawall has been rebuilt 18 times at public expense.

The U.S. West Coast is also subject to coastal damage. An interesting example involved Ellen Scripps, heiress to a publishing fortune and benefactress of, among other good works, the land and first buildings of what is now the Scripps Institution of Oceanography. After World War I, Scripps developed land along the San Diego shore into Sunset Cliffs Park, a property landscaped with gazebos, walkways, and gardens, which she donated to the citizens of California. Unfortunately the sandstone that formed the dramatic shore cliffs and adjacent park was so soft that a determined artist could scratch his initials into the rock using only his fingernails. The cliffs eroded at a rapid rate, and 55 years after its completion almost nothing remained of the once-beautiful park. The coast in the area continues to erode, threatening a highway and many expensive homes.

The long-term human relationship with both of these shores has been less than successful. Learning about coasts can prevent misplaced efforts and development.

Our personal experience with the ocean usually begins at the coast. These temporary, often beautiful junctions of land and sea are subject to rearrangement by waves and tides, by gradual changes in sea level, by biological processes, and by tectonic activity. Coasts may be classified in many ways, but one of the most useful schemes is based on the degree to which a coast has been modified by marine processes. In this chapter we survey the natural processes that shape coasts and the impact of human interference with these processes.

CLASSIFYING COASTS

A good classification scheme allows us to group things into categories determined by common origins or the influence of common processes. The *process* of categorizing things often makes processes easier to comprehend. Coasts have been hard to classify because, as we will see, many factors combine to shape them.

As we saw in Chapter 3, no area of geology has been left undisturbed by the revelations of plate tectonics. In the 1960s geologists began to classify coasts according to tectonic position. *Active* coasts, near the leading edge of moving continental masses, were found to be fundamentally different from the more *passive* shores near trailing edges. The ages and characteristics of coasts are better understood by taking plate movements into account, as are the coasts' composition and physical shapes. But the slow forces of plate movement are frequently obscured by the more rapid action of waves, by the erosion of land, and by the transport of sediments.

Another important factor that must be considered in the classification of coasts is long-term change in sea level. Five factors can cause sea level to change. Three of these factors are responsible for **eustatic change**—variations in sea level that can be measured all over the world ocean:

- The amount of water in the world ocean can vary. Sea level is lower during periods of global glaciation (ice ages) because there is less water in the ocean; it is higher during warm periods, when the glaciers are smaller. Periods of abundant volcanic outgassing or an increase in the number of icy comets striking the Earth can also add water to the ocean and raise sea level.

- The volume of the ocean's "container" may vary. As noted in Chapter 4, high rates of seafloor spreading are associated with the expansion in volume of the mid-ocean ridges. This displaces the ocean's water, which climbs higher on the edges of the continents. Sediments shed by the continents during periods of rapid erosion can also decrease the volume of ocean basins and raise sea level.

- The water itself may occupy more or less volume as its temperature and salinity vary. During times of global warming, seawater expands and occupies more volume, raising sea level.

Of course the continents rarely stay still as sea level rises and falls. Local changes are bound to occur, and two other factors produce variations in local sea level:

- Tectonic motions and isostatic adjustment can change the height and shape of the coast.

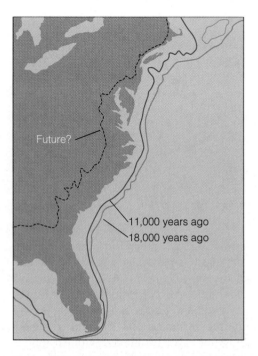

Figure 12.1 Because of lower sea level within the last 18,000 years, the position of the coast in the eastern United States has been as much as 200 kilometers (125 miles) seaward from the present shoreline, leaving much of the continental shelf exposed. In the distant future, if the ocean were to expand and the polar ice caps were to melt because of global warming, sea level could rise perhaps 100 meters (330 feet), driving the coast inland as much as 250 kilometers (160 miles).

- Wind and currents, seiches, storm surges, and other effects of water in motion can force water against the shore or draw it away.

Sea level has been at its current elevation (give or take 0.5 meter, 1.5 feet) for only about 3,500 years. Over the past million years worldwide sea level has varied from about 120 meters (400 feet) below to about 100 meters (330 feet) above its present position. (Sea level differences over the geologically short time of the last 35,000 years have also been substantial, as indicated in Figure 4.9.)[1] These changes in sea level have produced major differences in the position and nature of coastlines, especially in areas where the edge of the continent slopes gradually or where the coast is rising or sinking. **Figure 12.1** shows an estimate of previous shore positions along the U.S. East Coast in the geologically recent past, and a prediction for the distant future should the present warming trend cause more of the polar ice to melt.

[1]Consider the implications that arise for coastal civilizations. During the time between 18,000 years ago and 8,000 years ago, sea level rose over 100 meters (330 feet) at a rate of about 1 centimeter (½ inch) a year; over 0.5 meter in a human lifetime. Could this have given rise to the flood legends common to many religions?

Figure 12.2 A false-color photograph of Chesapeake Bay taken from space. The complex bay is an example of a drowned river valley.

Figure 12.3 Trollfjord, a deep fjord in the Lofoten Islands, northern Norway. The ship is a coastal steamer from Bergen, one of 11 ships that carry passengers and cargo to villages otherwise isolated by the steep terrain.

What coastal classification scheme best combines these elements? One of the most useful is the one popularized by Francis Shepard and used by many geological oceanographers: *primary and secondary coasts.* If a coast is essentially in almost the same condition as it was when sea level stabilized after the last ice age, Shepard called it a **primary coast.** Primary coasts are young coasts in which terrestrial influences (that is, processes that occur at the boundary between land and air) dominate. If the coast has been significantly changed by wave action and other marine processes since sea level stabilized, he termed it a **secondary coast.** Secondary coasts are usually older than primary coasts; that is, they have been exposed to marine action for a longer time. Secondary coasts retain little (if any) evidence of the nonmarine processes that produced them. Some coasts show both characteristics: A single long shoreline can consist of both primary and secondary coasts.

PRIMARY COASTS

Primary coasts are often rough and irregular. The ocean has not had time to modify the terrestrial features provided by changes in sea level, the scouring of glaciers, deposition of sediment at the mouths of rivers, volcanic eruptions, or the movement of the Earth along faults.

Land Erosion Coasts

When sea level was lower during the last ice age, rivers cut across the land and eroded sediment to form a coastal river valley. When higher sea level returned, the valley was flooded, or *drowned,* with seawater. Chesapeake Bay (**Figure 12.2**) and the Hudson River valley are examples of drowned river mouths. Coasts dominated by these sunken river valleys are sometimes called *ria coasts,* from the Spanish word for river.

Glaciers sometimes form in river valleys when rivers cut through the edges of continents at high latitudes. Deep, narrow embayments known as **fjords** are formed by tectonic forces and later modified by glaciers eroding valleys into deep, U-shaped troughs. Fjords (**Figure 12.3**) are found in British Columbia, Greenland, Alaska, Norway, New Zealand, and other cold mountainous places.

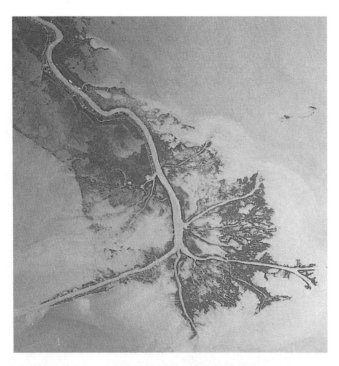

Figure 12.4 River deltas form at places where sediment-laden rivers enter enclosed or semienclosed seas, where wave energy is limited. The bird's-foot shape of the Mississippi Delta is seen clearly in this photograph. Submerged overlapping lobes of the Mississippi Delta extend completely across a wide continental shelf, and sediment from the river is now being deposited directly on the continental slope. Lobed and bird's-foot deltas form where deposition overwhelms the processes of erosion and sediment transportation.

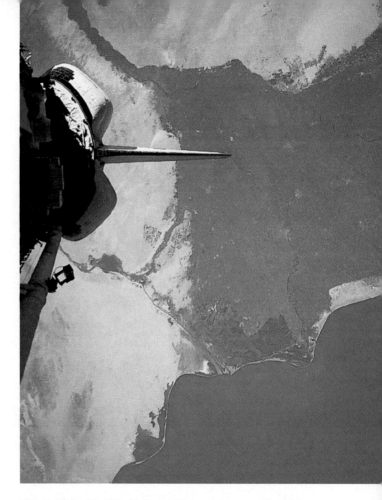

Figure 12.5 The Nile Delta, which has been smoothed into a more rounded shape by post-depositional erosion and sediment transportation. (The tail of the space shuttle can be seen in the upper left of the photo.)

Coasts Built Out by Land Processes

In a few places, sediments washing off the land have built out the coasts extensively. The shoreline in such places is much different from its configuration at the end of the last ice age. The most important of these costal features are **deltas.** The term, first used by Herodotus in the fifth century B.C. to describe the triangular mass of sediment at the terminus of the Nile, is derived from the triangular shape of the capital Greek letter delta (Δ).

Deltas do not form at the mouth of every sediment-laden river. A broad continental shelf must be present to provide a platform on which sediment can accumulate. In addition (as befits a primary coast), marine processes must not dominate: Tidal range should be low, and waves and currents generally mild. There are no large deltas along the east coast of the United States because sediments arriving at the coast are deposited in the sunken river mouths or dispersed by tides and currents. Also, there are no large deltas along the western margins of North and South America because these coasts are converging margins, where an oceanic plate is being sub-

ducted and the continental shelf is very narrow; sediment that would form a delta is swept down the continental slope or dispersed along the coast by waves. Deltas are most common on the low-energy shores of enclosed seas (where the tidal range is not extreme) and along the tectonically stable trailing edges of some continents. The largest deltas are those of the Gulf of Mexico (the Mississippi, **Figure 12.4**), the Mediterranean Sea (the Nile in **Figure 12.5**, and the Rhône, Po, and Ebro), the Ganges–Brahmaputra river system in the Bay of Bengal, and the huge deltas formed by the rivers of China that empty into the South China Sea.

The shape of a delta represents a balance between the accumulation of sediments and their removal by the ocean. For a delta to maintain its size or grow, the river must carry enough sediment to keep marine processes in check. The combined effects of waves, tides, and river flow determine the shape of a delta. In 1975, William Galloway, a geologist at the University of Texas, classified deltas by the relative influence of those three factors. (See **Figure 12.6**. Note that any delta strongly affected by only one of the three factors is placed at an apex of the

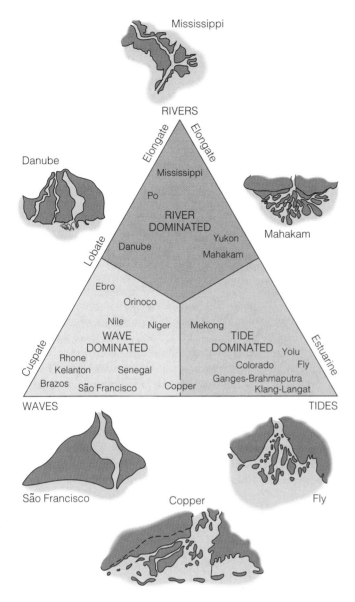

Figure 12.6 Each delta reflects the balance between the rate of sediment delivery at the river mouth and the mechanisms of its dispersal. This triangular diagram classifies river deltas according to the influence of the three major factors affecting their development: the river, waves, and tides.

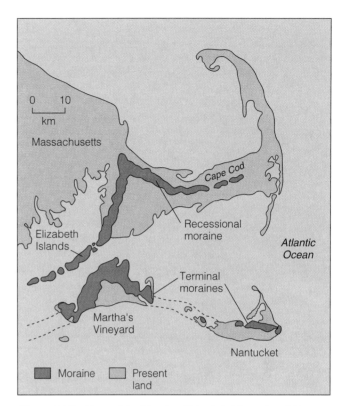

Figure 12.7 Most of the islands of Martha's Vinyard and Nantucket was formed from debris bulldozed into place by an advancing glacier. The spine of Cape Cod is built of material dropped as the glaciers retreated.

triangle.) A delta with a balance of forces is placed near the middle of the triangle. *River-dominated deltas* are fed by a strong flow of fresh water and continental sediments, and they form in protected marginal seas. They terminate in a well-developed set of *distributaries*—the split ends of the river—in a characteristic bird's-foot shape (as shown in the Mississippi, Figure 12.4). In *tide-dominated deltas,* freshwater discharge is overpowered by tidal currents that mold sediments into long islands parallel to the river flow and perpendicular to the trend of the coast. The largest tide-dominated delta has formed at

the mouths of the Ganges–Brahmaputra river system on the Bay of Bengal. *Wave-dominated deltas* are generally smaller than either tide- or river-dominated deltas and have a smooth shoreline punctuated by beaches and sand dunes. Instead of a bird's-foot pattern of distributaries, a wave-dominated delta will have one primary exit channel.

Deltas are not the only types of coasts built out by the land. The glaciers that covered the poleward parts of the continents during the ice age bulldozed great quantities of sediments and rocks to their outer margins. When the glaciers retreated, they left **moraines**—hills and ridges of sediments—some of which still stand above sea level. Part of Long Island, New York, is a glacial moraine, as are the oval-shaped hills of Boston and sections of the Puget Sound coast at Seattle. Perhaps the most famous glacial moraine in the U.S. is the area around Cape Cod, Massachusetts. The cores of the islands of Martha's Vinyard and Nantucket represent the farthest advance of the glacial debris, and the spine of the Cape itself is a remnant of material dropped by the glacier as it retreated northward when the climate warmed (see **Figure 12.7**).

Figure 12.8 Lava builds the Hawaiian coast during a 1987 eruption of Kilauea volcano. The typical lobe shape of a volcanic coast is visible in the background. In the foreground, a black sand beach is forming as hot lava meets cool seawater.

Volcanic Coasts

As we saw in Chapter 4, most islands that rise from the deep ocean are of volcanic origin. If the volcanism has been recent, the coasts of a volcanic island will consist of lobed lava flows extending seaward (**Figure 12.8**). Concave shorelines (as in **Figure 12.9**) result when a small coastal volcano collapses and seawater fills the crater.

Coasts Shaped by Earth Movements

Coasts can coincide with places where the Earth's crust is being warped or faulted. When the seabed on the seaward side of a coastal fault moves *downward,* a steep escarpment that continues to a greater depth than a wave-cut cliff can result. When the landward side of the fault moves *upward,* the part of the coast previously submerged during high tides can be left high and dry. The 1964 Alaska earthquake (described in Chapter 3) caused parts the of Prince William Sound shore to rise more than 3.5 meters (12 feet). As can be seen in **Figure 12.10**, in some places nearly half a kilometer (one-fourth mile) of the old sea-cut bench was exposed, even at high tide.

Fault coasts sometimes occur along transform faults. The Pacific coast of North America provides two striking examples: the Gulf of California (located along the San Andreas Fault between Baja California and the mainland of Mexico) and the area around Point Reyes, north of San Francisco. Baja California was once a part of the North American continent, but movement at the boundary between the Pacific and North American plates has torn the finger of land on the west side of the fault from the land mass to the east. The young, narrow, straight gulf has formed as seawater intruded. As can be seen in **Figure 12.11**, a photo taken near Point Reyes, fault coasts can be startlingly straight.

SECONDARY COASTS

Secondary coasts are those coasts that have been significantly changed by wave action and other marine processes after sea level has stabilized. Land erosion and marine erosion both work to change a primary coast to a secondary coast. Secondary coasts are shaped and attacked from the land by stream erosion, the abrasion of wind-driven grit, the alternate freezing and thawing of water in rock cracks, the probing of plant roots, glacial activity, rainfall, dissolution by acids from soil, and slumping. (As any mountaineer will confirm, these factors are not limited to coasts.)

Attack from the sea is by waves and currents. The continual onslaught of waves does most of the work, with currents distributing the results of the waves' labor.

Figure 12.9 Concave shapes may also occur on a volcanic coast. The explosion and collapse of a volcano close to shore caused this indentation on the coast of Isla Encantada, Gulf of California, Mexico.

Figure 12.10 This former seafloor at Prince William Sound, Alaska, was raised 8 meters (26 feet) above sea level by tectonic uplift during the great earthquake of 27 March 1964. The exposed surface, which slopes gently from the base of the sea cliffs to the water, is about 400 meters (about one-quarter of a mile) wide. The light-colored coating on the rocks consists mainly of the dried remains of small marine organisms. The photo was taken at about 0.0 tide, 30 May 1964.

Figure 12.11 A characteristically straight fault coast near Point Reyes, California.

Figure 12.12 Waves from a large Pacific storm strike the north coast of the Hawaiian island of Oahu.

Some indication of the violence of this activity may be implied from **Figure 12.12**.

Large storm surf routinely generates pressures of 27,000 kilograms per square meter (25 tons per square yard). The crashing waves push air and water into tiny rock crevices. The repeated buildup and release of pressure within these crevices can weaken and fracture the rock. But it is not the hydraulic pressure of moving water alone that abrades the coasts. Tiny pieces of sand, bits of gravel, or stones hurled by waves toward the shore are even more effective at eroding the shore. **Dissolution,** the dissolving of minerals in the rocks by water, contributes to the erosion of easily soluble coastal rocks such as limestone. Even the digging and scraping activities of marine organisms have an effect.

The rate at which a shore erodes is proportional to the hardness and resistance of the rock, the violence of the wave shock to which it is exposed, and the local range of tides. Hard rock resists wear. Coasts made of granite or basalt may retreat an insignificant amount over a human lifetime; the granite coast of Maine erodes only a few centimeters per decade. Coasts of soft sandstone or other weak (or soluble) materials, however, may disappear at a rate of a few meters per year. In an extreme case, soft shore cliffs facing the North Sea in East Suffolk, England, eroded more than 11 meters (36 feet) during two hours of a severe storm!

Marine erosion is usually most rapid on **high-energy coasts,** areas frequently battered by large waves. High-energy coasts are most common adjacent to stormy ocean areas of great fetch and along the eastern edges of continents exposed to tropical storms. The coasts of Maine and British Columbia and the southern tips of South America and South Africa are typical high-energy coasts. **Low-energy coasts** are only infrequently attacked by large waves. Because of their generally protected location in the Gulf of Mexico, the U.S. Gulf states share a low-energy coast—at least between hurricanes!

Waves can only affect the coast where they strike, so erosion is concentrated near average sea level. A shore with little tidal variation erodes quickly because the wave action is concentrated near one level for longer times. Low-energy coasts protected by offshore islands usually erode slowly, as do areas below the low tide line. Some erosion does occur below the surface because of the orbital motion of water in waves, but even the largest waves have little erosive effect at depths greater than about 15 meters (50 feet) below average sea level. Cliffs above shore are subject to pounding either directly from waves or by rocks hurled by waves. In one case, windows of a Scottish lighthouse 100 meters (328 feet) above sea level were broken by wave-thrown rocks during an Atlantic winter storm!

Some Features of Secondary Coasts

Erosive forces can produce a wave-cut shore showing some or all of the features illustrated in **Figure 12.13**. Note the complex, small-scale irregularities of this rocky erosional coastline. **Sea cliffs** slope abruptly from land into the ocean, their steepness usually resulting from the collapse of undercut notches. The position of the sea cliffs marks the shoreward limit of marine erosion on a coast. The parade of waves cuts **sea caves** into the cliffs at local zones of weakness in the rocks. Most sea caves are accessible only at low tide. A blowhole can form if erosion follows a zone of weakness upward to the top of the cliff. When the tide is at just the right height, spray can blast from the fissure as waves crash into the cliff. Offshore features of rocky coasts can include natural arches; the stacks that remain when the arch falls; and a smooth, nearly level **wave-cut platform** just offshore,

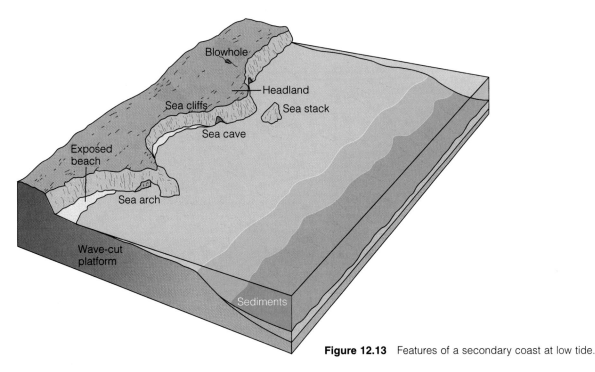

Figure 12.13 Features of a secondary coast at low tide.

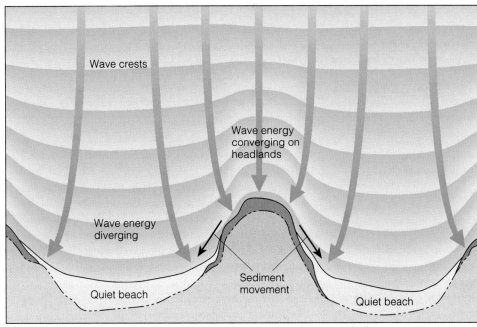

Figure 12.14 Wave energy converges on headlands and diverges in the adjoining bays. The accumulation of sediment in the tranquil bays eventually smooths the contours of the shore.

which marks the submerged limit of rapid marine erosion. Much of the debris removed from cliffs during the formation of these structures is deposited in the quieter water farther offshore, but some can rest at the bottom of the cliffs as exposed beaches. Indeed, as we shall soon see, broad beaches are often features of secondary coasts.

Shore Straightening

The first effect of marine erosion on a newly exposed coast is to intensify the irregularity of the coastline. This happens because coastal rocks are usually not uniform in composition over long horizontal distances. Some hard rocks will resist erosion well, while softer rocks on the same coast may disappear almost overnight. (This explains the uneven character of the stacks, arches, and sea cliffs described above.)

Eventually, however, coastal erosion tends to produce a smooth shoreline. Wave energy is focused onto headlands and away from bays by wave refraction (**Figure 12.14**). Sediment eroded from the headlands tends to collect as beaches in the relatively calm bays. As erosion continues, the deposits may eventually protect the base of the shore cliffs from the waves. Coastal irregularities

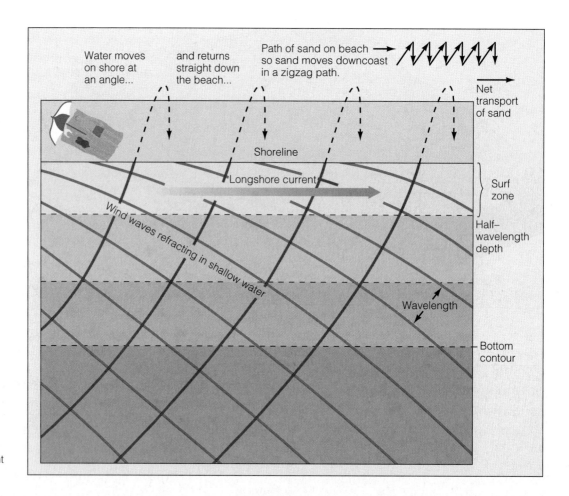

Figure 12.15 The wind wave refraction and longshore current components of longshore drift.

are thus smoothed with the passage of time. As you might expect, straightening occurs most rapidly on high-energy coasts.

The Accumulation of Beaches

If the submerged slope of the seafloor is steep, eroded sediments will soon drain to deeper waters. If the slope is not too steep, sediments will be transported along the coast by wave and current action. The net amount of sediment (usually sand) that moves along the coast, driven by wave action, is referred to as **longshore drift.** Longshore drift occurs in two ways: the wave-driven movement of sand along the exposed beach, and the current-driven movement of sand in the surf zone just offshore (**Figure 12.15**).

As we have seen, wind waves arriving from distant storms do not always approach the coast straight on. Most wind waves approach at an angle and then refract in shallow water to break almost parallel to shore. Refraction is usually incomplete, however, and some angle remains when the waves break. If sediments have accumulated to form a beach, water from the breaking wave will rush up the beach at a slight angle but return

to the ocean by running straight downhill under the influence of gravity. The sand grains disturbed by the wave will follow the water's path, moving up the beach at an angle but retreating down the beach straight down the slope. Net transport of the grains is *longshore,* parallel to the coast, away from the direction of the approaching waves. Net sand flow along the U.S. East and West coasts is usually to the south because the waves that drive the transport system usually approach from the north, where storms most commonly occur.

Sediments are also transported in the surf zone in a **longshore current.** The waves breaking at a slight angle distribute a portion of their energy away from their direction of approach. This energy propels a narrow current in which sediment already suspended by wave action can be transported downcoast. The speed of the longshore current sometimes reaches almost 4 kilometers (about 2½ miles) per hour.

Sand moving in the wash of waves along the beach and sediments propelled in the longshore current just offshore are often joined by much greater loads of sediments brought to the coast by rivers. Net southward transport of all this material along the central California coast exceeds 230,000 cubic meters (300,000 cubic yards) per year. Typical East Coast figures are about two-thirds of this value. Transport is to the south because southern-moving waves from northern storms provide most of the beach energy on the East and West coasts.

BEACHES

The most familiar feature of a secondary coast is the beach. A **beach** is a zone of unconsolidated (loose) particles that covers part or all of a shore. The landward limit of a beach may be vegetation, a sea cliff, relatively permanent sand dunes, or construction such as a seawall. The seaward limit occurs where sediment movement on- and offshore ceases—a depth of about 10 meters (33 feet) at low tide. The continental United States has 17,762 kilometers (10,983 miles) of beaches, about 30% of the total shoreline. Beaches result when sediment, usually sand, is transported to places suitable for deposition. Such places include the calm spots between headlands, shores sheltered by offshore islands, and regions with usually quiet surf. Sometimes the sediment is transported a very short distance—grit may simply fall from the cliff above and accumulate at the shoreline—but more often the sediment on a beach has been moved for long distances to its present location.

Wherever they are found, beaches are in a constant state of change. They may be thought of as rivers of sand—zones of continuous sediment transport.

The Composition and Slope of Beaches

The material of beaches can range from boulders, through cobbles and pebbles and gravel, to very fine silt. The rare black sand beaches of Hawaii are made of finely fragmented lava. Some beaches consist of shells and shell debris, or fragments of coral. Unfortunately some also include large quantities of human junk: Glass or plastic beaches are not unknown. Cobble beaches can be very steep (occasionally with slopes in excess of 20°), but wide beaches of fine sand are sometimes nearly as flat as a parking lot.

In general, the flatter the beach, the finer the material from which it is made (**Table 12.1**). The relation between particle size and beach slope depends on wave energy, particle shape, and the porosity of the packed sediments. Water from waves washing onto a beach—the **swash**—carries particles onshore, increasing the beach's slope. If water returning to the ocean—the **backwash**—carries back the *same* amount of material as it delivered, the beach slope will be in equilibrium; so the beach will not become larger or steeper.

On fine-grain beaches, the ability of small, sharp-edged particles to interlock discourages water from per-

Table 12.1 The Relationship Between the Particle Size of Beach Material and the Average Slope of the Beach

Type of Beach Material	Size (mm)	Average Slope of Beach
Very fine sand	.0625–.125	1°
Fine sand	.125–.25	3°
Medium sand	.25–.50	5°
Coarse sand	.50–1.0	7°
Very coarse sand	1–2	9°
Granules	2–4	11°
Pebbles	4–64	17°
Cobbles	64–256	24°

Source: Shepard, 1973.

colating down into the beach itself; so water from waves runs quickly back down the beach, carrying surface particles toward the ocean. This process results in a very gradual slope. Broad flat beaches also have a large area on which to dissipate wave energy, and they can provide a calm environment for the settling of fine sediment particles. In contrast, coarse particles (gravel, pebbles) do not fit together well and readily allow water to drain between them. Onrushing water disappears *into* a beach made of coarse particles, so little water is left to rush down the slope, thereby minimizing the transport of sediments back to the ocean. Thus larger particles tend to build up at the back of the beach, increasing its steepness.

Beach Shape

Figure 12.16 shows a beach profile, or cross section, affected by small-to-moderate wave and tidal action. The scale is exaggerated vertically to show detail. The key feature of any beach is the **berm,** an accumulation of sediment that runs parallel to shore and marks the normal limit of sand deposition by wave action. The peaked top of the berm, called the **berm crest,** is usually the highest point on a beach. It corresponds to the shoreward limit of wave action during most high tides. Inland of the berm crest, extending to the farthest point where beach sand has been deposited, is the **backshore.** The backshore is the relatively inactive portion of the beach, which may include windblown dunes and grasses. The **foreshore,** seaward of the berm crest, is the active zone of the beach, washed by waves during the daily rise and fall of the tides. It extends from the base of the berm—

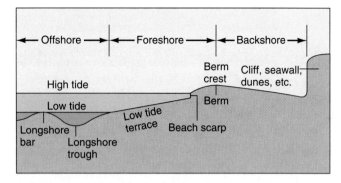

Figure 12.16 A typical beach profile.

where a **beach scarp** (a vertical wall of variable height) is often carved by wave action at high tide—to the low tide mark where the offshore zone begins. The **low tide terrace** is the smooth area between the beach scarp and low tide mark, on which waves expend most of their energy. The low tide terrace is the part of the beach favored by runners because of its constant slope and hard, water-packed sand. Below the low tide mark, wave action, turbulent backwash, and longshore currents excavate a **longshore trough** parallel to shore. Irregular **longshore bars** (submerged or exposed accumulations of sand) complete the seaward profile.

This beach profile is only temporary, generated by the interplay of sediments, waves, and tides. Great storm waves can rearrange a beach in a day, transporting thousands of tons of sediment from the beach to hidden sand bars offshore. **Figure 12.17** shows this transition in progress during a period of high wave action on a southern California beach. The beach scarp has grown to enormous proportions as sand is gouged from a thick berm with each new wave on the incoming tide. Later in the year smaller waves will move the sand back onto the beach.

Indeed, most temperate-climate beaches undergo a seasonal transformation. Beaches are cut to a lower level in winter than in summer because higher waves accompany winter storms. **Figure 12.18** traces the changes in a beach profile through most of a year. Notice that the beach builds by deposition during the summer from its low point in April to its height in December and then is torn down again by winter storms. Changes from summer to winter on a southern California beach are shown in **Figure 12.19**.

Minor Beach Features

Small features are among the things that make beaches such interesting places. *Ripples* in the sand (**Figure 12.20a**) are caused by rushing currents; similar structures can be seen in dry river bottoms or stream beds. *Rills* (**Figure 12.20b**) are small branching surface depressions that channel water back to the ocean from a saturated beach during a falling tide. Diamond-shaped *backwash marks* (**Figure 12.20c**) form when projecting shells, pebbles, or animals interrupt the backwash along the low tide terrace and cause uneven deposition of very fine sediment of a contrasting color. Another result of uneven

Figure 12.17 A high beach scarp in southern California. Rising tides and large late-summer waves have removed huge amounts of sand from a thick berm. The sand was deposited through the spring and summer months when wave energy was lower.

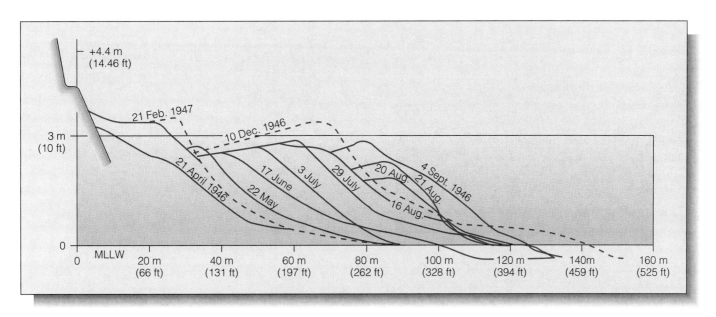

Figure 12.18 Seasonal changes in a beach profile. The berm at Carmel, California, builds 60 meters (200 feet) seaward during summer when the waves are small. The beach then almost completely disappears when struck by the large waves of winter storms (dashed lines).

a

b

Figure 12.19 The movement of sand on Carlsbad State Beach, California. (a) The beach in the summer of 1978. (b) The same beach in the spring of 1983 after a winter of severe storms. The sand has moved offshore, and coastal structures have been damaged.

deposition can be seen in beach layering. A vertical slice into a calm beach will often reveal layer upon layer of sediments with different colors and textures (**Figure 12.20d**).

On a somewhat larger scale, the beach shoreline is sometimes scalloped by *cusps* (*cuspis* = point) (**Figure 12.21**). These repeating points, interrupted by miniature bays, are evenly spaced every several meters. Cusps tend to form at neap tide periods and are destroyed during the greater range of spring tides. Cusp size appears to be correlated with wave energy. Their origin is not well understood, but they are probably caused by interference between approaching waves. Offshore sand bars and troughs are also more pronounced during periods of moderate tides. The variability of cusps and bars reflects the transitory nature of beach processes.

Figure 12.20 Minor features of beach sand. (a) Ripple marks. (b) Backwash marks. (c) The streamlike pattern of rill marks. (d) Layers of sediments are clearly seen in this newly eroded beach face. The alternating layers were deposited by small waves during many ebb and flood tide cycles. Larger storm-driven waves have exposed the layers to view.

Figure 12.21 Beach cusps at San Simeon, California, formed in a gently sloping fine-sand beach. The horizontal beach berm can be seen to the right.

Coastal Cells

As we have seen, most new sand on a coast is brought in by rivers or originates from the erosion of cliffs. The sand is moved parallel to the beach by longshore drift, and it is moved on- and offshore at right angles to the beach as the seasons change. If a beach is stable in size, neither growing nor shrinking, the amount of new sand entering must be balanced by the amount of old sand being removed. Sand that drifts below the reach of wave action is lost from the coast and may migrate farther out on the continental shelf. Some sand is driven by longshore currents into the nearshore heads of submarine canyons. Sand moving away from shore in these canyons sometimes forms impressive sandfalls (see Figure 4.13 for an example) and is lost from the beaches above.

The natural sector of a coastline in which sand *input* and sand *outflow* are balanced may be thought of as a **coastal cell.** The main features of such a cell are illustrated in **Figure 12.22a**. The size of coastal cells varies greatly. They are often very large along the relatively smooth, tectonically passive trailing edges of continents; coastal cells along the southeast coast of the United States, for example, are hundreds of kilometers long. On the active leading edge of the continent, they are smaller. Four cells exist in the 360 kilometers (225 miles) between southern California's Point Conception and the Mexican border. Each terminates in a submarine canyon at the downcoast end (see **Figure 12.22b**).

LARGE-SCALE FEATURES OF SECONDARY COASTS

We have already noted the presence of small longshore bars off beaches (see Figure 12.16), but secondary coasts, especially the coasts along a subsiding irregular conti-

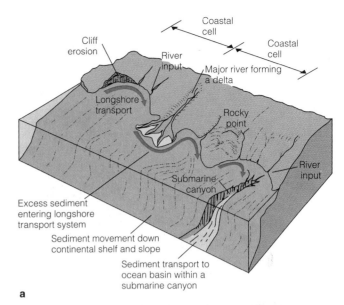

Figure 12.22 Coastal cells. (a) The general features of coastal cells, in which sand is introduced by rivers, transported southward by the longshore drift, and trapped within the nearshore heads of submarine canyons. (b) A series of coastal cells in southern California. The white arrows indicate the flow of sand in rivers toward the beaches; the red arrows show sand flowing into submarine canyons.

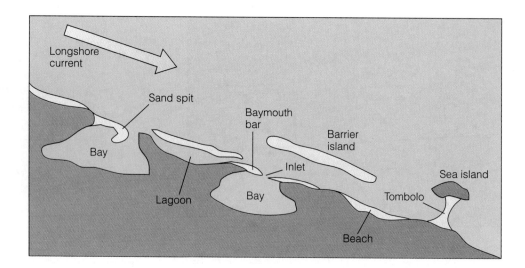

Figure 12.23 A composite diagram of the large-scale features of an imaginary secondary coast. Not all these features would be found in such close proximity on a real coast.

nental margin, also exhibit much larger features. Some of these features are illustrated in **Figure 12.23**.

Sand Spits and Bay Mouth Bars

Sand spits are among the most common of these. A sand spit forms where the longshore current slows as it clears a headland and approaches a quiet bay. The slower current in the mouth of the bay is unable to carry as much sediment, so sand and gravel are deposited in a line downcurrent of the headland (**Figure 12.24**). As can be seen in Figures 12.23 and 12.24, sand spits often have a curl at the tip. This is caused by the current-generating waves being refracted around the tip of the spit.

A **bay mouth bar** forms when a sand spit closes off a bay by attaching to a headland adjacent to the bay. The bay mouth bar protects the bay from waves and turbulence and encourages the accumulation of sediments there. An **inlet**—a passage to the ocean—may be cut through a bay mouth bar by tidal action, by water flowing from a river emptying into the bay, or by heavy storm rains. A bay mouth bar is shown in **Figure 12.25**.

Barrier Islands and Sea Islands

Secondary coasts can also develop narrow, exposed sandbars that are parallel to but separated from land. These are known as **barrier islands** (see **Figure 12.26**). About 13% of the world's coasts are fringed with barrier islands.

Barrier islands can form when sediments accumulate on submerged rises paralleling the shoreline. Some islands off the Mississippi–Alabama coast developed in this way. Larger barrier islands are thought to form in a different way, however. Near the end of the last major rise in sea level, about 6,000 years ago, coastal plains

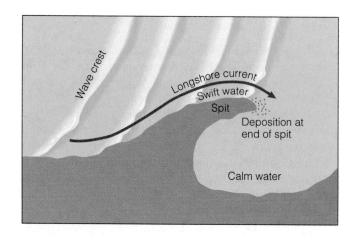

Figure 12.24 A sand spit forms where the longshore current slows as it clears a headland and approaches a quiet bay.

near the edge of the continental shelf were fronted by lines of sand dunes. Rising sea level caused the ocean to break through the dunes and form a **lagoon**—a long, shallow body of seawater isolated from the ocean. The high lines of coastal dunes became islands. As sea level continued to rise, wave action caused the islands and lagoons to migrate landward. Most of the barrier islands off the southeast coast of the United States originated in this way. They are still migrating slowly landward as sea level continues to rise.[2]

There are 295 barrier islands along the East and Gulf coasts of the United States, with a combined length of 2,591 kilometers (1,610 miles). Every year severe storms generate waves intense enough to erode barrier island

[2]This type of coast superficially resembles the drowned valleys seen on some primary coasts. Here, however, indentations are minor. Barrier coasts lack long embayments such as Chesapeake Bay.

Figure 12.25 A bay mouth bar. The inlet is now closed, but increased river flow (from inland rainfall) or large waves combined with very high and low tides could break the bar.

Figure 12.26 Barrier islands off the Texas Gulf Coast. (This photo, taken from space, is on a much larger scale than Figure 12.25.)

beaches. The largest of these storms can generate waves that overwash the low islands. Runoff from rivers swollen by rains, coupled with water driven by wind waves and storm surge, can rapidly flood a lagoon and cut new inlets through barrier islands.

Despite these dangers, about 70 of the barrier islands have been commercially developed, and millions of people live on them. The most famous barrier islands include Atlantic City, New Jersey; Ocean City, Maryland; Miami Beach and Palm Beach, Florida; and Galveston, Texas. Roughly once every 100 years a winter storm has

catastrophic effects on populated areas of Atlantic barrier islands, and the southeastern Atlantic and Gulf coasts must contend with occasional large hurricanes. The continuing subsidence of these passive coasts (combined with changes caused by commercial development and the ongoing rise in sea level) will undoubtedly cost lives and destroy property. **Figure 12.27** suggests the extent of the threat.

Unlike barrier islands, **sea islands** contain a firm central core that was part of the mainland when sea level was lower. The rising ocean separated these high points

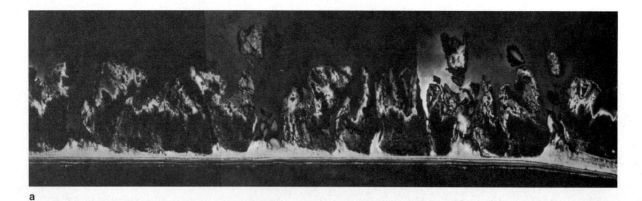

a

b

c

Figure 12.27 Barrier island modification: actual and potential. (a) The beach extending along the Matagorda Peninsula (Texas) barrier in September 1960. (b) The same area six days after the passage of hurricane Carla in September 1961. The beach and island have been breached, and washover deltas are clearly seen. (c) Ocean City, Maryland, a developed barrier island. Host to 8 million visitors a year, this city (and others similarly situated) has no effective protection against flooding and damage from severe storms.

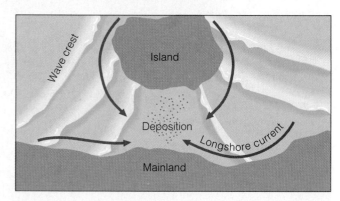

Figure 12.28 Formation of a tombolo. The island acts as a barrier to wave energy, allowing sediments to accumulate behind it. Eventually a sand bar connects the island to the mainland.

Figure 12.29 A small section of the Great Barrier Reef, Queensland, Australia. This coast has been extensively modified by biological activity.

Figure 12.30 A mangrove coast in Florida. Mangrove trees trap sediments, building and stabilizing the coast.

from land, and sedimentary processes surrounded them with beaches. Hilton Head, South Carolina, and Cumberland Island, Georgia, are sea islands (see again Figure 12.23). If the island is close to shore, a bridge of sediments called a **tombolo** may accumulate to connect the island to the mainland (see **Figure 12.28**). Tombolos can also connect offshore rocky outcrops or volcanoes to the mainland.

COASTS FORMED BY BIOLOGICAL ACTIVITY

Coasts can be extensively modified by the activities of animals and plants. The most dramatic modifications occur in the tropics, where coral polyps form reefs around volcanic islands or along the margin of a continent. The greatest of all reefs is the Australian Great Barrier Reef (**Figure 12.29**), which begins in the Torres Strait separating New Guinea and Australia and runs down the northeastern coast of Australia for 2,500 kilometers (1,500 miles). The reef is not a single object, but a composite of more than 3,000 individual coral reefs covering 350,000 square kilometers (135,000 square miles)—collectively the largest structure made by living organisms on Earth.

Corals animals (about which you will learn more in Chapter 18) are related to the familiar sea anemones found along American coasts. Reef-building coral polyps secrete a cup-shaped calcium carbonate skeleton, which remains behind after the animal dies; the accumulation of skeletons gradually forms the reef. These corals grow best in brightly lighted water about 5 to 10 meters (16 to 33 feet) deep, and in ideal conditions they grow at a rate of about 1 centimeter (½ inch) per year.

The Florida Keys—a series of low islands extending south of the tip of Florida—are an excellent example of a coral reef coast in the continental U.S. How can a coral coast extend above sea level? The Keys are relatively high because they were formed during a time between ice ages when sea level stood about 20 feet (6 meters) higher than it does today. The coral reef islands of the Pacific and Indian oceans are much lower—a great storm can submerge and fracture them. Those that extend above sea level do so because chunks of reef margin are thrown toward the center of these islands by storm waves and winds. The accumulated blocks are cemented together by the limestone (calcium carbonate) they contain.

Other coasts have been formed by mangroves, trees that can grow in salt water. The coast of southwestern Florida has been extended and shaped by the activity of mangroves, whose root systems trap and hold sediments around the plant (**Figure 12.30**). The root complex forms an impenetrable barrier and safe haven for organisms around the base of the trees. You will learn more about mangroves in the section on marine plants in Chapter 14.

ESTUARIES

An **estuary** (*œstus* = tide) is a body of water partially surrounded by land, where fresh water from a river mixes with ocean water. Estuaries are areas of remarkable biological productivity and diversity. Chesapeake Bay, San Francisco Bay, and Puget Sound are all estuaries.

Three factors determine the characteristics of estuaries: the shape of the estuary, the volume of river flow at the head of the estuary, and the range of tides at the estuary's mouth. The mingling of waters of different densities, the rise and fall of the tide, and the variations in river flow—along with the actions of wind, ice, and the Coriolis effect—guarantee that patterns of water circulation in an estuary will be complex.

Estuaries are categorized by their circulation patterns. The simplest circulation patterns are found in **salt wedge estuaries,** which form where a rapidly flowing large river enters the ocean in an area where tidal range is low or moderate. The exiting fresh water holds back a wedge of intruding seawater (**Figure 12.31a**). Note that density differences cause fresh water to flow over salt water. The seawater wedge retreats seaward at times of low tide or strong river flow, and it returns landward as the tide rises or when river flow diminishes. Some seawater from the wedge joins the seaward-flowing fresh water at the steeply sloped upper boundary of the wedge, and new seawater from the ocean replaces it. Nutrients and sediments from the ocean can enter the estuary in this way. Examples of salt wedge estuaries are the mouths of the Hudson and Mississippi rivers.

A different pattern occurs where the river flows more slowly and tidal range is moderate to high. As their name implies, **well-mixed estuaries** contain differing mixtures of fresh and salt water through most of their length. Tidal turbulence stirs the waters together as river runoff pushes the mixtures to sea. A well-mixed estuary is illustrated in **Figure 12.31b**. The mouth of the Columbia River is an example.

Deeper estuaries exposed to similar tidal conditions but greater river flow become **partially mixed estuaries.** Partially mixed estuaries share some of the properties of salt wedge and well-mixed estuaries. Note in **Figure 12.31c** the influx of seawater beneath a surface layer of fresh water flowing seaward; mixing occurs along the junction. Energy for mixing comes from both tidal turbulence and river flow. England's Thames River, San Francisco Bay, and Chesapeake Bay are examples.

In well-mixed and partially mixed estuaries in the Northern Hemisphere, the incoming seawater will press against the right side of the estuary because of the Coriolis effect. Outflowing river water will also trend to the right of its direction of travel. This rightward drift can be

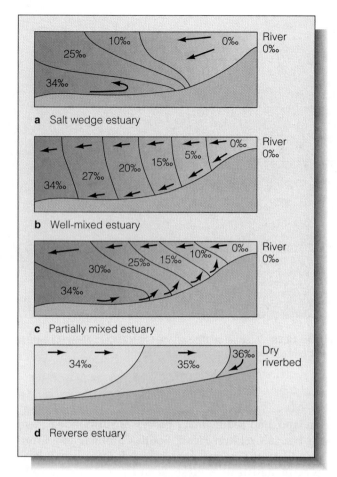

Figure 12.31 Types of estuaries in vertical cross sections. The salinity values show the amount of mixing between fresh water (0‰) and seawater (34‰) in the various types. (a) Salt wedge estuary. (b) Well-mixed estuary. (c) Partially mixed estuary. (d) Reverse estuary, in which evaporation plays a major role.

seen in the contour of lines representing surface salinity in Chesapeake Bay (**Figure 12.32**).

Reverse estuaries can form along arid coasts when rivers cease to flow. The evaporation of seawater in the uppermost reaches of these estuaries will cause water to flow from the ocean into the estuary, producing a gradient of *increasing* salinity from the ocean to the estuary's upper reaches (see **Figure 12.31d**). Reverse estuaries, sometimes called *lagoons*, are common on the Pacific coast of Mexico's Baja peninsula and along the U.S. Gulf Coast.

The coasts of the United States contain about 15,150 square kilometers (5,850 square miles) of estuarine waters. Many of the estuaries along the East Coast have been formed by the combined effects of rising sea level and tectonic subsidence. Chesapeake Bay and the Hudson River valley are examples of drowned river mouths. Some West Coast estuaries were formed by tectonic

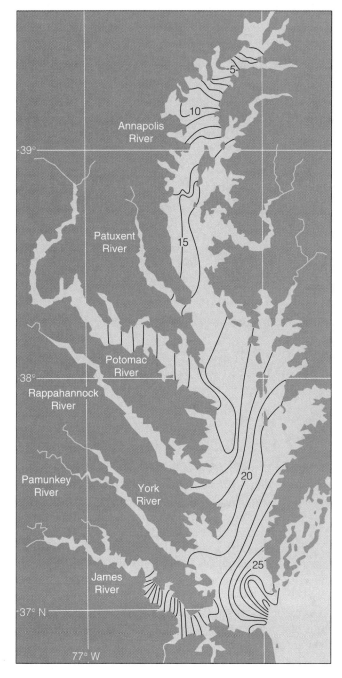

Figure 12.32 Cheasapeake Bay, an example of a partially mixed estuary. The typical distribution of surface salinity in the estuary ranges from 28‰ at the mouth to 1‰ near the upper reaches. The Coriolis effect forces the inflowing salt water against the right bank. (Notice how the 20‰ contour lines trend toward the right bank.) Compare this diagram to the photograph in Figure 12.2.

forces; San Francisco Bay has filled with water because of faulting and local subsidence. Estuaries can also form in river-fed bays behind bay mouth bars or within river deltas. Estuaries have become much more common as sea level has risen over the last 6,000 years.

As we will see in Chapter 17, estuaries often support a tremendous number of living organisms. The easy availability of nutrients and sunlight, protection from wave shock, and the presence of many habitats permit the growth of many species and individuals. Estuaries are frequently nurseries for marine animals; several species of perch, anchovy, and Pacific herring take advantage of the abundant food in estuaries during their first weeks of life. Unfortunately for their inhabitants, estuaries are in high demand for development into recreational resources and harbors. Estuaries have become the most polluted of all marine environments.

CHARACTERISTICS OF UNITED STATES COASTS

Plate tectonic forces have had immense influence on the margins of continents, and the edges of the United States are no exception. The results of plate movement on the West Coast differ greatly from those on the East and Gulf coasts, primarily because the West Coast is near an active plate margin while the East and Gulf coasts are not.

The West Coast

The West Coast is an actively rising margin on which volcanoes, earthquakes, and other indications of recent tectonic activity are easily observed. West Coast beaches are typically interrupted by jagged rocky headlands, volcanic intrusions, or the effects of submarine canyons. Wave-cut terraces (**Figure 12.33**) are found as much as 400 meters (1,300 feet) above sea level in a number of places, evidence that tectonic uplift has exceeded the general rise in sea level through the past million years.

Most of the sediments on the West Coast originated from erosion of relatively young granitic or volcanic rocks of the coastal mountains. The particles of quartz and feldspar that constitute most of the sand were transported to the shore by flowing rivers. The volume of sedimentary material transported to West Coast beaches from inland areas greatly exceeds the amount originating at the coastal cliffs. Deltas tend not to form at West Coast river mouths because the continental shelf is narrow, river flow is generally low (except for the Columbia River), and beaches are usually high in wave energy. The predominant direction of longshore drift is to the south because northern storms provide most of the wave energy.

Figure 12.33 Wave-cut terraces on San Clemente Island.

The East Coast

The East Coast is a passive margin, tectonically calm and subsiding because of its trailing position on the North American Plate. Subsidence along the coast has been considerable—3,000 meters (10,000 feet) over the last 150 million years. A deep layer of sediment has built up offshore, material that helped produce today's barrier islands. Relatively recent subsidence has been more important in shaping the present coast, however. Coastal sinking and rising sea level have combined to submerge some parts of the East Coast at a rate of about 0.3 meter (1 foot) per century. This process has formed the huge flooded valleys of Chesapeake and Delaware bays, the landward-migrating barrier islands, and the shrinking lowlands of Florida and Georgia.

Rocks to the north (in Maine, for example) are among the hardest and most resistant to erosion of any on the continent, so beaches are uncommon in Maine. But from New Jersey southward the rocks are more easily fragmented and weathered, and beaches are much more common. As on the West Coast, sediments are transported coastward by rivers from eroding inland mountains, but the transported material is trapped in estuaries and therefore plays a less important role on beaches.

Eastern beaches are typically formed of sediments from shores eroding nearby, or from the shoreward movement of offshore deposits laid down when the sea level was lower. The amount of sand in an area thus depends in part on the resistance or susceptibility of nearby shores to erosion. Sand moves generally south on these beaches just as it does on the West Coast, but the volume of moving sand in the East is less.

As we have seen, glaciers have also contributed to the shaping of the northern part of the East Coast: large portions of Long Island and all of Cape Cod are remnants of debris deposited by glaciers.

The Gulf Coast

The Gulf Coast experiences a smaller tidal range and—hurricanes excepted—a smaller average wave size than either the West or East coasts. Reduced longshore drift and an absence of interrupting submarine canyons allow the great volume of accumulated sediments from the Mississippi and other rivers to form large deltas, barrier islands, and a long raised "super berm" that prevents the ocean from inundating much of this sinking coast.

These are fortunate conditions because the rate of subsidence in the Gulf Coast is greater than that for most of

breakwater

Figure 12.34 Growth of a beach protected by a breakwater: Santa Monica, California.

the East Coast. Human activity can make the subsidence even worse. Pumping from oil and water wells, combined with sediment starvation and dredging, have exacerbated the situation around some large cities. At Galveston, Texas, for example, sea level is nearly 64 centimeters (25 inches) higher than it was a century ago, and most of New Orleans is now 2 meters (6.6 feet) below sea level. As we have seen, the results of hurricanes at such places can be tragic. The protective natural berm can easily be breached, and flood waters can surge far inland.

HUMAN INTERFERENCE IN COASTAL PROCESSES

Beaches exist in a tenuous balance between accumulation and destruction. Human activity can tip the balance one way or the other. For example, consider the rocky **breakwater** shown in **Figure 12.34**. The breakwater interrupts the progress of waves to the beach, weakening the longshore current and allowing sand to accumulate there. Without dredging, the beach will eventually reach the breakwater and fill the small boat anchorage the breakwater was built to provide. This is a minor example of human alteration of a beach, yet it serves to introduce the growing problem of human influences on coastal processes.

We often divert or dam rivers, build harbors, and

develop property with surprisingly little understanding of the impact our actions will have on the adjacent coast. Our role then becomes that of powerless observers. Residents of eroding coasts can only accept the inevitable loss of their property to the attack of natural forces, but residents of coasts in which deposition exceeds erosion are sometimes presented with alternatives. The choices are almost never simple. For example, should rivers be dammed to control devastating floods? If the dams are built, they will trap sediments on their way from mountains to coast. Beaches within the coastal cell fed by the dammed river will shrink because the sand on which they depend (to replenish losses at the shore) is blocked. Alarmed coastal residents will then take steps to hang onto whatever sand remains. They may try to trap "their" beaches by erecting **groins:** short extensions of rock or other material placed at right angles to longshore drift, to stop the longshore transport of sediments. This temporary expedient usually accelerates erosion downcoast (**Figure 12.35a**). Diminished beaches then expose shore cliffs to accelerated erosion. Wind wave energy that would have harmlessly churned sand grains now speeds the destruction of natural and artificial structures. Seawalls don't help, either (**Figure 12.35b**). They increase beach erosion by deflecting wave energy onto the sand. Churning by this increased energy eventually undermines the seawall, causing it to collapse. The importation of sand trapped behind dams (or from other sources) is also only a temporary—and very expensive—expedient (**Figure 12.35c**).

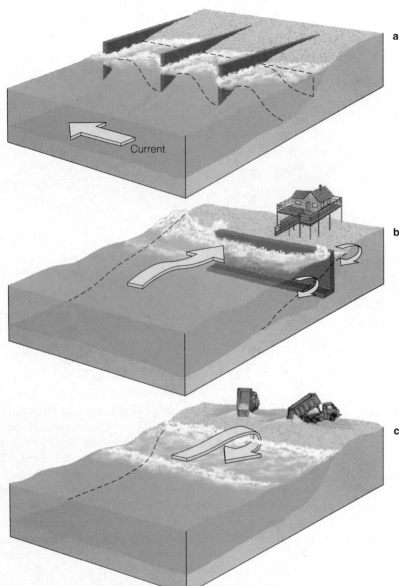

a Groin
Groins are structures that extend from the beach into the water. They help counter erosion by trapping sand from the current. Groins accumulate sand on their updrift side, but erosion is worse on the downdrift side, which is deprived of sand.

b Seawall
Seawalls protect property temporarily, but they also increase beach erosion by deflecting wave energy onto the sand in front of and beside them. High waves can wash over seawalls and destroy them and property.

c Importing sand
Importing sand to a beach is considered the best response to erosion. The new sand is often dredged from offshore and can cost tens of millions of dollars. Because it is often finer than beach sand, dredged sand erodes more quickly.

Figure 12.35 Some measures taken to slow beach erosion, and why these measures are largely ineffective.

Southern California provides many case studies. Beaches in the coastal cells there are generally shrinking because most sediment-carrying rivers have been dammed. Any structure that interferes with longshore drift will destabilize the beach because new sand is not readily available to take the place of sand trapped by the structure. **Figure 12.36** shows how the beach changed after a breakwater was constructed early in this century to protect the entrance to San Diego Bay. The Hotel del Coronado, a beautiful resort built in 1888 in nearby Coronado, lost its broad protecting beach because the new breakwater blocked sand replenishment from the mouth of San Diego Bay, to the north. The hotel was nearly lost to wave action as the beach shrank. Fortunately, sand driven from the south by waves diffracting off the tip of nearby Point Loma and by waves from tropical cyclones off Mexico have slowly filled in much of the

area south of the breakwater. Some of the beach has returned, but not to its original width. The hotel is not out of danger.

Less frequently, the deposition of sand is an unwanted result of human constructions. Interruption of longshore transport caused a new beach to form against what once were bare cliffs upcurrent from the breakwater at Santa Barbara. Downcurrent, calm water at the breakwater's end interfered with longshore drift, and sand accumulated in the harbor (**Figure 12.37**). Without continuous dredging of about 600 cubic meters (800 cubic yards) per day, the harbor would soon fill with sand. The dredging cost to taxpayers of Santa Barbara is considerable.

What are the implications of these unlooked-for sand movements? Douglas Inman, director of the Center for Coastal Studies at Scripps Institution of Oceanography,

Figure 12.36 The beach and the road north of the Hotel del Coronado were destroyed because sand replenishment was blocked after the building of the San Diego breakwater. In 1905, more than 30,000 200-pound sandbags were placed around the hotel to save it from destruction by the waves.

a

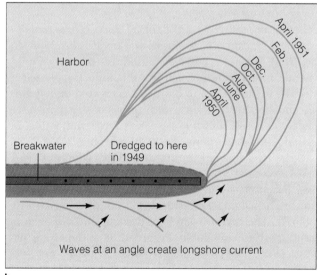

b

Figure 12.37 (a) A sand spit grows into the small boat anchorage at Santa Barbara, California. The area within the white square is enlarged in (b). Sand accumulates at the tip of the breakwater when longshore transport is interrupted.

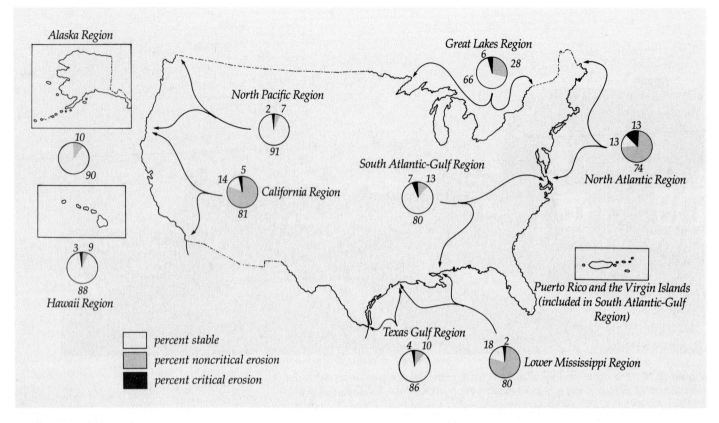

Figure 12.38 Shore erosion by region.

feels that *at least 20% of the United States' beach-bounded coastline is in danger of serious or catastrophic alteration.* On the West Coast a 30-year period of relatively mild weather may be ending. During this time, people felt it was safe to build close to the shore. Increased dam building and breakwater, jetty, and groin construction have made southern California's beaches more vulnerable. In coastal California alone, 3,666 homes and 1,020 businesses were damaged during the stormy winter of 1983; losses exceeded $100 million. A large storm in January 1988 did similar damage along the same coast, a clear indication that coastal residents do not always learn by example. The barrier islands of the East and Gulf coasts are at least as vulnerable. **Figure 12.38** summarizes the progress of shore erosion by regions.

Shores that look permanent through the short perspective of a human lifetime are in fact among the most temporary of all marine structures. Let's enjoy them in their present stages.

Q-AND-A

1. I've noticed currents moving seaward through the surf on high-energy beaches. What causes these so-called rip tides?

These small currents are caused by the movement offshore of a large amount of water at one time. Because they're not caused by gravitational forces, rip tide *isn't a good term for them; they're properly called* **rip currents.**

*Rip currents form when a group of incoming waves piles an excess of water on the landward side of a submerged sand bar faster than the longshore current can carry it. The water breaks through the bar in a few places and flows rapidly through the surf back to sea. Rip currents are often made visible by the muddy color of their suspended sediments contrasting with the cleaner water just offshore (**Figure 12.39**). A strong swimmer can use the rip current to his or her advantage, hitching a ride seaward through the churning surf to where the body surfing is best. To escape this narrow ocean-going band, however, a swimmer is advised to swim slowly parallel to shore and then return to shallow water. The higher the surf, the greater the probability of rip currents.*

Rip currents are sometimes called **undertows,** *a word as deceptive and inaccurate as* rip tide. *There aren't any small-scale, nearshore features that suck swimmers beneath the surface; even the legendary whirlpools would have trouble accomplishing that task (see, for example, Box 11.2).*

2. My foot tends to sink whenever I stand on the beach and let water from a wave run over it. The sand

moves away from the edges of my foot, and I sink in. Why?

This is a good example of water's ability to carry more sediment as its speed of flow increases. Your foot interrupts the flow of water up or down the beach after a wave breaks, and the water must speed up to get around your foot. Fast running water moves sand more effectively than slow running water; so the sand immediately next to your foot is removed. This process, termed scouring, *becomes a problem when structures are placed in shallow water.*

3. How can any beach accumulate material when it slopes *seaward*? Why doesn't all the sediment making up the beach just flow downhill into the ocean and disappear?

Wave action and beach permeability combine to build beaches. Small and moderate waves tend to predominate in places where there are beaches. These waves keep sand on or near shore because the orbital motion of water within them before they break moves sand more easily toward *shore than* away *(see* **Figure 12.40**). *After the waves break, a certain amount of the water that rushes onto the beach soaks into the sand. That much less water is therefore available to push sand grains downhill and out to sea. Coarse-grain beaches are proportionally steeper because a higher percentage of the uprushing water percolates into the beach, and so less is left to rush down the slope and "unpack" the beach.*

4. I hope someday to live near the ocean. What should I look for in buying property there?

Firm ground! Coastal Maine would be an ideal bet. The dense metamorphic rock of much of the Maine shore is stable (within the human time frame) and hard—ideal footings for a house. Make sure the site is far enough inland to avoid high surf, storm surge, and tides. If the winters in Maine don't appeal to you, coastal Florida might make a good choice—if your children aren't hoping to inherit the property. If you insist on building on a barrier island, make sure your home is on the mainland side of the southern end! Parts of the West Coast are all right, but local variability on that active margin makes some knowledge of the geological history of the area very valuable. For example, some parts of the San Diego shoreline are eroding about 3 meters (10 feet) per year—hardly a solid investment.

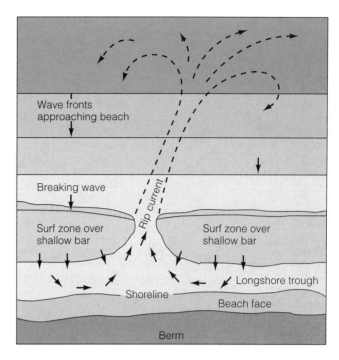

Figure 12.39 Rip currents. Rip currents are created when waves breaking on a submerged bar cause water to pile up shoreward of the bar. The excess water then flows seaward as a swift and narrow rip current in a channel of its own making, ending in turbulence outside the surf.

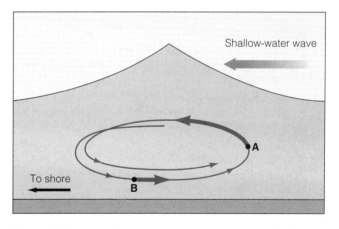

Figure 12.40 In small or moderate waves, a sand grain at point A moves easily *toward* shore, free of most bottom friction. Movement *offshore* by the sand grain at point B is impeded by friction with the bottom. New flow may be reversed from onshore to offshore by high waves because of strong backwash.

Terms and Concepts to Remember

backshore
backwash
barrier island
bay mouth bar
beach
beach scarp
berm
berm crest
breakwater
coast
coastal cell
delta
dissolution
erosion
estuary
eustatic change

fjord
foreshore
groin
high-energy coast
inlet
lagoon
longshore bar
longshore current
longshore drift
longshore trough
low-energy coast
low tide terrace
moraine
partially mixed
    estuary

primary coast
reverse estuary
rip current
salt wedge estuary
sand spit
sea cave
sea cliff
sea island
secondary coast
shore
swash
tombolo
wave-cut platform
well-mixed estuary

Study Questions

1. How is a primary coast different from a secondary coast?

2. In what other ways can coasts be classified? What are the advantages of classification? Are there disadvantages?

3. What features would you expect to see along a primary coast? A secondary coast? How long would you expect the features to last?

4. What two processes contribute to longshore drift? What powers longshore drift? What is the predominant direction of drift on U.S. coasts? Why?

5. What are some of the features of a sandy beach? Are they temporary or permanent? Is there a relationship between wave energy on a coast and the size (or slope, or grain size) of beaches found there?

6. How are deltas classified? Why are there deltas at the mouths of the Mississippi and Nile rivers, but not at the mouth of the Columbia River?

7. What is a coastal cell? Where does sand in a coastal cell come from? Where does it go?

8. How are estuaries classified? Upon what does the classification depend? Why are estuaries important?

9. Compare and contrast the U.S. West, East, and Gulf coasts.

10. How do human activities interfere with coastal processes? What steps can be taken to minimize loss of life and property along U.S. coasts?

For Further Study

Bascom, W. 1980. *Waves and Beaches*. Rev. ed. Anchor/Doubleday. A valuable reference, well and clearly written, nicely illustrated.

Boicourt, W. C. 1993. "Estuaries: Where River Meets the Sea." *Oceanus* 36 (no. 2): 29–37. An overview of the main estuaries of the continental U.S., the processes that shape them, and the human impact on them.

Broecker, W. S., and G. H. Denton. 1990. "What Drives Glacial Cycles?" *Scientific American,* January, 48–56. Sea level change is an important factor in coastal morphology. The authors think the variation in the heat of Northern Hemisphere summers forces ice ages and therefore changes in sea level.

Burk, K. 1979. "The Edges of the Ocean: An Introduction." *Oceanus* 22 (no. 3, Fall): 2–9. Terminology and overview.

Davis, R. A. 1994. *The Evolving Coast*. New York: Scientific American Library. A clearly written and very well illustrated overview of coastal processes and problems.

Dolan, R., and H. Lins. 1987. "Beaches and Barrier Islands." *Scientific American*, July, 68–77. Excellent summary of the evolution of these ephemeral structures.

Emery, K. O. 1969. "The Continental Shelves." *Scientific American,* September, 106–22. The size of the shelf changes as sea level varies.

Fairbridge, R. W. 1960. "The Changing Level of the Sea." *Scientific American,* May, 70–79. Formation and melting of glaciers and tectonic changes are mostly to blame. (See also article by Broecker, above.)

Flanagan, R. 1993. "Beaches on the Brink." *Earth,* November, 24–33.

Giese, G. S., and A. Aubrey. 1987. "Losing Coastal Upland to Relative Sea Level Rise: Three Scenarios for Massachusetts." *Oceanus* 30 (no. 3, Fall): 16–22. An East Coast version of Kuhn (see below).

Gornits, V., et al. 1982. "Global Sea Level Trend in the Past Century." *Science* 215 (no. 4540): 1611–14. Mean sea level rise has been 12 centimeters (4.7 inches) in the past century. Most of this rise has resulted from ocean warming and the subsidence of coasts.

Holman, R. A., et al. 1993. "The Application of Video Image Processing to the Study of Nearshore Processes." *Oceanography* 6 (no. 3): 78–85. New imaging technology can be used to analyze coastal processes.

Inman, D. L., and R. A. Bagnold. 1963. "Littoral Processes." In *The Sea*, vol. 3, edited by M. N. Hill. New York: Wiley.

Inman, D. L., and B. M. Brush. 1973. "The Coastal Challenge." *Science* 181 (no. 4094): 20–32. Discusses the dangers inherent in coastal development.

Kennett, J. 1982. *Marine Geology*. Englewood Cliffs, NJ: Prentice-Hall. Summary of causes and degrees of sea level changes.

King, C. A. M. 1972. *Beaches and Coasts*. London: Edward Arnold. This comprehensive text uses a competing system of coastal classification, that of H. Valentin. An interesting contrast to Shepard's scheme.

Kuhn, G. G., and F. P. Shepard. 1984. *Sea cliffs, Beaches, and Coastal Valleys of San Diego County.* Berkeley: University of California Press. Some disturbing histories and unsettling implications about the rate at which the coastline is changing.

Pethick, J. 1984. *An Introduction to Coastal Geomorphology.* Baltimore, MD: Edward Arnold. An introduction to the study of coastal landforms, readable and thorough.

Pilkey, O. H. 1983. *Coastal Design: A Guide for Builders, Planners, and Home Owners.* New York: Van Nostrand-Reinhold. Pilkey is a leading force for responsible coastal development in the Southeast. His guidelines hold for any coast.

Sackett, R. 1983. *The Edge of the Sea.* Planet Earth series. Chicago: Time–Life Books. Excellent photos of coasts and coastal processes.

Shepard, F. P. 1969. *The Earth Beneath the Sea.* New York: Atheneum. One of the best general references on marine geology for the layman.

Shepard, F. P. 1973. *Submarine Geology.* 3d ed. New York: Harper & Row. *The Earth Beneath the Sea* in greater depth, with a more scholarly emphasis.

Shepard, F. P. 1977. *Geological Oceanography: Evolution of Coasts, Continental Margins, and the Deep Ocean Floor.* New York: Crane, Russack.

Shepard, F. P., and H. R. Wanless. 1971. *Our Changing Coastlines.* New York: McGraw-Hill.

Walker, J. 1982. "Walking on the Shore, Watching the Waves and Thinking on How They Shape the Beach." *Scientific American,* August.

Oceanus 36 (nos. 1 and 2, Spring and Summer 1993). Both issues were dedicated to coastal science and policy.

13 LIFE IN THE OCEAN

The Breath of the Planet

Most biologists agree that life on Earth arose in the ocean (see Chapter 1). The chemical reactions of this planet's living things proceed only in water that is slightly saline, and life, whether terrestrial or aquatic, depends on substances dissolved in water. All life on Earth thus reflects its marine origin.

Life and the ocean and atmosphere influence each other through a large and complex set of interactions. In recent years, two contrasting views of these interactions have been put forward. One view holds that life has prospered in spite of many obstacles because it adapts to conditions as it finds them. The competing view, suggested in 1979 by James Lovelock, a British chemist and inventor, holds that all life on Earth actually defines and exerts control over the conditions necessary for its survival. Lovelock's original hypothesis, which envisions the surface of the Earth as a single huge organism intentionally creating an optimum environment for itself, is called the Gaia hypothesis. (*Gaia*—pronounced *guy*'a— was the early Greek Earth goddess.) A less dramatic version of this hypothesis, accepted today by many scientists, acknowledges that life influences the environment but that it does so without intentional control.

What evidence do we have that life influences the environment? Consider a couple of the many examples, involving atmospheric gases. As we saw in Chapter 7 (see Figure 7.6), about 19% of the light approaching Earth from the sun is absorbed by carbon dioxide (CO_2) and water vapor in the atmosphere. This is a near-optimal amount of energy to maintain atmospheric temperature at levels ideal for life: Too little light absorbance and the ocean would freeze, too much and it would boil. As we will see shortly, plants consume CO_2 as they grow. Moreover, too much CO_2 in the atmosphere would lead to runaway atmospheric heating. However, plants consume CO_2 as they grow, and the excess CO_2 would stimulate plant growth; this would in turn cause the CO_2 level to fall, cooling the atmosphere. Too cold an atmosphere would cause plant growth to slow, allowing CO_2 levels to rebuild until temperature was again optimal. Thus, the CO_2-plant-energy system acts as a global thermostat.

Dimethyl sulfide (DMS), a chemical used by small marine plants to balance water pressure within their cells, may be another atmospheric temperature regulator. More marine plants would mean more DMS. Increased quantities of DMS in the ocean would lead to more gaseous DMS in the atmosphere over the ocean. Marine air is usually so clean that the cloud cover over the ocean is limited by the supply of condensation nuclei needed to form clouds; but DMS (and its oxidation by-products) can act as condensation nuclei and help to form dense—and highly reflective—cloud layers. Thus more plant activity again would cause the Earth to cool and would decrease the solar energy available for photosynthesis. Less plant activity would cause it to warm, encouraging plant growth. Similar feedback loops involving marine plants have been proposed as the mechanisms that modulate seawater salinity, the quantities of atmospheric oxygen and nitrogen, and the ocean's acid–base balance.

Are these delicate interactions proof of *intentional* control? Most researchers say no. But the fact of the interactions is clear: Life, atmosphere, ocean, and land do profoundly affect each other. Rifting lithospheric plates and volcanic eruptions spewing carbon dioxide, sulfurous gases, water vapor, and other substances are thus more than geological phenomena. These and other physical processes affect life in hundreds of ways. Geology and biology are different faces of the same coin.

In this chapter, we study the special interrelationship of living things and energy—which led to the proposal of the Gaia hypothesis—by examining the basic life processes that have evolved to capture, transform, store, and transmit energy. The physical characteristics of the ocean can facilitate or impede this flow of energy and, therefore, the success of marine organisms. The chapter also discusses the process through which organisms change and prosper: evolution by natural selection. But to begin with, we will consider one of the fundamental problems of science: defining life.

A coral reef in the Red Sea.

Life on Earth is notable for both unity and diversity: *diversity* because there are at least 50 million different species (kinds) of living things on Earth; *unity* because each species shares the same underlying mechanisms for capturing and storing energy, manufacturing proteins, and transmitting information between generations. The atoms in living things are no different from the atoms in nonliving things, and the energy that powers living things is the same energy found in inanimate objects. This makes life difficult to define. Our working definition of life recognizes the highly organized nature of living material and the complex ways living things manipulate energy.

Because of its watery nature and origin, in a sense all life on Earth is marine. The oceanic environment is a relatively easy place for cells to capture, transform, and store the energy they require to maintain their complex organization and to grow and reproduce. In part at least, life in the ocean is so successful—total marine productivity is so high—because of the ocean's benign physical characteristics, but these characteristics may also be limiting factors for an organism.

Evidence suggests that life has diversified and adapted to varying environments by the process of natural selection. A great revolution in biology has sprung from the understanding that life has changed through time, and that life and the Earth have grown old together.

BEING ALIVE

Differences between living and nonliving or once-living things seem obvious until we try to describe them in detail. We cannot always see the difference in a quick visual inspection, as **Figure 13.1** points out. Chemically, there are no well-defined differences, no convenient signs to help us gauge the difference between life and nonlife. The same atoms move continuously in and out of living and nonliving systems. With your last breath you exhaled millions of carbon atoms that entered your body as food in your last meal. Before the day is done some of those atoms may be incorporated into a nearby house plant. These carbon atoms have moved from life to nonlife and back again, all in the space of an afternoon. As the Russian researcher A. I. Oparin observed early in this century, there is nothing special about the atoms or energy of life, nothing to distinguish them from their nonliving counterparts. The free exchange of identical components between life and nonlife complicates any attempts at formal definition. Yet intuitively, we can all make this distinction. How? By considering the familiar things living organisms do.

Figure 13.1 Living or nonliving? The brightly colored "rock" in this picture is a colony of small marine plants. The distinction between life and nonlife does not lie in composition or appearance, but in the ability to manipulate energy.

Energy, Entropy, and Life

Living organisms require **energy,** the capacity to do work. Living things cannot create new energy, but they can transform one kind of energy to a different kind. A plant can transform light energy into chemical energy; an animal can transform chemical energy into energy of movement by the muscles, can transform energy of movement into heat, and so on. In one way or another, all life activity is involved, directly or indirectly, in energy transformation and transfer. Energy must therefore be central to anyone's definition of life.

Living organisms are complex assemblies; they need energy to build that complexity from large numbers of simple molecules. The **second law of thermodynamics** shows, however, that disorder inevitably tends to increase in the universe as time passes. Things run down, become disorganized, and break. **Entropy** is a measure of this disorder.

Living things are not exempt from the second law of thermodynamics, but they can delay that inevitable descent into entropy because the transformation of energy in living things allows a temporary and local remission of the second law. Thus, living organisms might be defined as localized regions where the flow of energy results in *increased* order; that is, areas of great complexity and low entropy. Car engines use the energy in gasoline only to move. Organisms use energy in food to move, to maintain their highly complex organization, and to grow. Cars and word processors don't fix themselves; fish and people do. This sophisticated use of energy seems to be a basic attribute of living things.

The main source of energy for living things on Earth is the sun. Life prospers, becomes ever more complex, and evolves into millions of forms by accepting sunlight and radiating waste heat to the cold of space. With only a few exceptions, organisms get their power directly or indirectly through the capture, storage, and transmission of energy from sunlight. Light energy is transformed into chemical energy and finally into heat as organisms temporarily forestall the disorderly fate decreed by the second law.

In the long run, however, this fate is inescapable. The sun powers the biological systems of Earth, and the sun is "winding down." As the sun dies, so will life on Earth. Death is marked by an irreversible increase in the entropy of an organism. Without an unbroken flow of energy, the complexity of living things returns to the simplicity of atoms and heat.

A Working Definition of Life

Here, then, is a definition of *life* that incorporates most of these ideas: Living things contain matter in a highly ordered, low-entropy state; they can capture, store, and transmit energy. Organisms are also capable of reproduction and change through time, and they adapt to their

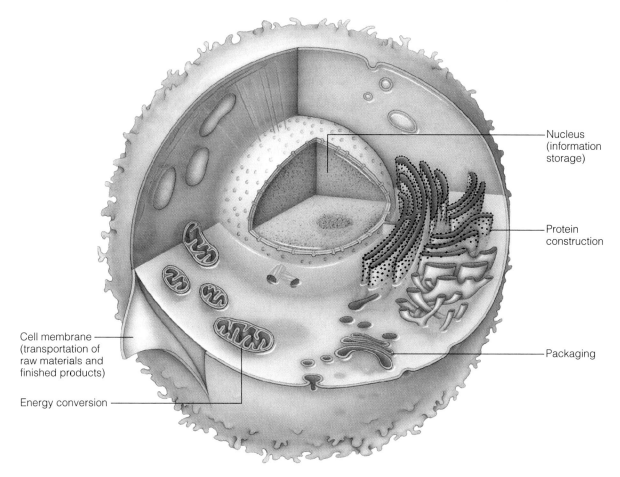

Cell membrane
(transportation of
raw materials and
finished products)

Energy conversion

Nucleus
(information
storage)

Protein
construction

Packaging

Figure 13.2 A cell and a few of its functions. A living cell is a self-contained factory that absorbs raw materials from its surroundings and uses them to maintain itself and manufacture finished products for the use of the organism as a whole.

environment. The essential differences between living and nonliving systems appear to be in the way living things use energy and in the extent to which component molecules are organized within the system.

THE CAPTURE AND FLOW OF ENERGY

The basic organizational unit of life is the **cell,** a compartmented package of complex chemistry encompassed by a semipermeable **membrane** that regulates the passage of many molecules into and out of the cell (**Figure 13.2**). Virtually all organisms are composed of one or more cells. Most cells are smaller than the width of a human hair. The flow of energy and cycling of molecules—the business of life—are conducted within and between cells. Nearly all of the energy used by cells (and living organisms) is derived directly or indirectly from nuclear reactions in the sun.

The sun produces enormous quantities of energy, some in the form of visible light, a tiny portion of which strikes the Earth's surface. Only about 1 part in 2,000 of this light is captured by organisms capable of binding its energy into molecules, but that "small" input of energy powers all the growth and complexity of living systems. Light energy from the sun is trapped by chlorophyll in organisms called *producers* (certain bacteria, algae, and green plants) and changed into chemical energy. The chemical energy is used to build simple carbohydrates and other organic molecules—**food**—which is then used by the plant or eaten by animals (or other organisms) called *consumers.* The energy is released when food is used for growth, repair, movement, reproduction, and the other functions of organisms. The metabolism of food eventually produces waste heat, which flows away from Earth into the coldness of space. This one-way flow of energy through living systems is shown in **Figure 13.3**.

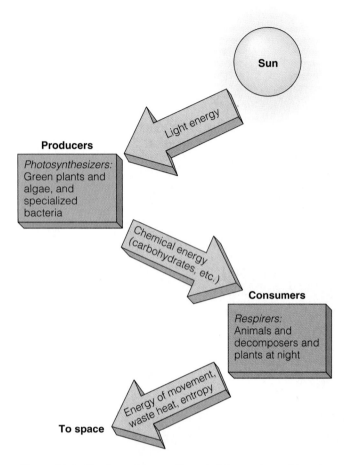

Figure 13.3 The flow of energy through living systems. At each step, energy is degraded (transformed into a less useful form).

Capturing Energy: Photosynthesis and Primary Productivity

Living things cannot use the energy of sunlight until it is transformed into the energy of chemical bonds. A method for transforming light energy into chemical bond energy (food energy) evolved early in the history of life on Earth. Central to the process is the chlorophyll molecule. Light energy is transported in photons, tiny packets of electromagnetic radiation. **Chlorophyll** is a pigment that collects the energy of many photons and transfers it to electrons. The energetic electrons then trickle through a series of biochemical reactions that gently release the stored energy in controlled ways. Some of this energy is used directly to assemble molecules of ATP (adenosine triphosphate), which is the immediate energy source for all living things. Most of the energy, however, is used to manufacture the carbohydrate **glucose,** a small food molecule. The energy that binds together the atoms in glucose is thus derived from sunlight. Later, the glucose is used to produce more ATP, needed by the plant to do such work as chemical synthesis, growth, and reproduc-

tion. Because light energy is used to synthesize molecules rich in stored energy, the process is called **photosynthesis** (*photos* = light, *syn* + *tithenai* = to place together).

Glucose is made up of atoms of carbon, hydrogen, and oxygen. Where did those atoms come from? Carbon arrives in the cell from surrounding water or air in the form of carbon dioxide. Water itself is also involved in making glucose. Here is a general formula for photosynthesis:

$$6CO_2 \; + \; 6H_2O \; \xrightarrow{\text{sunlight}} \; C_6H_{12}O_6 \; + \; 6O_2$$

carbon + water (yields) glucose + oxygen
dioxide

Notice in this formula that a large molecule is formed from small ones. Light energy is used to bind atoms into a glucose molecule. Oxygen is a by-product of these reactions.

Directly or indirectly, photosynthesis produces the food needed by nearly all organisms. The production of new organic matter out of inorganic matter is called **primary productivity,** *primary* because this is the initial trapping of energy and binding of individual carbon atoms into carbohydrate (food) molecules. (More on primary productivity will be found in Chapter 14's discussion of marine plants.)

Photosynthesis is the usual method of primary productivity, but there is another. **Chemosynthesis,** employed by a few relatively simple forms of life, is the production of usable energy directly from energy-rich inorganic molecules in the environment. As we shall see in Chapter 17, some unusual forms of marine life depend on chemosynthesis. Chemosynthetic production of food is very small, however, in comparison to photosynthetic production.

Harvesting Energy by Respiration: Obtaining Energy from Food Molecules

ATP is the immediate source of usable energy for all life on Earth. Some ATP is directly assembled during photosynthesis, but not nearly enough to satisfy the energy demands of most living things. As you have just seen, plants store a great deal of energy as food. Taking apart food molecules yields a substantial energy harvest, and this energy is used to assemble most of the ATP needed by plants and animals. Plants make enough food through photosynthesis to meet their own needs, but animals must eat the photosynthesizers (or organisms that have consumed photosynthesizers) to gain an adequate supply of food.

The process in which food is disassembled and energy is harvested is called cellular **respiration.** (Respiration is a biochemical process and should not be confused with the mechanical process of breathing.) A food molecule—usually glucose—accompanied by oxygen is broken down into carbon dioxide and water. The process is shown here:

$$C_6H_{12}O_6 + 6O_2 \rightarrow 6CO_2 + 6H_2O + \text{enough energy to assemble 36 ATP molecules}$$

$$\text{glucose} + \text{oxygen (yields) carbon} + \text{water} + \text{chemical} \\ \text{dioxide} \qquad \text{energy}$$

Notice in this formula that oxygen is being consumed and that small molecules are formed from a large one—the opposite of photosynthesis. Respiration liberates large amounts of energy, energy that is packed into ATP for use by the cell. Having started as light, this energy now resides as chemical energy within the bonds of ATP. Waste heat is released as the ATP is used (see again Figure 13.3). The biochemical processes used to liberate energy from food seem to have evolved even earlier in the history of life than photosynthesis.

Now look at **Figure 13.4.** *The beginning products for photosynthesis are the end products for respiration, and vice versa.* The only loose ends are those of energy flow. The individual molecules of carbon dioxide, glucose, oxygen, and water move back and forth in an endless cycle as energy flows through the system. Photosynthesis and respiration are opposite and complementary processes. All the free oxygen in the atmosphere and ocean passes through organisms—in by respiration, out by photosynthesis—about every 2,000 years, and all the available carbon dioxide cycles through organisms in the reverse direction every 300 years. All the water in the ocean is broken down and re-formed by the chemistry of photosynthesis and respiration every 2 million years!

Feeding (Trophic) Relationships

Photosynthetic and chemosynthetic organisms can be called either **primary producers** or **autotrophs** (*auto* = self + *trophe* = nourishment) because they make their own food. The bodies of autotrophs are rich sources of chemical energy for any organisms capable of consuming them. **Heterotrophs** (*hetero* = other, different) are organisms, such as animals, that must consume other organisms because they are unable to synthesize their own food molecules. Some heterotrophs consume autotrophs, and some consume other heterotrophs.

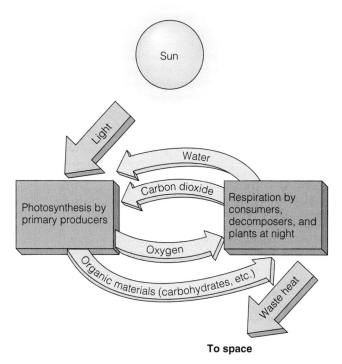

Figure 13.4 The flow of matter through living systems. Note that the beginning products for photosynthesis are the end products for respiration, and vice versa.

We can label organisms by their positions in a who-eats-whom feeding hierarchy called a **trophic pyramid** (*trophos* = one who feeds). The primary producers shown at the bottom of the pyramid in **Figure 13.5** are mostly chlorophyll-containing photosynthesizers. The animal heterotrophs that eat them are called **primary consumers** (or herbivores), the animals that eat primary consumers are called secondary consumers, and so on to the **top consumer** (or top carnivore).

Note that the mass of the consumers becomes smaller as energy flows toward the top of the pyramid. There are many small primary producers at the base, but only a very few large top consumers at the apex. Only about 10% of the energy from the organisms consumed is stored in the consumers as flesh; so each **trophic level** is about 1/10 the mass of the level directly below. The rest of the energy is lost as waste heat as organisms live and work to maintain themselves.

Pyramids such as these can lead to the misconception that one kind of fish eats only one other kind of fish, and so on—an idea used to humorous effect in **Figure 13.6.** Real communities are more accurately described as food *webs* (**Figure 13.7**). A **food web** is a group of organisms linked by complex feeding relationships in which the flow of energy can be followed from primary producers

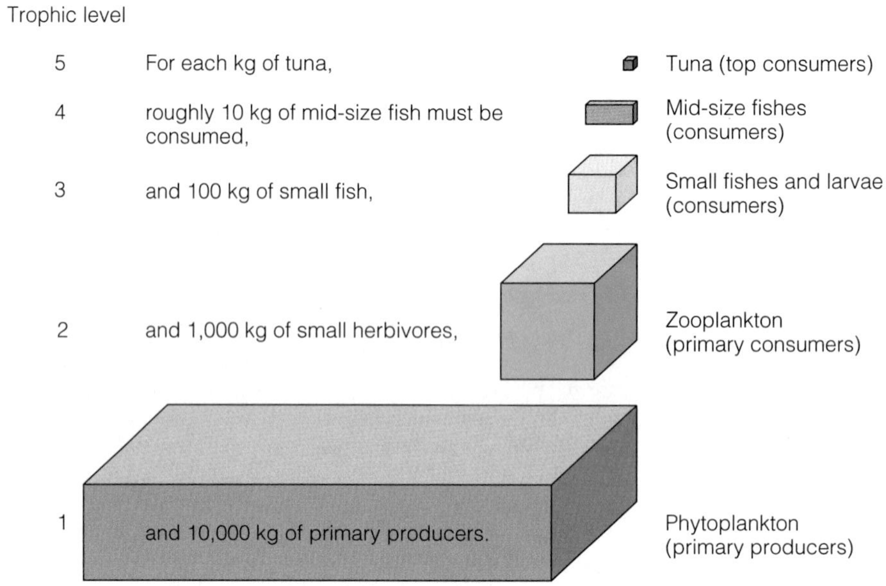

Trophic level

5	For each kg of tuna,	Tuna (top consumers)
4	roughly 10 kg of mid-size fish must be consumed,	Mid-size fishes (consumers)
3	and 100 kg of small fish,	Small fishes and larvae (consumers)
2	and 1,000 kg of small herbivores,	Zooplankton (primary consumers)
1	and 10,000 kg of primary producers.	Phytoplankton (primary producers)

Figure 13.5 A generalized trophic pyramid. The tuna is only one example of a top consumer in open-water food webs.

Figure 13.6 Food webs are never this simple!

through consumers. Organisms in a food web almost always have some choices of food species.

So organisms interact with each other, feed on one another, and transfer energy as food from producing autotrophs through a web of consuming heterotrophs. Nearly all are ultimately dependent on sunlight and photosynthesis. What is the physical role of the ocean in all this?

PHYSICAL FACTORS AFFECTING MARINE LIFE

Marine organisms depend on the ocean's chemical composition and physical characteristics for life support. Any aspect of the physical environment that affects living organisms is called a **physical factor.** Living in the ocean may have advantages over living on land: Physical con-

ditions in the sea are usually more benign and less variable than physical conditions on land. The most important physical factors for marine organisms are water transparency, dissolved nutrients, temperature, salinity, dissolved gases, acid–base balance, and hydrostatic pressure.

Transparency

On land, most photosynthesis proceeds at or just above ground level. But seawater is relatively transparent, which allows photosynthesis to proceed for some distance below the ocean surface. Incoming sunlight must run a gauntlet of difficulties, however, before it can be absorbed by the chlorophyll in marine autotrophs.

One important factor is that water is more transparent to some colors of light than others. In clear water, blue light penetrates to the greatest depth, whereas red light

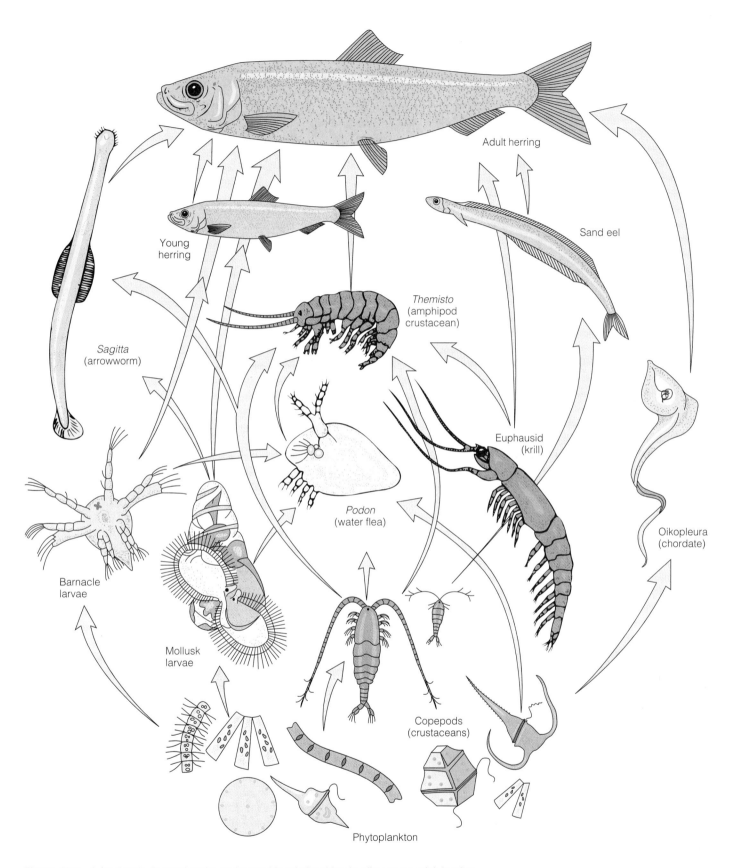

Figure 13.7 A food web, illustrating the major trophic relationships leading to an adult herring. The arrows show the direction of energy flow.

Adult herring

Young herring

Sand eel

Sagitta (arrowworm)

Themisto (amphipod crustacean)

Euphausid (krill)

Podon (water flea)

Oikopleura (chordate)

Barnacle larvae

Mollusk larvae

Copepods (crustaceans)

Phytoplankton

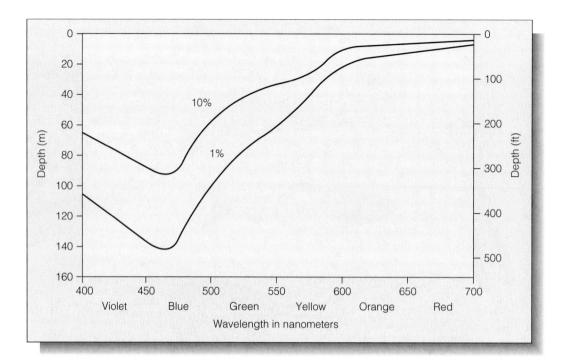

Figure 13.8 Depths at which the surface radiation of light is reduced to 10% and 1% for various colors in clear ocean water.

is absorbed near the surface. **Figure 13.8** shows the depths attained by light of various wavelengths (colors) in clear ocean water. Light energy absorbed by water turns to heat.

The depth to which light penetrates is also limited by the number and characteristics of particles in the water. These particles, which may include suspended sediments, dustlike bits of once-living tissue, or the organisms themselves, scatter and absorb light. High concentrations of particles quickly absorb most blue and ultraviolet light. This absorption, combined with the reflection of green light by chlorophyll within the producers, changes the color of productive coastal waters to green.

How far down does light penetrate? Near the coasts, conditions become uncomfortably dark for divers at depths around 30 to 40 meters (100 to 130 feet). The human eye is most sensitive to the blue wavelengths, and in very clear tropical waters with a smooth surface and high solar angle, human observers in submersibles have seen dark blue light at about 200 meters (660 feet). There is evidence to suggest that some deep-water fish use even this dim light for body orientation, feeding, and predator avoidance.

Photosynthesis proceeds slowly at low light levels. As you may recall from Chapter 7, the sunlit layer of water at the ocean's surface is called the **photic zone** (*photos* = light). Most of the biological productivity of the ocean occurs in the upper half of the photic zone, an area near the surface called the **euphotic zone** (*eu* = good + *photos*

= light). This is where marine plants trap more energy than they use. Though it is difficult to generalize for the ocean as a whole, the euphotic zone typically extends to a depth of approximately 40 meters (130 feet) in mid-latitudes. The upper productive layer of ocean is a very thin skin indeed; the water within this zone amounts to less than 2% of world ocean volume. Nearly all marine life depends on this fine illuminated band.

Dissolved Nutrients

The photosynthesis formula given earlier shows only the raw materials and the final products; it omits the many intermediate steps needed to convert energy from light to chemical form. The chemistry of life depends on the molecules that support it—especially substances containing nitrogen, calcium, phosphorus, potassium, silicon, sulfur, and sodium; some substances also contain trace elements such as manganese, zinc, copper, cobalt, and iodine. These molecules enter the cell as **nutrients,** compounds that autotrophs require for the production of organic matter.[1] Some nutrients help form the structural parts of organisms, some make up the chemicals that directly manipulate energy, and some have other functions. A few of these necessary nutrients are always present in seawater, but most are not.

[1]A broader definition of a *nutrient* includes all needed substances an organism obtains from its environment except oxygen, carbon dioxide, and water.

The main inorganic nutrients required in primary productivity include nitrogen (as nitrate, NO_3^-) and phosphorus (as phosphate, PO_4^{3-}). As any gardener knows, plants require fertilizer—mainly nitrates and phosphates—for success. Ocean gardeners would have more trouble raising crops than their terrestrial counterparts, though, because the most fertile ocean water contains only about 1/10,000 the available nitrogen of topsoil. Phosphorus is even scarcer in the ocean. Fortunately, less of it is required by living things, which contain about 1 atom of phosphorus for every 16 atoms of nitrogen.

Nitrogen and phosphorus are often depleted by autotrophs during times of high productivity and rapid reproduction. Also in short supply during rapid growth are dissolved silicates and calcium compounds (used for shells and other hard parts) as well as trace elements such as iron, copper, and magnesium (used in enzymes, vitamins, and other large molecules). Marine plants have no choice but to recycle these nutrients; nutrient cycles move nutrient molecules rapidly and repeatedly into and out of living systems. Though primary productivity may be very high when light is available, the total mass of living material cannot increase until more nutrients are made available by recycling, upwelling, runoff from land, or other means.

Temperature

The rate at which complex chemical reactions occur in a living organism is largely dependent on the molecular vibration we call *heat*. Living chemistry happens in water; so keeping the temperature of the reactions between water's freezing and boiling points might seem to be enough. That is too broad a range, however. Increased heat agitates molecules, and too much heat shakes and distorts the large molecules of life—literally cooking them. Too little heat slows molecular vibration so much that the molecules no longer react rapidly enough to keep the organism alive. Freezing water locks reacting chemicals within an unmoving lattice, and ice crystals destroy cells by shattering their membranes.

Since agitation brings reactants together, warmer temperatures increase the rate at which chemical reactions occur. Thus, an organism's **metabolic rate,** the rate at which energy-releasing reactions proceed within an organism, increases with temperature. The metabolic rate approximately doubles with a 10°C (18°F) temperature rise. The interior temperature of an organism is directly related to the rate at which it generates ATP, moves, reacts, and lives.

The great majority of marine organisms are "cold-blooded," or **poikilothermic** (*poikilos* = various + *therm* = heat), having an internal temperature very close to that of their surroundings. A few complex animals—the mammals and the birds—are "warm-blooded," or **homeothermic** (*homoio* = same); they have a stable high internal temperature.

In general, the warmer the environment of a poikilotherm within its tolerance range, the more rapidly its metabolic processes will proceed. Tropical fish in a heated aquarium will therefore eat more food and require more oxygen than goldfish of the same size living in an unheated but otherwise identical aquarium. The tropical fish will generally grow faster, have a faster heartbeat, reproduce more rapidly, swim more swiftly, and live shorter lives. But you can't just crank the heater up another notch for even faster fish—eventually the little fellows will cook as their proteins distort. The upper limit of temperature that a poikilotherm can tolerate is often not much higher than its optimum temperature. The lower limit is usually more forgiving because molecules are merely slowed.

Do homeotherms have narrow temperature requirements? Yes and no. Homeotherms can tolerate a tremendous range of *external* temperature compared to poikilotherms—think of a whale migrating from polar waters to tropics, or an emperor penguin incubating an egg at −51°C (−60°F)—but their *internal* temperatures vary only slightly. In our own case, consider the temperatures of places inhabited by humans in contrast to the narrow internal temperature range considered normal. Sophisticated thermal regulation mechanisms make it possible for homeotherms to live in a variety of habitats, but they pay a high price. Their high metabolic rates make proportionally high demands on food supply and gas transport, but the benefit of having a biochemistry fine-tuned to a single efficient temperature is worth the regulatory difficulties involved.

Ocean temperature varies with depth and latitude. The average temperature of the world ocean is only a few degrees above freezing, with warmer water found only in the lighted surface zones of the temperate and tropical ocean and in rare warm, deep chemosynthetic communities. Though temperature ranges of the ocean are considerable (**Figure 13.9**), they are much narrower than comparable ranges on land. Water's great ability to accept heat, yet rise little in temperature (and vice versa), is an important moderating force on the rate of marine biochemical reactions. Marine organisms are almost never exposed to sustained temperatures above 30°C (86°F), but some terrestrial species must tolerate long periods when temperature reaches 60°C (140°F).

Salinity

Cell membranes are greatly affected by the salinity of surrounding water. As we saw in Chapter 6, the salinity

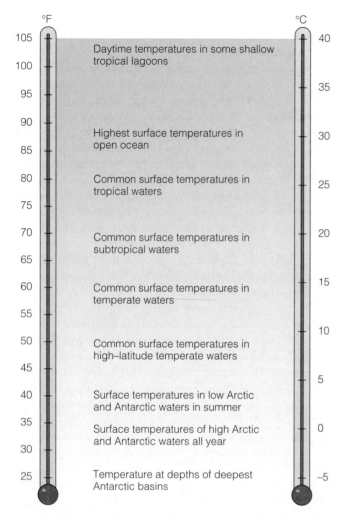

Figure 13.9 Temperatures of marine waters.

°F

105
100
95
90
85
80
75
70
65
60
55
50
45
40
35
30
25

°C

40
35
30
25
20
15
10
5
0
-5

Daytime temperatures in some shallow tropical lagoons

Highest surface temperatures in open ocean

Common surface temperatures in tropical waters

Common surface temperatures in subtropical waters

Common surface temperatures in temperate waters

Common surface temperatures in high–latitude temperate waters

Surface temperatures in low Arctic and Antarctic waters in summer

Surface temperatures of high Arctic and Antarctic waters all year

Temperature at depths of deepest Antarctic basins

Dissolved Gases

Virtually all marine organisms require dissolved gases—in particular carbon dioxide and oxygen—to stay alive. Oxygen does not dissolve easily in water, and as a result there is about 100 times more gaseous oxygen in the atmosphere than in the ocean. But CO_2, essential to primary productivity, is much more soluble and reactive in seawater than oxygen is (as **Table 13.1** shows). Although as much as 1,000 times more carbon dioxide than oxygen can dissolve in water, normal values at the ocean surface average around 50 milliliters per liter for CO_2 and around 6 milliliters per liter for oxygen. At present the ocean holds about 60 times as much carbon dioxide as the atmosphere. Because of this abundance, marine plants almost never run out of CO_2.

Deep water tends to contain more carbon dioxide than surface water. Why should this be? Table 13.1 also shows the relationship between water's temperature and its ability to dissolve gases. Note that colder water contains more gas at saturation. You may recall that the deepest and densest seawater masses are formed at the surface in the cold polar regions, and, as we have seen, more CO_2 can dissolve in that low-temperature environment. The dense water sinks, taking its large load of CO_2 to the bottom, and the pressure at depth helps to keep it in solution. CO_2 also builds in deep water because only heterotrophs (animals) live and metabolize there, and because CO_2 is produced as decomposers consume falling organic matter. No photosynthetic primary producers are present in the dark depths to use this excess CO_2.

Rapid photosynthesis at the surface lowers CO_2 concentrations and increases the quantity of dissolved oxygen. Oxygen is least plentiful just below the limit of photosynthesis because of respiration by many small animals at middle depths. These relationships were shown in Figure 6.10.

Low oxygen levels can sometimes be a problem at the ocean surface. Plants produce more oxygen than they use, but they produce it only during daylight hours. The continuing respiration of plants at night will sometimes remove much of the oxygen from the surrounding water. In extreme cases this oxygen depletion may lead to the death of the plants and animals in the area, a phenomenon most noticeable in enclosed coastal waters during spring and fall plankton blooms.

The greatest variability in levels of dissolved gas is found at the surface near shore. Less dramatic changes occur in the open sea.

pH

The acidity or alkalinity of a solution is expressed in terms of a *pH scale,* a logarithmic measure of the concen-

of seawater can vary in places because of rainfall, evaporation, runoff of water and salts from land, and other factors. Surface salinity varies most, from lows of 6‰ or less (along the coast of the outer Baltic Sea in early summer) to year-round highs exceeding 40‰ in the Red Sea. Salinity is less variable with increasing depth, with the ocean typically becoming slightly saltier with depth.

Fluctuating salinity can physically damage membranes, and concentrated salts can alter protein structure. Salinity can affect specific gravity, density, and therefore the buoyancy of an organism. Salinity is also important because it can cause water to enter or leave a cell through the membrane, changing the cell's overall water balance. Seawater is nearly identical in salinity to the interior of all but the most advanced forms of marine life; this means that maintaining salt balance, and therefore water balance, is easy for most marine species.

Table 13.1 Solubility of Gases in Seawater as a Function of Temperature (Salinity = 33‰)			
	Solubility (ml/l at atmospheric pressure)[a]		
Temperature	N_2	O_2	CO_2
0°C (32°F)	14.47	8.14	8,700.0
10°C (50°F)	11.59	6.42	8,030.0
20°C (68°F)	9.65	5.26	7,350.0
30°C (86°F)	8.26	4.41	6,660.0

Source: F. G. Walton-Smith. CRC Handbook of Marine Science (Cleveland, OH: CRC Press, 1974).

[a]Figures are given at saturation, the maximum amount of gas held in solution before bubbling begins.

tration of hydrogen ions in a solution. Recall that 7 on the pH scale is neutral, with smaller numbers indicating greater acidity and larger numbers indicating greater alkalinity. Figure 6.11 shows a pH scale and the pH of some familiar solutions.

Average seawater is slightly alkaline (at about pH 8), its pH maintained within a narrow range by the carbonate buffer system described in Chapter 6. This is fortunate for Earth's life-forms, whose complex chemistry invariably depends on precisely shaped enzymes. Like heat, strong acids or bases would distort the shapes of these vital proteins and they would lose their ability to function in chemical reactions. The normal pH range of seawater is much less variable than that of soil. Terrestrial organisms are sometimes limited by the presence of harsh alkali soils that damage cell components. Here again the general advantage goes to marine organisms.

Small pH changes in seawater can be important, however. When dissolved in water, CO_2 combines with a water molecule to form, among other things, carbonic acid, H_2CO_3. In areas of high primary productivity, pH will rise slightly (the water will become more alkaline) because CO_2 is being removed from solution for use in photosynthesis. And because surface temperatures are generally warmer in such areas, less CO_2 can dissolve to replace what is consumed. One of the reasons why coral reefs are not found in cold water is that the greater solubility of CO_2 at low water temperature lowers pH (makes the water more acidic) and impedes the deposition of calcium carbonate.

As we have seen, more CO_2 may be present in deeper water. With higher pressures, colder temperatures, and no photosynthetic activity to remove it, the CO_2 will form carbonic acid and lower the pH of water, making it slightly more acid. Thus seawater below the calcium car-

bonate compensation depth, about 4,500 meters (14,800 feet), has a relatively low pH of around 7.6. As noted in Chapter 5, this lower pH has a profound effect on calcium-containing marine sediments, which can be dissolved in this more acidic environment.

Human activity since the beginning of the Industrial Revolution has greatly increased the amount of fossil fuel burned and therefore the amount of CO_2 being injected into the atmosphere. Yet until quite recently the amount of CO_2 in the air has remained relatively constant. Where has the excess CO_2 been going? Because of its solubility, much of the CO_2 has accumulated in the ocean. Not all of it has, however, and the amount of CO_2 in the atmosphere has begun to climb.

Hydrostatic Pressure

Marine organisms are often subject to great pressure from the constant weight of water above them, but this **hydrostatic pressure** presents very little difficulty to them. In fact, the situation in the ocean is parallel to that on land. Land animals live in air pressurized by the weight of the atmosphere above them (1 kilogram per square centimeter, or 14.7 pounds per square inch, at sea level) without experiencing any problems. Indeed, atmospheric pressure is necessary for breathing, flight, and some other physical necessities of life.

Pressures inside and outside an organism are virtually the same, both in the ocean and at the bottom of the atmosphere. Thus marine organisms do not need heavy shells to keep from being crushed by hydrostatic pressure. Great pressure does have some chemical effects: Gases become more soluble at high pressure, some enzymes are inactivated, and metabolic rates for a given temperature tend to be slightly higher. These effects are felt only at great depth, however. Unless marine organisms have gas-filled spaces in their bodies, a moderate change in pressure has little effect.

The Interplay of Factors

The physical factors we have surveyed work in concert to provide the physical environment for oceanic life. A change in one factor usually produces a change in others. For example, the ocean absorbs more solar radiation in spring, increasing surface temperature. The resulting evaporation increases surface salinity. Increasing salinity causes an increase in water density. If density becomes very high at the surface, the warm saline water will sink. It will be replaced by nutrient-rich water from below, stimulating plant growth. The growing plants will absorb CO_2, causing the pH of the water to rise. And so it goes.

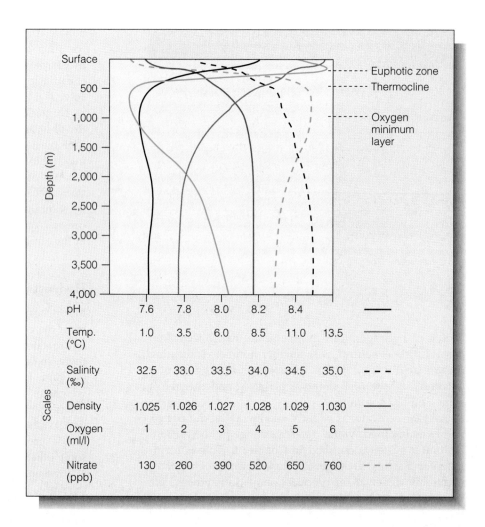

pH	7.6	7.8	8.0	8.2	8.4		——
Temp. (°C)	1.0	3.5	6.0	8.5	11.0	13.5	——
Salinity (‰)	32.5	33.0	33.5	34.0	34.5	35.0	- - -
Density	1.025	1.026	1.027	1.028	1.029	1.030	——
Oxygen (ml/l)	1	2	3	4	5	6	——
Nitrate (ppb)	130	260	390	520	650	760	- - -

Figure 13.10 The variability of physical factors with depth, illustrated by a sampling off southern California.

Figure 13.10 charts a number of physical factors at one location off southern California. The intermix of related factors, all changing with depth, all influencing (and being influenced by) living organisms, gives us a glimpse of how intricate is the relationship of marine life to its physical environment.

Often too much or too little of a single physical factor can adversely affect the function of an organism. We call that factor a **limiting factor,** a physical or biological necessity whose presence in inappropriate amounts limits the normal action of the organism. Try to imagine, for example, an ocean area in which everything is perfect for photosynthesis—warmth, nutrients, proper pH, adequate CO_2—but there is *no light*. In that circumstance no photosynthesis would occur; light is the limiting factor. If light were present but nitrates were absent, nitrate nutrients would be the limiting factor. Sometimes *too much* of something can be limiting—heat, for instance.

ORGANISMS AND OCEAN TOGETHER

To the list of physical factors discussed above we can add the concept of biological factors. A **biological factor** is a biologically generated aspect of the environment that affects living organisms. The most important biological factors for most marine organisms are the feeding (trophic) relationships we have discussed. Some other factors include crowding, metabolic wastes in the water, and the defense of territory. Biological factors operate in association with purely physical factors. We now turn to the biological action on this physical stage, looking at some additional influences on the lives of organisms.

Diffusion, Osmosis, and Active Transport

According to the second law of thermodynamics, unless energy is supplied to maintain the organization of a system, the system tends to become disorganized. Among

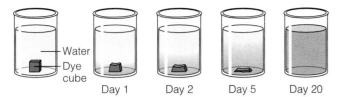

Figure 13.11 An example of diffusion. Molecules of dye gradually diffuse through the surrounding water as a dye cube dissolves. Random molecular movement spreads the dye away from the region of high concentration at the cube's surface. At the same time, water moves from its own region of high concentration (next to the dye cube) to areas of lower concentration (within the disintegrating dye cube itself). After many days, dye will be distributed evenly throughout the water. Stirring would greatly accelerate the mixing process, but this input of mechanical energy is not required if you don't mind waiting for diffusion alone to accomplish the task.

other things, this statement implies that a concentration of one substance within another (an organized system) will disperse (become less organized) if each substance is free to move. For example, a cube of soluble dye in water represents a concentration of dye. If the cube is not disturbed, the dye molecules will eventually diffuse evenly throughout the water, though this process of **diffusion** might take weeks (**Figure 13.11**). The energy to distribute the dye comes from heat, the random vibration of molecules: The warmer the water, the faster the diffusion. The net transfer of material in diffusion occurs from a region of high concentration to regions of lower concentration.

Diffusion is an important marine process. For example, minerals dissolve, and their components tend to diffuse randomly throughout a liquid environment. Liquids and gases can also diffuse through water from zones of high concentration to zones of low concentration. Mass transport (the movement of substances in currents, for example) is more important than diffusion in moving dissolved substances over large distances. Diffusion is most important over small distances, particularly the distances within and between living cells.

Cells are bounded by membranes, complex films through which certain selected substances can passively diffuse. Molecules will tend to diffuse across membranes from areas where they are highly concentrated to areas where they are less concentrated. For example, oxygen resulting from photosynthesis will diffuse through a membrane from inside a plant cell (a region of high oxygen concentration) to the outside (a region of lower oxygen concentration). Because biological membranes are selective and allow only certain kinds of small molecules to pass, they are considered *semipermeable.*

Diffusion of water through a membrane is called **osmosis** (*osmos* = a thrusting). In osmosis, water moves between two solutions of different water concentrations

through a membrane permeable to water but not to salts. If the water outside a cell membrane contains less dissolved salt than the water inside, it will diffuse from the region of higher water concentration (outside the cell) to the region of lower water concentration (inside the cell), causing the cell to swell (as in Figure 13.13b). Salt is prevented by the nature of the membrane from moving outside to balance the situation. This is why human swimmers avoid getting fresh water in nasal membranes: Our body fluids are more saline than fresh water; so cells in our sinuses take up the water and swell painfully.

Most simple marine organisms have nearly the same concentration of dissolved substances in their body fluids as seawater does. They are almost **isotonic** to their fluid environment (*isos* = equal + *tonos* = strength) and so experience little net flow of water through their outer membranes. In fresh water a marine animal would be **hypertonic** (*hyper* = over) to its surroundings; water would move *into* the animal through its cell membranes. If the animal had no way to eliminate the water, its cells would burst. The same kind of marine animal moved to Utah's highly saline Great Salt Lake would be **hypotonic** (*hypo* = under), and water would flow *out of* its cells, resulting in dehydration and collapse. **Figure 13.12** summarizes these relationships. Because they are nearly isotonic to their surroundings, simple marine organisms have a big advantage over their freshwater counterparts; freshwater animals must expend large amounts of energy in excretory systems designed to transport water from their tissues back to the outside. Because of these specialized excretory mechanisms, very few freshwater and marine organisms can successfully change places.

There are some interesting exceptions to the near-isotonicity of marine organisms, mostly in mobile vertebrates that encounter changing osmotic environments. Most ocean fish, for example, have body fluids only about one-third as saline as seawater. These fish are therefore hypotonic to their surroundings and are constantly losing water through their gill membranes. To counteract dehydration, these fishes drink seawater and excrete salts though special *chloride cells* in their gills. Sea birds have salt-excreting glands in the skull above their eyes that do the same job. Salmon, which migrate from fresh water to the sea and back again, have large kidneys to export excess water during the freshwater phase of their life cycle. They also recover salts from their urine and absorb salts from their food, energy-expensive adaptations to maintain a stable internal environment in either fresh water or the ocean.

Active transport is the reverse of passive diffusion. In active transport dissolved substances are "pumped" through a membrane "uphill" from a region of low concentration to a region of high concentration. Active transport requires energy because this "uphill" movement

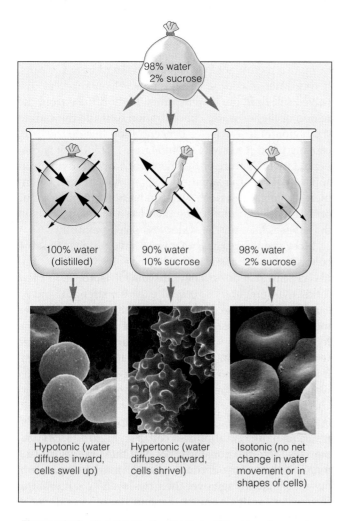

Figure 13.12 The effects of osmosis in different environments. The semipermeable bag at the top of the figure represents the concentration of water and dissolved solids typical in the cells of higher forms of life. If the bag is placed into the solutions indicated, it will swell, shrink, or stay the same size. The cells pictured here are human red blood cells immersed in distilled water, concentrated seawater, and water of the same tonicity as human blood, respectively.

(Top bag label) 98% water / 2% sucrose

(Beaker labels) 100% water (distilled) — 90% water 10% sucrose — 98% water 2% sucrose

(Bottom labels) Hypotonic (water diffuses inward, cells swell up) — Hypertonic (water diffuses outward, cells shrivel) — Isotonic (no net change in water movement or in shapes of cells)

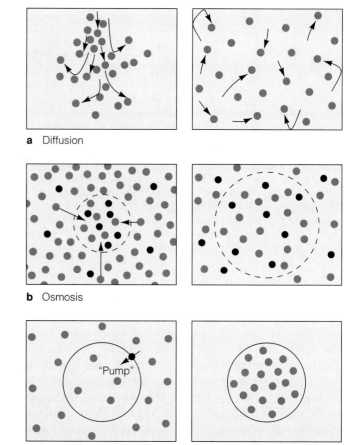

a Diffusion

b Osmosis

c Active transport ("Pump")

Figure 13.13 A comparison of the three main ways by which substances move into and out of cells. (a) In diffusion, molecules introduced into a container (left) become evenly distributed after a period of time (right). (b) In osmosis, the diffusion of water may cause a cell to swell. The blue dots represent water molecules, the black dots represent dissolved particles, and the arrows indicate the direction of water movement into the original cell. Under different conditions, water may move out of the cell, causing it to shrink. (c) Active transport enables a cell to accumulate molecules even when there are more inside the cell than outside. Cells may expel other molecules by the same process.

defies the normal "downhill" functioning of the second law of thermodynamics. When a cell exports a finished product (the glucose made in photosynthesis, for example) through a membrane to a storage area in which this product is concentrated, it uses active transport. Active transport is a common process in living things, and much of any organism's energy is expended on facilitated movement against the normal concentration gradient.

Diffusion, osmosis, and active transport are all temperature-dependent, proceeding at a more rapid rate as temperature rises. These three processes are compared in **Figure 13.13**.

Surface-to-Volume Ratio

Though individual organisms are often large, the cells that comprise them are always small. The smaller the

cell, the more efficiently materials can cross the outer cell membrane to be distributed through the interior. A small cell has enough surface area to permit the passage of oxygen (or carbon dioxide, or glucose, or wastes) at a rate sufficient to supply the metabolic needs of its internal machinery.

If a round cell grew only by expanding its volume, the surface area of its outer membrane would not increase at an adequate rate. The cell would soon be unable to shuttle materials through the membrane in large enough quantities to keep itself alive. Note in **Figure 13.14** that the volume of a spherical cell increases with the *cube* of its diameter, while its surface area increases only with

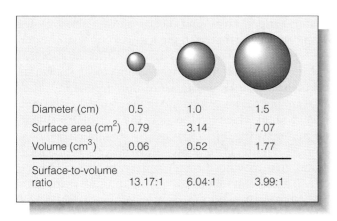

Diameter (cm)	0.5	1.0	1.5
Surface area (cm^2)	0.79	3.14	7.07
Volume (cm^3)	0.06	0.52	1.77
Surface-to-volume ratio	13.17:1	6.04:1	3.99:1

Figure 13.14 Surface-to-volume ratio: the relationship between surface area and volume when a sphere is enlarged. Notice that as the diameter increases, the volume is increasing more quickly than the surface area. The surface-to-volume ratio thus *decreases* as the cell increases in size.

the *square* of its diameter. If the cell grew to 4 times its original diameter, its volume would increase 64 times, but its surface area would increase only 16 times. Each square unit of outer membrane would have to serve four times as much interior volume as before. Past a certain point, the inward flow of nutrients and gases (and the outward flow of wastes) would not be fast enough to supply the cell's metabolic needs, and the cell would die. Large cells have an inefficient ratio of surface area to internal volume.

Instead, cells grow by dividing, not by limitless expansion. Diffusion, osmosis, and active transport can be accomplished more efficiently by cells with a high **surface-to-volume ratio**—that is, by small cells with a relatively large surface area and a relatively small volume.

Gravity and Buoyancy

The density of seawater is nearly the same as the average density of living material; so the weight of a marine organism is usually counterbalanced by its buoyancy. Seawater has a density of about 1.025 (pure water = 1.000). Some parts of a fish are heavier than seawater and some parts lighter, but overall a marine fish has a density of around 1.07—about 5% denser than the same volume of seawater. To counteract the weight of heavy muscle and bone, many swimming fishes have gas-filled bladders. Gas is much lighter (less dense) than seawater; so only a small volume of gas is required. Fishes from very deep water usually don't have gas bladders, but they do have lightweight bones and oily, watery flesh to lighten their weight and make swimming easier. At the other end of the scale, most members of the mackerel family are such strong swimmers that the heavy weight of their muscular bodies is of no apparent concern (see Figure 13.16d).

Adaptations for balancing weight and buoyancy can become quite sophisticated. The skeleton of a marine organism, if present, often plays a minor role in physical support. The skeleton may be reduced in size, strength, and weight because its major function is only to provide stiffness for swimming, predation, or defense. A few marine animals further lighten themselves by substituting solutions rich in ammonium chloride (NH_4Cl) for seawater inside their bodies. This ammonium chloride solution has the same osmotic characteristics as seawater but weighs less per unit volume. Some jellyfish and other floating invertebrates actively transport heavy sulfate ions to the outside, allowing lighter chloride ions to take their place within. Many planktonic (drifting) forms of life store food energy in lightweight waxes and oils. These also provide weight balance or flotation. Indeed, this trick contributes to the diatoms' great success (news of which awaits you in the next chapter).

Viscosity and Movement

Viscosity is a fluid's internal resistance to flow. Molasses flows much more slowly than water; so molasses has a much higher viscosity than water. Temperature has an effect on viscosity, and molasses in January flows very slowly indeed. Water has a much lower viscosity than molasses (and is much easier to swim through), but the viscosity of water is also affected by temperature. Seawater at 0°C (32°F) is about 25% more viscous than seawater at 20°C (68°F), a significant difference. Salinity also affects viscosity. Salt water is more viscous than fresh water. Air is much less viscous than seawater, but animals that move rapidly through air must still contend with its viscosity. Note that birds and other flying species are streamlined.

For marine organisms dealing with the viscosity of seawater, the main problem is **drag,** the resistance to movement of an organism induced by the fluid through which it swims. The amount of drag depends on the viscosity of the water and the speed, shape, and size of the moving organism.

Very small marine organisms are more directly affected by viscosity than larger ones. Swimming is difficult for tiny animals; the viscosity of water impedes their efforts at movement. Gradual sinking, however, can be a problem, and very small drifting organisms have plumes, hairs, ribbons, spines, and other extensions to increase their friction-producing area. Since warm water is less viscous than cold, warm-water species bear more of this ornamentation, as **Figure 13.15** indicates. Tropical species also tend to be smaller than related polar species. Why? Because a larger surface-to-volume ratio provides

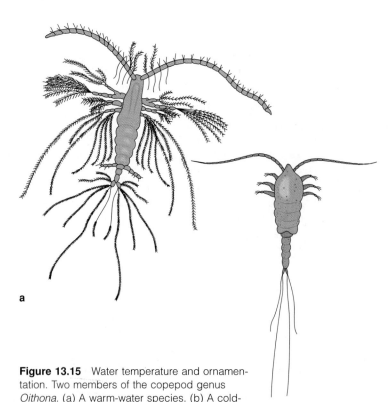

Figure 13.15 Water temperature and ornamentation. Two members of the copepod genus *Oithona*. (a) A warm-water species. (b) A cold-water form.

greater surface area—and hence slower sinking—for small drifting organisms in the less viscous warm water.

Large swimming animals have a different problem with viscosity: the chaotic water movement known as **turbulence.** Water's resistance to flow results in turbulence around the swimmer and in the swimmer's trailing wake, both of which tend to slow the animal. An animal can minimize this frictional loss by streamlining. As an example, **Figure 13.16** shows three objects of identical frontal area moving through water. Streamlined shape (c) will have only about 6% of the drag of shape (a)'s disk. Note the teardrop shape of the fast-swimming tuna in Figure 13.16d.

Other unique adaptations have evolved in swimming species. Some fish secrete a small amount of friction-reducing mucus or oil onto their surface to minimize the formation of eddies, and some can tuck maneuvering fins into body recesses. Some fast-moving marine mammals have skins that passively indent to smooth out eddies as they form. In a sense, any reduction of drag is food saved, prey not needed, a greater ability to escape being eaten, and energy that can go to other concerns (such as reproduction).

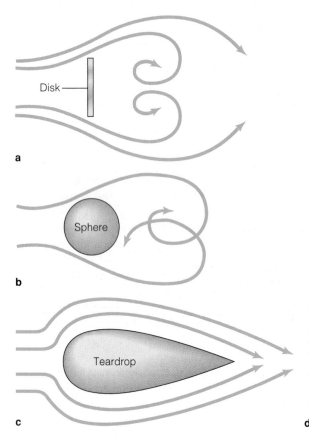

Figure 13.16 Turbulence and drag. At the same speed, with the same frontal diameter, shape (a) will have about 15 times more drag than shape (c). Shape (b) shows only a small improvement in drag over the disk. The fast-swimming yellowfin tuna seen in (d) is shaped like a teardrop to minimize turbulence and drag.

Water Movement

The role of ocean currents in the horizontal distribution of marine organisms or their larvae is obvious, but there are more subtle advantages of drifting with the currents. For example, some kinds of small drifting animals are known to swim each day through considerable *vertical* distances. Many reasons have been proposed to explain this behavior. One advantage to the organisms may be predator avoidance; they feed near the surface only during the relatively safe, dark, night hours. Other benefits may include increased opportunity for feeding during the movement itself or for trace nutrient uptake in deeper water. Another possibility is that these animals may be taking advantage of the differential speeds of currents with depth. Geostrophic currents move fastest at the surface and less rapidly as depth increases. Drifting animals grazing on primary producers at the surface could descend, drift at a different rate, and then reappear in a different surface water mass a day later. This tactic could prevent being "stuck" in one surface mass with few opportunities for feeding.

Other kinds of water movement that influence marine life are associated with the rise and fall of tides, wave action and wave shock, and density differences. In each case, organisms are directly influenced by movement that, in cases such as tidal height and wave shock, can be limiting factors.

CLASSIFICATIONS OF THE MARINE ENVIRONMENT

Scientists have found it useful to divide the marine environment into **zones,** areas with homogeneous physical features. Attempts to classify the oceanic realm began in France in the 1830s but were hampered by lack of knowledge about the physical and biological features of the ocean. The situation changed rapidly near the end of the 1800s with the invention of tools for measuring physical factors and taking samples at various depths. Now these divisions can be made on the basis of light, temperature, salinity, depth, latitude, water density, or almost any of the other physical dimensions we have discussed. Some classifications are more useful than others, however, and we will survey those classifications in this section.

Classification by Light

The photic zone, the sunlit top layer of the ocean, extends in the tropics to a depth of approximately 200 meters (660 feet) and in productive mid-latitude water down to about 100 meters (330 feet). The upper half of the photic zone—the layer in which most biological pro-

ductivity occurs—is called the euphotic zone (*eu* = good). Below the photic zone is the **aphotic zone** (*a* = without), the dark zone that extends to the bottom.

Classification by Location

When the great English and American oceanographic institutions were founded, the need for a common, detailed nomenclature of position became urgent. Using a model from Harald Sverdrup as a point of departure, the scheme shown in **Figure 13.17** was devised by the National Academy of Sciences in the mid-1950s. It has withstood the test of time and is in common use today.

The primary division is between water and ocean bottom. Open water is called the **pelagic zone** (*pelagius* = of the sea) and is divided into two subsections: the **neritic zone** (*neritos* = shallow), near shore over the continental shelf, and the deep-water **oceanic zone,** beyond the continental shelf. The oceanic zone is further divided by depth into zones. The *epipelagic zone* (*epi* = atop) corresponds to the lighted photic zone. In the aphotic depths are layered the *mesopelagic* (*mesos* = in the middle), *bathypelagic* (*bathos* = depth), and *abyssopelagic* (*a* = without + *byssos* = bottom) zones. Abyssopelagic water is the water in the deep trenches.

Divisions of the bottom are labeled **benthic** (*benthos* = bottom) and begin with the intertidal **littoral zone** (*litoral* = of shore), the band of coast alternately covered and uncovered by tidal action. (The *supralittoral zone,* the splash zone *above* the high intertidal, is not technically part of the ocean bottom.) Past the littoral is the **sublittoral zone** (*sub* = below), which is further divided into inner and outer segments: The *inner sublittoral* is ocean bottom near shore, and the *outer sublittoral* is ocean floor out to the edge of the continental shelf.[2] The **bathyal zone** covers seabed on the slopes and down to great depths, where the **abyssal zone** begins. The **hadal zone** (*Hades* = underworld) is the deepest seabed of all, the trench walls and floors.

EVOLUTION AND LIFE IN THE OCEAN

Life on Earth seems very well suited to the physical conditions found here. The composition of the atmosphere and ocean seems ideal, as do the average surface temperature, amount of sunlight, seasons, tides, winds and currents, nutrients, and dissolved gases. All organisms circulate seawater-like fluids within themselves, their cells bathed in life-maintaining saline fluids even if they are hundreds of kilometers from the sea. Life captures,

---

[2]Many workers use the words *supertidal, intertidal,* and *subtidal* instead of the *littoral* terms.

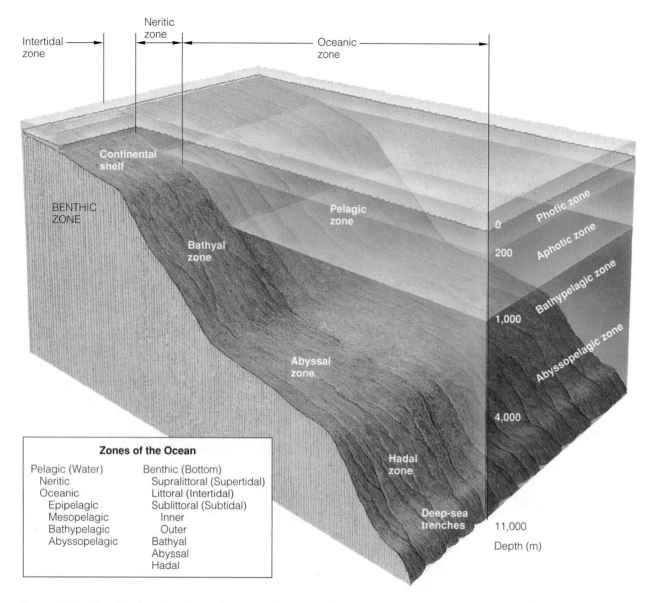

Figure 13.17 Classification of marine environments. This diagram is designed to show divisions and is somewhat exaggerated in indicating the proper proportions. (Compare the hypsographic curve in Figure 4.3.)

transforms, and stores a nearby star's energy, which is then transferred from one living thing to another in a cascade of common molecules. Everything seems close to perfect. Why?

Marine life has not accidentally capitalized on its rich fluid home. Earth's organisms did not arise in just a few thousand years. They have changed, generation by generation, over almost 4 billion years. The ability of living things to change through time to fit the physical and chemical environment, to become ever more efficient at extracting energy from their surroundings, to colonize virtually every location capable of sustaining them, and finally to investigate themselves by scientific logic has come about through **evolution.**

The Concept of Evolution

Evolution means change. Clothing styles evolve, systems of government evolve, our perceptions of the world evolve; *things change.* The preponderance of evidence suggests that living things change, too. That animals and plants might be capable of change with the passage of time was not a popular idea in the nineteenth century. Indeed, the proposal that organisms do change with time began what has since been called a revolution in biology. The unlikely revolutionary was Charles Darwin, a quiet and thoughtful English scholar (**Figure 13.18**).

In the early nineteenth century, most people in the Western world—including scientists—accepted Irish

Figure 13.18 Charles Darwin as a young man.

a

b

Figure 13.19 The *Beagle* expedition. (a) A replica of HMS *Beagle* in the Straits of Magellan. (b) The route of *Beagle*'s five-year voyage around the world. A stop at the Galápagos Islands, about 1,000 kilometers (1,600 miles) off the coast of Ecuador, particularly influenced Darwin's thinking because these isolated islands support a number of unique plant and animal species.

EQUATOR

Galápagos Islands

bishop James Ussher's calculation that Earth was only about 6,000 years old (see Chapter 3). But railroad cuts being made through the hills near Cambridge, where Darwin was studying for the ministry, exposed many layers of rock in the Earth and suggested that the planet was much older. Geologist Charles Lyell had studied these cuts, as well as mine shafts, wells, and many other exposed areas. His book *Principles of Geology* strongly suggested that the geological processes of mountain building and erosion have always been extremely slow. If that were so, the presence of great mountain ranges and canyons also served as evidence for an Earth of great age. Some radical thinkers believed the planet could be as much as 1 million years old. Few people took them seriously.

Darwin was a skilled amateur naturalist and observer. In 1831, after completing his theological studies at Cambridge, he was invited to be the staff naturalist on a South American mapping expedition aboard HMS *Beagle*. He accepted the unpaid position over his father's strenuous objections. His shipboard reading included Lyell's book. During *Beagle*'s five-year expedition around the world (**Figure 13.19**), Darwin saw sights that led him to accept Lyell's ideas about Earth's great age. He also began collecting observations on what was to become his most important work, contemplation of what he called the "species problem."

Most scientists trained in the Western tradition believed the Earth's animals and plants had been made by God in their present form and had remained unchanged in the 6,000 years since the Creation. Indeed, if the Earth were only 6,000 years old, organisms would not have had time to change naturally. But scientists already knew that organisms could change. Controlled breeding—of plants, pigeons, hunting dogs, and race horses, for example—could produce many distinct

BOX 13.1 ● *Sea Serpents*

On 4 October 1848, the British frigate *Daedalus* arrived at Plymouth from the East Indies. Rumors immediately swept the port that the crew had sighted a 100-foot-long sea monster at 4 o'clock in the afternoon of 6 August between the Cape of Good Hope and St. Helena. The captain and most of the ship's officers and crew observed it for some 20 minutes (**Figure a**).

When this startling news reached the Lord Commissioners of the Admiralty in London, they immediately requested a formal report from Peter M'Quahe, captain of *Daedalus*. He wrote:

> The object was discovered to be an enormous serpent, with head and shoulders kept about four feet constantly above the surface of the sea, and as nearly as we could approximate by comparing it with the length of what our main-topsail yard would show in the water, there was at the very least 60 feet of the animal [above the water], no portion of which was, to our perception, used in propelling it through the water. . . . It passed rapidly, but so close under our lee quarter, that had it been a man of my acquaintance, I should easily have recognised his features with the naked eye; and it did not, either in approaching the ship or after it had passed our wake, deviate in the slightest degree from its course to the [southwest], which it held on at the pace of from 12 to 15 miles per hour, apparently on some determined purpose.
>
> The diameter of the serpent was about 15 or 16 inches behind the head, which was, without any doubt, that of a snake, and it was never, during the 20 minutes that it continued in sight of our glasses, once below the surface of the water; its colour a dark brown, with yellowish white about the throat. It had no fins, but something like a mane of a horse, or rather a bunch of seaweed, washed about its back.

This letter was published in *The Times* of 13 October. The British public were greatly excited by the report. After all, naval officers do not tell lies to the Lords of the Admiralty!

Nearly 60 years later, Sir Arthur Rostron, commodore of the Cunard Line and, as captain of *Carpathia*, the man responsible for rescuing more than 700 survivors of *Titanic*, reported a similar experience. Rostron was chief officer on the liner *Campania* when, on 26 April 1907, he saw something curious. "It was a sea monster!" he wrote in his memoirs.

a "Appearance of the sea serpent when first seen from HBM ship *Daedalus*," according to the engraving that was published in *The Times* along with Captain M'Quahe's letter.

It was no more than fifty feet from the ship's side when we passed it and so both I and the junior officer of the watch had a good sight of it. So strange an animal was it that I remember crying out: "It's alive!" One has heard such yarns about these monsters and cocked a speculative eye at the teller, that I wished as never before that I had a camera in my hands. . . . The thing was turning its head from side to side for all the world as a bird will on a lawn between its pecks.

I was unable to get a clear view of the monster's "features" but we were close enough to realize its head rose eight or nine feet out of the water while the trunk of the neck was fully twelve inches wide [**Figure b**].

What were these competent, sea-wise men seeing? Almost certainly, they were seeing something, an object rare enough to excite comment and disbelief. Perhaps they sighted the giant squid, a huge mollusk known to reach a total length of 17 meters (57 feet). Its two longest tentacles could conceivably extend from the water, periscope fashion, in a passable imitation of a serpent. But squid tentacles (and the squid to which they are attached) couldn't move consistently along the ocean surface "at the pace of from 12 to 15 miles per hour." A large seal might, but even the largest seals and sea lions do not approach the sizes reported in these and other sightings. More than one large seal could be mistaken for a single individual in poor observing conditions, however. Whale sharks certainly qualify in size (they can reach lengths to 18 meters, or 60 feet), and although they don't often stick parts of their bodies out of water, they can move purposefully in one direction for long periods of time. And what of masses of seaweed caught in a current? Mariners have reported possible "sea monsters" that upon closer inspection turn out to be clumps of seaweed rolled into curious shapes by storm waves; it's interesting that Captain M'Quahe reported "something like a mane of a horse, or rather a bunch of seaweed, washed about its back." Lines of dolphins rising regularly to take air could also produce a long serpentlike effect. And don't forget upright-floating tree trunks, flocks of birds flying close to the water, or flotsam and other debris.

As we have seen, scientists never say never. Many new species await discovery in the open ocean, and large oceanic reptiles may be among them. Beautiful plesiosaur-like animals now thought to be extinct may someday be found, but the odds are overwhelmingly against such a dramatic discovery. No reptile could live undiscovered in the greatest depths because it would have to come to the surface to feed and breathe. Vertical movement from the surface to the abyss would expose the animal to terrific environmental stresses; no known animal is adapted to survive at all depths. Except for a few deep-diving marine mammals, air-breathers are limited to the uppermost layers. Any true sea serpents would probably live near the surface, and we would almost certainly be familiar with them. And no fresh plesiosaur carcasses or bones have washed ashore anywhere.

Does this mean that there are no sea serpents in the ocean? Not necessarily, but it does suggest that such monsters are much more often found in the popular press and public imagination than in the ocean itself.

b An artist's impression of a sea monster.

Source: J. B. Sweeney, *A Pictorial History of Sea Monsters and Other Dangerous Marine Life* (New York: Crown, 1972).

breeds within a few generations. Simply put, the "species problem" was this: If breeders could change animals and plants so quickly, what could nature do in great spans of time? If Lyell was on the right track—if Earth was very old—species might have had enough time to change by themselves in response to their environment. Darwin wondered how such changes might occur and what their nature would be.

For more than 20 years Darwin amassed a wide variety of detailed evidence in support of his developing theory of species formation, but he put off publishing it in definitive form. In June 1858, he received a letter from Alfred Wallace, a young scientist working in Indonesia. Wallace outlined the main points of a theory on species that coincided with Darwin's own. Wallace had thought through the ideas in less than a week during a feverish illness. Encouraged by friends and colleagues, Darwin prepared a short summary of his researches, and the two men's theories were announced simultaneously at a meeting of the Linnaean Society, one of the world's foremost natural science associations. Darwin's famous and controversial book-length version of his paper, *On the Origin of Species by Means of Natural Selection,* was published the next year. It contained an idea whose time had come.

The Theory of Evolution by Natural Selection

And what was this remarkable idea? Why was evolution by **natural selection** controversial then, and why is it capable of stirring passions even today? Darwin had discovered a mechanism for how species might change with time. Here are Darwin's main points:

1. In any group of organisms, more offspring are produced than can survive to reproductive age.

2. Random variations occur in all organisms. Some of these variations are inheritable; that is, they can be passed on to the offspring.

3. Some inheritable traits increase the probability that the organisms possessing them will survive. These are favorable traits.

4. Because bearers of favorable traits are more likely to survive, they are also more likely to reproduce successfully than bearers of unfavorable traits. Thus, favorable traits tend to accumulate in the population; they are *selected.*

5. The physical and biological (*natural*) environment itself does the selection. Favorable traits are retained because they contribute to the organism's success in its environment. These traits show up more often in succeeding generations if the environment stays the same. If the environment changes, other traits become favorable—and the

organisms with those traits live most effectively in the environment.

It is easy to see how random variations are selected for or selected against by environmental pressures, but how do entirely new traits arise? They come about by spontaneous **mutation**, an inheritable change in an organism's genes (the structures that contain its assembly instructions). The vast majority of mutations are unfavorable; so the organisms possessing them are eliminated by other organisms or by the physical environment. For example, a tuna born with no eyes could not see to feed; so it would not live to reproductive age. But a tuna born with extraordinarily good eyesight might have more to eat than its cohorts, and being better nourished it might be especially effective in its reproductive efforts, spreading its genes far and wide. It is the accumulation of these favorable traits, either variations or mutations—and the elimination of unfavorable traits— that makes life possible in changing conditions on Earth.

Note that although mutations occur randomly, evolution by natural selection is anything but random. The natural environment winnows favorable mutations from unfavorable ones—hence the origin of the term *natural selection.* The process takes a great deal of time, but time is in great supply now that geologists have shown the Earth to be about 4.6 billion years old.

This evolutionary viewpoint provided a new way of looking at life: An organism is a vessel holding a particular combination of traits, testing that combination. If the organism is successful it will reproduce, and the genes responsible for its good traits will continue in the population. If the organism is not successful, that combination will be eliminated. There is no biological predeterminism. Organisms don't *want* to evolve, and individual organisms *don't* evolve. Rather, generation after generation, groups of individuals respond to environmental pressures by change. The changes can be in shape, size, color, biochemistry, behavior, or any other aspect of the organism. Evolution by natural selection is the accumulation of these beneficial inheritable structural or behavioral traits, known as favorable **adaptations.** Organisms with favorable adaptations have more reproductive success than less well adapted organisms.

Evolution, then, is the maintenance of life under changing conditions by continuous adaptation of successive generations of a species to its environment. It is a remarkable, beautiful, productive theory.

What Are Species?
How Do New Species Arise?

A **species** is a group of actually (or potentially) interbreeding organisms that is reproductively isolated from

Figure 13.20 The marine iguana, genus *Amblyrhyncus*, found only in the Galápagos Islands, is believed to have evolved through reproductive isolation from species on the South American mainland. The land iguana, genus *Iguana*, is a modern representative.

all other forms of living things. Variations can occur within a species, a situation easily observed in domestic dogs. The fact that all breeds of domestic dogs are interfertile (capable of giving birth to puppies that will grow up and reproduce) is evidence that they are all members of a single species.

The process of species formation is known as **speciation.** One way for new species to form is by physical isolation, such as that created when land animals or birds are rafted or blown from a mainland shore to an isolated oceanic island. Because the number of breeding animals within a species on an island may be small, evolutionary change may be rapid; that is, favorable traits may accumulate quickly in the population, and the species will change relatively rapidly to suit its new habitat. In general, the smaller the reproducing population, the more rapid the rate of evolutionary change.

For example, on the Galápagos Islands off the coast of Ecuador, Darwin observed finches and marine lizards, most of which closely resembled their mainland South American ancestors (see **Figure 13.20**). They were not the same species of birds and reptiles that occurred in mainland South America, however. This suggested to Darwin that isolation was a driving force of evolution.

An ideal example of the role of isolation in the formation of new species is that of Australian mammals. Driven by the forces of plate tectonics, Australia drifted away from the supercontinent of Pangaea as it broke up during the Cretaceous period between 63 and 130 million years ago, when marsupial (pouched) mammals were the dominant mammalian form. Many terrestrial organisms were stranded on this great land mass, iso-

lated from populations elsewhere in the world. Away from Australia, after the split, the placental mammals became dominant. More efficient at reproduction than marsupials, the placentals displaced most of the marsupials everywhere they were in competition. But the Australian marsupials were isolated from this interference, and they radiated into many different "occupations." Kangaroos in Australia are the marsupial counterparts of deer and antelope, Tasmanian wolves the counterparts of dogs, koalas the counterparts of monkeys. This is an instance of **divergent evolution,** one group radiating into many different species and life-styles.

Once a new species has been formed, it will remain reproductively isolated from its ancestral relatives even if the cause of the original isolation is removed. That is, a Galápagos marine iguana and a mainland tree iguana will not produce fertile young even if they could be induced to mate. Why not? Because a new species, by definition, is genetically different from its precursor population. If viable, fertile offspring are produced, the mating animals are not separate species.

Evolution in the Marine Environment

Evolutionary theory contains important implications for marine science. The ocean has a much larger inhabitable volume than dry land; and as we've seen in our discussion of physical factors, it is often an easier place to live than either the terrestrial or freshwater realms. The ocean contains only a fifth of the species known to science, but some scientists suggest that counting species is not a valid way of assessing the success of an environment. All major animal groups (known as *phyla*) are found in the sea, and one-third of them are exclusively marine. If plants and single-celled organisms are included, at least 80% of all phyla include marine species. Perhaps an examination of life-styles (what an organism *does*) is more important in judging the biological diversity of an environment than what an organism *is*. There are more ways of "making a living" in the ocean than on land. Indeed, some important oceanic life-styles have no terrestrial counterparts. For example, filter feeding—practiced by clams, barnacles, and some whales—relies on a relatively dense fluid matrix and is therefore not seen on land. Also, marine food chains tend to be more complex than terrestrial ones and contain more trophic levels. Marine life has evolved in countless ways to take advantage of nearly every scrap of energy, nearly every nook and cranny of space.

Since physical conditions in the open ocean are relatively uniform, large marine animals with similar lifestyles but different heritages eventually tend to look much the same. That is, similar conditions may result in coincidentally similar organisms. The shark (a fish), the ichthyosaur (an extinct marine reptile), the penguin (a bird), and the porpoise (a mammal) resemble each other in shape even though they are only remotely related, because the physics of rapid movement through water requires a similar streamlined shape (see **Figure 13.21**). Traits leading to this shape were independently selected by environmental conditions. These accumulated adaptations resulted in superficially similar animals, each derived from different and diverse stock. The process is known as **convergent evolution.**

Evolution, or change by natural selection, seems to explain much of the biological world we see today. Darwin's theory has lifted biology from a mere catalog of differences among living organisms to a strong statement of common descent and underlying unity. It has given us a tool with which to investigate the living world and, by implication and extension, to answer some very interesting questions. True, many questions remain to be answered. We don't know if evolution progresses continuously and slowly, as Darwin suggested, or in punctuated fits and starts interspersed with long, slow periods of little change. We don't yet know if the Earth is old enough to allow for the great diversity of biological form we see to have evolved by natural selection as we understand the process. Biologists and geologists continue to work on these issues.

We do know that life and the Earth change together. Life is tenacious; it has survived catastrophe and calm. In every instance, however, *the environment isn't right for the organisms; rather, the organisms are right for the environment.* One is reminded of author Pär Lagerkvist's observation, "Things need be as they are." Living things have adapted to the physical conditions of their environment, and they continue to increase in numbers, complexity, and efficiency.

Q-AND-A

1. Why are nitrates often a limiting factor in photosynthesis? If air is 78% nitrogen, there should be lots of nitrogen around even if the solubility of nitrogen in the ocean is low.

   Yes, but marine plants cannot use dissolved atmospheric nitrogen (N_2) directly; so it is not a nutrient. The preferred forms of nitrogen for uptake are ammonia and nitrate (NO_3^-), an ion formed by the oxidation of ammonia and nitrites. Nitrate runoff from soil is an especially rich source of this often limiting nutrient, which is why coastal water tends to support greater plankton populations than oceanic water. There is an important exception, however. Some kinds of blue-green algae (cyanobacteria) can use dissolved atmospheric nitrogen; so they take N_2 directly from seawater. Their contribution to total oceanic productivity is relatively low, however.

2. What proportion of total world productivity is achieved by chemosynthetic means?

   An interesting and controversial question! Chemosynthesis occurs in the marine environment primarily at hydrothermal vents associated with the 65,000-kilometer (41,000-mile) network of mid-ocean ridges. Specialized communities of organisms have evolved in the zones of warm or hot mineral-rich water percolating through the seafloor in these places. Once thought to be rare and isolated, vent communities have now been discovered in broadly separate locations and may be quite common to mid-ocean ridges. The trophic pyramid in these deep communities is based on chemosynthetic bacteria. Some biological oceanographers have suggested that if vent communities are abundant on ridges, chemosynthesis there may account for a much larger proportion of total oceanic productivity than was previously thought.

3. If humans have a fluid much like seawater bathing their cells, why can't we drink seawater and survive?

   Human cells function in an environment hypotonic to seawater; that is, blood plasma is less saline than seawater. Drinking seawater therefore causes water to leave the intestinal walls, flood the intestine, and leave the body. There is a net loss via intestine or kidneys even if the seawater is diluted with fresh water before drinking. Moral: Never drink seawater in a survival situation at sea, and never dilute fresh water with seawater (in any proportion) to stretch your supply.

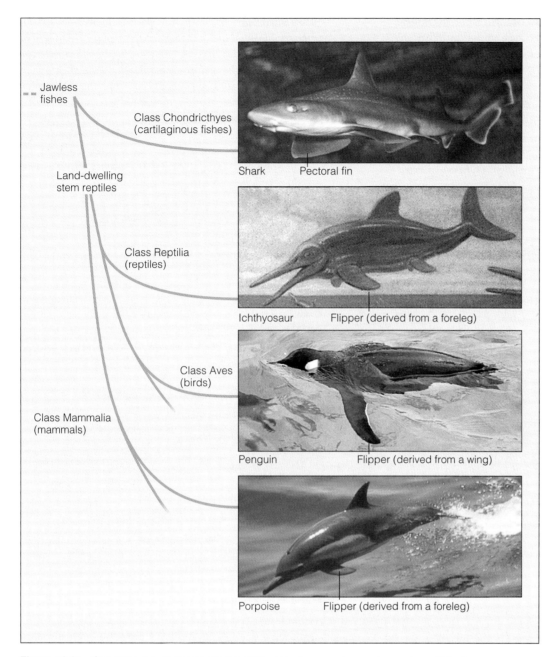

Jawless fishes

Class Chondricthyes (cartilaginous fishes)

Land-dwelling stem reptiles

Shark   Pectoral fin

Class Reptilia (reptiles)

Ichthyosaur   Flipper (derived from a foreleg)

Class Aves (birds)

Class Mammalia (mammals)

Penguin   Flipper (derived from a wing)

Porpoise   Flipper (derived from a foreleg)

Figure 13.21 Convergent evolution in sharks, ichthyosaurs, penguins, and porpoises. Selection for adaptations that permitted rapid swimming resulted in superficially similar shapes among these four kinds of vertebrates, even though they are only remotely related.

4. Perfectly smooth body surfaces are best for attaining high speeds, right?

Not always. Animals that move rapidly through fluids (water or air) are almost never perfectly smooth; their skin usually contains small projections such as hair or feathers or scales. Even small whales that appear glassy smooth have flexible skins that indent to smooth out flow irregularities. A small degree of surface roughness and "give" makes it more difficult for the fluid to adhere to the moving surface, thereby cutting drag.

This finding has been put to good use: As a result, golf balls are covered by little dimples instead of being smooth, America's Cup yacht builders now incise patterns of tiny lines into their previously flawless hulls, and competitive swimmers no longer need to shave their bodies for maximum speed.

5. Chlorophyll is green. That must mean that it reflects green light while absorbing red and violet light (otherwise it would appear some other color). Energy in

red and violet light is mainly used in photosynthesis, but there's virtually no red and violet light in the ocean past about 5 meters of depth. How can photosynthetic autotrophs in the ocean be so productive?

Trickery, that's how. There are accessory pigments *surrounding chlorophyll that absorb the available blue and green light and transfer its energy to chlorophyll. These accessory pigments cause many marine plants to look tan, or red, or some color other than pure, unmasked green. More on this in Chapter 14's discussion of marine plants.*

6. How important was the revolution in biology sparked by Darwin's work?

Very important. According to Ernst Mayr, a famous evolutionary biologist and philosopher of biology, "The Darwinian revolution was the most fundamental of all intellectual revolutions in the history of mankind. While such revolutions as those brought about by Copernicus, Newton, Lavoisier, or Einstein affected only one particular branch of science, or the methodology of science as such, the Darwinian revolution affected every thinking man. A world view developed by anyone after 1859 was by necessity quite different from any world view formed prior to 1859." How could it be otherwise? (See Mayr, 1988.)

Terms and Concepts to Remember

abyssal zone
active transport
adaptation
aphotic zone
autotroph
bathyal zone
benthic zone
biological factor
cell
chemosynthesis
chlorophyll
convergent evolution
diffusion
divergent evolution
drag
energy
entropy
euphotic zone
evolution
food
food web

glucose
hadal zone
heterotroph
homeotherm
hydrostatic pressure
hypertonic
hypotonic
isotonic
limiting factor
littoral zone
membrane
metabolic rate
mutation
natural selection
neritic zone
nutrient
oceanic zone
osmosis
pelagic zone
photic zone
photosynthesis

physical factor
poikilotherm
primary consumer
primary producer
primary productivity
respiration
second law of thermodynamics
speciation
species
sublittoral zone
surface-to-volume ratio
top consumer
trophic level
trophic pyramid
turbulence
viscosity
zone

Study Questions

1. What evidence suggests that life on Earth originated in the ocean?

2. How are living things different from nonliving things? How can you differentiate between them?

3. The second law of thermodynamics states that entropy (disorganization) tends to increase with time. Living things tend to become *more* complex with time (embryos grow to adults, populations evolve). How can that be?

4. Look at the general formulas for photosynthesis and respiration. What similarities do you see? Differences? How are the formulas related? Where does solar energy go? How does chemosynthesis differ from photosynthesis?

5. What is a trophic pyramid? What is the relationship of organisms in a trophic pyramid? Does this have anything to do with food webs?

6. Name and discuss five physical factors of the marine environment. How is each different in the ocean from those on the land?

7. What is a biological factor? Give examples.

8. How is a poikilotherm different from a homeotherm? Which are you? What are the advantages of a high, stable internal body temperature? The disadvantages?

9. What is a limiting factor? Can you think of some examples not given in the text?

10. How are diffusion, osmosis, and active transport related? Use these concepts to explain the problems a freshwater goldfish would have if it suddenly found itself in the ocean.

11. What problems do pelagic organisms encounter in maintaining their position in the water column and in swimming? How are these difficulties overcome?

12. How is the marine environment classified? Which scheme is most useful? Why, and to whom?

13. How would you define biological success? Does success depend on the size of an organism? Its beauty? The amount of space it controls? Its numbers?

14. How is evolution by natural selection thought to work?

15. Why did the concept of biological evolution have to wait until Earth was shown to be very old? What does time have to do with it?

16. How do new species arise? Are there thought to be more terrestrial species or marine species? Why?

For Further Study

Alvarez, L. W., et al. 1980. "Extraterrestrial Cause for the Cretaceous–Tertiary Extinction." *Science* 208: 1095–1100. Conditions for life on Earth have not always been stable. Occasional collisions with comets and asteroids may have caused the extinction of many (or most) marine species.

Bent, H. 1965. *The Second Law.* New York: Oxford University Press. A curious and interesting blend of highly technical material and information for the interested reader.

Briggs, J. C. 1974. *Marine Zoogeography.* New York: McGraw-Hill. A useful chapter on the evolution of marine life and the integration of continental drift with the distribution of marine forms. Also a good (though now dated) literature review on the origin of life in the ocean.

Cloud, P. 1983. "The Biosphere." *Scientific American,* September, 176–89. Earth and life have evolved together.

Darwin, C. 1859. *On the Origin of Species by Means of Natural Selection.* Reprint. New York: Random House, 1963. Chapter heads and summaries are eminently readable, but the great wealth of examples in the text itself makes for heavy going. Still, it's worth a look at the original.

Denton, E. 1960. "The Buoyancy of Marine Animals." *Scientific American,* July, 118–28. Who floats, how, and why.

Devlin, K. M., and L. Barker. 1976. *Photosynthesis.* New York: Van Nostrand-Reinhold. Excellent technical reference on photosynthetic biochemistry.

Eiseley, L. 1956. "Charles Darwin." *Scientific American,* February, 62–72. A wonderfully written short biography.

Friedrich, H. 1969. *Marine Biology: An Introduction to Its Problems and Results.* Translated from German by G. Vevers. London: Sedgwick & Jackson. Fine exposition on the influence of physical factors on life in the sea.

Grobstein, C. 1974. *The Strategy of Life.* 2d ed. New York: Freeman. Thought-provoking discussion of the difficulty of defining life. Unfortunately now out of print.

Hedgpeth, J. W., ed. 1957. *Treatise on Marine Ecology and Paleoecology.* Vol. 1, *Marine Ecology.* Geological Society of America Memoir 67. Washington, DC: Geological Society of America. A classic compendium of review articles on marine ecology by the field's top researchers. Invaluable, but, again, unfortunately out of print.

Herring, P. J., and M. R. Clarke, eds. 1971. *Deep Oceans.* New York: Praeger. Outstanding general reference aimed at the technically literate lay reader. Diverse collection of papers by eminent European scientists on a great variety of marine topics; not limited to abyssal depths.

Isaacs, J. 1969. "The Nature of Oceanic Life." *Scientific American,* September, 146–62. A valuable overview, part of the now-classic issue of *Scientific American* dedicated to the revolution in marine science.

Jerlov, N. G. 1951. "Optical Studies of Ocean Waters." *Report of the Swedish Deep-Sea Expedition, 1947–8* 3:1–59. An often excerpted classic paper on the subject.

Levinton, J. S. 1982. *Marine Ecology.* Englewood Cliffs, NJ: Prentice-Hall. Excellent standard text.

May, R. M. 1988. "How Many Species Are There on Earth?" *Science* 241 (no. 4872): 1441–49. A great many, it seems—possibly 50 million! Most are terrestrial insects.

Mayr, E. 1988. *Toward a New Philosophy of Biology: Observations of an Evolutionist.* Cambridge, MA: Harvard University Press, Bel-

knap Press. One of the twentieth century's greatest biologists contemplates evolution.

McConnaughey, B., and R. Zottoli. 1983. *Introduction to Marine Biology.* 4th ed. St. Louis: Mosby. Eclectic general reference.

Milne, D. H. 1995. *Marine Life and the Sea.* Belmont, CA: Wadsworth. A complete treatment, well written and up to date.

Moore, H. B. 1958. *Marine Ecology.* New York: Wiley. Comprehensive look at the interaction of physical factors and organisms.

Nicol, J. A. C. 1967. *The Biology of Marine Animals.* 2d ed. New York: Wiley. A classic. Many examples and explanations.

Nybakken, J. W. 1988. *Marine Biology: An Ecological Approach.* 2d ed. New York: Harper & Row. One of the best and most readable general texts available.

Russell-Hunter, W. D. 1970. *Aquatic Productivity: An Introduction to Some Basic Aspects of Biological Oceanography and Limnology.* New York: Macmillan.

Sagan, Carl. 1980. *Cosmos.* New York: Random House. This is the best-selling science book in history, and not without good reason. The sections on life are wonderful to read. Very highly recommended.

Sagan, Carl. 1988. "Life." In *Encyclopædia Brittanica,* vol. 22. Chicago: Encyclopædia Brittanica. Perhaps the best single reference on the subject. Written in a brilliant style.

Schmidt-Nielsen, K. 1972. *How Animals Work.* London: Cambridge University Press. A simplified look at structure, function, and biochemistry.

Schopf, J. W., ed. 1991. *Major Events in the History of Life.* Boston: Jones & Bartlett. Symposium volume. Recent thoughts by experts on the Burgess Shale, Precambrian microfossils, primordial soup, the earliest plants, and other events of interest.

Stanley, S. M. 1986. *Earth and Life Through Time.* New York: Freeman. Excellent standard text.

Starr, C., and R. Taggart. 1995. *Biology: The Unity and Diversity of Life.* 7th ed. Belmont, CA: Wadsworth. Perhaps the best general biology text available. The principles of unity and evolution serve as organizers. Gracefully written, thorough.

Steadman, D. W., and S. Zousmer. 1988. *Galápagos—Discovery of Darwin's Islands.* Washington, DC: Smithsonian Institution Press.

Sverdrup, H., M. Johnson, and R. Fleming. 1942. *The Oceans: Their Physics, Chemistry, and General Biology.* Englewood Cliffs, NJ: Prentice-Hall. Technical discussion of salinity and other physical factors as they affect marine life.

Thorne-Miller, B. L. 1991. *The Living Ocean: Understanding and Protecting Biodiversity.* Washington, DC: Island Press. Though only a fifth of known species are marine, there are more orders and phyla in the sea than on land. The author wonders if species number is the proper delineator of biodiversity.

Volpe, E. P. 1985. *Understanding Evolution.* 5th ed. Dubuque, IA: W. C. Brown. Thorough, technical, comprehensive survey of biological evolution.

Weiner, J. 1995. "Evolution Made Visible." *Science* 267 (no. 5194): 30–31. Charles Darwin didn't think we could watch evolution in action, but modern biologists are getting a good look at processes such as natural selection and the origin of species.

14

PRIMARY PRODUCTIVITY, PLANKTON, AND PLANTS

The Forest Beneath the Waves

Diving in a kelp forest is a joy for any good swimmer interested in the marine environment. The experience can be pleasing for anyone, but for marine scientists it can be rapturous. The ocean often moves placidly in a kelp forest—the long seaweeds interfere with the circular movement of water molecules as the waves pass. Because the plants secrete a lubricant that smooths the diver's way, visitors can glide easily between the closely entwined seaweeds, gently nudging the strands from their path. Sunlight filters between the fronds, sending illuminating shafts flickering into the deeper water below. Schools of fishes often move among the kelp, and countless animals, many too small to be seen, nestle around their dark bases. The feeling is like that of being in a great grove of redwoods, but the diver's relative weightlessness allows for exploration of nearly all parts of the area. It is usually quiet here, peaceful and calm.

A biologist sees even more beauty in the scene and notes the rapid growth: The seaweeds are longer on this visit than a few weeks ago, and their stalks are thicker. Some worms have begun to make spiral tracings on the kelp surfaces. The animals on the seabed are different now and are arrayed in new patterns. A family of otters has moved to the area, attracted by the nutritious sea urchins scattered across the gravel-covered bottom. Schools of anchovetta flash overhead, swimming between the kelp blades in long, follow-the-leader schools, efficiently sieving the ocean for food. Importantly, there are many more plantlike organisms living here than the big seaweeds swaying in the currents: A

Sunlight penetrates a kelp forest, one of the ocean's most beautiful and productive habitats.

gentle living haze crowds the water immediately ahead of the diver's faceplate. A flashlight beam reveals swarms of dust-sized organisms in every direction. Countless millions of organisms drift unobserved, too small to reflect the light. Life abounds.

The ocean here fairly hums with productivity. Food is being produced rapidly and in great quantity. Big plants and microscopic ones are harnessing light from the sun to assemble carbohydrates as their forbears have done since the ocean was young. The temperate coastal ocean brims with life, much of it dependent on the subtle, light-driven biochemistry proceeding in this lovely, gently moving place.

The photosynthesizing organisms at the base of marine food webs—organisms like the seaweeds and the drifting microscopic cells—are the primary producers, creatures able to make complex food molecules out of water and carbon dioxide. You learned about photosynthesis, as this transformation is called, in Chapter 13; in this chapter we examine primary productivity and then turn our attention to the producers themselves.

CHAPTER OVERVIEW

Primary producers are organisms that synthesize food from inorganic substances. Small, drifting, plantlike organisms—the phytoplankton—are responsible for most of the ocean's primary productivity. Larger marine plants, usually attached to the ocean bottom, account for much of the rest. Many physical and biological factors influence primary productivity, the most important being light and the availability of inorganic nutrients.

Phytoplankton—and zooplankton, the small, drifting or weakly swimming animals that consume them—are usually the first links in oceanic food webs. Plankton are most common along the coasts, in the upper sunlit layers of the temperate zone, in areas of equatorial upwelling, and in the southern subpolar ocean. Marine scientists have been inspired by the beauty and variety of plankton since first observing them under the microscope in the nineteenth century.

Most of the larger marine plants, the ones informally known as seaweeds, are classified by their pigments into three large groups: green, brown, and red algae. Some forms of brown algae, which we call *kelp*, grow in great underwater forests. Not all large marine plants are seaweeds; some are sea grasses and mangroves.

Some marine autotrophs are commercially important. The worldwide value of products containing living and fossilized marine plant substances exceeded $36 billion in 1993.

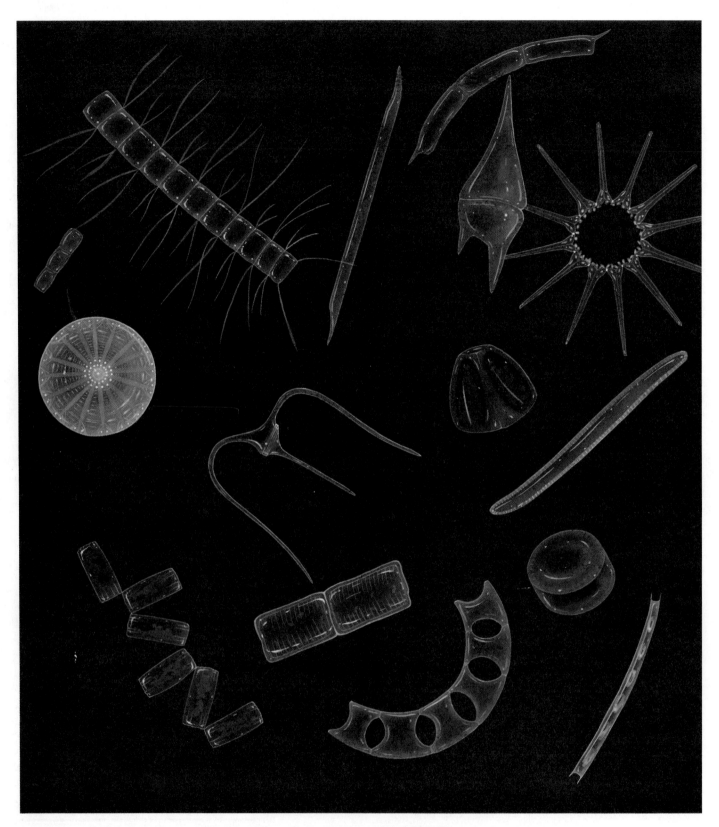

Figure 14.1 Assorted phytoplankton, the ocean's most prolific producers. To the unaided eye these organisms would barely be visible as specks in the water. Under a low-power light microscope their diversity and intricacy can be appreciated. For an indication of their true size, see Figure 14.5c.

PRIMARY PRODUCTIVITY

The synthesis of organic materials from inorganic substances by photosynthesis or chemosynthesis is called **primary productivity.** Primary productivity is expressed in grams of carbon bound into organic material per square meter of ocean surface area per year ($gC/m^2/yr$). The organic material produced is usually glucose, a carbohydrate. The source of carbon for glucose is dissolved CO_2. **Phytoplankton**—single-celled, plantlike organisms that drift near the ocean surface (**Figure 14.1**)—account for between 90% and 98% of oceanic carbohydrate production. **Seaweeds**—larger marine plants—may be more obvious to us, but they contribute only 2% to 10% of the ocean's primary productivity. Chemosynthetic organisms may account for around 1% of the total. Though estimates vary widely, recent studies suggest that total ocean productivity ranges from 50 to 100 grams of carbon bound into carbohydrates per square meter of ocean surface per year (50–100 $gC/m^2/yr$).

The total weight of a primary producer is assumed to be about 10 times the mass of the carbon it has bound into carbohydrates. Thus, a primary productivity of 100 $gC/m^2/yr$ represents the yearly growth of about 1,000 grams of primary producers for each square meter of ocean surface, on a worldwide average. Between 20 and 25 billion metric tons (22 to 28 billion tons) of carbon is believed to be bound into carbohydrates in the ocean each year; so between 200 and 250 billion metric tons (220 to 280 billion tons) of marine autotrophs are produced annually! Each year this vast bulk is consumed by the respiration of the producers themselves and by the consumers that graze on them. The component atoms are then reassembled by photosynthesis into carbohydrates in a continuous cycle.

Measuring Primary Productivity

The rate at which autotrophs form organic material is not easy to measure. At first glance the problem might seem simple to solve: Collect all the organisms—large and small, autotrophs and heterotrophs—in an area and analyze their organic content and weight. Since all organisms are dependent on primary productivity, the mass of living tissue (or **biomass**) in the area would seem directly proportional to productivity.

This approach has drawbacks. For example, a dense population of phytoplankton (a high biomass) would interfere with light penetration; so the autotrophs would manufacture carbohydrates slowly, and productivity would be low. In contrast, a sparse population of phytoplankton (a low biomass) might encounter ideal conditions for photosynthesis and produce carbohydrates at a rapid rate. Small animals might immediately consume

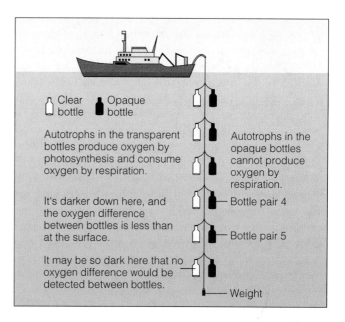

Figure 14.2 The light bottle/dark bottle technique for estimating marine primary productivity. Clear and opaque bottles are filled with water from the area to be studied. Pairs of the bottles are lowered to various depths, and photosynthesis and respiration are allowed to occur. The difference in the quantity of oxygen in each pair of bottles over time gives an indication of the net productivity at each depth.

this production and keep the biomass at low levels, but the productivity would be high.

What researchers need is a method of measuring productivity directly. Since we know the formula for photosynthesis (see Chapter 13), measuring any one component of photosynthesis will tell us about the others. For example, from a measurement of oxygen production by primary producers we can calculate the rate of carbohydrate production. This measurement is complicated by the fact that the primary producers *use* oxygen as well as *make* it. We must somehow isolate production from consumption if we are to estimate productivity.

One way to do this is to place identical water samples (containing identical populations of organisms) into pairs of identical transparent and opaque bottles, and then suspend chains of the bottles from the side of a ship from dawn until noon (**Figure 14.2**). The transparent bottles admit light; the opaque bottles block it. Autotrophs within the transparent bottles both produce and consume oxygen; autotrophs within the opaque bottles consume oxygen but are unable to generate it; and heterotrophs consume oxygen in both bottles. The difference in the quantity of oxygen in each pair of bottles over time gives an indication of the net rate of productivity at each depth. Consider, for example, bottle pair #4 in Figure 14.2. If we measure a net gain of 4 milliliters of oxygen in the transparent bottle but a drop of 3 milliliters in the opaque bottle, we know that 7 milliliters of oxygen was actually produced in the transparent bottle. (The difference—3 milliliters—was consumed by the respiration of the autotrophs and heterotrophs in the transparent bottle.) Light-dark bottle experiments may be conducted for various species and concentrations of producers at different light, nutrient, and temperature levels. Extrapolation from these experiments to other oceanic areas can then yield data on overall productivity.

A more sensitive way to measure productivity is by tagging atoms of carbon in CO_2 with radioactive tracers. The rate at which the marked carbon is incorporated into carbohydrates is a direct indication of primary productivity. This laboratory procedure, unlike the at-sea oxygen balance method, requires costly and delicate equipment, but it provides researchers with the most accurate estimates of productivity in controlled situations.

Ocean conditions are complex and variable, however—rarely controlled. A new and promising method of gauging productivity may be the most effective and useful of all. Recent advances in remote sensing have made it possible to estimate the chlorophyll content of ocean water from orbiting satellites (**Figure 14.3**). Because the amount of chlorophyll present is directly related to the rate of photosynthesis, chlorophyll content is a good indicator of productivity.

Factors That Limit Productivity

As you may recall, a limiting factor is a physical or biological necessity whose presence in inappropriate amounts limits the normal actions of an organism—in this case, production of carbohydrates. Photosynthetic autotrophs require four ingredients to produce carbohydrates: water, carbon dioxide, inorganic nutrients, and sunlight. Obviously, water is not a limiting factor in the ocean. Carbon dioxide is almost never a limiting factor either, because of its high solubility in water and because of the large quantity of carbon dioxide dissolved in the ocean. So the two potential limiting factors in primary productivity are nutrients and light.

Autotrophs require inorganic nutrients for two purposes: to construct the large organic molecules that make primary productivity possible and to construct their skeletons or protective shells. **Nonconservative nutrients** are nutrients that change in concentration with biological activity. After a period of rapid phytoplankton growth—a phenomenon known as a **plankton bloom**—the ocean surface is often drained of nonconservative nutrients such as nitrates, phosphates, and silicates. These nutrients become parts of the producers or of the animals that have eaten the producers. When either dies, their bodies tend to fall below the productive sunlit

layer, and the nutrients they contain are lost to immediate recycling. The lost nutrients may be replaced by upwelling of nutrient-rich deep water, runoff of nutrients from land, and additions from the atmosphere.

Lack of nutrients is the most common factor limiting primary productivity. Adding a few hundred kilograms of nitrate- and phosphate-rich lawn fertilizer to a calm, warm, sunlit ocean area depleted of nutrients will usually turn the place into a rich phytoplankton soup in a matter of hours or days.

If adequate nutrients are present, productivity depends on illumination. Too little light is obviously limiting for photosynthesizers; very little photosynthesis proceeds below 100 meters (330 feet), and no photosynthesizers are known to function below 268 meters (879 feet). Too much light can also be limiting, however. You might think that productivity would be highest right at the ocean surface, where light is brightest. It isn't. Light there is often strong enough to overwhelm the photosynthetic chemistry of some photosynthesizers, especially phytoplankton.

Quantity is one important aspect of the light received by marine autotrophs. Quality, or color, is another. Chlorophyll is a green pigment and thus absorbs best in the red and violet wavelengths. Chlorophyll looks green because it reflects green light. But red light is effectively absorbed and converted into heat near the ocean surface; very little red light penetrates past three meters (10 feet).

How, then, can a few species of tropical marine plants live at depths of over 250 meters (820 feet)? Because most marine autotrophs have evolved specialized accessory pigments. **Accessory pigments** (also called *masking pigments*) are light-absorbing compounds closely associated with chlorophyll molecules. They don't resemble chlorophyll chemically, but they bind loosely with it. Their presence in autotrophs greatly enhances photosynthesis because they absorb the dim blue light at depth and transfer its energy to the adjacent chlorophyll molecules. Accessory pigments may be brown, tan, olive green, or red; they give most marine autotrophs, especially seaweeds, their characteristic color. The masking effect of accessory pigments is often so complete that an observer is unaware that the plant contains any chlorophyll whatever, even when the seaweed is transparent. The absence of accessory pigments allows the bright green of chlorophyll to shine through.

Phytoplankton species near the surface may or may not contain accessory pigments, but many seaweeds (especially those adapted to deep water) are completely dependent on them. Deep-water seaweeds do not grow rapidly because the transfer of energy from weak blue light is relatively inefficient—but neither do they have much competition from other plants at those depths.

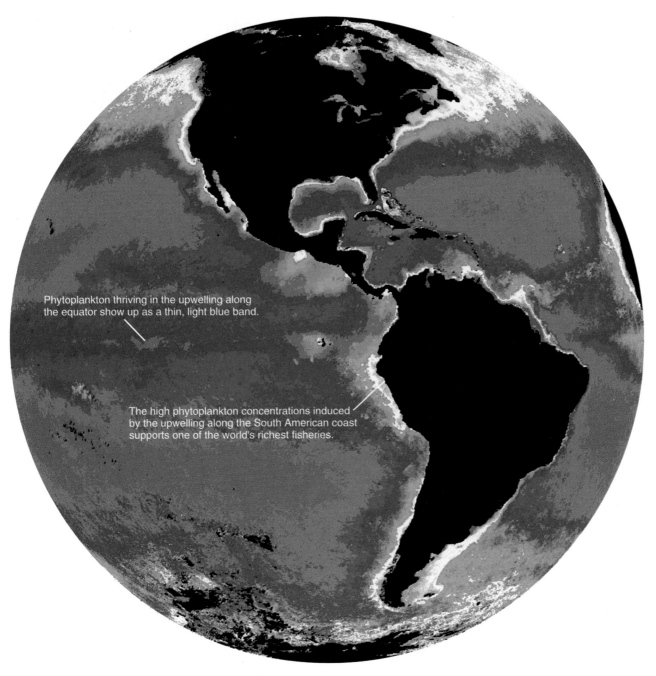

Phytoplankton thriving in the upwelling along the equator show up as a thin, light blue band.

The high phytoplankton concentrations induced by the upwelling along the South American coast supports one of the world's richest fisheries.

Figure 14.3 Measuring the concentration of chlorophyll in ocean water by satellite. This image from the Coastal Zone Color Scanner aboard the satellite *Nimbus 7* represents average conditions over several years. It shows the concéentration of chlorophyll in the upper layer of the ocean, with higher amounts indicated by green, orange, and red colors. Note the high phytoplankton concentrations induced by increased nutrient availability along the coasts and the thin, light blue band representing upwelling and productivity along the equator west of South America. The centers of the oceanic gyres contain relatively few phytoplankton, as shown by their purple hue.

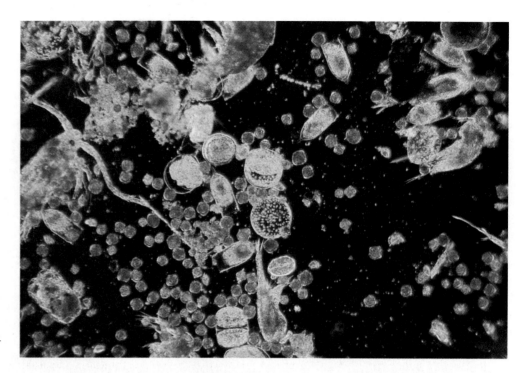

Figure 14.4 Concentrated plankton collected with a plankton net. Both plant and animal species are shown. Magnification here is about 100×; the round diatoms are about 100 micrometers (0.004 inch) in diameter.

PLANKTON

The phytoplankton we have discussed are members of the plankton community. The organisms that constitute **plankton** are as important as they are inconspicuous. The word is derived from the Greek word *planktos*, meaning "wandering." Plankton drift or swim weakly, going where the ocean goes, unable to move consistently against waves or current flow.

The diversity of planktonic organisms is astonishing. There are giant, drifting jellyfish with tentacles 8 meters (25 feet) long; small but voracious arrowworms; many single-celled creatures that glow brightly when disturbed; mollusks with slowly beating flaps that resemble butterfly wings; crustaceans that look like microscopic shrimp; miniature, jet-propelled animals that live in jellylike houses and filter food from water; and shimmering, crystal-shelled algae. The only feature common to all plankton is their inability to move consistently laterally through the ocean. However, many can and do move vertically in the water column. **Figure 14.4** provides a photograph of concentrated plankton.

The plankton contain many different plantlike species and virtually every major group of animals. Thus, the term is not a collective natural category like mollusks or algae, which would imply an ancestral relationship among the organisms; instead it describes a common ecological connection. Members of the plankton community, informally referred to as **plankters,** can and do interact with one another. There is grazing, predation, parasitism, and competition among members of this dynamic group.

Collecting and Studying Plankton

The first researcher to appreciate the uniqueness of plankton was the Danish biologist Anders S. Oersted. He was prompted to look for tiny drifters in the ocean after noticing microscopic algae that discolored water in an aquarium. His observations, made during a voyage from Denmark to the West Indies in 1847, revealed that the open ocean was not devoid of plant life. He correctly concluded that these single-celled autotrophs were the primary food source for most animal life in the ocean. His brief report, written in Danish, was generally overlooked by the scientific community for almost 50 years. The first large-scale, systematic study of plankton was carried out by biologists aboard the research vessel *Meteor* during the German Atlantic Oceanographic Expedition of 1925–26. Many of the tools and techniques they pioneered are still in use today.

Plankton nets (Figure 14.5) of the type perfected for the *Meteor* Expedition are essential to plankton studies. These conical nets are customarily made of nylon or dacron cloth woven in a fine interlocking pattern to assure consistent spacing between threads. The net is hauled slowly for a known distance behind a ship, or it is cast to a set depth; then it is reeled in. Trapped organisms are flushed to the net's pointed end and carefully removed for analysis. Quantitative analysis of plankton

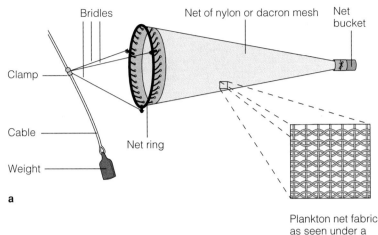

Figure 14.5 Plankton nets. (a) The standard conical net is made of fine mesh and has a mouth up to 1 meter (3.3 feet) in diameter. The net is towed behind a ship for a set distance. The number of organisms present in the water can be estimated if the trapped organisms are counted and the volume of sampled water is known. A plankton net's filtering efficiency (the effectiveness with which it removes organisms from the intended water sample) must be taken into account when calculating the volume of water sampled. An abundance of large organisms or great numbers of individuals can sometimes clog the net, reducing its filtering efficiency.

(b) The net shown here has a somewhat coarser mesh because its target organisms, small shrimplike crustaceans known as krill, are relatively large.

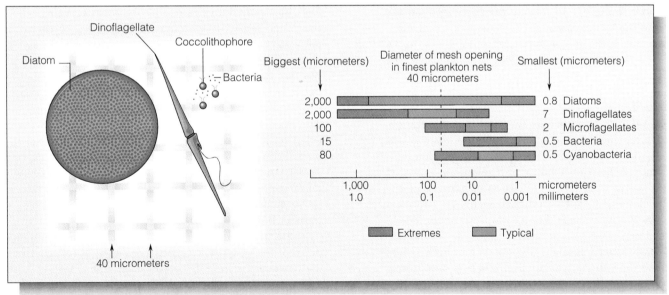

(c) Sizes of individual plankton relative to the mesh of a fine plankton net. Note that coccolithophores and bacteria (as well as viruses) will easily pass through the net.

requires both a count of the organisms and an estimate of the sampled volume of water.

Very small plankton can slip through a plankton net. Their capture and study requires either concentration by centrifuge or entrapment by a plankton filter through which water is drawn. The filter is later disassembled and the plankton studied in place.

Measurements of physical ocean conditions—such as dissolved carbon dioxide and oxygen content, pH, temperature, and light intensity at the time and place of sampling—are of special importance in interpreting the samples. **Synoptic sampling** (*syn* = together + *optikos* = to see), simultaneous sampling at many locations, can be useful in pinpointing the often subtle interplay between species and physical conditions that complicates our understanding of plankton biology.

Phytoplankton: The Autotrophs

Autotrophic plankton are generally called *phytoplankton*, a term derived from the Greek word *phyton*, meaning "plant." A huge, nearly invisible mass of phytoplankton drifts within the sunlit surface layer of the world ocean. Phytoplankton are critical to all life on Earth because of their great contribution to food webs and their generation of large amounts of atmospheric oxygen through photosynthesis. Planktonic autotrophs are thought to bind *at least* 20 billion metric tons of carbon into carbohydrates each year, at least 30% of the food made by photosynthesis on Earth! These easily overlooked, mostly single-celled, drifting photosynthesizers are much more important to marine productivity than the larger and more conspicuous seaweeds.

There are eight major types of phytoplankton, of which the most important are the dinoflagellates, the diatoms, and the coccolithophores.

Dinoflagellates As a child I remember walking along a dark California beach where the sand in my footprints glowed after each step. Handfuls of sand thrown out over the wet beach surface burst into fans of blue light strong enough to cast a shadow, and swimmers emerged from the warm water with thousands of glowing points of light clinging to their bodies. The light I saw was caused by dinoflagellates that had multiplied rapidly in the nearshore waters off southern California. In the late 1950s, wastewater treatment was not as thorough as it is today, and the effluent was not pumped as far out to sea. The combination of warm, calm water and abundant nutrients often triggered such *blooms*. These displays are rare today; but on warm September or October nights after the northeast winds have blown the surface water out to sea and upwelling has brought nutrients to the shore, the sand sometimes glows, and waves can flash blue as they break.

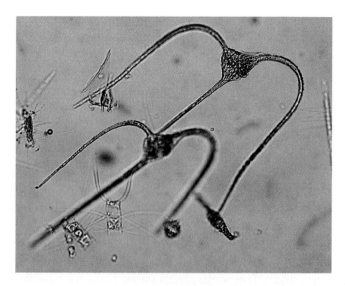

Figure 14.6 Two representatives of the genus *Ceratium*, a common dinoflagellate. Two flagella beat in opposing grooves in the armor-plated body; one groove is visible at the equator of the upper cell, between the two "horns." These specimens are about 0.5 millimeter (0.02 inch) across.

Most **dinoflagellates** are single-celled autotrophs (**Figure 14.6**). A few species live within the tissues of other organisms (the zooxanthellae of coral animals you will meet in Chapter 18, for example), but the great majority of dinoflagellates live free in the water. Most have two whiplike projections, called **flagella,** in channels grooved in their protective outer covering of cellulose. One flagellum drives the organism forward while the other causes it to rotate in the water. (Hence the name: *dino* = whirling, *flagellum* = whip.) Their flagella allow dinoflagellates to adjust orientation and vertical position to make the best photosynthetic use of available light.

Dinoflagellates are widely distributed, solitary organisms that reproduce by simple fission; they rarely form colonies. During reproduction, the cellulose covering that surrounds most species splits, and the single cell divides in half. Each daughter subsequently replaces the missing portion of covering. Under favorable conditions the organisms can reproduce once a day, growing in number but not in size.

Some species of dinoflagellates can become so numerous that the water turns a rusty red as light reflects from the accessory pigments within each cell. These species are responsible for the phenomenon of *red tide* (see **Box 14.1**). During times of such rapid growth (usually in springtime), concentration of these microscopic organisms may briefly reach 6 million per liter (23 million per gallon).

Some dinoflagellates are **bioluminescent** (*bios* = life, *luminis* = light), as I learned in my nighttime walks on

BOX 14.1 ● *Red Tides*

The somewhat misleading name *red tide* is used to describe patches of water turned a rusty red by the abundant growth of pigmented phytoplankton, usually certain dinoflagellate species (**Figure a**). Red tides occur when water runoff from land, coastal upwelling, or seasonal warming stimulates the growth and reproduction of these dinoflagellates. Though red tide is not a tidal phenomenon, the organisms constituting it can benefit from the stirring of nutrients that tidal flow can provide.

Although the dinoflagellates responsible for most red tides are comparatively simple organisms, they are unusual in their ability to synthesize potent toxins as by-products of metabolism. Among the most effective poisons known, these toxins (if ingested) may affect nearby marine life (**Figure b**) or even humans. Some of the toxins are similar in chemical structure to the muscle relaxant curare, but they are tens of times more powerful.

One of these compounds, a neurotoxin (nervous system poison), affects between 10,000 and 50,000 individuals annually—most of them on tropical and subtropical islands where herbivorous fishes make up much of the diet. The fishes graze on seaweeds, inadvertently ingesting the dinoflagellates attached to the surface of the fronds. The toxin is stored in fat and so is most concentrated in the oldest and fattest (and thus most desirable) fish. Victims complain of abdominal pain, muscular aches, dizziness, anxiety, tingling in the hands and feet, and a curious inversion of the senses (hot seems cold, and vice versa); many will die from the effects.

Clams, mussels, scallops, and oysters can also be dangerous; and shellfish poisonings are becoming increasingly common in North America. These illnesses result when people consume invertebrates that filter seawater to obtain food. Paralytic, diarrhetic, neurotoxic, and amnesic shellfish poisonings have been reported. A lethal dose of paralytic toxin kills by interfering with the normal function of heart muscle and affecting breathing. Neurotoxic compounds block communication between nerves and muscles; the first symptoms of intoxication are usually tingling lips and difficulty in swallowing. Diarrhetic poisoning causes diarrhea, nausea, and vomiting. Most mysterious is amnesic poisoning, a poorly understood disorder that results in the permanent loss of short-term memory. In 1987 three people in eastern Canada died, and 105 were affected by amnesic poisoning. A rare species of diatom has been implicated in this and similar incidents.

a

b

Red tides and toxic dinoflagellates. (a) Scanning electron micrograph of *Gymnodinium breve*, the dinoflagellate responsible for the outbreaks of red tides along the Florida coast. Other dinoflagellate species produce red tides in other parts of the world. (b) Portion of a fish kill that resulted from a dinoflagellate bloom.

Humans should avoid eating filter feeders during summer months, when toxin-producing phytoplankton are most abundant. If shellfish from a particular area are unsafe, a governmental agency will issue an advisory, which may remain in effect for six weeks or more until the danger is past. Such warnings should be observed; a single clam can accumulate enough toxin to kill a human. Whales and seabirds are also at risk from dinoflagellate toxins.

People can be affected even if they don't eat seafood. Beachgoers in North Carolina complained of asthma-like symptoms when they inhaled dinoflagellates that had dried on the beaches and been blown inland by strong winds.

The number and severity of red tide outbreaks and poisonings appears to be increasing. Perhaps this is not surprising: Coastal waters receive industrial, agricultural, and domestic wastes, rich in plant nutrients that stimulate dinoflagellate growth. Also, the long-distance transport of algal species in the ballast water of cargo vessels can introduce alien species into coastal waters, where they may thrive in the absence of the organisms that naturally consume them. Australia has recently issued strict guidelines for discharging ballast in the country's ports. Perhaps other nations should adopt similar restrictions. In any case, seafood consumers should be aware of the source of their meal and the nature of poisoning symptoms.

the beach. "Waste" oxygen, left over from photosynthetic reactions, combines with enzymes and an imaginatively named protein, *luciferin*, causing the organism to glow briefly. The release of light is not accompanied by a release of heat; so the organism does not overheat in the oxidation process.

Diatoms   Dinoflagellates are an old and successful group; they have lived in the ocean for more than a billion years. But the dominant photosynthetic organisms in plankton are the more biochemically sophisticated **diatoms.** Compared to the dinoflagellates, diatoms evolved comparatively recently. Diatoms began to dominate phytoplanktonic productivity in the Cretaceous period about 100 million years ago. Their abundance and photosynthetic efficiency further increased the proportion of free oxygen in Earth's atmosphere. More than 5,600 species of diatoms are known to exist. The larger species are barely visible to the unaided eye. Most are round, but some are elongated or branched or triangular.

Typical diatoms are shown in **Figures 14.7** and **14.8.** The name means "to cut through" (*dia* = through, *tomos* = to cut), a reference to the patterns of perforations through the diatom's rigid cell wall, or **frustule.** As much as 95% of the mass of the frustule consists of silica (SiO_2), giving this beautiful covering the optical, physical, and chemical characteristics of glass—an ideal protective window for a photosynthesizer. Magnification reveals that the frustule consists of two closely fitting halves, or **valves,** which fit together like a well-made gift box, the top valve adhering tightly over the lip of the bottom one. The pattern of perforations, slits, striations, dots, and lines on the surface of the valves is different for each diatom species (see Figure 14.8).

Inside the diatom's tailored valves lies the most nearly perfect photosynthetic machine yet to evolve on the planet. Fully 55% of the energy of sunlight absorbed by a diatom can be converted into the energy of carbohydrate chemical bonds, one of the most efficient energy conversion rates known. Excess oxygen not needed in the cell's respiration is released through the perforations in the frustule into the water. Some oxygen is absorbed by marine animals, some is incorporated into bottom sediments, and some diffuses into the atmosphere. Most of the oxygen we breathe has moved recently through the glistening pores of diatoms.

For more effective light absorption, chlorophyll is accompanied in diatoms by accessory pigments called **xanthophylls.** These yellow or brown pigments give most diatoms a yellow-green or tan appearance. Diatoms store energy as fatty acids and oils, compounds that are lighter than their equivalent volume of water and thus assist in flotation. As you might guess, flotation is a potential problem for diatoms because the weight of their heavy silica frustule seems at odds with their need to stay near the sunlit ocean surface. Oil floats, glass sinks, and a balanced amount of both produces neutral buoyancy. Not all diatoms need to float, however. Many nonplanktonic species lie on shallow bottoms, where light and nutrients are able to support photosynthesis. These benthic species are nearly always elongated (or *pennate*) in shape.

Like most single-celled organisms, diatoms reproduce by dividing in half and drifting apart (or, in the species of diatoms that form chains, remaining linked in long lines of cells). The cells may divide as rapidly as once each day. In most species the new valve is generated *within* the old one during division; so the average size of individuals in the population becomes smaller with time (see **Figure 14.9**). Individuals reach a minimum of about one-fourth the size of the original cell. When the cells become too small—or, more accurately, when the cells have too high a ratio of glass to living tissue—they become too heavy to float in spite of their buoyant oils. The problem is solved by sexual reproduction, which generates an **auxospore**, a naked cell without valves. If conditions are still favorable for growth, the auxospore will expand to the diatom's original size, form two thin new valves, and begin the cycle anew. If growing conditions are unsuitable, the auxospore will become dormant to await an opportunity for growth—which may be weeks or months away. When diatoms die, their valves can drift to the seafloor to accumulate as layers of siliceous ooze (see Chapter 5).

Coccolithophores and Other Phytoplankton   Most other types of phytoplankton are extraordinarily small and so are called **nanoplankton** (*nanus* = dwarf). The **coccolithophores,** for example, are tiny single cells covered with disks of calcium carbonate (coccoliths) fixed to the outside of their cell walls (**Figure 14.10**). Coccolithophores live near the ocean surface in brightly lighted areas. The translucent covering of coccoliths may act as a sunshade to prevent absorption of too much light. In areas of high coccolithophore productivity, most notably in the Mediterranean and Sargasso seas, their numbers occasionally become so great that the water appears milky or chalky. Coccoliths can also build seabed deposits of ooze. The famous White Cliffs of Dover in England consist largely of fossil coccolith ooze deposits uplifted by geological forces.

Other very small phytoplankters, the **silicoflagellates,** possess filamentous internal supporting structures of silica, the same material from which diatoms construct their valves. One or two flagella are always present. Compared to diatoms, however, the overall structure and biochemistry of silicoflagellates seem primitive. Lit-

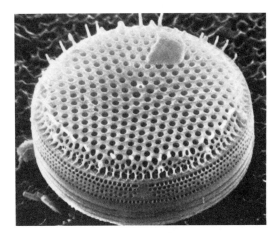

a

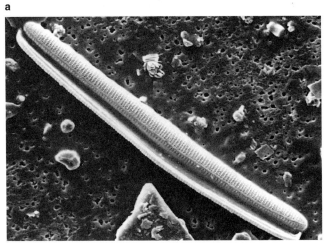

b

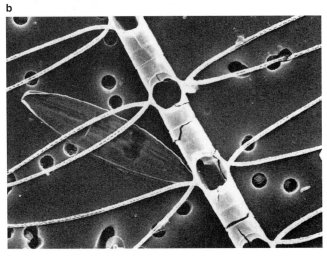

c

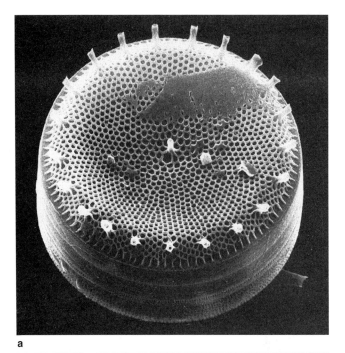

a

b

Figure 14.7 Diatoms. (a) Scanning electron micrograph of the upper valve ("shell") of *Thalassiosira,* a centric (round) diatom about 500 micrometers in diameter. Note the many fine holes in the silica-rich structure. (Figure 14.8b shows the valve perforations of a similar species in more detail.) (b) Scanning electron micrograph of *Nitzschia,* a pennate (elongated) diatom. (c) *Chaetoceros,* a chain-forming centric diatom. Each cell is about 15 micrometers (300 millionths of an inch) across.

Figure 14.8 The valve, or "shell," of the diatom *Coscinodiscus.* (a) A scanning electron micrograph of the beautiful silica frustule. The valve is about 750 micrometers (0.03 inch) in diameter. (b) A closer view, showing the intricate perforations (2340×).

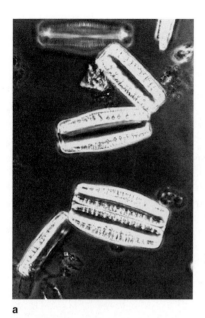

a

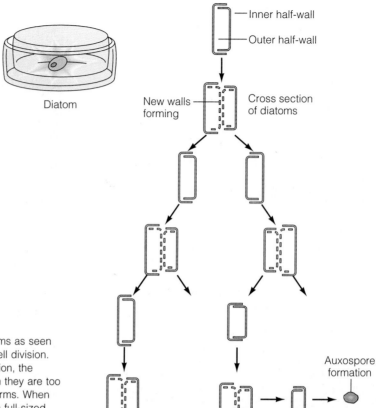

Inner half-wall
Outer half-wall

New walls forming
Cross section of diatoms

Auxospore formation

b

Figure 14.9 How diatoms divide. (a) Dividing diatoms as seen with a light microscope. (b) Diatoms reproduce by cell division. Because a new inner shell is formed after each division, the diatoms become smaller with each generation. When they are too small to permit further reproduction, an auxospore forms. When conditions are suitable, it will germinate to produce a full-sized diatom.

Figure 14.10 Calcium carbonate plates in place on *Umbilicosphaera*, a coccolithophore. The tiny plates are 6 micrometers (120 millionths of an inch) across, about four times larger than large ultraplankton.

tle is known about their worldwide distribution or abundance. As **Box 14.2** suggests, recent evidence suggests that silicoflagellates, coccolithophores, and other, even tinier photosynthesizers may contribute much more to overall plankton productivity than previously thought.

The Depth of Greatest Productivity

As we have seen, phytoplankton photosynthesis is inhibited by the bright light at the ocean surface. Phytoplankters are adapted to operate optimally using less light than is found at the surface; this permits greater productivity throughout the euphotic zone as a whole. Though it is difficult to generalize for all the ocean, the depth at which phytoplankton productivity is often greatest is typically around 20 meters (66 feet) at local noon, when illumination is greatest and the depth of maximum productivity the deepest. Figures averaged for a whole day indicate that the depth of greatest productivity is between 5 and 10 meters (17 and 33 feet) below the surface, about one-fifth the depth of the photic zone (see **Figure 14.11**).

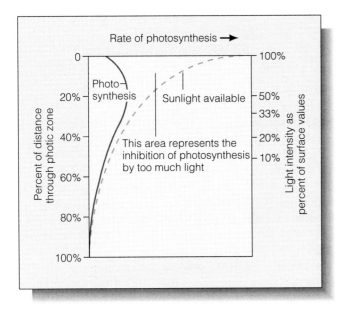

Figure 14.11 The rate of photosynthesis as a function of depth. The photic zone may extend to a depth of 100 meters (330 feet) in clear oceanic waters (as shown here) or to less than 20 meters (66 feet) in turbid nearshore waters. Note that too much light inhibits photosynthesis near the surface. As the depth increases, the available sunlight decreases, but the plants become more efficient at using it. Maximum productivity (the highest net rate of photosynthesis) takes place about 20% of the way through the photic zone.

Compensation Depth

Recall that autotrophs respire as they photosynthesize; thus, they use some of the carbohydrates and oxygen they produce. Carbohydrate production usually exceeds consumption, but not always. The deeper a phytoplankter's position, the less light it receives. At a certain depth, the production of carbohydrates and oxygen by photosynthesis through a day's time will exactly equal the consumption of carbohydrates and oxygen by respiration. This break-even depth is called the **compensation depth.** Compensation depth usually corresponds to the depth to which about 1% of surface light penetrates. If bottle pair #5 in Figure 14.2 were at compensation depth, the amount of oxygen produced in the transparent bottle would exactly equal the amount consumed (3 milliliters), an indication of zero net productivity.

Figure 14.12 shows compensation depth graphically. Like the depth of greatest productivity, compensation depth changes with sun angle, turbidity, surface turbulence, and other factors. Remember that the compensation depth is always below the depth of greatest productivity and that producers can still "make a profit" between the depth of greatest productivity and the compensation depth. If a producer slips below its compensation depth for more than a few days, however, it will consume its carbohydrate reserves and die. For this rea-

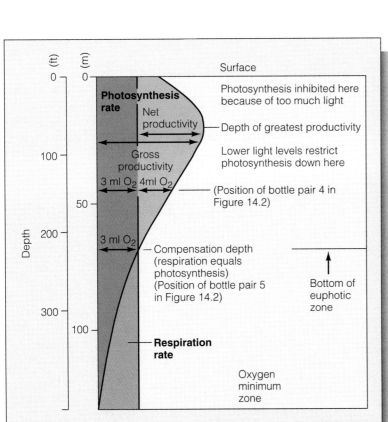

Figure 14.12 Compensation depth and its relationship to other aspects of productivity.

BOX 14.2 ● *Ultraplankton!*

In the mid-1970s marine biologists began to appreciate the contribution to oceanic productivity of a new class of phytoplankton they termed *ultraplankton*. **Ultraplankton** are the smallest of the small—diminutive nano-plankters so tiny that they were long mistaken for fine particles of suspended sediments. These organisms are too small to be resolved by light microscopes and slip undetected through all but the very finest filters. Their size, typically about 2 micrometers (40 millionths of an inch) across, is made up for by their abundance: an astonishing 100,000 cells per milliliter (3 million cells per ounce) in tropical and subtropical waters previously thought to be all but sterile.

Synechococcus is typical of this newly recognized type of organism. It was discovered when individual cells—little more than naked photosynthetic machines—fluoresced a bright orange when struck by ultraviolet light. Analysis of the fluorescence spectrum (living examples of *Synechococcus* are too small to study directly) revealed the presence of an odd chlorophyll variant previously known only in a rare mutant strain of corn. Structural details revealed by electron micrographs suggest that *Synechococcus* and its relatives may be cyanobacteria. Even smaller photosynthesizers have been discovered, some less than 1.5 micrometers (30 millionths of an inch) in diameter.

Recent estimates suggest that ultraplankton may account for up to 70% of all the photosynthetic activity of the world ocean! How could such a huge contribution to oceanic productivity go unnoticed for so long? In part because these autotrophs are so extremely small, and in part because the products of their photosynthetic activity are closely cycled to even smaller heterotrophic bacteria in the immediate vicinity! Here is a complete microecosystem—a community operating on the smallest possible scale—that manufactures and consumes carbohydrates and carbon dioxide in amounts almost beyond comprehension, a sort of ecological black market below the "official economy" of the relatively huge diatoms and dinoflagellates. As if they weren't busy enough, these bacteria also decompose organic material (spilled into the water when phytoplankton are eaten by zooplankton), clean up the liquid excretions of zooplankton dissolved in seawater, and process the small amounts of cytoplasm that phytoplankton exude into the ocean as they age. Many marine biologists now believe that the greatest fraction of organic particles in the water column of the open ocean is composed of these metabolically active bacterial cells—the ultraplankton.

And what happens to these bacteria? In an 1989 article in the British journal *Nature*, biologists announced the discovery of *still smaller* organisms, viruses less than 0.2 micrometer (4 millionths of an inch) in diameter. These viruses—a class of simple creatures known as *bacteriophages*—inject their genetic material into a bacterium and trick it into manufacturing more viruses. The bacteria die in the process. Researchers reported finding viruses in numbers that stagger the imagination: up to 100 million viruses per milliliter (nearly 3 billion per ounce). Ultraplankton, indeed!

Source: Bergh, O., K. Y. Borshem, F. Bratbak, and G. Heldal. 1989. "High Abundance of Viruses Found in Aquatic Environments." *Nature* 340 (no. 6233): 467–68.

Ultraplankton. (a) Bacteriophages. These viruses inject their genetic material into a bacterium and trick it into manufacturing more viruses. (b) Three bacteria—ellipsoidal, S-shaped, and C-shaped—from the Gulf of Bothnia (east of Sweden). The dark dots within the bacteria are viruses. The large hexagonal structure near the S-shaped bacterium is a relatively large virus with a diameter of 160 nanometers. The bacteria and viruses were photographed using a transmission electron microscope.

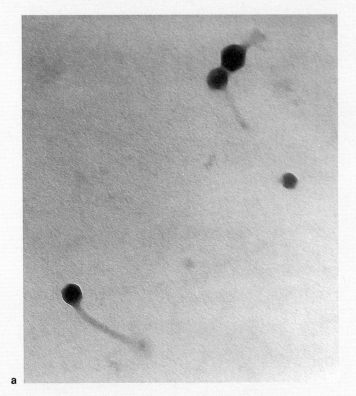

a

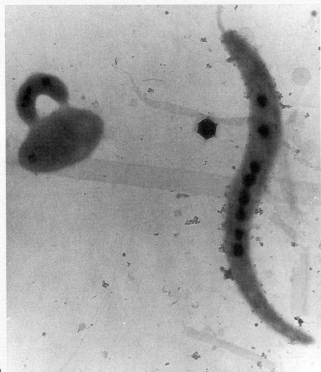

b

son diatoms contain a small quantity of lightweight oil that facilitates flotation, and dinoflagellates have whirling flagella to propel themselves toward optimal light.

Because of their greater efficiency, diatoms have a deeper compensation depth than dinoflagellates. Open tropical seas have the deepest potential compensation depths, but these regions are not generally productive because of nutrient deficiencies.

Global Distribution of Plankton Productivity

Where is phytoplankton productivity the greatest? This question is among the most important in marine biology. Since phytoplankton form the base of nearly all oceanic food webs, the biological characteristics of any ocean area will depend heavily on the presence and success of phytoplankton.

Because of coastal upwelling and land runoff, nutrient levels are highest near the continents. Plankton are most abundant there, and productivity is highest. The water above some continental shelves sustains productivity in excess of 1 $gC/m^2/day!$ But what of the open ocean? Where is productivity greatest away from land?

In the Tropics? The open tropical oceans have abundant sunlight and CO_2 but are generally deficient in surface nutrients because the strong thermocline discourages the vertical mixing necessary to bring nutrients from the lower depths. The tropical oceans away from land are therefore oceanic deserts nearly devoid of visible plankton. The typical clarity of tropical water underscores this point. In most of the tropics productivity rarely exceeds 30 $gC/m^2/yr$, and seasonal fluctuation in productivity is low.

Tropical coral reefs are exceptions to this general rule. Reef areas, which account for less than 2% of the tropical ocean surface, are productive places because autotrophic dinoflagellates live within the tissues of coral animals and don't drift randomly as plankton. Nutrients are made available by coastal upwelling and by the coral's own metabolism. These nutrients are cycled tightly through the reef and do not sink.

In the Polar Regions? At very high latitudes, the low sun angle, reduced light penetration due to ice cover, and weeks or months of darkness in winter severely limit productivity. At the height of summer, however, 24-hour daylight, a lack of surface ice, and the presence of upwelled nutrients can lead to spectacular plankton blooms. The surface of some sheltered bays can look like tomato soup because dinoflagellates and other plankton are so abundant. This bloom cannot last, however, because nutrients are not quickly recycled and because the

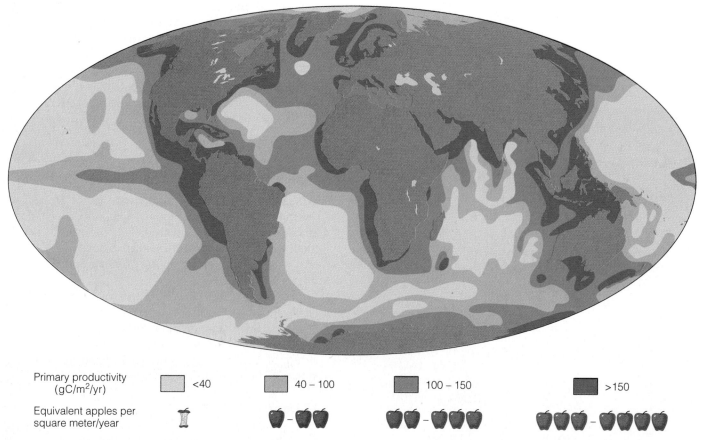

Primary productivity (gC/m²/yr): <40 | 40 – 100 | 100 – 150 | >150

Equivalent apples per square meter/year

Figure 14.13 Distribution of primary productivity in the world ocean.

sun is above the critical angle for a few weeks at best. The short-lived summer peak does not compensate for the long, unproductive winter months.

Average productivity at very high northern latitudes tends to be lower than at high southern latitudes, averaging less than 25 gC/m²/yr. The Arctic Ocean is almost surrounded by land masses that limit water circulation and, therefore, nutrient upwelling. The southern ocean, on the other hand, is enriched by water upwelling to replace sinking Antarctic Bottom Water. This rich mixture is stirred by the West Wind Drift. The Antarctic accounts for a much greater share of high-latitude production than the Arctic because more nutrients are available.

In the Temperate and Subpolar Zones? The tropics are generally out of the running because of nutrient deficiency, and the north polar ocean suffers from slow nutrient turnover and low illumination; so the overall productivity prize goes to the temperate and southern subpolar zones. Thanks to the dependable light and moderate nutrient supply, annual production in the nearshore temperate and southern subpolar ocean areas is the greatest of any open-ocean area. Typical productiv-

ity in the temperate zone is about 120 gC/m²/yr. In ideal conditions southern subpolar productivity can approach 250 gC/m²/yr!

Figure 14.13 shows the levels of productivity in tropical, temperate, and polar ocean areas. Note that *nearshore* productivity is nearly always higher than *open-ocean* productivity, even in the relatively productive temperate and south subpolar zones.

Curiously, the open-ocean area with the greatest annual productivity is an exception to the general picture developed in this section. The slender, cold finger of high productivity pointing west from South America along the equator is a result of wind-propelled upwelling due to Ekman transport on either side of the geographic equator (see again Figures 9.12 and 9.15). Look for this area in Figure 14.13.

Phytoplankton Productivity Through the Seasons

Figure 14.14 diagrams the relationship of phytoplankton biomass to season. The low, flat line representing annual tropical productivity contrasts with the high, thin peak representing the Arctic summer. The higher of the two

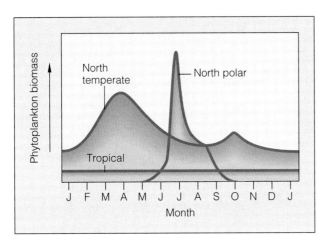

Figure 14.14 Variation in the biomass of phytoplankton by season and latitude. (Note that the total area under each curve represents productivity.)

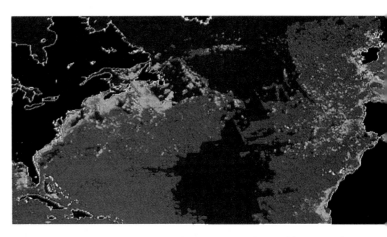

a Photosynthetic activity in winter. (In this color-enhanced image, red-orange shows where chlorophyll concentrations are greatest.)

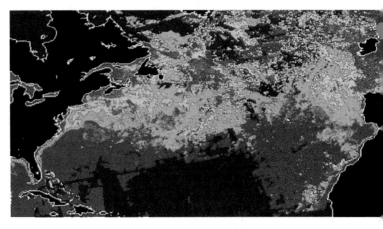

b Photosynthetic activity in spring.

Figure 14.15 Seasonal variation in phytoplanktonic photosynthetic activity in the North Atlantic, derived from Coastal Zone Color Scanner data acquired during the winter (January–March) and spring (April–June) of 1979 and 1980. (a) From the temperate regions to the poles, the nutrient supply of the upper ocean is replenished through the deep vertical mixing driven by surface cooling and intense winter storms. (b) As the seasons progress from winter to spring, the amount of sunlight available for photosynthesis increases, and the water stabilizes. The resultant period of exuberant growth is called the *spring bloom*. Red and orange colors indicate areas of high phytoplankton concentration; yellow and green represent areas of moderate concentration. Note the continuously high phytoplankton biomass in the temperate coastal zones and major fishing grounds. Black areas in the Atlantic represent areas where no data were available.

peaks for the temperate zone indicates the plankton bloom of northern spring, caused by increasing illumination; the smaller temperate zone peak, representing the northern fall, is caused by nutrients upwelling toward the surface.

Figure 14.15 dramatically shows the seasonal variation in phytoplankton biomass for the North Atlantic nearshore temperate zone. The spring bloom extends from North Carolina to Spain!

Estimates of World Productivity

Figure 14.13 showed marine primary productivity for various ocean areas. Estimates of total world primary productivity—and of the proportion of world productivity contributed by the ocean—have changed considerably over the years. The first closely reasoned estimates (made after the First World War but before *Meteor* scientists issued their reports on the nature and prevalence of marine plankton) suggested that land productivity greatly exceeded marine productivity. After *Meteor*, and continuing until the early 1970s, estimates of marine productivity steadily rose as scientists learned more about diatom efficiency, the rate of nutrient turnover, and the great biomass of phytoplankton. By 1970 W. D. Russell-Hunter, an expert in aquatic productivity, gave an upper estimate of world ocean productivity as 200 gC/m2/yr and considered marine phytoplankton to be responsible for "between 75% and 80% of the world's organic productivity." In 1974 V. J. Chapman, an algal physiologist, went even further in suggesting that "the marine planktonic algae . . . have been estimated to carry on about 90% of all photosynthetic activity on Earth."

These high estimates of oceanic productivity were later revised downward. An estimate compiled by Matthew Dring and published in 1982 reversed the land–sea proportion, attributing 71.6% of world productivity to terrestrial autotrophs. Dring's figures combine a much higher estimate for land productivity with a slightly lower estimate for marine productivity. The dramatic differences among these figures show the level of uncertainty surrounding such estimates.

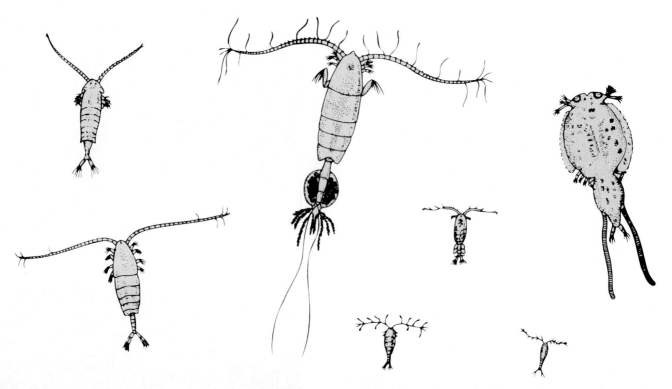

Figure 14.16 Zooplankton: assorted marine copepods. The largest of the animals pictured here is about 1 millimeter (1/25 inch) long.

More recent studies and satellite observations of ocean chlorophyll content, especially in nearshore areas, suggest that the pendulum may be swinging back toward the idea of oceanic dominance in productivity. The contribution of autotrophic nanoplankton to total productivity is now being reevaluated. As noted in Box 14.2, initial indications are that the role of cyanobacteria may have been seriously underestimated. And as we shall see in a moment, seaweeds may also account for a greater percentage of world primary productivity than previously considered. Many specialists now believe that total land and sea productivity are approximately equal.

Zooplankton: The Heterotrophs

Although this chapter emphasizes the producers, we will also spend a moment with the heterotrophic plankton, often collectively called **zooplankton** (*zoion* = animal). Zooplankters are the most numerous primary consumers of the ocean. They graze on the diatoms, dinoflagellates, and other phytoplankton at the bottom of the trophic pyramid in a way analogous to cows grazing on grass. The variety of zooplankton is surprising; nearly every major animal group is represented. Each is expert at the painstaking concentration of food from water. The most abundant zooplankters, accounting for about 70% of individuals, are tiny shrimplike animals called *copepods*

(**Figure 14.16**). Copepods are crustaceans, a group that also includes crabs, lobsters, and shrimp.

Not all zooplankters are small, however. Many are in the 1- to 2-centimeter (½- to 1-inch) size range. The largest drifters are giant jellyfish of genus *Cyanea;* their bells may be more than 3.5 meters (12 feet) in diameter! We have a special term for plankton larger than about 1 centimeter (½ inch) across: **macroplankton** (*makros* = large).

Most zooplankton spend their whole lives in the plankton community; so we call them **holoplankton** (*holos* = entirely). But some planktonic animals are the juvenile stages of crabs, barnacles, clams, sea stars, and other organisms that will later adopt a benthic or nektonic (swimming) lifestyle. These temporary visitors are **meroplankton** (*meros* = mixed). Most animal groups are represented in the meroplankton—even the powerful tuna serves a brief planktonic apprenticeship. These useful categories can be applied to phytoplankton as well as zooplankton. Holoplanktonic organisms are by far the more numerous forms of both phytoplankton and zooplankton.

Particularly lovely holoplanktonic organisms are the **radiolarians,** which are relatives of the amoeba. Like the amoeba they are able to move their protoplasm in streams, but unlike the common amoeba they possess spikelike silica projections on which the flowing occurs.

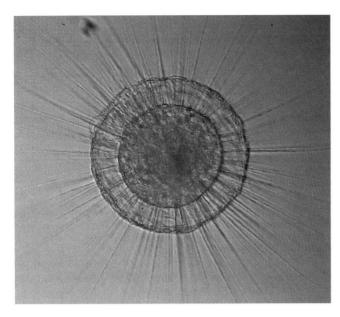

a

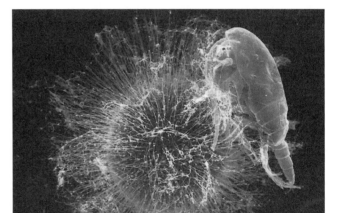

b

Figure 14.17 A type of radiolarian with needlelike pseudopods radiating from its body like rays of the sun.

Patterns of light refracting through the protoplasm and silica needles can be beautiful when viewed with a microscope (**Figure 14.17**).

Other forms related to amoebas are **foraminifera** (**Figure 14.18**), which also extend long protoplasmic filaments. Most foraminifera have calcium carbonate shells. As we saw in Chapter 5, extensive white deposits of calcareous ooze have been built on the seabed from their skeletons. As was the case with some phytoplankton sediments, some of these layers have been uplifted and can be found on land.

Plankton and Food Webs

Zooplankton and other animals eat phytoplankton, and still other animals eat these primary consumers. The mass of zooplankton is typically about 10% that of phytoplankton, which is reasonable because of the harvesting relationship that exists between them. The zooplankters depend on phytoplankton for both food and oxygen. The activity of zooplankton often creates an **oxygen minimum zone** below the well-lighted surface zone: Oxygen is depleted by the animals there and not replaced by phytoplankton (see Figure 14.12). Some species of zooplankton and a number of kinds of small swimming animals make nightly migrations from the oxygen minimum zone toward the darkened surface layer to feed on the smaller organisms drifting there.

It is interesting to note that the largest marine animals, such as whale sharks (fish) and baleen whales

Figure 14.18 Foraminiferans. The word means "bearers of windows." (a) Light streams through thin parts of the shells. (b) *Orbulina* feeding on a copepod (compare with Figure 14.16). The foram's spines are made of calcium carbonate and have a sticky layer on their surfaces. Zooplankton that bump into the spines stick to them and are subsequently digested.

(mammals), do not expend their energy tracking down and attacking big animals. Instead these largest of all feeders concentrate zooplankton from the water and consume it in vast quantity. The zooplankton they eat are not usually the primary consumers but the somewhat larger **secondary consumers,** usually crustaceans such as krill that have themselves fed on the microscopic primary consumers. In this way whales and other large filter feeders can harvest energy closer to the source, gaining the advantage of efficiency and quantity.

LARGER MARINE PLANTS

We now shift our attention from drifters to attached autotrophs. Attached plants account for between 2% and 10% of the ocean's total primary productivity.

Almost all marine autotrophs, large or small, are algae. **Algae** is a collective term for autotrophs possessing chlorophyll and capable of photosynthesis but lacking vessels to conduct sap. The single-celled diatoms and dinoflagellates discussed earlier are classified as **unicellular algae**. *Seaweed* is the informal name for large, marine **multicellular algae**. A few species of marine plants are not algae but **angiosperms** (*angios* = covered, *sperma* = seed); these are flowering plants. The main groups of marine angiosperms are sea grasses and mangrove trees. Marine angiosperms are not considered seaweeds.

Vascular Versus Nonvascular Plants

Anyone who has enjoyed maple syrup on pancakes knows something about sap. Land plants are loaded with sap, which they pump between roots and leaves through special vessels. Bundles of these vessels are visible, as the tubes found within most plant stems or as the veins in leaves. Nearly all land plants larger than a thumbtack have such conducting vessels and thus are known as **vascular plants.**

Why vessels and sap? In land plants, two of the four ingredients needed for photosynthesis are in the air (carbon dioxide and sunlight) and two are in the soil (water and inorganic nutrients). But for photosynthesis to proceed, all four must be present at the same place—in the leaves. A maple tree absorbs water and nutrients from the soil, and its vessels transport them to the leaves. There, in the presence of chlorophyll, they combine with carbon dioxide from the air and energy from the sun. After photosynthesis has bound carbon dioxide into sugars and other carbohydrates, separate vessels take some of the carbohydrates down to the roots to "pay" them for their efforts. These carbohydrates give maple syrup its sweetness.

Algae are **nonvascular plants;** that is, they do not have vessels. Algae require the same four ingredients for photosynthesis as vascular plants, but in their case the ingredients are already present in one location. Either the alga is very small (perhaps a single cell) and lives in a moist spot on land where conditions are ideal, or it lives in water.

Multicellular marine algae occur in a great variety of sizes and shapes. The largest can reach 62 meters (205 feet) in length; the smallest appear as smears of cells on the surface of rocks. Some types of algae form underwater forests; others grow in isolation. Lifeless seaweeds drying on shore communicate none of their natural beauty and grace to the beachcomber; but to a diver the marine forest from which they came reveals extraordinary grace and form. It discloses sheltering nurseries, conceals complex interrelationships, and even provides part of a sushi lunch. None of these plants grows below the euphotic zone because all depend on photosynthesis to produce the energy-rich compounds necessary for life. Nearly 7,000 species of multicellular marine algae have been identified.

The Problems of Large Marine Plants

At first glance marine plants appear to have an easy life, but living near the ocean surface is not without hazards. Intertidal plants may find themselves exposed to the drying effects of air and sunlight when the tide is out, and they may be lashed against the rocks by waves when the tide is in. The physical nature of the plants themselves provides some defense against these difficulties. Their bodies are flexible, easily able to absorb shock, resistant to abrasion, streamlined to reduce water drag, and very strong. Their surfaces are often covered by a slick, mucilaginous material that lubricates them as they move and that retards drying and deters grazing animals.

Marine algae have internal fluids nearly isotonic to seawater, but marine angiosperms contain fluids hypotonic to seawater; water tends to flow from their cells to the ocean. Sea grasses and mangroves must expend much energy to retain water in their tissues. Shallow-water species are often exposed to freshwater runoff from land, but most marine plants can cope with only moderate changes in salinity.

Thermal stress is more limiting. Higher temperatures lead to higher metabolic rates, and the oxygen level in warm seawater may not be high enough to support the respiratory needs of the plants at night. Warmer temperatures can also shatter the delicate accessory pigments and proteins required for algal photosynthesis and respiration. Small increases in temperature may prevent reproduction in some species of algae.

Nutrients are also limiting. Large algae are rare in warm, nutrient-poor waters, and divers visiting the tropics are usually surprised to find no sign of kelp forests. By contrast, chilly temperate and subpolar zones of nutrient upwelling often support thick algal mats and dense marine forests.

Yet another difficulty is the location of adequate anchorage or substrate for these large plants. Attached seaweeds require a stable footing. A sandy or muddy bottom is unsuitable for colonization by most large algae, and less than 2% of the ocean floor is shallow enough and solid enough to permit the growth of these attached plants. Mangroves and most sea grasses, on the other hand, require a fine, silty substrate.

The marine life-style also offers some advantages. Marine plants suffer no droughts and nearly always have enough carbon dioxide for photosynthesis. Assuming suitable nutrient levels and a good foothold, only sunlight is required for productivity. Being submerged in seawater brings the additional advantage of lightweight construction. A seaweed doesn't require strong support structures because it has nearly the same density as the surrounding seawater. More of its bulk can thus be dedicated to photosynthesis. Indeed, as we shall see, productivity in some seaweed beds may be the highest of *any* autotrophic community on Earth.

Structure of Seaweeds

Common terms like *leaf, stem,* and *root* are inappropriate for seaweeds because the definitions of those parts assume the presence of vessels. The structures in nonvascular plants that superficially resemble leaves are called **blades** (or fronds), the stemlike structures are termed **stipes,** and the root-shaped jumble at the plant's base is appropriately named a **holdfast. Gas bladders** assist many species in reaching strongly illuminated surface water. Blades, stipes, and holdfast comprise the body of the plant, the **thallus.** These parts are labeled in **Figure 14.19.**

A thallus may be large or small, branching or tufted, in sheet form or filamentous, encrusting or elongated, rounded or pointed—algal variety is tremendous. Algal blades are symmetrically equipped with photosynthesizing tissue; they absorb gases across their entire surface and even participate in reproduction. Stipes are strong, photosynthesizing, shock-absorbing links tying the blades (or sheets or filaments) into a unit. The holdfast does not take up water and nutrients from the substrate as does the vascular root it superficially resembles, but it does anchor the plant in place and may provide incidental shelter for a rich variety of animal life. The gas bladders range in size from tiny, grapelike bunches to single volleyball-sized floats. The flotation gas is usually nitrogen, oxygen, and argon in proportions similar to those in air; but the floats of one species of genus *Pelagophycus* contain more than 2% carbon monoxide, an amount sufficient to kill a mouse if it were placed within the structure. The metabolic pathways leading to this gas are presently unknown.

Classification of Seaweeds

The presence of accessory pigments was first used to classify large marine algae by William Harvey in 1836.

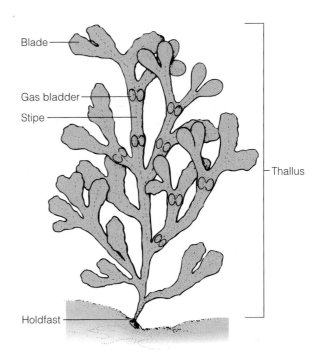

Figure 14.19 The parts of a multicellular alga.

Though group names have changed and other diagnostic characteristics have been added, Harvey's segregation of multicellular marine algae into three divisions based on their color is still employed today. The green algae, with their unmasked chlorophyll, are the **Chlorophyta** (*chloros* = green, *phyton* = plant); the brown algae **Phaeophyta** (*phae* = tan, dusky); and the red algae **Rhodophyta** (*rhodon* = rosy red).

The Chlorophytes The clear green color of chlorophytes is evidence of their lack of accessory pigments and suggests that they live at or near the surface, where red light is available. Land plants are thought to have evolved from chlorophyte ancestors. Only about 10% of the 7,000 species of Chlorophyta are marine, but some of those species are widely distributed and present in great numbers. Genus *Ulva* (**Figure 14.20**) is a familiar, delicate, lettucelike edible plant able to tolerate the often impure waters of urban coasts. *Ulva* makes quick use of concentrated nutrients near sewage outfalls to grow and multiply. Most other species of green algae prefer clean water and are branched or threadlike. Some encrust hard surfaces, and some live on sand, but none reaches the size of the large brown seaweeds, the phaeophytes.

The Phaeophytes Virtually all of the 1,500 living species of phaeophytes are marine. Some species of these largest of algae, which include the **kelps,** can reach lengths of 40 meters (132 feet); the record length exceeds

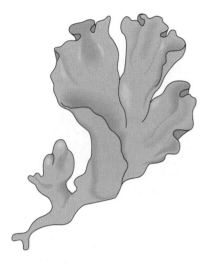

Figure 14.20 *Ulva*, a marine chlorophyte.

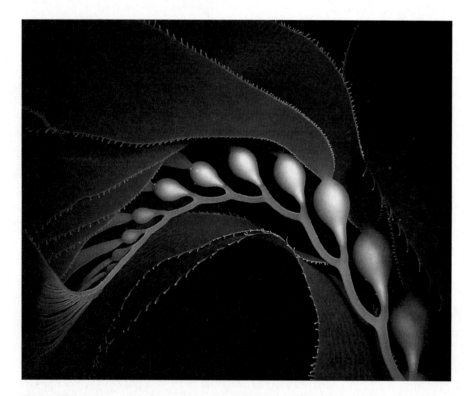

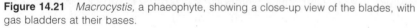

Figure 14.21 *Macrocystis*, a phaeophyte, showing a close-up view of the blades, with gas bladders at their bases.

60 meters (200 feet). To attain these dimensions the plant can grow at the spectacular rate of 50 centimeters (20 inches) per day! Part of the strategy of rapid growth is to reach bright surface water as soon as possible. Some brown algae are annuals; others live for up to seven years. The tan or brown color of phaeophytes comes from the accessory pigment **fucoxanthin,** which permits photosynthesis to proceed at greater depths than is pos-

sible for the unmasked chlorophytes. In ideal circumstances some larger brown algae can grow in water up to about 35 meters (115 feet) deep.

The Pacific's giant kelp forests, the world's largest, consist mostly of the magnificent genus *Macrocystis* (*macro* = large, *cyst* = bladder). The size and shape of the plant and its dense canopy are shown in **Figure 14.21.** Most brown algae live in temperate and polar habitats

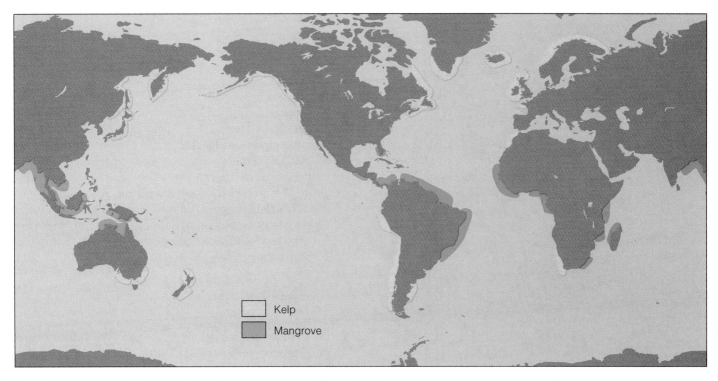

Figure 14.22 Distribution of kelp beds and mangrove communities worldwide.

poleward of the 30° latitude lines, but a few live in the tropics. The worldwide distribution of kelp is shown in **Figure 14.22.**

Not all seaweeds are attached by tenacious holdfasts to rocks on the bottom. Millions of tons of the genus *Sargassum,* named after the Portuguese word for the grapes its gas bladders resemble, grow in huge, free-floating masses in the central North Atlantic gyre—the so-named Sargasso Sea—where water is too deep for attachment. These weedy clumps are often home to a wide variety of dependent organisms. *Sargassum* may have changed history. Columbus's men were near mutiny in 1492 when clumps of these seaweeds were sighted. The mariners assumed that land must be nearby because all the seaweeds they knew were nearshore species, so they calmed down in anticipation of the end of the voyage. Little did they know they were still three weeks from landfall! Similar floating rafts of *Sargassum* are also found in the Pacific. Attached forms of *Sargassum* are common in harbors in Japan and California, where they secure themselves to rocks, pilings, or ship hulls (**Figure 14.23**).

The Rhodophytes Most of the world's seaweeds are red algae; there are more rhodophyte species than all other major groups of algae combined, about 4,000 species. Rhodophytes tend to be smaller and more anatomically and biochemically complex than phaeophytes. Red algae do inhabit high latitudes, but they thrive in warmer mid- and low-latitude areas, where they usually outnumber all other algal species. Rhodophytes excel in dim light because of their sophisticated accessory pigments, reddish proteins called **phycobilins.** These compounds absorb and transfer enough light energy to power photosynthetic activity at depths where human eyes cannot see light. The record depth for a photosynthesizer is held by a small rhodophyte discovered in 1984 at a depth of 268 meters (879 feet) on a previously undiscovered seamount in the clear, tropical Caribbean. The deepest rhodophytes grow very slowly and may be tens or even hundreds of years old.

Paradoxically, many red algae also live on rocks right at the water's surface. Surface light contains all colors of the spectrum; so the color-shifting capabilities of accessory pigments are not needed there. It is believed that the dark accessory pigments may act to shade the productive machinery from brilliant light. Most of the common surface-dwelling "reds" grow as a purple or pink film on rocks, shells, other seaweeds, and even glass or plastic. These coralline, or calcareous, algae look like a brilliant ceramic glaze, a bit of melting raspberry sherbet, or a knobby plastic membrane thrown carelessly over the surface. The name of one very common form,

Figure 14.23 *Sargassum*, a phaeophyte. Attached forms of *Sargassum* are common in temperate Pacific harbors. The fragments seen here are about 8 centimeters (3 inches) long.

genus *Lithophyllum*, comes from the Latin terms for "stony" and "leaf," an apt description of an alga able to deposit calcium carbonate within its tissues (**Figure 14.24a**). Others grow in erect branching or bushy forms (**Figure 14.24b**). The gritty calcium in these plants may deter grazing animals.

Another group of shallow rhodophytes is very important in the life of coral reefs. Like coral animals, encrusting coralline algae of genus *Lithothamnium* (**Figure 14.24c**) also remove large quantities of dissolved calcium carbonate from seawater and deposit it within their tissues. This activity helps cement the reef into a mass capable of resisting heavy surf. Some reefs in the Indian Ocean are not, as was once thought, of coral origin but rather were formed almost exclusively by the activity of coralline algae.

Seaweed Zonation

Zones or bands of dominant seaweeds are plainly visible on the sloping nearshore seafloor to the limit of light penetration. The vertical distribution of large marine plants depends in part on the pigments they possess. Chlorophytes lack accessory pigments and are rarely found more than 10 meters (33 feet) below the surface. Phaeophytes can operate from the surface down to about 30 meters (100 feet). Rhodophytes, as we have seen, can function from the surface to the photosynthetic record depth of 268 meters (879 feet).

Factors other than the response of pigments to light are also important in determining where these algae will

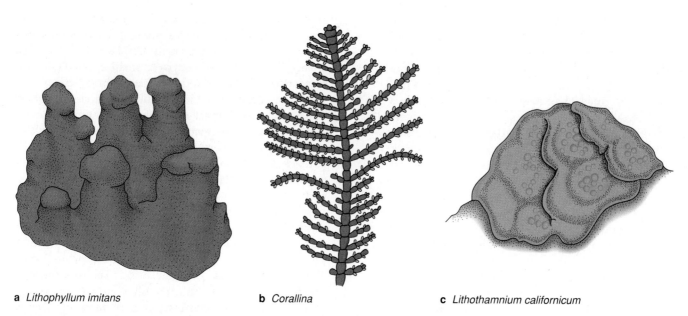

a *Lithophyllum imitans* **b** *Corallina* **c** *Lithothamnium californicum*

Figure 14.24 Various rhodophytes. *Lithophyllum* (a) and *Lithothamnium* (c) are encrusting species. *Corallina* (b) is an erect species.

prosper, especially in the intertidal zone. Wave shock, varying salinity and pH, the presence of grazers from land and sea, the drying effects of sun and wind, abrasion by waterborne particles of sand—all these factors and many more affect the location of intertidal plants. The resulting effects of these factors on marine plants (and animals) can be seen at low tide by any observant beachcomber. Look for bands of different marine algae (and angiosperms) the next time you go to the shore (see Figure 17.6).

Reproduction of Seaweeds

Any idea that seaweeds are simple plants evaporates when we study their life cycles. Many seaweeds reproduce in a cycle called **alternation of generations,** in which a sexual reproducing stage alternates with a separate asexual reproducing stage. The asexual reproducing plants possess a complete set of chromosomes. Cells on their blades divide to produce two kinds of spores, cells with half the normal number of chromosomes. These spores drift into the plankton, settle, and grow into two kinds of sexual reproducing plants, which in most species bear no resemblance to their parents. One of these kinds of plants then produces male gametes (equivalent to sperm cells); the other produces female gametes (equivalent to eggs). The gametes are released into the phytoplankton, where they meet and fuse at fertilization. The resulting cell grows into an asexual reproducing plant to complete the cycle.

In most chlorophytes the two generations look almost exactly alike. Most phaeophytes, however, including giant kelp, have a tiny sexual stage. Many brown algae were classified as two or three different species before researchers recognized the distinct plants as alternate forms of one species! Red algae have an even more complex reproductive cycle, which usually involves three distinct phases rather than two.

Marine Angiosperms

Angiosperms are advanced vascular plants that reproduce with flowers and seeds. Most large land plants are angiosperms. Relatively few angiosperms live in water; the advantages conferred by vessels and roots are largely unnecessary in an aquatic environment. However, a few species of angiosperms have recolonized the ocean. All of these have descended from land ancestors, and all live in shallow coastal water. Angiosperms live at the surface, where the red light required for photosynthesis is abundant; they have no need for accessory pigments, and their chlorophyll is unmasked. The most conspicuous marine angiosperms are the sea grasses and the mangroves.

Figure 14.25 The sea grass *Phyllospadix* in a tide pool.

Sea Grasses Many people lump **sea grasses** in the informal seaweed group, but their resemblance to large marine algae is only superficial. These plants are not true grasses, but they do have leaves and stems, as well as roots capable of extracting nutrients from the substrate. Extensive stands of sea grasses are found on the coasts of North America, on the Atlantic coast of Europe, in Eastern Asia, in temperate Australia, and in South Africa. They form broad gray or green submerged meadows, which support extraordinarily rich communities of heterotrophs (**Figure 14.25**). The life cycle of sea grasses is much like that of other angiosperms, but their stringy pollen is distributed by flowing water rather than by insects or wind. About 45 species of sea grasses are known.

The most common sea grass is eelgrass, genus *Zostera* (**Figure 14.26a**), a common inhabitant of the muddy shallows of calm bays and estuaries of the U.S. East and West coasts. Similar habitats along the Gulf Coast harbor stands of turtle grass (genus *Thalassia*) and manatee grass (genus *Syringodium*), angiosperms named after the animals that once shared their habitat and (in the case of the manatee) fed on them.

Perhaps the most beautiful sea grass is the vivid, emerald-green surf grass, genus *Phyllospadix*, with its seasonal flowers and fuzzy fruit (Figure 14.25 and **Figure 14.26b**). These hardy plants survive in the turbulent, wave-swept intertidal and subtidal zones of temperate East Asia and western North America.

Sea grasses can be remarkably productive. Some are capable of binding 1,000 gC/m^2/yr, three to five times the nearshore average for phytoplankton. One explanation for this efficiency is the advantage conferred by roots. Anaerobic bacteria in subtidal mud have been shown to bind dissolved nitrogen into nitrates easily available to these plants.

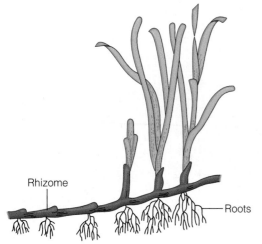

Rhizome

Roots

a *Zostera*

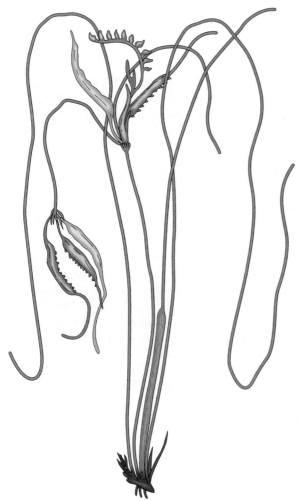

b *Phyllospadix*

Figure 14.26 Although they are called grasses and resemble seaweeds, sea grasses are neither. These flowering vascular plants are not true grasses. (a) *Zostera*, or eelgrass. (b) *Phyllospadix*, or surfgrass.

Mangroves Low, muddy coasts in tropical and some subtropical areas are often home to tangled masses of trees known as **mangroves.** These large, flowering plants are never completely submerged, but because of their intimate association with the ocean we consider them to be marine plants (**Figure 14.27**). They thrive in the sediment-rich lagoons, bays, and estuaries of the Indo-Pacific, tropical Africa, and the tropical Americas. Their distribution depends on temperature, currents, and rainfall (see again Figure 14.22).

The sediment in which mangrove trees live must be covered with brackish or salt water for part or all of the day. Many mangroves avoid taking up salt ions from seawater, or they selectively remove salt from sap with salt-excreting cells. The fine coastal mud they colonize doesn't provide firm footing for these substantial plants; so an intricate network of arching prop roots is required for support. The strutlike prop roots are supplemented by many smaller roots equipped with breathing pores and air passages. Atmospheric oxygen is conducted by these passages to the portions of the plant submerged in oxygen-deficient mud.

The root system also traps and holds sediments around the plant by interfering with the transport of suspended particles by currents. Mangrove forests thus assist in the stabilization and expansion of deltas and other coastal wetlands. The root complex also forms an impenetrable barrier and safe haven for organisms around the base of the trees.

The mangrove communities of south Florida, among the world's most widespread, consist primarily of the red mangrove (genus *Rhizophora*). The spreading leaves of the tree protect the residents from the tropical sun, and the tangled roots keep out large predators and also harbor huge numbers of fiddler crabs, worms, marine and terrestrial snails, fish, oysters, and red algae. Birds and insects inhabit the treetops, their droppings enriching the sediments below.

Mangrove seeds germinate on the trees. If the tide is out when the seed drops, the force of the fall will plant the seed in the muck, where it will continue to grow. If the tide is in, the seed will drop into the water, suspend growth, float to a new location, and resume growth when it gets a foothold in the mud. The trees mature in 20 to 30 years and can attain a height of 8 to 10 meters (26 to 33 feet). New mangroves colonize mud and sandbars away from shore; they stabilize them and eventually create new land.

Larger Marine Plants, Productivity, and Food Webs

What is the contribution of large marine plants to worldwide primary productivity? As we saw in our discussion of phytoplankton, estimates vary with the era and the

Figure 14.27 A mangrove (*Rhizophora*) growing in salt water in Everglades National Park, Florida. The tangled roots descending from the main branch are called *prop roots* or *stilt roots*. They provide anchorage for the mangrove, trap sediment, and provide protection for small organisms.

estimator. The proportion of marine productivity assigned to large marine plants has traditionally been about 2%, but K. H. Mann suggested in 1973 that the contribution of sea grasses and multicellular algae to oceanic productivity may have been underestimated. Mann reported productivity of 1,000 gC/m^2/yr in communities of Gulf Coast turtle grass (genus *Thalassia*), and his studies of a Nova Scotian bay indicated that seaweed productivity was three times that of the resident phytoplankton. Productivity within the seaweed zone itself averaged 1,750 gC/m^2/yr!

More recent studies have shown that intertidal kelp communities in Washington State may also be highly productive, fixing carbon at a rate of more than 1,000 gC/m^2/yr. Roger Lewin reported that these small, wave-swept algal communities have 10 times the productivity of a tropical rain forest and perhaps 20 times the productivity of nearshore phytoplankton on a unit area basis. Under ideal conditions, sea palms—a phaeophyte, genus *Postelsia* (**Figure 14.28**)—can bind an astonishing 14,600 gC/m^2/yr! Intense wave action in their habitat apparently facilitates nutrient flow, optimizes the distribution of light on photosynthetic tissue, speeds gas exchange, and deters grazers and predators. These data suggest that despite the relatively small coastal areas available to them, larger marine plants may account for up to 10% of total marine productivity. What to believe? Earlier we said 2% to 10%—probably as good a guess as any!

What happens to the organic material produced by large marine autotrophs? Some of the energy is transmitted to heterotrophs when the body of the plant is eaten directly by grazing animals, but most of the mass of attached marine plants enters the food web after the plant dies. The plant's decomposing mass is ground into fine particles by the action of surf and sediments. Bacteria and scavengers digest the bits and pieces and then pass the energy to other organisms when they themselves are eaten.

Organic molecules produced by marine algae can also enter the food web in a much less conventional way. Many species of large phaeophytes are surprisingly leaky; carbohydrates and other products of photosynthesis diffuse from their blades and stipes like tea from a teabag. Up to one-half of all organic matter produced by these large plants can be lost in this way. The foam visible in surf near kelp beds is produced in part by these substances. Sea urchins and other heterotrophs can absorb these molecules directly through their skin or outer membranes, "feeding" on kelp plants that may be tens or even thousands of meters away.

COMMERCIAL IMPORTANCE

Chances are good that you have had some recent contact with marine autotrophs even if you haven't been in the ocean. The mucilaginous material that is so effective in making algal blades slick, in lowering friction, and in deterring grazers is also harvested and made into an important commercial product called *algin*. When separated and purified, its long, intertwining molecules are used to stiffen fabrics, make adhesives, suspend water and oil together in salad dressings, prevent the formation of gritty crystals in ice cream, clarify beer and wine, and manufacture shoe stains, soaps, and shaving cream. Fast-food restaurants are using carageenan, a similar sea-

Figure 14.28 *Postelsia palmaeformis*, a deceptively delicate-looking, palm-shaped phaeophyte capable of extra-ordinarily high productivity in high-energy environments.

weed extract, to replace some of the fat in newly popular healthier hamburgers. These substances also prevent fire-extinguishing foams from dispersing; and they permit chocolate milk to remain on the refrigerator shelf without separating and keep the abrasives in liquid car waxes from settling to the bottom of the bottle. In biological laboratories bacteria are cultured on agar made from seaweed extracts. Very likely even the ink that forms the letters you are now reading has an algin or carageenan component!

Seaweeds are eaten directly, too. Animals raised for food or fur eat the plant material remaining after the mucilaginous components are removed. Seaweeds' mineral content and roughage are also useful in human nutrition, and human consumption of seaweed is especially high in East Asian countries; the Japanese consume more than 150,000 metric tons of *nori* annually, primarily in the form of sushi. Seaweed and seaweed extracts are also eaten in Britain, Ireland, and New Zealand.

Seaweeds are big business. Algal products were key substances in the production of over $36 billion worth of products in 1993. The future may be even brighter. The complex biochemistry of the brown and red algae holds much promise in pharmaceutical research. Seaweed-based drugs against parasitic infections, thyroid imbalance, kidney disease, high blood cholesterol, hypertension, and heavy metal poisoning have already been developed.

Large phaeophytes are harvested by bargelike ships equipped with a submerged array of hedge-trimmerlike blades (see Figure 19.19). Kelp is clipped about a meter (3.3 feet) below the surface and brought on deck by a moving conveyor. The plants grow rapidly back to their original length.

Plankton also has commercial importance. When geological forces lift ancient oozes above sea level, the beds can be mined as a source of diatomaceous earth for swimming pool filters, polishing abrasives, and paint filler. Production of Diatomite, the commercial material, was valued at $200 million in 1993.

Q-AND-A

1. Phytoplanktonic organisms account for a significant percentage of world primary productivity. Why are phytoplankton so successful? Is there something about the planktonic life-style that lends itself to high efficiency?

   Phytoplankton are successful because of their size. Because water supports them, small marine autotrophs don't require the elaborate support systems of large terrestrial plants. Their small size allows easy diffusion of required nutrients into their single cells and prompt transport of wastes out; vessels and sap are not needed. Nearly all of their tissue is photosynthetic.

   With transparent siliceous valves to protect them, diatoms are especially successful. The relatively recent evolution of diatoms was a high point in the development of marine autotrophs, and the additional oxygen these organisms contributed to the atmosphere has greatly influenced the success of life on land. More plants have led to more animals. After a few hundred million years of evolution, the planktonic life-style is elegantly tuned and interlocked into the complex and productive system we now observe in the world as a whole.

2. Do zooplankton have a compensation depth? If so, would it be above, below, or at the same level as the compensation depth of most phytoplankters?

   The concept of compensation depth is meaningless for zooplankton. Compensation depth applies only to autotrophs,

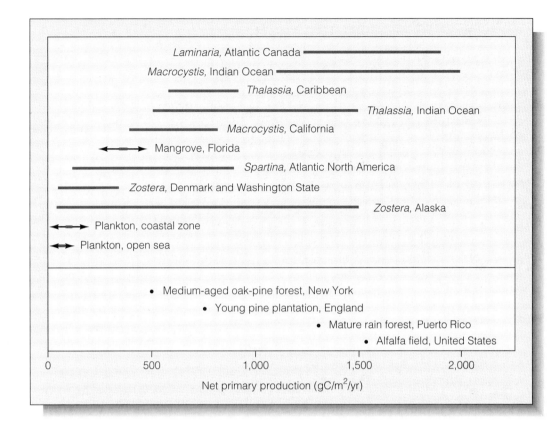

Figure 14.29 The productivity of various large marine plants compared with that of some terrestrial communities. Broken lines are estimates based on biomass data.

organisms capable of photosynthesis and *respiration. Since animals aren't autotrophic, productivity in them can never equal consumption.*

3. Why is the compensation depth for diatoms usually lower than that for dinoflagellates?

   Because diatoms are typically more efficient. That is, a diatom can usually accomplish the same amount of primary production with less light than a dinoflagellate can.

4. Does that mean diatoms are always found in a lower layer than dinoflagellates?

   Not always. When both are present they are usually mixed together. Diatoms tend not to tolerate bright light as well as dinoflagellates, however, and so may be absent at the surface.

5. If scientists have found as many as 100 million viruses in 1 milliliter (one-fifth of a teaspoon) of seawater from the open ocean, does this mean I'll get a bad cold—or worse—whenever I go swimming?

   No. These viruses specifically target the bacteria and cyanobacteria of the ultraplankton. They inject their own genetic material into these tiny cells and "trick" the cells into replicating viruses rather than cells. They cannot infect humans.

6. Isn't it too cold for seaweeds to survive at high latitudes?

   No. As with phytoplankton, seaweed productivity can be high regardless of water temperature. Indeed, kelp in some Canadian bays grows most rapidly when water is near 0°C (32°F). The important limiting factors are the few hours of daylight in winter months, the relative lack of nutrients in the north polar ocean, and the abrasive grinding of ice in nearshore areas.

7. Can algae to assist in the fight against world hunger?

   Nutritious, fast-growing foodstuffs are in great demand. It is more efficient to turn for food to productive organisms at the base of the food web than to top consumers. The most productive large plants in the world are phaeophytes. Remember, under ideal conditions sea palms (genus Postelsia) can bind an astonishing 14,600 gC/m^2/yr! Though marine algae are not particularly high in calories, it is possible that these plants, properly exploited, could relieve a great deal of human suffering.

8. How does the productivity of large marine plants compare with the productivity of large land plants?

   Look at Figure 14.29. Note that an alfalfa field, the most productive terrestrial community listed, is not as productive as Canadian Laminaria, a brown alga studied by Mann. Other contrasts in the chart are also interesting. The 14,600 gC/m^2/yr figure for Postelsia mentioned in the text was not known when this chart was drawn in 1973. That figure would be clear off the chart!

Terms and Concepts to Remember

accessory pigment
algae
alternation of
  generations
angiosperm
auxospore
bioluminescence
biomass
blade
Chlorophyta
coccolithophore
compensation
  depth
diatom
dinoflagellate
flagellum
foraminifera
frustule
fucoxanthin
gas bladder

holdfast
holoplankton
kelp
macroplankton
mangrove
meroplankton
multicellular algae
nanoplankton
nonconservative
  nutrient
nonvascular plant
oxygen minimum
  zone
Phaeophyta
phycobilin
phytoplankton
plankter
plankton
plankton bloom
plankton net

primary
  productivity
radiolarians
Rhodophyta
sea grass
seaweed
secondary
  consumer
silicoflagellate
stipe
synoptic sampling
thallus
ultraplankton
unicellular algae
valve
vascular plant
xanthophyll
zooplankton

Study Questions

1. What do primary producers produce? How is productivity expressed?

2. How can primary productivity be measured? Which method is considered the most accurate?

3. What factors limit productivity? What methods have marine producers evolved to cope with the lack of red light needed by chlorophyll for photosynthesis?

4. What are plankton? How are plankton collected? How are members of the plankton community differentiated? How are zooplankton different from phytoplankton?

5. Describe the most abundant and important phytoplankters. Which are most efficient in converting solar energy to energy in chemical bonds?

6. Is phytoplankton productivity highest at the ocean surface? What advantage would optimum productivity at a depth below the surface provide to phytoplankton?

7. What is compensation depth? What happens to phytoplankton below that depth? To zooplankton?

8. Where in the ocean is plankton productivity the greatest? Why?

9. How does a nonvascular plant differ from a vascular plant? Why are most marine plants nonvascular?

10. What are algae? Are all algae seaweeds? How are seaweeds classified? Which seaweeds live at the greatest depths? Why?

11. Give examples of marine angiosperms. Are they vascular or nonvascular plants? Are they considered seaweeds?

12. Of what commercial importance are marine plants?

For Further Study

On Plankton Biology and Estimates of General Productivity

Ackelson, S., W. M. Balch, and P. M. Holligan. 1988. "White Waters of the Gulf of Maine." *Oceanography* 1 (no. 2): 18–22. Coccolithophore blooms in the northwest Atlantic.

Anderson, D. M. 1994. "Red Tides." *Scientific American,* August, 62–68. Information about the organisms and the growing threat that red tides pose to marine and human health.

Banse, K. 1994. "Grazing and Zooplankton Production as Key Controls of Phytoplankton Production in the Open Ocean." *Oceanography* 7 (no. 1): 13–20. Phytoplankton–zooplankton interrelationships are more complex than generally assumed.

Chapman, V. J. 1974. "Algae." In *Encyclopædia Britannica,* vol. 1. Chicago: Encyclopædia Britannica. A very high estimate (90%) of phytoplankton's contribution to world productivity.

Devlin, R. M., and A. V. Barker. 1976. *Photosynthesis.* New York: Van Nostrand-Reinhold. Discussion of the structure and role of accessory pigments.

Dring, M. J. 1982. *The Biology of Marine Plants.* Contemporary Biology Series. London: Edward Arnold. Comprehensive review of estimates of worldwide primary productivity.

Grahame, J. 1987. *Plankton and Fisheries.* London: Edward Arnold. Summary of general plankton biology and the role of plankton in food webs.

Hardy, A. 1956. *The Open Sea.* Vol. 1, *The World of Plankton.* London: Collins. A classic, readable and enthusiastic. (American printing by Houghton Mifflin, Boston.)

Isaacs, J. D. 1969. "The Nature of Oceanic Life." *Scientific American,* September, 146–62. General discussion of marine trophic relationships.

Kirk, J. T. O. 1994. *Light and Photosynthesis in Aquatic Ecosystems.* 2d ed. Cambridge: Cambridge University Press.

Koblentz-Mishke, I. J., et al. 1970. "Plankton Primary Production of the World Ocean." In *Scientific Exploration of the South Pacific,* 183–93, edited by W. S. Wooster. Washington, DC: National Academy of Sciences.

Levinton, J. S. 1982. *Marine Ecology.* Englewood Cliffs, NJ: Prentice-Hall. Detailed discussion of phytoplankton ecology.

Malone, J. 1980. "Algal Size." In *The Physiological Ecology of Phytoplankton,* 433–63, edited by I. Morris. Berkeley: University of California Press. Suggests that nanoplankton make a much greater contribution to primary productivity than previously estimated.

Mays, B. 1976. "When the Sea Runs Red." *Oceans* 9 (no. 6): 52–56. Popular account of dinoflagellate blooms.

Milne, D. H. 1995. *Marine Life and the Sea.* Belmont, CA: Wadsworth. Complete general text in marine biology, with a complete section on plankton ecology.

Newell, F. E., and R. C. Newell. 1973. *Marine Plankton: A Practical Guide.* London: Hutchinson. Line drawings with aids to identification.

Platt, T., and S. Sathyendranath. 1988. "Oceanic Primary Production: Estimation by Remote Sensing at Local and Regional Scales." *Science* 241 (no. 4873): 1613–20. Satellite oceanography; remote sensing of chlorophyll.

Raymont, J. E. G. 1980. *Plankton and Productivity in the Oceans.* Vol. 1, *Phytoplankton.* 2d ed. Oxford: Pergamon Press. A definitive source, excellent bibliography and charts. Highly recommended for those wishing a deeper picture of planktonic concerns. Technical.

Round, F. E., R. M. Crawford, and D. G. Mann. 1990. *The Diatoms: Biology and Morphology of the Genera.* New York: Cambridge University Press. Technical treatise.

Russell-Hunter, W. D. 1970. *Aquatic Productivity.* New York: Macmillan. Probably the best general introduction to the logic and literature of this fascinating field. Lively writing and a genuine concern for productivity harnessed for human needs. Perhaps overly high estimates of phytoplankton's contribution to overall productivity.

Ryther, J. H. 1959. "Potential Productivity of the Sea." *Science* 130: 602–8; Reprinted in Hazen, W. E., ed. 1970. *Readings in Population and Community Ecology.* 2d ed. Philadelphia: Saunders. One of the classic papers in plankton biology.

Ryther, J. H. 1969. "Photosynthesis and Fish Production in the Sea." *Science* 166 (no. 3901): 72–76. An important historical paper giving initial estimates of worldwide productivity.

Smith, D. L. 1977. *A Guide to Marine Coastal Plankton and Marine Invertebrate Larvae.* Dubuque, IA: Kendall-Hunt. A laboratory guide to plankton.

Stanley, S. M. 1986. *Earth and Life Through Time.* New York: Freeman. Evolution of plankton, evidence in the geologic record.

Steeman-Nielsen, E. 1963. "Productivity: Definition and Measurement." In *The Seas,* vol. 2, edited by M. N. Hill. New York: Wiley.

Steeman-Nielsen, E. 1975. *Marine Photosynthesis.* New York: Elsevier.

Strickland, J. D. H., and L. D. B. Terhune. 1961. "The Study of *in situ* Marine Photosynthesis Using a Large Plastic Bag." *Limnology and Oceanography* 6: 93–96. There's nothing like going out and doing it, even though results are sometimes inconclusive.

Tait, R. V., and R. S. DeSanto. 1972. *Elements of Marine Ecology.* New York: Springer-Verlag.

Wickstead, J. H. 1965. *An Introduction to the Study of Tropical Plankton.* London: Hutchinson. Shows the highly ornamented tropical forms designed to retard sinking by elaborate projections and shapes.

Wickstead, J. H. 1976. *Marine Zooplankton.* London: Edward Arnold.

Wimpenny, R. S. 1966. *The Plankton of the Sea.* New York: Elsevier. Good companion volume to Hardy, and confined to plankton. A complete treatise for the layman. Highly recommended.

Oceanus 35 (no. 3, Fall 1992) is dedicated to the subject of biological oceanography and contains a number of important papers on the primary producers.

On Larger Marine Plants

Chapman, A. R. O. 1987. *Functional Diversity in Plants in the Sea and on Land.* Boston: James Bartlett.

Dawson, E. Y. 1966. *Marine Botany: An Introduction.* New York: Holt, Rinehart & Winston. The definitive reference by the acknowledged expert in the field.

Dawson, E. Y., and M. S. Foster. 1982. *Seashore Plants of California.* Berkeley: University of California Press.

Garrison, T. 1987. "Seaweeds." *California Diver,* September-October. General article for California divers.

Garrison, T. 1989. "Encrustations." *Pacific Diver,* May-June. The article includes information on encrusting rhodophytes.

Leigh, E. G., Jr., et al. 1987. "Wave Energy and Intertidal Productivity." *Proceedings of the National Academy of Sciences* 84: 1314. Turbulence increases nutrient and gas mixing, and therefore productivity.

Lerman, M. 1986. *Marine Biology: Environment, Diversity, and Ecology.* Menlo Park, CA: Benjamin/Cummings. Concise look at algal reproduction, information on marine angiosperms.

Lewin, R. 1987. "Life Thrives Under Breaking Ocean Waves." *Science* 235 (no. 4795): 1465–66.

Littler, M. M., et al. 1985. "Deepest Known Plant Life Discovered on an Uncharted Seamount." *Science* 227 (no. 4682): 57–59. A photosynthesizing rhodophyte at 268 meters!

Mann, K. H. 1973. "Seaweeds: Their Productivity and Strategy for Growth." *Science* 182 (no. 4116): 975–81. A rethinking of the contribution of large marine algae to coastal productivity. Includes comparisons of various terrestrial and marine productivity estimates.

McConnaughey, B., and R. Zottoli. 1983. *Introduction to Marine Biology.* 4th ed. St. Louis: Mosby.

Polunin, N. 1960. *Introduction to Plant Geography.* New York: McGraw-Hill. Distribution of mangroves and sea grasses.

Schmitz, K., and C. Lobban. 1976. "A Survey of Translocation in Laminariales (Phaeophyceae)." *Marine Biology* 36: 207–16. Movement of material through brown algae.

Sumich, J. L. 1988. *An Introduction to the Biology of Marine Life.* 4th ed. Dubuque, IA: W. C. Brown. Excellent general summary of marine plants.

The Times *Atlas of the World.* 1985. London: Times Books. Source of distribution of mangroves shown in Figure 14.22.

Zottoli, R. 1978. *Introduction to Marine Environments.* 2d ed. St. Louis: Mosby. Algal zonation.

This Old Shell

The mild, supportive ocean brims with animal life. **Animals** are active multicellular organisms incapable of synthesizing their own food. Animals must obtain food from primary producers—or from other animals that have consumed primary producers—to ensure their own survival.

All animal species encounter the same basic set of problems—to find food, avoid predators, and reproduce—and all have ways to solve these problems. It is the *variety* of survival strategies—adaptations to their environment—that makes marine animals so interesting. Each species, from sponge to whale, has a unique set of adaptations. It may sound like a circular argument, but by definition each living species is successful precisely because it *is* living: Its continuously evolving suite of adaptations has brought it through the rigors of

food finding, predator avoidance, and reproduction time and time again.

As a specific example, let's consider the everyday problems of one familiar marine animal, a hermit crab. These small, pleasant relatives of edible crabs and lobsters have engaged the attention of generations of intertidal visitors. The hermits rush around the floors and sides of tidal pools, withdrawing into their borrowed shells at the slightest sign of danger. Their activity appears random, but their fighting, snooping, hiding, probing, and scuffling is purposeful. Like all animals, hermit crabs must struggle to eat, avoid predators, and mate. Hundreds of structural and behavioral adaptations contribute to their success. Sensors on the hermit's antennae and mouthparts alert the animal to the presence of food; good eyesight, muscular coordination, and

A hermit crab surveys his domain.

a tough, formfitting covering usually foil fast-moving predators; and brilliant blue bands around the tips of the legs may signal the crab's availability for mating.

One unique behavioral adaptation shared by all hermit crabs involves the selection of a temporary home. The front parts of a hermit crab—mostly pincers, antennae, and mouthparts—are formidable, but its hindquarters are delicate and subject to attack. To protect his flank, a hermit searches for any enclosed portable object to climb into, usually an unoccupied snail shell. A hermit crab's borrowed or stolen shell is a source of both inordinate pride and perpetual difficulty. No shell is ever completely satisfactory. A hermit crab will carefully inspect any substitute dwelling, occupied or not, and consider whether to abandon his current digs in favor of the new candidate. House hopping proceeds fairly smoothly until the supply of suitably sized shells is exceeded by the number of potential occupants (as might happen when the number of crabs grows rapidly in times of abundant food). Then things get more serious. Snails can be evicted from their self-made homes even before they're through with them; in this case the hermit may get a house *and* a meal in a single transaction. Two crabs may fight for hours, even days, over one shell. Two others might simultaneously occupy the opposite ends of an abandoned worm tube, thus spending most of their time pulling each other in different directions. Sometimes two crabs swap shells at a moment's notice. Renter's remorse sets in almost immediately, and they're off to see if they can find something even more suitable. At times human observers can't resist laughing at the all-too-human goings-on.

But animals like the hermit crab teach us an important lesson: *No matter how improbable a structure or behavior seems, it must benefit the species in some way—or it would not be present.* In the long run, every knob, feeler, and movement helps an organism to succeed.

The history of animals begins far back in the history of the ocean. The path leading to today's formidable array of animals began with the availability of food.

CHAPTER OVERVIEW

Animals are heterotrophs; that is, they cannot synthesize their own food and must ultimately depend on primary producers (autotrophs) for nutrition. Single-celled, animal-like organisms evolved when increasing levels of free oxygen in the atmosphere permitted them to metabolize food obtained from autotrophs. True multicellular animals arose around 700 million years ago, near the end of the "oxygen revolution." More than 10 million species of animals are now thought to exist on Earth.

Biologists classify animals and other living things using a hierarchical system based on common ancestry. The invertebrates (animals without backbones) and vertebrates (animals with backbones) are grouped by similarities in external appearance and internal architecture into groups called *phyla* (singular, *phylum*). Porifera, Cnidaria, Platyhelminthes, Nematoda, Annelida, Mollusca, Arthropoda, and Echinodermata are the major phyla of marine invertebrates. Each phylum has characteristics that differentiate it from all others.

Figure 15.1 Ancient marine animals. (a, b) From Australia's Ediacara Hills, fossils of segmented marine animals about 600 million years old. (c) From the Burgess Shale of British Columbia, a marine worm about 530 million years old. In basic form, it resembles species alive today. (d) A 500-million-year-old trilobite fossil. Trilobites were once very abundant in the ocean, but all have died out.

THE ORIGIN OF ANIMALS

As you may recall from Chapter 1, the first organisms to evolve on Earth were probably tiny creatures adept at absorbing organic molecules that formed spontaneously in the ocean. Primitive life-forms could use the energy stored in these food molecules for growth and reproduction. Competition for food increased as these organisms grew more numerous. Had a photosynthetic mechanism not evolved, life would probably have died out when the supply of usable energy-rich molecules in the environment was exhausted. Using sunlight, the first simple autotrophs assembled their own food from inorganic molecules and then broke the food down to release energy. With the success and proliferation of simple autotrophs, and with the abundance of oxygen they provided, the way was clear for the evolution of animals.

The first animal-like creatures were single-celled organisms. They began to prosper in the ocean during **the oxygen revolution,** a time of radical change in the Earth's atmosphere. During the oxygen revolution, between about 2 billion and 400 million years ago, the activity of photosynthetic autotrophs changed the composition of the atmosphere from less than 1% free oxygen to its present oxygen-rich mixture of more than 20%. The growing abundance of free oxygen made it possible for heterotrophs to complete the disassembly of food molecules obtained by eating the autotrophs. Ozone derived from this oxygen blocked most of the sun's dangerous ultraviolet radiation from reaching the Earth's surface, permitting life to survive at the surface of the ocean and, later, on land.

Animals grew in complexity as they became more abundant. Instead of drifting apart after reproduction, some dividing cells stuck together and formed colonies. True animals evolved as these colonies distributed labor among specialized cells, eventually increasing the degree of interdependence among cells within the colony. The colonies ceased to be simple aggregations of individuals and began to take on specific architectures for specific tasks.

A group of animals that shares similar architecture, level of complexity, and evolutionary history is known as a **phylum** (plural, *phyla; phylon* = tribe). No one knows how many phyla of animals may have developed during the time of rapid animal proliferation that occurred near the end of the oxygen revolution. Small but fascinating "snapshots" of early marine life are preserved in fossils found in the Ediacara Hills of Australia, in the Burgess Shales of British Columbia, and in the Chengjiang beds of southwestern China. These sites were once parts of the

Figure 15.2 Diversity in marine invertebrates. (a) Gooseneck barnacles await the return of high tide to resume feeding. (b) A sea star prowls its domain. (c) A small crab inspects sediment for food. (d) An anemone's stinging cells are visible as light-colored dots on its tentacles.

warm, sediment-covered continental shelves of equatorial land masses. Animals inhabiting these places were abruptly buried—perhaps by turbidity currents—and their delicate features preserved. The animals in **Figures 15.1a** and **15.1b,** about 600 million years old, show evidence of segmented body plans; the 530-million-year-old worm in **Figure 15.1c** is similar in outline to species alive today. Not all groups of early marine animals survived. Once abundant, trilobites like the one shown in **Figure 15.1d** have all perished. Indeed, most of the animals in these ancient fossil beds are extinct and may represent failed experiments in animal evolution, unique designs that were not suited to later environmental conditions.

Our survey of the survivors follows the course of their evolution. There are thousands of marine animal species, and we will be able to touch only on the major groups. This chapter introduces the animals that do not have backbones, the **invertebrate** phyla (*invertebratus* = lacking a backbone). The great majority of animals are invertebrates, and **Figure 15.2** provides some inkling of their diversity. Chapter 16 discusses the more advanced vertebrate animals. Both chapters reflect the increasing complexity of animals, as we move from those groups whose basic structure seems to have solidified relatively early in the history of animals to those groups that seem to have evolved more recently.

CLASSIFICATION

Before looking at the animals themselves, we need to investigate an important question: How shall the multitudes of animals (and other kinds of organisms) be organized into more easily studied groups, based on similarity? Classification schemes have been around for as long as people have looked at living things, but devising a scheme of biological classification forces us to think carefully about two problems: (1) What do we mean by a *kind* of animal or plant, and (2) What attributes should we use to *organize* these kinds of organisms into larger categories? This study of biological classification is called **taxonomy** (*taxo* = to put in order, *onoma* = name).

Defining the Kinds of Animals and Plants

At first glance it might seem easy to decide on a definition for a particular kind of animal or plant, but the criteria are not always obvious. Consider mallard ducks, for instance. The male mallard is much more colorful and slightly larger than the female. Should male mallards be classified as a different species (a different kind of animal) from females? Of course not. If external appearance won't suffice, what criteria should be used? The problem was first addressed successfully by John Ray, a seventeenth-century English botanist who stressed internal structural likenesses as the basis of classification. Ray's use of the word *species* approximated current usage, but the modern definition of *species* had to await Charles Darwin's revolutionary insights (see Chapter 13). We now define a **species** as a population of organisms whose members interbreed under natural conditions and produce fertile offspring, but who are reproductively isolated from other such groups.

Systems of Classification

The second problem in taxonomy, the difficulty of classifying groups of species into larger categories, is not as easily solved. The Greek philosopher Aristotle proposed a system of classifying animals based on their exterior similarities, but his results were not very useful. Using his system we would place airline pilots, gliding squirrels, flying fish, and grasshoppers into the same group because each can fly! Such a system is an **artificial system of classification.** (Another artificial system of classification would be the arrangement of books by jacket color or page size or typeface.) By contrast, the **natural system of classification** for living organisms, which biologists use today, relies on an organism's structural and biochemical similarities to other organisms. We place all insects together regardless of their flying ability,

Figure 15.3 Carolus Linnaeus—the father of modern taxonomy—in Laplander costume. (He went on a scientific expedition to Lapland in 1732.)

just as we place all books by Melville together, all compositions of Mozart together, and all sea stars together—because each group has a common underlying natural origin. The groups are arranged *systematically*—that is, in some order that makes structural and evolutionary sense.

One of the first persons to classify groups of organisms into natural categories was the eighteenth-century Swedish naturalist Carl von Linné, or as he called himself, **Linnaeus (Figure 15.3).** In his zeal to classify every aspect of the natural world, Linnaeus invented three supreme categories, or **kingdoms:** animal, vegetable, and mineral. Today's biologists leave the mineral kingdom to the geologists and have expanded Linnaeus's two living kingdoms to five. The names and characteristics of these five kingdoms are listed in **Table 15.1.**

Determining the placement of an organism into a kingdom requires a fundamental understanding of the nature of the organism and its evolutionary relationship to other organisms. Placement is determined by science and by tradition. Four of the modern kingdoms are "natural"; that is, they include organisms of a similar fundamental nature: These are Monera, Fungi, **Plantae,** and **Animalia,** listed in Table 15.1. But one kingdom in Table 15.1, kingdom **Protista,** is a somewhat "unnatural" collection of diverse forms that do not clearly fit into any of the other four kingdoms. Protists are all unicellular organisms, but they include both autotrophs (like diatoms and most dinoflagellates) and heterotrophs (like

Table 15.1	Classification of Organisms into Five Kingdoms	
Kingdom	Characteristics	Examples
Monera	Single-celled organisms lacking a nucleus and other internal structural subdivisions; feed by absorption, photosynthesis, chemosynthesis	Bacteria, cyanobacteria (blue-green algae)
Protista	Single-celled organisms possessing a nucleus and other internal structural subdivisions; feed by absorption, photosynthesis, or ingestion of particles	Diatoms, dinoflagellates, protozoa (foraminiferans, radiolarians, amoebas, etc.)
Fungi	Multicellular organisms lacking photosynthetic ability, nutrition by absorption	Mushrooms, molds
Plantae	Multicellular photosynthetic autotrophs	Large algae, mosses, ferns, flowering plants
Animalia	Multicellular heterotrophs	Invertebrates, vertebrates

foraminiferans and radiolarians). Some microbiologists believe that the kingdom Protista will be split into many separate kingdoms as our understanding of these organisms advances.

Linnaeus's great contribution was a system of classification based on **hierarchy,** a grouping of objects by degrees of complexity, grade, or class. In this boxes-within-boxes approach, sets of small categories are nested within larger categories. Linnaeus devised names for the categories, starting with kingdom (the largest category) and passing down through phylum, class, order, family, and genus, to species (the smallest category). In 1758, he published a catalog of all animals then known, his monumental *Systema Naturae* (The System of Nature). **Figure 15.4** shows the classification of a familiar sea gull using the Linnaean method. Note the nested arrangement of category within category, each category becoming more specific with each downward step.

Scientific Names

Linnaeus also perfected the technique of naming animals. The *genus* and *species* names—the names of the last two nested categories—constitute an organism's **scientific name.** *Octopus bimaculatus* is the scientific name of a common West Coast octopus: *Octopus* is the generic name, *bimaculatus* the specific name. A closely related species, *Octopus dofleini,* is a larger animal that ranges to Alaska. *Octopus bimaculatus* and *Octopus dofleini* are not interfertile (they're not the same species), but as their shared generic name suggests, they are closely related.

Scientific names are an indispensable part of biology. They have three important characteristics:

1. Scientific names are usually *permanent.* An organism retains its original specific name forever (pro-

vided that certain agreed-upon naming procedures are followed), but generic names may occasionally be altered as additional information about relationships is acquired.

2. Scientific names traditionally describe the organism in *unchanging words,* usually derived from Latin or Greek. (*Octopus bimaculatus,* for example, literally means an "eight-footed, two-spotted" animal.)

3. Scientific names are *monitored* by an international agency to help prevent duplication and assure adherence to the rules.

The advantage of a scientific name over a common name is immediately apparent to anyone trying to identify a shell found on the beach. The same shell may have many different common names in many different languages, but it will have only one scientific name. When you discover that name in a good guide to shells, you can use it to find references that will tell you what is known about the animal, its life-style, its range, and its evolutionary history.

INTRODUCING THE INVERTEBRATES

More than 90% of all living and fossil animals are invertebrates. The category is convenient though somewhat artificial, containing as it does a vast variety of different organisms in many diverse phyla that have little in common. Invertebrates are generally soft-bodied animals lacking a rigid internal skeleton for the attachment of muscles; but many invertebrates possess some sort of hard, protective outer covering, which can be continuous (like a snail shell) or segmented (like a lobster shell). These animals range in size from microscopic worms to

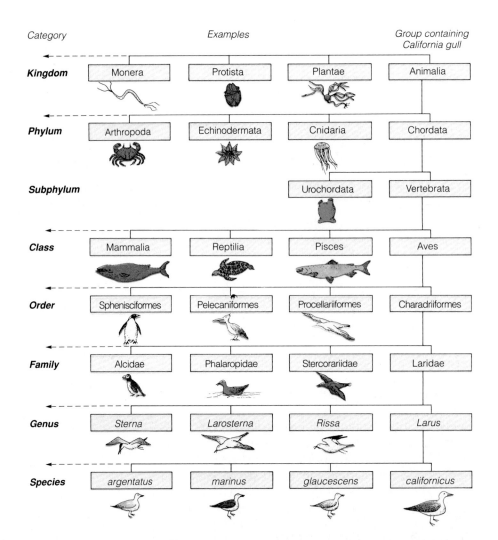

Category	Examples			Group containing California gull
Kingdom	Monera	Protista	Plantae	Animalia
Phylum	Arthropoda	Echinodermata	Cnidaria	Chordata
Subphylum			Urochordata	Vertebrata
Class	Mammalia	Reptilia	Pisces	Aves
Order	Sphenisciformes	Pelecaniformes	Procellariiformes	Charadriiformes
Family	Alcidae	Phalaropidae	Stercorariidae	Laridae
Genus	*Sterna*	*Larosterna*	*Rissa*	*Larus*
Species	*argentatus*	*marinus*	*glaucescens*	*californicus*

Figure 15.4 The modern system of biological classification, using the California gull (*Larus californicus*) as an example.

giant squids. At least 33 invertebrate phyla are currently recognized, and almost every one of them has marine representatives. The nine most conspicuous phyla of these are presented in this chapter in order of increasing complexity (for an overview, see **Table 15.2**). **Table 15.3** lists some of the less well known phyla that have living marine forms.

The Protozoa

Though they are heterotrophic, protozoa are single-celled and therefore technically not animals. The usually microscopic organisms constituting the many phyla of **protozoa** (*protos* = first, *zoon* = animal) are contained within the kingdom Protista. The amoeba and paramecium familiar to generations of high school biology students are protozoans.

Though some of the 50,000 known species of protozoans are parasitic (that is, they live on or within the body of a host), the great majority, including nearly all the marine species, are free-living. Many protozoans

Table 15.2   Major Animal Phyla with Marine Examples

Phylum	Marine Examples
Invertebrates	
Protozoa[a]	Foraminiferans, radiolarians, amoebas
Porifera	Sponges
Cnidaria	Coral, jellyfish, sea anemones, siphonophores
Platyhelminthes	Flatworms, flukes, tapeworms
Nematoda	Roundworms
Annelida	Segmented worms
Mollusca	Chitons, snails, bivalves, squid, octopuses
Arthropoda	Crabs, shrimp, barnacles, copepods, krill
Echinodermata	Sea stars, sea urchins, sea cucumbers
Chordata[b]	Tunicates, salps, *Amphioxus*
Vertebrates	
Chordata[b]	Fishes, reptiles, birds, mammals

[a]The protozoa are a multiphyletic group contained within the kingdom Protista. All other listed phyla are contained within the kingdom Animalia.

[b]Phylum Chordata includes both vertebrate and invertebrate classes.

move by bending rows of tiny, eyelashlike projections called *cilia*, or by flexing whiplike flagella. Others move by rippling their bodies in flowing waves, or by other means. Reproduction is usually by simple division, but many species reproduce sexually.

We have already met some marine protozoans in previous chapters. The deep-sea oozes discussed in Chapter 5 are often formed from the remains of planktonic foraminiferans (**Figure 15.5**), radiolarians, and heliozoans (relatives of the common amoeba). (Other protists are shown in Figures 5.6a, 14.17, and 14.18.) About a third of the deep-ocean floor (some 130 million square kilometers, or 50 million square miles) is covered with calcareous ooze derived from a single planktonic foraminiferan genus, *Globigerina*. Up to 50,000 of their shells may be found in a single gram of sediment. The relatively insoluble siliceous (glassy) remains of radiolarians also form deep-ocean oozes. The broken skeletons of these delicate and beautiful zooplankters cover up to 8 million square kilometers (3 million square miles) of seafloor in the Indian and Pacific oceans.

Bottom-dwelling protozoans are found by the millions hunting and scavenging in subtidal muds, in the coral sand and debris of tropical reefs, in tidal pools, and in most other benthic habitats.

Figure 15.5 A planktonic foraminiferan, a marine protist. Golden brown algae are dispersed throughout the foraminiferan's spines.

Phylum	Common Name and Examples	Characteristics
Ctenophora	Sea gooseberries, comb jellies	Pelagic, gelatinous, predatory, using sticky tentacles to entangle zooplankters
Nemertea	Ribbon worms	Flat, unsegmented worms with extensible "harpoons"; mostly marine, some pelagic and some bottom dwellers
Kinoryncha	(No common name)	Microscopic, spiny, segmented, wormlike; one of the several phyla that live between grains in marine sediments
Bryozoa	Moss animals	Common, encrusting, colonial marine forms, a few in fresh water; individuals are polyplike suspension feeders, equipped with a crown of tentacles known as a *lophophore*
Phoronida	(No common name)	Wormlike marine suspension feeders with lophophores; form leathery tubes often reinforced with sand or shell fragments
Brachiopoda	Lampshells	Bivalved animals, superficially like clams but with lophophore feeding apparatus; once widespread in the ocean, now rare; mainly in deep water
Priapulida	(No common name)	Small, rare, subtidal, predatory wormlike organisms
Sipuncula	Peanut worms	Extensible marine burrowing worms, many with tentacles
Echiura	Spoon worms	Burrowing marine worms
Tardigrada	Water bears	Tiny, eight-legged animals that live between sediment grains
Pogonophora	Beard worms	Deep-sea worms that form upright, chitinous tubes fixed to bottom sediments; found in vent colonies
Chaetognatha	Arrowworms	Stiff-bodied, planktonic predators, some of which migrate vertically through the water column
Hemichordata	Acorn worms	Unsegmented marine worms that form U-shaped burrows in sandy or muddy bottoms

Table 15.3 Other Phyla with Many Marine Representatives

Phylum Porifera

Sponges belong to the phylum **Porifera** (*porus* = holes, *ferre* = to bear), the most primitive true animals. Nearly all of the 10,000 species of these simple attached animals are marine. Sponges are widely distributed from intertidal zone to abyss and are found at all latitudes and in most benthic habitats.

Sponges range from the size of a bean to the size of a small automobile and come in a few basic shapes: branching, vaselike, and encrusting. Most of the sponges we see in the intertidal and subtidal zones of temperate waters are encrusting forms that adhere to the shaded underside of rocky ledges (**Figure 15.6**).

All sponges are **suspension feeders:** They strain plankton and tiny organic food particles from the surrounding water. A large sponge may filter more than 1,500 liters (400 gallons) of water each day. **Figure 15.7** shows a cutaway diagram of a simple upright sponge. Water carrying food and oxygen enters at the incurrent pores and is swept toward the excurrent openings by flagellated collar cells. The sticky collars snare food particles drifting past, and digestion begins. The captured nutrients are distributed to other cells of the organism by wandering cells in the body of the sponge. Sponges have no digestive system, only individual digestive cells; they have no circulatory, respiratory, or nervous systems. Excretion and the movement of gases into and out of the animal occur by simple diffusion.

The flagellated cells of more elaborate sponges are concentrated in hundreds of tiny chambers. Their interiors resemble a Swiss cheese. A skeletal network of spicules (needles) of calcium carbonate or glassy silica prevents the internal chambers and canals from collaps-

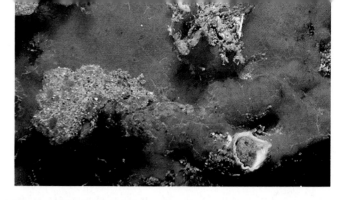

Figure 15.6   An encrusting sponge.

ing; a fibrous protein called *spongin* often serves the same purpose. Commercial natural bath sponges (not to be confused with the brightly colored synthetic sponges encrusting supermarket shelves) have been treated to retain only this spongin matrix; all the cells are gone.

Phylum Cnidaria

Jellyfish, sea anemones, and corals belong to the phylum **Cnidaria** (*knide* = nettle), which contains about 9,000 mostly marine species. (You may be more familiar with this phylum's old name, Coelenterata.) This group of carnivorous animals takes its name from the large, stinging cells called **cnidoblasts** (*knide* = nettle, *blastikos* = to shoot upward), deployed on tentacles that bend or retract toward the mouth. Each cnidoblast contains a capsule from which a coiled thread may be forcefully ejected (**Figure 15.8**). The thread can repel an aggressor or penetrate or entangle prey, often immobilizing the victim with a toxin. Then the prey—which may include the larger zooplankters and small fish—is drawn into the mouth, leading to a saclike digestive cavity. The digested food is absorbed by cells of the inner layer and trans-

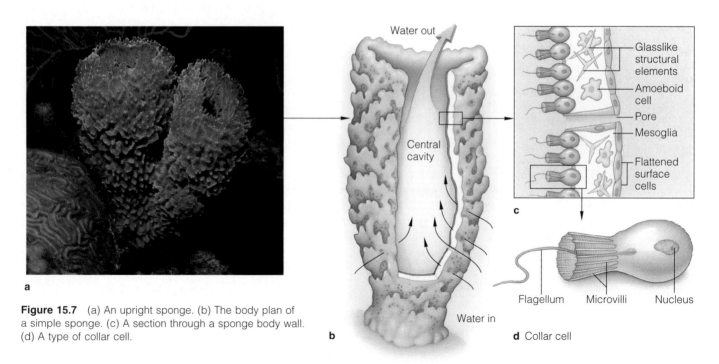

Figure 15.7   (a) An upright sponge. (b) The body plan of a simple sponge. (c) A section through a sponge body wall. (d) A type of collar cell.

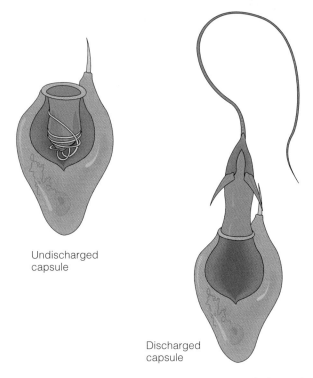

Figure 15.8 A cnidoblast, with stinging capsule undischarged (left) and discharged (right).

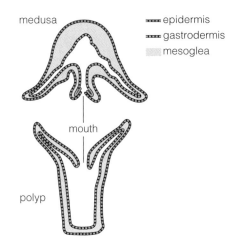

Figure 15.9 The medusa and polyp body plans of Cnidaria compared. The inner layer (the gastrodermis) is responsible for digestion and reproduction, the outer layer (the epidermis) for capture of prey and protection from attack. The layers are connected by a jellylike mesoglea.

medusa

epidermis
gastrodermis
mesoglea

mouth

polyp

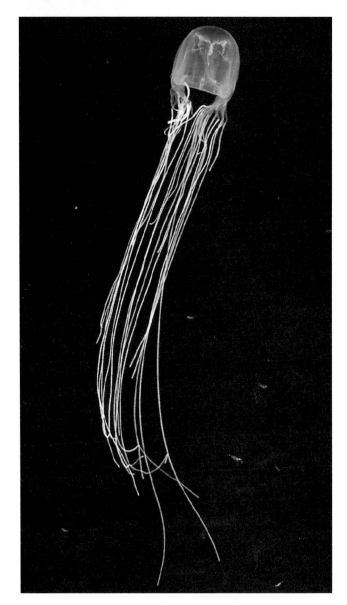

Figure 15.10 A sea wasp (*Chironix*), one of the most dangerous jellyfish. An inhabitant of tropical waters from Africa to northeastern Australia, it can kill a human within 3 minutes. The tentacles of a large specimen can be 15 meters (50 feet) long. *Chironix* has probably been responsible for more human deaths than sharks.

ported to other parts of the animal by migratory cells and by diffusion. Because the digestive cavity has only one opening, indigestible bones and other wastes are eliminated through the mouth.

Members of this group are built of two layers of cells. The inner layer (the gastrodermis) is responsible for digestion and reproduction, the outer layer (the epidermis) for capturing prey and protection from attack. The layers are connected by a jellylike mesoglea (*mesos* = middle, *glia* = glue). Cnidarians exhibit **radial symmetry;** that is, their body parts radiate from a central axis like the spokes of a wheel. Some, such as sea anemones and corals, attach to rocks or other objects; others, such as jellyfish, swim freely in the water. No cnidarian possesses a definite head or concentration of sensory recep-

tors, but a primitive network of nerves permits some species to respond to stimuli. These relatively simple animals depend on diffusion to move wastes and gases; they have no excretory or circulatory systems.

Cnidarians occur in two forms: medusae and polyps (**Figure 15.9**). Jellyfish are examples of the **medusa** body plan, named after a Greek mythological monster with a woman's face and hair that was a mass of writhing snakes. Medusae are predatory animals that swim by the rhythmic contraction of their bell-shaped bodies. They catch their prey with trailing tentacles armed with cnidoblasts. Some medusae are microscopically small, but the genus *Cyanea* can reach a bell diameter of 3.5 meters (12 feet), with tentacles 18 meters (60 feet) long! A smaller, more dangerous species is shown in **Figure 15.10.**

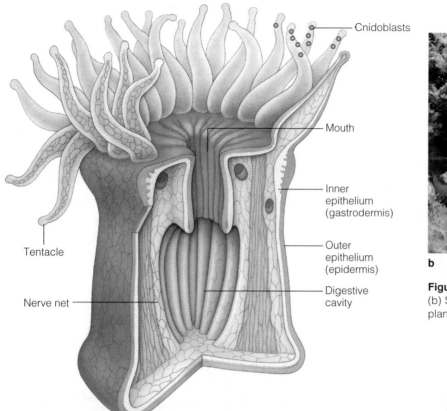

Cnidoblasts

Mouth

Inner
epithelium
(gastrodermis)

Tentacle

Outer
epithelium
(epidermis)

Digestive
cavity

Nerve net

a

b

Figure 15.11 (a) A cross section through a polyp.
(b) Sea anemones, cnidarians based on the polyp body
plan.

Sea anemones and corals are examples of the sedentary **polyp** (*polypous* = many-footed) body plan. Sea anemones (**Figure 15.11**) have no skeleton and attach firmly to the substratum or burrow into it with a sticky basal disc on which they can slide slowly, like a snail. Other species like the coral (which we will discuss in Chapter 18) lack a basal disc; they contain a calcareous skeleton covered by living tissue and are permanently cemented in place (see Figures 18.15, 18.16, and 18.18). The polyp's stout muscular body supports more substantial tentacles than those of the medusae, from which polyps presumably evolved. A few cnidarians alternate beween the polyp and medusa body plans at different times in their life cycles. Some begin life as polyps and later float free as medusae; others reverse the process.

An unusual group of cnidarians, the siphonophores, are colonial combinations of as many as 1,000 polyps and medusae that function as a single entity. One subindividual generates gas or oil and acts as a float for the colony; others perform capture and feeding duties, reproductive functions, or defensive actions. Stinging tentacles in the genus *Physalia*, the Portuguese man-of-war (**Figure 15.12**), can be 18 meters (60 feet) long and can cause intense pain, even death, to swimmers contacting them.

The Cnidaria are a very successful group, but their simple architecture would not serve more advanced organisms. Their blind-sac digestive system allows only one batch of food to be processed at a time; feeding

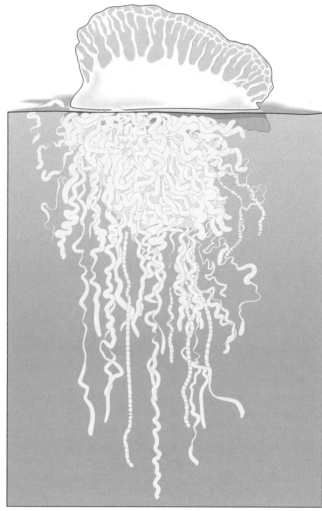

Figure 15.12 Portuguese man-of-war (*Physalia*), a dangerous siphonophore.

opportunities arising during digestion cannot easily be accommodated. The lack of a distinct head with a concentration of sense organs is a drawback, as is the absence of circulatory, respiratory, and excretory systems. Having only two cell layers limits the complexity of systems and structures that can form within the organisms.

THE WORM PHYLA

A transition from relatively simple to more advanced organisms is made in the worm phyla, three of which are discussed here. The worm body plan exhibits **bilateral symmetry** rather than radial symmetry; that is, each has a left side and a right side that are mirror images of one another. Nearly all worms have some concentration of sensory tissue in what may be termed a head, and many have flow-through digestive systems and systems to circulate fluids and eliminate waste. Some are efficient parasites, but most are free-living. A few burrow in cavities; others roam the seabed or lurk under rocks.

Phylum Platyhelminthes

The simplest worms are the well-named flatworms of phylum **Platyhelminthes** (*platys* = flat, *helmins* = worm) (**Figure 15.13**). More than 15,000 species of flatworms are known. Some are parasitic of vertebrates, such as the tapeworms of fish and marine mammals. However, most marine flatworms are free-living; they can be found on the shady underside of intertidal rocks or sharing colonies or burrows with other animals. Few examples exceed 3 centimeters (1¼ inches) in length.

Nearly all flatworms are carnivorous, and most feed upon prey or food particles sucked into the central digestive cavity through a mouth opening. Since these forms have no anus, tiny shells and other undigestible particles must be ejected by the same route.

Flatworms are the most primitive organisms with a central nervous system. In some species a complex of nerve cells serving as a rudimentary brain connects the animal's simple nervous system to a pair of light-sensitive eyespots. The eyespots are small pigmented cups that can sense only the presence or absence of light—certainly a critical factor for an animal only a few cells thick that must avoid light to prevent overheating or detection.

Larger flatworms are necessarily thin because they lack a true respiratory and excretory system. Gases must be exchanged and wastes eliminated by diffusion through the animal's surface; so no cell can be more than a few cell diameters from the outside. Flatworms have a metabolic rate about 10 times that of cnidarians; so their maximum thickness is limited by diffusion rate.

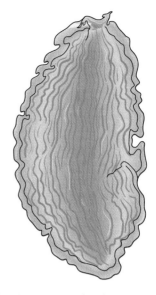

Figure 15.13 *Prostheceraeus*, a free-living tropical marine flatworm.

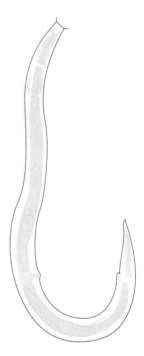

Figure 15.14 A common, free-living marine nematode of the type sometimes found in large numbers in nearshore marine sediments.

Phylum Nematoda

The first animals in our phylum survey that possess a flow-through digestive system (a digestive tract with mouth and anus rather than a digestive cavity) are members of the plentiful phylum **Nematoda** (*nematos* = thread). Also called *roundworms* (**Figure 15.14**), these most successful of the worm phyla are present in nearly every imaginable terrestrial, aquatic, and marine habitat. Most of the 12,000 known species are free-living and

BOX 15.1 ● *If It Moves, Don't Eat It!*

People who enjoy sushi and sashimi (raw fish) run a slight risk of contracting herring worm disease or some other malady caused by marine parasites. More than 1,200 cases of humans being infected by marine parasites have been reported, mostly from Japan but also from Europe and North America. About 50 cases of parasitic marine worm infections have been reported in the United States since 1980, probably because of an increase in the popularity of raw fish sushi, a Japanese delicacy.

The most dangerous parasites are nematodes of the genus *Anisakis*. These coiled worms—most often found in herring, red snapper, salmon, halibut, and some species of bass—are about 4 centimeters (1½ inches) long when unwound. They are usually coiled in balls the size of pinheads, either between groups of muscles or at the junction of muscles and organs. The worms look like tan or brown dots, often contrasting nicely with light-colored flesh.

Ingesting *Anisakis* can lead to much unpleasantness. Sometimes, the parasite can be felt in the throat within 24 hours of the meal; coughing may dislodge it. Worms that have been swallowed may cause intense abdominal pain, nausea, vomiting, fever, and diarrhea. Symptoms of an infestation may resemble those of a peptic ulcer. A person will become dangerously ill if the worm penetrates the intestine. Japanese physicians see these infections rather frequently and recognize them, but American doctors often mistake the signs for an ulcer or appendicitis and treat them accordingly. In a recent issue of the *New England Journal of Medicine,* a surprised surgeon reported removing a stricken patient's apparently normal appendix. Just as he was closing the incision he saw a 4.2-centimeter (1.65-inch) nematode crawl onto the surgical drapes.

Sushi lovers can greatly lessen the danger of *Anisakis* infection by patronizing specialty restaurants that employ experienced chefs. Sashimi from tuna or octopus is almost always free of the worms; salmon is potentially the most troublesome. Inspect each piece carefully, and *don't eat anything that moves!*

These parasites are killed by proper freezing or cooking. The fish must be frozen at −20°C (−4°F) for more than 60 hours or cooked for at least 5 minutes at 60°C (140°F). (Microwave cooks please note: Quick microwaving of fish does not raise all parts of the portion to the same high temperature.) You can safely enjoy a hot fish dinner if the fish has been uniformly heated.

Source: Wittner, M., et al. 1989. "Eustrongyliadiasis—A Parasitic Infection Acquired by Eating Sushi." *New England Journal of Medicine* 320 (no. 17): 1124–26.

microscopic, thriving in garden soil and marine sediments. Some make perfect parasites, however, and nearly all vertebrates and many invertebrates are parasitized by species of these long, thin worms. Readers who enjoy eating raw fish (sashimi) may wish to read the information on fish parasites in **Box 15.1**.

The nematodes' true claim to marine fame rests on their astonishing numbers in some soft sediments. A Dutch researcher investigating shallow subtidal mud in the North Sea reported 4,420,000 tiny individuals in a sample 1 meter square (10.8 square feet) by 4 centimeters (1.57 inches) deep! Samples of fine beach sand taken at low tide in southern California have yielded comparable numbers. I recall adding a protein stain to a wet sand sample in a bucket and watching thousands of transparent, previously invisible worms materialize by taking up the stain. Swimmers and gardeners have encountered these unobtrusive worms by the thousands without ill effects. Free-living worms cannot become parasitic.

Phylum Annelida

Members of the phylum **Annelida** (*annelus* = ring) are the most evolutionarily advanced worms. Their bodies are divided into a number of similar rings or segments. **Metamerism,** as this segmentation is called, is a convenient strategy for increasing the size of an animal simply by adding nearly identical units. We will see evidence of metamerism in most higher animals. Each segment of an annelid can have its own circulatory, excretory, nervous, muscular, and reproductive systems, but some segments (such as those forming the head) are specialized for specific tasks. The familiar garden earthworm is an annelid.

The 5,400 species of class **Polychaeta** (*poly* = many, *chaetae* = bristles), the largest and most diverse class of annelids, are the most important marine annelids (**Figure 15.15**). Polychaetes are often brightly colored or iridescent worms with pairs of bristly projections extending from each segment. They range in length from about 1 to 15 centimeters (½ to 6 inches). Some polychaetes burrow through and devour sediments or move freely over the bottom in search of food; others construct fixed parchmentlike or calcareous tubes from which only parts of their heads emerge. Mobile polychaetes have well-developed heads with prominent sense organs, and they can function as efficient predators. Some skewer prey with their sharp mouthparts. Tube-dwelling polychaetes are a common sight to divers on most continental shelves. The "feather duster" tops of some of them (**Figure 15.16**) exchange gases and remove food from the water. A few tropical forms have biting jaws at their tips, a surprise for the injudicious investigator! All can instantly retract their feathery gills when disturbed.

THE ADVANCED INVERTEBRATES

Phylum Mollusca

The conspicuous phylum **Mollusca** (*molluscus* = soft-bodied) contains 80,000 species, second in size only to the huge phylum Arthropoda. The phylum Mollusca includes such diverse members as clams, snails, octopuses, and squid. Most mollusks are marine, and most have an external or internal shell. A few molluscan species possess acute sight and even considerable intelligence.

Mollusks and annelids probably shared a common origin—possibly a distantly ancestral segmented worm—and therefore share a few basic characteristics. Like annelids, mollusks are bilaterally symmetrical and generally have obvious heads, flow-through digestive tracts, and well-developed nervous systems. A few mollusks are segmented. Unlike annelids, however, some mollusks achieve great size, secrete beautifully fitted shells in which to take refuge, and exhibit great structural diversity.

We will briefly discuss four molluskan classes here: The class **Polyplacophora,** the chitons; the class **Gastropoda,** the snails; the class **Bivalvia,** the clams, oysters, and mussels; and the class **Cephalopoda,** the nautiluses, octopuses, cuttlefish, and squid. Though greatly different in shape and habits, each class shows its link to a common ancient ancestor by sharing similar underlying parts. **Figure 15.17** shows some of the structural similarities in their body plans.

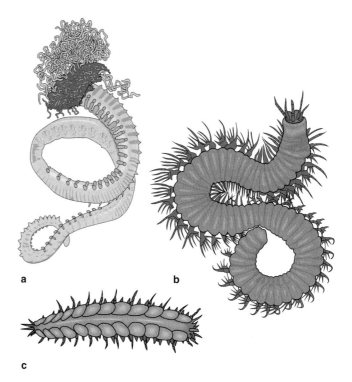

Figure 15.15 Marine annelids of the class Polychaeta. (a) *Amphitrite*, a sedentary tube worm that extends its sticky tentacles across the seafloor to capture food. (b) *Eunice*, an active predator. (c) *Serpula*, which lives in stony tubes on the sides of rocks and captures prey with its grasping mouthparts.

Figure 15.16 A tube-dwelling polychaete with its "feather duster" appendages deployed. A mucous coating on the "duster" catches food, and the appendage itself works as a gill for gas exchange.

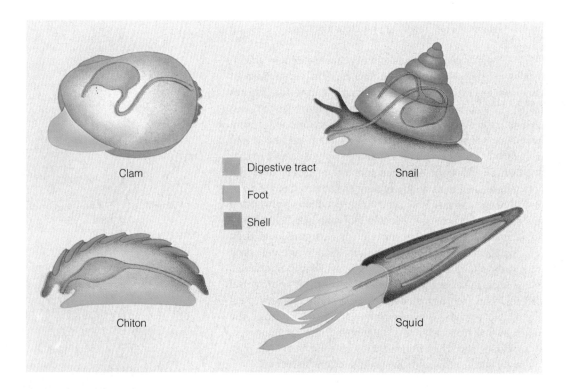

Clam

Snail

Digestive tract

Foot

Shell

Chiton

Squid

Figure 15.17 The molluscan body plan as it has been modified in the major groups.

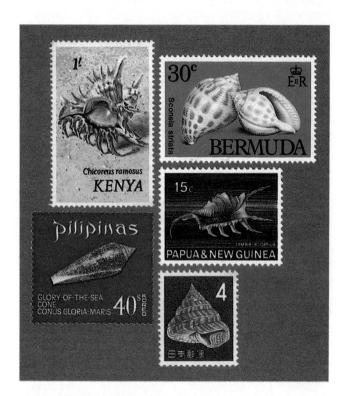

Figure 15.18 Gastropod shells are often aesthetically pleasing. They have been prized by collectors for centuries. Gastropod shells have been used to decorate many objects, postage stamps included.

Polyplacophorans (*poly* = many, *plakinos* = made of boards), oval animals usually 2 to 8 centimeters (1 to 3 inches) long, are found on intertidal or subtidal rocks. Also called **chitons,** they creep slowly on a sticky foot as their rasping mouthparts scrape algae from the surfaces around them. Their convex shells consist of eight articulated plates. The division of the shell into plates permits chitons to roll into a tight ball, their only defense if dislodged from a rock. They lack well-developed eyes or brains. (Note: *Chiton* rhymes with *Brighton*.)

Gastropods (*gaster* = stomach, *pod* = foot) are more familiar, including the abalone, conch, limpet, and garden snail. Members of this largest class of mollusks usually inhabit relatively large shells, where they can seek refuge in case of danger. Some gastropods are grazers, some are suspension feeders, and some are predators. A few marine species—the pteropods and heteropods—are planktonic, but most marine gastropods are found wandering on rocky bottoms or other firm substrates.

Gastropod shells are often structures of great beauty (**Figure 15.18**). The animal adds to and enlarges the opening as it grows. The shell is frequently coiled to compress its mass and allow for easier maneuverability. A foot and head protrude from the shell while the snail moves about. The shell itself is secreted in three principal layers: a fibrous outer covering that may serve to distribute shock; a strong, crystalline layer of calcium carbonate ($CaCO_3$) to provide strength; and an inner layer of smooth $CaCO_3$ to provide nonabrasive surroundings for

Figure 15.19 A translucent purple nudibranch (*Flabellina*) searches for food. The brilliant gill-like structures on its back assist in gas exchange. The carnivorous animal's vivid colors are derived in part from pigments in the small sedentary animals it eats. Its own terrible taste seems to discourage animals from eating it.

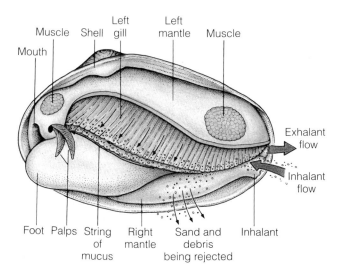

Figure 15.20 Bivalves are suspension feeders that make their living by filtering the water for edible particles. In this diagram (showing a bivalve with its left shell removed), water and tiny bits of food are swept into the animal by the movement of tracts of cilia on the gills. Food settles onto the gills and is then driven toward the mouth and swallowed.

the resident. Not all gastropods have shells. Nudibranchs (*nudus* = naked, *branchia* = gills), also called *sea slugs*, are lovely (or even bizarre-looking), shell-less gastropods (**Figure 15.19**).

Although some snails use their foot to burrow, a gastropod foot cannot attach to sand or mud. The shell-and-foot configuration of polyplacophorans and gastropods is therefore not well suited to life on most of the ocean floor, which is covered by sediments. Evolution of the bivalves (*bi* = two, *valv* = door) admitted mollusks to this rich sedimentary habitat. Animals enclosed in twin shells (clams, oysters, mussels, and scallops), bivalves surrender mobility for protection, and they gather food by suspension feeding rather than pursuit (see **Figure 15.20**). Some giant clams (*Tridacna*) have shells more than 1 meter (3.3 feet) wide and live at the surface of the sediment; but most bivalves are smaller and burrow into the sand or mud, extending their siphons to the surface to obtain water and eject wastes. Burrowing species dig with a strong muscular foot. In other species the foot has other functions: In mussels it secretes the tough threads that attach the organisms to wave-swept rocks—a neat turnabout trick by which bivalves can invade a gastropod habitat![1] An unusual use for these threads is described in **Box 15.2**.

[1]The biological adhesives secreted by mussels have been the subject of much research. Chemists would like to synthesize the material, a strong and flexible glue that doesn't degrade in water. For more information on these remarkable substances, see the article by Amato listed in this chapter's "For Further Study" bibliography.

The most highly evolved mollusks are the magnificent cephalopods (*cephalon* = head, *pod* = foot), a group of marine predators containing nautiluses, octopuses, and squid. These well-named animals have a head surrounded by a foot divided into tentacles. The nautiluses retain a large coiled external shell, but squid have only a thin vestige of the shell within their bodies—and octopuses have none at all. Cephalopods can move by creeping across the bottom, by swimming with special fins, or by squirting jets of water from an interior cavity.

Most cephalopods catch prey with stiff adhesive discs on their tentacles that function as suction cups, and they tear or bite the flesh with horny beaks. Nautiluses (**Figure 15.21**) are an open-ocean group that hunts at considerable depths, their strong shells buoyed with gas-filled chambers. They form a remnant population of a once large and important group of cephalopods prominent in the fossil record. Little is known of their natural history. Squid and octopuses are more advanced cephalopods. Their skin is embedded with tiny, extensible pigment cells called **chromatophores** (*chromos* = color), which can rapidly expand or contract to change the animal's color and pattern. Squid and octopuses can also confuse predators with clouds of ink. Some kinds of squid eject a kind of "dummy" of coagulated ink that's a rough duplicate of their size and shape. The squid is long gone by the time the attacker discovers the deception! At least one species of squid living below the euphotic zone produces a sparkling luminous ink instead of black ink (which would be ineffective in the darkness). Squid can

Figure 15.21 A chambered nautilus (*Nautilus*), from the only living cephalopod group with an external shell. The animal occupies the outermost chamber; the other chambers are partially filled with gas to provide buoyancy. Chambers are added as the animal grows.

Figure 15.22 A large squid (*Dosidicus*) and a diver inspect one another. The diver wrote, "I felt the cold embrace of tentacles with their sharp, toothed suction cups digging into my bare skin. It was like somebody was throwing a cactus on my neck." He returned to the boat minus his dive lights and decompression meter, with "nasty lesions" from sharp protrusions on the suction cups.

Figure 15.23 A large octopus—a fast, agile, and intelligent predator.

grow to surprising sizes (**Figure 15.22**): The record length, including tentacles, is 18 meters (59 feet)! Most are much smaller.

Squid may be the largest invertebrates, but octopuses are the most intelligent (**Figure 15.23**). About as smart as puppies and with even better eyesight, some nearshore species of octopuses kept in captivity soon learn to recognize their keepers and forage at night through adjacent aquariums for tidbits. Octopuses use their visual acuity and intelligence to good advantage in their intertidal or subtidal homes, memorizing the positions of hiding places, escape holes, and good hunting locations.

Only about 450 species of cephalopods live today, a small percentage of the number of species known from fossils. The small number of species suggests that sophistication is no guarantee of biological success.

Phylum Arthropoda

The phylum **Arthropoda** (*arthron* = joint, *pod* = foot)—a group that includes the lobsters, shrimp, crabs, krill, and barnacles—is a phylum of superlatives. Over a million species of arthropods are now known, but some experts suggest that more than 10 million species of insects—the most numerous members of this phylum—may exist on Earth! Arthropods are the most successful of Earth's animal phyla, occupying the greatest variety of habitats,

BOX 15.2 ● *The Golden Fleece*

The Golden Fleece of Greek mythology, sought by Jason and the Argonauts through many adventures, has a basis in fact. The name is believed to refer to a shining translucent fabric, also known as cloth-of-gold, which has been woven for centuries from the bysssal threads of the pen shell *Pinna nobilis.* These bivalves, which grow to a length of 60 centimeters (2 feet), live among rocks partially buried in Mediterranean mud and sand and anchor themselves to the substratum with long, fine filaments of a clear, gold-bronze color with a high metallic sheen. The bushlike clusters of byssal threads—named after the Greek word for flax—can be washed, dried, combed, carded, and spun into a fine yarn that can be woven into garments so sheer that a shoulder cape made from the material could be passed through a finger ring.

Fabrics made of *Pinna* silk were the most expensive of textiles, truly the cloth of kings. In medieval times, fishermen used unwieldy 20-foot tongs to snag pen shells with the byssal threads attached. It took a pound of threads and many weeks of skilled effort to produce 3 ounces of finished filaments. According to Procopius, a sixth-century Byzantine historian, a reigning emperor sent a complete robe of the precious stuff as a gift to the king of Armenia. (No report survives of the king's reaction!) Cloth-of-gold collars decorated the necks of kings and potentates depicted in illuminated manuscripts from the fourteenth and fifteenth centuries. In 1745, Pope Benedict was presented with a pair of sea-silk stockings, a gift so valuable that it was packaged in a solid silver box. Even Queen Victoria owned a pair, which she described as "wonderfully soft and warm."

Golden Fleece was processed in the Mediterranean until the early years of this century. The art was kept alive in Sicily, where the thread, mixed with conventional silk, was knitted into gloves, caps, stockings, and even coats. The industry died after World War I, but determined and patient searchers can still find fishing villages in southern Italy where *Pinna* silk scarves or gloves can be purchased. Imagine owning a garment made of the same legendary fabric that hurried Jason and his friends on their mythical quest!

---

consuming the greatest quantities of food, and existing in almost unimaginable numbers. The arthropod body plan is a variation on the basic annelid theme; bodies show clear segmentation with a pair (or pairs) of appendages per segment. All are bilaterally symmetrical.

Arthropods have not achieved the nervous system development of cephalopod mollusks, nor do they have the advantages of intelligence or extraordinary eyesight. They do, however, exhibit three remarkable evolutionary advances that have led to their great success:

1. An *exoskeleton,* a strong, lightweight, form-fitted external covering and support.

2. *Striated muscle,* a quick, strong, lightweight form of muscle that makes rapid movement and flight possible.

3. *Articulation,* the ability to bend appendages at specific points. The appendages of more primitive phyla can usually bend anywhere (like wet spaghetti), but arthropod appendages bend at a joint. There are no ball-and-socket joints in arthropods; instead, each joint along an appendage moves through a different plane to ensure a full range of motion.

Most important of these advances is the **exoskeleton.** Unlike the often cumbersome shell of a gastropod, the exoskeleton of an arthropod fits and articulates like a finely tailored suit of armor. It is made in part of a tough, nitrogen-rich carbohydrate called **chitin,** which may be strengthened by calcium carbonate. Its three layers serve to waterproof the covering, tint it a protective color, and make it resilient and strong. Muscles within the animal are attached to the exoskeleton to move the appendages.

Such an arrangement sounds ideal, but the difficulties encountered by an organism with an exoskeleton are profound. How can muscle leverage be obtained? How can encrusting organisms be discouraged? How can ducts and feeding passages remain unblocked? And, perhaps most critically, how can the organism grow? That each of these problems has been solved is obvious by the group's overwhelming success, but the growth issue deserves a closer look.

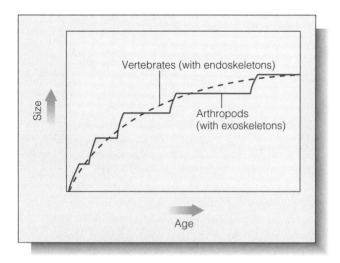

Figure 15.24 Generalized arthropod and vertebrate growth curves compared. The vertical segments of the arthropod curve represent molting periods.

a

b

Figure 15.25 A molting arthropod. (a) A Dungeness crab (*Cancer magister*) backing out of the exoskeleton (*right*) that it is abandoning. (b) Clear of the old exoskeleton, the soft-bodied crab takes in water and expands. It immediately begins to secrete a new exoskeleton. Note the obvious increase in the animal's size.

We vertebrates grow by steadily adding length to the bones of our *internal* skeleton and bulk to our bodies. An *external* skeleton obviously limits growth and must be shed, or **molted,** at regular intervals. Arthropods do not have a steady growth pattern; instead, their external growth progresses in a series of steplike jumps (**Figure 15.24**) as the animal molts and replaces its exoskeleton. The arthropod grows without getting bigger between these jumps in size. An aquatic arthropod slowly substitutes body mass for water held in the tissues between molts. When molting, it suddenly takes on water from outside the body, expanding its tissues without growing in muscle mass. The shell splits and falls away, and, through a magnificently orchestrated sequence of glandular secretions, the animal quickly regenerates a new exoskeleton one size larger (**Figure 15.25**).

The largest class of arthropods, the class Insecta, is poorly represented in the sea: There is only one marine genus and five known open-ocean species, all of which are water striders. The class **Crustacea** (*crustaceus* = having a shell or rind), however, includes 30,000 species of primarily marine, gill-breathing lobsters, crayfish, shrimp, crabs, water fleas, copepods, krill, amphipods, barnacles, and others. Their bodies usually have between 16 and 20 segments; the appendages may be specialized for sensing, food handling, walking, fighting, defense, and so forth. About 70% of all zooplankton are minute crustaceans like those in **Figure 15.26** (and in Figure 14.16), which graze on diatoms and dinoflagellates. (Thumb- or bean-sized planktonic crustaceans called *krill* constitute most of the food of the largest whales.) More familiar to most of us are the lobsters and crabs we esteem for food. The largest crustacean is the giant king crab, which can reach a legspan of 3.6 meters (12 feet). The heaviest individual, however, was a lobster caught off Chatham, Massachusetts, in 1949. (The beast weighed 22 kilograms, or 48 pounds, and reportedly served 10!)

Though they look rather like mollusks, barnacles are also crustaceans (**Figure 15.27**). They resemble shrimp attached to larger objects by their heads and encased in a heavy and highly modified calcareous exoskeleton. Barnacles feed by combing the water with feathery feet for plankton. Some forms of barnacles attach to rocks or boat bottoms by an elongated stalk, others attach to firm surfaces, and a few attach only to the skin of whales.

The activity and importance of billions of crustaceans as scavengers, predators, parasites, grazers, and general participants in oceanic biology are hard to overestimate. Large or small, obvious or retiring, crustaceans dominate the world of marine animals.

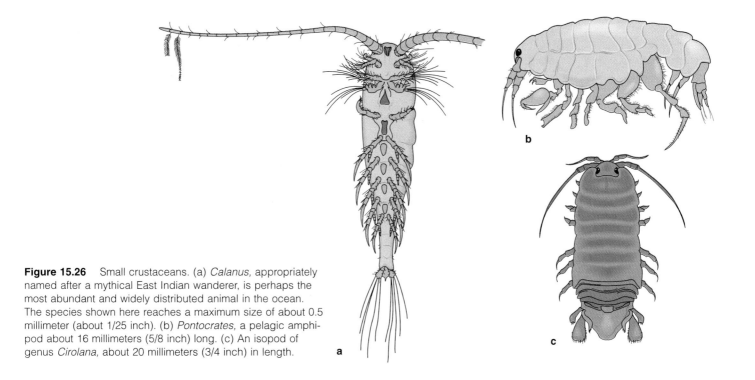

Figure 15.26 Small crustaceans. (a) *Calanus*, appropriately named after a mythical East Indian wanderer, is perhaps the most abundant and widely distributed animal in the ocean. The species shown here reaches a maximum size of about 0.5 millimeter (about 1/25 inch). (b) *Pontocrates*, a pelagic amphipod about 16 millimeters (5/8 inch) long. (c) An isopod of genus *Cirolana*, about 20 millimeters (3/4 inch) in length.

Figure 15.27 Gooseneck barnacles (genus *Lepas*) extend their feathery feet to comb the water for zooplankton. Though they superficially resemble mollusks, barnacles are crustaceans.

Phylum Echinodermata

The exclusively marine phylum **Echinodermata** (*echinos* = hedgehog, *derma* = skin) is an odd group sharply different from other members of the animal kingdom. The 6,000 species of echinoderms lack eyes or brains, have a radially symmetrical body plan based on five sections or projections (**Figure 15.28**), move slowly, and include only two known parasitic representatives.

Linnaeus classified echinoderms as primitive organisms, but study of their anatomy and life cycles indicates they possess some remarkably complex features. The relative position of organisms in the simple-to-advanced scheme we have been following is based not only on the appearance and capabilities of the adult organism, but also on that organism's embryonic development. The developmental stages through which an animal passes from fertilization to hatching (or birth) often provide clues to its evolutionary history. Echinoderms look primitive, but these radial animals have bilaterally symmetrical larvae. That is, they develop complex radial body plans after passing through some early stages shared by other advanced phyla. No other animals begin life as bilaterally symmetrical creatures and metamorphose into radial forms.

Living echinoderms are divided into five classes: The four most familiar ones are the class **Asteroidea,** the sea

Figure 15.28 Pentamerous (five-sided) symmetry in the four main classes of echinoderms. At some time in their lives, all members of this phylum possess a radially symmetrical body plan based on five sections or projections. (a) A sea star; (b) a brittle star (so named because its fragile arms will break off if grabbed by a predator); (c) a close-up of the five-part jaws centered on the underside of a sea urchin (see also Figure 15.31); (d) a sea cucumber.

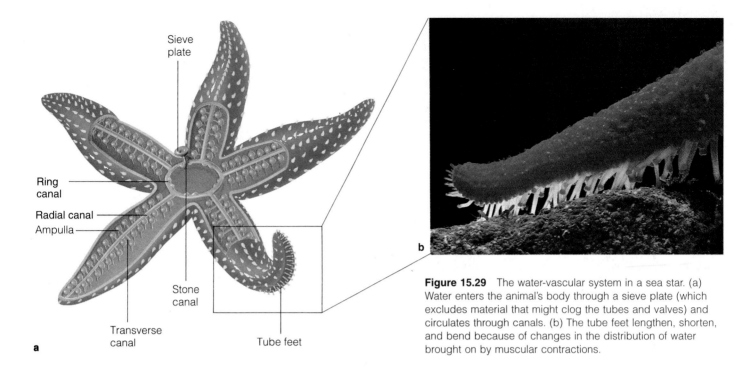

Figure 15.29 The water-vascular system in a sea star. (a) Water enters the animal's body through a sieve plate (which excludes material that might clog the tubes and valves) and circulates through canals. (b) The tube feet lengthen, shorten, and bend because of changes in the distribution of water brought on by muscular contractions.

Labels in figure (a): Sieve plate, Ring canal, Radial canal, Ampulla, Stone canal, Transverse canal, Tube feet

stars; the class **Ophiuroidea,** the brittle stars; the class **Echinoidea,** the sea urchins and sand dollars; and the class **Holothuroidea,** the sea cucumbers.

Nearly all asteroids (*aster* = star, *oidea* = resembling), or sea stars, are star-shaped echinoderms with arms not completely delineated from the central disc (see Figure 15.28a). The arms usually have spiny projections on top and delicate tube feet beneath. The tube feet work like suction cups and can grip objects; they also participate in gas exchange. Tube feet are part of a sea star's most striking feature: its unique **water-vascular system** (**Figure 15.29**), a complex of water-filled canals, valves, and projections used for locomotion and feeding. Operating like a hydraulic power system, the water-vascular system's plumbing can transmit forces generated by muscles at one side of the sea star to arms on the other side. Using this system, the tube feet of a sea star can grip a clam or mussel and exert a continuous pull (sometimes for hours) to force open its valves, even though one or more of the star's arms may tire. When the mussel finally opens, the sea star expels its stomach from its mouth, slips it between the mussel's shells, and digests the victim in place.

Delicate ophiuroids (*ophidion* = a snake) have long, slender arms (see Figure 15.28b). They are called *brittle stars* because of their unusual strategy for evading capture: If grasped by a predator, a brittle star will often detach its arm and escape. (The brittle star will later regenerate the arm.) Ophiuroids are perhaps the most widely distributed benthic marine animals; a few species

Figure 15.30 Ophiuroids (brittle stars) feed on edible particles in the surface layer of sediments on the continental slope off New England. Ophiuroids are among the most widely distributed of all benthic animals.

have been found in great numbers on the deep, sedimented seabeds of the world ocean (**Figure 15.30**). Many ophiuroid species also live beneath intertidal and subtidal rocks. Longitudinal grooves on the underside of each arm enable some species of brittle stars to locate food particles and transfer them by cilia to the mouth. Other

Figure 15.31 The prickly top of a sea urchin. Long, thin tube feet are visible between the spines. A close-up of the five-part jaws of the same sea urchin (centered on its underside) may be seen in Figure 15.28c.

species wave their arms through the water to capture plankton on sticky strands of mucus strung between adjacent arm spines.

Echinoids (*echinos* = hedgehog) are familiar to most coastal residents. The prickly appearance of a sea urchin (**Figure 15.31**) or the smooth velvety surface of a sand dollar do not at first suggest the phylum's five-sided symmetry. Close observation, however, reveals just such an arrangement, overlain by a few bilaterally symmetrical features. Urchins can feed either by taking bits of food into the mouth with complex grasping, chewing jaws (Figure 15.28c) or simply by absorbing food molecules into the mucus layer that covers their bodies and flows toward the mouth.

Holothuroids (*holos* = entirely, *thur* = tubelike), the well-named sea cucumbers, also do not look five-sided at first glance (see Figure 15.28d). Most sea cucumbers feed by thrusting small, sticky appendages into the surrounding sediments, retracting the appendages, and eating the catch. The main claim to fame of certain nearshore species is an exuberant and seemingly suicidal defense mechanism: When annoyed, these holothurans pressurize themselves and eviscerate violently, forcefully ejecting their digestive systems, respiratory apparatus, and gonads! The water around them churns with viscera, most of which are coated with the same sticky substance used to attach food to the feeding structures. A predator eating this material could glue his gill openings or esophagus shut. But what good is a defense mechanism that results in the death of the defender? None, of course. The holothuran does not die, but lies quietly for two or three weeks regenerating its insides from food reserves

in its intact outer covering. After this brief vacation, it's as good as new!

Phylum Chordata

All members of the phylum **Chordata**, the most advanced animal phylum, possess a stiffening **notochord** (*notus* = back, *chorda* = cord), a tubular dorsal nervous system, and gill slits behind the oral opening at some time in their development. The notochord was critical in evolution. It permitted a more complex embryonic development by providing a rigid "scaffold" on which the developing embryo could be constructed, and it provided an internal mechanical foundation for skeletal and muscle development. About 5% of the 45,000 species of chordates lose their notochord as they develop; these are called *invertebrate chordates*. The other 95% of chordates retain their notochord (or the vertebral column that forms around it) into adulthood; these are the familiar vertebrate chordates (such as fish, reptiles, birds, and mammals), our topics for the next chapter.

Two invertebrate chordates are of interest here. The **tunicates,** or sea squirts, are suspension feeders that superficially look and function like sponges. Their common name comes from an extraordinarily strong and flexible tunic (outer covering). Close investigation reveals a body plan much different from the primitive sponges, however (see **Figure 15.32**). Solitary or colonial, attached as adults or free-swimming, tunicates filter water with a special mucus plankton net capable of trapping a wide variety of microscopic food particles. The mucus is generated by a long glandular seam on one side of a basketlike interior structure (the pharynx); it is then driven by tiny cilia around to the other side and is collected and swallowed by an esophagus leading to a small stomach. Salps, related zooplanktonic forms, act almost like miniature jet engines—taking in water at one end, filtering it, and ejecting it from the other end to force themselves ahead. It stretches the imagination to consider these animals within the same phylum as seagulls or dolphins or people, but all chordate embryos share the same fundamental architecture.

Amphioxus (= sharp at both ends), another invertebrate chordate, is a small, semitransparent animal that buries itself in sand in shallow marine waters worldwide (**Figure 15.33**). *Amphioxus* swims by undulating its body in a fishlike motion and feeds by combing the water for microscopic organisms. The animal's importance lies in its well-developed dorsal tubular nerve cord, which is very similar to the spinal cord of a vertebrate. *Amphioxus* is a transitional form, an invertebrate with some vertebrate features. With this shining, 3-centimeter (1½-inch) animal we complete our cursory survey of the invertebrates and turn our attention to the vertebrates.

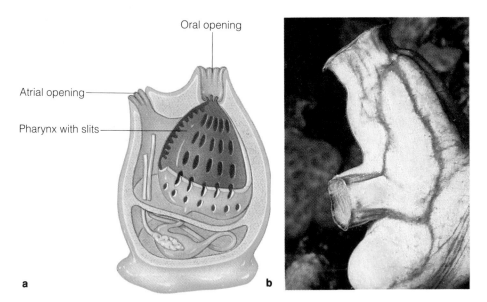

Oral opening

Atrial opening

Pharynx with slits

a   b

Figure 15.32 A tunicate. (a) Body plan showing the oral opening, through which water enters the animal's filtration system, and the atrial opening, from which it exits. (b) An adult tunicate.

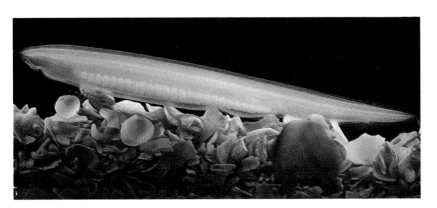

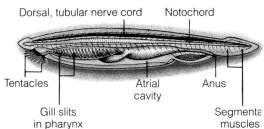

Dorsal, tubular nerve cord   Notochord

Tentacles   Atrial   Anus
           cavity

Gill slits   Segmenta
in pharynx   muscles

Figure 15.33 *Amphioxus*, an invertebrate chordate whose notochord and dorsal tubular nervous system make it a transitional species between invertebrates and vertebrates.

Q-AND-A

1. What's more important in classifying an organism—noting its similarities to other organisms or its differences?

All life-forms on Earth are variations on the same biochemical theme. All organisms synthesize proteins, bind energy into food, transfer information between generations, and move substances through membranes in basically the same way. No one in Linnaeus's time could have foreseen this modern message of biological unity. Linnaeus and his colleagues were deluged with new specimens from every corner of the world. They naturally concentrated on the diversity of the organisms they were classifying, and they arranged the taxonomic hierarchy accordingly.

Today's taxonomists are interested in both unity and diversity. They use a classification system based on five kingdoms (rather than the two kingdoms of living things used by Linnaeus) because they have discovered fundamental structural differences within organisms of the original groups. The number of phyla has increased for the same reasons. Yet biologists have also learned to prize unity. Seeking differences as a basis for including an organism in a category has given way to seeking similarities with other organisms already in the category. The difference may sound subtle, but the change in philosophy is profound. The old biology-as-catalog concept has been replaced by an appreciation for the underlying sameness of life on Earth.

2. **What do you mean when you say a phylum is *diverse*?**

Diversity is a measure of the structural or physiological variability within a group. Organisms within some groups at a certain taxonomic level (phylum, class, and so on) might look quite uniform, but organisms within another group at the same taxonomic level might hardly be recognizable as being related. For instance, members of phylum Cnidaria seem reasonably similar: People can generally see a relationship between the polyps or medusae of coral, anemones, and jellyfish. The phylum Chordata, in contrast, is hugely diverse. Elephants and tunicates in the same phylum? Only by careful observation of stages in the organisms' development can the researcher be certain that the underlying architectural similarities outweigh the eventual differences.

3. **Why are scientific names so complicated?**

They aren't. A scientific name usually describes an organism in a language that is stable. It applies to an exact species of animal or plant, and it allows you to search the scientific literature to discover what is known about the organism. Scientific names almost never change as common names can. Using the scientific name can prevent confusion and lessen the chance of accidentally duplicating research.

The first name of the binomial pair, the genus (generic) name, is equivalent to your surname (last name). The second name, the species (specific) name, is equivalent to your first name. Scientific names are italicized or underlined, and the first letter of the generic name is always capitalized. You use scientific names more often than you realize. For instance, Gorilla is both the common and the generic name of that formidable animal. The same goes for Hibiscus, Octopus, and many others.

Why use Latin and Greek names? When the Linnaean system was invented, every European scholar read Latin (and most read Greek) in addition to their native language. No matter where they lived, academics from all Western countries could communicate with each other in Latin up to about the beginning of this century. Another reason to use these classical languages in nomenclature is that word meanings don't change. To illustrate how important that stability is, consider the saying, "The exception proves the rule." It doesn't make much sense until you understand that our English word prove formerly meant "test." Try it that way, and this old saying is logical!

Finally, consider unfortunate common names like jellyfish or starfish. These animals are not fish at all. With scientific names, we don't have that problem.

4. **Representatives of the phylum Mollusca are both the largest and the most intelligent invertebrates. Yet mollusks are *not* considered the most successful nor the most highly evolved invertebrates. Why not?**

The most often used measure of success in biology is the number of species and individuals within a group. The number of arthropod species and individuals greatly exceeds the number of molluskan species and individuals; so arthropods are more successful.

The question of evolutionary position—that is, which group is most highly evolved—is a matter of some controversy, but consider this automotive analogy. Modern cars possess a number of sophisticated technological features, such as electronic fuel injection, turbochargers, independent suspension, high-speed radial tires, antilock brakes, aerodynamic enhancements, nifty stereo systems, and so forth. All of these innovations could be fitted to a 1959 Cadillac, but the older car's primitive chassis design would limit the performance of the total package. Cephalopod mollusks are an old chassis design. Excellent eyes and relatively high intelligence have been added to the ancient chassis, but the physical limitations of musculature and the lack of a jointed skeleton do not allow the class to exploit the full potential of these "options." Though well represented in the fossil record, fewer than 500 species of cephalopods exist today. They may represent an evolutionary dead end.

Terms and Concepts to Remember

animal	Cephalopoda	Echinoidea	metamerism	phylum	radial symmetry
Animalia	chitin	exoskeleton	Mollusca	Plantae	scientific name
Annelida	chiton	Gastropoda	molt	Platyhelminthes	species
Arthropoda	Chordata	hierarchy	natural system of	Polychaeta	suspension feeder
artificial system of	chromatophore	Holothuroidea	classification	polyp	taxonomy
classification	Cnidaria	invertebrate	Nematoda	Polyplacophora	tunicate
Asteroidea	cnidoblast	kingdom	notochord	Porifera	water-vascular
bilateral symmetry	Crustacea	Linnaeus, Carolus	Ophiuroidea	Protista	system
Bivalvia	Echinodermata	medusa	oxygen revolution	protozoa	

Study Questions

1. What is an animal? How is an animal different from an autotroph? From a protist?

2. When did the first true animals evolve? What atmospheric changes had to happen before animal life was possible? Are descendants of most of the early forms of animal life represented in the ocean today?

3. How many living kingdoms did Linnaeus invent? How many do biologists use today? What are they? What major characteristics determine an organism's placement within each of them?

4. How does a natural system of classification differ from an artificial system? Can you give an example of each? Was the hierarchy-based system invented by Linnaeus natural or artificial? What *is* a hierarchy-based system?

5. What are the advantages of a scientific name? What are the characteristics of a scientific name?

6. This chapter has touched on nine animal phyla (and the heterotrophic protists informally termed *protozoa*). List them and give an example of each.

7. Which animal phylum is most successful? How is success usually defined? What structural advances contribute most to that phylum's immense success?

8. Which phylum contains the largest representatives? The most intelligent?

9. Which phyla exhibit radial symmetry? Bilateral symmetry? No symmetry?

10. How can an arthropod grow within a "tailored" shell? How can an animal grow without getting bigger, or get bigger without growing?

For Further Study

Abbott, R. T. 1972. *Kingdom of the Seashell*. New York: Crown. Pictures illustrating the startling diversity of the phylum Mollusca.

Amato, I. 1991. "Stuck on Mussels: A Mollusk's Natural Grasp of Adhesive Science Captivates Biochemists." *Science News*, January.

Barnes, R. D. 1987. *Invertebrate Zoology*. 5th ed. Philadelphia: W. B. Saunders. A standard text.

Buchsbaum, M., R. Buchsbaum, R. Pearse, and J. Pearse. 1987. *Living Invertebrates*. Palo Alto, CA: Blackwell Scientific. Beautifully illustrated and readable; a classic.

Cloud, P. 1988. *Oasis in Space: Earth History from the Beginning*. New York: Norton. Well-argued history of the oxygen revolution. See especially his Figure 13.11.

Glaessner, M. F. 1984. *The Dawn of Animal Life: A Biohistorical Study*. Cambridge: Cambridge University Press.

Gould, S. J. 1989. *Wonderful Life: The Burgess Shale and the Nature of History*. New York: Norton. Information about the mid-Cambrian "snapshot" mentioned in the text.

Halstead, B. W., et al. 1990. *A Color Atlas of Dangerous Marine Animals*. Boca Raton, FL: CRC Press.

Hamner, W. 1994. "A Killer Down Under: Deadly Jellyfish of Australia." *National Geographic*, August, 116–30. Good advice on giving these lethal organisms plenty of room.

Jørgensen, C. B. 1966. *Biology of Suspension Feeding*. New York: Pergamon Press.

Kozloff, E. N. 1990. *Invertebrates*. Philadelphia: W. B. Saunders. Outstanding text, very well written by an expert marine biologist.

Margulis, L., and K. Schwarz. 1988. *Five Kingdoms*. 2d ed. San Francisco: Freeman. Excellent overview of the difficulties in classifying organisms into phyla.

May, R. M. 1988. "How Many Species Are There on Earth?" *Science* 241 (no. 4872): 1441–49. Perhaps as many as 50 million!

Mayr, E., E. Linsley, and R. Usinger. 1953. *Methods and Principles of Systematic Biology*. New York: McGraw-Hill. A classic text on taxonomic methods and philosophy.

Milne, D. H. 1995. *Marine Life and the Sea*. Belmont, CA: Wadsworth. A new and complete treatment of marine invertebrates and their ecological setting.

Ricketts, E., J. Calvin, and J. Hedgpeth. 1985. *Between Pacific Tides*. 5th ed. Stanford, CA: Stanford University Press. The most recent edition of an influential and beautifully written book that has stimulated generations of marine biologists.

Rossen, C. L., and S. W. Tolle. 1989. "Management of Marine Stings and Scrapes." *Western Journal of Medicine* 150 (January): 97–100. What to do next.

Russell-Hunter, W. D. 1968. *A Biology of Lower Invertebrates*. New York: Macmillan. This and the companion volume are fine introductions for the interested layman.

Russell-Hunter, W. D. 1969. *A Biology of Higher Invertebrates*. New York: Macmillan.

Wilson, E. O. 1992. *The Diversity of Life*. New York: Harvard University Press, Belknap Press. A distinguished biologist discusses the unity and diversity of life, and its history on the Earth.

MARINE ANIMALS II: THE VERTEBRATES

" . . . silver-shining, swift, strong, streamlined . . ."

Vertebrates are the members of the phylum Chordata that possess backbones. The word is derived from *verte-bratus*, meaning "jointed," a reference to the segments of the backbone. About 95% of all chordates are verte-brates—nearly 50,000 species. Here we find the familiar creatures drawn by generations of children—the fishes, frogs, lizards, chickens, cats, and dogs most of us first think of when we hear the word *animal*.

Vertebrates are some of the ocean's most conspicuous and intriguing organisms, with shapes and behaviors that can stretch the imagination. Among the vertebrates are huge whales that weigh 110 tons and adult fish so tiny that 4,500 of them weigh barely a pound. There are

birds—albatrosses—that can stay aloft over the water for more than three months without alighting. There are strange bioluminescent fishes with mouths so large that they can consume prey almost twice their own size. There are turtles that find their way unerringly back to an isolated stretch of breeding beach after a decade at sea. There are birds that fly through water and fish that fly through air, snakes more poisonous than any found on land, huge reptiles larger than the average dinosaur, sharks that passively comb the water for food and sharks that relentlessly pursue large prey.

Like all organisms, vertebrates are adapted for success within their particular environments. The nature of

A school of yellowfin tuna.

vertebrate adaptations often amazes us. Consider the members of the tuna family, the ocean's fastest and widest-ranging animals. The body of a tuna is dedicated to speed. Its fins retract into slots, its eyes form a smooth surface with the rest of the head, and it may consume as much as 25% of its weight in food each day. Indeed, a tuna uses so much energy that one of its greatest physiological problems is to avoid overheating! The biological equivalent of the legendary Flying Dutchman, these powerful fishes are fated to travel continuously. They depend on forward motion to pass oxygen-rich water over their gills. If they ever stopped they would suffocate, and their massive bodies would fall to the depths.

Tuna and their relatives swim enormous distances and exhibit astonishing bursts of speed. Studies have shown that albacore tuna regularly migrate from the coast of California to Japan and back, a one-way trip of 8,500 kilometers (5,300 miles), moving at an average speed of not less than 26 kilometers (16 miles) per day. Tagged bluefin tuna have traveled at least 7,770 kilometers (4,830 miles) across the North Atlantic in 119 days—that is, over 65 kilometers (40 miles) each day. In reality, the bluefins must have traveled much farther, continually detouring from a straight-line course to hunt for food. The fastest tuna can maintain speeds of over 75 kilometers (47 miles) per hour, and the sailfish, a close relative, can rocket to 110 kilometers (68 miles) per hour for a short time.

These magnificent fishes are valued for their meat, especially in Japan where they are highly prized for sashimi, the raw fish in sushi. In the Tokyo fish market, an old and fat bluefin tuna sold for $69,273 in 1992, a record for a single fish! In 1989 a school of huge bluefin appeared off the California Channel Islands—the largest weighed 458 kilograms (1,008 pounds). Many lucky fishermen paid off home mortgages and boat loans in a week of heroic fishing. But it is the living animal that provides the greatest inspiration: silver-shining, swift, strong, and streamlined, these silent nomads slip through the ocean more than a million miles in a lifetime.

CHAPTER OVERVIEW

Vertebrates (animals with backbones) have evolved intricate and effective adaptations for capturing prey, avoiding danger, maintaining thermal and osmotic balance with their surroundings, and competing for space.

There are six living classes of marine vertebrates. The three classes of fishes have been most successful in colonizing the marine environment, but reptiles, birds, and mammals have also established important populations in the ocean. Marine mammals represent the high point of the evolution of size and intelligence in marine life.

VERTEBRATE EVOLUTION AND CLASSIFICATION

Like other chordates, vertebrates had a remote marine ancestor. The first chordates had the stiffening notochord that gives the phylum its name, but they lacked the backbones of true vertebrates. The line to modern vertebrates probably passed through an *Amphioxus*-like predecessor that lived in the ocean more than 500 million years ago. Besides the backbone, vertebrate chordates differ from the invertebrate chordates discussed in Chapter 15 by having an internal skeleton of calcified bone or cartilage (or both). This scaffold allows uninterrupted support during growth; it also protects vital organs and provides a foundation to which muscles may attach to permit the strength and rapid responses characteristic of active animals. The vertebrate skull, a special unit of the skeleton, provides secure housing for the brain, eyes, and other sense organs that have made the evolution of intelligence possible. Only the simplest vertebrates lack jaws. The central nervous system is partially enclosed within the backbone, which extends from the skull, and the pairs of nerves passing between the vertebral segments allow rapid and efficient communication between brain and body. The most abundant and successful vertebrates are the fishes, which exist in an extraordinary variety of form and habitat. Least successful in the marine environment are the amphibians.

As is the case with any natural system of organization, vertebrate classification reflects our understanding of vertebrate evolution. **Figure 16.1** indicates the likely evolutionary relationship among vertebrate species. Note that all higher vertebrates appear to be derived from fishlike ancestors. In an architectural sense all higher vertebrates (including amphibians, reptiles, birds, and mammals) are highly modified, four-limbed, air-breathing fish!

Vertebrate taxonomy has itself evolved markedly over the last 25 years. In the past, taxonomists listed five classes within the chordate subphylum **Vertebrata:** fish, amphibians, reptiles, birds, and mammals. This scheme did not recognize the true importance and independence of the three major groups of fishes, however. Recent vertebrate taxonomies elevate the fish groups to the status of class, equal with amphibians, reptiles, birds, and mammals. A vertebrate taxonomy reflecting this view is included as **Table 16.1.** We will follow this scheme in our brief introduction to marine vertebrates.

FISHES

Fishes are vertebrates that usually live in water and that possess gills for breathing and fins for swimming. There are more species of fishes, and more individuals, than species and individuals of all other vertebrates *com-*

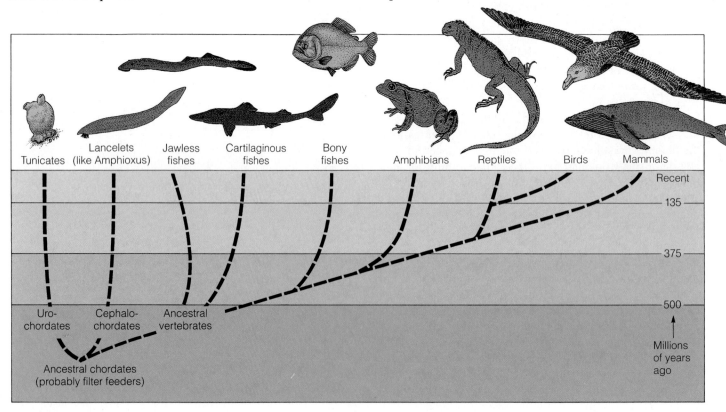

Figure 16.1 One proposed family tree for the vertebrates and their relatives, the invertebrate chordates.

Table 16.1 Classification of Major Groups Within Phylum Chordata

Group	Examples
Phylum Chordata	
Subphylum Urochordata	Tunicates[a]
Subphylum Cephalochordata	*Amphioxus*[a]
Subphylum Vertebrata	
Class Agnatha (jawless fishes)	Hagfishes, lampreys
Class Chondrichthyes (fishes with cartilage skeletons)	Sharks, skates, rays
Class Osteichthyes (fishes with bony skeletons)	Tuna, halibut, sea horse
Class Amphibia	Frogs, salamanders (no true marine representatives)
Class Reptilia	Crocodiles, turtles, sea snakes, and marine iguanas
Class Aves	Albatrosses and petrels, pelicans, gulls, penguins
Class Mammalia	Whales, seals and sea lions, otters, sirenians

[a]Invertebrate chordate discussed in Chapter 15.

bined—not a surprising fact considering the vast oceanic habitat the planet provides. Fishes range in adult length from less than 10 millimeters to over 20 meters (0.4 inch to 60 feet), and they weigh from about a 0.1 gram to about 41,000 kilograms (0.004 ounce to 45 tons). Some fish are capable of short bursts of speed in excess of 120 kilometers (75 miles) per hour; some species hardly ever move.

Fishes live near the surface and at great depth, in warm water and cold, even frozen within ice or dried in balls of mud. Like other cold-blooded organisms, or **poikilotherms,** the great majority of fishes are incapable of generating and maintaining a steady internal temperature from metabolic heat; thus the internal body temperature of a fish is usually the same as that of the surrounding environment. About 40% of the 30,000-plus fish species live all or part of their lives in fresh water; 60% live exclusively in seawater. Fishes have evolved to fit almost every conceivable watery habitat, but they are most numerous on the bottom or in productive seawater over the continental shelves. Some species have a "sixth sense," an ability to detect small changes in the electrical field surrounding their bodies, that assists in the detection of prey or avoidance of predators. Some electric eels, catfish, and rays can use internally generated electricity for defense and offense.

The first fishes probably evolved in the ocean around 500 million years ago. These jawless animals were little more than motile sucking digestive tubes, but they did have the structural advantages of the chordate body plan. The earliest jawed fishes, with their grasping and crushing mouths, were far more successful at feeding on invertebrates with shells or exoskeletons than their jawless predecessors had been. Early jawed fishes were also equipped with paired fins, which stabilized their movements and minimized pitching and rolling when they attacked their prey. The numbers and types of jawed fishes increased dramatically beginning about 410 million years ago. By the end of the Devonian period, the so-called Age of Fishes from 408 to 360 million years ago, jawed fishes had radiated into a vast number of aquatic and marine habitats in which their dominance has remained unchallenged. The ancestral jawed fishes gave rise to cartilaginous fishes (animals whose skeleton is made of stiff cartilage) and bony fishes. Only a few jawless forms, such as lampreys and hagfishes, have persisted to the present day.

The three living fish classes are as different from one another as they are different from amphibians, reptiles, birds, and mammals. We will look briefly at each in turn, beginning with the jawless fishes and ending with the advanced bony fishes.

Class Agnatha

Hagfishes and lampreys—members of the class **Agnatha** (*a* = lacking, *gnathos* = jaw)—lack jaws and have no paired appendages to aid in locomotion. Their thick, snakelike bodies are pierced by gill slits and (in some species) the openings of slime glands. Their round, sucking mouths are surrounded by organs sensitive to touch and smell. Agnathan eyes are degenerate and covered by thick skin. The body ends in a flattened tail that undulates to provide forward motion. The skin consists of resilient fibrous layers currently in high demand in South Korea for the manufacture of "eelskin" leather goods sold in upscale leather shops and department stores. Fewer than 50 species of agnathans are known.

The pinkish, soft hagfishes (**Figure 16.2a**) live in colonies on continental shelf sediments, where they burrow for polychaete worms or scavenge for weak or dead organisms. They prefer feeding on soft inner flesh and internal organs of their prey, which they abrade with a rasping tongue. If a piece of food is too large, a hagfish will tie its flexible body into a loose knot, pass the knot toward the head, and brace against the prey for a better tearing grip (**Figure 16.2b**). A hagfish defends itself primarily by producing copious quantities of clinging slime from glands along the side of its body. This slippery, odorous mass deters all but the most determined would-be diner. The hagfish passes a knot down its body to shed the slime soon after danger has passed.

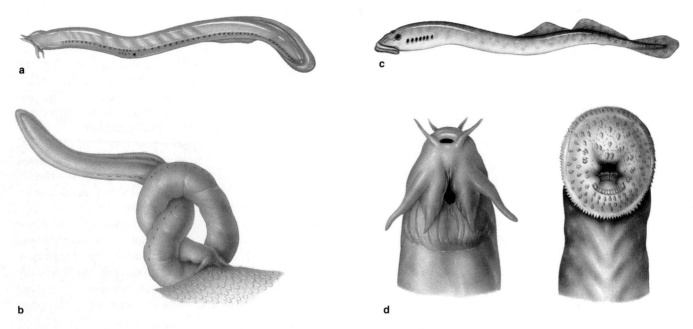

Figure 16.2 A comparison of hagfish and lamprey, jawless fishes of the class Agnatha. (a) The hagfish. (b) Hagfish knotting its body to tear off a chunk of prey. (c) The lamprey. (d) Mouths of hagfish (left) and lamprey (right). Lengths: hagfish, 75 centimeters (30 inches); lamprey, 90 centimeters (35 inches).

Lampreys (**Figure 16.2c**), a group closely allied to hagfishes, possess a toothed, funnel-shaped mouth with which they can wear a passage into another vertebrate. Continuous sucking on this seeping wound provides nourishment for the lamprey. Prey are usually bony fish, but lampreys have been found feeding on several species of whales and porpoises. A lamprey will usually detach before it has killed the host. The largest hagfishes and lampreys can reach 1 meter (3.3 feet) in length, but most are smaller.

Class Chondrichthyes

Members of the group that includes sharks, skates, rays, and chimaeras have prowled the world ocean for at least 280 million years. Sharks have a history twice as long as that of dinosaurs. Fossilized teeth, fin spines, and shark's eggs found in marine sediments tell us that today's sharks and rays are not much different from their ancient ancestors.

All members of the class **Chondrichthyes** (*chondros* = cartilage, *ichthys* = fish) have a skeleton made of a tough, elastic tissue called **cartilage,** the material that gives your ears and nose their shape. Though there is some calcification in the cartilaginous skeleton, true bone is entirely absent from this group. Unlike the more primitive agnathans, these fish have jaws with teeth, paired fins, and often active life-styles. Sharks and rays tend to

be larger than either agnathans or bony fishes, and except for some whales, sharks are the largest living vertebrates. Chimaeras, such as ratfishes, are comparatively rare forms found mostly at and below middle depths.

Only a small fraction of all fish species are members of class Chondrichthyes. About 350 species of sharks and 320 species of rays are known to exist. Nearly all are marine, though a few species inhabit estuaries and a very few are permanent inhabitants of fresh water. Although there are many exceptions, sharks tend to favor swimming through open water, whereas rays tend to be found on or near the bottom.

Skates and rays have a flattened appearance, with spreading pectoral fins attached to their bodies to form a triangular or rounded winglike shape. They glide through the water with a slow, rhythmic flapping of their fins. Neither sharks nor rays have gas bladders, and both are slightly negatively buoyant: They will slowly sink if they stop swimming. The skin of rays is usually smooth and greatly variable in color. (Sharks, in contrast, have a rough skin with thousands of tiny toothlike projections.) One family of rays carries a defensive barb at the base of the tail capable of inflicting a serious wound. Another ray family contains members 7 meters (22 feet) across and weighing 1,700 kilograms (1.9 tons), an example of which is the giant manta (**Figure 16.3**). Still another produces a violent electric shock capable of stunning prey or disabling a human diver. The largest rays feed on plank-

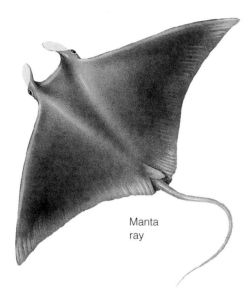

Figure 16.3 A manta ray, sometimes called a devilfish because of the cartilaginous protuberances of the head, which are used to guide plankton into the mouth.

ton, but most smaller species crush mollusks and arthropods with smooth calcified plates in the mouth.

Sharks have an undeservedly bad reputation. More than 80% of shark species are less than two meters (6.6 feet) long as adults, and only a few of the remaining 20% are aggressive toward humans. Like other cartilaginous fishes, sharks are not very intelligent and certainly don't hold grudges or behave in the malignant ways so vividly portrayed in recent popular novels and movies. Still, some sharks are indeed dangerous to humans, and the great white sharks in the genus *Carcharodon* (*karcharos* = sharp, *odontos* = tooth) are perhaps the most dangerous of all (**Figure 16.4a**). These swimmer's nightmares attain lengths of 7 meters (23 feet) and weigh up to 1,400 kilograms (3,000 pounds). (Reports of even larger great whites probably have resulted from misidentification of large, harmless species.) Great whites are not actually white, but rather grayish brown or blue above and creamy on the lower half. A dangerous relative, the mako shark (**Figure 16.4c**), reaches lengths of 4 meters (13 feet) and is even known to attack small boats. These and other predatory species, such as tiger sharks and hammerhead sharks, are attracted to prey by vibrations in the water,

a

b

Figure 16.4 (a) The great white shark, genus *Carcharodon*, a predator of seals, sea lions, and large fish. The great white shark is also one of the most dangerous of sharks encountered by human swimmers. (b) The jaws of a great white shark. (c) The mako shark, genus *Isurus*, is also dangerous to swimmers and is common to the Atlantic and Pacific oceans.

c

Figure 16.5 A diver hitches a ride on a whale shark. Unless he is struck by the fish's tail as he dismounts, the diver is in no danger. Divers are not part of the diet of this type of shark.

which they detect with sensitive organs arrayed in lines beneath the surface of their skin. Smell also plays an important role in hunting their prey, usually fish and marine mammals.

Though most famous, the so-called man eaters are not the largest of shark species. This honor goes to the immense, warm-water whale sharks in the genus *Rhineodon* (*rhine* = a rasp, a reference to the fish's rough skin), which reach sizes in excess of 18 meters (60 feet) and 41,000 kilograms (90,000 pounds) (**Figure 16.5**). Whale sharks and their somewhat smaller relatives, the basking sharks, are docile and present little threat to people. These greatest of fishes swim slowly near the surface with their huge mouths open, feeding on plankton. They may filter as much as 2,200 cubic meters (2,500 tons) of water per hour through a fine mesh of gill rakers. Accumulated plankton are periodically backflushed into the mouth, where they are concentrated for swallowing.

Class Osteichthyes

The 27,000-plus species of bony fishes, members of class **Osteichthyes** (*osteum* = bone, *ichthyes* = fish), owe much of their great success to the hard, strong, lightweight skeleton that supports them. These most numerous of fish—and most numerous and successful of all vertebrates—are found in almost every marine habitat, from tide pools to the abyssal depths. Their numbers include the air-breathing lungfishes and lobe-finned coelacanths, whose ancient relatives broke from the path of fish evolution to establish the dynasties of land vertebrates.

About 90% of all living fishes are contained within the osteichthyan order **Teleostei** (*teleos* = perfect, *osteon* = bone), which contains the cod, tuna, halibut, perch, and other familiar species (**Figure 16.6**). Within this large category are different fishes with gas-filled swim bladders to assist in maintaining neutral buoyancy, independently movable fins for well-controlled swimming and communication, great speed for pursuit or avoidance of predators, highly effective camouflage, social organization, the ability to cluster together in defensive schools, orderly patterns of migration, and other advanced features. Their economic importance is great: Some 70 million metric tons (77 million tons) of bony fishes are taken annually from the ocean to help satisfy the human demand for protein.

THE PROBLEMS OF FISHES

What problems are unique to a fish? Seawater may seem to be an ideal habitat, but living in it does present difficulties. Water is about 1,000 times denser and 100 times more viscous than air, and it impedes motion more effectively at low speeds. How can a fish best move through it? How can a fish maintain its vertical position

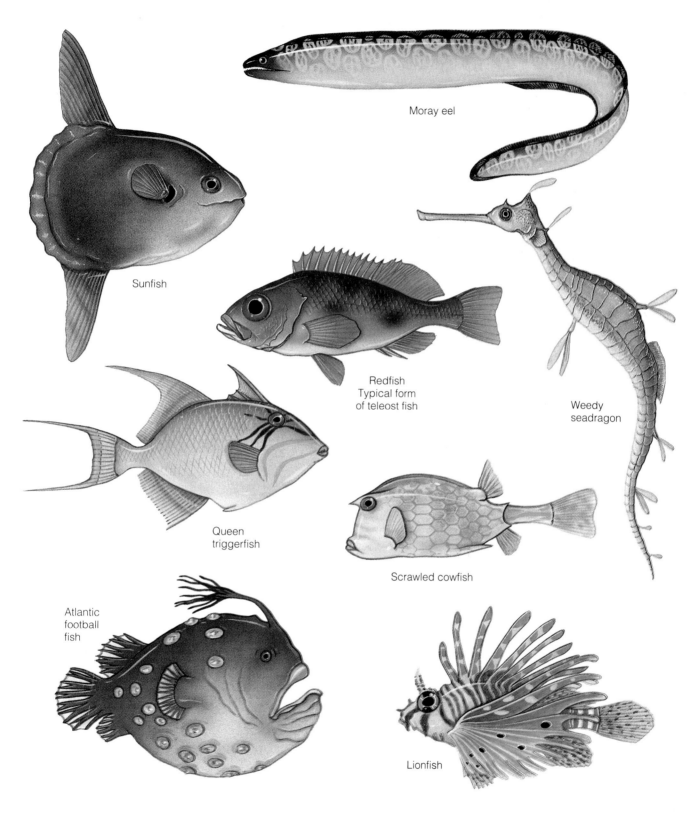

Moray eel

Sunfish

Redfish
Typical form
of teleost fish

Weedy
seadragon

Queen
triggerfish

Scrawled cowfish

Atlantic
football
fish

Lionfish

Figure 16.6 Some of the diversity exhibited by teleost (bony) fishes. These fishes are not drawn to the same scale.

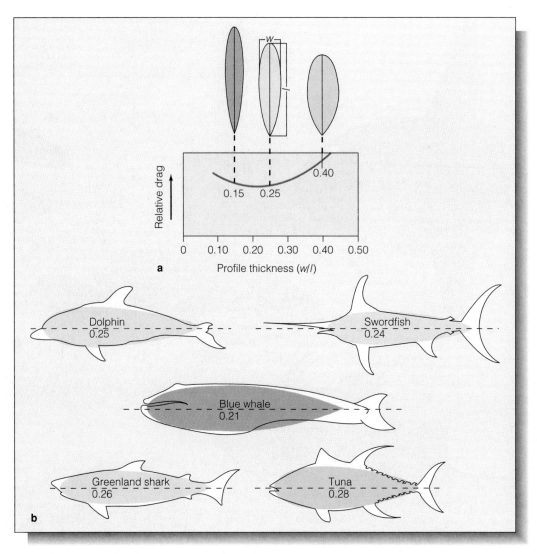

Figure 16.7 The effect of body shape on drag. (a) Streamlined profiles with width (*w*) equal to approximately one-fourth of length (*l*) minimize drag. (b) Width-to-length ratios (*w*/*l*) for several swimming vertebrates. Note that the optimum ratio is near 0.25.

in the water column? Must a fish swim constantly to offset the weight of muscle and bone? What about breathing? Can oxygen and carbon dioxide be exchanged efficiently underwater? What about osmotic balance? How can some species of fish move from fresh water to salt water? How is salt balance maintained across the thin gill membranes? How can predators be thwarted? Can group behavior or camouflage or subterfuge be employed? There would seem to be many problems, but these most successful vertebrates have structures and behaviors to cope.

Movement, Shape, and Propulsion

Active fish usually have streamlined shapes that make their propulsive efforts more effective. A fish's resistance to movement, or *drag*, is determined by frontal area, body contour, and surface texture. Drag increases geometrically with increasing speed. Faster-swimming fish are therefore more highly modified to minimize the slowing effects of the dense, relatively sticky medium in which they live.

The most effective antidrag shape is the tapering, torpedolike body plan shown in **Figure 16.7a.** Such shapes produce minimum drag when they are circular in cross section, when their greatest width is about one-fourth their length, and when the point of maximum width occurs about two-fifths of the distance from the leading tip. **Figure 16.7b** shows some very efficient swimming vertebrates along with their profile ratio (width to length). Rapidly swimming marine mammals have a similar shape; fish have no monopoly on streamlining.

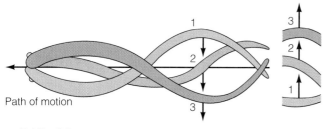

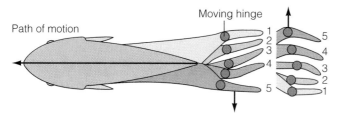

a Eel-like fishes

Moving hinge

b Advanced fishes

Figure 16.8 How a fish's body shape affects its efficiency of movement. The undulating movement of eel-like fishes (a) requires more energy to generate forward motion than the hinged-tail movement of the more advanced fishes (b).

Figure 16.9 The tunny, a pelagic teleost adapted for swimming at high sustained speeds.

A fish's forward thrust comes from the combined effort of body and fins. Muscles within slender, flexible fish (such as eels) cause the body to undulate in S-shaped waves that pass down the body from head to tail in a snakelike motion. The eel pushes forward against the water much as a snake pushes against the ground (**Figure 16.8a**). This type of movement is not very efficient, however: The body must wave back and forth across a considerable distance, exposing a large frontal area to the water; and the long body length (relative to width) needed to propagate the wave requires increased surface area, which increases drag. A more efficient swimming mechanism is shown in **Figure 16.8b**. More advanced fishes have a relatively inflexible body that undulates rapidly through a shorter distance; they also have a hinged, scythelike tail to couple muscular energy to the water. The fish's body can be shorter and can face more squarely in the direction of travel; so the drag losses are lower.

How efficient are the best swimmers, and how fast can they swim? Estimates vary, but it is thought that in the fastest fish between 60% and 80% of muscle force delivered to the tail fin results in forward motion. Swordfish and marlin can reach 120 kilometers (75 miles) per hour in short bursts! Some of the fastest tuna (and a few swift sharks) sustain high speeds by maintaining their internal body temperature a few degrees above that of the surrounding water. This adaptation permits them to oxidize food more rapidly and generates greater mus-

cle power per unit of weight. The muscular tunny shown in **Figure 16.9** incorporates all these sustained-speed advances; few other animals are capable of such prodigious power output for such a long time. The bluefin tuna, a close relative, swims at a more or less constant 14 kilometers (9 miles) per hour and may travel more than 1.6 million kilometers (1 million miles) in a lifetime!

Maintenance of Level

The density of fish tissue is typically greater than that of the surrounding water; so fishes will sink unless their weight is offset by propulsive forces or by buoyant gas- or fat-filled bladders. Cartilaginous fishes have no **swim bladders** and must swim continuously to maintain their position in the water column. Sharks generate lift with an asymmetrical tail and fins that act like airplane wings. Bony fishes that appear to hover motionless in the water usually have well-developed swim bladders just below their spinal columns. The volume of gas in these structures provides enough buoyancy to offset the animal's weight. The quantity of gas is controlled both by secretion and absorption of gas from the blood and by muscular contraction of the swim bladder to compensate for temporary changes in depth.

A swim bladder may make life easier for a slow-moving teleost, but the fastest bony fishes lack swim

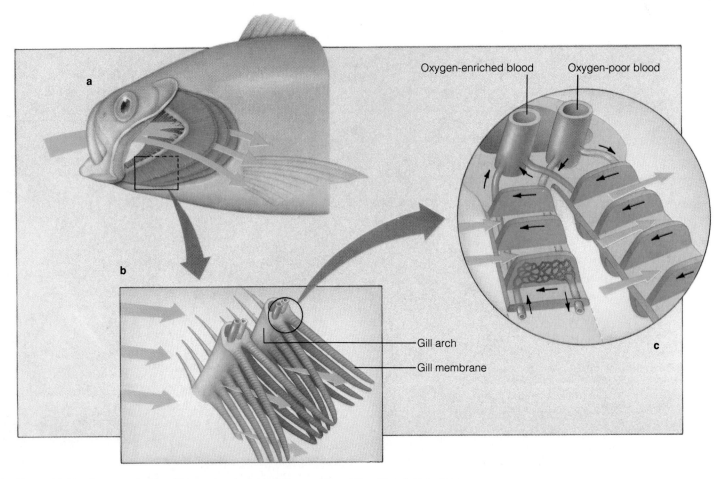

Oxygen-enriched blood   Oxygen-poor blood

Gill arch

Gill membrane

a

b

c

Figure 16.10 Cutaway drawing of a mackerel, showing the position of the gills (a). Broad arrows in (b) and (c) indicate the flow of water over the gill membranes of a single gill arch. Small arrows in (c) indicate the direction of blood flow through the capillaries of the gill filament in a direction opposite to that of the incoming water. This mechanism is called *countercurrent flow.*

bladders. Why? Fast, powerful predators such as tuna, mackerel, and swordfish must be able to chase prey between depths. The expansion and compression of gas in the bladder would vary rapidly with changing depth, and the risk of rupture would be too great. These speedy predators have power to spare and don't seem inconvenienced by their slight negative buoyancy.

Gas Exchange

How can fish breathe underwater? **Gas exchange,** the process of bringing oxygen into the body and eliminating carbon dioxide, is essential to all animals. At first glance the task may seem more difficult for water breathers than for air breathers, but air-breathing animals add an extra step. We air breathers must first dissolve gases in a thin film of water in our lungs before they can diffuse across a membrane. (The ease with which most gases dissolve in water was discussed in Chapter 7.)

Fish take in water containing dissolved oxygen at the mouth, pump it past fine **gill membranes,** and exhaust it through rear-facing gill slits. The higher concentration of free oxygen dissolved in the water causes oxygen to diffuse through the gill membranes into the animal; the higher concentration of CO_2 dissolved in the blood causes CO_2 to diffuse through the gill membranes to the outside. The gill membranes themselves are arranged in thin filaments and plates efficiently packaged into a very small space (**Figure 16.10**). Water and blood circulate in opposite directions—in a countercurrent flow—which increases transfer efficiency.

An active fish like a mackerel requires so much oxygen and generates so much waste CO_2 that its gill surface area must be 10 times its body surface area. (Sedentary fishes have proportionally less gill area.) With their large gill area and countercurrent flow, active fish extract about 85% of the dissolved oxygen in water flowing past their gills. Air-breathing vertebrates, by contrast, extract only about 25% of the oxygen from air entering their lungs.

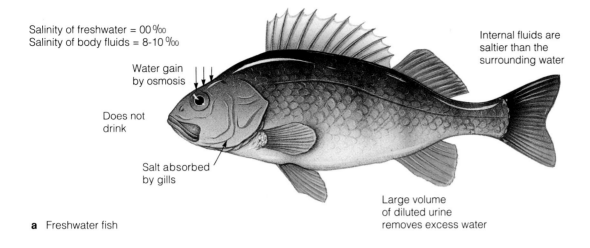

Salinity of freshwater = 00‰
Salinity of body fluids = 8-10‰

Water gain by osmosis

Does not drink

Salt absorbed by gills

Internal fluids are saltier than the surrounding water

Large volume of diluted urine removes excess water

a Freshwater fish

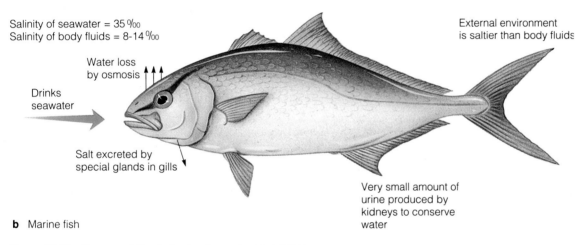

Salinity of seawater = 35‰
Salinity of body fluids = 8-14‰

Water loss by osmosis

Drinks seawater

Salt excreted by special glands in gills

External environment is saltier than body fluids

Very small amount of urine produced by kidneys to conserve water

b Marine fish

Figure 16.11 Osmoregulation in freshwater and marine fishes.

Osmotic Considerations

Marine invertebrates and primitive vertebrates such as agnathans have an internal salt concentration nearly identical to that of seawater. The body fluids of more advanced vertebrates generally are only about one-third as saline as seawater; they are **hypotonic** to their oceanic surroundings. The gill and intestinal membranes of fish are necessarily thin to allow passage of gases and food molecules, but this thinness also makes the passage of water, or **osmosis,** inevitable.

As we saw in Chapter 13, osmosis moves water across membranes from regions of high water concentration to regions of low water concentration. Marine teleosts (bony fishes), with their higher relative internal concentration of water (and lower concentration of salts) per unit of fluid volume, continuously *lose* water to their environment; freshwater teleosts, with their lower rela-

tive internal concentration of water, constantly *absorb* water from their environment. If these fishes were incapable of **osmoregulation**—that is, if they had no active way of adjusting their internal salt concentration—they would quickly die of fluid imbalance.

Figure 16.11 summarizes the ways bony fishes cope with osmoregulatory difficulties. The skin of both freshwater and marine teleosts is nearly impermeable to water and salts. A freshwater fish (**Figure 16.11a**) does not drink water. It uses large kidneys to generate copious quantities of dilute urine to export the invading water, and it actively absorbs salts through its gills from both the surrounding water and from its own urine. A marine fish (**Figure 16.11b**), on the other hand, makes only small quantities of urine, actively drinks seawater (some species drinking up to 25% of their body weight per day), and eliminates excess salts through special salt-secreting cells in the gills. Bony fish consume a substantial amount of energy in these essential osmoregulatory tasks.

Chondrichthyes (sharks, rays) employ a different strategy. Their internal fluids are supplemented with urea to produce a nearly neutral osmotic pressure across gill membranes. Although the proportion of dissolved solids in their body fluids is different from that in seawater, the quantity of dissolved solids is nearly identical, and the net flow of water into or out of the gills and intestinal membranes is minimal. Shark meat has an unpleasant, bitter taste (or tastes uncomfortably rich) unless it has been soaked in seawater to wash away the urea.

Feeding and Defense

Competitive pressure among the large number of fish species has caused a wonderful variety of feeding and defense tactics to evolve. Sight is very important to most fishes, enabling them to see their prey or avoid being eaten. Even some deep-water fishes that live below the photic zone have excellent eyesight for seeing luminous cues from potential mates or meals. Hearing is also well developed, as is the ability to detect low-frequency vibrations with the **lateral-line system (Figure 16.12)**. This mechanism consists of a series of small canals in the skin and bones around the eyes, over the head, and down the sides of the body. The canals are richly supplied with nerves and connect to the surface through tiny pores in the skin. The nerves report changes in current direction, water pressure, or sonic environment to the brain. Predatory sharks use their lateral-line systems to detect prey.

Smell and taste also play a role, with some fish having taste organs located on probelike extensions around the mouth. These assist in the search for food. The bizarre flattened crossbar that gives the hammerhead shark its name may provide a kind of stereo smell used to sense differing amounts of interesting substances in the water. Salmon smell their way to their home streams after years at sea by detecting faint chemical traces characteristic of the water from the stream in which they hatched.

Defense measures are well advanced in fishes. Some fishes such as sea horses (and their relatives the box fishes) depend on the simple expedient of armor plating for protection. Others, such as the puffer fish, inflate with water and erect bristly spines to become less attractive as a snack. More subtle means of offense or defense depend on trickery—looking like something you're not, or changing color to blend with the background. These kinds of **cryptic coloration** (*kryptos* = hidden) or camouflage may be active or passive. **Figure 16.13** shows an example of passive cryptic coloration: The kelp bass closely resembles its seaweed habitat. Small animals for-

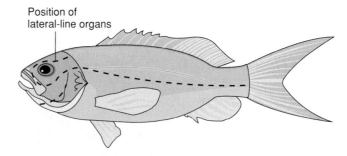

Figure 16.12 A fish's lateral-line system, which detects low-frequency vibrations. Lines of tiny pores at the fish's skin are external evidence of the lateral-line system.

aging nearby run the risk of being eaten, and the fish itself may escape the notice of a larger predator. An example of active cryptic coloration is shown in **Figure 16.14**. Here a turbot—a bottom dweller—has reproduced the pattern of the background on its own surface, and so blends inconspicuously into the surroundings. Actively swimming fish employ a different method of color blending called *countershading*. Their dark tops and silvery bottoms make the fish less obvious to predators above or below.

About a quarter of all bony fish species exhibit **schooling** behavior at some time during their life cycle. A fish school is a massed group of individuals of a single species and size class, packed closely together and moving as a unit. There is no leadership in fish schools, and the movement of fish within them seems to be controlled automatically by direct interaction between lateral-line sensors and the locomotor muscles themselves. I can personally attest to the effectiveness of schooling as a means of defense. On a few diving trips I've noticed a large moving mass just beyond the limit of clear visibility. Is it a fish school, or is it a single large animal? Many predators might not stay around long enough to find out! Schools have the added benefits of reducing chance detection by a predator, providing ready mates at the appropriate time, and increasing feeding efficiency.

Perhaps the most effective way to avoid being eaten is simply to disappear. The surface-feeding flying fishes do this by accelerating rapidly, leaping into the air, and spreading their fins to act as wings. Refractive differences between air and water prevent the surprised predator from seeing where its intended meal has gone!

AMPHIBIANS

Amphibians (frogs, salamanders, toads) are specialized vertebrates adapted to fresh water and to moist habitats on land. Only about 2,000 living species exist, and none is exclusively marine—although some large Southeast

a

b

Figure 16.13 Passive cryptic coloration in a kelp fish. (a) The fish resembles a seaweed blade in shape, color, and pattern. (b) The fish nestles within the blades, nearly disappearing from view, moving with the blades as they sway in the water.

Figure 16.14 Active cryptic coloration in a turbot. The fish is able to blend into its surroundings because of chromatophores (skin cells that can change color). Though the fish is in plain view, the only obvious features are its round, black eyes.

Asian frogs can tolerate water with a salinity of up to 28‰ for extended periods of time. Amphibians depend on the constant flow of water through their skins into their bodies to provide the fluid for the formation of urine to remove nitrogenous wastes. Placing an amphibian in seawater would cause water to flow through the skin in the opposite direction, dehydrating the animal. Their skin is too permeable to permit amphibians to colonize the marine environment.

MARINE REPTILES

Each of the three main groups within the class **Reptilia** has marine representatives: turtles, sea snakes and marine lizards (iguanas), and marine crocodiles. Like all reptiles, marine reptiles are poikilothermic, breathe air with lungs, are covered with scales and a relatively impermeable skin, and are equipped with special **salt glands** to concentrate and excrete excess salts from body fluids. Except for one widely ranging species of turtle, all marine reptiles require the warmth of tropical or subtropical waters.

Sea Turtles

The best-known and most successful living marine reptiles are the eight species of sea turtles. Unlike land turtles, from which they evolved, sea turtles have relatively small, streamlined shells without enough interior space to retract the head or limbs. The shell provides an effective passive defense, and adult sea turtles have no predators except humans. Their forelimbs are modified as flippers and provide propulsive power; their hind limbs act as rudders. The two species of green sea turtles (*Chelonia*) are the most abundant and widespread of living species (**Figure 16.15**). Green turtles range over great distances looking for the marine algae, turtle grass, and other plants on which they feed. The largest living turtle is the carnivorous Atlantic leatherback, a streamlined animal with a soft, skin-covered "shell"; it reaches

Figure 16.15 A green sea turtle.

Figure 16.16 A marine crocodile.

lengths in excess of 2 meters (6½ feet) and may weigh more than 600 kilograms (1,300 pounds).

Sea turtles are justly famous for their remarkable feats of navigation, which in some cases have been progressively increased (as you may recall from Chapter 3) by seafloor spreading. Sea turtles return at two-, three-, or four-year intervals to lay eggs on the beaches at which they were hatched. Homing behavior can be a great advantage to an animal; if the parent survived its earliest childhood at this location, it will probably be a suitable place for hatching the next generation. The navigation of green turtles to tiny Ascension Island, an emergent point of the Mid-Atlantic Ridge between Brazil and Africa, has been extensively studied. Researchers have found that the turtles use solar angle (to find latitude), wind wave direction, smell, and visual cues—first to find the island and then to discover the spot on the beach where they hatched perhaps 20 years before!

All marine turtles are in danger of extinction. Though they are protected and no longer used extensively for human food, their breeding beaches are being developed or invaded by noisy recreational pursuits. In addition, their eggs and shells are in great demand, they are often drowned in fishing and shrimping nets, and their feeding areas are increasingly damaged by pollutants. Much of the tropical turtle grass has disappeared because of human interference. In addition, some species mistake floating plastic bags and other debris for the jellyfish on which they normally feed; the plastic clogs their digestive tract, and they starve to death. The outlook for these creatures is not bright.

Marine Crocodiles

Fortunately, perhaps, there is only one living species of true marine crocodile (**Figure 16.16**). Marine crocodiles live in mangrove swamps and reefs islands and on isolated mainland shores in the tropical western Pacific. They hunt in packs and sometimes move ashore to consume any large, warm-blooded animals they can find. They are extremely aggressive, extraordinarily fast in attack, and responsible for a few human deaths each year. Male marine crocodiles of northern Australia and Indonesia have attained lengths of 7 meters (23 feet) and may weigh over a ton! (The much smaller and less dangerous Florida and Louisiana crocodilians live in fresh, brackish, and salt water.)

Marine Lizards

The only surviving marine lizard is the Galápagos marine iguana (**Figure 16.17**), which lives among the intertidal rocks and beaches of the Galápagos Islands off the Ecuadoran coast. These mottled gray to brownish-tan lizards grow to more than a meter (3.3 feet) in length and weigh about 9 kilograms (20 pounds). Although relatively graceless on land, they swim efficiently by undulatory movements of body and tail.

Galápagos iguanas live in large colonies at the water's edge and return to their settlements after foraging a short distance offshore for encrusting seaweeds. They gnaw the plants from subtidal rocks using small, spade-shaped teeth. Their ancestors presumably reached the islands from the coast of South America by riding on

Figure 16.17 A marine iguana.

Figure 16.18 An Australian sea-krait, a kind of sea snake, cruises a tropical reef. All sea snakes are venomous, and all must come to the surface to breathe air.

logs or rafts of vegetation drifting from mainland South America in the Humboldt Current.

Sea Snakes

Sea snakes are probably the most evolutionarily advanced of the marine reptiles. Fifty species of sea snakes are known; most are native to the warm Indian Ocean and western Pacific areas. None lives in the Atlantic. They hunt among the coral heads or uneven rocky bottoms and rise at intervals to breathe. With the exception of one egg-laying species, sea snakes give birth in the water to living young and never come onto land. All are highly venomous: Sea snake venoms are among the most powerful animal poisons known.

Sea snakes are effective swimmers, with flattened tails. These generally slow-moving snakes lunge at passing small fish, grasp them in their teeth, and wait as a saliva-borne neurotoxic venom seeps into the wound. (Sea snakes do not have hollow injection fangs.) The paralyzed fish is then swallowed head first.

Though not particularly aggressive, these snakes are very dangerous to humans. Members of genus *Pelamis* (**Figure 16.18**) are responsible for perhaps 20 fatalities every year, mainly among Asian fishermen who are scratched by a snake's teeth while removing it from a net. Sea snakes are most common in Southeast Asia, in the Gulf of California, and on the tropical Pacific coasts of Central and South America. There is concern that warming water will allow these unwelcome animals to extend their range to the southwestern coast of the United States.

MARINE BIRDS

Birds (class **Aves**) probably evolved from small, fast-running dinosaurs about 160 million years ago. Their reptilian heritage is clearly visible in their scaly legs and claws and in the configuration of their internal organs and skeletons. The success of the 8,600 living species of birds has resulted in large part from the evolution of feathers (derivatives of reptilian scales) used to insulate the body and to provide aerodynamic surfaces for flight. Birds (and mammals) are **homeotherms:** They generate and regulate metabolic heat to maintain a constant internal temperature that is generally higher than their surroundings.

Flying birds have light, thin, hollow bones without fatty insulation; they have forsaken the heavy teeth and jaws of reptiles for a lightweight beak. Their highly efficient respiratory systems can accept great quantities of oxygen, and their large, four-chambered heart circulates blood under high pressure. All birds lay eggs on land, and most incubate them and provide care for the young. Some seabirds may stay at sea for years, but all must eventually return to land to breed.

Only about 270 kinds of birds, about 3% of known bird species, qualify as seabirds. Most seabirds live in the Southern Hemisphere. Like the marine reptiles, seabirds have special salt-excreting glands in their heads to eliminate the excess salt taken in with their food. Salty brine from these glands may sometimes be seen dripping from the tip of their beaks. Marine birds are voracious feeders and are often found wherever the ocean teems with life.

True seabirds generally avoid land unless they are breeding; they obtain virtually all their food from the sea and seek isolated areas for reproduction.

Seabirds may be divided into four groups: tubenoses (albatrosses, petrels), pelicans and their relatives, the gull group, and penguins.

The Tubenoses

The 100 species of tubenoses of order Procellariiformes (*procella* = storm) are the world's most oceanic birds. Their prosaic common name does not convey any sense of their beauty and grace; it refers to the plumbing in their beak responsible for sensing air speed, detecting smells, and ducting saline water from the salt glands. Foraging for months across the ocean in conditions of strong winds and high waves, seeking no shelter during storms, enduring both tropical heat and polar sleet, soaring continually through air with a gliding efficiency exceeding that of the most perfectly built human sailplane—the beautiful albatrosses, petrels, and shearwaters are true masters of the sky.

The largest of the tubenoses are the magnificent wandering albatrosses (*Diomedea*), which reach a wingspan of 3.6 meters (12 feet) and a weight of 10 kilograms (22 pounds). (Ancestors of these great birds were even larger—some had wingspans in excess of 5.5 meters, or 18 feet!) The key to the albatross's success lies in its aerodynamically efficient wing (**Figure 16.19**), a very long, thin, narrow, cupped and pointed structure ideal for high-speed soaring and gliding. This beautiful wing allows albatrosses to cover great distances in search of food with very little expense of energy, flying continuously for weeks at a time. Using the uplift from wind deflected by ocean waves to stay aloft, they soar in long looping arcs. Albatrosses rarely flap their wings.

Satellite tracking data indicate that wandering albatrosses routinely cover 15,000 kilometers (9,300 miles) on foraging trips and reach speeds of 80 kilometers (50 miles) per hour. High speeds over long distances are their specialty. One bird was observed to travel 808 kilometers (502 miles) at an average speed of 56 kilometers (35 miles) per hour.

Albatrosses were once thought to locate food exclusively by sight, but recent research suggests that their extraordinarily acute sense of smell also plays a role in feeding. They can evidently find schools of fish by the odor of fish oil wafted tens of kilometers downwind. Tubenoses catch fish (and squid) by dipping their bill into the water during flight or during rare, brief stops on the surface when the water is calm. Albatrosses take shore leave only during breeding season. Chicks are hatched and raised on remote islands to which albatrosses regularly migrate from virtually any point over

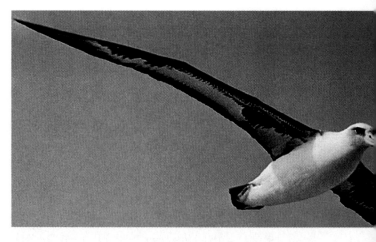

Figure 16.19 Wandering albatross (*Diomedea exulans*). This largest of all albatrosses can have a wingspan of 3.6 meters (12 feet) and may routinely cover 15,000 kilometers (9,300 miles) on foraging trips lasting a little over a month. They reach speeds of 80 kilometers (50 miles) per hour.

the world ocean. They are astonishing animals. No one who has seen one sweeping over the sea surface ever forgets the experience.

The Pelicans and Their Relatives

Birds in the order Pelecaniformes (*pelekan* = pelican)—which includes cormorants, frigate birds, and boobies in addition to pelicans—all have throat pouches and webbed feet. These large birds, commonly seen off tropical and subtropical coasts, don't spend as much time over the open sea as their tubenose relatives. Pelicans have broad, flat wings well adapted to slow flight (see **Figure 16.20**). The wings are folded back when the pelican crashes into the water in pursuit of prey. In contrast, the aerobatic frigate birds can neither walk nor swim and so must feed on flying fish caught during flight, on small squid or fish obtained while hovering, or on regurgitated food stolen from other seabirds as they return to their rookeries from a day of hunting. These highly maneuverable birds are extraordinarily light and delicate; a frigate bird's skeleton weighs less than its feathers. The most oceanic of this group are the relatively small tropic birds, which are among the few seabirds that regularly inhabit the relatively unproductive mid-oceanic gyres.

The Gulls

The order Charadriiformes (*charadra* = a cleft dweller) is a shorebird group with both marine and freshwater representatives. A few of the 115-plus species of gulls and

Figure 16.20 Brown pelican (*Pelecanus occidentalis*).

Figure 16.21 Common puffin with fish in its mouth.

terns are familiar to anyone who has spent time near the shore. Some gulls are found far inland at lakes or at dumps (where they can indulge their skill at scavenging), but most are found along the coasts eating nearly anything available. They are extraordinarily maneuverable and efficient flyers, buoyant swimmers, and—thanks to long legs at the middle of the body—surprisingly good runners.

Terns, smaller and finer in shape than gulls, are more oceanic and may travel to sea for prolonged periods in search of food. Most terns are excellent divers and plunge for their prey. The arctic tern has the most extensive migratory route of any bird, completing a 24,000-kilometer (15,000-mile) round-trip each year. It avoids winter altogether by moving from high northern latitude to high southern latitude at just the right times.

Gull relatives also include the endearing puffins (**Figure 16.21**). These birds don't fly well, and they must find rookeries on cliffs close to their feeding areas. They make up for this deficiency by being able to "fly" effectively through water, their stubby wings acting as high-speed paddles for pursuit of schooling fishes and other prey.

The Penguins

Penguins have completely lost the ability to fly, but they use their reduced wings to swim for long distances and with great maneuverability. The name of their order, Sphenisciformes (*spheniskos* = little wedge), refers to the shortness and shape of their wings. Their flightlessness makes it practical to have fatty insulation, greasy peglike feathers, stubby appendages, and large size and weight; indeed, such heat-conserving adaptations are critical to marine survival in very cold climates. Their neutral buoyancy is an advantage as they forage for food underwater. Emperor penguins, the largest of the living penguin species (see Figure 18.7), may dive to depths of 265 meters (875 feet) and stay submerged for 10 minutes or more. Penguins feed on fish, large zooplankters, bottom-dwelling mollusks or crustaceans, and squid. A few of the 18 species spend two uninterrupted years at sea between breedings.

Penguins are native only to the Southern Hemisphere and range from the size of a large duck (see **Figure 16.22**) to a height of more than a meter (3.3 feet) and a weight exceeding 36 kilograms (80 pounds). They are thought to consume about 86% of all food taken by birds in the southern ocean—about 34 million metric tons (37 million tons) per year, mostly larger zooplanktonic crustaceans. The small Galápagos penguin lives a comparatively easy life fishing the cold, nutrient-rich Humboldt Current at the equator, but its Antarctic relatives lead what must surely be the most rigorous existence of any seabird. For example, emperor penguins breed and incubate during the bitterly cold Antarctic winter. Unlike most other birds, emperors do not establish a territory, instead huddling together in tight crowds to conserve heat. The mass of penguins moves slowly around the breeding area as warm penguins from the center of the mob circulate to the outside to be replaced at the core by their chilled peripheral friends.

Figure 16.22 A rock hopper penguin, one of the smaller species. The largest penguins can be 1.2 meters (4 feet) tall.

MARINE MAMMALS

The class **Mammalia** (*mamma* = breast), to which humans belong, is the most advanced vertebrate group. About 4,300 species of mammals are known. The three living groups of marine mammals are the porpoises, dolphins, and whales of order **Cetacea;** the seals, sea lions, walruses, and sea otters of order **Carnivora;** and the manatees and dugongs of order **Sirenia.**

Each of these orders arose independently, from land ancestors. They exhibit the mammalian traits of being homeothermic, breathing air, giving birth to living young that they suckle with milk from mammary glands, and having hair at some time in their lives. Unlike other mammals, however, these extraordinary creatures have become adapted to life in the ocean and are dramatically different in appearance and function both from other modern forms and from the terrestrial precursors from which they branched only about 50 million years ago.

All marine mammals share four common features:

1. Their *streamlined body shape* with limbs adapted for swimming makes an aquatic life-style possible. Efficient locomotion depends on minimum drag and maximum ability to transfer propulsive energy from the muscles to the water. Thin, stiff flippers and tail flukes situated at the rear of the animal drive it forward, and similarly shaped forelimbs act as rudders for directional control. Drag is reduced by a slippery skin or hair covering.

2. They *generate internal body heat* from a high metabolic rate and conserve this heat with layers of insulating fat and, in some cases, fur. They also have relatively few surface capillaries, thus further reducing heat loss from the blood in the skin. Their large size gives them a favorable surface-to-volume ratio; with less surface area per unit of volume, they lose less heat through the skin. This is why there are no marine mammals smaller than a sea otter; a small mammal would lose body heat too rapidly. These adaptations are critical to animals living in cold water where food is readily available.

3. The *respiratory system is modified* to collect and retain large quantities of oxygen. The air duct "plumbing" of marine mammals is typically much different from that of land mammals, and the lungs can be more thoroughly emptied before drawing a fresh breath. The biochemistry of blood and muscle is optimized for the retention of oxygen during deep, prolonged dives. Some whales can stay submerged for 90 minutes.

4. A number of *osmotic adaptations* free marine mammals from any requirement for fresh water. Unlike other marine vertebrates, the marine mammals do not have salt-excreting glands or tissues. They swallow little water during feeding (or at any other time), and their skin is impervious to water. This minimal seawater intake, coupled with their kidneys' abilities to excrete a concentrated and highly saline urine, permit them to meet their water needs with the metabolic water derived from the oxidation of food.

Order Cetacea

The 90-plus living species of cetaceans (*ketos* = whale) are thought to have evolved from an early line of ungulates—hooved land mammals related to today's horses and sheep—whose descendants spent more and more time in productive, shallow waters searching for food.

BOX 16.1 ● *Deep Voices*

Humpback whale

The largest vertebrates make the loudest sounds. No one is sure just how whales produce the rumbles, clicks, whistles, and groans familiar to generations of sailors, but acoustical biologists are making progress.

The sound-making apparatus of toothed (odontocete) whales serves two purposes: communication and feeding. The squeals and whistles associated with communication appear to be made in the same way sounds are produced by the stretched neck of a deflating rubber balloon. These noises are usually accompanied by streams of bubbles from the blowhole. The extremely loud clicks, sounds important in feeding, are generated by an as-yet unknown mechanism. Researchers do know that oil-filled chambers within the head of a toothed whale focus and direct the sounds once they are produced. These sounds are so powerful that they can debilitate prey.

Like all whales, baleen (mysticete) whales lack vocal cords, but they do have a constriction between the back of their larynx and the tube that carries air to their lungs. One theory suggests that the tissues of this constriction vibrate at low frequencies as air is forced around or across it. A sac extending off the laryngeal cavity—a device that probably plays a role in pressure regulation during diving—may resonate with and amplify these vibrations and direct them outside. Baleen whales don't expel air from their blowholes during vocalization; so the air used in making the sounds must move back and forth within a closed interior system.

A competing theory holds that sounds are made by direct muscular action. Nerves activated by the brain could cause muscles to twitch at sonic frequencies, and the sound generated would move through the skin and into the water. Perhaps both methods are used; sounds are often heard to overlap antiphonally, sometimes producing unbroken melodies lasting a few minutes. No comparative analysis of the theories has been possible because of the difficulty of closely observing vocalizing baleen whales. Even obtaining fresh and undamaged whale specimens is difficult.

The patterns of communicating and socializing sounds of whales have been extensively studied. Humpbacks, a baleen species, are particularly melodious. Each of the populations of humpbacks—one in each of the major ocean basins—sings a unique song. Within each population, all humpbacks share the same song, a complex sequence of roars, ticks, ascending and descending scales, and booms. Each song lasts from 10 to 30 minutes and then repeats. Changes occur in the songs over time, but the population's shared song is still recognizable even after many years have passed. Significant changes in the song don't occur during the six silent months when the humpbacks are visiting their cold-water feeding grounds, but they do occur during the six months of breeding activity when the songs are in continuous use.

What do these repetitive songs tell us about the life of a humpback whale? So far, very little. But the complex syntax of the songs—complete with themes and variations, rhymes, and inverted phrases—suggests a rich shared heritage and extensive interactions among individuals in each population. One wonders how the noises of human civilization, from submarine sonar and seismic profilers to ships' propellers and engine exhausts, influence their symphonic display.

Modern whales range in size from 1.8 meters (6 feet) to 33 meters (110 feet) in length and weigh up to 100,000 kilograms (110 tons). Their paddle-shaped forelimbs are used primarily for steering, and their hind limbs are reduced to vestigial bones that do not protrude from the body. They are propelled mainly by horizontal tail flukes that are moved up and down by powerful muscles at the animal's posterior end. A thick layer of oily blubber provides insulation, buoyancy, and energy storage. One or two nostrils are located at the top of the head and have special valves to prevent intake of water when submerged. Whales have large, deeply convoluted brains and are thought to form complex family and social groupings.

Modern cetaceans are further divided into two suborders. **Figure 16.23** shows representative whales in each

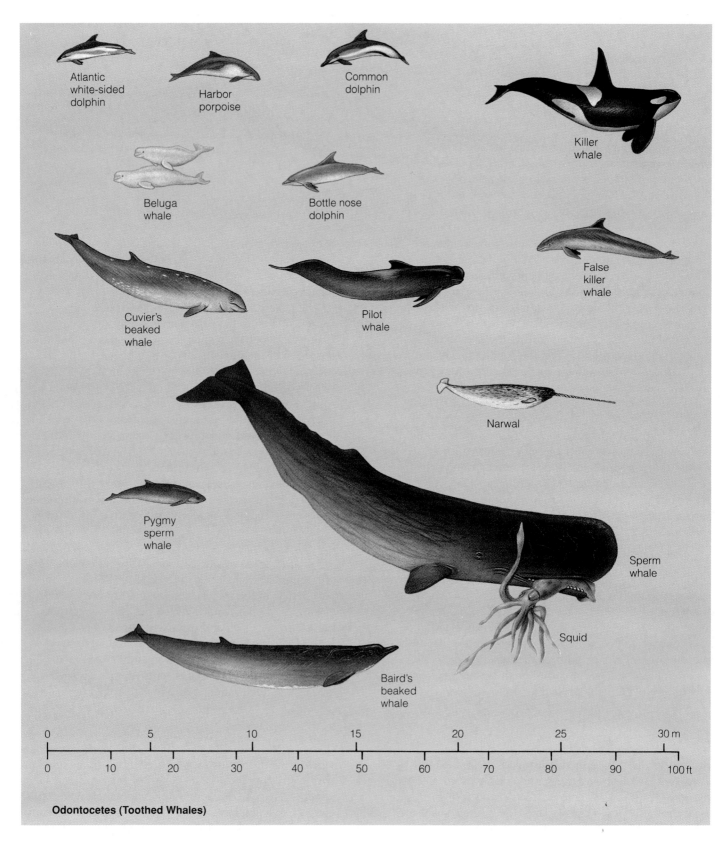

Atlantic white-sided dolphin

Harbor porpoise

Common dolphin

Killer whale

Beluga whale

Bottle nose dolphin

False killer whale

Cuvier's beaked whale

Pilot whale

Narwal

Pygmy sperm whale

Sperm whale

Squid

Baird's beaked whale

| 0 | 5 | 10 | 15 | 20 | 25 | 30 m |

| 0 | 10 | 20 | 30 | 40 | 50 | 60 | 70 | 80 | 90 | 100 ft |

Odontocetes (Toothed Whales)

Figure 16.23   Some representatives of the order Cetacea.

Humpback whale

Bowhead whale

Right whale

Minke whale

Blue whale

Feeding on krill

Fin whale

Sei whale

Gray whale

Mysticetes (Baleen Whales)

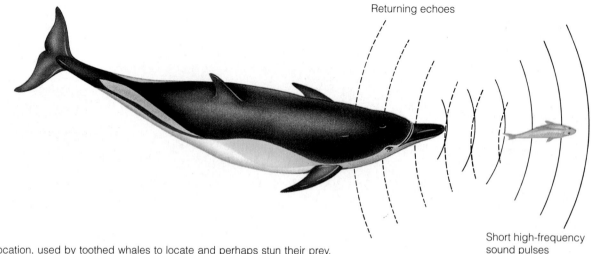

Figure 16.24 Echolocation, used by toothed whales to locate and perhaps stun their prey.

Returning echoes

Short high-frequency sound pulses

division. Members of suborder **Odontoceti** (*odontos* = tooth), the toothed whales, are active predators and possess teeth to subdue their prey. Toothed whales have a high brain-weight-to-body-weight ratio, and though much of their additional brain tissue is involved in formulating and receiving the sounds on which they depend for feeding and socializing, many researchers believe them to be quite intelligent. Smaller whales in this group include the orca (killer whale) and the familiar dolphins and porpoises of oceanarium shows. The largest toothed whale is the 18-meter (60-foot) sperm whale, which can dive to at least 1,140 meters (3,740 feet) in search of the large squid that provide much of its diet.

Toothed whales search for prey using **echolocation,** the biological equivalent of sonar; they generate sharp clicks and other sounds that bounce off prey species and return to be recognized (**Figure 16.24**). Reflected sound is also used to build a "picture" of the animal's environment and to avoid hitting obstacles while swimming at high speed. Odontocete whales are now thought to use sound offensively as well. Recent research indicates that some odontocetes can generate sounds loud enough to stun, debilitate, or even kill their prey. In one experiment, dolphins produced clicks as loud as 229 decibels, equivalent to a blasting cap exploding close to the target organism. Sperm whales, it has been calculated, may generate sounds exceeding 260 decibels! (The decibel scale is logarithmic; compare this figure to the 130-decibel noise of a military jet engine at full power 20 feet away!) How this prodigious noise is generated is not yet known, nor do we know how such energy is radiated from the whale without damaging the organs that produce and focus it.

Suborder **Mysticeti** (*mystidos* = unknowable), the whalebone or baleen whales, have no teeth and are thought to have branched from the line leading to toothed whales early in whale evolution. Filter feeders rather than active predators, these whales subsist primarily on krill, a relatively large, shrimplike crustacean zooplankter obtained in productive polar or subpolar waters. They do not dive deep but instead commonly feed a few meters below the surface. Their mouths contain interleaving triangular plates of bristly, hornlike **baleen** (**Figure 16.25**), used to filter the zooplankton from great mouthsful of water or from mud scooped from a shallow seabed. The plankton are concentrated as water is expelled, then swept from the baleen plates by the whale's tongue, compressed to wring out as much seawater as possible, and swallowed through a throat not much larger in diameter than a grapefruit. A great blue whale, largest of all animals, requires about 3 metric tons (6,600 pounds) of krill each day during the feeding season; about 1 million Calories per day. The short, efficient food chain from phytoplankton to zooplankton to whale provides the vast quantity of food required for their survival. A feeding blue whale with its mouth hugely distended with seawater and krill is shown in **Figure 16.26.**

Mysticeti is an excellent name for these odd and wonderful animals. We know comparatively little of their social structure, intelligence, sound-producing abilities, navigational skills, or physiology. We do know that humpback whales use complex songs in group communication and that blue whales may use very low-frequency sound to communicate over tremendous distances. Some species migrate annually from polar to tropical waters and back. Until recently our primary response to all whales has been to slaughter them in countless numbers for meat and oil, with little thought for their extraordinary abilities and assets. More of this

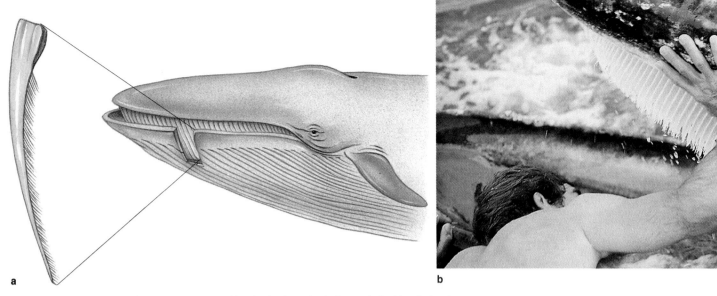

Figure 16.25 (a) A plate of baleen and its position in the jaw of a baleen whale. For clarity, the illustration shows an area of the mouth cut away. (b) A student and a live whale, up close and personal. This gray whale uses its stiff, coarse baleen plates to sieve crustaceans from shallow bottom mud.

Figure 16.26 A blue whale surfaces from a dive with throat pleats distended with water and food. The whale will filter the water in its mouth through plates of baleen, leaving behind krill, which it swallows.

depressing history may be found in the discussion of marine resources in Chapter 19.

Order Carnivora

The order Carnivora (*carnis* = flesh, *vorare* = to devour) includes land predators ranging from dogs and cats to bears and weasels, but the members of the carnivoran suborder **Pinnipedia**—the seals, sea lions, and walruses (see **Figure 16.27**)—are almost exclusively marine. Unlike the cetaceans, the gregarious pinnipeds (*pinna* = wing, *pedalis* = foot) leave the ocean for varying periods of time to mate and raise their young. They appear to have evolved from the same stock as modern bears.

True seals have a smooth head with no external ear flaps, the external part of the ear having been sacrificed to further streamline the body. They are covered with a short, coarse hair without soft underfur. Seals are graceful swimmers that pursue small fish (their usual prey) with powerful side-to-side strokes of their hind limbs. These rear appendages are partially fused and always point back from the hind end of the body; thus, they are of very little use for locomotion on land. The elephant seal, named for its large size and long snout, holds the diving depth record for all air-breathing vertebrates: 1,560 meters (5,120 feet).

Sea lions, familiar to many as the performers in "seal" shows, have hind limbs with a greater range of motion

a

b

c

Figure 16.27 Some representatives of the suborder Pinnipedia. (a) The California sea lion, the "seal" of seal shows. (b) Three juvenile harbor seals anxiously await lunch in a marine mammal care facility. (c) A walrus.

and are thus more mobile on land. They have a streamlined head with small external ears and a pelt with soft underfur; unlike seals, they use their front flippers for propulsion.

Walruses are much larger than either seals or sea lions and may reach weights of 1,800 kilograms (2 tons). Walruses use their tusks like sled runners to guide their sensitive "whiskers" just above the sediment surface, looking for clam siphons at depths up to 91 meters (300 feet). They dig up the clams with their mouths, crush the clam shell and remove the meat, then eject the inedible fragments. The large tusks are also useful for hauling their heavy bodies onto ice floes.

The suborder **Fissipedia** (*fissus* = split, *pedalis* = foot) has many members (including cats, dogs, raccoons, and

bears) and two truly marine representative: sea otters and polar bears, relative newcomers to the marine environment.

Sea otters, active and pleasing creatures (**Figure 16.28**), are the smallest of marine mammals. Human demand for their fur, the densest and warmest of any animal, caused their near-extermination. The modern population of the Pacific sea otter descends from a very few individuals accidentally overlooked by fur hunters of the late nineteenth and early twentieth centuries. Playful and intelligent, otters rarely exceed 120 centimeters (4 feet) in length; they eat voraciously, consuming up to 20% of their body weight in mollusks, crustaceans, and echinoderms each day. Sometimes they lie on their back in the water, balancing a rock on their chest. They ham-

Figure 16.28 Two sea otters (*Enhydra lutris*) in a California kelp bed.

Figure 16.29 A female polar bear with two cubs takes time out from seal hunting; the Arctic Ocean stretches beyond.

mer the shell of the prey against the rock until the shell cracks. Then, morsels of food are extracted with small nimble fingers, and rolling over in the water cleans away the debris.

Polar bears (**Figure 16.29**) spend most of their lives stalking seals and stranded whales on the frozen northern polar ocean. They swim between ice floes with large, oarlike forepaws and can cross 100 kilometers (62 miles) of open water. The world's largest bears, male polar bears may grow to 2.5 meters (8.2 feet) and weigh 800 kilograms (1,800 pounds). They wander over enormous distances in search of food. One tagged bear moved 3,200 kilometers (2,000 miles) in a year, and bears have been seen in the vicinity of the pole. Apart from humans, polar bears have no enemies. Once an object of exploitation for their thick white fur, their numbers appear to be increasing under protection. As many as 40,000 are thought to exist today.

Figure 16.30 A manatee, or sea cow.

Order Sirenia

The bulky, lethargic, small-brained dugongs and manatees, collectively called sirenians (*siricis* = a mermaid) (**Figure 16.30**), are the only herbivorous marine mammals. Like the cetaceans, they appear to have evolved from the same ancestors as modern ungulates. They make their living grazing on sea grasses, marine algae, and estuarine plants in coastal temperate and tropical waters of North America, Asia, and Africa. Some species live in fresh water. The largest sirenians reach 4.5 meters (15 feet) in length and weigh 680 kilograms (1,500 pounds). They were first compared to mermaids by early Greeks who noted the manatee's habit of resting in an upright position in the water and holding a suckling calf to her breast. Sirenians have been hunted extensively, and only about 10,000 individuals are now thought to exist worldwide. Even though protected now, many are killed or wounded each year in Florida by the propellers of powerboats.

The animals of Chapters 15 and 16 do not exist in isolation but interact with each other and with plants in complex marine communities. We turn our attention in the next chapter to these groupings.

Q-AND-A

1. What is the most widely distributed and successful marine vertebrate?

   Probably small pelagic bony fishes of genus Cyclothone. *These* bristlemouths *are found just below the euphotic zone in virtually all the world ocean away from shore. Bristlemouths feed on plankton and on organisms swimming into their visual field at the limit of light penetration. Some ichthyologists believe there are more living individuals of* Cyclothone microdon *than of any other single vertebrate species on Earth.*

2. How do birds like the arctic tern and wandering albatross navigate across the trackless ocean?

   No one is certain. Experiments done at Cornell University with homing pigeons suggest that homing birds use a combination of magnetic and optical cues to return to their starting points. Even polarized light and the positions of certain stars might be involved. However it works, the behavior is not learned but is instinctive to the bird.

3. What's the difference between a dolphin *fish* and a dolphin *mammal*?

   The name confusion began in Aristotle's time and has not abated. Dolphin mammals are small toothed whales; dolphin fish are teleosts (bony fish). The dolphin fish appears on restaurant menus and in seafood markets as mahi-mahi, dorado, and other names. Dolphin mammals are slaughtered in dismaying numbers in association with tuna fishing by some non-U.S. fishing fleets, but none of their meat appears on U.S. plates.

4. How is a porpoise different from a dolphin?

   Dolphin *and* porpoise *are common names of two subtly different groups of odontocetes.* Porpoise, *as a term, refers to the smaller members of the group, which have spade-shaped teeth, a triangular dorsal fin, and a smooth front end tapering to a point.* Dolphins *are usually larger and have an extended bottlelike jaw filled with sharp, round teeth. The small jumping whales in most oceanarium shows are dolphins, but killer whales are a species of large porpoise. To make matters even more complicated, the common dolphin seen in ocean-themed amusement parks,* Tursiops truncatus, *is often referred to as a porpoise, even by show announcers. This confusion between common names points out how useful scientific names can be. The real name of the animal,* Tursiops, *is clear and unambiguous.*

5. If whales are so intelligent, why haven't they learned to avoid the whalers' catcher boats?

   There are many theories, some of which are presented in an interesting and curious book called Mind in the Waters *by Joan McIntyre (see bibliography). It may be that they are not as intelligent as was once thought. It may be that they have no conception of violence—a paradoxical thought considering the way the odontocetes make their living. These animals should perhaps be considered intelligent for their native environment but not be evaluated in relation to terrestrial organisms in a radically different environment with radically different problems.*

6. If some seals and whales dive to great depths, why don't they get the bends?

The bends is a painful and occasionally fatal condition brought on by nitrogen gas—the most plentiful component of air—leaving solution and forming bubbles in the blood. The condition is analogous to what happens in a newly opened soda bottle. Human divers take a source of air with them to depth, breathe the air under pressure, and dissolve excess nitrogen gas in their blood. If they have been down long enough, or deep enough, the release of pressure upon surfacing will be like taking the cap off the soda bottle. Whales don't use scuba tanks—they have no source of supplemental air at depth. They "tank up" at the surface by oxygenating their blood and tissues, but they have no excess gas to bubble from the blood at the end of their dives.

7. Why did marine mammals evolve so quickly? Fifty million years doesn't seem like a very long time, relatively speaking, to derive such large and elaborately adapted creatures from small, herbivorous, land-dwelling precursors.

The different groups of marine mammals arose from different terrestrial stocks. The first forms spent only a little time in the sea, perhaps in enclosed bays and estuaries. The feeding situation in these locations was probably very favorable, and the generations of organisms that followed spent progressively more time in the water. The most successful organisms, those with the most effective aquatic feeding structures and behaviors, prospered and reproduced. Eventually, some of the animals changed sufficiently to abandon their land existence altogether. The fossil record of the evolution of whales is surprisingly clear; many of the intermediate forms have been identified and studied. One recently discovered intermediate form has forelegs. The rate at which these animals evolved was appropriate to the changing conditions they faced and to the enhanced feeding opportunities the ocean presented. The rate of change was neither fast nor slow, but suited to the development of the organisms themselves. For more information on this fascinating topic, please see Berta (1994), listed in this chapter's bibliography.

Terms and Concepts to Remember

Agnatha	gill membranes	poikilothermic
Aves	homeothermic	Reptilia
baleen	hypotonic	salt glands
Carnivora	lateral-line system	schooling
cartilage	Mammalia	Sirenia
Cetacea	Mysticeti	swim bladder
Chondrichthyes	Odontoceti	Teleostei
cryptic coloration	osmoregulation	Vertebrata
echolocation	osmosis	vertebrate
Fissipedia	Osteichthyes	
gas exchange	Pinnipedia	

Study Questions

1. Are all chordates vertebrates? Are all vertebrates chordates? What distinguishes a vertebrate?

2. There are seven living classes of vertebrates, but only six are marine. List the seven classes. Which class has no permanent marine representatives? Why not?

3. What are the classes of living fishes? Which is considered most primitive? Most advanced? Which class has the largest individuals? Which is the most economically important?

4. How do fishes swim? What problems are associated with swimming? How do fishes overcome these problems? How do fishes make defensive use of color, shape, and schooling behavior?

5. How do fishes "breathe" underwater? Do marine fishes experience any unusual problems by being immersed in seawater?

6. Compare and contrast the groups of living marine reptiles.

7. How are seabirds different from land birds? Are there more species of seabirds than land birds?

8. Compare and contrast the groups of living seabirds. Which group is responsible for the greatest food consumption?

9. What characteristics are shared by all marine mammals?

10. How are odontocete (toothed) whales different from mysticete (baleen) whales? Which are the better known and studied?

For Further Study

General References

Averbach, P. S. 1991. "Marine envenomations." *New England Journal of Medicine* 325 (no. 7): 486–93. If it stings, this discusses it and tells how to treat the wound and its effects.

Briggs, J. C. 1974. *Marine Zoogeography.* New York: McGraw-Hill. Distribution of vertebrates and environmental factors relating to distribution. Technical.

Jensen, A. C. 1978. *Wildlife of the Oceans.* New York: Abrams. A beautiful photographic treatment.

Pough, F. H., et al. 1989. *Vertebrate Life.* 3d ed. New York: Macmillan. Excellent and up-to-date general text containing a thorough discussion of the evolution of vertebrates.

Romer, A. S., and T. S. Parsons. 1977. *The Vertebrate Body*. 5th ed. Philadelphia: W. B. Saunders. A standard reference for vertebrate matters.

Sanderson, S. L., and R. Wassersug. 1990. "Suspension-Feeding Vertebrates." *Scientific American*, March. Animals that can eat plankton grow in huge numbers or to enormous size.

References for Fishes

Bray, R., and M. Hixon. 1974. "Night-Shocker: Predatory Behavior of the Pacific Electric Ray *Torpedo californica*." *Science* 200: 333–34.

Fierstine, H. L., and V. Walters. 1968. "Studies in Locomotion and Anatomy of Scombroid Fishes." *Memoirs of the Southern California Academy of Sciences* 6: 1–31. Propulsion in the mackerel family, the fastest of fishes.

Garrison, T. 1986. "Rays." *California Diver* 2 (no. 1, July/August): 17–20.

Garrison, T. 1988. "Sharks." *Pacific Diver* 1 (no. 3, December).

Gilbert, P. W., et al. 1982. "Sharks." *Oceanus* 24 (no. 4). Overview written for the general reader.

Gray, J. 1957. "How Fishes Swim." *Scientific American*, August, 48–54.

Hardy, A. 1965. "Fish and Fisheries." Part 2 of *The Open Sea: Its Natural History*. London and Boston: Houghton Mifflin.

Lagler, K. F., et al. 1977. *Ichthyology*. 2d ed. New York: Wiley. An excellent and readable standard text.

Love, R. M. 1995. *Probably more than you want to know about fishes of the Pacific Coast*. 2d ed. Santa Barbara: Really Big Press. An entertaining and informative book written by an expert.

Marshall, N. B. 1965. *The Life of Fishes*. London: Weidenfeld & Nicholson. Reprint. New York: Universe Books, 1970.

Nelson, J. S. 1984. *Fishes of the World*. 2d ed. New York: Wiley. Excellent systematic overview.

Partridge, B. L. 1982. "The Structure of Fish Schools." *Scientific American*, June, 114–23.

Shaw, E. 1962. "The Schooling of Fishes." *Scientific American*, June, 128–36.

Stevens, J. D., ed. 1987. *Sharks*. New York: Facts on File. Perhaps the best general reference on sharks that is available to the interested amateur.

Triantafyllou, M. S., and G. S. Triantafyllou. 1995. "An Efficient Swimming Machine." *Scientific American*, March, 64–70. Instinctive control of vortices lets fishes swim the way they do. The article describes a robotic tuna that has managed to swim in the same way.

References for Reptiles

Carr, A. 1973. *So Excellent A Fishe: A Natural History of Sea Turtles*. New York: Anchor/Doubleday. A beautifully written general account of sea turtles, with emphasis on navigational studies.

Carr, A., and P. J. Coleman. 1974. "Seafloor Spreading Theory and the Odyssey of the Green Turtle." *Nature* 249: 128–30.

Lohmann, K. J. 1992. "How Sea Turtles Navigate." *Scientific American*, January, 100–106. The most important cues are wave direction and Earth's magnetic field.

Minton, S. A., and H. Heatwole. 1978. "Snakes and the Sea." *Oceans* 11 (no. 2): 53–56. Human contact with sea snakes appears to be increasing.

References for Birds

Croxall, J. P., ed. 1987. *Seabirds: Feeding Ecology and Role in Marine Ecosystems*. Cambridge: Cambridge University Press.

Davis, L. S., and J. T. Darby, eds. 1990. *Penguin Biology*. San Diego: Academic Press. The first collection of biological and ecological studies of penguins since the mid-1970s.

Jouventin, P., and H. Weimerskirch. 1990. "Satellite Tracking of Wandering Albatrosses." *Nature* 343 (no. 6260): 746–68. Source of albatross travel data.

Lanting, F. 1986. "Riders on the Wind." *Oceans* 19 (September–October): 24–31. Excellent illustrated article on Pacific albatrosses.

Löfgren, L. 1984. *Ocean Birds*. New York: Knopf. Beautiful book combining pictures and text.

Nelson, B. 1979. *Seabirds*. New York: Addison-Wesley. A standard reference.

Peterson, R. T. 1990. *A Field Guide to the Western Birds*. Boston: Houghton Mifflin. A recent revision of a classic. An older companion volume is available for East Coast birds.

Welty, J. C. 1982. *The Life of Birds*. 3d ed. Philadelphia: W. B. Saunders. Standard text.

Whittow, G. C., and H. Rahn, eds. 1984. *Seabird Energetics*. New York: Plenum. What seabirds eat, where, and how much energy they obtain per unit of search effort.

Wilford, J. N. 1986. *The Riddle of the Dinosaur*. New York: Knopf. Part II discusses the evolution of birds.

Wilson, E. A. 1967. *Birds of the Antarctic*. Edited by B. Roberts. London: Blandford Press. A modern collection of Wilson's accurate and beautiful watercolors. Dr. Wilson perished with Scott on their return from the South Pole (see Chapter 18).

References for Mammals

Ackerman, D. 1990. "A Reporter at Large—Whales." *New Yorker*, 26 February, 51–86. A wonderfully written article on Roger Payne and his research into humpback whale biology in Argentina.

Amato, I. 1993. "A Sub Surveillance Network Becomes a Window on Whales." *Science* 261 (no. 5121): 549–50. Information on Integrated Undersea Surveillance System (IUSS) whale data, described as "the acoustic Rosetta stone for whales."

Anderson, H. T., ed. 1969. *The Biology of Marine Mammals*. New York: Academic Press.

Berta, A. 1994. "What Is a Whale?" *Science* 263 (14 January): 180–81. Recent summary of whale evolution.

Clark, M. R. 1979. "The Head of the Sperm Whale." *Scientific American*, January, 128–39. The author suggests that buoyancy control is the main function of the complex structures of the head. The article was written prior to the discovery of the stunning magnitude of sperm whale sounds.

Deep Voices. 1978. Capitol Records ST-11598. A phonograph record of whale sounds that will give your audio system a thorough low-frequency workout.

Ellis, R. 1982. *Dolphins and Porpoises.* New York: Knopf. Richard Ellis is one of the finest painters of cetaceans.

Kanwisher, J. W., and S. H. Ridgway. 1983. "The Physiological Ecology of Whales and Porpoises." *Scientific American,* June, 110–21.

Leatherwood, S., and R. R. Reeves, eds. 1990. *The Bottlenose Dolphin.* San Diego: Academic Press.

Maxwell, G. 1967. *Seals of the World.* London: Constable. The author of *Ring of Bright Water,* a popular book about otters, has written this excellent compendium.

McIntyre, J. 1974. *Mind in the Waters.* New York: Scribner. An unusual book concerning cetacean intelligence and conservation.

Minasian, S. S. 1984. *The World's Whales: The Complete Illustrated Guide.* Washington, DC: Smithsonian Books. The title is not hyperbole. Extraordinary photographs and clear text, taxonomically arranged.

Norris, K. S., and B. Möhl. 1983. "Can Odontocetes Debilitate Prey with Sound?" *American Naturalist* 122 (no. 1): 85–104.

Orr, R. 1972. *Marine Mammals of California.* Berkeley: University of California Press.

Payne, R., and S. McVay. 1971. "Songs of Humpback Whales." *Science* 173 (no. 3997): 585–97.

Reeves, R. R., S. Leatherwood, and B. Stewart. 1992. *The Sierra Club Handbook of Seals and Sirenians.* San Francisco: Sierra Club Books.

Riedman, M. 1990. *The Pinnipeds: Seals, Sea Lions, and Walruses.* Berkeley: University of California Press. Thorough, well-illustrated overview with an extensive bibliography.

Ryan, P. R. 1978. "Marine Mammals: A Guide for Readers." *Oceanus* 21 (Spring): 9–16.

Tangley, L. 1984. "A Whale of a Bang." *Science '84,* 73–74. A short article on the possible use of sonic bursts for hunting by odontocetes.

Whitehead, H. 1984. "The Unknown Giants: A Rare Look at Sperm and Blue Whales." *National Geographic,* December, 772–89. Remarkable photographs!

Würsig, Bernd. 1989. "Cetaceans." *Science* 244 (no. 4912): 1550–57. Excellent review article with extensive bibliography.

Zapol, W. M. 1987. "Diving Adaptation of the Weddell Seal." *Scientific American,* June, 100–105.

Note: Oceanus 21 (no. 2, 1978) devotes an entire issue to marine mammals.

17 MARINE COMMUNITIES

Job Opportunities in the Intertidal Zone

A visitor to a temperate rocky shore will surely be impressed by the variety of life there. Producers and consumers will be huddled against the waves, each specialized for existence in one of the Earth's most demanding locations. Before long, the visitor's attention will likely turn to the many snails in the area. Some snails can be seen living in tiny cracks high in the intertidal zone—much higher and they would be land animals. More live farther down the shore, among the rocks. A few burrow into kelp, some scrape algal films from hard surfaces, and some are active predators. Some are beautiful, many are drab. A few are active during the day, but most are busiest after dark. Some, like the abalone, can be almost as big as a catcher's mitt; others, the smallest limpets, are near-microscopic. Each has the same underlying anatomy, and each feeds using the same basic structures. But we may be curious about why are there so *many* species of snails? Wouldn't

competition for food eventually leave one species the winner?

In fact, each species has its own role in this intertidal zone. Each interacts with other organisms and with its physical surroundings in a unique way. One snail feeds by scraping algae off the *tops* of rocks, another by scraping algae off the *bottom* of the same rocks. One builds a curved tube underneath these rocks, then builds a mucous net at high tide, and ultimately reels in drifting bits of algae the others are unable to snare. Still another gastropod species, an agile carnivore, surprises small arthropods by harpooning them with a needlelike tooth, a specialized spear that has evolved from part of the same structure the other snails use for grazing. Any leftovers will be cleaned up by yet another kind of snail, a species of scavenger.

Like the other animals we have studied in the last two chapters, these snails don't exist in isolation. They live in groups, each group interacting with its surroundings. A group of organisms of the same species occupying a specific area is called a **population.** The many populations of organisms that interact with one another at a particular location constitute a **community.** There are many marine communities. Most are named for the dominant organism within them (such as a mussel bed community) or for their location (such as an intertidal community).

CHAPTER OVERVIEW

Organisms are distributed throughout the marine environment in specific communities—groups of interacting producers, consumers, and recyclers that share a common living space. The location of a community and the variety of organisms that constitute it depend on the physical and biological characteristics of that living space.

Any community is a dynamic place, waxing and waning as its residents respond to environmental fluctuations. Patterns of growth and distribution of organisms within communities depend on the often subtle interplay of physical and biological factors. The relative numbers of species and individuals in a community depend in part on whether its environment is relatively easy and free of stressors, or relatively hard and full of potential limiting factors. A few prominent marine communities are compared and contrasted in this chapter, and some representative adaptations of their residents are noted.

The chapter also discusses symbioses: intimate relationships of different species in mutualistic, commensalistic, and parasitic associations. Symbiotic relationships are very common in the ocean, and most forms of marine life are actively involved in them.

Periwinkle snails cluster along the edge of a tide pool.

THE DIVERSITY OF COMMUNITIES

Communities can be huge or surprisingly small. The largest marine community—which is also the most sparsely populated—lies within the uniform mass of permanently dark water between the sunlit surface and the deep bottom. Few animals live there because so little food is available, but those organisms that survive are among the strangest in the ocean. Opportunities for feeding in the deep open-ocean community are few and far between, and some animals are able to consume prey larger than themselves should the occasion arise. Because so few animals are present, mating is also a rare event; in a few species males and females become permanently bonded during their first encounter, the male burrowing into the female's body for a lifelong free ride.

In contrast, the smallest obvious marine communities may be those established against solitary rocks on an otherwise flat, featureless seabed. Drifting larvae will colonize the place; the established community can seem an oasis of life and activity in an otherwise static sedimentary desert. Seaweeds will grow, worms will burrow, snails will scrape food from the hard surfaces, and small fishes will nestle among crevices. Hundreds of small plants and animals can live their lives within a meter of each other, interacting in a compact solitary community with no similar environment available for thousands of meters. The larvae of the next generation drift away with little chance of finding a suitable place to carry on their lives. Microscopic communities also exist; in fact, an interacting set of populations can exist on a single grain of sand or on one decomposing fish scale.

ORGANISMS WITHIN COMMUNITIES

There are many different places to live and many different "jobs" for organisms within even a simple community. A **habitat** is an organism's "address" within its community, its physical *location*. Each habitat has a degree of environmental uniformity. An organism's **niche** (*nidus* = nest) is its "occupation" within that habitat, its relationship to food and enemies, an expression of what the organism is *doing*. For example, the small fishes living among the coral heads in a coral reef community share the same habitat, but each species has a slightly different niche. Each population in the community has a different "job" for which its shape, size, color, behavior, feeding habits, and other characteristics particularly suit it.

The Influence of Physical and Biological Factors

Physical and biological factors in the environment determine the location and composition of a community. As we saw in Chapter 13, physical factors such as temperature, pressure, and salinity affect the success of an organism: Seaweeds adapted to cold water usually cannot survive prolonged exposure to abnormally warm water, and lower-than-normal salinity can dangerously disrupt the fluid balance of most marine invertebrates. Biological factors are influences on an organism by members of its own population or other populations, in its own or other communities. Biological factors include crowding, predation, grazing, parasitism, shading from light, waste substances, and competition for limited oxygen.

A physical or biological factor that limits an organism's success in a community is a limiting factor for the organism, a limiting factor that prevents the organism from feeding, growing, reproducing successfully, defending itself, sensing danger, or otherwise functioning successfully. As you may remember from Chapter 13, some tropical fishes die in an unheated home aquarium because their physiology is optimized for a water temperature higher than the interior temperature of a typical house. If their natural tropical environment temporarily dropped to the temperature of a cool room, some of the fishes would become too sluggish to catch fast-moving food or avoid larger predators. Temperature would *limit* their success. Other fishes can adapt relatively easily to changes in temperature, however. Temperature would be a limiting factor to a sensitive tropical fish species, but not to species with a wider temperature tolerance.

The prefixes *steno-* (meaning "narrow") and *eury-* (meaning "wide, broad") are sometimes used to describe those species, populations, or individuals that have narrow or wide tolerance to specific factors. **Stenothermal** species of tropical fish function better in a heated aquarium, but **eurythermal** fish do not need this extra attention. Likewise, **stenohaline** (*halos* = salt) marine organisms require a stable saline environment; they are unable to withstand relatively small fluctuations in salinity. **Euryhaline** species, however, can withstand a wide range of salinity and perhaps even tolerate some exposure to fresh water. Temperature and salinity are only two of the many physical factors that vary in a marine organism's native community. Light and pressure are examples of other such factors.

Figure 17.1 provides an idealized look at the tolerance of organisms to a varying physical factor such as temperature. A eurythermal organism's optimal range is wide (represented by a broad curve), but a stenothermal organism's optimal range will be narrow (which would be represented by a steeper, tighter curve). The same reasoning applies to the salinity tolerance of stenohaline or euryhaline species, or to the pressure tolerance of stenobaric (*baros* = pressure) or eurybaric ones.

An animal or plant is almost never exposed to fluctuations of only one physical or biological factor in its environment. An organism may die if subjected to changes in

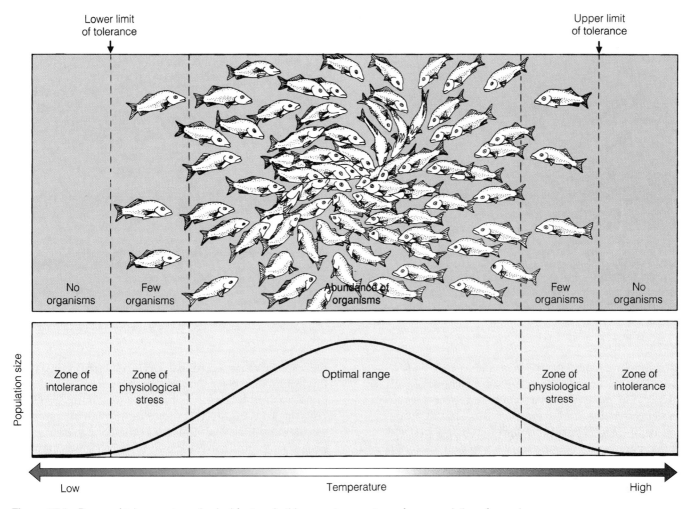

Figure 17.1 Range of tolerance to a physical factor—in this case, temperature—for a population of organisms.

several environmental factors at once, even though each individual change is within its range of tolerance. Even small changes in temperature or salinity, survivable in themselves, might prove lethal if a particular species of tropical fish were exposed simultaneously to both.

A favorable balance of physical and biological factors is critical to each individual organism's success—and therefore to community success and longevity. The study of this balance, and of the relationships of organisms and interactions within communities, is called **ecology** (*oikos* = house, *logos* = study of). Marine ecologists are concerned with the types of organisms within marine communities, as well as their habitats, niches, distribution, numbers, and reactions to variations in their environment.

Competition

The availability of resources such as food, light, and space within a community determines the number and composition of the populations of organisms within that community. Competition for the necessities of life may occur within the community between members of the *same* population or between members of *different* populations. Subtle swings in physical or biological factors may give one population the advantage for a time but then shift to favor another.

When members of the same population (all members of the same species) compete with each other, some individuals will be larger, stronger, or more adept at gathering food, avoiding enemies, or mating. These animals tend to prosper, forcing their less successful relatives to emigrate, fight, or die in the course of competition. The most successful organisms have the most surviving offspring; so useful inheritable variations are passed along in greater quantity to the next generation. As we saw in Chapter 13, this kind of competition continually fine-tunes a population to its environment.

When members of different populations compete, one population may be so successful in its "job" that it elim-

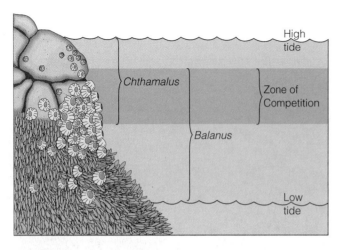

Figure 17.2 Competition between two species of barnacles prevents either from occupying as much of the intertidal zone as possible. The central zone of overlap indicates the area where *Chthamalus* (the smaller barnacle) and *Balanus* (the larger barnacle) compete for food and space.

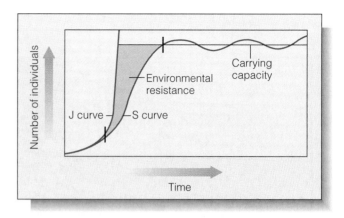

Figure 17.3 The J-shaped curve of population growth of a species is converted to an S-shaped curve when the population encounters environmental resistance. The physical or biological conditions responsible for the cessation of growth are called *limiting factors.*

inates competing populations. In a stable community, two populations cannot occupy the same niche for long. Eventually the more effective competitor overwhelms the less effective one. Extinction from this kind of head-to-head competition is probably uncommon, but restriction of a population because of competition between species is not. For example, the little barnacle *Chthamalus* lives on the uppermost rocks in many intertidal communities; the larger barnacle *Balanus* lives on the lower rocks (see **Figure 17.2**). Planktonic larvae of both species can attach themselves to rocks anywhere in the intertidal zone and begin to grow. In the lower zone the faster-growing *Balanus* push the weaker *Chthamalus* off the rocks, but at higher positions *Balanus* cannot survive because it is not as resistant to drying and exposure as the tough little *Chthamalus*. At the top and bottom of their distribution the two species do not compete for food or space. The competition at the intersection of their ranges prevents each species from occupying as much of the habitat as might otherwise be possible.

Growth Rate and Carrying Capacity in Communities

Organisms newly introduced into a favorable environment with no competitors for food or space will reproduce exponentially, tracing a J-shaped population growth curve. In nature very few populations reproduce at this maximal rate, however, because environmental conditions are rarely ideal and because limiting factors in the environment quickly slow the rate of population growth. The sum of the effects of these limiting factors in

the environment is called **environmental resistance.** Environmental resistance causes the actual population growth curve to be lower than the maximum potential growth curve. **Figure 17.3** shows the growth rate in number of individuals over time for both potential and actual situations. Note that the curve resulting when limiting factors intrude is S-shaped; it gradually flattens toward an upper limit of the number of individuals in the population. The final number of organisms oscillates around the **carrying capacity** of the environment for that species: the population size of each species that a community can support indefinitely under a stable set of environmental conditions.

The carrying capacity changes if environmental conditions change. A marine population could *crash* if up-welling ceased, if new predators were introduced, if climate varied, if food supplies dwindled, or if a new parasite infiltrated the population.

Distribution of Organisms in a Community

As we have seen, physical and biological factors affect the number and positions of organisms in a community. The number of individuals per unit area (or volume) is known as the **population density.** Rare individuals have a much lower population density than dominant ones. In general, more different species exist in benign habitats where physical factors stay near optimal values (like a coral reef or rain forest), and fewer species exist in rigorous habitats where physical factors range to extremes (like a beach or a desert). That is, "easy" habitats typically have high **species diversity**—they contain more

BOX 17.1 ● *Steinbeck, Ricketts, and Communities*

Many people know that John Steinbeck was awarded the 1962 Nobel Prize for literature for his 1939 novel *The Grapes of Wrath*. Few know that this famous writer was an avocational marine biologist with a deep interest in marine intertidal communities. He was introduced to this rich habitat by Ed Ricketts, a real person who became the fictional character "Doc" in Steinbeck's popular novels *Cannery Row, Tortilla Flat,* and *Sweet Thursday*. These novels were set in Monterey, then an important northern California fishing town. Ed Ricketts ran a small commercial biological supply business on Cannery Row. When Steinbeck and Ricketts first met (in a dentist's office), each had heard of the other: Steinbeck knew that Ricketts was a curious character interested in invertebrate zoology and classical music, and Ricketts knew that Steinbeck was a promising newspaper reporter and budding author. They became fast friends and went on many collecting expeditions together. They even wrote a book, *The Sea of Cortez*, about their trip to collect marine invertebrates on the shore of the Gulf of California; it combines travel adventures, humor, vivid scientific description, and personal philosophy.

Ed Ricketts was ahead of his time as a biologist. His approach to intertidal life was community based—what we would today call an ecological emphasis. He attempted to combine all his experiences and observations in an environment into an integrated picture of the whole. Today's biologists can appreciate Ricketts's urge to bring together the physical and biological factors affecting each species within a community, to decipher, as he once wrote, "the Zen of a segment of shore."

In 1939, Ricketts published a landmark book, *Between Pacific Tides*. Unlike previous guides to seashore life, *Between Pacific Tides* was organized by community and not by organism type. "The treatment," he wrote in the preface to the first edition, "is ecological and inductive; that is, the animals are treated according to their most characteristic habitat, and in the order of their commonness, conspicuousness, and interest." The graceful writing and accurate observations quickly made it a classic, a book that has deeply influenced generations of marine scientists (whose labs are, not surprisingly, often filled with classical music). *Between Pacific Tides* is now in its fifth edition.

Ed Ricketts died in an automobile accident in May 1948. Steinbeck wrote a foreword to the second edition, which Ricketts was preparing. The foreword concludes: "There are good things to see in the tide pools and

Ed Ricketts at the Great Tide Pool, Pacific Grove, California.

there are exciting and interesting thoughts to be generated from the seeing. Every new eye applied to the peep hole which looks out at the world may fish in some new beauty and some new pattern, and the world of the human mind must be enriched by such fishing."

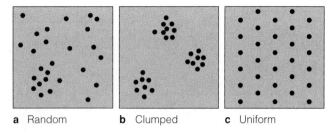

a Random   **b** Clumped   **c** Uniform

Figure 17.4 Random, clumped, and uniform population distribution patterns. The clumped pattern is most common in nature.

species in more niches within a given area—and harsh habitats usually have a lower species diversity. Relatively few species can cope with the stressful environment of the polar ocean surface, for example, but many species have adapted to the relatively benevolent environment of a tropical reef.

Individual organisms are almost never distributed randomly throughout their habitat. A **random distribution** implies that the position of one organism in a community in no way influences the position of other organisms in the same community. Further, a truly random distribution (such as that shown in **Figure 17.4a**) indicates that conditions are precisely the same throughout the habitat, an extremely unlikely situation except possibly in the unvarying benthic communities of abyssal plains.

The most common pattern for distribution of organisms is small patchy aggregations, or clumps. **Clumped distribution** (see **Figure 17.4b**) occurs when conditions for growth are optimal in small areas because of physical protection (in cracks in an intertidal rock), nutrient concentration (near a dead body on the bottom), initial dispersal (near the position of a parent), or social interaction.

Uniform distribution, with equal space between individuals (**Figure 17.4c**) (the arrangement of trees we see in orchards), is the rarest natural pattern of all. The distribution of some garden eels throughout their territories becomes almost uniform because each eel can extend from its burrow just far enough to hassle neighbor eels spaced at equal distance. But there is a break in the order of position. They don't line up row upon row like apple trees. The closest approximation to a uniform distribution in nature is probably that of breeding king penguins in an area bounded by a straight cliff. The birds nesting nearest the cliff arrange raised hillocks of stones one pecking-distance apart along the cliff face; the second row of birds is interspersed between the first; and so on. For a few rows away from the cliff the pattern sometimes resembles a checkerboard (**Figure 17.5**).

Figure 17.5 King penguins incubating their eggs on South Georgia Island, a rare example of uniform population distribution.

CHANGE IN MARINE COMMUNITIES

Like the organisms that constitute them, communities change through time, but marine communities do not evolve as rapidly as terrestrial communities. The slow changes associated with seafloor spreading, climate cycles, atmospheric composition, or newly evolved species have shaped this generally slow evolution. As on land, the species, community composition, and location of a marine community are changed by the environmental factors to which members of the community are exposed. Communities themselves can gradually modify the physical aspects of their environment. A coral reef is an extreme example: The massive accumulation of coral and sediments within the reef can alter current patterns, influence ocean temperature, and change the proportions of dissolved gases.

But rapid changes can occur in marine communities. A natural catastrophe—a volcano erupting, a landslide that blocks a river, the collision of an asteroid with Earth, for example—can disrupt a community. Similarly, human activities—such as the altering of an estuary by damming a river, dumping excess nutrients into a nearshore area, or stressing organisms with toxic wastes—can cause rapid, disruptive changes. Offshore communities change abruptly near new sewage outfalls, for example.

A stable, long-established community is known as a **climax community.** This self-perpetuating aggregation of species tends not to change unless disrupted by severe external forces, such as violent storms, significant changes in current patterns, epidemic diseases, or influx of great amounts of fresh water or pollutants. A disrupted climax community can be reestablished through the process of **succession,** the orderly changes of a community's species composition from temporary inhabitants to long-term inhabitants. Disruption makes the environment more hostile to the original species, but destruction of species in the original community leaves open habitats and niches. A few highly tolerant species will move into the area, eventually drawing in other species that depend on them. If the environment is permanently changed by the disruption, a different climax community will be established than was previously present.

EXAMPLES OF MARINE COMMUNITIES

There are many distinct marine communities. We briefly discussed the plankton community in Chapter 14, and coral reef and polar communities are included in the next chapter. Here we survey some other important examples.

Rocky Intertidal Communities

Anyone who spends time at the shore, especially a rocky shore, is soon struck by a curious contradiction. Although the rocky shore looks like a very difficult place for organisms to make a living, the **intertidal zone**—the band between the highest high tide and lowest low tide marks—is one of Earth's most densely populated areas. Hundreds of species and individuals crowd this junction of land and sea.

The problems of living in the intertidal zone are formidable. The tide rises and falls, alternately drenching and drying out the animals and plants. **Wave shock,** the powerful force of crashing waves, tears at the structures and underpinnings of the residents. Temperature can change rapidly as cold water hits warm shells, or as the sun shines directly on newly exposed organisms. In high latitudes, ice grinds against the shoreline; in the tropics, intense sunlight bakes the rocks. Predators and grazers from the ocean visit the area at high tide, and those from land have access at low tide. Too much fresh water can osmotically shock the occupants during storms. Annual movement of sediment onshore and offshore can cover and uncover habitats. Yet, astonishingly, the richness, productivity, and diversity of the intertidal rocky community— especially in the world's temperate zones—are matched by very few other places. There is intense competition for space. Life abounds.

One reason for the great diversity and success of organisms in the rocky intertidal zone is the large quantity of food available. The junction between land and ocean is a natural sink for living and once-living material. Minerals dissolved in water running off the land serve as nutrients for the inhabitants of the intertidal zone as well as for plankton in the area. The crashing surf and strong tidal currents keep nutrients stirred and ensure a high concentration of dissolved gases to support a rich population of autotrophs. Many of the larval forms and adult organisms of the intertidal community depend on plankton as their primary food source.

Another reason for the success of organisms here is the large number of habitats and niches to be occupied (see **Figure 17.6**). The habitats of intertidal animals and plants vary from hot, high, salty splash pools to cool, dark crevices. These spaces provide hiding places, quiet places to rest, attachment sites, jumping-off spots, cracks from which to peer to obtain a surprise meal, footing from which to launch a sneak attack, secluded mating nooks, or darkness to shield a retreat. The niches of the creatures in this community are varied and numerous— encrusting algae produce carbohydrates, snails scrape algae from rocks, hermit crabs scavenge for tidbits, and octopuses wait in ambush for likely meals.

The most obvious and important physical factor in intertidal communities is the rise and fall of the tides (see Chapter 11). Organisms living between the high and low tide marks experience very different conditions from those residing below the low tide line. Within the intertidal zone itself, organisms are exposed to varying amounts of emergence and submergence. For example, **Figure 17.7a** plots the number of hours of exposure to air in a California intertidal zone through six months' time versus the tidal height. Because some organisms can tolerate many hours of exposure while others are able to tolerate only a very few hours per week or month, the animals and plants sort themselves into three or more horizontal bands, or subzones, within the intertidal zone. Each distinct zone is an aggregation of animals and plants best adapted to the conditions within that particular narrow habitat. The zones are often strikingly different in appearance, even to a person unfamiliar with shoreline characteristics. This zonation is clearly evident in the rocky shore of **Figure 17.7b.**

For intertidal areas exposed to the open sea, wave shock is a formidable physical factor. Fist-sized rocks have been thrown 100 meters (330 feet) into the air by the force of breaking waves. Large intertidal plants must be immensely strong, elastic, and slippery to avoid being shredded by wave energy. **Motile** animals move to protecting overhangs and crevices, where they cower during intense wave activity. Attached, or **sessile** (*sessilis* = sitting), animals hang on tightly, often gaining assistance from rounded or low-profile shells, which deflect the violent forces of rushing water around their bodies. Some sessile animals have a flexible foot that wedges

a

Figure 17.6 A Pacific Coast tide pool and intertidal shore. (a) A diagrammatic view. (b) Key.

b

1. *bushy red algae*, Endocladia
2. *sea lettuce, green algae*, Ulva
3. *rockweed, brown algae*, Fucus
4. *iridescent red algae*, Iridea
5. *encrusting green algae*, Codium
6. *bladderlike red algae*, Halosaccion
7. *kelp, brown algae*, Laminaria
8. *Western gull*, Larus
9. *intrepid marine biologist*, Homo
10. *California mussels*, Mytilus
11. *acorn barnacles*, Balanus
12. *red barnacles*, Tetraclita
13. *goose barnacles*, Pollicipes
14. *fixed snails*, Aletes
15. *periwinkles*, Littorina
16. *black turban snails*, Tegula
17. *lined chiton*, Tonicella
18. *shield limpets*, Collisella pelta
19. *ribbed limpet*, Collisella scabra
20. *volcano shell limpet*, Fissurella
21. *black abalone*, Haliotis
22. *nudibranch*, Diaulula
23. *solitary coral*, Balanophyllia
24. *giant green anemones*, Anthopleura
25. *coralline algae*, Corallina
26. *red encrusting sponges*, Plocamia
27. *brittle star*, Amphiodia
28. *common starfish*, Pisaster
29. *purple sea urchins*, Strongylocentrotus
30. *purple shore crab*, Hemigrapsus
31. *isopod or pil bug*, Ligia
32. *transparent shrimp*, Spirontocaris
33. *hermit crab*, Pagurus, *in turban snail shell*
34. *tide pool sculpin*, Clinocottus

into small cracks to provide a good hold; others (like mussels) form shock-absorbing cables that attach to something solid.

Desiccation (drying) by exposure to air and sunlight is another source of intertidal stress. Again, motile organisms have an advantage because they can move toward water left in tidal pools or muddy depressions by the retreating ocean. Attached animals and plants must await the water's return, huddled in low spots, moist pockets, or cracks in the rocks—or in tightly closed shells. Water trapped within a shell can keep gills moist for the needed exchange of gases. A protective mucous coating can retard evaporative water loss from exposed soft animal body parts or blades of seaweed. When the weather is too warm—or when the tide is out for an unusually long time—a deceptive calm settles over the zone, to be relieved only by the returning ocean.

Sand Beach and Cobble Beach Communities

Some intertidal areas are sandy, some are muddy; others consist of gravel or cobbles. (A few shores combine all of these elements within a small area.) The usual rigors of the intertidal zone are intensified for organisms surviving on loose substrates. Indeed, it may surprise you to learn that in spite of its generally benign conditions, the ocean contains what may well be the most hostile, rigor-

ous, and dangerous environments for small living things on Earth: high-energy sand and cobble beaches.

As environments go, sand beaches don't seem particularly nasty places to us; many people consider the beach to be about the finest habitat around. Seals and sea lions spend a lot of time at the beach and seem to enjoy the experience as much as people do. In short, for organisms of about our size, the problems of living on a beach are manageable.

But for smaller organisms a beach is a forbidding place. Sand itself is the key problem. Many sand grains have sharp, pointed edges; so rushing water turns the beach surface into a blizzard of abrasive particles. Jagged grit works its way into soft tissues and wears away protective shells. A small organism's only real protection is to burrow below the surface, but burrowing is difficult without a firm footing. When the grain size of the beach is small, capillary forces can pin down small animals and prevent them from moving at all. If these organisms are trapped near the sand surface, they may be exposed to predation, to overheating or freezing, to osmotic shock from rain, or to crushing as heavy animals walk or slide on the beach.

As if this weren't enough, those that survive must contend with the difficulty of separating food from swirling sand and the dangers of leaving telltale signs of their position for predators, or being excavated by crash-

Figure 17.7 The relationship between amount of exposure and vertical zonation in a rocky intertidal community. (a) A graph showing intertidal height versus hours of exposure. The 0.0 point on the graph, the tidal datum, is the height of mean lower low water. (b) Vertical zonation, showing four distinct zones. The uppermost zone (I) is darkened by lichens and cyanobacteria; the middle zone (II) is dominated by a dark band of the red alga *Endocladia;* the low zone (III) contains mussels and gooseneck barnacles; and the bottom zone (IV) is home to sea stars (*Pisaster*) and anemones (*Anthopleura*). *Note:* The bands in the photograph (b) correspond approximately to the heights shown in the graph.

ing waves. A few can run for their lives; some larger beach-dwelling crabs depend on their good eyesight and sprinting ability to outrace onrushing waves!

To these horrors must be added the usual problems of intertidal life discussed earlier. Not surprisingly, very few species have adapted to wave-swept sandy beaches! Some of the successful ones are shown in **Figure 17.8.** The few that have done so—mostly small, fast-burrowing clams, sand crabs, sturdy polychaetes, and other minute worms—consume a rich harvest of plankton and organic particles washed onto the beach and filtered from the water by the uppermost layer of sand.

Cobble beaches are even more uninviting (and they're murder on bare feet). The rounded rocks clack and bump together as waves pound the shore; most small animals are crushed. Except for nimble, insectlike "beach hoppers" and a few species of scavenging terrestrial insects, most loose, rock-strewn shores are understandably void of anything much larger than microscopic organisms.

Surely the most difficult of all are the black sand beaches, derived from pulverized lava on tropical volcanic islands such as Hawaii. Besides all the other difficulties mentioned, we must add the lava's ability to store solar heat until temperatures approach 71°C (160°F) just below the surface of the sand. Almost nothing can tolerate these beaches for more than just a few minutes— including human feet!

Salt Marshes and Estuaries

Muddy-bottomed salt marshes are among the most interesting intertidal shores. Much of the high primary productivity of a salt marsh comes from sea grasses, mangroves, and other vascular plants that can prosper in a marine (or partly marine) environment.

As you may recall from Chapter 12, salt marshes often form in an **estuary,** a broad, shallow, river mouth where fresh water and salt water mix (see Figures 12.31 and

Figure 17.8 Sand beach organisms. (a) Dime-sized coquina clams (*Donax*) lie at the surface awaiting a ride up the beach on an incoming wave. They will bury themselves in the loose sediment, push up their siphons, and filter the water for food. When the tide retreats, they will again pop to the surface and allow the waves to take them back down the beach. (b) A sand crab (*Emerita*), beloved of all beach-going children, attempts to bury itself in anticipation of an onrushing wave. It gleans food from passing water with its feathery antennae. (c–f) Examples of interstitial animals. These organisms are tiny enough to live in the spaces (or interstices) between sand grains, too small to be seen by the unaided eye. (c) A polychaete worm; (d) a crustacean; (e) a tardigrade; (f) a brachiopod.

12.32). A characteristic of estuaries is the reduction of wave shock: Surf is blocked from estuaries by longshore bars or by twisting passages connecting to the ocean. The salinity of water within an estuary may vary with tidal fluctuations, from seawater through **brackish** water (mixed salt and fresh water) to fresh water. Many of the organisms living in estuaries are necessarily euryhaline. In areas near the river entrance, however, the water may be almost fresh, whereas near the outlet it may be of oceanic salinity. These different salinities often lead to a distinct horizontal zonation of organisms. Temperature range is also potentially extreme, especially in the tropics or during the temperate-zone summer when a receding tide abandons residents to the heat of the sun. Strong

currents may move in estuaries as the tide rises and falls and the river flows. Flowing water takes the place of waves in mixing nutrients and gases in the intertidal estuary community.

Estuarine marshes (such as the one shown in **Figure 17.9**) are richer and exhibit greater species diversity than marshes exposed only to seawater. Primary productivity in estuaries is often extraordinarily high because of the availability of nutrients, the great variety of organisms present, strong sunlight, and the large number of niches. Decomposition of fast-growing, salt-tolerant plants provides the raw material for the large and complex food webs and rapid nutrient turnover characteristic of these communities. The standing biomass (mass of living mat-

Figure 17.9 An estuarine marsh. Urban developers often destroy coastal marshes to build marinas or homes, but citizens near this marsh in Orange County, California, have recognized its natural value and have set it aside as a marine preserve.

ter per unit area or volume) in a typical estuary is among the highest per unit of surface area of any marine community.

Estuarine organisms show unique adaptations to their rich and variable environment. Some estuarine plants trap fine silt particles at their roots, thus countering the erosive action of current flow. Small plants are often filamentous, bristling with tiny projections used to anchor themselves to the substratum. Larger plants have extensive root systems to hold themselves in place and to colonize new areas. Most of the resident animals burrow into the muck, scurry rapidly across the surface, or hide in the vegetation. Clams and snails work their way through the substratum, obtaining food and shelter at the same time. Polychaete worms dig for targets of opportunity, and crabs dart for any interesting morsels. Since planktonic larvae would be washed out to sea, most estuarine organisms produce nonplanktonic larvae, lay eggs on firm objects, or carry eggs on their bodies.

Estuaries are sometimes called marine nurseries because so many juvenile organisms are found there. This is especially true for fish. Many pelagic species spend their larval lives in the protective confines of an estuary, taking advantage of the many feeding opportunities available. Most of the commercially exploited fish species on the U.S. Atlantic coast utilize estuaries as juvenile feeding grounds. The human pressures of development and pollution are thus doubly stressful in estuaries, affecting both permanent residents and the sensitive larval stages of open-water animals.

You may also recall that estuaries are not permanent features—they are very sensitive to changes in sea level. Most East Coast estuaries probably formed during the sea level rise of the last 3,000 to 10,000 years. Estuaries are probably more common today than they were at the height of the last ice age, when sea level was lower.

The Open Ocean

About 83% of the ocean's total biomass is concentrated in its uppermost 200 meters (660 feet); only 0.8% is found below 3,000 meters (10,000 feet). Nearly all deep-ocean habitats are sparsely populated, but the few species of animals that have adapted to this impoverished place range through virtually all oceanic latitudes. There are no photosynthetic autotrophs in the deep ocean because there is no light. With the exception of rift communities (more on this in a moment), consumers at great depths

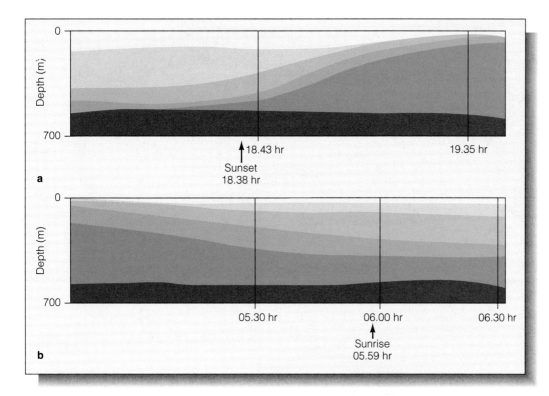

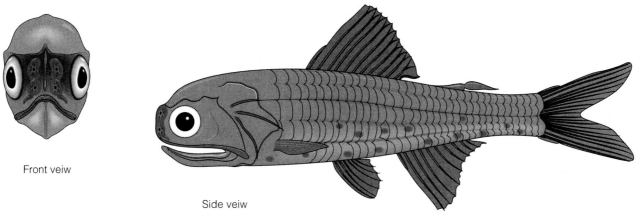

Figure 17.10 The movements of scattering layers, as recorded by an echo sounder. (a) Three distinct layers move toward the surface at sunset. (b) Before sunrise the layers move down again. This phenomenon is caused by organisms that migrate up and down with changing amounts of sunlight. (In both traces another scattering layer remains at a constant depth.)

Front veiw

Side veiw

Figure 17.11 *Diaphus elucens,* a common mesopelagic fish with large light organs (here shaded blue) between the eyes and on the body. The fish is 5.5 centimeters (2 inches) long.

must depend on the productivity of the water column above.

A peculiar pelagic community lives at the uppermost limits of the permanent darkness. Named after its ability to reflect sound pulses and appear to echo sounders as a false bottom, the **deep scattering layer (DSL)** is a relatively dense aggregate of fishes, squid, and other animals that usually migrate up and down in synchrony with daylight (**Figure 17.10**). Deep scattering layers—there is often more than one—are found in all ocean areas except the Arctic, and they are best developed in places of high surface productivity. The DSL is most pronounced during daylight hours, when members of the community congregate at the lowest limit of light penetration. At nightfall many of the organisms migrate to the surface to feed on plankton. Most residents of the deep scattering layer have large, sensitive eyes, which permit them to feed by detecting the faint shadows of prey above. Some members of the community have built-in luminescent organs that cast dim blue light downward; this light masks their own shadows; so they have less chance of being detected and eaten (see **Figure 17.11**).

Between the deep scattering layer and the bottom, in the bathypelagic zone, the ocean is nearly devoid of life. Very little food is available in this zone; tiny crumbs of organic material are usually broken down by microbes

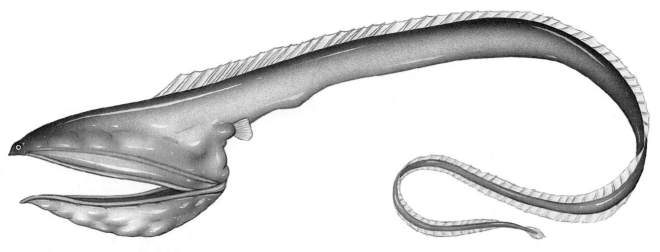

Figure 17.12 The deep-sea gulper (*Eurypharynx pelecanoides*), a bathypelagic species with a worldwide distribution beneath tropical waters. Its length is about 60 centimeters (24 inches).

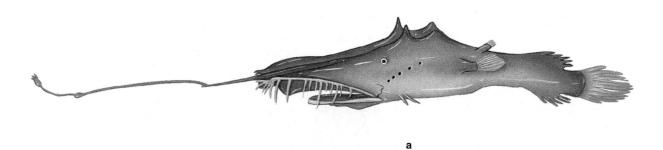

a

before reaching mid-depths, and the bodies of large organisms continue their fall to the seafloor.

The few animals in this vast middle volume of ocean below the DSL are among the Earth's most bizarre. Gulper eels (**Figure 17.12**) have extendable jaws and a stomach capable of consuming prey larger than the eels themselves, an adaptation of great importance when one considers that a gulper eel may not encounter a feeding opportunity more often than once or twice a year! Bioluminescence, the biological production of light, is important here in both feeding and mate attraction. Some deep-swimming organisms attract their infrequent meals with a luminous lure (**Figure 17.13**). These animals also use patterns of glowing spots or lines to identify themselves to members of the same species, a necessary first step in mating. Some use flashes of light to dazzle or frighten potential predators.

The Deep-Sea Floor

Most of the deep-ocean floor is an area of endless sameness. It is eternally dark, almost always very cold,

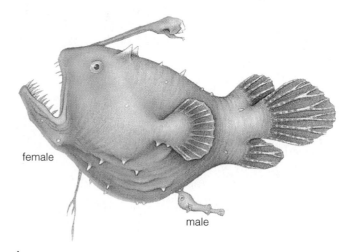

female

male

b

Figure 17.13 Two species of deep-sea anglers with bioluminescent lures. (a) *Lasiognathus saccostoma*, known only from the western Atlantic, has fishing equipment that consists of an extensible rod, filament, "float," and illuminated lure with little hooks. Its body is about 20 centimeters (8 inches) long. (b) A female *Melanocoetus johnsoni*, smaller at 10 centimeters (4 inches), carries a permanently attached male.

slightly hypersaline (to 36‰), and highly pressurized. Life here is more plentiful and obvious than in the bathypelagic water above, but as bottoms go, the density of life at great depths is extremely low. For example, 5,000 grams of living organisms can be found on a square meter of nearshore seafloor, and 200 grams might be recovered from the same area of continental shelf, but less than 1 milligram per square meter is typical for deep-ocean benthic communities! Currents are usually so weak that they impose no strain on residents, but they do transport the few drifting organisms and waft the smell of food. Most bottom dwellers are blind.

The feeding strategies of animals living on the deep-ocean floor are often bizarre. Tripod fish (**Figure 17.14**) use sensitive extensions of their fins and gill coverings to detect the movement of prey many meters away. Some organisms whose mouths blend with the natural contours of the ooze act as living caves into which small creatures crawl for protection. The predator need not even swallow to get the prey into its gut—back-pointing spines direct the victim along a one-way path to the stomach! Other species are capable of smelling large, sunken dead animals for miles down-current and then spending weeks or months slowly following the scent to its source. The metabolic rate of organisms in cold water tends to be low; so most deep animals require relatively little food, move slowly, and live very long lives. Some may feed less than once in a year and may live to be hundreds of years old. Common deep benthic representatives are seen in **Figure 17.15**.

This deep uniform environment, though rigorous, holds benefits for those few species adapted to it. Perhaps the best-adapted organisms are the ubiquitous ophiuroids (brittle stars) that inhabit sedimentary bottoms, sometimes at arm-tip to arm-tip densities, at nearly all latitudes. They feed by collecting tiny nutritive particles—some of which will have fallen through the water for many months—along grooves beneath their arms and transporting them to their mouths. As you may recall from Chapter 15, brittle stars are among the most widely distributed of Earth's animals (see Figure 15.30).

The organisms within deep pelagic and benthic communities share some curious adaptations. Gigantism is a common characteristic: Individuals of representative families tend to be much larger in deep water than related individuals in the shallow ocean. Fragility is also common in the depths. Not only are heavy support structures unnecessary in the calm deep environment, but the relatively low water pH and high pressure discourage the deposition of calcium—and thus skeletal development. Some animals have slender legs or stalks to raise themselves above the sediment, and some come apart like warm gelatin at the slightest touch. Except for its influence on enzyme activity, hydrostatic pressure is

Figure 17.14 A blind tripod fish, an abyssal benthic species. The long, curved projections are thought to aid in sensing the distant vibrations of prospective prey.

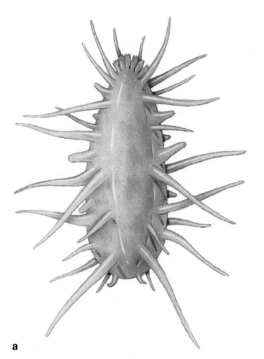

a

b

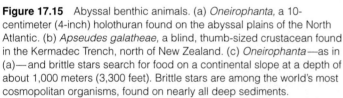

c

Figure 17.15 Abyssal benthic animals. (a) *Oneirophanta*, a 10-centimeter (4-inch) holothuran found on the abyssal plains of the North Atlantic. (b) *Apseudes galatheae,* a blind, thumb-sized crustacean found in the Kermadec Trench, north of New Zealand. (c) *Oneirophanta*—as in (a)—and brittle stars search for food on a continental slope at a depth of about 1,000 meters (3,300 feet). Brittle stars are among the world's most cosmopolitan organisms, found on nearly all deep sediments.

not a problem for these animals. They live in balance with the great pressure; their internal pressure is precisely the same as that outside their bodies.

Recent research has shown that seabed communities are not confined to the uppermost layer of sediments. Populations of bacteria have been found as deep as 500 meters (1,640 feet) below the Pacific seafloor. The bacteria depend for energy on the metabolism of sulfur and methane compounds laid down in the sediments more than 4 million years ago.

Vent Communities

Biologists around the world were excited in 1977 when scientists in Woods Hole Oceanographic Institution's submersible *Alvin* (Box 4.1c) discovered an entirely new type of marine community over 3,000 meters (10,000 feet) below the surface.[1] They were searching the near-

---

[1]Ironically there wasn't a single marine biologist on this research cruise—only geologists, geophysicists, chemists, physicists, and a science writer!

freezing bottom for the source of some unusually warm water that had been detected by remote probes. What they found were jets of superheated water (to 350°C, 650°F) blasting from rift vents in the young oceanic ridge 350 kilometers (220 miles) north and east of the Galápagos Islands. Clustered around the vents were dense aggregations of large, previously unknown animals. Bottom water in the area was laden with hydrogen sulfide, carbon dioxide, and oxygen, on which specialized bacteria were found to live. These bacteria evidently form the base of a food chain that extends to the animals. Large crabs, clams, sea anemones, shrimp, and unusual worms were found in this warm oasis.

Some of the *tube worms,* contained in their own long parchmentlike tubes, measured 3 to 4 meters (10 to 13 feet) in length and were the diameter of a human arm. These strange animals have been tentatively identified as pogonophorans, members of a small phylum of invertebrates also found in fairly shallow water. Three species of the newly and appropriately named genus *Riftia* (**Figure 17.16**) have been identified so far. The tubes of these pogonophorans are flexible and capable of housing the

Figure 17.16 *Riftia*, large tube worms found around hydrothermal vents on oceanic ridges.

Figure 17.17 A vent field dominated by the giant white clam *Calyptogena magnifica*.

length of the animal when it retracts. The animals extend tufts of tentacles from the openings of their tubes. Feeding was something of a puzzle because these animals have no mouth, digestive tract, or anus. The trunks of the worms were found to contain large *feeding bodies* tightly packed with bacteria similar to those seen in the water and on the bottom near the geothermal vents. The worms' tentacles absorb hydrogen sulfide from the water and transport it to the bacteria, which then use the hydrogen sulfide as an energy source to convert carbon dioxide to organic molecules. The ultimate source of the worms' energy (and the energy of most other residents in this community) is this energy-binding process, called *chemosynthesis,* which replaces photosynthesis in the world of darkness.

The clams and shrimp of these vent communities are equally unusual. For example, the large, white clam *Calyptogena* grows among uneven basaltic mounds (**Figure 17.17**). Each the size of a shoe, the clams shelter the same kinds of bacteria as *Riftia.* Though the clam retains its filter-feeding structures, it too derives nutrition from the bacteria. Small shrimp discovered at the vents in 1985 have been found to possess special organs that may allow them to sense heat from the vents. Such an adaptation would permit them to range away from the vents for food, yet return to the warmth and richness of the home community.

Vent communities have now been found off Florida and Louisiana, off California and Oregon, in the North Sea, east of Japan, and in several other locations. Not all

BOX 17.2 ● *Deep-Sea Clambake*

In April 1991, observers were startled to find the freshly roasted remains of a large tube worm community at a depth of 2,500 meters (8,200 feet) on the East Pacific Rise west of Central America. Oceanographers from the University of California at Santa Barbara and Columbia University's Lamont–Doherty Earth Observatory, who photographed the scene, dubbed it a "tube worm barbecue." Members of the deep community—mostly pogonophoran worms and snails—had been baked by a surge of rapidly moving superheated water. The force of the water flow was enough to tear many of the cooked bodies apart. Bits of flesh and debris hung in the water around the site.

Because scavengers had not yet gathered at the location, the scientists believed the violence had occurred only hours before their arrival. Since fresh lava flows had partially buried some of the scorched creatures, they concluded that they had just missed witnessing a mid-ocean ridge volcanic eruption!

Other expeditions have seen evidence of similar activity. During 1978, just one year after the discovery of vent communities, a joint French-American-Mexican expedition visiting a section of the East Pacific Rise south of Baja California came upon broken columnar chimneys blasting extremely hot water and surrounded by thou-

sands of boiled giant clams. A similar pattern of devastation was seen in 1989 on the seabed west of Oregon.

No one has yet witnessed a ridge eruption and the subsequent formation of oceanic crust. But these frequent occurrences must certainly be responsible for the destruction of countless vent communities along the mid-ocean ridges. Organisms in these communities have evolved several strategies for coping with catastrophe. Members of the vent communities often grow very rapidly: Giant clams can grow up to 500 times faster than deep-sea clam species not associated with vents, and the pogonophorans (giant tube worms) have been found to regenerate with remarkable speed. Individual members of one population of pogonophorans recovered from a near-disaster by growing 1.2 meters (4 feet) in little more than two years. Also, most species of vent organisms have long-lived planktonic larvae that are borne by the slow abyssal currents on lengthy recolonization journeys; thus, some youngsters will survive the holocaust simply by drifting away.

Communities everywhere must change with ambient conditions. Even when those conditions lead to a particular community's destruction, the characteristics of the organisms can contribute to their eventual success somewhere else.

are on oceanic ridges, and not all are in areas spouting hot water. The slow upward percolation of cool, mineral-rich water seeping from beneath sediments encourages the growth of mats of chemosynthetic bacteria that are able to metabolize sulfur-containing compounds or methane. The methane, where present, appears to come from the decomposition of organic material in the sediment or underlying sedimentary rocks. Cool or cold vent communities dependent on these bacteria may be more widespread than those near areas of superheated water, and they may be longer lived (see **Box 17.2**).

Studies of vent communities suggest many questions. Do hot vent communities occupy the active central rift valleys of a significant percentage of the 65,000 kilometers (41,000 miles) of Earth's oceanic ridges? Did the hot or cold deep vents serve as the birthplace of life on Earth? Perhaps the deep vent communities will prove to

be more important to overall marine productivity than has been previously supposed. Marine biologists are eager to continue their explorations.

SYMBIOTIC INTERACTIONS AND DEPENDENCIES

Newcomers to biology are often surprised to learn that more than half of the animal species known to science are not free-living. Most are actively involved in close symbiotic relationships with at least one other life-form in their community. These relationships are often intricate and sometimes quite strange.

Symbiosis (*sym* = with, *bios* = life) is the biologist's term for the co-occurrence of two species in which the life of one is closely interwoven with the life of the other.

Figure 17.18 Mutualism. Some species of sea anemones have a symbiotic relationship with anemone fish, in which the fish receives protection from predators and the anemone receives scraps of food from the fish.

The symbiotic bond is often so strong that one organism is totally dependent on the other.

Symbioses

There are three general types of symbioses: mutualism, commensalism, and parasitism.

In **mutualism,** as the name implies, both the symbiont and the host—the larger organism with which the symbiont lives—benefit from the relationship. True mutualism is rare among marine organisms, but a few examples have been observed. One is the relationship between an anemone fish and its sea anemone. In this symbiosis a small, brightly colored anemone fish nestles within the tentacles of a sea anemone (**Figure 17.18**). The mechanism that permits the little fish to do this without being stung is not well understood, but biologists believe that the fish gradually desensitizes the stinging cells of the anemone by using mucous secretions from its own skin. In return for the anemone's protection the fish feeds the

anemone scraps of food and may even lure prey within the anemone's reach.

Another example of mutualism is the relationship between certain cnidarians, such as coral animals, and the specialized dinoflagellates (known collectively as *zooxanthellae*) that live within their tissues. Both organisms benefit: The autotrophic dinoflagellates have a safe home and a ready source of carbon dioxide, and the animals have a handy, built-in source of carbohydrates. Without their resident dinoflagellates, the reef corals (about which we will learn more in Chapter 18) would be unable to deposit calcium, and the rich tropical reef communities we know today would not exist.

Perhaps the most striking examples of mutualism involve cleaning symbioses. In these relationships small organisms (usually fish or shrimp) establish a "cleaning station" at which they remove dead tissue and troublesome surface parasites from the skin, mouth, and gill coverings of larger animals (usually fish) visiting the site. The cleaner eats the dead tissue and parasites; so again both animals benefit. The cleaned animals frequently defend the cleaning station and its cleaners from attack by would-be predators.

In **commensalism,** the symbiont benefits from the association while its host neither benefits nor is harmed. For example, biologists once believed that the association between pilot fish and shark was mutualistic; the pilot fish was thought to guide the shark to a meal and in turn be permitted to dine on the scraps. We now know the pilot fish is only an opportunistic commensal, taking what scraps he can from the shark's meal. (On the other hand, the mutualistic anemone fish/anemone partnership was thought to be commensal before the anemone fish was observed feeding the anemone.)

Some of the most curious commensals are those that enter their partner's habitat and, after a period of growth, cannot escape. Small pea crabs of genus *Fabia,* for example, live inside mussel shells, eating food particles that are brought inside by the normal feeding and respiratory actions of the mussel. After a time, the crab grows so large that it cannot leave the mussel because it is larger than the gap between the mussel's shells.

Parasitism is the most highly evolved and by far the most common symbiotic relationship. The parasite lives in (or on) the host for at least part of its life cycle and obtains food at the host's expense. For obvious reasons, parasites do not usually kill their hosts, but they can seriously affect the host organism by reducing its feeding efficiency, depleting its food reserves, reducing its reproductive potential, lowering its resistance to disease, or otherwise sapping its energy. The host–parasite relationship is finely balanced and extraordinarily delicate. The parasite must in some way be aware of the host's physical condition in order to avoid weakening the host so much that it dies. On the other hand, the parasite must take as much energy from the host as possible in order to ensure its own success.

All major phyla have parasitic marine representatives. However, the most widely distributed and successful group of parasitic marine animals are the roundworms of the phylum Nematoda. Like nearly all parasites, nematodes have a species-specific relationship with a host. A **species-specific relationship** is an exclusive relationship between two species; parasites can usually parasitize only one species of host. The reason for this interdependency is the delicacy of the biochemical feedback mechanisms that inform the parasite that its activity may be overstressing the host. The feedback responses are, by necessity, tailored to specific host–parasite pairs. The parasite will not usually survive if it settles in or on a host for which it is not specifically "programmed."

More than one species of parasite *can* infect a single host, however. A significant percentage of the weight of many fishes may be in nematode worms and other para-sites. The parasitic burden of a normal, apparently healthy sea lion may exceed 2.3 kilograms (5 pounds) and 20 species! Parasites are usually small, but one species of nematode parasite of the fin whale reaches 7 meters (23 feet) in length and is the diameter of a pencil. And, by the way, parasites can have their own parasites!

These three categories (mutualism, commensalism, parasitism) form a continuum in nature. There are few clearly mutualistic relationships that do not suggest at least a touch of commensalism, and few commensalistic relationships that do not hint of parasitism. The whale barnacles pictured in **Figure 17.19** glean most of their food directly from the ocean, but some of their nutrition is derived from the flesh and circulating fluids of their host. It's sometimes hard to tell where one kind of relationship stops and another begins.

Dependencies

Marine animals that can capture and eat only a single prey species are said to be *dependent* on that species. Though it is a controversial stance, some marine biologists consider such a specific feeding relationship to be a variety of symbiosis. These relationships are called **dependencies.** Dependencies differ from true symbioses because the lives of the species involved are not interwoven for extended periods of time.

One of the most extreme dependencies is that of a recently discovered species of copepod (a planktonic arthropod), which can eat only one particular size of one species of diatom because of the configuration of the copepod's mouthparts. Such a specific feeding dependency is unusual; most organisms are able to eat a broader range of food species.

Q-AND-A

1. If the rocky, sandy, or muddy intertidal zones represent such a challenging mix of environmental factors, why do so many organisms live there?

   Difficulty *in biology is a relative term. It may seem a circular argument, but wherever organisms live, conditions for life at that place are biologically tolerable, food is available, and environmental conditions are not so extreme as to preclude success. Organisms live in abundance where energy is available. Where there is food, or sunlight, or biodegradable compounds, there is life. Natural selection has sorted out the ways that work in this zone from the ways that do not, and the adaptations that work give the organisms living in the intertidal zone's many niches access to a rich harvest of nutrients.*

Figure 17.19 Whale barnacles. These arthropods can be buried up to 3 centimeters (1¼ inches) deep in the skin of certain whales. They derive part of their nutrition from the flesh and circulating fluids of the whale. Some individuals are up to 7.6 centimeters (3 inches) across.

2. Is the species-specificity rule of parasitism ever broken?

Yes, sometimes with catastrophic results. As an example, the lung flukes that inhabit the respiratory tract of most sea lions will "abandon ship" if the sea lion is weakened or dying. A sea lion in this condition sometimes comes out of the water onto a beach to rest. If your pet dog should discover the animal there and sniff at its nose, some of the parasites could transfer from sea lion to dog and establish themselves in the dog's lungs. Because the dog is not the species-specific host for the lung parasites, the biochemical machinery that tells the parasite that it is weakening its host is missing. The dog may die a painful death in a few weeks from an overinfestation with lung flukes. The parasites, will, of course, also die. The interaction will have been a failure in all respects.

3. Could there be any huge undiscovered Godzilla-type sea monsters in the deep ocean?

Probably not, unless they can extract energy directly from water molecules! The deep pelagic feeding situation is simply not rich enough to support the energy needs of an active population of violent, aggressive, city-eating (metrophagous?) reptiles. Scientists never say never, but classic science fiction films aside, it doesn't look promising.

4. What are the historical foundations of ecology?

Many students confuse ecology with conservation efforts such as recycling or with environmental activism such as picketing large industrial polluters. But neither of these activities is ecology

Ecology as a subdivision of natural science has a long history, which began with the writings of Pliny and other Romans who took an interest in a unified view of nature. Some European Renaissance scholars developed a similar outlook. In more recent times the French naturalists René-Antoine Réaumur (1663–1757) and Georges-Louis de Buffon (1707–88)—along with the great German zoologist and Darwinian Ernst Häckel, who coined the word oekologie *in 1869—all helped to build the foundations of the modern science. Population studies were advanced by Elton in his 1927 book* Animal Ecology, *and modern biometric analyses of populations and communities were pioneered by the*

Australian H. G. Andrewartha and his colleagues and contemporaries. The important concept of primary productivity was developed by two American freshwater biologists—E. Birge and C. Juday—in the 1930s.

Modern ecology came of age in 1942 when Robert L.

Lindeman, an American, began the study of ecological energetics. His work detailing the flow of energy through ecosystems was among the first to consider an ecosystem as a unified whole. Most modern ecological studies can be traced to his insight.

Terms and Concepts to Remember

brackish	estuary	species diversity
carrying capacity	euryhaline	species-specific
climax community	eurythermal	relationship
clumped distri-	habitat	stenohaline
bution	intertidal zone	stenothermal
commensalism	motile	succession
community	mutualism	symbiosis
deep scattering	niche	uniform distri-
layer (DSL)	parasitism	bution
dependency	population	wave shock
desiccation	population density	
ecology	random distri-	
environmental	bution	
resistance	sessile	

Study Questions

1. How does a population differ from a community? A niche from a habitat?

2. What is a limiting factor? Give a few examples.

3. In what ways can members of the same population compete with one another? How might members of different populations compete? Contrast the results of these kinds of competition.

4. Describe a typical growth curve for a population. What factors influence the shape and eventual height of the curve?

5. What factors influence the distribution of organisms within a community? How are these distributions described? Why is random distribution so rare?

6. What problems confront the inhabitants of the intertidal zone? How do you explain the richness of the intertidal zone in spite of these rigors?

7. Which marine habitat is the most sparsely populated? Why?

8. Describe the residents of the deep ocean. How do inhabitants of the vent communities differ from other deep-floor organisms? What is the source of nutrition for vent communities?

9. What are the three types of symbioses? Give an example of each.

10. Why must the host–parasite relationship be so finely balanced? What would be the result of an imbalance?

For Further Study

General Ecology and Marine Ecology

Andrewartha, H. G. 1961. *Introduction to the Study of Animal Populations.* Chicago: University of Chicago Press. An important classic.

Briggs, J. C. 1974. *Marine Zoogeography.* New York: McGraw-Hill.

Hazen, W., ed. 1970. *Readings in Population and Community Ecology.* Philadelphia: W. B. Saunders. Compilation of important original papers in general ecology. General principles; some marine examples.

Hedgpeth, J., ed. 1957. *Treatise on Marine Ecology and Paleoecology.* Vol. 1, *Ecology* (Memoir 67). Washington, DC: Geological Society of America. An extraordinary compilation of material on biological and physical factors, habitats, and whole communities, written by experts in each area.

Levinton, J. S. 1982. *Marine Ecology.* New York: Prentice-Hall. An excellent technical introduction to marine ecological theory. Chapter 5 summarizes marine community diversity and evolution.

MacGinitie, G., and N. MacGinitie. 1968. *The Natural History of Marine Animals.* Chapters 4, 5, and 8 cover animal groupings, relationships, and habitats.

Miller, G. T. 1992. *Living in the Environment.* 7th ed. Belmont, CA: Wadsworth. Part 2, on basic ecological concepts, is especially recommended.

Milne, D. H. 1995. *Marine Life and the Sea.* Belmont, CA: Wadsworth. An excellent ecological approach, readable and complete.

Moore, H. B. 1958. *Marine Ecology.* New York: Wiley. An especially thorough introduction to individual habitats, with an excellent presentation of physical and biological factors.

Nybakken, J. W. 1971. *Readings in Marine Ecology.* New York: Harper & Row. Reprints of important original papers in marine ecology.

Nybakken, J. W. 1988. *Marine Biology: An Ecological Approach.* 2d ed. New York: Harper & Row. Excellent general marine biology text, with an ecological emphasis.

Starr, C., and R. Taggart. 1995. *Biology: The Unity and Diversity of Life.* 7th ed. Belmont, CA: Wadsworth. Chapters 45 and 46 provide a concise general introduction to the principles of ecology.

Note: Oceanus 35 (no. 3, Fall 1992) was devoted to the topic of marine biology.

Specific Habitats and Communities

Adler, T. 1994. "Bacteria Found Deep Below Ocean Floor." *Science News* 146 (no. 14): 215. Specialized bacteria can live up to 500 meters below the surface. They break down energy-rich methane and sulfur compounds dissolved in upward-percolating water. The compounds are derived from sedimentary rock.

Ballard, R. D. 1977. "Notes on a Major Oceanographic Find." *Oceanus* 20 (no. 3, Summer): 35–44. News of the startling discovery of the rift community.

Ballard, R. D., and F. J. Grassle. 1979. "Return to Oases of the Deep." *National Geographic* 156 (no. 5): 686–703. Wonderful pictures of *Alvin*, the vents themselves, and the odd organisms of the vent communities.

Childress, J. J., H. Felbeck, and G. Somero. 1987. "Symbiosis in the Deep Sea." *Scientific American*, May, 115–20. The high density of life near hydrothermal vents is explained by mutually beneficial symbioses.

Dietz, R. S. 1962. "The Sea's Deep Scattering Layer." *Scientific American*, August, 44–50.

Gage, J. D., and P. A. Tyler. 1991. *A Natural History of Organisms at the Deep-Sea Floor.* New York: Cambridge University Press. A scholarly celebration of the growth of our knowledge about biological systems on the seafloor.

Garrison, T. 1989. "Marine Parasites." *Pacific Diver*, October, 19–20.

Heezen, B. C., and C. D. Hollister. 1971. *The Face of the Deep.* New York: Oxford University Press.

Hinton, S. 1969. *Seashore Life in Southern California.* Berkeley: University of California Press. Excellent nontechnical field guide, with fine drawings and photographs.

Horn, M. H., and R. N. Gibson. 1988. "Intertidal Fishes." *Scientific American*, January, 64–70. Fishes in the intertidal zone have special adaptations for clinging, and some species can breathe air.

Ingle, R. M. 1954. "The Life of an Estuary." *Scientific American*, May, 64–68. A key paper on estuarine adaptation.

Isaacs, J. D., and R. A. Schwarzlose. 1975. "Active Animals of the Deep-Sea Floor." *Scientific American*, October, 84–91.

Koehl, M. A. R. 1982. "The Interaction of Moving Water and Sessile Organisms." *Scientific American,* December, 124–34. Attached intertidal organisms show a remarkable array of adaptations to the forces exerted by strong currents and crashing waves.

Kozloff, E. 1983. *Seashore Life of the Northern Pacific Coast.* Seattle: University of Washington Press. A thorough, well-written treatment.

Limbaugh, C. 1961. "Cleaning Symbioses." *Scientific American,* August, 42–49. Cleaner wrasses and shrimp are discussed.

Lutz, R. 1991. "The Biology of Deep Sea Vents and Seeps." *Oceanus* 34 (no. 4, Winter 1991/92): 75–83. Information on hot and cold vents and seeps.

Monastersky, R. 1989. "Deep-See [sic] Shrimp" *Science News* 155 (11 February). The 1985 discovery of *Rimicaris exoculata* is discussed. These small shrimp possess special structures that may sense heat (or very dim light) issuing from hydrothermal vents.

Ricketts, E. F., J. Calvin, J. Hedgpeth, and D. W. Phillips. 1985. *Between Pacific Tides.* 5th ed. Palo Alto, CA: Stanford University Press. A gracefully written classic applicable especially to the western American coast.

Robison, Bruce H. 1995. "Light in the Ocean's Midwaters." *Scientific American*, July, 60–65. Between the sunlit waters near the surface and the pitch darkness at the seafloor is our planet's largest and most fantastic community, illuminated only by the chilly radiance of its luminous natives.

Yonge, C. M. 1949. *The Sea Shore.* London: Collins. A classic, applicable to both sides of the North Atlantic.

18

POLES AND TROPICS: OCEANOGRAPHY OF THE EXTREMES

Half the World in Half a Day

It's possible to fly from the polar regions to the tropics in about half a day, and passengers who select their takeoff time properly may spend all of the flight in daylight. If they forgo in-flight movies and fitful sleep, commuters along the shortest route from London to Hawaii (via Anchorage, Alaska) can experience the dramatic contrasts of a quick passage from frozen oceanic wastes to vibrant tropical expanses. After flying northwest across the Atlantic, after passing the smoldering gray mass of Iceland (a visible bit of the Mid-Atlantic Ridge), our travelers see the white shoulders of Greenland rise from a cold, cloudy, ice-flecked sea. Another hour's flight—over a white, mostly featureless expanse—reveals the deep fjords of western Greenland, the gashed valleys that calve icebergs like the one that sank the RMS *Titanic* in the days of slower (but not always safer) intercontinental travel. Its flight path takes the airliner to still higher latitudes, across the stormy, ice-choked seas far above the Arctic Circle and over the barren northernmost islands beyond mainland Canada—islands named for Axel Heiberg and Harald Sverdrup, pioneers of turn-of-the-century oceanographic exploration.

After the plane turns southward and refuels at Anchorage, observers cannot fail to notice a rapidly changing sea: The storms and ice fall behind, and the horizon-to-horizon ocean takes on a different cast. As time passes, the low gray light turns sharper, bluer, more benign. The smooth sea brightens, reflecting more light from its calmer surface. Six hours later: the volcanic tips and surf-ringed coasts of the islands of Hawaii. Turquoise water over reefs—a color last seen at the base of great icebergs near Greenland—now signifies warmth, life in boundless variety, the promise of journey's end. A voyage that occupied nearly two years for Captain Cook, an enthusiastic connoisseur of such oceanic differences, takes only 12 hours today. Clouds, ice, currents, eroded mountains, active volcanoes, changes in sun angle, marine and terrestrial biology—all are visible from cruising altitude for those willing to look outside. The informed eye can draw much from such a journey. Demand a window seat!

The jagged eastern coast of Greenland from 40,000 feet. A 747 jetliner allows very comfortable Arctic exploration. Figure 18.13 shows where this photo was taken.

Exposure to very different amounts of sunlight causes most of the differences in ocean conditions between polar and tropical regions. Polar ocean areas tend to be colder, less stratified by density, and richer in dissolved nutrients and gases than tropical ocean areas. The polar oceans also contain greater populations (but of fewer species) than their tropical counterparts.

In only a few places does the brilliant blue tropical ocean live up to its popular reputation as an environment that teems with vividly colored animals. These productive locations, usually associated with shallow coral reefs, occupy less than 2% of the tropical ocean surface but account for more than half of the tropical ocean's total biological productivity. (The temperate ocean, the bands of water between the polar and tropical extremes, are the most biologically productive areas of all.)

The physical and biological differences between the polar and tropical oceans are largely confined to the uppermost 2,000 meters (6,500 feet); conditions at great depth are similar in all areas of the world ocean.

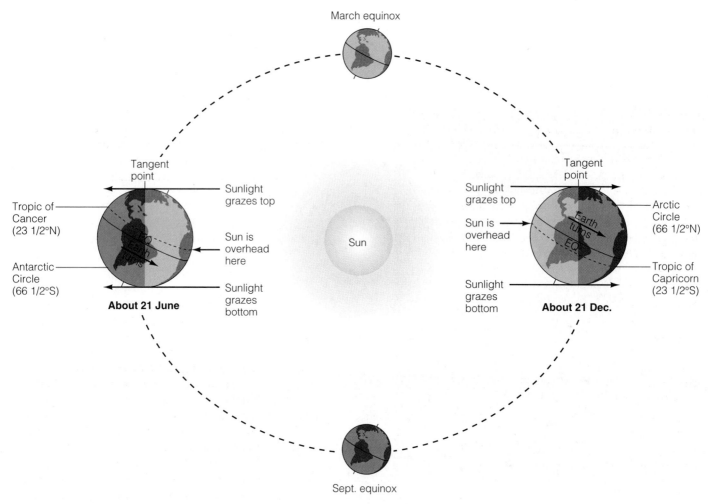

Figure 18.1 The astronomical definition of tropical and polar regions. At the winter solstice (about 21 December), the sun is directly overhead at 23½°S, marking the Tropic of Capricorn, and it does not shine at all north of 66½°N, the Arctic Circle. At the summer solstice (about 21 June), the sun is directly overhead at 23½°N, marking the Tropic of Cancer, and it does not shine at all south of 66½°S, the Antarctic Circle. (The solstice is the extreme of northward or southward travel of the subsolar point—the point on the Earth directly beneath the sun—during a year.)

DEFINING THE POLAR AND TROPICAL REGIONS

The extraordinary variety of light, ocean surface, and cloud cover exposed by the flight just described is more meaningful if we consider the geography involved. As you may recall from Chapter 8, the Earth is tilted on its axis about 23½° from a line perpendicular to the plane of its orbit around the sun (see **Figure 18.1**). Because of this tilt, the sun shines most directly on the Northern Hemisphere for one-half of the year and on the Southern Hemisphere for the other half. Because the tilt is 23½°, the southernmost point at which the sun appears directly overhead is 23½°S (south latitude). The sun reaches this latitude, called the **Tropic of Capricorn,** on about 21 December. Six months later—on about 21 June—movement of the Earth in its orbit has brought the overhead position of the sun to 23½°N. This latitude is called the **Tropic of Cancer.** The **tropics** are astronomically defined as the area between the Tropic of Cancer and the Tropic of Capricorn, the area in which the sun shines directly overhead at some time during the year. We refer to the tropics as *low latitudes* because this area includes latitudes from 0° (the equator) to 23½°N and 23½°S. The tropics of Cancer and Capricorn lie 2,570 kilometers (1,597 miles) north and south of the equator, respectively. The warm area of the planet between them is 91% ocean.

The polar regions may also be defined by the position of the sun. Figure 18.1 shows sunlight grazing the Earth's polar regions on about 21 December, when the sun is over the Tropic of Capricorn. Notice that the zone between the northern tangent point (the point at which the sun's rays just graze the Earth's surface) and the North Pole is in continuous darkness, while the zone between the southern tangent point and the South Pole is in continuous sunlight. The circles these tangent points inscribe on the Earth as the Earth rotates on about 21 December (and about 21 June) are the **Arctic Circle** and the **Antarctic Circle.** Knowing the tilt of the Earth relative to the plane of its orbit allows us to calculate the latitude of the Arctic and Antarctic circles: $90° - 23\frac{1}{2}° = 66\frac{1}{2}°$ north and south latitudes. The **polar regions** are astronomically defined as those areas lying between the Arctic and Antarctic circles and the respective pole. We refer to the polar regions as *high latitudes* because they include the latitudes from $66\frac{1}{2}°$ (north or south) to $90°$ (the poles). The Earth's surface within the Arctic Circle is mostly ocean; within the Antarctic Circle it is mostly land. The cold areas within the polar circles are usually covered with ice. The polar circles are 2,623 kilometers (1,630 miles) from the poles.

The polar regions are cold because they radiate more heat into space than they receive from the sun. The tropics receive more heat from the sun than they radiate into space, making their land and water surfaces warm. Differences in light and heat are the main reason why the polar and tropical ocean areas—and the organisms and communities within them—are so dissimilar. Temperature differences also drive the large-scale atmospheric and oceanic circulation described in Chapters 8 and 9.

The polar and tropical regions stand in dramatic contrast to the temperate zones, where most of the world's human population lives. The **temperate zones** are the mid-latitude areas between the polar and tropical regions; that is, between $23\frac{1}{2}°$N and $66\frac{1}{2}°$N and between $23\frac{1}{2}°$S and $66\frac{1}{2}°$S. The area of the temperate zones is greater than the area of the tropical and polar regions combined. As the name implies, temperatures tend to be temperate—that is, not extreme—at these in-between latitudes.

Of course, the turbulent ocean and atmosphere are not bounded by imaginary lines drawn on the Earth's surface. Because of the many variables induced by the circulation of water and air, the oceanographic and biological boundaries between the regions cannot be drawn with geometric precision. Temperature rather than sun position is the most useful way to delineate the zones. Tropical oceans nearly always have surface temperatures above 20°C (69°F), and the warmest open surface water is found in the nearly current-free calm north of Australia, where surface temperatures can exceed 30°C (86°F). Polar oceans rarely get warmer than 10°C (50°F), and the coldest water is found where sea ice is forming at −2°C (28°F). Water at the surface in the temperate zone averages about 10°C (50°F).

HOW POLAR AND TROPICAL OCEAN AREAS DIFFER

A ship moving south from the equator encounters progressively colder water, but surface water does not become steadily colder simply as a function of latitude. An observer aboard might notice very little change in temperature for days and then suddenly record an abrupt 10°C (16°F) drop in temperature within a few miles. Salinity might also fall abruptly, and the resident organisms would probably change. The ship has crossed a **convergence zone,** an area where waters from different regions meet. **Figure 18.2** shows the positions of the major surface water convergence zones and surface water masses of the world ocean.

Convergence Zones: The Boundaries

The most pronounced convergence zone is the **Antarctic Convergence,** which encircles the continent of Antarctica between about 50°S and 60°S. South of the Antarctic Convergence is cold Antarctic Circumpolar Water, in which flows the West Wind Drift, greatest of all geostrophic currents (see Chapter 9). To the north is Subantarctic Surface Water, warmer and more saline than the water south of the Antarctic Convergence. The **Subtropical Convergence,** between about 40°S and 50°S, marks the northern boundary of Subantarctic Surface Water. The position of the Subtropical Convergence varies with the time of year, the position of currents, and other factors. The surface water equatorward of the Subtropical Convergence—appropriately known as Central Water—is the most abundant, warmest, and saltiest of the three surface water masses.

The organization of surface water masses in the Northern Hemisphere is less orderly because of the different configuration of continents there. As in the south, Central Water is most abundant, warmest, and saltiest. The transition from Central Water to Subarctic Surface Water occurs at about 45°N in the Pacific and at about 60°N in the Atlantic, at the Northern Hemisphere's Subtropical Convergence. The transition to Arctic Water is a subtle one, and the **Arctic Convergence** is often poorly defined. A circumpolar current like that of the southern polar ocean does not form in the north because continents interrupt its flow.

At last we can venture an oceanographic definition of the polar, tropical, and temperate oceanic regions. The water masses of each region have distinctive temperature, density, salinity, and organisms to differentiate them. Oceanographers define the **polar ocean areas** as the zones to the south of the Antarctic Convergence and to the north of the Arctic Convergence. The **tropical ocean area** is the warm central zone equatorward of the

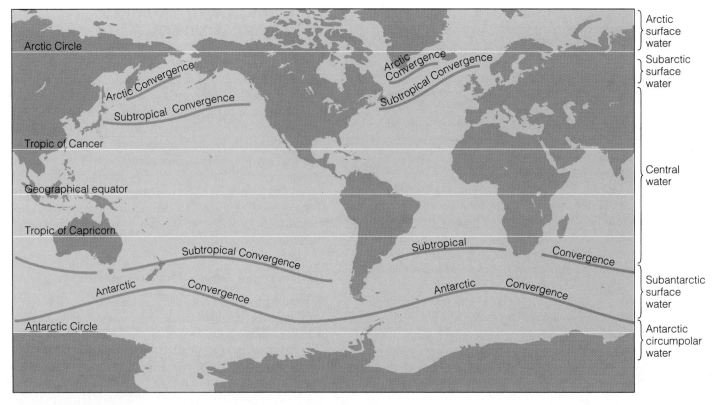

Figure 18.2 Major convergence zones and surface water masses, marking the oceanographic boundaries of the polar and tropical zones, compared to the astronomical boundaries marked by circles of latitude.

Subtropical convergences, where surface temperatures are nearly always above 20°C (69°F). **Figure 18.3** shows global climatic divisions, including subpolar and subtropical areas. Note that these climatic divisions correspond more closely with the boundaries of convergence zones than with astronomical latitude circles.

Temperature, Salinity, and Density

Surface temperature is proportional to available sunlight. More solar energy is available in the tropics than in the polar regions; so the water there is warmer. The ocean's sunlit upper layer is also thicker in the tropics, both because the solar angle there is more nearly vertical and also because water in the open tropical ocean contains fewer suspended particles (and is therefore clearer than water in open temperate or polar regions). Because the ocean is heated to a greater depth, the tropical thermocline is deeper than thermoclines at higher latitudes. It is also much more pronounced—the transition to the colder, denser water below is more abrupt in the tropics than at high latitudes. Usually there is no distinct thermocline near the poles because surface water in the polar regions is nearly as cold as water at great depths. (Sea-

sonal changes in surface temperature are greatest in the temperate zones.) You may wish to review the thermal cross section of the ocean shown in Figures 7.11 and 7.13.

Changes in the salinity of the open ocean are caused mainly by changes in the balance of precipitation and evaporation. As you may recall from Chapter 7, salinity tends to be high in areas where evaporation exceeds precipitation. The areas of excess evaporation, and consequently of highest surface salinity, lie near the edges of the tropics in zones of high atmospheric pressure (see Figure 7.15). Where precipitation exceeds evaporation, salinity will be low. Rainfall is highest near the meteorological equator (about 5° north of the geographic equator), and the ocean surface there tends to be relatively low in salinity. Salinity in the polar regions is influenced by water runoff from land and by processes associated with the freezing or melting of ocean water.

Temperature and salinity influence water density. **Figure 18.4** shows the average variation of salinity, temperature, and density with latitude. Note the approximate symmetry around the meteorological equator at about 5°N.

Variations in temperature, salinity, and density between high and low latitudes are usually confined to the

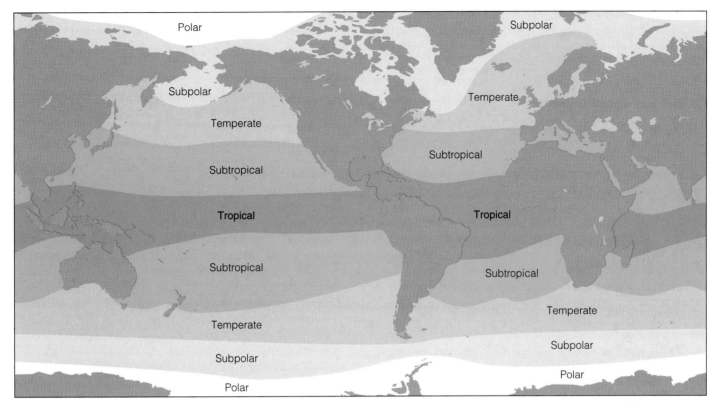

Figure 18.3 The climate zones of the open ocean. Note that the tropical and subtropical zones are the largest.

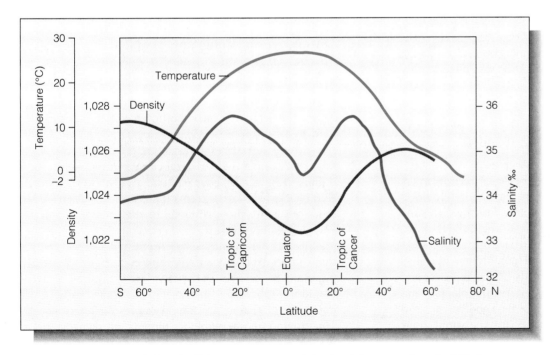

Figure 18.4 Average variation of salinity, temperature, and density at the sea surface, by latitude.

uppermost 2,000 meters (6,500 feet) of water. Below this depth conditions are similar at all latitudes. The deep ocean is nearly as cold, salty, and dense in the equatorial Pacific as it is off the Siberian coast.

Vertical Water Movement

The vertical movement of large volumes of water from the surface to great depths (and vice versa) is possible only where surface water density is similar to deep water density. The great difference in temperature, and therefore density, between surface water and deep water in the tropics makes the water column very stable and prevents an exchange of surface and deep water. This stability is maintained despite the fact that the surface of the tropical ocean is in constant horizontal motion, churned by tropical cyclones and stirred by currents.

Vertical movement of water in the northern polar ocean is also limited. There, however, the stratification is caused largely by a salinity difference between surface water and water at great depths. The surface of the Arctic Ocean receives a large volume of freshwater runoff from Siberian and Canadian rivers. The proximity of continental masses blocks the formation of large currents, and the landlocked northern ocean communicates sluggishly with other ocean areas; so the surface water tends not to mix with deeper water or to flow to lower latitudes.

In contrast, the southern polar ocean is only weakly stratified. The cold temperature of southern ocean surface water closely matches that of deep water; so no thermocline divides surface water from deep water. The absence of confining continental margins and mixing at the boundaries of the West Wind Drift minimize salinity differences. Turbulence and weak stratification encourage a huge volume of deep-water upwelling and therefore high surface nutrient levels.

Productivity

Primary productivity (see Chapter 13) tells us how much life a marine area can support by providing us with a measure of food production by photosynthetic plants. As you may remember, photosynthesis requires sunlight, carbon dioxide, and inorganic nutrients such as nitrates, phosphates, and trace elements. The inorganic nutrients are recycled from plants to animals to decomposers and back to plants again. Unfortunately for the plants, valuable nutrients incorporated into the bodies of dead plants and animals tend to sink below the sunlit zone. The resulting low nutrient concentrations are the most important factors limiting the growth of marine plants. Plant productivity cannot continue unless the upwelling of deep water returns these nutrients to the surface.

As we have just seen, the opportunity for exchange of nutrient-depleted surface water and nutrient-rich deep water is relatively high in the Antarctic. The return of sunlight in the southern summer triggers tremendous phytoplankton blooms; Antarctic water sometimes becomes a rich planktonic soup as millions of diatoms compete for nutrients and sunlight. Indeed, because of the high nutrient levels, the summer waters between the Antarctic Convergence and the mainland of Antarctica are among the most productive on Earth. Phytoplankton blooms in the northern polar ocean are not usually as exuberant because the volume of upwelling water is much less there.

Because of the stability of the horizontal layers, upwelling is uncommon in the tropical ocean, except in the equatorial Pacific or in areas where currents impinge on interrupting islands or continents. The clear blue of the tropical ocean is the sure signature of an oceanic desert in which deep upwelling rarely occurs. When nutrients are available or are tightly recycled, as in shallow coral reefs, the tropical ocean explodes into vigorous productivity.

Where is the *best* combination of factors to promote productivity? In the higher latitudes of the temperate zones. Here sunlight is available year-round, and nutrients are present from upwelling or current transport. Over a year's time the average total productivity of the temperate zones is greater than the productivity of the polar and tropical zones combined.

Differences in Polar and Tropical Marine Organisms

Relatively few species have been able to adapt to the stern polar conditions. The harshness of the polar environment has favored the evolution of marine species that tend to move more slowly, develop and grow at slower rates, live longer lives, have fewer offspring, and attain larger sizes than their tropical counterparts. Small polar organisms generally have less elaborate and ornamented shapes, produce larval forms that more closely resemble the adult, store proportionally greater reserves of food, and evolve more slowly.

When growing conditions are optimal, the few species that can withstand polar environmental rigor frequently populate an area in extremely large numbers. Phytoplankton, the small drifting plants that begin most marine food chains, are fine examples of this principle. The brief, intense plankton blooms of polar summer usually consist of only 1 or 2 species of primary producers, but a similar bloom in temperate waters may comprise 10 or 20 species. Penguins are another example: Breeding grounds are often jammed with thousands of individuals of one species nesting as closely together as possible.

The situation is reversed in the tropics. Environmental conditions at low latitudes are not as intimidating, and a great many kinds of animals and plants have adapted to life in the warm zone. There are thousands of species of tropical fish, tropical birds, tropical plants, tropical insects, and tropical plankton, but you usually won't see many examples of any one species in a small area. Try remembering the idea this way: In the tropics, if many organisms are present there will be *few* individuals of *many* species. In the polar regions, if many organisms are present there will be *many* individuals of *few* species.

The overall appearance of the resident species often differs with latitude. Whereas polar species are typically dull in color or pattern, tropical organisms are frequently conspicuous in coloration and behavior. Their bright colors and patterns are probably related to the need to dif-ferentiate among many different species for feeding and mating competition.

THE POLAR OCEANS

The two polar ocean areas are nearly complete opposites; they are "poles apart" in their characteristics as well as their locations. The Earth's North Pole lies in the middle of an ocean—the **Arctic Ocean**—an area covered by a layer of ice and bounded by the continents of Eurasia and North America. The South Pole is located near the center of a continent—Antarctica—a large land mass at the center of the Antarctic Ocean. The **Antarctic Ocean** surrounds the continent and is bounded not by land but by the Antarctic Convergence. The Arctic Ocean is about one-fourth the size of the Antarctic Ocean. The polar ocean areas are shown in **Figure 18.5.**

Figure 18.5 (a) The far north: an ocean ringed by land. (b) The far south: land surrounded by ocean.

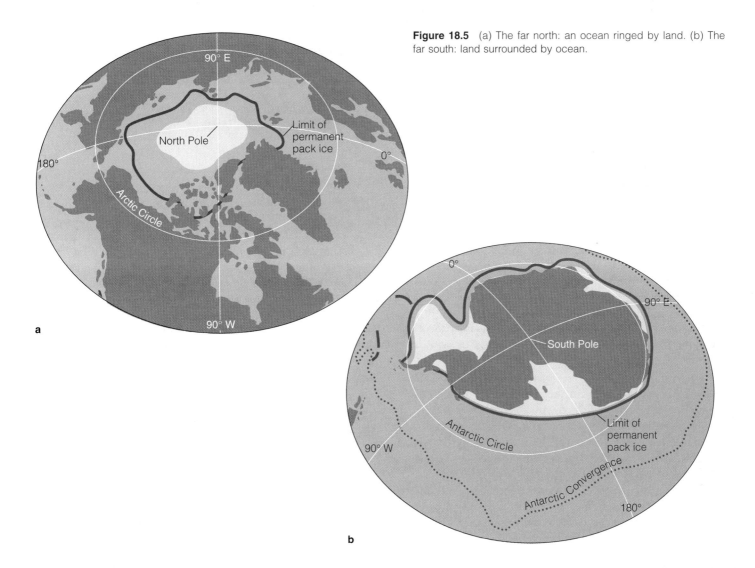

Figure 18.6 A mosaic of 40 images of Antarctica obtained by the *Galileo* spacecraft in December 1990. Three oceans may be seen surrounding the icy continent: the Pacific to the lower right, the Indian to the upper right, and a small section of the Atlantic at the upper left. Nearly the entire continent is illuminated, just two weeks before Antarctic midsummer. The faint blue line along the curved limb of the Earth, at the bottom, marks the atmosphere.

The Southern Polar Ocean

Although it encompasses nearly one quarter of the world ocean, the southern polar ocean contains only one-tenth of the ocean's heat. With an average surface temperature of about 2°C (36°F), its cold waters act as a vast heat sink for energy transported from low latitudes by atmospheric circulation and ocean currents. During the southern winter, as much as 20 million square kilometers (7.7 million square miles) of southern ocean—an area about twice the size of the continental United States—is covered by ice.

The Antarctic continent is an extraordinary and extreme place. You can sense its isolation and desolation in **Figure 18.6.** The world's fifth largest continent, Antarctica is 14 million square kilometers (5.4 million square miles), almost twice the size of Australia. About 98% of the continent is covered by ice, which averages about 2,000 meters (6,500 feet) in depth but is over 4,000 meters (13,000 feet) deep in places. It is also the highest of all continents. Though 80% of Earth's fresh water is frozen in its ice cap, the continent is paradoxically the largest of all deserts—because the average precipitation is equivalent

to only about 5 centimeters (2 inches) of water per year. The world's lowest temperature—a bone-chilling −89.2°C (−128.6°F)—was recorded in Antarctica by Soviet scientists on 21 July 1983! The great weight of the continent's 31.2-million-cubic-kilometer (7.5-million-cubic-mile) shroud of ice has forced Antarctica down into the mantle to maintain isostatic equilibrium. This explains the great depths of the seaward boundary of continental shelves found there, about 600 meters (2,000 feet) rather than the 150 meters (500 feet) observed around most other continents.

The southern polar ocean is bounded on the south by these deep continental shelves, and its floor stretches north to blend into the ridges and abyssal plains of the Indian, Atlantic, and Pacific oceans. The densest water of any in the open ocean, Antarctic Bottom Water is formed as seawater freezes near Antarctica. Salts are not incorporated into ice crystals; so sea ice forming over the continental shelf sheds a frigid brine, which sinks and spreads across the seabed all the way into the Northern Hemisphere. Between 20 and 50 million cubic meters (26 to 65 million cubic yards) of this water is formed every second!

Figure 18.7 Emperor penguins in the colony at Cape Crozier, in a watercolor by Edward A. Wilson. Emperor chicks are born toward the end of the Antarctic winter. The parents take turns guarding the chicks, warming them between their legs until the chicks are large enough to fend for themselves.

Antarctica is an exceptionally inhospitable environment for life. On the continent itself, no food is available inward from the shore; so the several species of birds and mammals that venture onto land depend entirely on marine food for survival. Moreover, almost all of the Antarctic bird species spend the winter in warmer climates. Only one species, the emperor penguin (**Figure 18.7**), can withstand the dark and bitterly cold Antarctic winters. Emperors somehow manage to mate, lay eggs, and begin rearing their young during this stressful time. Emperor penguins go to sea for a season of feeding when spring comes, leaving only a few species of flightless insects as permanent residents.

Thanks to the moderating effects of seawater, marine organisms are much more numerous. The keystone of the Antarctic ecosystem is *Euphausia superba*, a thumb-sized, shrimplike crustacean commonly called **krill** (**Figure 18.8**). Krill, primary consumer extraordinaire, grazes on the abundant diatoms that prosper in the nutrient-rich open water of the southern polar ocean. In turn, krill are eaten in tremendous numbers by seabirds, squid, fishes, and whales. The important position of krill in Antarctic food chains is shown in **Figure 18.9**.

Some 500 to 750 million metric tons (450 to 680 million tons) of krill inhabit the Antarctic Ocean, with the greatest concentrations in the productive upwelling cur-

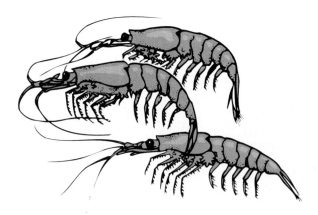

Figure 18.8 Krill (*Euphausia superba*). These shrimplike crustaceans, shown here approximately life-size, occur throughout the world ocean; they are particularly numerous in Antarctic seas.

rents of the Weddell Sea. Krill travel in great schools that can extend over several square kilometers. They behave more like schooling fish than pelagic crustaceans; their primary swimming mode is horizontal, not vertical. Japanese researchers, using two ships equipped with side-scan sonar, tracked a large school of swimming krill for 14 days across 278 kilometers (172 miles). This new finding threatens krill's usual classification as zooplankton (animals unable to move consistently in one direction against waves or current flow).

The Northern Polar Ocean

About 80% of the ocean area north of the continental masses of Eurasia and North America, some 12.2 million square kilometers (4.7 million square miles), is mostly covered by a jumbled layer of permanent ice called the **ice cap.** In winter, the ice cap is typically about 3 meters (10 feet) thick, but wind and ice expansion can shove and fold the ice into long ridges that may push 10 meters (33 feet) above sea level and 40 meters (130 feet) below. Beneath the Arctic Ocean's frozen cover lie the Earth's widest continental shelves and an ocean basin with an average depth of about 3,700 meters (12,000 feet). Water at the bottom of the Arctic Ocean is the coldest of all seawater. (Antarctic Bottom Water is slightly warmer; but being saltier, it is also denser.) The Arctic Ocean has lower average salinity because fresh water from Siberian and Canadian rivers flows into the area during the summer. This lower surface salinity means that Arctic Ocean water freezes at a slightly higher temperature than Antarctic Ocean water.

Weak currents move this water and its floating ice pack, a large floating expanse of broken ice masses pressed and frozen together. Circulation within and around the isolated Arctic Ocean basin is not as thorough as in the southern polar ocean; there is no north-

ern equivalent of the West Wind Drift. Indeed, some oceanographers consider the Arctic Ocean to be an estuary of the Atlantic; the major circulation into and out of the Arctic basin is through a single deep channel east and north of Greenland. Some seawater enters the circulating system at the Bering Strait between Alaska and Siberia, crosses the top of the world, and exits along the west coast of Greenland.

Arctic pack ice supports many more species of large animals than are found on the pack ice or mainland shores of Antarctica. Because of the proximity of liquid water, temperatures in the far north are not as extreme as those of the far south. Some species of seals and an occasional polar bear have been seen as far north as the pole. The seals rise for air at **polynyas,** holes in the ice opened by wind and currents. Killer whales and narwahls also use these holes to breathe. The polar bears and killer whales prey on the seals. These animals share the ice pack with a few kinds of birds, one species of hare, and a transient human population.

ICE ON THE OCEAN

Ice, which covers about 6% of the world ocean, influences the weather and climate of the whole Earth. The presence of an ice covering greatly increases the ocean's reflectivity and insulates the ocean from the atmosphere. The seasonal freezing and thawing of ice is also important; water's latent heat of fusion moderates polar temperatures.

Insulation

The insulating property of ice increases worldwide atmospheric circulation. A 1-meter (3.3-foot) layer of ice reduces the exchange of heat energy between ocean and atmosphere by about 100 times. Deprived of heat from the ocean, air over the polar ice grows colder, increasing the temperature difference between the polar and tropical atmospheres. Global circulation becomes more vigorous as the atmosphere carries more heat toward the poles to redress the imbalance. Southern Hemisphere winds are stronger because the south polar region is colder than the north polar region.

On the other hand, the insulating layer of sea ice *prevents* wide swings in the temperature of the underlying liquid water. The relatively warm days of summer or cold days of winter have no significant influence on temperatures beneath the layer of ice.

Latent Heat of Fusion

As was discussed in Chapter 7, removing a calorie of heat from freezing pure water at 0°C (32°F) won't change

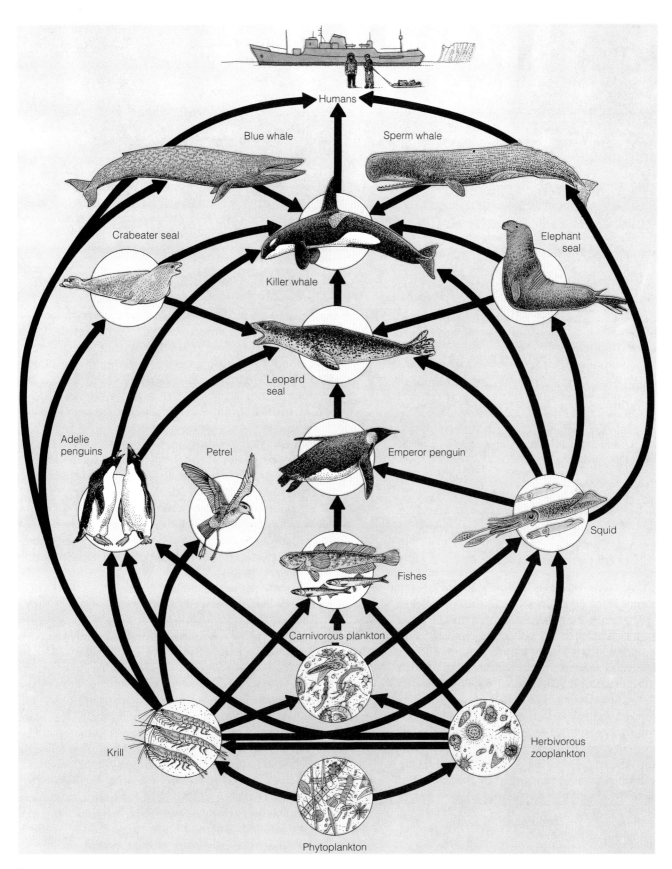

Figure 18.9 Greatly simplified food web in the Antarctic. There are many more participants, including an array of decomposer organisms.

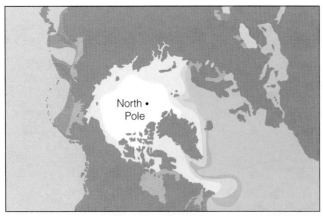

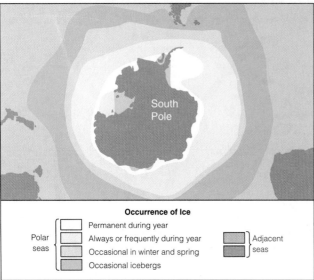

Occurrence of Ice

Polar seas
- Permanent during year
- Always or frequently during year
- Occasional in winter and spring
- Occasional icebergs

Adjacent seas

Figure 18.10 Seasonal ice limits in the Northern Hemisphere and the Southern Hemisphere.

Figure 18.11 Pancake ice near Cape Evans, Antarctica.

its temperature at all; 80 calories of heat energy must be removed per gram to form ice. More than 18,000 cubic kilometers (4,300 cubic miles) of polar ice, covering some 16 million square kilometers (6 million square miles) of surface, thaws and refreezes in the Southern Hemisphere each year (**Figure 18.10**). The annual change in sea ice cover is less in the Arctic, averaging about 5 million square kilometers (2 million square miles). The autumn cooling of the atmosphere in each hemisphere is delayed because heat energy is released as masses of water turn to ice. Heat is absorbed during ice melt in the spring. Seasonal extremes are thus moderated by the absorption and release of heat energy as ice thaws and refreezes.

The Formation of Sea Ice

Sea ice is a collective term for ice formed by the freezing of seawater. Seawater of normal salinity begins to freeze at about −1.8°C (28.7°F). When enough heat has been

removed from the surface water, clouds of needlelike ice crystals start to form. The ocean surface becomes calm and cloudy, and it no longer reflects objects in the sky. The crystals exclude most of seawater's salt and expand slightly as they grow. They float to the surface to form slushy mats. If the air is cold enough, the mats jostle into flat rounded shapes with raised edges called *pancake ice* (**Figure 18.11**). If the air temperature continues to fall, the pancakes coalesce to form sea ice, which in the absence of strong wave action gradually hardens to a firm opaque surface called an **ice floe** (*flo* = layer, expanse). Most ice floes range from about 10 meters to 10 kilometers (33 feet to 6 miles) in diameter. Seawater will freeze to the bottom of the floe; but because ice is a good insulator, its growing thickness makes the movement of heat from ocean through ice to atmosphere more difficult. The ice mass will not become thicker than 3 or 4 meters (10 or 13 feet) unless other factors intervene.

One of those factors is precipitation. Though annual snowfall is sparse within the polar circles, most of what does fall never thaws. Thus, the accumulation of snow on polar ice slowly increases its thickness. Other factors are currents and winds, which can break ice floes or jam them together into an ice pack. Currents and winds can cause sections of the ice pack to pull away from each other to form a polynya, through which a mammal or a submarine might surface; at other times they raft onto one another to form pressure ridges—lines of jagged, house-sized chunks of ice. In the Arctic, pressure ridges formed by converging floes may extend for many kilometers. Arctic pack ice is typically between three and seven years old; Antarctic pack ice is usually younger because a greater proportion of the southern pack thaws and re-forms each year.

Away from the main pack, isolated bits of ice (called *growlers* because of the sound they make when they scrape against a hull) may be encountered. The extent of pack ice in each hemisphere varies considerably from year to year depending on currents, insolation, winds and weather, and other factors.

Icebergs

Only a minute fraction of ocean ice is contained in the often fantastic structures of **icebergs.** Unlike sea ice, icebergs always originate from snow falling on or adjacent to land. Two types of icebergs are recognized: *pinnacled* (or *castellated*, after their castle-turret shapes), and *tabular.* The classic iceberg, with its craggy profile and towering projections, is a pinnacled iceberg.

Pinnacled icebergs, more characteristic of the northern polar ocean, usually begin life as glaciers in western Greenland. These slowly flowing ice rivers form at the edge of the great Greenland ice sheet, squeeze between the sharp-crested ridges of the mountains, and eventually reach the sea. Changing tides at the end of the deep, glacially cut fjords break off huge tongues of ice, which drift from shore as icebergs. The great weight of ice in a glacier compacts these glistening masses and makes them very dense; only about one-seventh of their bulk floats above sea level (see **Figure 18.12**). An average Greenland-born iceberg may weigh about 1 million metric tons (1.1 million tons) and last for two years. Wave action often forms submerged shoulders along icebergs' waterlines as they age.

The jagged underwater edge of a pinnacled iceberg broke the hull of the "unsinkable" White Star liner RMS *Titanic* on the night of 14 April 1912; 1,517 people lost their lives. The iceberg had reached the Atlantic shipping lanes after separating from its glacier, being swept down the west coast of Greenland, and then moving south toward the Gulf Stream and North Atlantic Current. A typical path is shown in **Figure 18.13.** Icebergs sometimes go even farther south—a small pinnacled iceberg was sighted near Bermuda in June 1926. Such icebergs are rarely a threat in the Pacific. Alaska has only a few fjord glaciers, and the icebergs they calve are small and generally drift slowly northward into the permanent pack.

Antarctic icebergs have a different origin and a different shape. The great weight of ice atop the Antarctic continent has depressed the center of the land mass; so most of the ice in Antarctica tends to remain inland rather than flow toward the coast. Some glacial and land ice accumulates into sheets and shelves along the sloping coastal edges of the continent, however, and it creeps seaward at a rate of perhaps 50 to 100 meters (up to 330 feet) each year. Parts of these huge ice sheets occasionally break off, creating the flat, tabular (table-shaped) icebergs most characteristic of the southern ocean (**Figure 18.14**).

Antarctic icebergs are notable for their tremendous

a

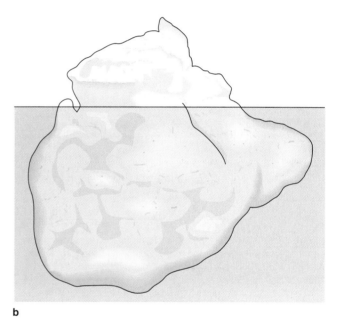

b

Figure 18.12 (a) A pinnacled iceberg. Most pinnacled icebergs are formed in the Arctic. Because they are calved from glaciers of compressed ice, only about one-seventh of an iceberg's mass is above the surface, as the sketch (b) shows.

size. Lengths of up to 8 kilometers (5 miles) are not unusual, with ice 45 meters (150 feet) above sea level. In 1927 a frozen section of 26,000 square kilometers (10,000 square miles)—about eight times the size of Rhode Island—broke from the Antarctic shore and floated north along the coast of Argentina! The giant iceberg carried an overheated load of Adélie penguins. Favorable currents on the west coast of South America have driven tabular icebergs (and itinerant penguins) as far north as the tropics. The ancestors of today's Galápagos penguins were almost certainly transported to their equatorial home in this way.

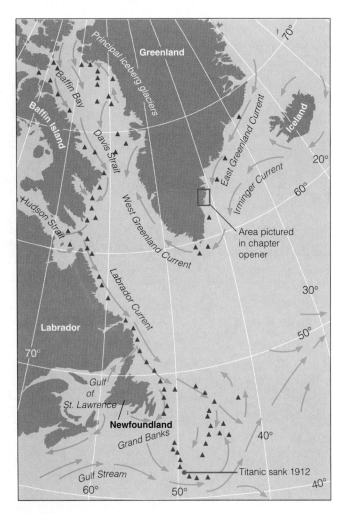

Figure 18.13 The drift path of icebergs from western Greenland into the North Atlantic. Of the many thousands of icebergs (▲) that break off the glaciers each year, about 400 ultimately drift south of Newfoundland.

Figure 18.14 High wind blows snow from the top of a tabular iceberg grounded at the edge of the southern polar ocean.

THE TROPICAL OCEAN

The tropical ocean occupies about 50% of the total area of open water and 30% of the area above Earth's continental shelves. The surface of the tropical ocean brims with light and heat energy. It is blue, brilliantly transparent, relatively high in salinity, and notable for a deep and abrupt thermocline, strong geostrophic currents, and relatively low concentrations of dissolved nutrients and gases. Storms and currents sweep across its surface, but except in some places near the equator, vertical circulation is rare. It supports surprisingly little life.

But what of the travel-poster view of the tropical ocean? What about the thousands of brightly colored fish, strange invertebrates, and breathtaking scenes of divers swimming through living reef formations—a vision of great productivity and activity? Such scenes,

photographed on reefs near islands and continents, are actually exceptional in their *rarity*; less than 2% of the tropical ocean could be included in such a description. The key to the difference between the open tropical ocean and the tropical reefs lies in the productivity of the reefs. As we will see, the source of that productivity lives within the reef itself.

Reefs

The word **reef** has many meanings. Sailors consider any shallow hazard to navigation a reef, and fishermen use the term to indicate a mass of fish or an object on which their nets may snag. Marine biologists often use the term to describe a wave-resistant structure dominated by a strong and rigid mass of living (or once-living) organisms. Not all reefs are built of coral; other reef builders include red and green algae, cyanobacteria, worms, and even oysters. But we think first of coral reefs when the words *reef* and *tropics* are mentioned together.

Coral

Although they look like flowers, **corals** are classified as cnidarians, along with sea anemones and jellyfish. Some corals are solitary animals with bodies up to 30 centimeters (12 inches) in diameter, but most of the more than 500 species are ant-sized organisms crowded into colonies called **coral reefs** (**Figure 18.15**). The coral animals themselves construct the reefs by secreting hard skeletons of aragonite (a fibrous, crystalline form of calcium carbonate). The matrix of cup-shaped individual skeletons

Figure 18.15 Close-up of hermatypic coral, showing expanded polyps.

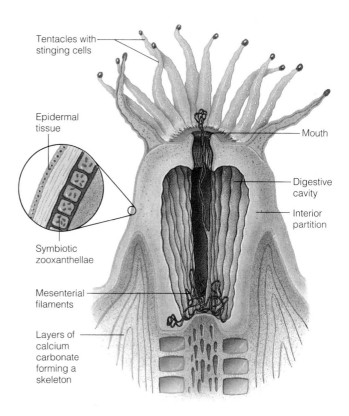

Tentacles with stinging cells

Epidermal tissue

Mouth

Digestive cavity

Interior partition

Symbiotic zooxanthellae

Mesenterial filaments

Layers of calcium carbonate forming a skeleton

Figure 18.16 Anatomy of a reef coral polyp, with a blowup detail showing a cross section of the outer covering and tissue.

secreted by coral animals gives the colony its characteristic shape.

An individual coral animal, or **polyp,** feeds by capturing and eating plankton that drift within reach of its rosette of tentacles. Victims are entrapped by stinging cells on each tentacle, transported to a central gastric cavity, and rapidly digested. Tropical corals feed at night; at dawn the polyps retract into their skeletal cups to escape strong sunlight and predators. Coral polyps can also feed by direct absorption, a process in which they simply transport dissolved food molecules through their body walls. The anatomy of a coral polyp is shown in **Figure 18.16.**

Tropical, reef-building corals are **hermatypic,** a term derived from the Greek word *hermatos*, meaning "mound-builder." Their bodies contain masses of tiny, symbiotic dinoflagellates called **zooxanthellae** (*xanthos* = yellow). These single-celled, plantlike organisms facilitate the rapid biochemical deposition of calcium carbonate into the coral skeleton. Coral's success in the nutrient-poor water of the tropics depends upon its intimate biological partnership with zooxanthellae. The microscopic zooxanthellae carry on photosynthesis, absorb waste products, grow, and divide within their

coral host. The coral animals provide a safe and stable environment and a source of carbon dioxide and nutrients; the zooxanthellae reciprocate by providing oxygen, carbohydrates, and the alkaline pH necessary to enhance the rate of calcium carbonate deposition. The coral occasionally absorbs one of the resident dinoflagellates, "harvesting" the organic compounds for its own use. The zooxanthellae are captive within the coral; so none of their nutrients are lost as they would be if the zooxanthellae were planktonic plants that could drift away from the reef. Instead, nutrients are used directly by the coral for its own needs. The cycling of materials is short, direct, quick, and very efficient. Even the shallow depth of the reef aids in nutrient use: Nutrients cannot sink out of the productive sunlit zone.

Because of the needs of its zooxanthellae, hermatypic corals depend on light and warmth. Reef corals grow best in brightly lighted water about 5 to 10 meters (16 to 33 feet) deep. Coral reefs can form to depths of 90 meters (300 feet), but growth rates decline rapidly past the optimum 5–10-meter depth. In ideal conditions coral animals grow at a rate of about 1 centimeter (½ inch) per year. They prefer clear water because turbidity prevents light penetration and because suspended inorganic particles interfere with feeding. The animals are protected from the harmful effects of bright sunlight by a mucous coating that contains an ultraviolet-blocking "suntan lotion."

Hermatypic corals also prefer water of normal or slightly elevated salinity. Coral animals are highly susceptible to osmotic shock, and exposure to fresh water is rapidly fatal; thus reefs growing in shallow water have a flat upper surface because rain is lethal. Fresh water and suspended sediments prevent reefs from forming near the mouths of rivers or in areas adjacent to islands or continents where rainfall is abundant. Reef corals are also susceptible to potentially devastating diseases such as bleaching, about which scientists as yet know little (see **Box 18.1**).

Corals reproduce both asexually and sexually. Colonies grow by budding: One animal splits from a parent and extends the skeleton outward to enclose itself. Colonies can become quite large in this way, sometimes reaching the size of small automobiles. Fragments of a large colony might be transported to a new location by storm waves and grow to large size by budding, but new colonies are usually formed by sexual reproduction. In the tropical Pacific, mature coral animals spawn in late October or November after a full moon. The spawning season is longer in the Caribbean, lasting from June through early October. Clouds of eggs and sperm litter the ocean surface during a spawn. Fertilized eggs form larvae, which drift as plankton for a few days and then settle to the bottom. If conditions are suitable, they begin to grow and bud into a new colony.

Nearly all of the reef-building corals are found within the 21°C (70°F) isotherm indicated in **Figure 18.17**, a zone that corresponds roughly to the 25° latitude lines in both hemispheres. Poleward of this area, water is too cold for the zooxanthellae to survive; temperatures below 18°C (65°F) cause their death. But as Figure 18.17 also shows, coral organisms are found in some cold, high-latitude waters as well. These corals lack zooxanthellae; so they deposit calcium carbonate much more slowly, and the structures they build do not resemble those found in the tropics. Instead, these deep-water corals, known as **ahermatypic** corals (*a* = without, *hermatos* = mound builder), build smooth banks on the cold, dark outer edges of temperate continental shelves from Norway to the Cape Verde Islands and off New Zealand and Japan. The rarest corals of all are large solitary organisms living on the abyssal floors and outer continental shelves of the Antarctic.

Other Reef Residents

Tropical coral reefs typically form in areas of high wave energy; indeed, reef organisms preferentially grow into high-energy environments because those places tend to be richer in nutrients and suspended organic material. In most reefs there is an approximate balance between con-

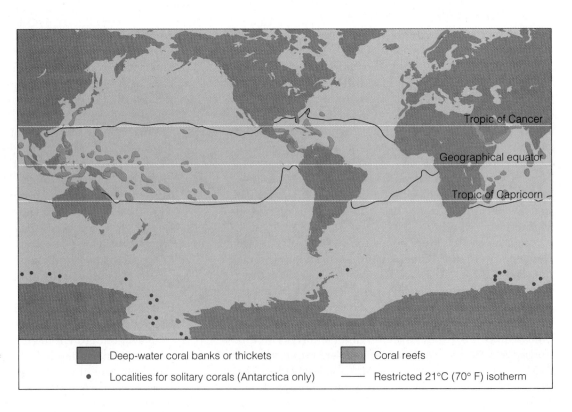

Figure 18.17 Distribution of coral reefs and their relation to sea-surface temperature.

Deep-water coral banks or thickets   Coral reefs

•   Localities for solitary corals (Antarctica only)   — Restricted 21°C (70° F) isotherm

BOX 18.1 ● *Coral Bleaching*

In 1983, marine scientists first encountered a baffling new disease of coral reefs. Because it causes coral animals to expel their brownish zooxanthellae and turn an uncharacteristic creamy color, the phenomenon is known as *coral bleaching*. Without zooxanthellae, the coral cannot secrete calcium carbonate and generate a skeleton or reef. After one bout of bleaching, corals usually recover; but while they are bleached they stop growing, leaving the reef vulnerable to erosion.

Some 60 species of coral are known to be affected. After the first outbreak in eastern Pacific reefs, the mysterious disease reached the Caribbean in 1987, attacking reefs from the Florida Keys to Jamaica and as far east as the Virgin Islands.

The causes for bleaching are unknown. Initial suspicions centered on global warming. If seawater were becoming warmer, the supersensitive corals might respond by changing in ways incompatible with needs of their zooxanthellae. The 1987 episode was widespread, occurring at distant sites at about the same time, leading scientists to believe that local conditions of pollution might not be to blame. But other researchers believe that a combination of temperature increase, sediments, and chemical pollutants may be the cause. Experiments in Australia have shown that excess phosphorus inhibits calcification of the reef and upsets the biological balance of its smallest inhabitants. Bleaching follows. Agricultural runoff and sewage outfalls are the apparent sources of the extra phosphorus.

Corals are subject to other unusual diseases as well. Elkhorn coral, the principal reef-building coral in the Caribbean, is affected by lethal *white-band disease*, and Pacific species are developing other peculiar maladies. Whether these problems are natural cyclical events or have been triggered by human activity remains to be discovered.

Most corals recover from these diseases, but by 1990 the recovery rate had slowed markedly. This trend raises fears for the future of the reefs themselves and lends urgency to the quest for a cause.

Bleached coral. The coral *Montastrea annularis* on Grecian Rocks Reef in the Key Largo National Marine Sanctuary during the "bleaching event," October 1987 (left). After the recovery of normal coloration in August of 1990 (right).

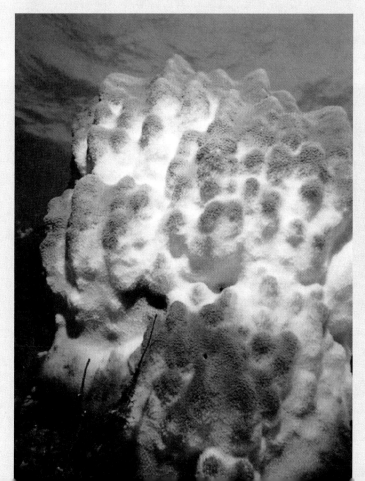

a

Figure 18.18 (a) The coral reef habitat. (b) Key. (c) A coral reef.

1 black-capped petrel
2 sea nettle
3 angelfish
4 lobed corals
5 sea whips and soft corals
6 triggerfish
7 sea fans
8 tube anemone
9 orange stone coral
10 bryozoans
11 brain coral
12 butterfly fish
13 moray eel
14 cleaner fish
15 tube corals

16 muricid snail
17 nudibranch
18 sponges
19 colonial tunicate
20 giant clam
21 purple pseudochromid fish
22 cobalt sea star
23 soft corals
24 barber pole shrimp
25 sea anemones
26 clown fish
27 worm tubes
28 cowry
29 sea fan

b

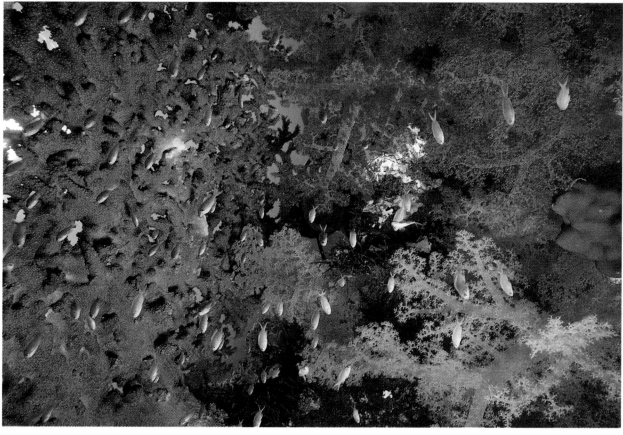

c

struction and destruction. The reef consists of actively growing coral colonies and fragments of material of different sizes, from worn coral boulders down to fine sand.

Corals are by no means the only participants in reef life, however; they may account for only about half of the biomass in these areas. Other reef residents include calcareous algae (whose secretions help "cement" the reef together), as well as a bewildering array of encrusting, burrowing, producing, and consuming creatures ranging upward in size from the microscopic. Some may tunnel into the coral or shatter it in search of food, contributing to the erosion of the reef. Fierce competition exists among reef organisms for food, living space, protection from predators, and mates. The bright colors, protective camouflage, spines, and various toxins and venoms common to tropical organisms are probably related to the intense struggle for existence that goes on in these beautiful but deceptively calm-looking places. A typical reef scene is shown in **Figure 18.18.**

Reef Types

In 1842, Charles Darwin classified tropical reef structures into three types: Fringing reefs, barrier reefs, and atolls (**Figure 18.19**). We still use his classification today.

Fringing Reefs   As their name implies, **fringing reefs** cling to the margin of land. As can be seen in **Figure 18.19a,** a fringing reef connects to shore near the water surface. Fringing reefs form in areas of low rainfall runoff primarily on the leeward (downwind side) of tropical islands. The greatest concentration of living material will be at the reef's seaward edge, where plankton and clear water of normal salinity are dependably available. Most new islands anywhere in the tropics have fringing reefs as their first reef form. Permanent fringing reefs are common in the Hawaiian Islands and in similar areas near the boundaries of the tropics.

Barrier Reefs   **Barrier reefs** are separated from land by a lagoon (**Figure 18.19b**). They tend to occur at lower latitudes than fringing reefs and can form around islands or in lines parallel to continental shores. The outer edge—the barrier—is raised because the seaward part of the reef is supplied with more food and is able to grow more rapidly than the shore side. The lagoon may be anywhere from a few meters to 60 meters (200 feet) deep, and it may separate the barrier from shore by only tens of meters or by as much as 300 kilometers (190 miles). Coral grows slowly within the lagoon because fewer nutrients are available and because sediments and fresh water run off from shore. As you would expect, conditions and species within the lagoon are much different from those of the wave-swept barrier. The calm lagoon is

often littered with eroded coral debris moved from the barrier by storms.

The greatest of all barrier reefs, the Great Barrier Reef of Australia, is the largest biological construction on the planet. It extends along the northeast coast of Queensland for 2,500 kilometers (1,560 miles) and is up to 150 kilometers (95 miles) wide. It isn't a single reef, but a conglomeration of thousands of interlinked segments. The segments present a steep outer wall to the prevailing currents and trade winds. At a growth rate of 1 centimeter (½ inch) per year, the structure is obviously of great age and astonishing volume. The huge reef is younger and thinner at its southern end; the slow northward movement of the Indian-Australian lithospheric plate in which Australia is embedded is thought responsible for this.

The variety of organisms within the Australian Barrier Reef staggers the imagination. About 500 species of hermatypic coral live there, nearly 10 times the number found in the western Atlantic. Over 1,000 species of bivalve mollusks, 3,000 species of shore fishes, and 40 species of sea snakes live in the area.

Atolls   An **atoll** (**Figure 18.19c**) is a ring-shaped island of coral reefs and coral debris enclosing, or almost enclosing, a shallow lagoon from which no land protrudes. Coral debris may be driven onto the reef by waves and wind to form an emergent arc on which coconut palms and other land plants take root. These plants stabilize the sand and lead to colonization by birds and other species. This is the tropical island of the travel posters.

Though an atoll's central lagoon connects to the deep water outside through a series of channels or grooves, coral does not usually thrive in the lagoon, both because the water may become too fresh during rains or too hot and because feeding opportunities for the coral are limited there. Despite these drawbacks, some lagoons have enough resources within to sustain small *patch reefs,* miniatures of the larger reef. A cross section of an atoll is shown in **Figure 18.20.**

Some atolls are isolated, but most occur in loose groups (**Figure 18.21**) in shallow continental shelf areas or in the deep open ocean. More than 300 atolls exist—most in the Pacific. They range in size from a few kilometers in diameter to Kwajalein in the Marshall Islands, whose slender, 280-kilometer (176-mile) ring of coral encloses a lagoon of 2,850 square kilometers (1,100 square miles).

How do atolls form? Scientists began speculating on the cause of their ring shape soon after the first scientific voyages published their reports. Charles Darwin imagined a volcanic island growing from the sea, accumulating a skirt of coral around its shore, and then slowly sub-

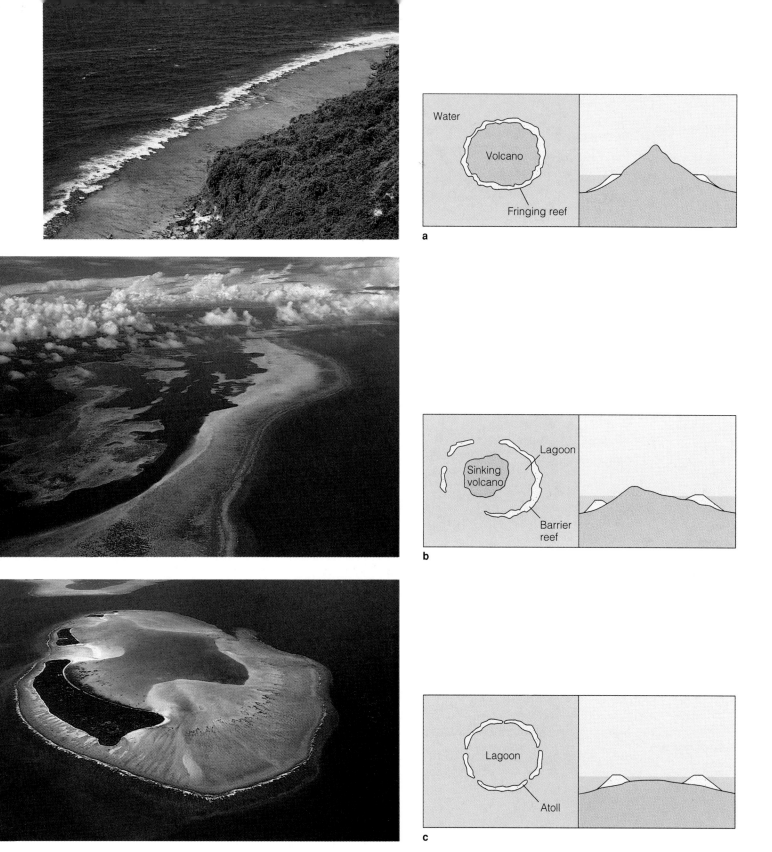

Figure 18.19 The three types of coral reefs. (a) The structure of a fringing reef like that on Moorea, in the Society Islands. (b) The structure of a barrier reef, an example of which is Bora Bora, also in the Society Islands. (c) An atoll, such as Aratika in the Tuamotu Archipelago.

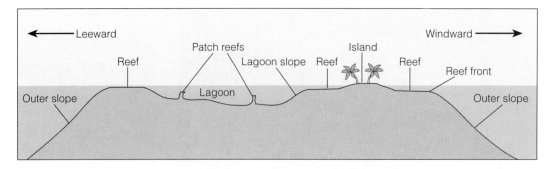

Figure 18.20 Cross section through an atoll. The windward side of an atoll is often wider and steeper, a result of more rapid coral growth from the increased availability of food.

Figure 18.21 A group of atolls in the Tuamotu Archipelago of the South Pacific, looking southeast as seen from *Apollo 7* at an altitude of 153 kilometers (96 miles).

siding at a rate equal to the growth rate of the coral. The central volcanic island would eventually sink from view, but the coral could grow continuously atop skeletons of past generations to maintain a living presence near the surface. **Figure 18.22** shows the progression. Note that the island begins with a fringing reef, passes through a barrier reef stage as it sinks, and eventually becomes an atoll as the peak disappears beneath the ocean surface. Should the island subside faster than about 1 centimeter per year (the growth rate of coral), all trace of both the island and the reefs would disappear. (The submerged island might become a guyot.)

This theory seemed reasonable, but Darwin couldn't explain what would cause volcanic islands to subside

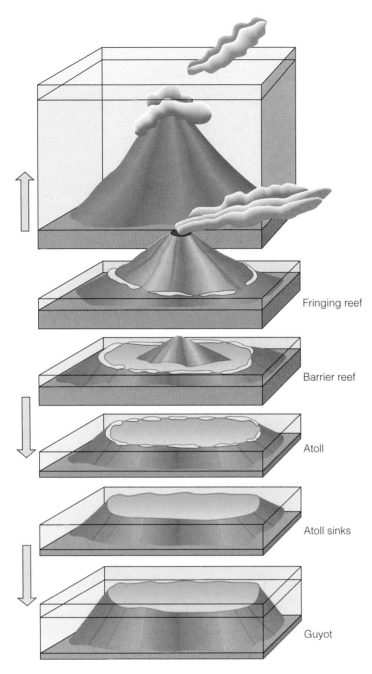

Figure 18.22 The history of an atoll. Volcanic activity at a spreading center builds the island, which acquires a fringing reef. As the island moves away from the spreading center, the volcano becomes inactive, the island slowly subsides, and the coral animals continue to build, forming first a barrier reef and then an atoll. If the subsidence rate increases above about 1 centimeter (½ inch) per year, the coral dies and the atoll becomes a guyot.

Fringing reef

Barrier reef

Atoll

Atoll sinks

Guyot

because he didn't know about plate tectonics. Now we know that volcanoes can form near spreading centers, ride outward and downward from their birthplaces, cease to be active as they leave their source of mantle heat, and sink as they are carried into deeper water—just slowly enough to permit coral growth to continue as they go.

An alternate hypothesis suggested that a slow but continuous rise in sea level could cause the growth of atolls. But this explanation could be correct only if the thickness of the accumulated coral was less than about 200 meters (660 feet), the minimum sea level of comparatively recent geologic time.

Drilling at Eniwetok atoll in 1952 in preparation for thermonuclear bomb testing settled the issue. The strength and composition of the island was probed in order to calibrate the yield of the bombs. The drills hit volcanic rock beneath the coral at depths of 1,267 meters (4,156 feet) and 1,458 meters (4,782 feet), much deeper than the lowest sea level of any known ice age. Darwin's hypothesis appears to be correct.

A BRIEF HISTORY OF EARLY TROPICAL AND POLAR EXPLORATION

Traveling in polar and tropical areas can be dangerous. Take your pick: polar storms or tropical cyclones; months of frigid darkness or day after day of blazing heat; being becalmed in the doldrums or being trapped in the ice. The ocean extremes did not give up their secrets easily, but the persistence of explorers and scientists has been—and continues to be—rewarded.

The Tropics

The tropical Atlantic was haphazardly investigated by European cultures from the dawn of voyaging, but the Pacific was intentionally opened to scientific exploration (and exploitation) by the major European powers in the mid-1700s. The earliest European explorers—from Spain, Portugal, and England—were interested in claiming territory and discovering faster, safer trade routes from the Orient to Europe. They were secretive about their discoveries, but information about prevailing winds and currents was slowly collated onto charts and eventually made its way to centers of commerce. Some of these observations were of scientific interest: For example, finding the fastest way across the Pacific led to location of the prevailing westerlies.

In the latter half of the eighteenth century, the first voyages with specific scientific goals and with scientific staff aboard were launched. The French Admiral Louis-Antoine de Bougainville was first on the scene, sent by

Figure 18.23 On his third Pacific voyage, Captain James Cook sailed north from Tahiti with orders to explore the northwestern coast of North America. On 19 January 1778 he became the first European to reach Hawaii. He wrote in his journal, "How do we account for this Nation spreading itself so far over this Vast ocean?"

his government in 1766 to circle the Earth on a voyage of exploration. Indeed, Bougainville's expedition was the first genuinely scientific voyage of circumnavigation. His primary goals were to make reliable charts of the parts of the central Pacific through which he passed and to collect and categorize plants that might prove commercially valuable. Part of his mission was to befriend the native peoples and to scout reprovisioning and repair locations for the French navy; Tahiti—with its abundant food, water, and lumber—was most important for this. Bougainville sighted the Great Barrier Reef (but did not appreciate its true extent) and turned north without discovering the Australian mainland. Unfortunately, without a chronometer Bougainville had no way to estimate longitude accurately, making his Pacific charts too inaccurate to use in navigating the maze of Pacific islands.

Captain James Cook's more successful voyages, described in Chapter 2, had similar scientific and logistical goals. Cook first arrived in the tropics two years after Bougainville left the Pacific. Cook's middle voyage (1772–75) was the most important of the three he made to the tropics. Extending his explorations east of Aus-

tralia and New Zealand (places he had discovered on his first voyage), Cook disproved the then-prevalent theory of a great continent at mid-latitude in the South Pacific. He explored and accurately charted many atolls and islands and reasserted the British presence in the waters around Tahiti. On his third voyage, he encountered Hawaii (**Figure 18.23**). On all three voyages, naturalists aboard collected biological specimens of great interest to researchers at home.

Sailing around the world sent people to fairly high latitudes even when they were interested in the tropics (see, for example, Figure 2.15). Though Cook explored the South Pacific past 70°S he did not discover Antarctica. He came within 75 miles of the Antarctic coast, but 47 more years passed before anyone actually saw that forbidding shore.

Charles Darwin's five-year voyage into the tropics aboard HMS *Beagle* (1831–36) may be one of the most fruitful scientific expeditions ever launched (see Figure 13.19). Not only did his first ideas on the origin of species form during that time, but his keen powers of observation were turned to the tropical seas on which *Beagle*

Figure 18.24 Fridtjof Nansen (1861–1930), pioneering oceanographer and polar explorer, looking every inch the Viking.

Figure 18.25 Roald Amundsen and his sturdy ship *Fram*. On 3 June 1910, one day after this picture was taken, *Fram* sailed from Norway to take on the rest of its cargo, embark its crew, and depart for the South. Amundsen and his team went on to become the first to stand at the South Pole.

sailed. Darwin became the first tropical marine biologist, publishing detailed monographs on coral reefs and organisms. Much modern work may be traced to his skill and encouragement.

The Poles

Arctic oceanography began with the pioneering efforts of Fridtjof Nansen (**Figure 18.24**). Nansen's interest in the Arctic began with a voyage to Greenland in 1882. In 1888 he returned to Greenland and with five companions made the first crossing of Greenland's great ice cap, a trek that consumed nine arduous months. In 1892 he proposed an even more hazardous expedition to prove

that surface currents flowed beneath the north polar ice pack. Nansen courageously allowed his specially designed ship *Fram* to be trapped in the Arctic ice, where he and his crew of 13 drifted with the pack for nearly four years (1893–96), exploring to 85°57′N, a record for the time. The 1,650-kilometer (1,025-mile) drift of *Fram* proved that no Arctic continent existed. Nansen's studies of the drift, of meteorological and oceanographic conditions, of life at high latitudes, and of deep sounding and sampling techniques form the underpinnings of modern polar science.

Living up to her name—*Fram* means "forward" in Norwegian—Nansen's ship continued to play a pivotal role in exploration (**Figure 18.25**). In 1910 Roald Amundsen, a student of Nansen's, set out in the sturdy little ves-

Figure 18.26 Dr. Edward Adrian Wilson, physician and ornithologist, chief scientist of the 1910–12 British Antarctic Expedition. Wilson and three other scientists conducted the first extensive scientific investigations at high latitude. In this photograph, he is completing a watercolor showing the positions of ice crystal halos around the sun.

sel for the coast of Antarctica, the first leg of a journey to the South Pole. Nansen himself settled down to a long and distinguished career as an oceanographer, inventor, zoologist, artist, statesman, and professor. He was awarded the Nobel Peace Prize in 1922 for his unstinting work in worldwide humanitarian causes.

Scientific curiosity, national pride, new ideas in ship-building, advances in nutrition, and great personal courage led in the early years of this century to the golden age of polar exploration. After some heroic attempts by a number of explorers to reach the poles, an American naval officer, Robert E. Peary, accompanied by his African-American assistant Matthew Henson and four Inuit (Eskimos), reached the vicinity of the North

Pole in April 1909. A party of five men led by Roald Amundsen of Norway achieved the South Pole in December 1911.

Peary and Amundsen's scientific contributions were limited to a few meteorological observations, but around the same time another Antarctic explorer, Captain Robert Falcon Scott of the British Royal Navy, led a scientific expedition to the South Pole. Chief scientist of the British expedition was Dr. Edward A. Wilson (**Figure 18.26**), a Cambridge-trained physician and naturalist with a special interest in ornithology. In 1911, before the assault on the pole, Wilson and three other scientists conducted the first extensive scientific investigations at high latitude. Laboring in a cramped expedition hut at Cape Evans, the researchers tended continuous-reading barometers and temperature gauges. They noted the direction of winds at the surface (and aloft by observing the smoke plume from Mt. Erebus, a nearby active volcano), measured atmospheric radioactivity, and recorded auroral displays. They also made weekly observations of the strength and direction of Earth's magnetic field, towed for plankton, trapped fish beneath the ice, made soundings through cracks in the pack ice, analyzed water samples from various depths, collected samples of life from various freshwater lakes and pools ashore, explored nearby dry valleys and glaciers, and analyzed the ecological relationships among species. Ice received their full attention; using methods pioneered by Nansen, the team made the first detailed on-site studies of the formation, deposition, temperature, and thermal conductivity of sea ice.

A scientific high point of the expedition was a five-week sledge journey led by Wilson to Cape Crozier to visit a breeding colony of emperor penguins. The birds' nesting habits were noted, and a few eggs were returned to base for embryological studies. The trip, made in the dead of winter, nearly cost the scientists their lives.

Although the expedition was a scientific success, the polar journey itself was a tragic failure. Under Scott's leadership, members of the expedition arrived at the pole four weeks after Amundsen, but they died on the return trip to base camp. There were many reasons for the polar party's inability to complete the 2,000-kilometer (1,300-mile) round-trip (Scott, Wilson, and a companion died just 18 kilometers—11 miles—from their final food depot). Among these reasons must surely be counted the team's unwillingness to abandon their scientific notes, their measurements and records, and 35 pounds of geological specimens collected en route. It was not geographical victory, but courage in adversity that captures the imagination.

Modern technology has eased the burden of high-latitude travel. In 1958, under the command of Capt. William Anderson, the U.S. nuclear submarine *Nautilus* sailed beneath the North Pole during a submerged transit beneath the Arctic pack from Point Barrow, Alaska, to the Norwegian Sea. As is apparent in Box 4.1e, a warm, strong, and stable nuclear submarine is a nearly ideal platform for conducting oceanographic research at high latitudes. A staff of seven scientists spent 38 days operating in the Arctic Ocean during a 1993 research cruise aboard USS *Pargo,* a fast and powerful *Sturgeon*-class submarine hardened for surfacing through ice (**Figure 18.27**, page 472). More than 9,000 kilometers (5,600 miles) of underway data—including bathymetry, gravity anomaly, temperature, salinity, ice draft, and images of the underside of the ice—were collected; 1,500 water samples were taken for biological and chemical analysis; and

four long-term observation buoys were placed. The Navy has agreed to host annual 45- to 60-day research cruises aboard *Pargo* until 1997. If the $200 million in conversion costs (and $10 million in annual upkeep) were forthcoming, the Navy might be persuaded to keep at least one *Sturgeon*-class submarine in oceanographic service past the vessels' turn-of-the-century decommissioning date. Polar science would be immeasurably richer for it.

Q-AND-A

1. What does the word *tropic* mean? Why are the tropic lines called the Tropic of Cancer and the Tropic of Capricorn?

   Tropic *derives from the Greek word* tropikos, *which refers to the equatorward turn of the sun at the time of the solstice. The* **solstice** *is the extreme of northward or southward travel of the subsolar point (the point on the Earth directly beneath the sun) during a year.*

   The names of the tropics of Cancer and Capricorn are derived from the sun's ancient position in the zodiacal constellations of Cancer and Capricorn at the time of the solstice. The sun no longer appears in these constellations at the time of the solstice, however, because of complex, long-period motions of the Earth relative to the sun.

2. Why are days longer in the summer and shorter in the winter? Why is this effect more pronounced when I travel farther north? In Canada in the summer it seems to stay light until very late.

   Yes, and Canadian winter days are correspondingly short. A look at **Figure 18.28** *(page 473) can help explain the phenomenon. Imagine yourself standing at 60°N, moving eastward with the rotation of the Earth through one day. If that day were 21 June, the day of summer solstice, very little of your time would be spent in the Earth's shadow. North of the Arctic Circle (66½°N), you could spend all of your day in the sun. How do you think the sun would appear to move through the day?*

3. Does the great diversity of marine species seen at the surface of some tropical ocean areas occur in the deep sea as well?

   No. The cold, unchanging regions of the deep ocean are populated by the same kinds of specialized organisms all over the world. Down there water is cold, food is scarce, and only a few species have adapted. A deep bottom sample off Tahiti would yield the same sorts of brittle stars, sea cucumbers, and unusual cnidarians as a sample from similar depth in the Arctic Ocean. Conditions—and species—below about 3,500 meters (12,000 feet) are similar anywhere in the world ocean.

BOX 18.2 ● *Shackleton's Miracle*

Perhaps the most arresting saga of polar exploration, and surely the one with the most surprising outcome, is the all-but-unbelievable 1914–16 adventure of Ernest Shackleton, a British merchant navy officer (**Figure a**). Shackleton's goal was to cross the Antarctic continent from sea to sea. Before his expedition could even reach its base, however, his ship, HMS *Endurance,* was immovably gripped by ice in Antarctica's Weddell Sea. *Endurance* was not built like *Fram;* it was slowly crushed by the ice and sank in November 1915 (**Figure b**). The party of 28 explorers rescued three small boats from the remains of their ship, along with a little food and some essential supplies. They camped beneath the overturned boats, drifted on the frozen sea for five months, and then slowly and laboriously dragged the boats to open water. After a brief time at sea, the party was cast ashore on desolate Elephant Island at the tip of Antarctica's Palmer Peninsula. This unexplored island was so isolated that no hope was held for rescue. Someone had to go for help.

Shackleton's largest boat was only 6.85 meters (22.5 feet) long, and ahead lay the most violent ocean area on Earth. Where could they go for help? There was only one chance: to attempt a landfall on South Georgia Island, site of a whaling station, over 1,300 kilometers (800 miles) away. With the force of the westerlies and the thrust of the West Wind Drift, there would be no hope of return if they were blown past the small island.

A crew of six set out on the morning of 24 April 1916 (**Figure c**). No one can imagine that voyage. The skill, fortitude, strength, and courage of these men against the bitter southern ocean with its mountainous waves and freezing spray is wonderful to consider. The wet discomfort and cold must have been all but unbearable.

After 16 days on howling seas, they sighted South Georgia Island and made a harrowing landing. Unfortunately, the whaling station was on the other side of the island; between them and safety lay a mountain range that had never been traversed! Three incapacitated sailors waited in a shore cave while Shackleton, Frank Worsley (captain of *Endurance*), and Petty Officer Crean set off across glaciers and snow-choked passes. They explored as they trudged onward, only to be frequently turned back by impassable terrain and towering cliffs. After three days of unremitting effort, exhausted and deeply chilled, they became lost. In fact, however, their overland navigation had been excellent, and the desper-

a A silhouette of Ernest Shackleton by Edward A. Wilson.

ate party was saved when they heard the sound of the whaling station's quit-work whistle in the distance. They stumbled into the whaling station at Stromness, only to have the workers reject their story as an impossibility and the ravings of madmen. The mountains were considered impassable (and indeed were not traversed again until 1963 by climbers equipped with modern gear). As for their wild story of a voyage over the towering seas of the West Wind Drift in a 22-foot boat—well, that was too ridiculous to be taken seriously!

Eventually, the whalers were convinced, and the men on the western side of South Georgia were rescued without incident. The first three attempts to reach the men back on Elephant Island were unsuccessful, however. The rescuers traveled to Argentina to obtain a larger ship, and finally, on 30 August 1916, Shackleton broke through the pack ice and found all 22 of his crew safe and well. Not a single man had died, and no member of the expedition sustained any serious injuries. Worsley's account of the boat journey to South Georgia Island makes splendid reading (see Worsley, 1977), especially beneath a sturdy Scottish wool blanket near a warm fireplace. Shackleton's boat, *James Caird,* is preserved in the British National Maritime Museum in Greenwich. It's well worth a visit.

c The boat *James Caird* puts to sea from Elephant Island in a desperate attempt at rescue.

b Shackleton's ship *Endurance* crushed, November 1915.

Figure 18.27 USS *Pargo* surfaced at the edge of the Arctic ice pack during the 38-day 1993 SCICEX cruise. A scientific staff of 7 was embarked along with a full operating crew; 39 shore-side researchers participated in the analysis of samples and data taken during the cruise. A strong, fast, quiet, powerful (and pleasantly warm) nuclear submarine makes an ideal platform for oceanographic research at high northern latitudes.

4. Can zooxanthellae live outside of a coral animal?

In laboratory cultures, yes. They change from the spherical shape seen within the coral to the typical biflagellate form characteristic of motile dinoflagellates. Researchers are uncertain whether all corals are host to the same species of zooxanthellae or whether several species exist. As far as we know, they do not normally live free in the ocean.

5. Will Antarctica continue to be set aside as a continent for scientific research?

Antarctica is the only continent without recognized sovereign countries. An Antarctic Treaty signed by twelve nations in 1959 was concerned primarily with demilitarization of the region south of 60°S, cooperative scientific investigation, protection of the polar environment, and prohibition of nuclear testing and waste disposal. The treaty was supplemented in 1972 and 1980 by agreements on biological conservation. It expired in 1991; its successor was signed by 39 nations on 4 October of that year.

Negotiating a new treaty was a daunting task. Explo-ration for mineral resources has already begun: Concentrations of copper and manganese have been discovered, as have traces of gold, silver, and platinum. The Ross, Bellingshausen, and Weddell seas may overlie substantial oil and gas deposits. Chile is planning to expand its tourist facilities. The original signatories seemed more interested in economic exploitation than in science, and many feared that scientific research might not continue to be the primary human activity on Antarctica. Unless people pressed for voluntary restraint, the continent's fragile ecosystems could quickly be destroyed. In recognition of this fact, the new international accord calls for a 50-year ban on drilling and mining, which may be lifted only by a three-fourths vote of all 26 consultative members of the treaty—those nations with full voting rights. In agreeing to sign the treaty, U.S. president George Bush said that the mining compromise "addresses our concerns and provides effective protection for Antarctica without foreclosing options for future generations." One can only hope that this pristine environment will be spared the human insults visited on all the other continents.

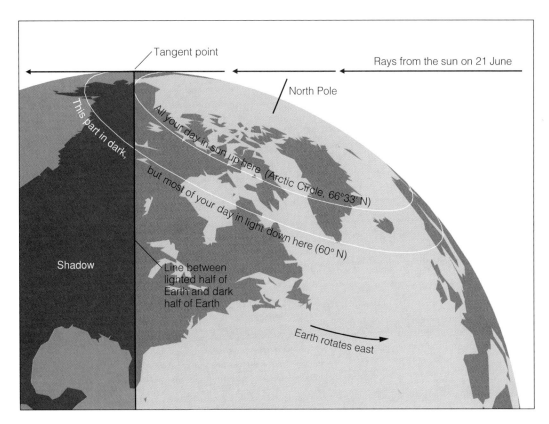

Figure 18.28 Why summer days are long at high latitudes.

Terms and Concepts to Remember

ahermatypic	coral reef	reef
Antarctic Circle	fringing reef	sea ice
Antarctic	hermatypic	solstice
Convergence	iceberg	Subtropical
Antarctic Ocean	ice cap	Convergence
Arctic Circle	ice floe	temperate zone
Arctic Convergence	krill	tropical ocean area
Arctic Ocean	polar ocean areas	Tropic of Cancer
atoll	polar regions	Tropic of Capricorn
barrier reef	polynya	tropics
convergence zone	polyp	zooxanthellae
coral		

Study Questions

1. How are the polar and tropical regions defined? Why do we use an astronomical definition *and* a biological/oceanographic definition?

2. What happens at convergence zones? Why are the convergence zones more pronounced in the Southern Hemisphere than in the Northern Hemisphere?

3. List and briefly explain how physical conditions in the polar ocean areas differ from those typical of the tropics.

4. How do marine organisms and communities differ in polar and tropical areas? What can you say about the relative number of species and individuals in each area?

5. In which major oceanic zone is primary productivity highest over a year's time? At any *one* time?

6. How does the northern polar ocean differ from the southern polar ocean? How do these differences influence the weather and climate of their respective hemispheres?

7. What is the relationship between coral animals and their symbiotic zooxanthellae? Are zooxanthellae always necessary for the mass accumulation of calcium carbonate?

8. How are tropical coral reefs classified? How do the reef types differ from each other? Who proposed the classification?

9. Are all tropical ocean areas biologically productive? Why or why not?

10. Knowing what you know about the Earth's oceanic extremes, where would you prefer to have been an explorer?

For Further Study

Bates, M. 1964. *Where Winter Never Comes.* New York: Scribner. An outstanding book on tropical biology for the general reader.

Batten, M. 1989. "Krill: Keystone of the Antarctic Ecosystem." *Calypso Log* 16 (no. 2, April). Horizontal swimming for considerable distances threatens krill's usual classification as zooplankton.

Brent, P. 1974. *Captain Scott and the Antarctic Tragedy.* New York: Saturday Review Press. Excellent general overview of polar exploration in general and of Captain Robert Scott's last expedition in particular.

Briggs, J. C. 1974. *Marine Zoogeography.* New York: McGraw-Hill. Who lives where, and why.

Burke, M. 1994. "Phosphorus Fingered as Coral Killer." *Science* 263 (25 February): 1086. Phosphate-rich fertilizers and sewage are implicated in eutrophic changes leading to the demise of coral reefs.

Cherry-Garrard, A. 1922. *The Worst Journey in the World.* Reprint. New York: Carroll & Graf, 1989. An inspiring and beautifully written account of Captain Scott's last expedition and the rescue attempt that followed its unsuccessful conclusion.

Culliney, J. L. 1989. *Islands in a Far Sea: Nature and Man in Hawaii.* San Francisco: Sierra Club Books. A beautiful book about tropical islands visited by millions of tourists.

Darwin, C. 1842. *The Structure and Distribution of Coral Reefs.* Reprint. New York: Dover, 1962. A classic.

Davies, P. J., et al. 1987. "Horizontal Plate Motion: A Key Allocyclic Factor in the Evolution of the Great Barrier Reef." *Science* 238 (18 December): 1697–99. The Great Reef becomes younger from north to south. The northward movement of the Indian-Australian plate is thought responsible.

Deacon, G. 1984. *The Antarctic Circumpolar Ocean.* Cambridge: Cambridge University Press. Excellent brief and nontechnical overview of most aspects of southern polar oceanography.

Fagerstrom, J. A. 1987. *The Evolution of Reef Communities.* New York: Wiley-Interscience. Interdisciplinary volume including biological and geographical information.

Fogg, G. E. 1992. *A History of Antarctic Science.* New York: Cambridge University Press. A new and comprehensive treatment.

Gordon, A. L., and J. C. Comiso. 1988. "Polynyas in the Southern Ocean." *Scientific American,* June, 90–97. Large gaps in the sea ice around Antarctica expose seawater to frigid air.

Goreau, T. F., et al. 1979. "Coral and Coral Reefs." *Scientific American,* August, 124–36.

Grigg, R. W., and D. Epp. 1989. "Critical Depth for the Survival of Coral Islands: Effects in the Hawaiian Archipelago." *Science* 243 (no. 4891):638–41. During the first half of the rapid rise in sea level at the end of the last ice age (18,000 to 9,000 years ago), the rising sea outpaced the ability of all reefs in the world submerged below critical depth to build upward.

Headland, R. K. 1989. *Chronological List of Antarctic Expeditions and Related Historical Events.* Cambridge: Cambridge University Press. Complete.

Hill, D. 1974. "Coral Islands, Coral Reefs, and Atolls." In *Encyclopædia Britannica.* Vol. 5. Chicago: Encyclopædia Britannica.

Huntford, R. 1986. *Shackleton.* New York: Atheneum. Excellent recent biography of Sir Ernest Shackleton, warts and all.

Huntford, R. 1987. *The Amundsen Photographs.* London: Hodder & Stoughton. The lantern slides used by Amundsen in lecture tours after his return from the South Pole were discovered in 1986 in an old malted milk box in an attic in Oslo. They are reproduced here.

Kaiser, J. 1994. "Nuclear Sub Is Researchers' Dream Boat." *Science* 265 (30 September): 2003. The USS *Pargo* expedition is discussed, and hopes of turning a Navy nuclear submarine into an oceanographic research vessel are outlined.

King, C. A. M. 1966. *An Introduction to Oceanography.* New York: McGraw-Hill (also London: Mackay, 1962). Thorough compendium of statistics for polar and tropical ocean.

King, H. G. R. 1969. *The Antarctic.* London: Blandford Press. Fine overview from one of Britain's modern polar explorers.

Levinton, J. S. 1982. *Marine Ecology.* Englewood Cliffs, NJ: Prentice-Hall. Perhaps the best general marine ecology text available. Detailed information about tropical marine biology.

Ley, W. 1962. *The Poles.* Chicago: Time–Life Books. Well-illustrated overview of the polar regions. An especially good review of the history of exploration of the far north.

Long, E. J., ed. 1964. *Ocean Sciences.* Annapolis, MD: United States Naval Institute Press. Chapter on polar oceanography by Walter Wittmann is especially useful.

Longhurst, A. R., and D. Pauly. 1987. *Ecology of Tropical Oceans.* Orlando, FL: Academic Press. Area, distribution, and general characteristics of the tropical ocean.

Lopez, B. 1986. *Arctic Dreams: Imagination and Desire in a Northern Landscape.* New York: Scribner. An evocative and beautifully written book.

Menard, H. 1986. *Islands.* New York: Freeman.

Murphy, R. C. 1962. "The Oceanic Life of the Antarctic." *Scientific American,* September, 186–210. A brief review article.

Neshyba, S. 1987. *Oceanography: Perspectives on a Fluid Earth.* New York: Wiley. Formation and influence of sea ice.

Pickard, G. L. 1964. *Descriptive Physical Oceanography.* London: Pergamon Press. Information on density stratification in the polar ocean.

Pyne, S. J. 1988. *The Ice: A Journey into Antarctica.* New York: Ballantine Books. A wonderfully written essay on the art, icescape, and history of the most alien landscape of our planet. Highly recommended.

Roberts, L. 1990. "Warm Water, Bleached Corals." *Science* 250 (no. 4978): 213. Information on the bleaching phenomenon.

Smith, R. C., et al. 1992. "Ozone Depletion: Ultraviolet Radiation and Phytoplankton Biology in Antarctic Waters." *Science* 255 (no. 5047): 952–59. Reduced concentration of atmospheric ozone, and consequently higher ultraviolet irradiance, has reduced phytoplanktonic productivity in the far south by as much as 6% to 12%.

Smith, W. O. 1990. *Polar Oceanography.* San Diego: Academic Press. Volume A covers physical science; volume B covers chemistry, geology, and biology.

Weller, G., et al. 1987. "Laboratory Antarctica: Research Contributions to Global Problems." *Science* 238 (4 December): 1361–67. Excellent review article on research in Antarctica, including information on the newsworthy ozone layer decrease and global warming.

Wilson, D. 1989. *The Circumnavigators*. New York: M. Evans. More information on Cook and Bougainville.

Wilson, E. A. 1972. *Diary of the* Terra Nova *Expedition to the Antarctic, 1910–1912*. Edited by H. G. R. King. London: Blandford Press. Diary of the chief scientist of Scott's last Antarctic expedition.

Worsley, F. A. *Shackleton's Boat Journey*. Reprint. New York: Norton, 1977. Worsley's account of the Shackleton expedition's most remarkable achievement, the open boat journey from the Palmer Peninsula to South Georgia Island. Very highly recommended.

Oceanus 31 (no. 2, Summer 1988) is devoted to Antarctica.

Charts

Antarctica. Supplement to *National Geographic*, April 1987. Washington, DC: National Geographic Society.

Arctic Ocean. Supplement to *National Geographic*, January 1990. Washington, DC: National Geographic Society.

19 MARINE RESOURCES

X *Marks the Spot*

World economies began a period of unprecedented growth in the decade following World War II. Leaders began to look to the sea to provide a greater proportion of the mineral and biological resources needed to sustain their economies and feed their citizens. It was comforting to think that as we learned more about the ocean, we would surely discover a vast treasure trove available for the taking.

The hidden treasure analogy is a good one. Popular folklore is full of stories about a person who finds a treasure map (the location of the treasure is always marked with a big *X*), follows its mysterious clues, and strikes it rich. Luck and a small amount of effort quickly lead to a life of ease. But with very few exceptions, the persistent prediction that the ocean will provide an increasing percentage of the material needs of a growing human population—of oceanic riches for everyone—has not been realized.

Because they are easier to exploit, solid mineral resources on the land have many economic advantages over mineral resources from the sea. On land, valuable elements have been concentrated into ores by sedimentation, weathering, and other natural processes, and the variety of minerals that can be extracted far exceeds anything we have seen in the ocean. Fluid resources such as petroleum (oil) and natural gas are probably as abundant in the continental shelves as on dry land, but obtaining these substances is very expensive and often dangerous. The land also provides much more food for humans than the ocean; worldwide only 3% to 5% of our food comes from the sea. Sea farming on a scale that would end world hunger, an activity dear to the hearts of science fiction writers in the 1950s, has failed to materialize. In fact, commercial marine farming has been successful only with a few expensive, high-protein species such as salmon, oysters, and abalone. Contrary to expectation, the ocean's natural treasures have not been easy to exploit.

Encrusted coins salvaged from the wreck of *Nuestra Señora de Atocha*, a Spanish treasure galleon that sank in shallow water off the Florida Keys in 1622. Marine resources are rarely this obvious or spectacular.

Marine resources include physical resources such as oil, natural gas, building materials, and chemicals; marine energy; biological resources such as seafood and kelp; and nonextractive resources like transportation and recreation. The contribution of marine resources to the world economy has become so large that international laws now govern their allocation.

In spite of their abundance, marine resources provide only a fraction of the worldwide demand for raw materials, human food, and energy. Similar resources on land can usually be obtained more safely and at lower cost. The management of marine resources—especially biological resources—for long-term benefit has been largely unsuccessful.

TYPES OF MARINE RESOURCES

We will discuss four groups of marine resources in this chapter. **Physical resources** result from the deposition, precipitation, or accumulation of useful substances in the ocean or seabed. Most physical resources are mineral deposits, but petroleum and natural gas, mostly remnants of once-living organisms, are included in this category. Fresh water obtained from the ocean is also a physical resource. **Marine energy resources** result from the extraction of energy directly from the heat or motion of ocean water. **Biological resources** are living animals and plants collected for human use. As the term suggests, **nonextractive resources** are uses of the ocean in place; transportation of people and commodities by sea, recreation, and waste disposal are examples.

Marine resources can be classified as either renewable or nonrenewable. **Renewable resources** are naturally replaced on a seasonal basis by the growth of marine organisms or by other natural processes. **Nonrenewable resources** such as oil, gas, and solid mineral deposits are present in the ocean in fixed amounts and cannot be replenished.

PHYSICAL RESOURCES

Petroleum and Natural Gas

Offshore petroleum and natural gas generated nearly $110 billion in revenues in 1992. Here the ocean makes a significant contribution to present world needs: About 28% of the crude oil and 21% of the natural gas produced in 1992 came from the seabed. About a third of known world reserves of oil and natural gas lie along the continental margins. Major U.S. marine reserves are located on the continental shelf of southern California, off the Texas and Louisiana Gulf Coast, and along the North Slope of Alaska. The deep-sea floors probably contain little or no oil or natural gas.

Oil, one of the few naturally occurring liquids, is a complex chemical soup containing perhaps a thousand compounds, mostly hydrocarbons. Petroleum is almost always associated with marine sediments, suggesting that the organic substances from which it was formed were once marine. Planktonic organisms or soft-bodied benthic marine animals are the most likely candidates. Their bodies apparently accumulated in quiet basins where the supply of oxygen was low and there were few bottom scavengers. The action of anaerobic bacteria converted the original tissues into simpler, relatively insoluble organic compounds that were probably buried—possibly first by turbidity currents, then later by the continuous fall of sediments from the ocean above. Fur-

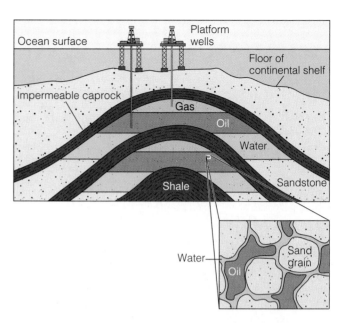

Figure 19.1 Oil and natural gas are often found together beneath a dome of impermeable caprock. Oil and gas are not found in great hollow reservoirs, but within pore spaces in rock (inset).

ther conversion of the hydrocarbons by high temperatures and pressures must have taken place at considerable depth, probably 2 kilometers (1.2 miles) or more beneath the surface of the ocean floor. Slow cooking under this thick sedimentary blanket for millions of years completed the chemical changes that produce oil.

If the organic material cooked too long, or at too high a temperature, the mixture turned to methane, the dominant component of natural gas. Deep sedimentary layers are older and hotter than shallow ones and have higher proportions of natural gas to oil. Very few oil deposits have been found below a depth of 3 kilometers (1.8 miles). Below about 7 kilometers (4.4 miles), only natural gas is found.

Oil is less dense than the surrounding sediments; so it can migrate from its source rock through porous overlying formations. It collects in the pore spaces of reservoir rocks when an impermeable overlying layer prevents further upward migration of the oil (**Figure 19.1**). When searching for oil, geologists use sound reflected off subsurface structures to look for the signature combination of layered sediments, depth, and reservoir structure before they drill.

Drilling for oil offshore is far more costly than drilling on land because special drilling equipment and transport systems are required. Most marine oil deposits are tapped from offshore platforms resting in water less than 100 meters (330 feet) deep (**Figure 19.2**). As oil demand

Figure 19.2 An oil drilling platform in the relatively shallow water off southern California. Compare this platform to the ones shown in Box 19.1.

(and therefore price) continues to rise, however, deeper deposits farther offshore will be exploited from larger platforms (see **Box 19.1**). The largest and heaviest platform is *Statfjord-B,* in position since 1981 northeast of the Shetland Islands in the North Sea. The tallest drilling platform is Platform *Auger,* a tension leg structure deployed by the Shell Oil Company in 1993. *Auger* is held in position 341 kilometers (214 miles) off the Louisiana coast by steel cables anchored into the seabed and pulling against the semisubmerged platform's buoyancy. It is anchored in 872 meters (2,860 feet) of water and—with its cables—is nearly a thousand meters (3,275 feet) tall! Some 32 wells on *Auger* are scheduled to produce 46,000 barrels of oil and 125 million cubic feet of natural gas per day in the year 2001. Total cost of the platform: $1.3 billion. Even larger tension leg platforms are being planned.

Sand and Gravel

Sand and gravel are not very glamorous marine resources, but they are second in dollar value only to oil and natural gas. More than 112 million metric tons (123 million tons) of sand and gravel, valued at about a third of a billion dollars, were mined offshore in 1987. Only about 1% of the world's total sand and gravel production is scraped and dredged from continental shelves each year, but the seafloor supplies about 20% of the sand and gravel used in the island nations of Japan and the United Kingdom. The world's largest single mining operation is the extraction of aragonite sands at Ocean Cay in the Bahamas. Sand is suction-dredged onto an artificial island and then shipped on specially designed vessels. This sand, about 97% calcium carbonate, is used in Portland cement, glass, and animal feed supplements, and also in the reduction of soil acidity.

Most of the exploitable U.S. deposits of marine sand and gravel are found off the coasts of Alaska, California, Washington, and the East Coast states from Virginia to Maine. Nearshore deposits are widespread, easily accessible, and used extensively in the building industry to manufacture concrete and to surface highways. Offshore oil wells in Alaska are built on huge man-made gravel platforms; the large quantities of gravel available at those locations make offshore drilling practical there.

Magnesium and Magnesium Compounds

Magnesium, the third most abundant dissolved element, precipitates from seawater, mainly in the form of magnesium chloride ($MgCl_2$) and magnesium sulfate ($MgSO_4$) salts. Magnesium metal (a strong, lightweight material used in aircraft and structural applications) can be extracted by chemical and electrical means from a concentrated brine of these salts. Worldwide, about half the production of metallic magnesium is derived from seawater; about 60% of U.S. production comes from a single seawater processing facility in Texas. The value of this metal to the U.S. economy was $275 million in 1992.

Magnesium compounds are also valuable. Magnesium salts are used in chemical processes, in foods and medicines, as soil conditioners, and in the lining of high-temperature furnaces. About 25% of the magnesium compounds produced in the United States are derived from seawater. In 1992 they were worth about $55 million to the U.S. economy.

Salts

As you may remember from Chapter 6, the ocean's salinity varies from about 3.3% to 3.7% by weight. When seawater evaporates, the remaining major constituent ions

BOX 19.1 ● *Drilling Platforms*

In August 1981, *Statfjord-B*, the heaviest manufactured object ever moved, was pulled by five tugs through a narrow Norwegian fjord toward the North Sea. Five days later, the giant oil drilling platform was on station 100 miles northeast of the Shetland Islands. Water was pumped into tanks at the bottom of the structure, and it settled to the bottom within 15 meters (50 feet) of its planned position. It rises 270 meters (890 feet) from the seabed and weighs 740,000 metric tons (824,000 tons).

The huge platform had been built in two parts. Its concrete base, consisting of 24 reinforced concrete ballast cells, was built in a drydock in Stavanger, Norway. Four hollow legs, also of concrete, rose from the cells. A separate steel superstructure supporting all the equipment needed for producing 150,000 barrels of oil a day—along with helicopter pads, generators, emergency equipment, and a 200-bed hotel to house the workers—was constructed nearby. The base and superstructure were joined in a delicate operation in a sheltered deep-water fjord (**Figure a**). The completed platform is nearly as tall as the Eiffel Tower and weighs nine times as much as the world's greatest warships, the U.S. Navy's *Nimitz*-class carriers. It cost $1.84 billion to build.

Statfjord-B is a gravity platform, one that is held on the seabed by its tremendous weight (**Figure b**). After the platform reached its final position in the North Sea, seawater was pumped into the ballast cells to sink the bottom of the platform into the sediment. Water was then pumped from beneath a steel skirt around the base to draw the platform into the seafloor. Concrete was forced into the gaps between the bottom of the base and the compressed seabed. The platform is designed to withstand waves more than 30 meters (100 feet) high and winds of more than 87 knots (100 miles per hour) without shifting more than a centimeter (half an inch).

The first wells were drilled a few months after the platform was positioned. Sixteen pipes running down two of the four legs carry the equipment needed to drill the wells, which do not go straight down but rather arc out in sweeping parabolas to reach the farthest and deepest parts of the oil field, 6,000 meters (19,700 feet) beneath the surface.

The Shell Oil Company's platform *Auger* is built in a different way. *Auger* is a tension leg platform, held in position by steel cables anchored into the seabed and pulling against the semisubmerged platform's buoy-ancy. Eight lateral cables anchored to the seabed at a radius of 2.7 kilometers (1.7 miles) from the base prevent sideways movement. The platform is deployed in 872 meters (2,860 feet) of water on the Garden Banks of the Gulf of Mexico; it began production in 1994. Peak production is expected to be about 46,000 barrels of oil per day and 125 million cubic feet of natural gas per day in the year 2001. Shell's even larger tension leg platform *Mars* (**Figure c**), due for deployment off Louisiana in 913 meters (2,933 feet) of water in 1996, is pictured here in this imaginary view in relation to the Houston skyline. Both platforms are designed to withstand hurricane-force waves of 22 meters (71 feet) and winds of 225 kilometers (140 miles) per hour.

a (left) *Statfjord-B* in Yrkjefjorden during assembly. The poet A. Alvarez has written: "As if from the bottom of one of Piranesi's imaginary prisons—a vast enclosed shadowy place, with gangways and galleries and ominous, purposeful machinery, all of it disproportionate to the human scale."

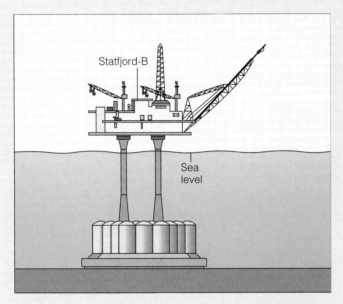

b A drawing of *Statfjord-B* in place on the seabed. The whole structure is 270 meters (890 feet) high.

c Platform *Mars,* due for deployment off Louisiana in 1996, is pictured in this imaginary view in relation to the Houston skyline. Tension leg platforms like this are held in position by steel cables anchored into the seabed, pulling against the semisubmerged platform's buoyancy. Lateral cables anchored to the seabed (not shown) prevent sideways movement. *Mars* is designed to withstand hurricane-force waves of 22 meters (71 feet) and winds of 225 kilometers (140 miles) per hour.

481

Figure 19.3 Salt evaporation ponds at the southern end of San Francisco Bay in California. Operators can segregate the various salts from one another by shifting the residual brine from pond to pond at just the right time during the evaporation process. The colors in the ponds are imparted by algae and other microorganisms that thrive at varying levels of salinity. In general, the highest salinity ponds have a reddish cast.

(see Table 6.1) combine to form various salts, including calcium carbonate ($CaCO_3$), gypsum ($CaSO_4$), table salt (NaCl), and a complex mixture of magnesium and potassium salts. Table salt makes up slightly more than 78% of the total salt residue.

Seawater is evaporated in large salt ponds in arid parts of the world (**Figure 19.3**). Operators can segregate the various salts from one another by shifting the residual brine from pond to pond at just the right time during the evaporation process. As we've seen, the magnesium salts are used as a source of magnesium metal and magnesium compounds. The potassium salts are processed into chemicals and fertilizers. Bromine (a useful component of certain medicines, chemical processes, and antiknock gasolines) is also extracted from the residue. Gypsum is an important component of wallboard and other building materials. About a third of the world's table salt is currently produced from seawater by evaporation. In North America, some of this salt is used for snow and ice removal. Salt is also used in water softeners, agriculture, and food processing. In 1992 the United States produced by evaporation about 4.2 million metric tons (4.6 million tons) of table salt with a value of about $110 million.

Manganese Nodules

In Chapter 5 you read about manganese nodules, rounded black objects that litter the abyssal plains, particularly in the Pacific (see Figure 5.9). These slow-growing lumps were first seen in bottom samples taken by scientists aboard HMS *Challenger* in 1874. The iron, manganese, copper, nickel, and cobalt content of the nodules makes them particularly attractive to industrial nations lacking onshore sources of these crucial materials. The U.S. Bureau of Mines estimates that the richest deposits in the Pacific exceed 16 billion metric tons (17.6 billion tons), about 20 times all known terrestrial reserves and more than 2,000 years' worth of production at present rates of use.

Manganese nodules have been dredged from the seabed in small-scale trials (see **Figure 19.4**), but no commercial mining ventures exist. Various recovery schemes have been proposed, including a system resembling a vacuum cleaner, but the difficulties of collecting large numbers of nodules from abyssal depths in excess of 4,000 meters (12,000 feet) has rendered these plans uneconomical—at least until prices for manganese, copper, and nickel rise as terrestrial sources are consumed.

Figure 19.4 Emptying a dredge of manganese nodules from a demonstration mining program.

Phosphorite Deposits

Also discovered during the *Challenger* Expedition were irregular chunks of phosphorite, first collected from the continental rise off South Africa. Sedimentary phosphorite deposits, from which industrial chemicals and phosphate-rich agricultural fertilizers may be made, formed from the decaying remains of marine organisms that lived in areas of extensive upwelling. The richest deposits occur at depths between 30 and 300 meters (100 to 1,000 feet). Post-*Challenger* investigations have revealed rich deposits of this important material off the coasts of Florida, California, western South America, and western Africa. Even though phosphorite deposits occur in much shallower water than manganese nodules do, the costs of recovering the resource from the ocean greatly exceeds that of recovery from land. Much of the U.S. supply of phosphorus and phosphates comes from the strip mining of ancient, uplifted shallow-water phosphorite deposits in central Florida.

Metallic Sulfides and Muds

The recent discovery of metal-rich sulfides around hydrothermal vents has spurred interest among economists as well as oceanographers. Heated seawater carrying large quantities of metals and sulfur leached from the newly formed crust pours out through vents and fractures. The metals—mainly zinc, iron, copper, lead, silver, and cadmium—combine with the sulfur and precipitate from the cooler surrounding water as mounds, coatings, and chimneys. While these deposits are certainly commercial-grade ores, they are neither large nor extensive. Also, they are subject to solution and oxidation on the seafloor and are not likely to be preserved in thick layers for long periods of time. Nevertheless, scientists are planning mineralogical studies of the fast-spreading rift zones along the East Pacific Rise near the mouth of the Gulf of California, the Galápagos Rift, and the Juan de Fuca Ridge off the coast of Oregon.

The Red Sea is another area where lithospheric plates are diverging and where molten material is close to the surface. Seawater seeping in through deep faults and fractures comes into contact with fresh, hot basalts and dissolves metals and salts from the rock. The recycled water emerges at temperatures of around 100°C (212°F) and is extremely saline, about 250‰ to 300‰ (average seawater is about 34‰). Though the solutions are hot, their great density causes them to stay on the floor of the Red Sea in deep, fault-bounded basins. Three great pools of hot water have been discovered at a depth of about 2,000 meters (6,600 feet). The metals precipitating within these basins produce muds rich in metal sulfides, silicates, and oxides of commercial concentration. Recovery of these ores is expected to begin in the late 1990s.

Fresh Water

Only 0.017% of Earth's water is liquid, fresh, and available at the surface for easy use by humans. Another 0.6% is available as groundwater within half a mile of the surface. Unfortunately, much of this is polluted or otherwise unfit for human consumption. The fact that fresh, pure water often costs more per gallon than gasoline emphasizes its scarcity and importance. More than any other factor in nature, the availability of **potable water** (water suitable for drinking) determines the number of people who can inhabit any geographic area, their use of other natural resources, and their life-style.

Fresh water is becoming an important marine resource. Exploitation of that resource by **desalination,** the separation of pure water from seawater, is already under way, mainly in the Middle East, West Africa, Peru, Florida, Texas, and California. More than 1,500 desalination plants are currently operating worldwide, producing a total of about 13.3 billion liters (3.5 billion gallons) of fresh water per day. The largest desalination plant, in Saudi Arabia, produces about 114 million liters (30 million gallons) daily!

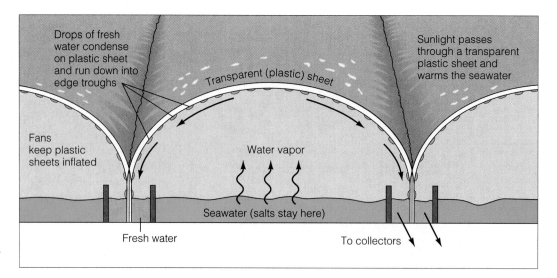

Figure 19.5 A low-energy, low-tech scheme for desalination.

Several desalination methods are currently in use. *Distillation* by boiling is the most familiar; about three-fourths of the world's desalinated water is produced in this way. Distillation uses a great deal of energy, making it a very expensive process. *Freezing* is another effective but costly method of desalination; ice crystals exclude salt as they form, and the ice can be "harvested" and melted for use. Solar or geothermal power may bring down the cost of distillation or freezing, but more efficient, less energy-intensive mechanisms are being developed. Among these is *reverse osmosis desalination*. In this process seawater is forced against a semipermeable membrane at high pressure. Fresh water seeps through the membrane's pores while the salts stay behind. About a quarter of desalinated water is produced in this way. Reverse osmosis uses less energy per unit of fresh water produced than distillation or freezing, but the necessary membranes are fragile and costly.

One of the most elegant ideas for inexpensive desalination is illustrated in **Figure 19.5.** Other than the sunlight needed to make water vapor, this passive system requires no energy except that required by pumps to move water through plastic-covered troughs and small fans to keep the plastic sheets inflated.

Another low-tech proposal for providing fresh water involves capturing and towing a huge iceberg to an area where water is in demand. In 1974 delegates to a Saudi Arabian–sponsored conference studied the economics of transporting a 100-million-metric-ton iceberg from Antarctica to that desert country. The plan was to wrap the iceberg in insulating plastic and tow it with several ships. If the iceberg lost only 20% of its mass during the eight-month trip, 80 million cubic meters of water could be recovered—enough to give nearly 600,000 people 380 liters (100 gallons) a day for a year. And the cost is potentially cheaper than water produced by desalination. Logistical problems and uncertainties over the calculations have prevented implementation of the idea, however. For the moment, desalination is a more certain alternative.

Desalination, water conservation, and perhaps iceberg harvesting will become more common as water becomes more polluted, scarcer, and more valuable.

MARINE ENERGY

The energy crises of 1973 and 1979 focused public attention on the need for unconventional sources of power. Sources of energy that are not consumed in use—solar power or wind power, for example—are preferable to nonrenewable sources such as fossil fuels. Anyone who has watched the ocean knows that so restless a place must surely be rich in energy. The energy is certainly there, but extracting it in useful form is not easy. **Table 19.1** lists four marine energy sources and the estimated power theoretically available from each.

Waves and Currents

Waves are the most obvious manifestation of oceanic energy—ask any surfer about the energy in a wave. As you may remember from Chapter 10, wind waves store wind energy and transport it toward shore. A wave 3 meters (10 feet) high transmits energy at a rate of 100 kilowatts per linear meter of crest. This is the power equivalent of a line of automobiles, packed as closely together as possible, rushing at full throttle toward the shore. Larger waves are proportionally more powerful.

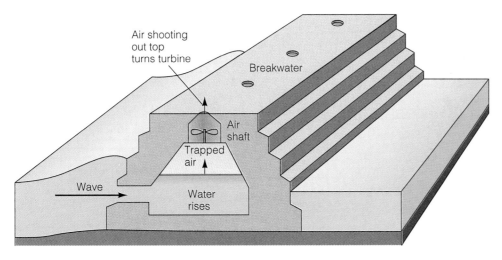

Figure 19.6 A device for using wave energy to generate power, one of several being tested around the world.

Table 19.1	Estimated Power Theoretically Available from Marine Energy Sources

Energy Source	Estimated Power (millions of megawatts)
Waves	2.5
Tides[a]	2.7
Currents	25.0
Thermal gradient	40,000

Source: Constans, 1979.

Note: Estimated world power needs by the year 2000 are 35 million megawatts.

[a]Tidal power is discussed in Chapter 11.

Many devices have been proposed to harness this energy; Japan, Norway, Great Britain, Sweden, the United States, and Russia have built small experimental plants to evaluate their effectiveness. One of these devices (shown in **Figure 19.6**) uses the rush of air trapped by waves entering breakwater caissons to power a generator. So far none of the plants has produced power at a competitive price, but some designs show promise. There is, however, no free lunch. Extensive use of wave power generators along a shore could deprive that shore of its natural wave energy. Changing the wave patterns could alter longshore transport or disrupt the life cycles of marine organisms. (And some schemes would ruin waves for surfers and beachcombers.)

Ocean currents might also be harnessed. Huge, slowly turning turbines immersed in the Gulf Stream have been proposed, but their necessary size and complexity make them prohibitively expensive.

Thermal Gradient

As can be seen in Table 19.1, the greatest potential for energy generation in the ocean lies in exploiting the thermal gradient between warm surface water and cold, deep water. As you may remember from Chapter 7, 1 calorie of heat will raise the temperature of 1 cubic centimeter of seawater by about 1°C (27°F). If we assume a temperature difference of 15°C between surface water and deep water, the total heat energy in each cubic meter of surface water (relative to deep water) is about *15 million calories.* If this energy were released in 1 second, it would generate 60 megawatts of power. Extracting the heat energy from 1,600 cubic meters of warm seawater per second would provide power equivalent to the full generating capacity of all the nuclear power plants in the United States! Note in Table 19.1 that the estimated power from this source is more than 1,000 times that projected for all human needs by the year 2000.

How might the immense potential of thermal gradients be harnessed? The earliest proposal was made in the 1880s by a French physicist and inventor, Jacques d'Arsonval. His idea for generating electricity is shown in **Figure 19.7.** Warm seawater would be pumped into the plant through openings near the ocean surface. This water would pass through heat exchangers and boil liquid ammonia—a liquid with a very low boiling point—into a pressurized vapor. This gas would turn a turbine, which would spin an electrical generator, and then pass into another heat exchanger cooled by water pumped from the depths. Here the ammonia would condense into a liquid, creating a vacuum that would draw more ammonia vapor through the turbine. The liquid ammonia would be pumped back to the first heat exchanger to repeat the cycle. A small plant was built on this model in

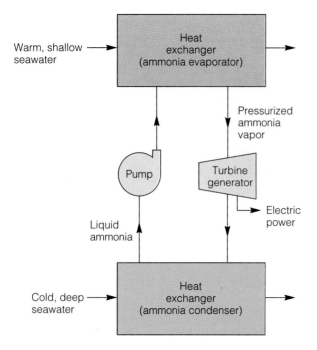

Figure 19.7 Basic aspects of the thermal energy conversion system proposed by Jacques d'Arsonval.

Figure 19.8 Design for a large-scale ocean thermal energy conversion (OTEC) plant for generating electricity from the temperature gradient in a tropical area of ocean. The platform would be about 181 meters (594 feet) in length and 75 meters (246 feet) in diameter. The cold-water inlet pipe would extend another 305 meters (1,000 feet) below the surface.

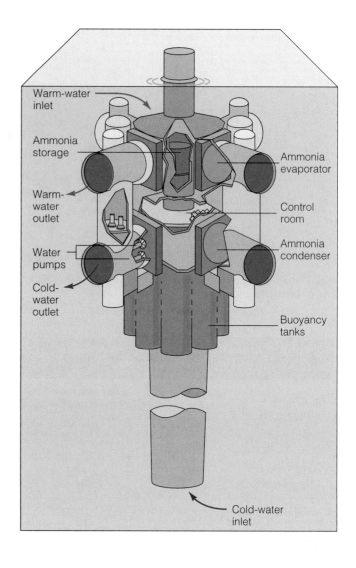

Cuba in 1930. It operated successfully for a short time before being destroyed in a tropical storm. **Figure 19.8** shows a modern design for an electrical generator, dubbed OTEC for *Ocean Thermal Energy Conversion*, that would use thermal gradients for power.

Why has this promising technology not been exploited? Mainly because of the low efficiency of the OTEC process. The efficiency of heat-driven power generators depends on the *difference* in temperature between the hottest part of the system and the coldest. A fossil-fueled generating plant can be highly efficient because of the great difference in temperature between the flame and the water (or air) cooling the heat exchanger. A nuclear plant is less efficient because the reactor core cannot be as hot as a flame. The efficiency of an OTEC plant—with only a 15°C temperature difference between "hot side" and "cold side"—would be only about 2%. Therefore, huge amounts of warm and cold water would

have to circulate through the plant. An OTEC plant with the same generating capacity as a single large nuclear power plant would need to process a continuous flow of water equal to five times the average flow of the Mississippi River—and that estimate doesn't include the power necessary to operate the OTEC plant's massive internal pumps.

There are other problems. An OTEC plant would need to be sited in the tropics, where warm water is layered over cold. Tropical cyclones common in these areas would wreak havoc with the plant's long pipes and delicate generators. Marine life stimulated by the upwelled, nutrient-rich cold water would grow in the surrounding ocean and foul the heat exchangers. Construction and maintenance costs would be astronomical, and the transmission of power to a distant shore presents special difficulties. Still, as energy becomes more and more costly, the OTEC option may prove practical.

BIOLOGICAL RESOURCES

Ancient kitchen middens (garbage dumps of bones and shells) found in many coastal regions demonstrate that humans have used the sea for thousands of years as a source of food and medicines. Now the human population threatens to outgrow its food supply. Contemporary food production and distribution practices are unable to satisfy the nutritional needs of all the world's 5.5 billion people, and starvation and malnutrition are major problems in many nations. Can the ocean help?

At least 12% of the caloric intake in an adequate human diet should be in the form of proteins, essential structural components of living cells. **Table 19.2** shows the high protein content of fish, crustaceans, and mollusks in comparison to other common foods of various cultures.

Compared to the production from land-based agriculture, the contribution of marine animals and plants to the human intake of all protein is small, probably around 4%. Although most of that protein comes from fish, marine sources account for only about 18% of the total *animal* protein consumed by humans. Fish, crustaceans, and mollusks contribute about 14.5% of the total; fish meal and by-products included in the diets of animals raised for food account for another 3.5%. About 85% of the annual catch of fish, crustaceans, and mollusks comes from the ocean, and the rest from fresh water.

The sea will probably not be able to provide substantially more food to help alleviate future problems of malnutrition and starvation caused by human overpopulation; indeed, population growth would likely absorb any resource increase. Nevertheless, these resources currently sustain a great many people.

Fish, Crustaceans, and Mollusks

Fish, crustaceans, and mollusks are the most valuable living marine resources. Commercial fishermen took nearly 87 million metric tons (96 million tons) of these animals in 1992. The largest share of the global catch of marine fish, crustaceans, and mollusks—over 68%—comes from the temperate waters over the continental shelves in the northwest, southeast, and west central Pacific, and the northeast Atlantic (see **Figure 19.9**).

Of the thousands of species of marine fishes, crustaceans, and mollusks, fewer than 500 species are regularly caught and processed. The 10 major groups listed in **Table 19.3** supply more than 95% of the commercial marine catch. Note that the largest commercial harvest is of the herring and its relatives, which accounts for nearly a quarter of the live weight of all living marine resources caught each year. **Figure 19.10** shows some of the most important commercial species.

Table 19.2 Amount of Protein Provided by Commonly Consumed Foods

Food	Protein by Weight (%)
Vegetable Protein	
Cassava	0.9
Plantain	1.4
Tapioca (cassava product)	1.5
Sweet potato	1.8
Taro	1.8
Chinese cabbage	1.8
White potato	2.0
Yam	2.1
Rice (polished)	5–8
Beans (dry)	19
Animal Protein	
Pork	16
Beef	17
Veal	19
Poultry	21
Oysters	**10**
Shrimp	**21**
Fish (flesh)	**18–25**

Source: *Encyclopædia Britannica*, 1974, vol. 7, 345.

Table 19.3 World Commercial Catch of Marine Fishes, Crustaceans, and Mollusks, by Species Group, 1992

Species Group	Millions of Metric Tons, Live Weight
Herring, sardines, anchovies	20.4
Cods, hakes, haddocks	10.5
Jacks, mullets, sauries	10.5
Mollusks	8.9
Redfish, basses, conger eels	5.9
Crustaceans	5.4
Tunas, bonitos, billfishes	4.4
Mackerel, snooks, cutlass fishes	3.3
Flounders, halibuts, soles	1.2
Miscellaneous marine fishes	16.1
Total for all marine sources, excluding marine mammals and aquatic plants	**86.6**

Source: U.S. Department of Commerce. *Fisheries of the United States,* 1993.

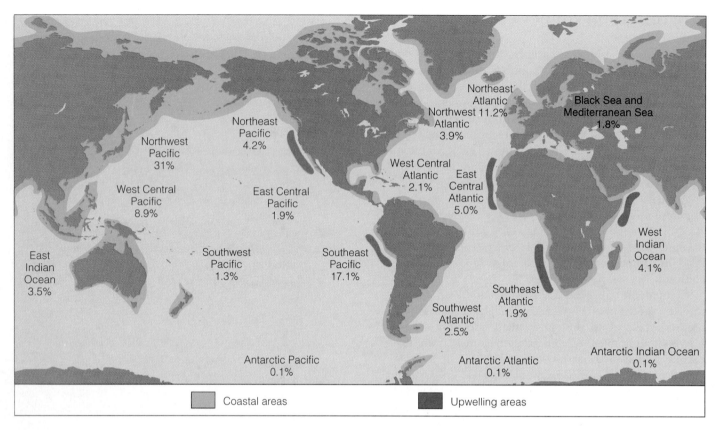

Figure 19.9 Location of the world's major commercial marine fisheries and distribution of the catch by fishery for 1990. Fishing is best in coastal areas over the continental shelves and in areas of upwelling.

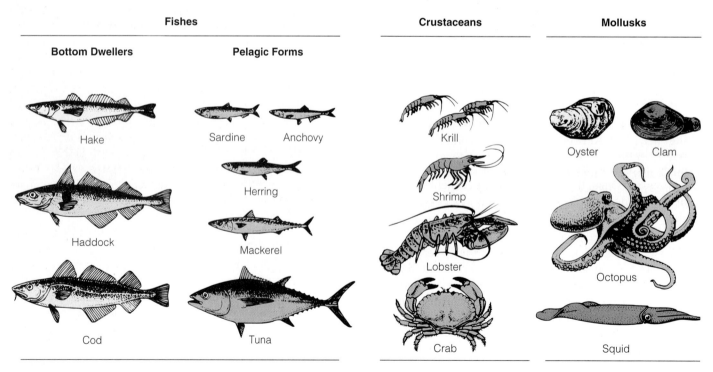

Figure 19.10 Some major types of commercially harvested fishes, crustaceans, and mollusks.

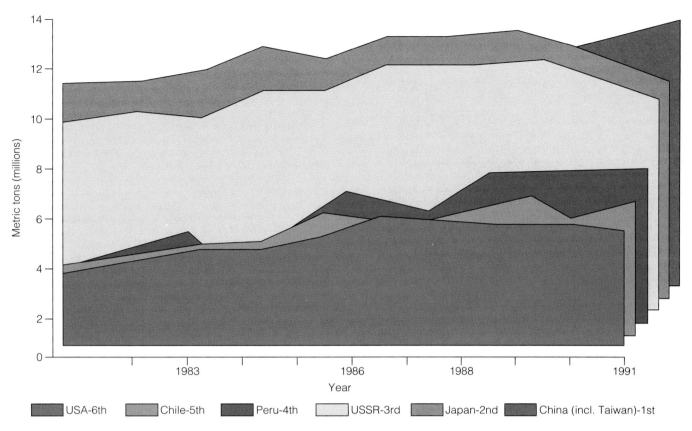

Figure 19.11 World commercial catch by leading countries, 1981–91. China, now the world's largest harvester of marine resources, will have a human population of about 1.4 billion in the year 2000.

The fastest growing fishery is that for the small crustaceans called *krill* (see Figure 18.8). The ocean around Antarctica supports tremendous numbers of krill, which eat the abundant diatoms of the surface waters and are in turn eaten by baleen whales and penguins. Krill have proliferated as the whale population has declined. The Soviets and Japanese pioneered krill harvesting and processing in the late 1960s. Today Russia fields about 100 processing ships; they produce krill "butter" and krill "cheese" spread, as well as large quantities of feed for poultry, cattle, and fur-bearing animals. Japan has about 14 processing ships; the Japanese are experimenting with krill cakes and a type of krill sausage thickened with algin. Krill is also sold raw and fresh-frozen. About 500,000 metric tons (550,000 tons) of krill are now being taken each year, making the krill fishery the 15th largest in landed catch. Some experts feel that the krill harvest might ultimately match the total production of the rest of the global marine fishery.

Fishing is a big business, employing more than 15 million people worldwide. Though estimates vary widely, the value in 1992 for the worldwide marine catch was thought to be around $55 billion! About half of the world's commercial marine catch is taken by only five

countries: In 1992 the ships of China led the world with 15% of the catch, followed by Japan (9%), Peru (7%), Chile (7%), Russia (6%), and the United States (6%) (**Figure 19.11**). About 75% of the annual harvest is taken by commercial fishermen who operate vast fleets working year-round, using satellite sensors, aerial photography, scouting vessels, and sonar to pinpoint the location of fish schools. Huge factory ships often follow the largest fleets to process, can, or freeze the animals on the run. Catching methods no longer depend on hooks and lines but rather on large trawl nets (**Figure 19.12**), purse seines, or gill nets. The living resources of the ocean are under furious assault: Between 1950 and 1990 the commercial marine fish catch increased fourfold.

The cost per unit of seafood has risen dramatically in spite of all this high-tech assistance. The increasing expense of fuel for the fishing fleets and processing plants, the cost of wages for the crews, and the greater distances that boats must cover to catch each ton of fish have all helped drive up the cost of seafood. In spite of greater efforts, the total marine catch leveled off in about 1970 and remained surprisingly stable until 1980, when greater demand and increasing prices began to drive the tonnage upward again. Annual harvests are now declin-

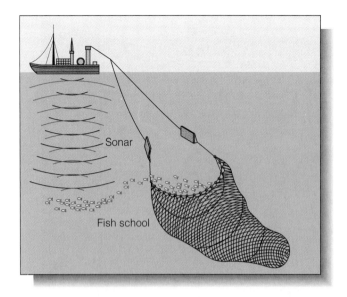

Figure 19.12 Stern trawler fishing. After sonar on the trawler finds the fish, they're captured by a trawl net towed along the bottom. Boards angled to the water flow keep the net's mouth open.

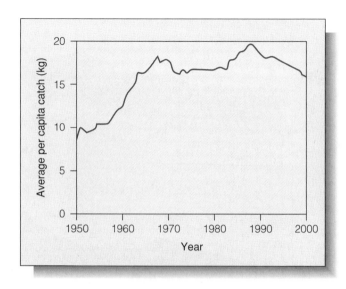

Figure 19.13 Average per capita world fish catch reached a peak in the late 1980s and has declined since. The decline is projected to continue past the end of the century.

ing; preliminary estimates from the United Nations suggest a drop of 7% between 1992 and 1994. And since 1970 the world population has grown; so the average *per capita* world fish catch has fallen alarmingly (see **Figure 19.13**).

Fishery Mismanagement

Can we continue to take this much food from the ocean year after year? The **maximum sustainable yield,** the maximum amount of each type of fish, crustaceans, and mollusks that can be caught without impairing future populations, probably lies between 100 and 135 million metric tons (110 and 150 million tons) annually. As can be seen in Table 19.3, current yield is nudging the lower figure. Fleets are obtaining fewer tons per unit of effort and are ranging farther afield in their urgent search for food. We may be perilously close to the catastrophic collapse of some fisheries. The National Marine Fisheries Service estimates that 45% of the fish stocks whose status is known are now overfished, and that populations of some species have fallen to less than 10% of the optimal level (the level that yields the largest sustainable catch).

The experience of the Peruvian anchovy fishery illustrates the effects of **overfishing**—harvesting so many fish that there is not enough breeding stock left to replenish the species. In 1960 the anchoveta catch in Peru was about 4 million metric tons (4.4 million tons); only 10 years later it had reached nearly 13 million metric tons a year (see **Figure 19.14**). The combination of overfishing

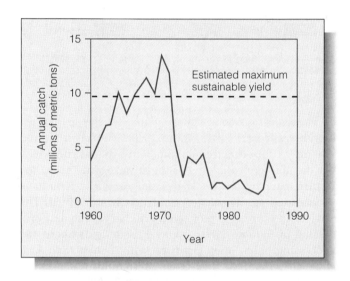

Figure 19.14 The Peruvian anchovy catch, showing the combined effects of overfishing and El Niño.

and an El Niño that began in 1971 dropped the 1972 catch to about 4.8 million metric tons. By 1980 less than 1 million metric tons were recovered, and the 1985 catch, further decimated by the 1983–84 El Niño, was less than 100,000 metric tons. Peru no longer has a commercial anchovy fishery, and in terms of total tonnage of fish caught, the country slipped from first place to fourth.

By the early 1980s overfishing had depleted the stocks of 24 other valuable fisheries. Cod and herring had been

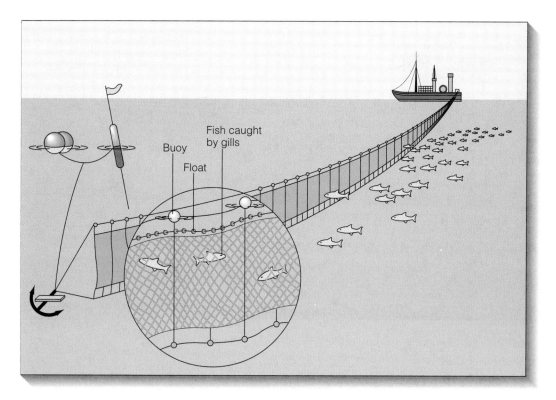

Figure 19.15 Drift net fishing. Typically, a fleet of huge factory ships uses sophisticated electronic detection devices and aircraft to find large schools of fish. Each ship then launches 20 to 50 fast, small catcher boats to set hundreds to thousands of miles of drift nets, weighted to stay at the desired depth. After drifting overnight the nets are hauled in by the factory ships, and the catch is processed on board by canning or freezing. Large-scale drift net fishing was banned in 1993, but it continues in the Pacific and has begun in the Atlantic.

overfished in the North Atlantic, and salmon and Alaskan king crab were becoming scarce in the Pacific Northwest. The Pacific sardine had been reduced to **commercial extinction**—depletion of a resource species to a point where it is no longer profitable to harvest.

Even when faced with evidence of overfishing, the fishing industry seldom follows a rational course. The industry's dominant motivating force is usually quick financial return, even if it means depleting a stock and disrupting the equilibrium of a fragile ecosystem. Long-term stability is forsaken for short-term profit. When the catch begins to drop, the industry increases the number of boats and develops more efficient techniques for capturing the animals. When the impending catastrophe is obvious, governments will sometimes intervene to set limits or close a fishery altogether. In Canada, 18,000 Newfoundlanders were put out of work in 1993 when a moratorium was placed on cod fishing in an effort to save the remaining stocks. The fishery has also collapsed off New England, where cod, haddock, and flounder populations are at or near historic lows.

A particularly disruptive fishing technique employs

drift nets—fine, vertically suspended nylon nets as much as 7 meters (25 feet) high and 80 kilometers (50 miles) long. A deployed drift net is shown in **Figure 19.15.** Drift net technology was pioneered by a United Nations agency to help impoverished Asian nations turn a profit from what had been subsistence fishing. Until 1993, Taiwanese, Korean, and Japanese vessels deployed some 48,000 kilometers (30,000 miles) of these "walls of death" each night—more than enough to encircle the Earth! Drift nets caught the fish and squid for which they were designed, but they also entangled everything else that touched them, including turtles, birds, and marine mammals (**Figure 19.16**). The process has been compared to stripmining—the ocean is literally sieved of its contents. An estimated 30 kilometers (18 miles) of these fine, nearly invisible nets were lost each night—about 1,600 kilometers (1,000 miles) per season. These remnants, made of nonbiodegradable plastic, become *ghost nets*, continuing to entangle fish and other animals for decades.

Though large-scale drift net fishing is now banned, pirate netting continues in the Pacific and has begun

Figure 19.16 An unintentional but unavoidable consequence of drift net fishing.

in the Atlantic. Completely unrestrained by regulation, these outlaw fishermen operate where their activities cause maximum damage to valuable reproductive stock.

Whaling

Since the 1880s, whales have been hunted to provide meat for human and animal consumption; oil for lubrication, illumination, industrial products, cosmetics, and margarine; bones for fertilizers and food supplements; and baleen for corset stays. **Figure 19.17** shows whales being butchered for food and oil. An estimated 4.4 million large whales existed in 1900; today slightly more than 1 million remain. **Table 19.4** indicates the extent of their slaughter. Eight of the 11 species of large whales once hunted by the whaling industry are commercially extinct. Again, a short-sighted industry is willing to take immediate profits despite obvious signs that the "fishery" is exhausted (see **Figure 19.18**).

Substitutes exist for all whale products, but the harvest of most commercial species did not stop until whaling became uneconomical. In 1986 the International Whaling Commission, an organization of whaling countries established to manage whale stocks, placed a moratorium on the slaughter of large whales. Except for a suspiciously large harvest of whales taken by the Japanese for "scientific purposes," commercial whaling ceased in 1987. Fewer than 700 large whales were taken in 1988 (see **Table 19.5**). Now, however, the number being taken appears to be rising: The American Cetacean Society estimates that about 1,400 large whales were taken in 1994.

Table 19.4	Worldwide Whale Catches, 1868–1984		
Year	Number of Whales	Year	Number of Whales
1868[a]	30	1963–64	63,001
1873	36	1964–65	64,680
1878	116	1965–66	57,891
1883	569	1966–67	52,238
1888	709	1967–68	46,645
1893	1,607	1968–69	42,126
1898	1,993	1969–70	42,480
1903	3,867	1970–71	38,771
1908	5,509	1971–72	32,133
1913	25,673	1972–73[d]	32,605
1918	9,468	1973–74	31,629
1923	18,120	1974–75	29,961
1928	23,593	1975–76	22,049
1933	28,907	1976–77	16,309
1938[b]	54,835	1977–78	13,638
1943	8,372	1978–79	10,668
1948	44,002	1979–80	3,542
1953	53,642	1980–81	2,928
1958	64,075	1981–82	2,050
1960–61	65,641	1982–83	1,683
1961–62[c]	66,090	1983–84	1,683
1962–63	63,579		

Source: Committee for Whaling Statistics, Oslo.

[a]The year the harpoon gun was invented.
[b]World catch record, in tonnage.
[c]Season catch record in individuals taken.
[d]U.S. ceases to import whale products.

Figure 19.17 The whaling industry has pushed most of the dozen or so species of great whales to the brink of extinction through overharvesting. This photograph shows a blue whale being butchered in the Coal Harbour Whaling Station, British Columbia, Canada, in 1963. Commercial whaling of great whales was banned in 1985 but began again in 1993.

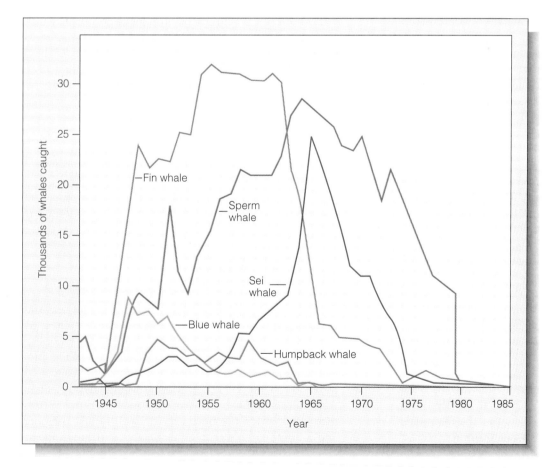

Figure 19.18 All five species included in this graph are commercially extinct, and the blue whale is still in danger of total extinction even though commercial blue whale hunting ended in 1964. At that time only about 1,000 blue whales were left, and experts fear that the population is too small to recover.

Table 19.5 Total Value of Whaling in 1988

Iceland agreed to stop all whaling in 1990, but, along with Norway, resumed commercial whaling in 1993. Under the guise of "scientific whaling," Japan took about 900 whales in 1994. Subsistence whaling is practiced by aboriginal peoples to provide whale products for their own use.

Number of Whales	Species	Nation	Estimated Value per Whale ($ millions)	Total Value ($ millions)
276	Minke	Japan	0.2	55.2
130	Minke	Greenland	0.2	26
68	Finback	Iceland	1.0	68
10	Sei	Iceland	0.5	5
168	Gray	USSR		Subsistence
3	Humpback	Bequia (Eastern Caribbean)		Subsistence
29	Bowhead	USA		Subsistence
684				**$154.2**

Source: Kraus, 1989.

What protection there is may have come too late to save some species from extinction—and protection may be only temporary. In June of 1992, under intense pressure from their major fishing industries, Iceland and Norway withdrew from the International Whaling Commission and made preparations to resume whaling. Norway resumed whaling in 1993. Their target, the small minke whale, is the smallest and most numerous of the great whale species (see Figure 16.23). The minke whale population has been estimated at 860,000, but the accuracy of this figure is in doubt. The decision to return to whaling may doom the minke to the same fate as most other whale species.

There is a glimmer of hope. In 1994 the International Whaling Commission voted overwhelmingly to ban whaling in about 21 million square kilometers (8 million square miles) around Antarctica, thus protecting most of the remaining large whales, which feed in those waters. Conservation efforts can do some good. Although it was hunted nearly to extinction, the California gray whale has long been off limits to hunters. Its numbers have grown, and it was removed from the endangered list in 1993.

Many more small whales have been killed than large ones, but not for food or raw materials. For reasons that are not well understood, porpoises (which are small whales) gather above schools of yellowfin tuna in the open sea. Fishermen have learned to find the tuna by spotting the porpoises. Nets cast to catch the tuna also entangle the air-breathing porpoises, and the mammals drown. Nearly all the dead porpoises are simply pitched over the side as waste. More than 6 million small whales have been killed in association with tuna fishing since 1971!

Passage of the U.S. Marine Mammals Protection Act in 1972 brought a drop in the number of porpoise deaths. However, from 1977 to 1987 the number of tuna boats in the U.S. fleet declined from more than 100 boats to 34 boats, while the foreign fleet rose from fewer than 10 to more than 70 boats. Foreign fishermen do not always abide by the porpoise-saving provisions of the act, but the U.S. Congress voted in 1988 to ban importation of tuna not caught in accordance with new methods designed to reduce the kill. In 1990, American tuna processing companies agreed to buy tuna only from fishermen whose methods do not result in the deaths of porpoises. American commercial fishermen have agreed to comply, and conservationists hope that the foreign fleets will follow their lead.

Fur-Bearing Mammals

Worldwide, between 400,000 and 500,000 seals and sea lions are taken annually for fur. Eight species of seal and one species of sea lion are of economic importance. By the turn of this century a few of these species had been hunted nearly to extinction, but a system of hunting quotas has allowed most of the populations to recover. A provision of the 1972 U.S. Marine Mammals Protection Act protects all fur-bearing marine mammals in U.S. territory except for the northern fur seal. Public pressure and an unfavorable economic climate contributed to the collapse of the Alaskan northern fur seal industry in 1986.

The harp seal, a species found in relatively great abundance on ice floes off the coasts of Canada in the Labrador and Barents seas, has attracted much popular attention in the last decade. Newborn harp seal pups are covered with dense, pure white fur. Each spring, several thousand are killed for their pelts. Because these pups are so attractive and helpless, much effort has been made to stop the slaughter. Harp seals are not an endangered species, however; the rate at which harp seals are harvested is considerably below their maximum sustainable yield, and the population continues to grow.

Botanical Resources

Marine plants are also commercially exploited. The most important commercial product is **algin,** made from the mucus that slickens seaweeds. When separated and purified, algin's long, intertwining molecules are used to stiffen fabrics; to form emulsions such as salad dressings, paint, and printer's ink; to prevent the formation of large crystals in ice cream; to clarify beer and wine; and to suspend abrasives. The U.S. seaweed gel industry produces more than $120 million worth of algin each year, and the annual worldwide value of products containing algin (and other seaweed substances) was estimated to be more than $35 billion in 1993. A harvesting barge is shown in **Figure 19.19.**

Seaweeds are also eaten directly. The Japanese consume 150,000 metric tons (165,000 tons) of *nori* each year; seaweed and seaweed extracts are also eaten in the United States, Britain, Ireland, New Zealand, and Australia. Their mineral content and fiber are useful in human nutrition.

Aquaculture

Aquaculture is the growing or farming of plants and animals in any water environment under controlled conditions. Aquaculture accounts for about 15% of the world's fish catch and 14% of the commercial harvest. About 10 million metric tons (14 million tons) of food are produced annually by freshwater aquaculture, mostly by China and the other countries of Asia. United States aquaculture data for 1991 are summarized in **Table 19.6.**

Mariculture is the farming of *marine* organisms, usually in estuaries, bays, or nearshore environments, or in

Figure 19.19 A kelp cutter harvesting kelp. Metal shears about a meter below the surface trim the kelp, and a moving ramp brings it aboard.

Figure 19.20 Oyster mariculture in Puget Sound. Oyster larvae attached to old oyster shells are put in screen bags, which are placed on intertidal mud flats. The bags must be lifted monthly to clean out excess mud. Oysters are ready for harvesting in about three years.

Species	Metric Tons[a]	Thousands of Pounds[a]	Thousands of Dollars
Finfish (FW)[b]			
Baitfish	9,608	21,182	55,948
Catfish	177,297	390,870	246,639
Salmon	7,599	16,753	44,156
Trout	26,954	59,422	59,142
Shellfish			
Clams (SW)[b]	1,716	3,784	11,133
Crawfish (FW)	27,481	60,585	33,285
Mussels (SW)	76	167	902
Oysters (SW)	9,241	20,373	61,851
Shrimp (FW)	184	406	2,407
Shrimp (SW)	1,600	3,527	14,110
Miscellaneous	**12,127**	**26,735**	**104,998**
Totals	**273,883**	**603,804**	**634,571**

Table 19.6 Estimated U.S. Aquaculture Production, 1991

Source: Fisheries Statistics Division, F/RE1, National Marine Fishery Service.

[a]Clams, oysters, and mussels are reported as meat weights (excluding shells); other species such as shrimp are reported as whole (live) weights.

[b]FW = freshwater, SW = seawater.

specially designed structures using circulated seawater (see **Box 19.2**). Mariculture facilities are sometimes placed near power plants to take advantage of the warm seawater flowing from their cooling condensers.

Worldwide mariculture production is thought to be about one-tenth that of freshwater aquaculture. Several species of fish, including plaice and salmon, have been grown commercially, and marine and brackish-water fish account for 67% of the total production. Several kinds of edible seaweeds are grown, generating 17% of mariculture production. The balance—16%—comes from crustaceans and mollusks, including shrimp, mussels, oysters, and abalone.

In the United States, annual revenue from mariculture is approaching $100 million, with most of the revenue generated from oyster mariculture (**Figure 19.20**). About half of the oysters consumed in North America are cultured. Not all mariculture produces food; cultured pearls are an important industry in Japan, China, and northern Australia. Japan leads the world in mariculture.

Ranching, or open-ocean mariculture, is an interesting variation. Juveniles are grown in a certain area, released into the ocean, and expected to return when mature. The natural homing instinct of salmon makes salmon ranching quite successful—about 20% of the world supply of salmon is ranched. Yet only about 1 in 50 of ranched juveniles return to the area of release. In Japan, recent attempts to extend ranching techniques to yellowtail and tuna have met with only limited success.

Mariculture is an expanding industry; it is growing at 8% annually compared to a slight decline for the world marine fishery as a whole. Mariculture produces mostly luxury seafoods such as oysters and abalone, and it uses fish meal from so-called *trash fish* as feed. (Trash fish are edible but considered unappealing because they are bony or bad tasting.) As world population increases and the demand for protein grows among the world's millions of undernourished people, however, today's trash fish may become more acceptable human food.

BOX 19.2 ● *Mariculture in Action: Ab Lab*

Abalone is a marine gastropod long prized as a delicacy, particularly in California and Japan. Divers and commercial fishermen have nearly exhausted the natural supply on both sides of the Pacific, however, and prices have risen sharply. Ab Lab, located on the California coast at Port Hueneme, was founded in 1972 to supply small abalone by mariculture. A few abalone are reintroduced into the wild, but most of Ab Lab's output is purchased by restaurants in the U.S. and in Japan. The live snails are shipped chilled in special packing containers to ensure freshness. Current production capacity is 400,000 100-gram, 82-millimeter (3¼-inch) red abalone per year.

Culture begins when eggs and sperm shed by captive breeding animals are combined. The resulting larvae are placed in shallow holding tanks for a short time and supplied with flowing seawater.

Soon they are transferred to nursery tanks (**Figure a**). Ab Lab uses 700 230-liter (60-gallon) fiberglass tanks to hold the baby abalone during the four to six months they require to reach a size of 6 millimeters (¼ inch). The snails grow on diatoms gleaned from the inner surface of the tank.

Finally, the abalone are moved to *raceways* (**Figure b**), shallow trays covered with seaweed taken from nearby kelp-growing areas leased from the State of California. The abalone stay in these raceways, eating seaweed and enjoying a life of ease in an aerated, predator-free, circulating seawater environment for another three years until they have grown to the market size of 8.2 centimeters (3¼ inches) (**Figure c**). Restaurants pay about $30.00 per shipped kilo for the animals—about $3.00 each.

a Nursery tanks holding baby abalone.

b Abalone raceways.

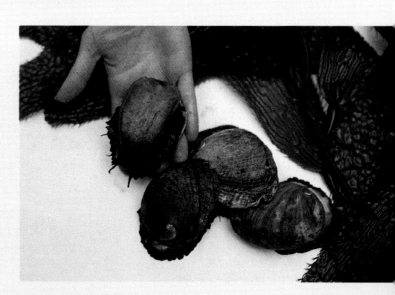

c Market-size red abalone.

Drugs

The earliest recorded use of medicines derived from marine organisms appears in the *Materia Medica* of the emperor Shen Nung of China, 2700 B.C. Modern medical researchers estimate that perhaps 10% of all marine organisms are likely to yield clinically useful compounds. One such medicine is derived from a Caribbean sponge and is already in use: Acyclovir, the first antiviral compound approved for humans, has been fighting herpes infections of the skin and nervous system since 1982.

Newly discovered compounds are also being tested. A common bryozoan—a small encrusting invertebrate—has been found to produce a potent anticancer chemical that is being prepared for human testing. Extracts from 30% of all tunicate species investigated show antiviral and antitumor activity; one of these extracts, Didemnin-B, shows promise as a treatment for malignant melanoma, the deadliest form of skin cancer. A South Pacific sponge makes a chemical that can kill *Candida,* a fungus that causes vaginitis and the throat infection known as *thrush.* A species of coral produces a powerful anti-inflammatory painkiller, a promising and nonaddictive drug for arthritis. Another compound derived from cyanobacteria stimulated the immune system of test animals by 225% and cells in culture by 2,000%; the drug may be useful in treating AIDS. Vidabarine, another antiviral drug developed from sponges, may attack the AIDS virus directly. Human trials of some of these drugs are already under way. In the not-too-distant future, drugs from marine organisms may be among our most valuable marine resources.

Other useful chemicals derived from marine organisms include Padan, a powerful new insecticide derived from an annelid worm, and a new family of steroids that may prove to be useful in the regulation of human growth.

NONEXTRACTIVE RESOURCES

Transportation and recreation are the main nonextractive resources the oceans provide. People have been using the ocean for transportation for thousands of years. Through most of this time the transport of cargo has produced far more revenue than the movement of passengers. At present, oil tankers ship the greatest gross tonnage of any type of cargo (over 260 million metric tons—287 tons—annually). Oil accounts for about 51% of the total value of world trade transported by sea; iron, coal, and grain make up 25% of the rest.

Nearly half of the world's crude oil production is transported to market by ships. Tankers are needed because very few of the major oil-drilling sites are close to areas where the demand for refined oil products is highest. The largest tankers (**Figure 19.21**) are more than 430 meters (1,300 feet) long and 66 meters (206 feet) wide, and they carry more than 500,000 metric tons (3.5 million barrels) of oil. The complexity of international crude oil trade routes (and volumes transported) is shown in **Figure 19.22**.

Modern harbors are essential to transportation. Cargoes are no longer loaded and off-loaded piece by piece by teams of longshoremen. Today's harbors bristle with automated bulk terminals, high-volume tanker terminals (both offshore and dockside), containership facilities, roll-on–roll-off ports for automobiles and trucks, and passenger facilities required by the growing popularity of cruising. Most of this specialized construction has occurred since 1960. New Orleans is now the greatest U.S. port; nearly 180 million tons of cargo—most of it grain—passed through its docks in 1992.

Transportation is sometimes combined with recreation. In the last decade, the cruise industry has experienced spectacular growth. Passengers on luxurious liners (**Figure 19.23**) can enjoy a few relaxing days on the

Figure 19.21 *Murex*, first of a series of five VLCCs (very large crude carriers) being built for the Shell Oil Company by Daewoo Heavy Industries, South Korea. Each ship is of double-hulled design and will carry 2.15 million barrels of crude oil at a service speed of 28 kilometers (18 miles) per hour. Somewhat smaller than the largest tankers, *Murex* is 332 meters (1,089 feet) long and extends 22 meters (72 feet) below the surface when fully loaded. The ship, built to high standards of reliability and safety, is seen here unballasted on the day of its naming, 17 January 1995.

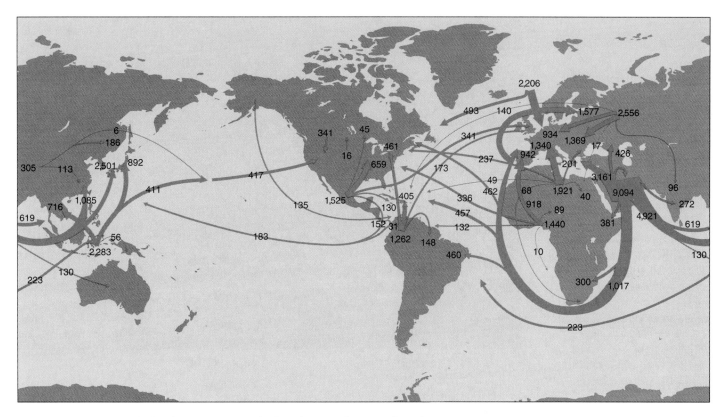

Figure 19.22 The complexity of international crude oil trade routes in 1984; all numbers are in thousands of barrels per day. Arrows indicate origins and destinations of the major shipments, but not necessarily specific routes.

Figure 19.23 The Cunard liner *Queen Elizabeth 2,* one of the largest ocean liners ever built and the only passenger ship in regular transatlantic service.

Figure 19.24 Public aquariums and marine parks are important resources.

ocean crossing the North Atlantic, visiting tropical islands, or touring places accessible to the public only by ship. Indeed, tourism is now the world's largest industry. Ocean-related leisure pursuits, including sport fishing, surfing, diving, day cruising, sunbathing, dining in seaside restaurants, and just plain relaxing contribute to the economy. In addition to being important producers of revenue, public aquariums and marine parks (like Sea World) are centers of education, research, and captive breeding programs (**Figure 19.24**). Even public interest and curiosity about whales are a source of recreational revenue. Whale-watching trips generate an annual revenue of $60 million; dolphin displays, whale artwork and books, and conservation group donations amount to another $50 million.

THE LAW OF THE SEA

Prehistoric peoples living near the shore were the earliest users of marine resources. With the rise of nation-states and military establishments, the fight for control of marine resources began. Some nations assumed that the ocean belonged to all and endeavored to guarantee free right to passage and resources. Others decided that it belonged to none and tried to control access to ports and resources by force. In 1604 Hugo Grotius, a learned Dutch jurist, wrote *De Jure Praedae* (On the law of prize and booty), a treatise justifying the action of a Dutch admiral who had successfully defended Dutch trading

rights in a dispute with Portugal. One chapter of this work, in which Grotius defended free ocean access for all nations, was reprinted in 1609 under the title *Mare Liberum* (A free ocean). *Mare Liberum* formed the basis for all modern international laws of the sea.

About a century later, in 1703, the concept of territorial seas adjacent to land was recognized. A country's seaward boundary was set at about 5 kilometers (3 miles)— the distance a cannonball could be fired from shore. This 5-kilometer (3-mile) limit stood until 1945.

The United Nations and the International Law of the Sea

After World War II the technology became available to search for oil and natural gas on continental shelves. After U.S. oil companies found rich deposits beyond the 5-kilometer (3-mile) limit off Louisiana, President Harry Truman issued a proclamation annexing the physical and biological resources of the continental shelf contiguous to the United States. Other nations rushed to make similar claims.

The United Nations then became involved. A committee of the General Assembly began to formulate policy, which was later presented at the First United Nations Conference on Law of the Sea in 1958 in New York. Twenty-four years of effort by delegates from many interested nations resulted in the 1982 Draft Convention on the Law of the Sea. In April 1982 the United Nations adopted the convention by a vote of 130 to 4, with 17 abstentions. (The United States, Turkey, Venezuela, and Israel voted against the convention.) By 1988 more than 140 countries had signed all or most parts of the treaty. It is now legally binding, but signatories have selectively chosen to respect or ignore its individual provisions.

Here are some important features of the 1982 Draft Convention:

- **Territorial waters** are defined as extending 12 miles (19.2 kilometers) from shore. A nation has the right to jurisdiction within its territorial waters. Straits used for international navigation are excluded from a nation's territorial waters in that any vessel has the right to innocent passage.

- The 200 nautical miles (370 kilometers) from a nation's shoreline constitute its **exclusive economic zone (EEZ).** Nations hold sovereignty over resources, economic activity, and environmental protection within their EEZs.

- All ocean areas outside the EEZs are considered the **high seas.** In tradition, the high seas are common property to be shared by the citizens of the world. An International Seabed Authority was established to oversee the extraction of mineral resources from the deep sea.

- The values of protecting the ocean and preventing marine pollution were endorsed.

- Subject to some conditions, the freedom of scientific research in the ocean was encouraged.

The convention places about 40% of the world ocean under the control of the coastal countries, within the EEZs. The resources of the remaining 60%, the **high seas,** are to be shared by the citizens of the world.

The United States Exclusive Economic Zone

The United States did not sign the 1982 Draft Convention for a variety of reasons. Among these was concern that private enterprise would be deprived of profits if it were made to share high seas resources with other countries. Instead, the United States unilaterally claimed sovereign rights and jurisdiction over all marine resources within its own 200-nautical-mile region, which it called the **United States Exclusive Economic Zone.** The proclamation—similar in most ways to the 1982 United Nations Treaty but lacking the provision of shared high seas resources—was signed by President Ronald Reagan on 10 March 1983. The United States EEZ brings within national domain over 10.3 million square kilometers (4 million square miles) of continental margins, an area 30% larger than the land area of the United States and a region of diverse geological and oceanographic settings (see **Figure 19.25**).

The first step in exploring this new region has been to map the surface of the seafloor, a project that is still going on. The first bottom surveys were conducted off the West Coast of the United States because of the energy and mineral resources known to exist there. Massive deposits of various metallic sulfides are being explored

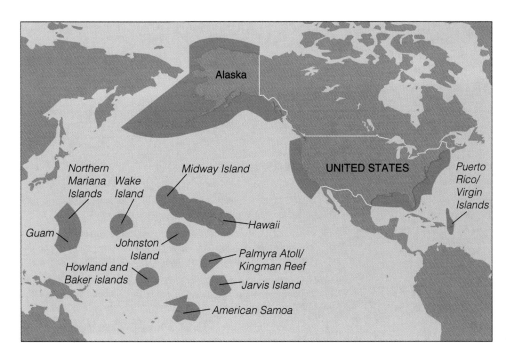

Figure 19.25 The U.S. Exclusive Economic Zone (EEZ), shown here in dark blue, covers a vast area of the continental margin. The edge of the continent as it extends out under the ocean contains a wealth of resources.

at the hydrothermal vents on the Gorda and Juan de Fuca ridges. Over 100 previously unknown volcanoes have been mapped within the U.S. EEZ off our West Coast. Huge faults, submarine landslides, seamounts, and details of the spreading crest of the oceanic ridges are a few of the features that have been discovered so far. Knowledge of the modern tectonic setting for ore formation has been valuable in locating new ore deposits on land.

Manganese nodules and crusts have been discovered within the United States EEZ off the East and West coasts, Hawaii, and the Pacific Island territories. Cobalt, present in the manganese crusts, is being studied to determine how the deposit is formed and how it can be retrieved economically.

The EEZ project also includes research into meteorology, accurate weather forecasting, environmental studies, effects of plate movement on the ocean floor, effects of mining on seafloor organisms, and geohazards such as submarine landslides and earthquakes. It is hoped that this new era of marine exploration and research will lead to informed decisions about offshore activities that will benefit not only the United States but other maritime nations. Maybe the treasure trove is really there, waiting for these new developments in marine technology and international cooperation to tap the riches of the sea.

Q-AND-A

1. When will we run out of oil?

   By the time the world's crude oil supplies have been substantially depleted, the total amount of oil extracted is expected to range from 1.6 to 2.4 trillion barrels. If this estimate is correct, consumption could continue at the present level until some time between 2025 and 2040, at which time it would drop quite rapidly. But, in fact, we will never run completely out of oil—there will always be some oil within the Earth to reward great effort at extraction. The days of unlimited burning of so valuable a commodity are nearing an end, however. Future civilizations will surely look back in horror at the fact that their ancestors actually burned something as valuable as lubricating oil.

2. How does the sea salt I see in the health food stores differ from regular table salt?

   Unlike the producers of regular table salt, the producers of commercial sea salt do not move the evaporating brine from pond to pond to isolate the different precipitates. Sea salt therefore contains the ocean's salts in their natural proportions, with NaCl making up 78% of the mix. Sea salt has a slightly bitter taste from the potassium and magnesium salts present. Some people believe that the variety of minerals in sea salt is beneficial, but most people obtain adequate amounts of these minerals from a normal diet.

3. You mentioned treasure at the beginning of the chapter. What about treasure as a marine resource?

   Florida entrepreneur Mel Fisher has recovered nearly half a billion dollars in gold and archaeological treasures from sunken ships along the U.S. southeast coast. (He believes there's more where that came from.) In 1981, a British salvage company recovered $80 million in gold ingots from a British warship sunk in 1942 off Murmansk. Private consortia are currently considering plundering the hulks of the RMS Titanic *and the German battleship* Bismarck. *Even priceless archaeological treasures are being excavated and sold to the highest bidder. Treasure is not always economic—the scholarly value of an ancient shipwreck is beyond monetary calculation.*

Terms and Concepts to Remember

algin	mariculture	physical resources
aquaculture	marine energy	potable water
biological resources	resources	renewable
commercial	maximum sustain-	resources
extinction	able yield	territorial waters
desalination	nonextractive	United States
drift net	resources	Exclusive Eco-
exclusive economic	nonrenewable	nomic Zone
zone (EEZ)	resources	
high seas	overfishing	

Study Questions

1. Distinguish between physical and biological resources, and between renewable and nonrenewable resources.

2. What are the three most valuable physical resources? How does the contribution of each to the world economy compare to the contribution of that resource derived from land?

3. How are oil and natural gas thought to be formed? How can these substances be extracted from the seabed? Why are the physical characteristics of the surrounding rock important?

4. What are the sources of metals mined or extracted (or potentially mined or extracted) from the sea?

5. What method of ocean energy extraction has been most practical? Which has the greatest potential? Why has the latter not been exploited on a large scale?

6. Does the ocean provide a substantial percentage of *all* protein needed in human nutrition? Of all *animal* protein? What is the most valuable biological resource? The fastest growing fishery?

7. What are the signs of overfishing? How does the fishing industry often respond to these signs? What is the usual result?

8. Distinguish between mariculture and aquaculture. Does freshwater or seawater aquaculture provide more human food, worldwide? What are the prospects for mariculture?

9. What is a nonextractive resource? Give some examples.

10. What are the advantages and disadvantages of a proclaimed EEZ? Do you feel the United States was justified in proclaiming its own EEZ separate from the provisions of the 1982 United Nations Convention?

For Further Study

Avery, W. H., and C. Wu. 1994. *Renewable Energy from the Ocean.* New York: Oxford University Press.

Batten, M. 1989. "Krill: Keystone of the Antarctic Ecosystem." *Calypso Log* 16 (no. 2, April): 17–18.

Borgese, E. M. 1983. "The Law of the Sea." *Scientific American,* March, 42. Historical viewpoint.

Broadus, J. M. 1987. "Seabed Materials." *Science* 235 (no. 4791): 853–60. Thorough summary of material resources.

Constans, J. 1979. *Marine Sources of Energy.* United Nations Department of International Economic and Social Affairs, OST. Elmsford, NY: Pergamon Press. Source of Table 19.1.

Cousteau, J. P. 1988. *Whales.* New York: Abrams. Source of some of the statistics on whaling.

Craig, J. R., D. J. Vaughan, and B. J. Skinner. 1988. *Resources of the Earth.* Englewood Cliffs, NJ: Prentice-Hall.

Faulkner, D. J. 1979. "The Search for Drugs from the Sea." *Oceanus* 22 (no. 2): 44–50.

Hawkes, N. 1990. *Structures.* New York: Macmillan. Information on construction of large oil platforms.

Holmes, B. 1994. "Biologists Sort the Lessons of Fisheries Collapse." *Science* 264 (27 May): 1252–53.

Hunt, J. M. 1981. "The Origin of Petroleum." *Oceanus* 24 (no. 2): 53–57.

Isaacs, J. D., and R. J. Seymour. 1973. "The Ocean as a Power Resource." *International Journal of Environmental Studies* 4: 201–5. Wave energy devices are proposed.

Kraus, S. D. 1989. "Whales for Profit." *Whalewatcher* 23 (no. 2, Summer). Total value of whaling, 1988.

Mangone, G. L. 1988. *Marine Policy in America.* 2d ed. New York: Taylor & Francis.

Mangone, G. L. 1991. *Concise Marine Almanac.* 2d ed., rev. New York: Taylor & Francis.

Miller, G. T. 1990. *Resource Conservation and Management.* Belmont, CA: Wadsworth.

Milne, D. 1995. *Marine Life and the Sea.* Belmont, CA: Wadsworth. A particularly thorough discussion of the present state of world fisheries.

O'Bannon, B. 1993. *Fisheries of the United States, 1992.* Washington, DC: U.S. Department of Commerce.

Penney, T. R., and D. Bharathan. 1987. "Power from the Sea." *Scientific American,* January, 86–93. A thorough discussion of the potential of warm seawater as a source of electrical energy.

Richards, W. E., and J. R. Vadus. 1980. "Ocean Thermal Energy Conversion: Technology Development." *Marine Technology Society Journal* 14 (no. 1, February–March): 3–14.

Ross, D. A. 1980. *Opportunities and Uses of the Ocean.* New York: Springer-Verlag. Excellent and concise coverage of marine resources.

Scheuer, P. J. 1990. "Some Marine Ecological Phenomena: Chemical Basis and Biomedical Potential." *Science* 248 (no. 4952): 173–77. Information on recent drugs and other useful chemicals derived from marine organisms.

United States Department of the Interior. 1993. *Minerals Yearbook, 1992.* Washington, DC: U.S. Government Printing Office.

World Resources Institute. 1986. *World Resources 1986.* New York: Basic Books.

Special Issues of Journals and Magazines

Oceanus ("Deep Sea Mining"), 1982, vol. 25, no. 3. Though now somewhat dated, it is highly recommended for the excellent photographs and articles on many phases of deep-ocean mining.

Oceanus ("Industry and the Oceans"), 1984, vol. 27, no. 1. Covers topics such as salmon ranching, kelp forest management, genetic engineering in the marine sciences, and more.

Oceanus ("The Exclusive Economic Zone"), 1984–85, vol. 27, no. 4. Contains articles on both living and mineral resources of the EEZ.

Scientific American ("Meeting the Challenge of Sustainable Development"), June 1992.

20 ENVIRONMENTAL CONCERNS

Fouling Our Own Nest

Why do we enjoy being near the ocean? Perhaps for the same reasons we keep aquariums or hang pictures of waves or dolphins on the walls of our homes and offices. We perceive the ocean as a pure, unspoiled, restlessly natural place; a refuge of calm in an increasingly complex, noisy, odorous, dangerous existence; a breath of crisp air, a school of fish near a diver's mask, a symbol of more natural times. It is sunsets and fog, waves to ride, and cold, clean spray in our faces. We love the ocean's freedom and strength, its feel on our skin, its changing moods.

How, then, can we explain our treatment of this revered place? We take solace from the ocean, but we rarely consider the effects of our behavior on the ocean and atmosphere. We seldom think of trash after it is thrown away, yet soon the highest point on the coast between Maine and the tip of Florida will be the growing mountain of refuse on Staten Island, which receives 28,000 tons of solid New York City waste each day. We accept the benefits of air conditioning and the convenience of plastic foam containers, yet chemical compounds associated with these luxuries appear to be responsible for a reduction in stratospheric ozone, which is responsible for blocking most of the dangerous ultraviolet radiation from the sun. We don't think twice about driving our cars or powering our industries with fossil fuels, yet their burning has caused the carbon dioxide content of the atmosphere to increase rapidly, which may be raising the Earth's surface temperature. Some of us pour used crankcase oil into sewers. We grumble at the inconvenience of pollution controls on gas pumps or automobile engines and think recycling aluminum cans or newspapers is just too much trouble. We vote against slow-growth initiatives, wind- and solar-power incentives, and bond issues for sewage treatment plants, cogeneration facilities, and mass transit. We willingly visit the ocean for recreation, relaxation, and spiritual renewal, yet by our actions or by our neglect we are helping to damage or destroy it.

In this chapter we will survey the pollutants, the habitat destruction, the mismanagement of living resources, and the global changes that are currently stressing the marine environment. It's a depressing list, but at the end you'll find a glimmer of hope.

Bits of plastic and oily debris litter the approach to a salt marsh.

Our species has always exercised its capacity to consume resources and pollute its surroundings, but only in the last few generations have our efforts affected the ocean and atmosphere on a planetary scale. The introduction into the biosphere of unnatural compounds (or natural compounds in unnatural quantities) has had and will continue to have unexpected detrimental effects. The destruction of marine habitats and the uncontrolled harvesting of the ocean's living resources have also disturbed delicate ecological balances. We find ourselves in difficult situations, for which solutions do not come easily.

MARINE POLLUTION

We define **marine pollution** as the introduction into the ocean by humans of substances or energy that changes the quality of the water or affect the physical and biological environment. It is not always easy to identify a pollutant; some materials labeled as pollutants are produced in large quantities by natural processes. For example, a volcanic eruption can produce immense quantities of carbon dioxide, methane, sulfur compounds, and oxides of nitrogen. Excess amounts of these substances produced by human activity may cause global warming and acid rain. For this reason we need to distinguish between *natural pollutants* and *human-generated* ones.

We have long used the sea as a dump for our wastes. About three-quarters of the pollution entering the ocean comes from human activities on land (**Table 20.1**). The ocean's great volume and relentless motion dissipate and distribute natural and synthetic substances, but its ability to absorb is not inexhaustible. No one knows to what extent we have contaminated the ocean. By the time the first oceanographers began widespread testing, the Industrial Revolution was well under way and changes had already occurred. Traces of synthetic compounds have now found their way into every oceanic corner. It is sad to consider that we will never know what the natural ocean was like nor what remarkable plants and animals may have vanished as a result of human activity. Our limited knowledge of pristine conditions is gleaned from small seawater samples recovered from deep within the polar ice pack and from tiny air bubbles trapped in glaciers. There are few undisturbed habitats left to study, and few marine organisms are completely free of the effects of ocean pollutants.

Characteristics of a Pollutant

A **pollutant** causes damage by interfering directly or indirectly with the biochemical processes of an organism. Some pollution-induced changes may be instantly lethal; other changes may weaken an organism over weeks or months, or alter the dynamics of the population of which it is a part, or gradually unbalance the entire community.

In most cases an organism's response to a particular pollutant will depend on its sensitivity to the combination of *quantity* and *toxicity* of that pollutant. Some pollutants are toxic to organisms in tiny concentrations. For example, the photosynthetic ability of some species of diatoms is diminished when chlorinated hydrocarbon compounds are present in parts-per-trillion quantities. Other pollutants may seem harmless, as when fertilizers flowing from agricultural land stimulate plant growth in estuaries. Still other pollutants may be hazardous to some organisms but not to others. For example, crude oil

Table 20.1 Sources of Marine Pollution

Source	Share of Total (percent)
Runoff and discharges from land	44
Airborne emissions from land	33
Shipping and accidental spills	12
Ocean dumping	10
Offshore mining, oil and gas drilling	1
All sources	100

Source: Joint Group of Experts on the Scientific Aspects of the Marine Environment, *The State of the Marine Environment,* UNEP Regional Seas Reports and Studies No. 115 (Nairobi: U.N. Environment Programme, 1990).

interferes with the delicate feeding structures of zooplankton and coats the feathers of birds, but it simultaneously serves as a feast for certain bacteria.

Pollutants also vary in their *persistence;* some reside in the environment for thousands of years, while others last only a few minutes. Some pollutants break down into harmless substances spontaneously or through physical processes (like the shattering of large molecules by sunlight). Sometimes pollutants are removed from the environment through biological activity. For example, some marine organisms escape permanent damage by metabolizing hazardous substances to harmless ones. Indeed, many pollutants are ultimately **biodegradable**—that is, able to be broken down by natural processes into simpler compounds. Most pollutants resist attack by water, air, sunlight, or living organisms, however, because the synthetic compounds of which they are composed resemble nothing in nature.

The ways in which pollutants are changing the ocean and the atmosphere are often difficult for researchers to determine. Environmental impact cannot always be predicted or explained. As a result, marine scientists vary widely in their opinions about what pollutants are doing to the ocean and atmosphere and what to do about it. Environmental issues are frequently emotional, and media reports tend to sensationalize short-term incidents (like oil spills) rather than more serious, long-term problems (like atmospheric changes or the effects of long-lived chlorinated hydrocarbon compounds).

Oil

Oil is a natural part of the marine environment. Oil seeps have been leaking large quantities of oil into the sea for millions of years. The amount of oil entering the ocean has increased greatly in recent years, however, because

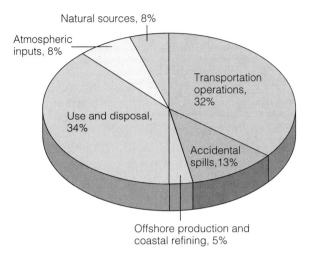

Figure 20.1 Sources of ocean oil pollution.

Natural sources, 8%

Atmospheric inputs, 8%

Transportation operations, 32%

Use and disposal, 34%

Accidental spills, 13%

Offshore production and coastal refining, 5%

Figure 20.2 A bird coated with crude oil. Such birds die unless the oil is removed with a detergent solution. Many will die of cold or toxic effects even when the oil is removed. About half a million seabirds die each year from the effects of oil pollution.

of our growing dependence on marine transportation for petroleum products, offshore drilling, nearshore refining, and street runoff carrying waste oil from automobiles (**Figure 20.1**).

In the early 1990s people were using nearly 5.3 billion metric tons (5.8 billion tons) of crude oil each year, about half of which was transported to market in large tankers. In 1993 about 3 million metric tons (3.3 million tons) of oil entered the world ocean. Natural seeps accounted for only about 8% of this annual input, about a quarter of a million metric tons. Some 45% of the total was associated with marine transportation. Most of this oil—32%—was not spilled in well-publicized tanker accidents but was released intentionally, quietly, and routinely, during the loading, discharging, and flushing of tanker ships. This oil is particularly harmful to seabirds: Between 150,000 and 450,000 marine birds are killed in the North Sea and North Atlantic regions each year by the routine release of oil from tankers (**Figure 20.2**). It's no wonder that a sea surface completely free of an oil film is quite rare.

It is difficult to generalize about the effects that a concentrated release of oil—an oil spill from a tanker, coastal storage facility, or well—will have in the marine environment. The consequences of a spill vary according to several factors: its location and proximity to shore; the quantity and composition of the oil; the season of the year, currents, and weather conditions at the time of release; and the composition and diversity of the affected communities. Intertidal and shallow-water subtidal communities are most sensitive to the effects of an oil spill.

Spills of *crude* oil are generally larger in volume and more frequent than spills of refined oil. Most components of crude oil do not dissolve easily in water, but those that do can harm the delicate juvenile forms of marine organisms even in minute concentrations. The remaining insoluble components form sticky layers on the surface that prevent free diffusion of gases, clog adult organisms' feeding structures, kill larvae, and decrease the sunlight available for photosynthesis. Even so, crude oil is not highly toxic, and it is biodegradable. Though crude oil spills look terrible and generate great media attention, most forms of marine life in an area recover from the effects of a moderate spill within about five years. For example, the 126 million gallons of light crude oil released into the Persian Gulf during the 1991 Gulf War (see **Table 20.2**) have dissipated relatively quickly and will probably cause little long-term biological damage.

Spills of *refined* oil, especially near shore where marine life is abundant, can be more disruptive for longer periods of time. The refining process removes and breaks up the heavier components of crude oil and concentrates the remaining lighter, more biologically active ones. Components added to oil during the refining process also make it more deadly. Spills of refined oil are of growing concern because the amount of refined oil transported to the U.S. has risen dramatically through the 1980s and 1990s. A 1969 spill of refined oil at West Falmouth, Massachusetts, killed countless marine organisms and caused grave environmental damage that is still apparent in the

Table 20.2 The 23 Worst Oil Spills in History

The *Exxon Valdez* disaster of March 1989 was the second worst in U.S. history. But when ranked with other massive oil spills involving both shipping and well accidents worldwide, it falls to the bottom of the top 23.

Date	Incident and Location	Size of Spill (millions of gallons)
3 June 1979	Well spill in Bay of Campeche off coast of Mexico	140
19–28 Jan. 1991	Discharged into Persian Gulf during Gulf War	126
4 Feb. 1983	Well spill at Nowruz, Iran, that dumped oil into Persian Gulf	80
6 Aug. 1978	Fire aboard *Castillo de Vellver*, off Cape Town, South Africa	78.5
16 March 1978	Tanker *Amoco Cadiz* ran aground off coast of France	68.7
19 July 1979	Collision of two ships off Trinidad and Tobago, *Atlantic Empress* and *Aegean Captain*	48.8
Aug. 1980	Leak at Well No. D103, Libya	42
2 Aug. 1979	Wreck of *Atlantic Empress*, Barbados	41.5
23 Feb. 1980	Wreck of *Irenes Serenade*, Greece	36.6
20 Aug. 1981	Leak at Kuwait National Petrol Tank, Kuwait	31.2
15 Nov. 1979	Wreck of *Independence*, Istanbul, Turkey	28.9
25 May 1978	Leak at No. 126 well/pipe, Iran	28
Jan. 1993	Wreck, *Braer,* Northern Scotland	25
6 July 1979	Leak at British Petroleum storage tank, Nigeria	23.9
15 Aug. 1985	Wreck of *Nova*, Kharg Island, Persian Gulf	21.4
11 Dec. 1978	Leak at BP-Shell fuel depot, Zimbabwe	20
4 Dec. 1992	Wreck of *Aegean Sea*, Spain	18
7 Jan. 1983	Wreck of *Assimi*, off Oman	15.8
12 June 1978	Tohoku Oil Co., Japan	15
31 Dec. 1978	Wreck of *Andros Patria*, Spain	14.6
10 Dec. 1983	Wreck of *Peracles*, Qatar	14
Jan. 1975	Spill from *Corinthos*, Delaware River	11.2
March 1989	Wreck of *Exxon Valdez*, Alaska	10.8

Sources: Oil Spill Intelligence Report; National Public Radio; *Los Angeles Times,* 28 Jan. 1991, 24 March 1994.

form of reduced species diversity and fertility. In a similar accident in Mexico, the tanker *Tampico* went aground and dumped her load of refined gasoline into an unspoiled Baja California bay. The area still shows biological scars after nearly three decades.

The volatile components of any oil spill eventually evaporate into the air, leaving the heavier tars behind. Wave action causes the tar to form into balls of varying sizes. Some of the tar balls fall to the bottom, where they may be assimilated by bottom organisms or incorporated into sediments. Bacteria will eventually decompose these spheres, but the process may take years to complete, especially in cold polar waters. This oil residue—especially if derived from refined oil—can have long-lasting effects on seafloor communities. The fate of spilled oil is summarized in **Figure 20.3.**

The methods used to contain and clean up an oil spill sometimes cause more damage than the oil itself. Deter-

gents used to disperse oil are especially harmful to living things. Cleanup of the 1969 *Torrey Canyon* accident off the southern coast of England, one of the first large tanker accidents, did much more environmental damage than the 100,000 metric tons (110,000 tons) of crude oil released. Some resort beaches in the south of England were closed for two seasons, not because of oil residue but because of the stench of decaying marine life killed by the chemicals used to make the shore look clean.

Even the more sophisticated methods that were used in dealing with the *Exxon Valdez* disaster, the second worst oil spill in U.S. history (and the 23rd worst spill ever; see Table 20.1), seem to have done more harm. The supertanker *Exxon Valdez* ran aground in Alaska's Prince William Sound on 24 March 1989. More than 40 million liters (almost 10.8 million gallons; 29,000 metric tons) of Alaskan crude oil—about 22% of her cargo—escaped from the crippled hull (see **Figure 20.4**). Only about 17%

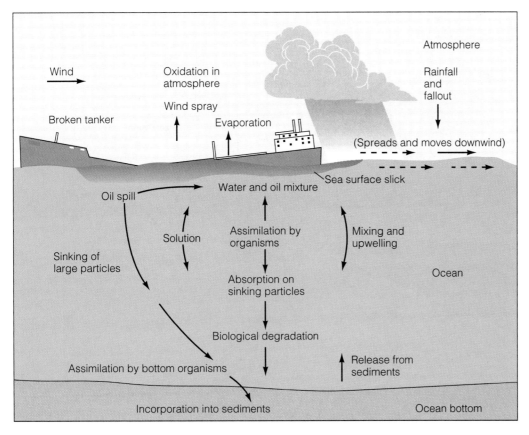

Figure 20.3 Some of the possible pathways that an oil spill can travel.

Figure 20.4 Cleaning up the March 1989 *Exxon Valdez* oil spill, Prince William Sound, Alaska.

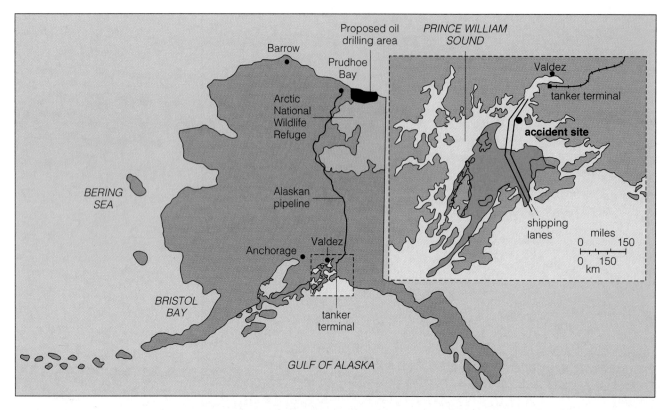

Figure 20.5 Site of the *Exxon Valdez* spill. About 10.8 million gallons of crude oil fouled 450 kilometers (300 miles) of Alaskan coastline.

of this oil was recovered by a work crew of more than 10,000 people using containment booms, skimmer ships, bottom scrapers, and absorbent sheets. About 35% of the oil evaporated, 8% was burned, 5% was dispersed by strong detergents, and 5% biodegraded in the first five months. The rest of the oil, some 30% of the spill, formed oil slicks on Prince William Sound and fouled more than 450 kilometers (300 miles) of coastline (**Figure 20.5**).

Recent analysis of the affected parts of Prince William Sound shows the cleaned areas to be in generally worse shape than areas left alone. Most of the small animals that make up the base of the food chain in these areas were cooked by the 65°C (150°F) water used to blast oil from between the rocks. Others were smothered when the high-pressure jets rearranged sand and mud. It appears that an overambitious cleanup program can be counterproductive. Sylvia Earle, chief on-site scientist of the National Oceanic and Atmospheric Administration (NOAA), has said, "Sometimes the best, and ironically the most difficult, thing to do in the face of an ecological disaster is to do nothing." The biological cost of the spill will not be known for a decade. The cost of the cleanup has exceeded $3.5 billion. The legal costs could be even higher: In 1994 a federal jury awarded $5 billion in puni-

tive damages to 14,000 plaintiffs including fishermen, native villagers, and fish processors. Exxon has appealed the verdict.

Of course the best way to deal with oil pollution is to prevent it from happening in the first place. Tanker design is being modified to limit the amount of oil intentionally released in transport. Legislation is being considered that would limit new tanker construction to stronger, double-hull designs (although shipping experts are uncertain that even a double-hull design could have prevented the *Exxon Valdez* spill). Perhaps most important, crew testing and training are being upgraded.

Heavy Metals

Small quantities of heavy metals are capable of causing damage to organisms by interfering with normal cell metabolism. Among the dangerous heavy metals being introduced into the ocean are mercury and lead. Human activity releases about 5 times as much mercury and 17 times as much lead as is derived from natural sources, and incidents of mercury and lead poisoning—major causes of brain damage and behavioral disturbances in children—have increased dramatically over the last two decades. Lead particles from industrial wastes, landfills,

Table 20.3	Synthetic Organic Chemicals That Have Been Detected in the Ocean
Name	**Major Health Effects**
Aldicarb (Temik)	High toxicity to the nervous system
Benzene	Chromosomal damage, anemia, blood disorders, and leukemia
Carbon tetrachloride	Cancer; liver, kidney, lung, and central nervous system damage
Chloroform	Liver and kidney damage; suspected cancer
Dioxin	Skin disorders, cancer, genetic mutations
Ethylene dibromide (EDB)	Cancer and male sterility
Polychlorinated biphenyls (PCBs)	Liver, kidney, and lung damage
Trichloroethylene (TCE)	In high concentrations, liver and kidney damage, central nervous system depression, skin problems, and suspected cancer and mutations
Vinyl chloride	Liver, kidney, and lung damage; lung, cardiovascular, and gastrointestinal problems; cancer and suspected mutations

Source: Miller, 1994.

and gasoline residue reach the ocean through runoff from land during rains, and the lead concentration in some shallow-water, bottom-feeding species is increasing at an alarming rate. Consumers should be wary of seafoods taken near shore in industrialized regions.

Copper, another heavy metal, is so effective in killing marine organisms that it has long been used in marine antifouling paints. The ship *Pac Baroness,* a freighter carrying 21,000 metric tons (23,000 tons) of finely powdered copper, sank in 448 meters (1,480 feet) of water off the coast of central California after a collision in 1987. Her toxic cargo was scattered over the seabed, and a plume of copper-tainted water has been detected 40 kilometers (24 miles) down-current from the wreck. Marine life has been significantly disrupted in the area, which is a major fishing zone for Dover sole and rock cod. A similar incident off Holland in 1965 killed more than 100,000 fish and destroyed commercial mussel beds.

Canadian and Norwegian researchers have found that coal combustion, electric utilities, steel and iron manufacturing, fuel oils, fuel additives, and incineration of urban refuse are the major sources of oceanic and atmospheric contamination by heavy metals. They now believe that the toxicity of heavy metals is a greater health problem than the toxicity of all organic and radioactive wastes. In the wellness-conscious 1990s, fish are seen as a safe and healthful food. But with the ocean still receiving heavy metal–contaminated runoff from the land, a rain of pollutants from the air, and the fallout from shipwrecks, we can only wonder how much longer most seafood will be safe to eat.

Synthetic Organic Chemicals

Heavy metals are not the only poisons that may contaminate seafood. Many different synthetic organic chemicals also enter the ocean and become incorporated into its organisms. Some of these synthetic organic substances are listed in **Table 20.3**. Ingestion of even small amounts of these compounds can cause illness or even death.

Halogenated hydrocarbons—a class of synthetic hydrocarbon compounds that contain chlorine, bromine, or iodine—are used in pesticides, flame retardants, industrial solvents, and cleaning fluids. The concentration of **chlorinated hydrocarbons**—the most abundant and dangerous halogenated hydrocarbons—is so high in the water off New York State that officials have warned women of childbearing age and children under 15 not to consume more than half a pound of local bluefish a week. (They are told *never* to eat striped bass caught in the area.) One administrator for the U.S. Environmental Protection Agency has written, "Anyone who eats the liver from a lobster taken from an urban area is living dangerously."

The level of synthetic organic chemicals in seawater is usually very low, but some organisms at higher levels in the food chain can concentrate these toxic substances in their flesh. This **biological amplification** is especially hazardous to top carnivores in a food web.

The damage caused by biological amplification of DDT, a chlorinated hydrocarbon pesticide, is particularly instructive. In the early 1960s, brown pelicans began producing eggs with thin shells containing less than normal amounts of calcium carbonate. The eggs broke easily, no

BOX 20.1 ● *Minamata's Tragedy*

For the people of Minamata, Japan, who were poisoned by mercury released into the ocean from a nearby factory, heavy metal pollution is a continuing horror story. Between 1953 and 1960 more than 100 people who ate shellfish taken from Minamata Bay were afflicted by a form of mercury poisoning now called Minamata disease (**Figure a**). Their symptoms included kidney damage, neuromuscular deterioration, birth defects, insanity, and eventually death.

The source of the mercury was a plastics plant that was dumping waste mercuric chloride into the bay. Bac-terial action changed this chemical into a form that could be taken up and concentrated by benthic organisms. Villagers who ate clams and oysters, especially pregnant women and young children, were seriously affected.

The discharge of mercury has been stopped, but the bay (which had been a source of food for the poor) is still completely unusable for fishing and clamming, even after more than 30 years. This situation will not change for generations; mercury is not easily removed from sediments and has a long residence time.

a A victim of Minamata disease.

chicks were hatched, and the nests were eventually abandoned. The pelicans were disappearing. The trail led investigators to DDT. Plankton absorbed DDT from the water; fishes that fed on these microscopic organisms accumulated DDT in their tissues; and the birds that fed on the fishes ingested it, too (**Figure 20.6**). The whole food chain was contaminated, but because of biological amplification the top carnivores were most strongly affected. A chemical interaction between DDT and the birds' calcium-depositing tissues prevented the formation of proper eggshells. DDT was eventually banned for use in the United States, and the pelican and osprey populations are recovering.

Biological amplification of other chlorinated hydrocarbons has also affected other species. **Polychlorinated biphenyls (PCBs),** fluids once widely used to cool and insulate electrical devices and to strengthen wood or concrete, may be responsible for the behavior changes and declining fertility of some populations of seals and sea lions on islands off the California coast. PCBs have also been implicated in a deadly viral epidemic among dolphins in the western Mediterranean. Ingestion of the chemical may have severely weakened the dolphins' immune system and made it impossible for them to fight the infection—up to 10,000 dolphins died during the summer of 1990. Even more alarming to biologists is the

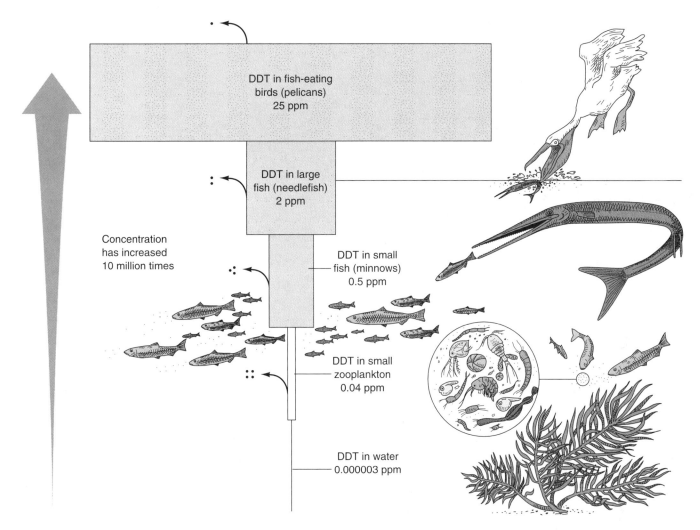

Figure 20.6 The concentration of the pesticide DDT in the fatty tissues of organisms was biologically amplified approximately 10 million times in this food chain of an estuary adjacent to Long Island Sound, near New York City. Dots represent DDT, and arrows show small losses of DDT through respiration and excretion.

recent discovery that the nearshore dolphins off U.S. coasts are intensely contaminated. The concentration of chlorinated hydrocarbons in these animals exceeds 6,900 parts per million (ppm), concentrations high enough to disrupt the dolphins' immune systems, hormone production, reproductive success, neural function, and ability to stave off cancers.[1] These levels vastly exceed the 50 ppm limit the U.S. government considers hazardous for animals, and the 5 ppm considered the maximum acceptable level for humans! Investigations are continuing, as are probes into the effects of dioxin and other synthetic organic poisons accumulating in the oceanic sink.

[1]If you drag a beached porpoise into the ocean, you could theoretically receive a $10,000 fine for improper disposal of polluted materials!

Figure 20.7 indicates the locations of U.S. coastal areas degraded by synthetic organic chemicals. Production of such chemicals subject to biological amplification in food chains currently exceeds 91 million metric tons (100 million tons) each year. Even vaster expanses of ocean will be affected if the predicted 1% of that production reaches the sea.

Eutrophication

Not all pollutants kill organisms. Some dissolved organic substances act as nutrients or fertilizers that speed the growth of marine autotrophs, causing eutrophication. **Eutrophication** (*eu* = good, well; *trophos* = feeding) is a set of physical, chemical, and biological changes that take place when excessive nutrients are released into the

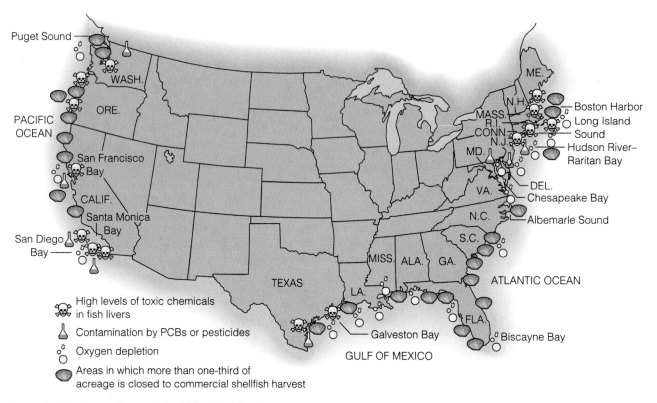

Puget Sound

WASH.

ORE.

PACIFIC
OCEAN

San Francisco
Bay

CALIF.

Santa Monica
Bay

San Diego
Bay

ME.

N.H.

MASS.
R.I.
CONN.
N.J.

MD.

DEL.

VA.
Chesapeake Bay

N.C.
Albemarle Sound

S.C.

MISS. ALA. GA.

LA.

TEXAS

Boston Harbor
Long Island
Sound
Hudson River–
Raritan Bay

ATLANTIC OCEAN

FLA.
Biscayne Bay

Galveston Bay

GULF OF MEXICO

High levels of toxic chemicals
in fish livers

Contamination by PCBs or pesticides

Oxygen depletion

Areas in which more than one-third of
acreage is closed to commercial shellfish harvest

Figure 20.7 Chemical assault on major U.S. coastal areas.

water. Too much fertility can be as destructive as too little. Eutrophication stimulates the growth of some species to the detriment of others, destroying the natural biological balance of an ocean area. The extra nutrients come from wastewater treatment plants, factory effluent, accelerated soil erosion, or fertilizers spread on land. They usually enter the ocean from river runoff and are particularly prevalent in estuaries. Eutrophication is occurring at the mouths of almost all the world's rivers.

The most visible manifestations of eutrophication are the red tides, yellow foams, and thick green slimes of vigorous plankton blooms (see **Figure 20.8**). These blooms typically consist of one dominant phytoplankter that grows explosively, overwhelming other organisms. Huge numbers of algal cells can choke the gills of some animals, and (at night, when sunlight is unavailable for photosynthetic oxygen production) deplete the free oxygen content of surface water (note the symbols for oxygen depletion in Figure 20.7). Toxic substances released from the abundant algae can sicken or kill other species. In 1978 a phytoplankton bloom was linked to the catastrophic die-off of fish near the German coast of the North Sea. In 1988 a similar fate befell dolphins off the U.S. East Coast. As many as 1 million fluke and flounder were killed in the summer of 1989 when an algal bloom

depleted the oxygen in New Jersey's Raritan Bay. During some of these blooms, beaches were covered with foul-smelling foam more than 1 meter (3.3 feet) thick.

These exceptional algae blooms appear to be increasing in number and intensity. There is little mention of foam events before about 1930, but since 1978 there has been at least one every year. A similar pattern has been reported for red tides.

Solid Waste

Not all pollutants enter the ocean in a dissolved state; much of the burden arrives in solid form. Some solid waste is ultimately biodegradable, but plastic—which now makes up almost 7% of the solid waste stream—is not. Scientists estimate that some kinds of synthetic materials—plastic six-pack holders, for example—will not decompose for about 400 years!

Americans generate 120 million metric tons of plastic waste each year, about 500 kilograms (1,100 pounds) per person. By the year 2000 plastics will account for more than 10% of all solid waste. Since the ocean is treated as the ultimate dump, some of this waste plastic finds its way to the sea. In September 1987, volunteers collected 307 tons of litter along the Texas Gulf Coast, most of it

Figure 20.8 A toxic algae bloom, seen from below the surface.

plastic. The haul included 31,733 bags, 30,295 bottles, and 15,631 six-pack yokes. In isolated areas the dunes of plastic debris can build to surreal proportions (see **Figure 20.9**).

A 1987 survey by the Woods Hole Oceanographic Institution found that each square mile of ocean surface off the northeast coast of the United States has more than 46,000 pieces of plastic floating on the surface. This material includes ropes, fishing line and nets, plastic sheeting and bags, and granules of broken plastic cups. A staggering 100,000 marine mammals and 2 million seabirds die *each year* after ingesting or being caught in plastic debris! Sea turtles mistake plastic bags for their jellyfish prey and die from intestinal blockages. Seals and sea lions starve after becoming entangled in nets (see **Figure 20.10**) or muzzled by six-pack rings. The same kinds of rings strangle fish and seabirds. Adding ingredients to

Figure 20.10 Sea lions (seen here) and seals die by the hundreds each year after becoming entangled in plastic debris, especially discarded and broken fishing nets.

Figure 20.9 Plastic mounds on an isolated shore of Niihau, one of the Hawaiian islands.

plastics that would hasten their decomposition would add only 5% to 7% to their cost, but this price increase is currently unacceptable to industry.

What should we do with plastic and other solid wastes such as glass and paper, disposable diapers, scrap metal, building debris, and all the rest? Dumping it into the ocean is clearly unacceptable, yet places to deposit this material are becoming scarce. In 1988 the average New Yorker threw away 1.13 tons of waste annually, up 14% in only two years. California's Los Angeles and Orange counties generate enough solid waste to fill Dodger Stadium every nine days. Transportation of waste to sanitary landfills becomes more expensive as nearby landfills reach capacity.

Is incineration the answer? We currently burn about 7% of our trash, a figure projected to reach 30% by the year 2000. By contrast, Sweden burns half its solid wastes. Unfortunately, even with air pollution control devices, incinerators still emit great quantities of tiny particles, heavy metal residue, and immense amounts of carbon dioxide. Dumping of the residual ash is also a problem; ash from Philadelphia's garbage incinerators has been turned away by states as far away as South Carolina, and one shipload of the stuff was refused entry at an African port!

Is recycling the answer? The Japanese currently recycle about 50% of their solid waste and are importing even more; scrap metal and waste paper headed for Japan are the two biggest exports from the Port of New York. Americans are buying back their own refuse in the form of appliances, automobiles, and the cardboard boxes that hold their televisions and compact disc players. Massachusetts and California have set a goal of recycling 25% of their waste; the city of Seattle is now approaching 30%. The direct savings to consumers, as well as the environmental rewards to ocean and air, will be significant.

The *best* solution is a combination of recycling and reducing the amount of debris we generate by our daily activities. We will soon have no other choice.

Sediment

Runoff from mining, farming, forestry, and other land uses often contains large amounts of sediment. This material can cloud the water, impede photosynthesis, and clog the gills of marine organisms. Coastal ecosystems have been smothered and buried by soils and sands washed into the ocean after strong rains. The problem is particularly apparent in areas of rapidly growing human populations and in the vicinity of coastal mining and dredging operations. In Louisiana, coastal marshes are vanishing as sediments from upstream bury them at the rate of 130 to 155 square kilometers (50 to 60 square miles) a year. The sediments come from poor agricultural practices and uncontrolled commercial development. Sediments washing from local rivers have killed 75% of Costa Rica's Caribbean reefs. Philippine reefs are similarly endangered. By the early 1980s about 75% of the reefs surrounding the Philippines had been damaged beyond recovery by the effects of sedimentation and erosion. The economic loss of Philippine reefs is tremendous; more than 100,000 jobs and $80 million in fish catches are lost each year. Because of lowered reef productivity, 5 million Filipinos are unable to catch the fish and crustaceans they require for proper nutrition. Half of all children living in coastal regions of the Philippines are malnourished.

Sewage

About 98% of sewage is water. Sewage treatment separates the fluid component from the solids, treats the water to kill disease organisms and reduce the levels of nutrients, and releases it into a river or the ocean. The remaining **sewage sludge** is a semisolid mixture of organic matter containing bacteria and viruses, toxic metal compounds, synthetic organic chemicals, and other debris. It is digested, thickened, dried, and shipped to landfills, burned to generate electricity, or dumped into the ocean. The liquid effluent wanders from its outlet and circulates with currents, but sludge and other insoluble residues may stay near the outfall or dump site for years. The amount of wastewater and sewage sludge increased by 60% in the 1990s.

Until 1992 almost 2 billion liters (half a billion gallons) of partially treated sewage poured into Boston Harbor every day. A new sewage treatment plant opened in that year, 22 years after the deadline mandated by the U.S. Clean Water Act. About the same amount of treated sewage pours from outfall pipes off Los Angeles. Treatment plants in southern California are sometimes overwhelmed after heavy rainstorms, and raw sewage enters the ocean in large quantity. Rain or shine, areas around the entrance to San Diego harbor are often so contaminated with sewage from the city of Tijuana, Mexico, that anyone who enters the water runs the risk of bacterial or viral infection.

Sludge may be an even greater problem than raw sewage. The water just south of Long Island, New York, has been one of the most intensely used ocean dumping sites in the world. The area is covered with sludge, which creates an oxygen-poor environment in which few animals can survive. During storms some of this material washes up on local beaches, contaminates shellfish beds, and routinely causes disease outbreaks among people consuming raw oysters and clams from the area. After a public outcry, a new dumping site was selected

Figure 20.11 Thermal effluent from power plants and sewage outfalls, shown in a thermograph of the Quinault River as it enters the Pacific at Taholah, Washington State. In this infrared photograph, white areas indicate warmer water.

beyond the edge of the continental shelf. Ten million tons of wet sludge produced by treatment plants in New York and New Jersey since March 1986 have been dropped there by huge barges. The plume from this sludge now contaminates the Gulf Stream.

Liquid effluent and sludge also contribute to eutrophication. As we've seen, some forms of algae thrive in nutrient-rich wastewater, where they multiply to prodigious numbers, secrete toxins, or deplete oxygen in the surrounding water and cause the death of fish and crustaceans. A huge anoxic zone in the Gulf of Mexico appears to have been triggered by wastewater.

Waste Heat

Shoreside electrical generating plants use seawater to cool and condense steam. The seawater is returned to the ocean about 6°C (10°F) warmer, a difference that may overstress marine organisms in the effluent area. **Figure** 20.11 shows how widely the heated water spreads out around a coastal power installation. Some recent power plant designs minimize environmental impact by pumping colder water from farther offshore, warming it to the temperature of the seawater surrounding the plant site, and then releasing it. This method minimizes impact on the surrounding communities, but it still shocks those eggs, larvae, plankton, and other organisms that are sucked through the power plant with the cooling water.

Introduced Species

Several thousand species are in transit every day in the ballast water of tankers and other ships. Juvenile forms of marine organisms can easily hitch rides across otherwise insurmountable oceanic barriers and set up housekeeping at distant shores. These foreign organisms sometimes outcompete native species and reduce biological diversity in their new habitats. New marine diseases can also be introduced in this way. Even canals and fishery enhancement projects can introduce potentially destabilizing new species.

The Costs of Pollution

In 1985 government and industry in the United States spent about $70 billion on the control of atmospheric, terrestrial, and marine pollution—an average of $289 for each American. This figure was equivalent to about 1.6% of the gross national product, or 2.7% of capital expenditures by U.S. business. That same year the U.S. lost 4% of its gross national product through environmental damage. Clearly the financial costs of pollution will continue to increase.

But there are other costs. Failure to control pollution will eventually threaten our food supply (marine and terrestrial), destroy whole industries, produce a greater disparity between have and have-not nations, and cause a decline in the health of all the planet's citizens. To these costs must be added the aesthetic costs of an ocean despoiled by pollution; few of us look forward to sharing the beach with oiled birds, jettisoned diapers, or clumps of medical waste.

HABITAT DESTRUCTION

The pollution processes we have discussed don't affect individual organisms alone. They influence whole habitats, especially the most complex and biologically sensitive shallow-water habitats.

Bays and Estuaries

The hardest hit habitats are estuaries, the hugely productive coastal areas at the mouths of rivers where fresh water and seawater meet. Pollutants washing down rivers enter the ocean at estuaries, and estuaries often contain harbors, with their potentials for oil spills. As little as 1 part of oil for every 10 million parts of water is enough to seriously affect the reproduction and growth of the most sensitive bay and estuarine species. Some of the estuaries along Alaska's Prince William Sound, site of the 1989 *Exxon Valdez* accident, were covered with oil to a depth of 1 meter (3.3 feet) in places. The spill's effects on the $150-million-a-year salmon, herring, and shrimp fishery will be felt for years to come.

Other West Coast habitats have been polluted in different ways. Bottom sediments in Seattle's Elliott Bay are contaminated with a poisonous mix of lead, arsenic, zinc, cadmium, copper, and PCBs. Tumors on the livers of English sole, which dwell on the sediments, have been linked to these compounds. Pollutants are so abundant in southern San Francisco Bay that clams and mussels contain near-lethal concentrations of heavy metals. Birds migrating from Central America to the Arctic Circle run a risk of being poisoned by stopping there to feed. There is also a risk to humans who eat ducks shot in this area because of the high concentrations of pollutants they contain. Similar warnings apply to all fish caught in and near southern California estuaries between Santa Barbara and Ensenada, Mexico.

Estuaries and bays along the U.S. Gulf Coast, one of the most polluted bodies of water on Earth, are also being severely stressed. About 40% of the nation's most productive fishing grounds, including its most valuable shrimp beds, are found in the Gulf. Nearly 60% of the Gulf's oyster and shrimp harvesting areas—about 13,800 square kilometers (5,300 square miles)—are either permanently closed or have restrictions placed on them because of rising concentrations of toxic chemicals and sewage. Half of Galveston Bay, once classed as the second most productive estuary in the United States, is off limits to oyster fishers because of sewage discharges.

Estuaries along the East Coast are also threatened. From southern Florida to central Georgia, more than 325 square kilometers (125 square miles) of sea grasses (which act as nurseries for a great variety of marine life) have been killed by a virus. Scientists speculate that pollutants from urban and agricultural runoff have weakened the plants' resistance. Fishermen to the north in Chesapeake Bay have been mystified by a sudden decline in the abundance of fish and crustaceans, a change marine scientists attribute to increasing pollution in the area. Lobstermen in New England have noticed an alarming increase in the incidence of tumors in lobsters' tail and leg joints. Could these changes also be caused by toxic wastes? The beluga whale population of Canada's St. Lawrence estuary collapsed in the early 1980s. High levels of PCBs, DDT, and heavy metals—substances biologically amplified in the whales' food—were blamed for the tumors, ulcers, respiratory ailments, and failed immune systems discovered during the autopsies of 72 dead whales.

People also "develop" estuaries into harbors and marinas. In the 1960s and 1970s California led the world in the acreage of bays and estuaries filled for recreational marinas. Harbors grew smaller as more of their area was filled for docks and storage facilities. One hundred and fifty years ago, San Francisco Bay covered 1,131 square kilometers (437 square miles). Today only 463 square kilometers (179 square miles) remains—the rest has been filled in. Filling of estuaries is just as threatening to the natural reproductive cycles of shrimp and fish as poisoning by toxic wastes.

Some states control the development of coastal regions. A citizens' initiative passed by Californians in

1972 limited development of that state's coastal zone. Massachusetts laws make it illegal to fill any marsh or estuarine region, even areas that are privately owned. Similar legislation is pending in a few other coastal states.

Coral Reefs

Pollution may also be damaging coral reefs. Marine biologists have been baffled by recent incidents of coral bleaching in the Caribbean and tropical Pacific (see Box 18.1). The 1989 bleaching event in the Caribbean is the most recent ecological disruption in a series that includes a massive fish kill in 1980, a die-off of about 95% of the individuals of one sea urchin species in 1983–84, and a slowly spreading coral malady known as white-band disease. Biologists do not know for certain what caused these disturbances, but increasing pollution is thought to be partially responsible.

Some chemical pollution is intentional. Especially damaging to tropical reefs has been the practice of using cyanide to collect tropical fish. Fishermen squirt a solution of sodium cyanide over the reef to stun valuable species. Many fish die; those that survive are sent to collectors all over the world. At the same time the invertebrate populations of the sensitive coral reef communities are decimated.

Not all coral reef pollutants are chemicals. Fishermen in Indonesia and Kenya dynamite the reefs to kill fish that hide among the coral branches. Reefs throughout the world are mined for construction material, for ornamental pieces, or for their calcium carbonate (to make plaster and concrete).

Other Habitats

Between 1963 and 1977, about half of India's extensive mangrove forests were cut down. About a third of Ecuador's mangroves have been converted to ponds used in shrimp mariculture. Agricultural expansion is expected to wipe out all Philippine mangroves in 10 years. Not even the calm, cold communities of the abyssal plains are safe from disruption. Imagine the effect manganese nodule mining will have on the delicate organisms of the deep bottom.

GLOBAL CHANGES

The ocean and the atmosphere are extensions of each other, and human activity has changed the atmosphere as it has changed the ocean. Pollutants injected into the air can have global consequences for the ocean and for all the Earth's inhabitants. Potentially the most destructive atmospheric problems are depletion of the ozone layer, global warming, and acid rain.

Ozone Layer Depletion

Ozone is a molecule formed of three atoms of oxygen. Ozone forms naturally when lightning strikes through air; larger quantities are generated spontaneously in the stratosphere (an upper layer of atmosphere). A diffuse layer of ozone mixed with other gases—the **ozone layer**—surrounds the world at a height of about 20 to 40 kilometers (12 to 25 miles).

Seemingly harmless synthetic chemicals released into the atmosphere—primarily **chlorofluorocarbons (CFCs)** used as cleaning agents, refrigerants, fire-extinguishing fluids, spray-can propellants, and insulating foams—are converted by the energy of sunlight into compounds that attack and partially deplete the Earth's atmospheric ozone. Ozone levels in the stratosphere have decreased by about 3% over some of the United States since 1969 (see **Figure 20.12**). A 4% drop has been noted over Australia and New Zealand, and a 50% decrease has been observed near the North and South poles (**Figure 20.13**). The amount of depletion varies with latitude (and with the seasons) because of variations in the intensity of sunlight.

This decline in ozone alarms scientists because stratospheric ozone intercepts some of the high-energy ultraviolet radiation coming from the sun. Ultraviolet radiation injures living things by breaking strands of DNA and unfolding protein molecules. Species normally exposed to sunlight have evolved defenses against average amounts of ultraviolet radiation, but increased amounts could overwhelm those defenses. Land plants such as soybeans and rice would be subjected to sunburn that decreases their yields. Even plankton in the uppermost 2 meters of ocean would be affected; recent research in fact indicates an alarming decrease in phytoplankton primary productivity of between 6% and 12% in the coastal waters around Antarctica.[2]

Our own species would not escape: A 1% decrease in atmospheric ozone would probably be accompanied by a 5% to 7% increase in human skin cancer (see **Figure 20.14**). According to medical researchers' estimates, the 4% ozone depletion over Australia and New Zealand will cause at least a 20% increase in human skin cancers over the next two decades. Strong ultraviolet light can also suppress the immune system and cause eye cataracts. The quantity of photochemical smog shrouding our cities will also increase as ozone levels fall.

In June 1990, representatives of 53 nations agreed to

[2]See the 1992 article by Smith, et al., listed in this chapter's bibliography.

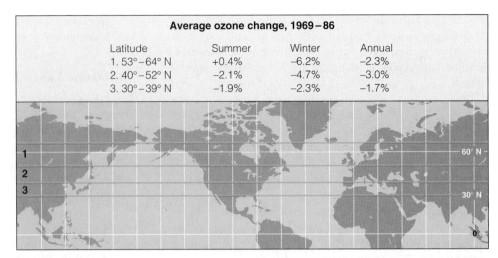

Average ozone change, 1969–86			
Latitude	Summer	Winter	Annual
1. 53°–64° N	+0.4%	–6.2%	–2.3%
2. 40°–52° N	–2.1%	–4.7%	–3.0%
3. 30°–39° N	–1.9%	–2.3%	–1.7%

Figure 20.12 Ozone depletion in the stratosphere above various latitudes of the Northern Hemisphere between 1969 and 1986. A 1991 NASA study showed that between 1978 and mid-1990 the decreases in ozone were about twice the percentages shown here.

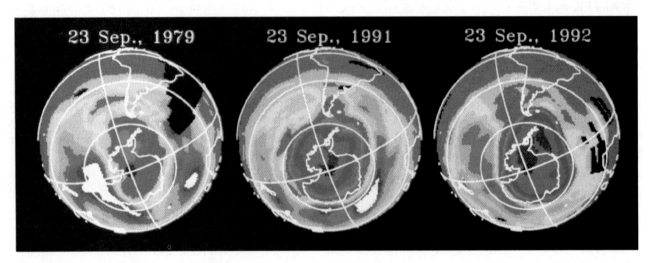

Figure 20.13 Expansion of the seasonal ozone hole over Antarctica, as recorded by the *Nimbus 7* satellite. The lowest ozone values (the "hole") are indicated by magenta and purple. Droplets of sulfuric acid, released during the 1991 volcanic eruption of Mount Pinatubo, contributed to the thinning.

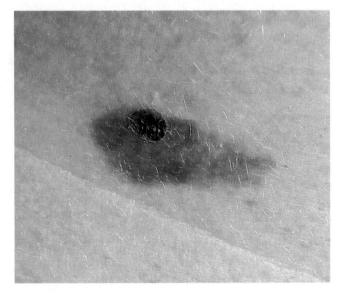

Figure 20.14 A malignant melanoma in an early stage. Left untreated, this most dangerous form of skin cancer could spread and cause the death of the patient. Skin cancer appears to be associated with exposure to the sun. If you spend much time in the sun you increase your risk of skin cancer. Be alert for changes in moles or for any pigmented lesion that is asymmetrical, has an irregular border, is an unusual color, grows in size, or begins to itch. If you notice any of these characteristics, see a physician.

ban major production and use of ozone-destroying chemicals by the year 2000. Legislation is being proposed to phase out CFC production in the United States, and a sense of great urgency surrounds ongoing research to find safe substitutes for these substances. Yet if their production were completely halted tomorrow, chlorine levels in the stratosphere would continue to rise for another 20 to 30 years, according to NASA sources. Moreover, CFCs stay in the atmosphere for up to 110 years; so their damaging effect will continue.

Figure 20.15 If ozone depletion continues, the "perfect tan" may have to be none at all.

Global Warming?

The surface temperature of the Earth fluctuates slowly over time. The global temperature trend has been generally upward in the 18,000 years since the last ice age, but the *rate* of increase has recently accelerated. This rapid warming may be the result of an enhanced **greenhouse effect,** the trapping of heat by the atmosphere. Glass in a greenhouse is transparent to light but not to heat. The light is absorbed by objects inside the greenhouse, and its energy is converted into heat. The temperature inside a greenhouse rises because the heat is unable to escape. On Earth **greenhouse gases**—carbon dioxide, water vapor, methane, CFCs, and others—take the place of glass. Heat that would otherwise radiate away from the planet is absorbed and trapped by these gases, causing a rise in surface temperature. **Figure 20.16** shows this mechanism.

A certain amount of greenhouse effect is necessary for life; without it, Earth's average atmospheric temperature would be about −18°C (0°F). Earth has been kept warm by natural greenhouse gases. The sources of these gases are volcanic and geothermal processes, the decay and burning of organic matter, and respiration and other biological sources. The removal of these gases by photosynthesis and absorption by seawater appears to prevent the planet from overheating. But the human demand for quick energy to fuel industrial growth, especially since the beginning of the Industrial Revolution, has injected unnatural amounts of new carbon dioxide into the atmosphere from the combustion of fossil fuels. Burning of forests and jungles exacerbates the problem. Carbon

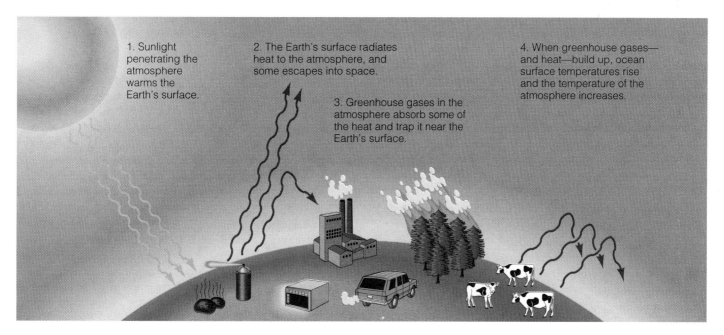

1. Sunlight penetrating the atmosphere warms the Earth's surface.

2. The Earth's surface radiates heat to the atmosphere, and some escapes into space.

3. Greenhouse gases in the atmosphere absorb some of the heat and trap it near the Earth's surface.

4. When greenhouse gases—and heat—build up, ocean surface temperatures rise and the temperature of the atmosphere increases.

Figure 20.16 How the greenhouse effect works.

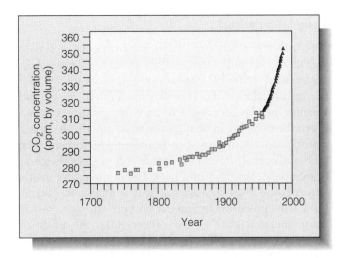

Figure 20.17 The rise in atmospheric carbon dioxide since 1750. Measurements up to about 1960 (■) came from samples of air trapped in the Antarctic ice; later measurements (▲) were made directly at Mauna Loa, Hawaii (after Watson et al., 1990).

dioxide is now being produced at a greater rate than it can be absorbed by the ocean. **Figure 20.17** shows how much carbon dioxide has increased in the atmosphere since about 1750. The atmosphere's carbon dioxide content now rises at the rate of 0.4% each year. CFCs and methane are increasing even more rapidly (see **Table 20.4**), and one molecule of CFC is equivalent in greenhouse heating effect to 10,000 carbon dioxide molecules.

There has been a 4°C (7°F) rise in global temperature from the end of the last ice age until today. Carbon dioxide and other human-generated greenhouse gases produced since 1880 are thought to be responsible for about 1°C (1.8°F) of that increase (see **Figure 20.18**). If current models of greenhouse warming are correct, we can expect global temperature to rise another 2.5°C (4.5°F) by the year 2030. A temperature increase of this magnitude would cause water in the ocean to expand; average sea level would rise between 8 and 29 centimeters (3 and 11 inches). (One European model predicts a 4°C—7°F—rise by 2070, with an ocean rise of 0.5 to 1.5 meters, or 1.6 to 5 feet.) Imagine the effect on the harbors, coastal cities, river deltas, and wetlands where one-third of the world's people now live. The costs to society would be very large.

But scientists are still uncertain about the magnitude of human-induced global warming. Unpredictable natural processes, such as fluctuations in the energy output of the sun or the eruption of Mount Pinatubo in the Philippines in 1991, can greatly influence temperature predictions. Researchers estimated that the huge volume of dust and gas injected by Mount Pinatubo into the stratosphere will more than compensate for greenhouse heating and will bring Earth's atmospheric temperature

down by about 1°C by the year 2000. So far, the predictions are on track, but Mount Pinatubo's effects will not last for long. Most researchers are counseling caution—and a reduction in the generation of greenhouse gases—until more data can be accumulated.

Unfortunately, it will be exceedingly difficult to curtail our production of carbon dioxide. In the last hundred years, industrial production has increased 50-fold; we have burned 1 billion barrels of oil, 1 billion metric tons of coal, and 10 billion cubic meters of natural gas. Carbon dioxide is a major product of combustion for these hydrocarbon compounds. The world's energy demand is projected to increase 3.5 times between now and the year 2025, with carbon dioxide emissions 65% higher than today. Some environmentalists have proposed the planting of millions of trees to take up the extra carbon dioxide, but these efforts would be futile in the face of such massive increases.

Alternatives to fossil fuel must be found if we are to maintain world economies and prevent an increase in global temperature with all its uncertainties. The only alternate source of energy that currently produces significant amounts of power is **nuclear energy,** which now generates about 17% of the electricity produced in the United States. Despite much publicity to the contrary, these pressurized water reactors have good records of dependable power production and safety.[3] The problem with nuclear power lies not so much in the everyday operation of the reactors but in disposing of the nuclear wastes they produce. By 1992 about 55,000 highly radioactive spent fuel assemblies were in temporary storage in deep pools of cooled water; they must be stored for 10,000 years before their levels of radioactivity will be low enough to pose no environmental hazard. Radioactive substances emit **ionizing radiation,** a form of energy able to penetrate and permanently damage cells. It is essential that artificial sources of ionizing radiation be isolated from the environment.

Of course the ocean has been investigated as a potential dumping site. Some low-level medical and industrial nuclear wastes have already been jettisoned at sea, and the results have not been promising. A few barrels of radioactive waste dropped to the seabed near the Farallon Islands off San Francisco have broken open. Drums of radioactive waste from another source have washed ashore in Oregon. Small amounts of radioactive waste now enter the ocean at the mouth of the Columbia River after having leaked from storage containers near Hanford, Washington. Fishermen far from the North Atlantic coast have snagged containers of radioactive residue in trawl nets. In 1993 a Russian warship was found dump-

---

[3] The Soviet reactor at Chernobyl that exploded in 1986 was of a much different design.

Table 20.4 Concentrations and Trends for Greenhouse Gases in the Atmosphere

Gas	Concentration Pre-1850	Concentration 1985	Trends 1975–85	Projections for Mid-21st Century
Carbon dioxide (CO$_2$)	275 ppm[a]	345 ppm	+4.6%	400–600 ppm
Methane (CH$_4$)	0.7 ppm	1.7 ppm	+11.0%	2.1–4 ppm
Chlorofluorocarbons (CFCs)	0	0.6 ppb[b]	+102%	2.7–7.8 ppb

Source: Adapted from Ramanathan, 1988.

[a]ppm = parts per million by volume

[b]ppb = parts per billion by volume

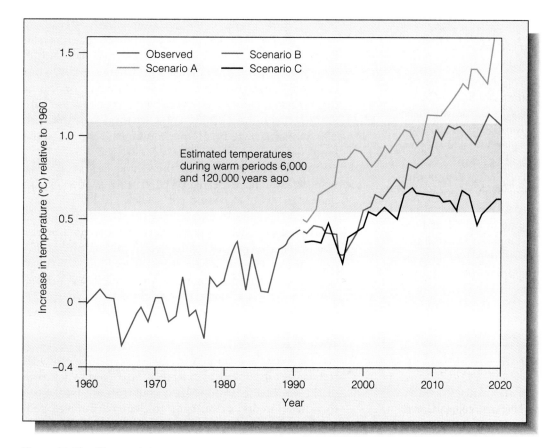

Figure 20.18 Observed changes in global temperature since 1960 and projected changes through the year 2020 as a result of the greenhouse effect. Scenario A assumes continued growth in the rate of release of carbon dioxide and similar gases. Scenario B assumes the implementation of some controls on the release of greenhouse gases. Scenario C assumes a complete halt to their release. Even under Scenario C, temperatures would still climb into the shaded area, which represents average temperatures during extremely warm periods about 6,000 and 120,000 years ago.

ing liquid nuclear wastes into the Sea of Japan. The countries of the CIS have long disposed of contaminated reactors and radioactive wastes by abandoning them at sea.

Will citizens of industrial countries (and countries wishing to become industrialized) agree to lessen the danger of increased global warming by slowing their economic growth, decreasing their dependence on fossil fuel combustion, and developing safe alternate sources of energy? Some insight may be gained from the behavior of ranchers and industrialists in the rain forests of New Guinea, the Philippines, and Brazil. The Amazon rain forest of Brazil is being burned at a rate of about 12 square kilometers *per hour* (almost 5 square miles *each hour*), acreage equivalent to the area of West Virginia every year. Huge stands of trees that should be nurtured to absorb excess carbon dioxide are being destroyed. The cleared land is used for farms, cattle ranches, roads, and cities. The priorities are obvious.

Acid Rain

Levels of sulfur dioxide and oxides of nitrogen, normally present in the atmosphere in small amounts, are also being increased by human activity. Excess quantities of these gases dissolve in water droplets to form acids, which precipitate to the surface as **acid rain.** Lakes and soil are particularly susceptible to damage by acid buildup.

Scientists generally believe that the natural buffering of seawater protects the open ocean from these effects, and indeed oceanic calcium reserves seem adequate to prevent any significant overall change in pH. Nearshore regions, however, appear to be threatened by the nitrates in acid rain, which can act as fertilizers and lead to eutrophication. Research in Chesapeake Bay suggests that about a quarter of the nitrate load in the bay is derived from acid rain, a figure double that of previous estimates. Again, delicate marine balances are threatened in unpredictable ways.

WHAT CAN BE DONE?

In a pivotal paper published in 1968, biologist Garret Hardin examined what he termed "The Tragedy of the Commons." Hardin's title was suggested by his study of societies in which some agricultural areas were held *in common*—that is, were jointly owned by all residents. Citizens of these societies owned small homes, plots of land, and perhaps a cow that was put to pasture on the commons. Each farmer *kept* the milk and cheese given by his cow but *distributed* the costs of cow ownership—overgrazing of the commons, cow excrement, fouled drinking water, and so on—among all the citizens. This arrangement worked well for centuries because wars, diseases, and poaching kept the numbers of people and cows well below the carrying capacity of the land. But

eventually political stability and relative freedom from disease allowed the human (and cow) population to increase. Farmers pastured more cows on the commons and gained more benefits. Soon the overstressed commons could no longer sustain the growing numbers of cows, and the area held in common was ruined. Eventually no cows could survive there.

The lesson applies to our present situation. Hardin noted that in our social system each individual tends to act in ways that maximize his or her material gain. Each of us gladly keeps the *positive* benefit of work but willingly distributes the *costs* among all. For example, this morning I drove to my college office; the benefit to me was one trip to my office. A cost of this short drive was the air pollution generated by the fuel combustion in my car's engine. Did I route the exhaust fumes through a hose to a mask held tightly over my nose and mouth? That is, did I reserve the environmental costs of my actions for my own use, just as I had reserved for myself the benefit of my ride to work? No. I shared those fumes with my fellow Californians, just as you shared your morning's sewage with your fellow citizens, or just as the factory down the road shared its carbon dioxide with all the world. Indeed, the world itself is our commons. The modern tragedy of the commons rests on these kinds of actions.

The carrying capacity of the whole Earth-commons may already have been exceeded. Births now exceed deaths by about 3 people per second, 10,400 per hour. *Each year* there are 95 million more of us, a total equal to nearly one-third of the population of the United States. The number of people has tripled in this century and is expected to double again before reaching a plateau sometime in the next century. Another billion humans will join the world population in the next 10 years, 92% of them in third world countries (see **Figure 20.19**).[4] One-fifth of the world's people already suffer from abject poverty and hunger.

This exploding population is not content with using the same proportion of resources used today. Citizens of the world's least developed countries are influenced by education and advertising to demand a developed-world standard of living. They look with justifiable envy on the United States, a country with 5% of the planet's population that consumes 35% of its raw material resources and 25% of its energy, while generating 30% of industry-related carbon dioxide. Can the world support a population whose expectations are rising as rapidly as their numbers? In Garrett Hardin's words, "We can maximize the number of humans living at the lowest possible

---

[4]In the United States alone, the population is growing by the equivalent of four Washington D.C.s every year, another New Jersey every 3 years, another California every 12. By the year 2000 another 40 million people will reside within U.S. boundaries.

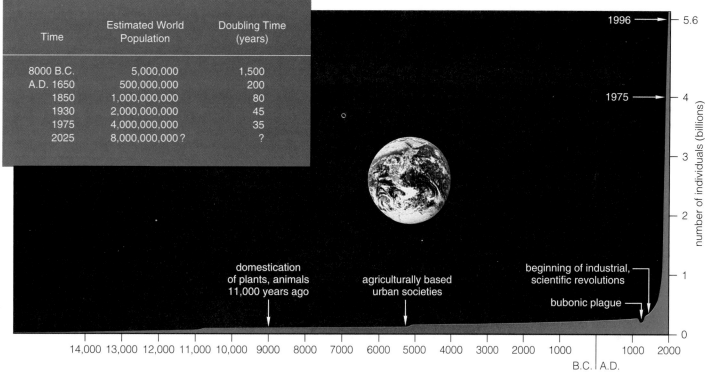

Time	Estimated World Population	Doubling Time (years)
8000 B.C.	5,000,000	1,500
A.D. 1650	500,000,000	200
1850	1,000,000,000	80
1930	2,000,000,000	45
1975	4,000,000,000	35
2025	8,000,000,000 ?	?

Figure 20.19 Growth curve for the human population. The diagram's vertical axis represents world population, in billions. (The slight dip between the years 1347 and 1351 represents the 25 million deaths in Europe from the bubonic plague.) The growth pattern over the past two centuries has been exponential, sustained by agricultural revolutions, industrialization, and improvements in health care. The list in the blue box tells us how long it took for the human population to double in size at different times in history. The number of people on Earth exceeds 5.6 billion.

level of comfort, or we can try to optimize the quality of life for a smaller population." The burgeoning human population is the greatest environmental problem of all.

We cannot expect science to solve the problem for us. Most of the decisions and necessary actions fall outside pure science in the areas of values, ethics, morality, and philosophy. The solution to environmental problems, if one exists, *lies in education and action*. Each of us is obliged to become informed on issues that affect the Earth, its ocean, and its air—to learn the arguments and weigh the evidence. Once informed, we must act in rational ways. Chaining yourself to an oil tanker is not rational, but selecting well-designed, long-lasting, recyclable products made by responsible companies with minimal environmental impact (and encouraging others to do so) certainly is. (Other things you can do are listed in **Box 20.2**.)

Obvious answers and quick solutions are often misleading; a great deal of research and work are needed to give reliable insight into the many difficult questions that confront us. The present trade-off between financial and ecological considerations is often strongly tilted in favor of immediate gain, short-term profit, and immediate convenience. Education may be the only way to modify these destructive behaviors. Garrett Hardin suggests that absolute freedom in a commons brings ruin to all.

Humanity is part of the natural world, not its master. We may be able to learn to live in harmony with this small, beautiful, blue world. We need not relive the Tragedy of the Commons on a planetary scale. True convenience and true progress depend on several things: the preservation of open space, serious and sustained attention to population control, conversion to a steady-state economy (instead of one that must grow to stay alive), business incentives for preservation, the use of renewable resources, and, above all, *public education in environmental issues*. We must ask ourselves difficult questions: "What is the optimal quality of life?" "How can I achieve balance between my material needs and the needs of the Earth?" "What do I want to leave for my children?" "How can I reserve the quiet, renewing ocean for myself, for all species, and for the future?" Our cities are crowded and our tempers are short. Times of turbulent change lie before us. The trials ahead will be severe.

Each of us, individually, needs to take a stand. We must preserve the sunsets and fog, the waves to ride, the cold, clean spray on our faces. Margaret Mead summarized our potential for making a difference: "Never doubt that a small group of thoughtful, committed citizens can change the world. Indeed it is the only thing that ever has." We need to start now.

BOX 20.2 ● *Some Things You Can Do for the Environment*

Each of us can make daily decisions that collectively make a difference. Some things we can do are listed here.

A. Energy

1. Drive fewer miles. Ride your bike or walk. Use public transportation. Carpool.
2. Use smaller wattage lightbulbs in lamps. Replace incandescent bulbs with compact 18-watt fluorescent ones. Turn off lights in rooms not being used. Switch to low-voltage exterior lighting.
3. Purchase energy-efficient appliances and cars. Government-mandated stickers on cars and appliances allow you to compare the amounts of energy each will use. Spend a bit more now to save a lot later.
4. Check insulation and weatherstripping around your house, apartment, or dorm. Keep thermostats relatively high in summer, low in winter.
5. Check into alternative energy sources such as solar and wind power units.
6. Buy and use rechargeable batteries. The disposable ones contain heavy metals that can pollute.

B. Food

7. Try to avoid foods from areas where pesticides have been used. Never eat fish caught in developed harbors or bays.
8. Consider how your food choices affect the environment. Eat foods low on the food chain. Avoid foods raised in places where endangered envronments (such as rain forests) were destroyed.
9. Raise a garden.
10. Buy bulk or unpackaged goods, or buy food in reusable containers.

C. Toxic Materials and Pollutants

11. Check for toxic materials (old cans of paint, pesticides, engine oil, pool chemicals, and so on) you may have around the house. Take them to an approved toxic waste receiving station.
12. Read labels. Buy the least toxic materials available.
13. Swat flies—don't spray them.
14. Buy clothing that does not require dry cleaning. Dry clean only when necessary.

D. Water

15. Use water-saving showerheads, sink faucet aerators, and toilets.
16. Sweep walks and driveways; don't hose them down.
17. Water lawns infrequently but thoroughly. Better yet, replace them with drought-tolerant ground covers.
18. Take shorter showers. Shower with a friend.
19. Run only full dishwasher and clothes washer loads.

E. Recycling and Waste Reduction

20. Recycle newspapers, cans, glass, and plastics.
21. Buy products that are refillable and recyclable.
22. Volunteer your time in a beach cleanup campaign.
23. Use cloth rags and diapers instead of paper diapers and paper towels.
24. Take your grocery bags back to the market for another trip home.

F. Preservation of Life and the Environment

25. Plant trees.
26. Preserve open space. Become involved in parkland and open space movements.
27. Avoid CFCs. Check air conditioners to see if they leak.
28. Boycott organizations that violate sound environmental practices.
29. Don't litter. Take your trash out of campsites.

G. Other Things

30. Write your legislators. Better yet, visit them!
31. Get involved in an environmental group such as the Sierra Club or Greenpeace.
32. Talk to your friends about environmental matters.
33. Buy quality. Purchase a well-built item, maintain it, repair it, keep it forever.
34. Sit quietly on a beach at sunset or sunrise.
35. Read. Listen.
36. Be positive about results. Be encouraged.

Source: Thanks to Jack Mertz.

Q-AND-A

1. How sure are scientists of the data and predictions discussed in this chapter?

It's difficult to gather and interpret data fast enough to answer the important questions. The debate on global warming, for example, has distinguished proponents on both sides: One group of experts is sure that global warming has begun, the other is confident it has not. "Warmers" point to rises in sea level since the mid-1800s; the thinning of pack ice near both poles; the expanding range of flowering plants at high latitudes; the 3-millimeter (⅛-inch) rise in average sea level detected by the TOPEX/Poseidon satellite; and the fact that the 7 warmest years of the past 110 years were 1990, 1991, 1988, 1987, 1983, 1981, and 1980, in descending order.[5] "Nonwarmers" note that the Greenland ice sheet has been growing thicker at a rate of about 2.5 centimeters (0.9 inch) per year since 1975; that based on satellite measurements, the global average temperature did not rise in the 10-year period from 1979 through 1988; and that a small rise in warmth will generate more high clouds, which will in turn reflect more sunlight and cause net global cooling. They also point out that two of the coldest winters ever recorded occurred in the past decade. For an insight into the controversy see the article by Jones and Wigley (1990) listed in this chapter's bibliography.

Most scientists, however, are convinced that global warming is or will become a reality. In a recent survey of 330 atmospheric and marine scientists in 41 countries, 96% said they believe the basic greenhouse effect thesis. Nearly two-thirds (65%) said they believe that the odds are better than 50–50 that global temperatures will rise 3°C (5.4°F) over the next century; 71% said they believe that a global temperature rise of a few degrees would cause "massive human disruption" by the end of the next century. It is important to realize, however, that the jury is still out on these issues.

2. Other than oil spills and tanker operations, how does oil enter the ocean?

*Every year more than 900 million liters (about 240 million gallons) of used motor oil—about 19 times the volume of the Exxon Valdez spill—is dumped down drains (**Figure 20.20**), poured into dirt, or concealed in trash headed for landfills. This oil is much more toxic than crude or newly refined oil because it has developed carcinogenic and metallic components from the heat and pressure within internal combustion engines. Some of this material seeps into groundwater and contaminates wells. Much of it, of course, ends up in the ocean.*

Figure 20.20 Storm drains empty directly into rivers, bays, and ultimately the ocean.

3. What can an individual do to minimize his or her impact on the ocean and atmosphere?

Remember that the Earth and all life-forms are interconnected. There are no true consumers, only users: Nothing can truly be thrown away (there is no "away"). We must abandon the pollute-and-move-on ethic that has guided the actions of most humans for thousands of years, and we must work toward a society more in harmony with the fundamental rhythms of life that sustain us. Our task is not to multiply and subdue the Earth. It may not be too late to change our ways. We need to act individually to effect change. We should think globally and act locally.

4. Is pollution always a bad thing?

Some forms of pollution bring temporary benefits. For example, some of Florida's once-endangered manatees are thriving at the warm outfalls of coastal power stations. On a larger scale, if increased greenhouse warming does develop, some computer models indicate increased rainfall, longer growing seasons, and increased crop yields over broad latitude bands in the temperate zones of both hemispheres. On the whole, though, the less human intervention in complex natural systems, the better.

5. What are the most dangerous threats to the environment, overall?

The underlying causes of the problems discussed in this chapter are (1) human population growth and (2) a growth-dependent economy. Stanford professor Paul Ehrlich said recently, "Arresting global population growth should be second in importance on humanity's agenda only to avoiding nuclear war." The present world population, now above 5.6 billion, seems doomed to reach at least 10 billion before leveling off. And what if everybody wants the same number of cows?

---

[5]Ash and gases from the eruption of Mount Pinatubo in the summer of 1991 kept that year from breaking the record; the first six months of 1991, however, were the warmest ever recorded.

6. What role does public perception play in pollution issues?

A large role, indeed! Until recently, relatively small-scale but highly visible insults have claimed most of the public's attention and have driven us to action. The messy breakup of the oil tanker Torrey Canyon *off the southern coast of England in 1969 galvanized world opinion and set the stage for the present environmental movement. More recently, in the summer of 1988 beachgoers were horrified to discover that more than 80 kilometers (50 miles) of northern New Jersey and Long Island beaches had been temporarily closed because of medical debris littering the shore. Some of the dozens of vials of blood, syringes, stained bandages, and surgical sutures tested positive for the viruses that cause AIDS and hepatitis B. Similar incidents occurred in Rhode Island and Massachusetts.*

Appalling and visible though such incidents are, their long-term effects on the ocean as a whole are negligible. Public attention has recently turned to issues of larger consequence. The drought and heat of the past few summers brought terms like greenhouse effect *and* ozone layer *to local newspapers and dinner table conversation. Weekend fishermen worry about eating their catch. They wonder about the unseen threats as much as the obvious ones. Perceptions are changing.*

Terms and Concepts to Remember

acid rain	eutrophication	pollutant
biodegradable	greenhouse effect	polychlorinated
biological	greenhouse gases	biphenyls (PCBs)
amplification	ionizing radiation	sewage sludge
chlorinated	marine pollution	
hydrocarbons	nuclear energy	
chlorofluorocarbons	ozone	
(CFCs)	ozone layer	

Study Questions

1. What is pollution? What factors determine how dangerous a pollutant is?

2. Why is refined oil more hazardous to the marine environment than crude oil? Which is spilled more often? What happens to oil after it enters the marine environment?

3. What heavy metals are most toxic? How do these substances enter the ocean? How do they move from the ocean to marine organisms and people?

4. Few synthetic organic chemicals are dangerous in the very low concentrations in which they enter the ocean. How are these concentrations increased? What can be the outcome when these substances are ingested by organisms in a marine food chain?

5. What is eutrophication? How can "good eating" be hazardous to marine life?

6. What parts of the marine environment are hardest hit by human activities?

7. What synthetic chemicals appear to be causing depletion of the Earth's protective ozone layer? What is the likely result?

8. What is the greenhouse effect? Is it always detrimental? What gases contribute to the greenhouse effect? Why do most scientists believe that the Earth's average surface temperature will increase over the next few decades? What may result?

9. What is the tragedy of the commons? Do you think Garrett Hardin was right in applying the old idea to modern times? What will you do to minimize your negative impact on the ocean and atmosphere?

10. How might global warming or a decrease in stratospheric ozone *directly* affect the ocean?

For Further Study

Anderson, D. M. 1994. "Red Tides." *Scientific American,* August, 62–68. Information about the organisms and the growing threat that red tides pose to marine and human health.

Beardsley, T. M. 1989. "Not So Hot: New Studies Question Estimates of Global Warming." *Scientific American,* November.

Beaty, J., and R. Spector. 1990. "The Dark Side of Worshiping the Sun." *Time,* 23 July. More than 600,000 new cases of skin malignancies were diagnosed in 1990; 27,600 will be potentially lethal melanomas.

Broadus, J. M. 1991. "The Sea Environment: Good News, Bad News." *Proceedings of the United States Naval Institute* 117 (no. 10, October): 50–55. Presentation of a "dirty dozen" environmental problems requiring attention.

Brown, L. 1994. *State of the World.* New York: Norton. An annual updating of the state of our relationship with the environment.

Carson, R. 1962. *Silent Spring.* New York: Houghton Mifflin. This elegantly written book is said to have begun the era of environmental awareness in the United States.

Cherfas, J. 1990. "The Fringe of the Ocean Under Siege from Land." *Science* 248 (no. 4952). Excellent summary of the present assault on the ecology of ocean margins.

Dickson, D. 1988. "Mystery Disease Strikes Europe's Seals." *Science* 241 (19 August): 893–95. Are toxic wastes suppressing the seals' immune response?

Duedall, I. 1990. "A History of Ocean Disposal." *Oceanus* 33 (no. 2, Summer): 29–37. Comprehensive and unpleasant to contemplate.

Ehrlich, A., and P. Ehrlich. 1987. *Earth.* New York: Franklin Watts. An excellent overview of environmental issues from two veterans in the field. Beautifully illustrated.

Farrington, J. W., et al. 1983. "Ocean Dumping." *Oceanus* 25 (no. 4): 39–50. A good overview of the problem.

Geraci, J. R., and D. J. St. Aubin. 1990. *Sea Mammals and Oil.* San Diego: Academic Press. Apart from sea otters, oil has had little discernible direct impact on marine mammals, this despite the pronouncements of the popular press. The species most at risk are sea otters, polar bears, fur seals, and ice seals.

Hardin, G. 1968. "The Tragedy of the Commons." *Science* 162 (13 December): 1243–48. Hardin makes the compelling case that freedom in a commons brings ruin to all.

Houghton, R. A., and G. M. Woodwell. 1989. "Global Climatic Change." *Scientific American,* April, 36–44. Evidence suggests that production of carbon dioxide and methane from human activities has already begun to change the climate and that radical steps must be taken to halt any further change.

Jackson, J. B. C., et al. 1989. "Ecological Effects of a Major Oil Spill on Panamanian Coastal Marine Communities." *Science* 243 (no. 4887): 37–43. The effect of a crude oil spill in a complex region of mangroves, sea grasses, and coral reefs is discussed.

Jones, P. D., and T. M. W. Wigley. 1990. "Global Warming Trends." *Scientific American,* August, 84–91. Part of a special issue entitled "Energy for Planet Earth."

Kerr, R. A. 1988a. "Is the Greenhouse Here?" *Science* 239 (5 February): 559–61.

Kerr, R. A. 1988b. "Stratospheric Ozone Is Decreasing." *Science* 239 (25 March): 1489–91. Some decreases greatly exceed predictions.

Kerr, R. A. 1988c. "Weather in the Wake of El Niño." *Science* 240 (13 May): 883. Warming trend graphed.

Kerr, R. A. 1988d. "Evidence of Arctic Ozone Destruction." *Science* 240 (27 May): 1144–45.

Kerr, R. A. 1988e. "Report Urges Greenhouse Action Now." *Science* 241 (1 July): 23–24.

Kerr, R. A. 1991. "A Lesson Learned, Again, at Valdez." *Science* 252 (no. 5004): 371. The cleanup at Prince William Sound was more damaging than the effects of the crude oil spilled.

Kerr, R. A. 1995. "Is the World Warming or Not?" *Science* 267 (no. 5198): 612. Climate researchers are not betting on a quick resolution to the global warming debate.

Koshland, D. 1989. "Oil Spills." (Editorial). *Science* 244 (no. 4905): 629. Source of data for the fate of *Exxon Valdez* oil.

Lenssen, N. 1989. "The Ocean Blues." *World Watch* 2 (no. 4, July–August).

Loughlin, T. R., ed. 1994. *Marine Mammals and the Exxon Valdez.* San Diego, CA: Academic Press. The series of studies described here began within days of the spill and continued through 1993.

Mangone, G. J. 1986. *Concise Marine Almanac.* New York: Van Nostrand-Reinhold. Excellent summary chapter on marine pollution.

McKibben, Bill. 1989. *The End of Nature.* New York: Random House. A beautifully written plea for rationality and change.

Miller, G. T. 1994. *Living in the Environment.* 8th ed. Belmont, CA: Wadsworth. Perhaps the best general text on environmental science available.

National Academy of Sciences. 1985. *Oil in the Sea.* Washington, DC.

Ramade, F. 1987. *Ecotoxicology.* 2d ed. New York: Wiley. Treatise on the effects of pollutants on marine ecosystems.

Ramanathan, V. 1988. "The Greenhouse Theory of Climate Change: A Test by Inadvertent Global Experiment." *Science* 240 (15 April): 293–99. Excellent and comprehensive review article.

Reese, H. C. 1991. "Trouble on the Wide Common." *Proceedings of the United States Naval Institute* 117 (no. 10, October): 60–64. Difficulties with leaking tankers on the oceanic "common."

Roberts, L. 1989. "*Valdez:* The Predicted Oil Spill." *Science* 244 (no. 4900): 20–21. The potential disaster of a major spill in Prince William Sound was forecast in a 1972 environmental impact statement.

Sagan, C. 1980. *Cosmos.* New York: Random House. The last chapter, "Who Speaks for Earth?", is especially recommended.

Schafer, H. A., et al. 1984. "Chlorinated Hydrocarbons in Marine Mammals." In *Southern California Coastal Water Research Project Biennial Report, 1983–1984,* 109–14.

Scotto, J., et al. 1988. "Biologically Effective Ultraviolet Radiation: Surface Measurements in the United States, 1974–1985." *Science* 239 (12 February): 762–64.

Smith, R. C., et al. 1992. "Ozone Depletion: Ultraviolet Radiation and Phytoplankton Biology in Antarctic Waters." *Science* 255 (no. 5047): 952–59. Reduced concentration of atmospheric ozone and consequently higher ultraviolet irradiance have reduced phytoplanktonic productivity in the far south by as much as 6% to 12%.

Sun, M. 1988. "Acid Rain Said to Threaten Bay." *Science* 240 (29 April): 601.

Toufexis, A. 1988. "The Dirty Seas." *Time,* 1 August. A fine summary of the situation.

Trefil, J. 1990. "Global Warming and a Scientific Free-for-All." *Smithsonian,* December. Predictions of nature's behavior are based on knowns and unknowns—thus the international policy debate.

Watson, R. T., H. Rodhe, H. Oeschger, and U. Siegenthaler. 1990. "Greenhouse gases and aerosols." *Climate Change: The IPCC Scientific Assessment,* ed. J. T. Houghton, G. J. Jenkins, and J. J. Ephraums, 1–40. New York: Cambridge University Press.

Weisskopf, M. 1988. "Plastic Reaps a Grim Harvest in the Oceans of the World." *Smithsonian,* March, 59–66. Excellent review of the growing crisis of plastic in the ocean.

White, R. M. 1990. "The Great Climate Debate." *Scientific American,* July, 36–43. The author feels we should take steps to limit the extent of global warming.

Issues of *Oceanus*

Oceanus 20 (no. 4, 1977). Entitled "Oil in Coastal Waters."

Oceanus 24 (no. 1, 1981). Devoted to pollution and entitled "The Oceans as Waste Space."

Oceanus 27 (no. 1, 1984). Deals with the relationships between industry and the ocean.

Oceanus 32 (no. 2, 1989). Entitled "The Oceans and Global Warming."

Oceanus 33 (no. 2, 1990). Deals with ocean disposal.

Special Issues of *Scientific American*

Scientific American 261 (no. 3, September 1989). This special issue is subtitled "Managing Planet Earth."

Scientific American 266 (no. 6, June 1992). This special issue is subtitled "Meeting the Challenge of Sustainable Development."

AFTERWORD

The marine sciences are at the threshold of a new age. The recent revolutions in biology and geology are being assimilated, and the road ahead seems clearer. A revolution in the design of sampling devices, robot submersible vehicles, and data processing has brought new vigor to oceanography. Satellite-borne sensors can provide data in an instant that would have taken years to collect using surface ships. Shipboard technology has become so sophisticated that Wyville Thomson or Fridtjof Nansen would hardly recognize our sensors or sampling devices.

The tools may be different, but the spirits of those who use them remain the same. Today's marine scientists are like all the men and women who have gone before: *We want to know about the ocean.* We haunt our mailboxes for journals bearing the latest research news, search television listings for any new ocean shows, inspect new samples with the enthusiasm of little kids, and share our insights with anyone at the drop of a hat. I am personally delighted that you have traveled with me this far. Those of us who enjoy an oceanographic background (and this now includes you) look at the Earth with greater understanding than we did before we began. The whole concept of an ocean world appeals to us, gives us profound pleasure, and sobers us with a deep sense of responsibility. In no other field of science do so many ideas interweave to form so rich a tapestry.

Our journey together is over, but before we go our separate ways, I have three last ideas to share:

- Change has been a recurrent theme of this book. The Earth's climate has changed with time, as has its atmospheric composition, its ocean chemistry, the size and positions of its continents, and its life-forms. The Earth may seem a calm and stable home, but it is really a violent place for inhabitation by such seemingly delicate objects as living things. Even so, life and the ocean have grown old together. The story of the Earth is the story of change and chance; its history is written in the rocks, the water, and the genes of the millions of organisms that have evolved here. We are survivors.

  But that survival may now be in question. Change is now progressing at an unnatural rate, and these human-induced changes are imposing stress on natural systems. What we do *with* and *to* the ocean is literally of planetary consequence. In the last century we have developed the physical, chemical, and biological machinery to destroy or rejuvenate the world ocean and all of its life. A painful time of inadvertent global experimentation lies just ahead.

  All of us who love the colors and textures of this small wet world need to act to moderate the negative effects of the looming environmental crisis. In Chinese, the written character for the word *crisis* has two components: danger and opportunity. Informed citizens will express their concern, discuss this concern with others, and act whenever possible to minimize the threats and take advantage of new opportunities. Intelligence and beauty must triumph; we have no other rational alternative.

- Appreciation of the ocean doesn't come exclusively from the realm of science. Philosophers, artists, composers, and poets have had much to say about the sea. Read Homer's description of the ocean in *The Odyssey* (try books iv, x, and xi). See how Lord Byron's feeling for the ocean colors his poetry (see, for instance, *Childe Harold's Pilgrimage,* stanzas 183 and 184). Read modern poet Robinson Jeffers's powerful *Continent's End.* Share Prospero's marine magic in Shakespeare's *The Tempest.* Find some of the evocative woodcuts of Rockwell Kent and the impressionistic ocean paintings of English artist J. M. W. Turner. Listen to Benjamin Britten's *Four*

Sea Interludes from *Peter Grimes,* and Ralph Vaughan-Williams's *Sea Symphony* and *Sinfonia Antartica* (but take care not to blow out your sound system). Read the ocean novels of Herman Melville and Jack London, and try reading the journals and accounts of the famous explorers and scientists you have met in this book. Sit on a quiet beach at night with the stars of the Milky Way shining softly overhead. The pervasive inspiration of the wave-breathing ocean is never far away.

- Don't let your involvement stop here. Lifelong learning is the truest joy, a pleasure that does not diminish with age, a source of wisdom and calm. We can learn much about patience, hope, and optimism from the ocean. We can learn much about the world—and about ourselves—by looking for the oceanic connections among things. I hope your interest in learning about the ocean has just been kindled. There is much good in the world. Go and add to it.

Measurements and Conversions

Other than the United States, only three countries in the world—Burma, Tonga, and South Yemen—do not use metric measurements. The metric system, a contribution of the French Revolution, conquered Europe along with Napoleon. It is based on a decimal system, a system familiar to Americans because of our decimal money system: 10 cents to a dime, 10 dimes to the dollar.

The first move toward a rational system of measurement was made in 1670 by Gabriel Mouton, the vicar of St. Paul's Church in Lyon, France. Instead of the then-prevalent measurement system based on the width of the king's hand, or the length of his outstretched right arm, or the weight of a particular basket of stones kept in the palace, Mouton suggested a length measure based on the arc of 1 minute of longitude, to be subdivided decimally. Other measurements would follow from this unit of length. His proposal contained the three major characteristics of the metric system: using the Earth itself as a basis for measurement, subdividing decimally (by 10s), and using standard prefixes (*kilo, centi, milli*, and so on). These ideas were debated for 125 years before being implemented by a commission appointed by Louis XVI in one of his last official acts before being imprisoned during the French Revolution. One ten-millionth of the distance from the North Pole to the equator (on the line of longitude passing through Paris) was selected as the standard unit of length, the meter. A new unit of weight was derived from the weight of 1 cubic meter of pure water. Temperature was to be based on pure water's boiling and freezing points. A list of prefixes for decimal multiples and submultiples was proposed. In 1795, a firm decision was made to establish the system throughout France, and in 1799, the metric system was implemented "for all people, for all time."

At first, people objected to the changes, but the government insisted that old measurements be included side by side with the equivalent new (metric) ones. In everyday competition, the advantages of the metric system proved decisive; in 1840, it was declared a legal monopoly in France. The French public had been won over to the new, simple, rational system of measurement. All of Europe followed, and eventually, virtually all other countries.

Not the United States, however. Though Ben Franklin proposed that the country convert in the eighteenth century, the people of the United States have continued to insist that the metric system—now known as the Système International (SI)—is too difficult to learn and work with. The federal government has urged conversion to metric units to increase opportunities for international trade. In August 1988, President Ronald Reagan signed the Omnibus Trade and Competitiveness Act. This act amended the 1975 Metric Conversion Act, stating that by 1992, all federal agencies must, wherever feasible, use the metric (SI) system in their purchases, grants, and other business.

The government may be making the change, but the public clings tenaciously to inches, pints, and pounds. Why? Is it really simpler to add $\frac{1}{16}$ of an inch, $\frac{1}{32}$ of an inch, and $\frac{3}{8}$ of an inch to cut a bookshelf to length? Can you remember how many cups to a quart? How many pints to a gallon? How many miles to a league?[1] The reason we continue to use the English system (which, of course, the English have long since abandoned) is because it is familiar to us. We know how long 5 inches is, and how much a quart is, and what 72° Fahrenheit represents. Perhaps by following the French example—by having measurements expressed everywhere in English *and* metric measurements—we may be able to make a complete conversion within a generation or two. That's why English and metric measurement are used together throughout this book. The process has already begun, of course: You use 35mm film, 2-liter soft drink containers, 750-milliliter wine bottles, 100-watt light bulbs—and you might run a 10-K (10-kilometer) race on Saturday.

The conversion factors listed here will give you an idea of how English and metric (SI) units are equivalent. Don't panic—the system is as rational and logical as it has always been. Note that 1 meter equals 100 centimeters and that 1 centimeter equals 10 millimeters. (See the table showing multiples and submultiples for an explanation of the relationship between prefixes like *centi* and *milli*.) Note that 2.54 centimeters equals 1 inch. (See the table of conversion factors if you wish to convert from one system to another.) Some numerical oceanographic data are included in supplemental tables.

---

[1]A mile is 5,280 feet, and a league is 5,280 yards. Captain Nemo, in his fictional submarine *Nautilus*, traveled a distance greater than twice the circumference of the Earth in Jules Verne's fantasy *20,000 Leagues Under the Sea*, a task made even more formidable by his total dependence on English units of measurement!

Scientific Notation

Multiples and Submultiples

	Name	Common Prefixes
10^{12} = 1,000,000,000,000	trillion	tera
10^{9} = 1,000,000,000	billion	giga
10^{6} = 1,000,000	million	mega
10^{3} = 1,000	thousand	kilo
10^{2} = 100	hundred	hecto
10^{1} = 10	ten	deka
10^{-1} = 0.1	tenth	deci
10^{-2} = 0.01	hundredth	centi
10^{-3} = 0.001	thousandth	milli
10^{-6} = 0.000001	millionth	micro
10^{-9} = 0.000000001	billionth	nano
10^{-12} = 0.000000000001	trillionth	pico

Conversion Factors

Area

1 square inch (in.^{2})	6.45 square centimeters
1 square foot (ft^2)	144 square inches
1 square centimeter (cm^2)	0.155 square inch
	100 square millimeters
1 square meter (m^2)	10^4 square centimeters
	10.8 square feet
1 square kilometer (km^2)	247.1 acres
	0.386 square mile
	0.292 square nautical mile

Mass

1 kilogram (kg)	2.2 pounds
	1,000 grams
1 metric ton	2,205 pounds
	1,000 kilograms
	1.1 tons
1 pound	16 ounces
	454 grams
	0.45 kilogram
1 ton	2,000 pounds
	907.2 kilograms
	0.91 metric ton

Length

1 micrometer (μm)	0.001 millimeter
1 millimeter (mm)	1,000 micrometers
	0.1 centimeter
	0.001 meter
1 centimeter (cm)	10 millimeters
	0.394 inch
	10,000 micrometers
1 meter (m)	100 centimeters
	39.4 inches
	3.28 feet
	1.09 yards
1 kilometer (km)	1,000 meters
	1,093 yards
	3,280 feet
	0.62 statute mile
	0.54 nautical mile
1 inch (in.)	25.4 millimeters
	2.54 centimeters
1 foot (ft)	30.5 centimeters
	0.305 meter
1 yard	3 feet
	0.91 meter
1 fathom	6 feet
	2 yards
	1.83 meters
1 statute mile	5,280 feet
	1,760 yards
	1,609 meters
	1.609 kilometers
	0.87 nautical mile
1 nautical mile	6,076 feet
	2,025 yards
	1,852 meters
	1.15 statute miles
1 league	15,840 feet
	5,280 yards
	4,804.8 meters
	3 statute miles
	2.61 nautical miles

Pressure

1 atmosphere (sea level)	760 millimeters of mercury at 0°C
	14.7 pounds per square inch
	33.9 feet of water (fresh)
	29.9 inches of mercury
	33 feet of seawater

Temperature

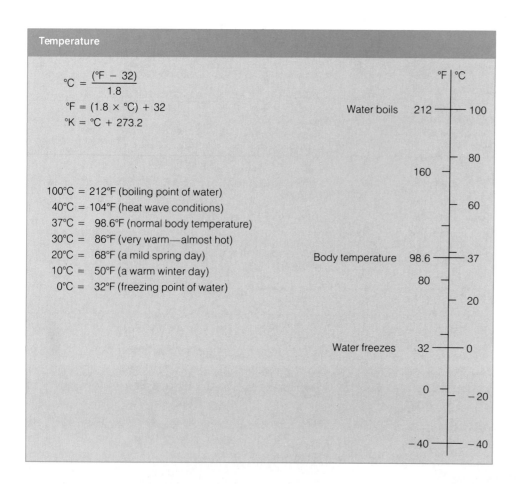

$$°C = \frac{(°F - 32)}{1.8}$$

$$°F = (1.8 \times °C) + 32$$

$$°K = °C + 273.2$$

100°C = 212°F (boiling point of water)
40°C = 104°F (heat wave conditions)
37°C = 98.6°F (normal body temperature)
30°C = 86°F (very warm—almost hot)
20°C = 68°F (a mild spring day)
10°C = 50°F (a warm winter day)
0°C = 32°F (freezing point of water)

Volume

1 cubic inch (in.³)	16.4 cubic centimeters
1 cubic foot (ft³)	1,728 cubic inches
	28.32 liters
	7.48 gallons
1 cubic centimeter (cc; cm³)	1 milliliter
	0.061 cubic inch
1 liter	1,000 cubic centimeters
	61 cubic inches
	1.06 quarts
	0.264 gallon
1 cubic meter (m³)	10^6 cubic centimeters
	264.2 gallons
	1,000 liters
1 cubic kilometer (km³)	10^9 cubic meters
	10^{15} cubic centimeters
	0.24 cubic mile

Time

1 hour	3,600 seconds
1 day	24 hours
	1,440 minutes
	86,400 seconds
1 calendar year	31,536,000 seconds
	525,600 minutes
	8,760 hours
	365 days

Speed

1 statute mile per hour	1.61 kilometers per hour
	0.87 knot
1 knot (nautical mile per hour)	51.5 centimeters per second
	1.15 miles per hour
	1.85 kilometers per hour
1 kilometer per hour	27.8 centimeters per second
	0.62 mile per hour
	0.54 knot

Some Familiar Metric Approximations

Measurement	Metric Unit	Approximate Size of Unit
Length	millimeter	diameter of a paper clip wire
	centimeter	a little more than the width of a paper clip (about 0.4 inch)
	meter	a little longer than a yard (about 1.1 yards)
	kilometer	somewhat further than $\frac{1}{2}$ mile (about 0.6 mile)
Mass (Weight)	gram	a little more than the mass (weight) of a paper clip
	kilogram	a little more than 2 pounds (about 2.2 pounds)
	metric ton	a little more than a short ton (about 2,200 pounds)
Volume	milliliter	five of them make a teaspoon
	liter	a little larger than a quart (about 1.06 quarts)
Pressure	kilopascal	atmospheric pressure is about 100 kilopascals

Source: U.S. Metric Board Report.

Numerical Oceanographic Data

Equivalences in Concentration of Seawater

Seawater with 35 grams of salt per kilogram of seawater	3.5 percent
	35 parts per thousand (‰)
	35,000 parts per million (ppm)

Speed of Sound

Velocity of sound in seawater at 34.85 parts per thousand (‰)	4,945 feet per second
	1,507 meters per second
	824 fathoms per second

Area, Volume, and Depth of the World Ocean

Body of Water	Area (10^6 km^2)	Volume (10^6 km^3)	Mean Depth (m)
Atlantic Ocean	82.4	323.6	3,926
Pacific Ocean	165.2	707.6	4,282
Indian Ocean	73.4	291.0	3,963
All oceans and seas	361	1,370	3,796

APPENDIX II

Geological Time

As we saw in Chapter 2, astronomers and geologists have determined that the Earth originated about 4.6 billion years ago. They have divided the Earth's age into eras, roughly corresponding to major geological and evolutionary changes that have taken place, as shown in the chart in **Figure 1**. Note that the time spans of the different eras are not shown to scale; if they were, the chart would run off the page.

Recall that life began fairly soon after the Earth's crust, atmosphere, and ocean formed. One way to conceive of the time span over which life evolved is to imagine it as a 24-hour clock, with life originating at midnight (**Figure 2**). In this scheme, invertebrates with hard parts (which make good fossils) became abundant about 4:30 P.M., and animals began to leave the ocean for land about 8 P.M. Our closest human ancestors (*Homo sapiens*) appeared about 2 seconds before midnight, agriculture began only ¼ second before midnight, and the industrial revolution has been around for ⁷⁄₁₀₀₀ of a second.

Era	Period	Epoch	Millions of Years Ago (mya)
CENOZOIC	Quaternary	Recent	0.01
		Pleistocene	1.65
	Tertiary	Pliocene	5
		Miocene	24
		Oligocene	37
		Eocene	58
		Paleocene	66
MESOZOIC	Cretaceous	Late	
			98
		Early	
			144
	Jurassic		
			208
	Triassic		
			245
PALEOZOIC	Permian		
			286
	Carboniferous		
			360
	Devonian		
			408
	Silurian		
			438
	Ordovician		
			505
	Cambrian		
			570
PROTEROZOIC			
			2,500
ARCHEAN			

Source: *Geological Time Scale, Decade of North American Geology*, Geological Society of America, 1983.

Figure 1   A Geological Time Scale

Times of Major Geological and Biological Events

1.65 mya to present. Major glaciations. Modern humans emerge and begin what may be greatest mass extinction of all time on land, starting with Ice Age hunters.

65–1.65 mya. Unprecedented mountain building as continents rupture, drift, collide. Major climatic shifts; vast grasslands emerge. Major radiations of flowering plants, insects, birds, mammals. Origin of earliest human forms.

65 mya. Asteroid impact? Mass extinction of all dinosaurs and many marine organisms.

135–65 mya. Pangea breakup continues, broad inland seas form. Major radiations of marine invertebrates, fishes, insects, dinosaurs. Origin of flowering plants.

181–135 mya. Pangea breakup begins. Rich marine communities. Major radiations of dinosaurs.

205 mya. Asteroid impact? Mass extinction of many organisms in seas, some on land; dinosaurs, mammals survive.

240–205 mya. Recovery, radiations of marine invertebrates, fishes, dinosaurs. Gymnosperms the dominant land plants. Origin of mammals.

240 mya. Mass extinction. Nearly all species in seas and on land perish.

280–240 mya. Pangea, worldwide ocean forms; shallow seas squeezed out. Major radiations of reptiles, gymnosperms.

360–280 mya. Tethys Sea forms. Recurring glaciations. Major radiations of insects, amphibians. Spore-bearing plants dominate. Gymnosperms present. Origin of reptiles.

370 mya. Mass extinction of many marine invertebrates, most fishes.

435–360 mya. Laurasia forms, Gondwana moves north. Vast swamplands, early vascular plants. Radiations of fishes continue. Origin of amphibians.

435 mya. Glaciations as Gondwana crosses South Pole. Mass extinction of many marine organisms.

500–435 mya. Gondwana moves south. Major radiations of marine invertebrates, early fishes.

550–500 mya. Landmasses dispersed near equator. Simple marine communities. Origin of animals with hard parts.

700–550 mya. Supercontinent Laurentia breaks up; widespread glaciations.

2,500–570 mya. Oxygen present in atmosphere. Origin of aerobic metabolism. Origin of protistans, algae, fungi, animals.

3,800–2,500 mya. Origin of photosynthetic bacteria.

4,600–3,800 mya. Formation of Earth's crust, early atmosphere, oceans. Chemical evolution leading to origin of life (anaerobic bacteria).

4,600 mya. Origin of Earth.

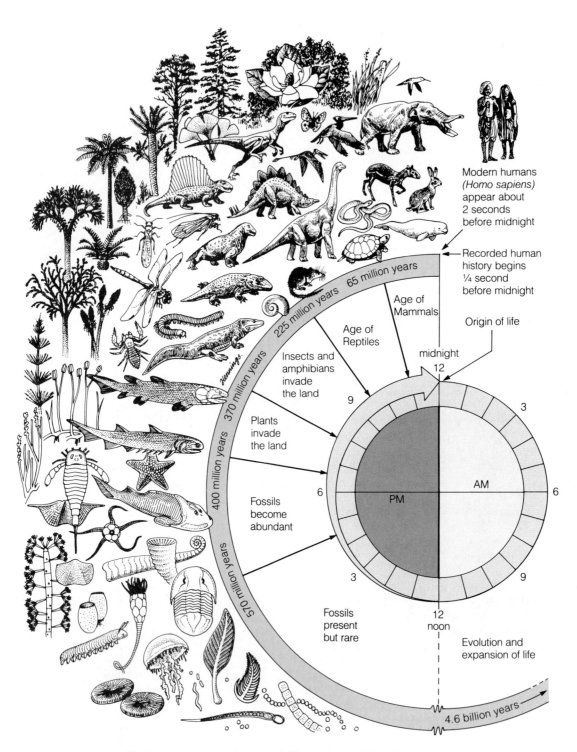

Figure 2 Greatly simplified history of the development of different forms of life on Earth through biological evolution, compressed to a 24-hour time scale.

Latitude and Longitude, Time, and Navigation

The ocean is large and easy to get lost in. A backyard, like that shown in **Figure 1**, is smaller, but we can still be lost in it if we don't have a frame of reference. Note that the yard is framed by a fence. We can refer to this frame to establish our position—in this case, at the intersection of perpendicular lines drawn from fence posts 2 and C. Many towns are arranged in this way: Fourth and D streets intersect at a precise spot based on the municipal frame of reference.

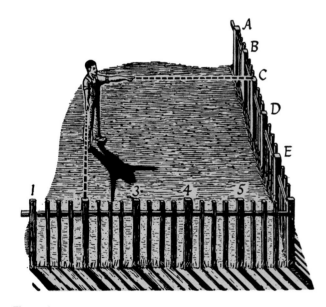

Figure 1

But the World Is Round: Spherical Coordinates

If the world were flat, a simple scheme of rectangular coordinates would serve all mapping purposes—a rectangle, like the yard in Figure 1, has four sides from which to measure. A sphere has no edges, no beginnings or ends, so what shall we use as a frame of reference for the Earth? Since the Earth turns, the poles, the axis of rotation, are the only absolute points of reference. We can draw an imaginary line equidistant from the North and South poles, a line that *equates* the globe into north-

ern and southern halves: the equator. Other lines, drawn parallel to the equator, further divide the sphere north and south of the equator. These lines, or parallels, are lines of **latitude** (**Figure 2**).

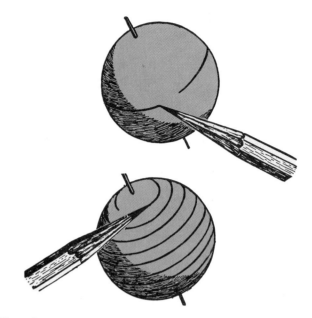

Figure 2

We can further subdivide the Earth by drawing lines at regular intervals through both poles. Note that unlike the parallels, these lines, called meridians, are all equally long. Meridians are lines of **longitude** (**Figure 3**).

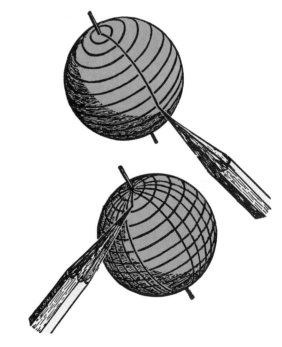

Figure 3

If you travel north from the equator, you can count the parallels (lines of latitude) that cross your path to find out how far you have gone. Likewise, if you travel east from a reference meridian, you can count the meridians (lines of longitude) that cross your path to find out how far you have gone. Just as a football player on the field knows his distance from the goal by the yard lines that cross his run, so you know how far north or east you have gone by the lines that have crossed your path.

Since there are no continuous lines of fence posts on the spherical Earth, our reference frame for latitude and longitude must be marked from the equator and poles by some other means. This is done by degrees.

Why Degrees?

Degrees measure fractions of a circle. We need to know what fraction of the Earth's circumference separates us from the equator and from the reference meridian to have a definite idea of our location.

Babylonian astronomers first divided the circle into 360 degrees (°). Why 360? The moon cycles around the Earth every 30 days. It takes about 12 months ("moonths") to make a year. Thus, $30 \times 12 = 360$, the number of days they supposed was in a year. Circles were divided the same way. As we saw in Chapter 2, the Greek librarian Hipparchus applied this division to the surface of the Earth.

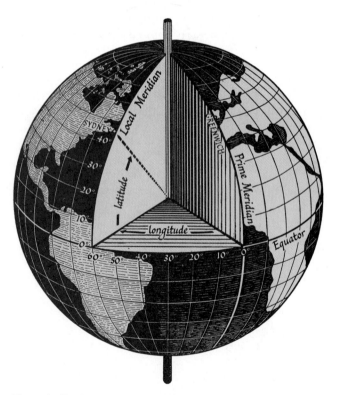

Figure 4 The latitude and longitude of Sydney, Canada.

In **Figure 4**, we have marked the position of Sydney, Canada. A line drawn from Sydney to the center of the Earth intersects the plane of the equator at an angle of 46° to the north. That is its latitude.

The reference meridian, the meridian from which all others are marked, is known as the prime meridian. Unlike the equator, there is no earthly reason why the prime meridian should pass through any particular place. It passes through Greenwich, England, because an international agreement signed in 1884 decreed it so. The meridian on which Sydney, Canada, lies intersects the plane of the prime meridian at an angle of 60°. The angular distance of Sydney from the prime meridian is 60° to the west. That is its longitude. So its position is 46°N 60°W.

We can do this for each hemisphere. A line drawn to the center of the Earth from Sydney, *Australia*, intersects the plane of the equator at an angle of 34° south latitude. Sydney, Australia, lies 151° east of the prime meridian. Thus, its position is 34°S 151°E. (Note that the greatest possible longitude is 180°; once you pass 180°, the line opposite the prime meridian, you begin to come around the other side of the Earth, and the angle to Greenwich decreases.)

What Does Time Have to Do with This?

Meridians are often numbered from the prime meridian in 15s. The Earth takes 24 hours to complete a 360° rotation. Divide 360° by 24 hours and you get 15, the number of degrees the sun moves across the sky in 1 hour. Meridians on a globe are often spaced to represent 1 hour's turning of the Earth toward or away from the sun, toward or away from noon.

You can use this fact to find your east-west position, your longitude. Imagine that you have a radio that can tell you the precise time of noon at Greenwich.[1] If your local noon comes *before* Greenwich noon, you are east of Greenwich. For instance, if the sun is highest in your sky at 10 A.M. Greenwich time, you are 2 hours before—30° east—of Greenwich. The Earth must turn 2 more hours before the sun will shine directly above the Greenwich meridian. If your local noon is *after* Greenwich noon, you are west of Greenwich. Suppose that the sun is at high noon and your chronometer, set at Greenwich time, says 6 P.M. That means that the Earth has been turning 6 hours since noon at Greenwich, and 6 hours times 15° per hour is 90°. That's your longitude relative to Greenwich: 90°W.

---

[1] Any shortwave radio will do. Tune it to 2.5, 5, 10, 15, or 20 mHz for radio stations WWV (Colorado) or WWVH (Hawaii). These stations broadcast time signals giving a measure of coordinated universal time, an international time standard based on the time at Greenwich. For a telephone report of coordinated universal time, call WWV at (303) 499-7111.

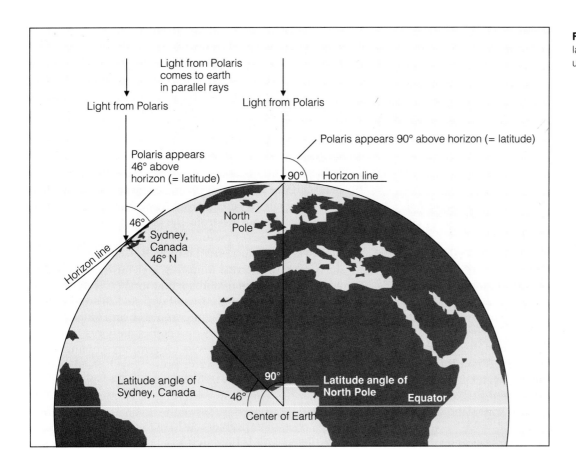

Navigation

Longitude is half the problem. To find latitude and obtain a position, we need to measure the angle north or south of the equator. But we can't use the time difference between local noon and Greenwich noon to determine longitude because the sun moves from east to west and we want to measure north-south position. Instead, we use the angle of the North Star above the horizon. Polaris, the current North Star, lies almost exactly above the North Pole. If we were standing at the North Pole, the North Star would appear almost directly overhead; ideally, the angle from the horizon to the star would be 90°, the same as the latitude of the North Pole (**Figure 5**). At Sydney, Canada, the angle from the horizon to the star would be about 46°—again, the same as the latitude. What would the angle of Polaris be at the equator, 0° latitude? (If you enjoyed your high school geometry course, you might try to prove that the angle from the horizon to Polaris is equal to the latitude at any position in the Northern Hemisphere.)

Polaris is not visible in the Southern Hemisphere, so how can we find south latitude? By finding the angle above the horizon of other stars. In practice, navigators in both hemispheres use a sextant to measure angles from the horizon to selected stars, planets, the moon, and the sun. The time of the observation is carefully noted. The navigator takes these readings to his or her stateroom, consults a series of mathematical tables, does some relatively simple calculations to compensate for observational errors, and comes up to the pilothouse with the vessel's latitude and longitude, accurate (in the best of circumstances) to within $\frac{1}{2}$ mile, marked on a small slip of paper. The daily results are always entered into the ship's log.

New Tricks

Discovering position by measuring the angular positions of heavenly bodies—celestial navigation—is a dying art. Global positioning satellites, loran-C, inertial platforms, radar, and other electronic wonders have largely replaced the romance of a navigator standing on the bridge squinting through a sextant. The slip of paper has been supplanted by the glow of back-lit liquid-crystal readouts or a chart with an X marking the ship's position, accurate to within 50 feet, feeding out of a slot. Still, when the power fails, the human navigator becomes the most popular person on board.

APPENDIX IV
Maps and Charts

It is easier to draw a diagram to show someone how to get to a place than to describe the process in words. For centuries, travelers have made special diagrams—maps and charts—to jog their own memories and to show others how to reach distant destinations. A **map** is a representation of some part of the Earth's surface, showing political boundaries, physical features, cities and towns, and other geographical information. A **chart** is also a representation of the Earth's surface, but it has been specially designed for convenient use in navigation. It is intended to be worked on, not merely looked at. A **nautical chart** is primarily concerned with navigable water areas. It includes information such as coastlines and harbors, channels, obstructions, currents, depths of water, and the positions of aids to navigation.

Any flat map or chart is necessarily a distortion of the spherical Earth. If we roll a flat sheet of paper around a globe to form a cylinder, the paper will contact the globe only along one curve. Let's assume that it's the equator. If the lines of latitude and longitude on the globe are covered with ink, only the equator will contact the paper and print an exact replica of itself. Unroll the cylinder, and that part of the new map will be a perfect representation of the Earth. To include areas north and south of the equator, we will have to "throw them forward" onto the paper; we need to *project* them in some way.

Now imagine our globe to be a translucent sphere. If we place a bright light at its center, we can project the lines of latitude and longitude onto the rolled paper cylinder (**Figure 1**). Careful tracing of these lines will result in a map, but the areas away from the equator will be distorted: The farther from the equator, the greater the distortion. A useful modification of this projection—one that does not distort high latitudes as dramatically—was devised by Gerhardus Mercator, a Flemish cartographer who published a map of the world in 1569. Though landmasses and ocean areas are not depicted as accurately in a Mercator projection as they would be on a globe, such a map is still useful because it enables mariners to steer a course over long distances by plotting straight lines.

The distortion in Mercator projections has led generations of school children to believe that Greenland is the same size as South America (**Figure 2**). Mercator charts can distort our perceptions of the ocean as well: The area of the continental shelves at high latitudes, the amount of primary productivity in the polar regions, and the importance of ocean currents at the northerly or southerly extremes of an ocean basin may be exaggerated if presented in Mercator projection. The projection used in this

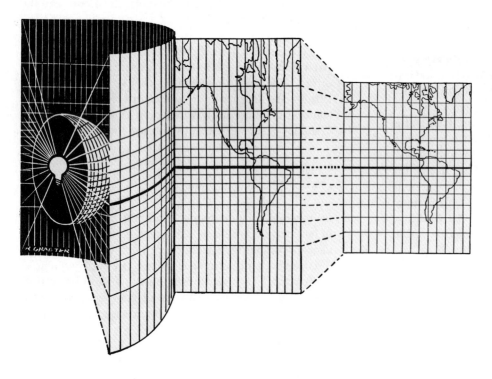

Figure 1 Central projection of a globe upon a cylinder, and a modified map structure, the Mercator, made to the same scale along the equator.

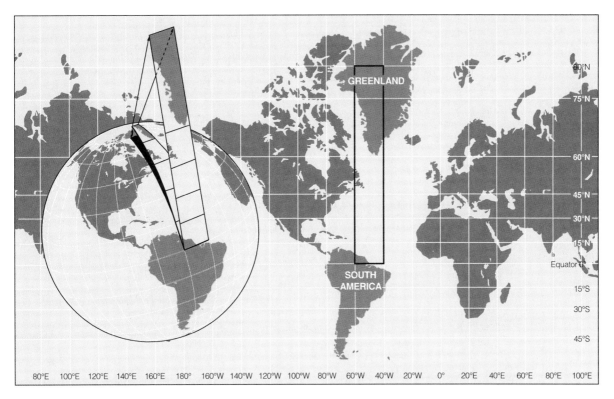

Figure 2 A gore of the globe peeled and projected according to the scheme devised by Gerhardus Mercator. This is the projection used on modern sailing charts. Note that this projection's distortion at high latitudes makes Greenland and South America appear about the same size. Next time you're near a globe, check their real sizes.

book—a further modification of the mercator projection known as the Miller projection—was chosen for its more accurate representation of surface area at high latitudes.

Mapmakers have invented other projections, each with advantages and disadvantages for particular uses. Some are conical projections: a flat sheet of paper wrapped into a cone with its edge touching the globe at a line of latitude north (or south) of the equator and the point of the cone above the North (or South) Pole. Conical projections do not distort high-latitude areas in the same way a Mercator projection does, and if drawn for the ocean area in which a mariner is sailing, can be used to draw great circle routes as straight lines. However, the distortions inherent in a conical projection prevent it from being used to represent more than about one-third of the globe on a single sheet of paper. Other projections, like the point-contact projection shown in **Figure 3**, try to minimize distortion around a specific location. All map and chart projections are distorted in some way; a sphere cannot be flattened onto a plane without deformation. Marine scientists necessarily become familiar with various chart projections and are careful to use the proper chart for its intended purpose.

Figure 4 is a Mercator projection of the world. On it are indicated areas of interest discussed in this book.

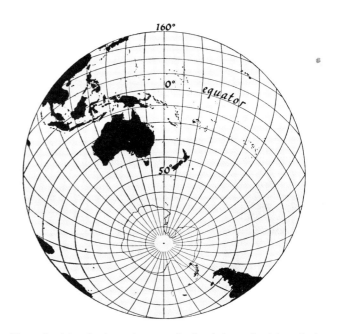

Figure 3 A Lambert equal-area projection (a type of point-contact projection), centered at 50°S, 160°E.

Figure 4 A map of the world, with many oceanic features labeled.

For Further Study

Krause, G., and M. Tomczak. 1995. "Do Marine Scientists Have a Scientific View of the Earth?" *Oceanography* 8 (no. 1): 11–16. Chart distortions often distort our interpretation of data, as this well-illustrated paper demonstrates.

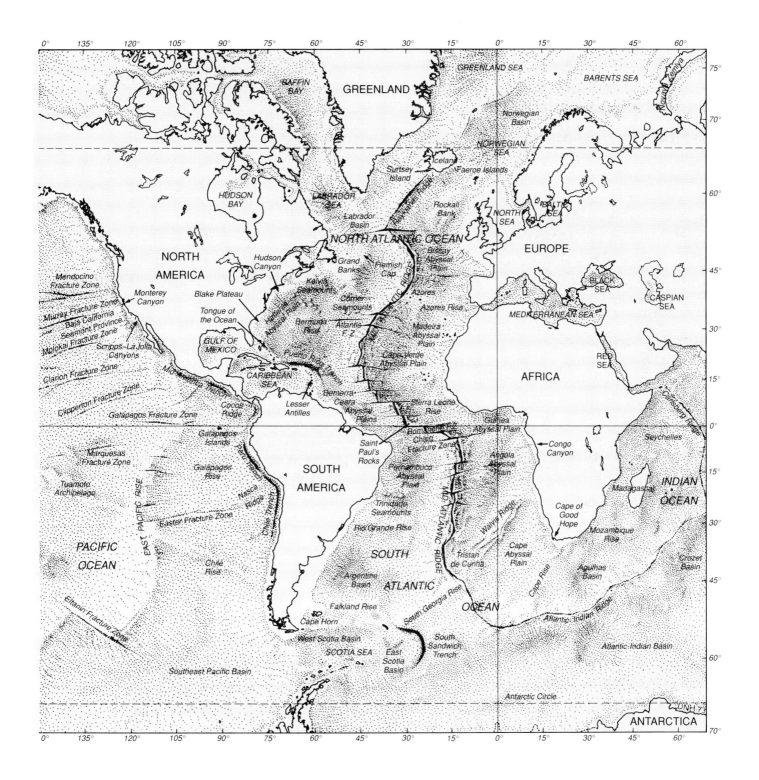

Taxonomic Classification of Marine Organisms

Exclusively nonmarine phyla generally have been omitted, along with most extinct phyla and classes.

KINGDOM MONERA. Bacteria and cyanobacteria, either single cells or simple associations of cells; autotrophic and heterotrophic forms.

KINGDOM PROTISTA. Eukaryotic single-celled forms.

PHYLUM CHRYSOPHYTA. Diatoms, coccolithophores, silicoflagellates.

PHYLUM PYRROPHYTA. Dinoflagellates, zooxanthellae.

PHYLUM CRYPTOPHYTA. Some "microflagellates"; cryptomonads.

PHYLUM EUGLENOPHYTA. A few "microflagellates"; mostly freshwater.

PHYLUM ZOOMASTIGINA. Nonphotosynthesizing flagellated protozoa.

PHYLUM SARCODINA. Amoebas and their relatives.
Class Rhizopodea. Foraminiferans.
Class Actinopodea. Radiolarians.

PHYLUM CILIOPHORA. Ciliated protozoa.

KINGDOM FUNGI. Fungi, mushrooms, molds, lichens; mostly land, freshwater, or highest supratidal organisms; heterotrophic.

KINGDOM PLANTAE. Photosynthetic autotrophs.

DIVISION CHLOROPHYTA. Multicellular green algae.

DIVISION PHAEOPHYTA. Brown algae, kelps.

DIVISION RHODOPHYTA. Red algae, encrusting and coralline forms.

DIVISION ANTHOPHYTA. Flowering plants (angiosperms). Most species are freshwater or terrestrial. Marine eelgrass, manatee grass, surfgrass, turtle grass, salt marsh grasses, mangroves.

KINGDOM ANIMALIA. Multicellular heterotrophs.

PHYLUM PLACOZOA. Amoeba-like multicellular animals.

PHYLUM MESOZOA. Wormlike parasites of cephalopods.

PHYLUM PORIFERA. Sponges.

PHYLUM CNIDARIA. Jellyfish and their kin; all are equipped with stinging cells.
Class Hydrozoa. Polyplike animals that often have a medusalike stage in their life cycle, such as Portuguese man-of-war.
Class Scyphozoa. Jellyfish with no (or reduced) polyp stage in life cycle.
Class Anthozoa. Sea anemones, coral.

PHYLUM CTENOPHORA. "Sea gooseberries," comb jellies; round, gelatinous, predatory, common.

PHYLUM PLATYHELMINTHES. Flatworms, tapeworms, flukes; many free-living predatory forms, many parasites.

PHYLUM NEMERTEA. Ribbon worms.

PHYLUM GNATHOSTOMULIDA. Microscopic, wormlike; live between grains in marine sediments.

PHYLUM GASTROTRICHA. Microscopic, ciliated; live between grains in marine sediments.

PHYLUM ROTIFERA. Ciliated; common in fresh water, plankton, and attached to benthic objects.

PHYLUM KINORYNCHA. Small, spiny, segmented, wormlike; live between grains in marine sediments; all marine.

PHYLUM ACANTHOCEPHALA. Spiny-headed worms; all parasitic in vertebrate intestines.

PHYLUM ENTOPROCTA. Polyplike, small, benthic suspension feeders.

PHYLUM NEMATODA. Roundworms. Common, free-living, parasitic.

PHYLUM BRYOZOA. Common, small, encrusting colonial marine forms.

PHYLUM PHORONIDA. Shallow-water tube worms; suspension feeders; a few centimeters long; all marine.

PHYLUM BRACHIOPODA. Lampshells; bivalved animals, superficially like clams; scarce, mainly in deep water.

PHYLUM MOLLUSCA. Mollusks.
Class Monoplacophora. Rare deep-water forms with limpetlike shells.
Class Polyplacophora. Chitons.
Class Aplacophora. Tusk shells, sand burrowing.
Class Gastropoda. Snails, limpets, abalones, sea slugs, pteropods.
Class Bivalvia. Clams, oysters, scallops, mussels.
Class Cephalopoda. Squid, octopuses.

PHYLUM ARTHROPODA.
Subphylum Crustacea. Copepods, barnacles, krill, isopods, amphipods, shrimp, lobsters, crabs.
Subphylum Chelicerata. Horseshoe crabs, sea spiders.
Subphylum Uniramia. Insects, centipedes, millipedes; one genus and five species in the ocean.

PHYLUM PRIAPULIDA. Small, rare, wormlike, subtidal.

PHYLUM SIPUNCULA. Peanut worms; all marine.

PHYLUM ECHIURA. Spoon worms.

PHYLUM ANNELIDA. Segmented worms; includes polychaetes such as feather duster worms.

PHYLUM TARDIGRADA. "Water bears"; tiny, eight-legged animals with the ability to survive long periods of hibernation.

PHYLUM PENTASTOMA. Tongue worms; parasites of vertebrates.

PHYLUM POGONOPHORA. Beard worms; no digestive system; deep-water tube worms; all marine.

PHYLUM ECHINODERMATA. Spiny-skinned, benthic, radially symmetrical, most with a water-vascular system.
- Class Asteroidea. Sea stars.
- Class Ophiuroidea. Brittle stars, basket stars.
- Class Echinoidea. Sea urchins, sand dollars, sea biscuits.
- Class Holothuroidea. Sea cucumbers.
- Class Crinoidea. Sea lilies.

PHYLUM CHAETOGNATHA. Arrowworms; stiff-bodied, planktonic, predaceous, common.

PHYLUM HEMICHORDATA. Acorn worms; unsegmented burrowers.

PHYLUM CHORDATA.
- Subphylum Urochordata. Sea squirts, tunicates, salps.
- Subphylum Cephalochordata. Lancelets, *Amphioxus*.
- Subphylum Vertebrata.
  - Class Agnatha. Jawless fishes: lampreys, hagfishes; cartilaginous skeleton.

Class Chondrichthyes. Sharks, skates, rays, sawfish, chimeras; cartilaginous skeleton.
Class Osteichthyes. Bony fishes.
Class Amphibia. Frogs, toads, salamanders; no marine species.
Class Reptilia. Sea snakes, turtles, one species of crocodile.
Class Aves. The birds.
- Order Sphenisciformes. Penguins.
- Order Procellariiformes. Albatrosses, petrels.
- Order Charadriiformes. The gulls.
- Order Pelecaniformes. The pelicans.

Class Mammalia. Warm-blooded, with hair and mammary glands.
- Order Cetacea. Whales, porpoises, dolphins.
- Order Sirenia. Manatees.
- Order Carnivora. Two marine families.
  - Suborder Pinnipedia. Seals, sea lions, walruses.
  - Suborder Fissipedia. Sea Otters.
- Order Primates. One family that regularly enters the ocean.
  - Family Hominidae. Humans.

Working in Marine Science

Working in the marine sciences is wonderfully appealing to many people. They sometimes envision a life of diving in warm, clear water surrounded by tropical fish, or descending to the seabed in an exotic submersible outfitted like Captain Nemo's fictional submarine in *20,000 Leagues Under the Sea*, or living with intelligent dolphins in a marine life park. Then reality sets in. There are rewards from working in the marine sciences, but they tend to be less spectacular than the first dreams of students looking to the ocean for a life's work.

A marine science worker is paid to bring a specific skill to a problem. If that problem lies in warm, tropical water or in a marine park, fine. But more likely, the problem will yield only to prolonged study in an uncomfortable, cold, or dangerous environment. The intangible rewards can be great; the physical rewards are often slim. Having said that, let me add that no endeavor is more interesting or exciting, and few are more intellectually stimulating. Doing marine science is its own reward.

Training for a Job in Marine Science

Marine science is, of course, science. And science requires mathematics—you need math to do the chemistry, physics, measurements, and statistics that lie at the heart of science. Your first step in college should be to take a math placement test, enroll in an appropriate math class, and spend time doing math. *Math is the key to further progress in any area of marine science.*

With your math skills polished, start classes in chemistry, physics, and basic biology. Surprisingly, except for one or two introductory marine science classes, you probably won't take many marine science courses until your junior year. These introductory classes will be especially valuable because a balanced survey of the marine sciences can aid you in selecting an appealing specialty. Then, with a good foundation in basic science, you can begin to concentrate in that specialty.

Other skills are important too. The ability to write and speak well is crucial in any science job. Also critical is computer literacy—preferably DOS, not Macintosh, by the way. Expertise in photography or foreign languages

or the ability to field-strip and rebuild a diesel engine or hydraulic winch will put you a step above the competition at hiring time. Certification as a scuba diver is almost mandatory; you can never have too much diving experience. (Remember, though, diving is only a tool, a way to deliver an informed set of eyes and an educated brain to a work site.) You should be in good health. Indeed, good aerobic fitness is essential in most marine science jobs; stamina is often a crucial factor in long experiments under difficult conditions at sea. It is also desirable to be physically strong—marine equipment is heavy and often bunglesome. And it helps greatly if you are not prone to seasickness.

Deciding what school to attend will depend on your skills. Readers of this book will probably be enrolled in a general oceanography course in a college or university. The first step would be to discuss your interests with your professor (or his or her teaching assistants). You'll need to attend a four-year college or university to complete the first phase of your training. If you're attending a two-year institution, picking a specific transfer institution can come later, but keep a few things in mind: No matter where you take your first two years of training, you need thorough preparation in basic science. You should attend an institution with strengths in the area of your specialty (such as geology, biology, and marine chemistry). And you should be reasonable in your expectations of acceptance if you're a transfer student (that is, don't try for Stanford or Yale with a B average).

Another thing: Most marine scientists have completed a graduate degree (a master's degree or doctorate). Most graduate students hold teaching or research assistantships (that is, they get paid for being grad students). In all, progress to a final degree is a long road, but the journey is itself a pleasure.

If the thought of four or more years of higher education doesn't appeal, does that mean there's no hope? Not at all. Many students begin a program with the goal of becoming a marine technician, animal trainer at a marine life park, marina or boatyard employee or manager, or crew member on a private yacht. Those jobs don't always require a bachelor's degree. Jobs at Sea World and other marine theme parks do require athletic ability, extreme patience, public speaking skills, a love of animals, and, usually, diving experience. Few positions are available, but there is some turnover in the ranks of junior trainers, and being hired is certainly possible.

Becoming a marine technician is an especially attractive alternative to the all-out chemistry-physics-math academic route. For every highly trained marine scientist, there are perhaps five technical assistants who actually do the experiments, maintain the equipment, work daily with organisms, and build special apparatus. Marine technicians tend to spend more time at hands-on tasks than marine scientists. Most of these folks (including the author of the letter that ends this appendix) have the equivalent of a two-year technical degree, usually from a community college.

Don't quit your job, burn your bridges, leave your family, sell your possessions, and dedicate yourself monklike to marine science. Do some investigation. Nothing is as valuable as *actually going out and talking to people who do things that you'd like to do*. Ask them if they enjoy their work. Is the pay okay? Would they start down the same road if they had it to do all over again? You may decide to expand your involvement in marine science in a more informal way, by becoming a volunteer; joining the Sierra Club, Audubon Society, Greenpeace, or other environmental group; working for your state's fish and game office as a seasonal aide; or attending lectures at local colleges and universities.

If you decide to continue your education, don't be discouraged by the time it will take. Have a general view of the big picture, but proceed one semester at a time. Again, remember that the educational journey is itself a great pleasure. Don Quixote reminds us of the joys of the road, not the inn.

The Job Market

Marine science is very attractive to the general public. People are naturally drawn to thoughts of working in the field. Unfortunately, there aren't a great many jobs in the marine sciences. But there will always be some jobs, and people will fill them. Those people will be the best prepared, most versatile, and most highly motivated of those who apply. Perhaps not surprisingly, marine biology is the most popular marine science specialty. Unfortunately, it is also the area with the smallest number of nonacademic jobs. Museums, aquariums, and marine theme parks employ biologists to care for animals and oversee interpretive programs for the public. A few marine biologists are employed as monitoring specialists by water management agencies like sanitation districts, which discharge waste into the ocean. Electrical utilities that use seawater to cool the condensers in power generating plants almost always have a handful of marine biologists on staff to watch the effects of discharged heat on local marine life and to write the reports required by watchdog agencies. State and federal agencies employ marine biologists to read and interpret those documents and to set standards. Relatively small businesses, like private shipyards, agricultural concerns, and chemical plants, can't afford their own staff biologists, so private consulting firms staffed by marine biologists and other specialists have arisen to assist in the preparation of the environmental impact reports required of businesses under various legislation.

There are more jobs in physical oceanography: marine geology, ocean engineering, and marine chemistry and physics. Thousands of marine geologists work for oil and mineral companies; indeed, with the increasing emphasis on offshore resources, the market for these people may be increasing. Marine engineers are needed to design, construct, and maintain offshore oil rigs, ships, and harbor structures. Marine chemists are hard at work figuring ways to stop corrosion and to extract chemicals from seawater. Physicists are vitally interested in the transmission of underwater sound and light, in the movement of the ocean, and in the role the ocean plays in global weather and climate. Economists, lawyers, writers, and mathematicians also work in the marine science field.

Many biological and physical oceanographers are teachers and professors. Indeed, there are nearly as many marine scientists employed in the academic world as there are in private industry and government. If you like the idea of teaching, you might consider this avenue. The demand for science teachers at all educational levels is already great and is expected to increase.

Four factors will be significant in influencing your employability:

1. *Experience*. Employers are favorably impressed by experience, especially work experience related to the duties of the position for which you are applying. Volunteer work counts.

2. *Grades*. Good grades are important, especially for positions in government agencies. A grade point average of 3.0 or higher in all college work increases your chances of employment and should give you a higher starting salary.

3. *Geographical availability*. Don't restrict yourself geographically. Not everyone can work in Hawaii or California, but four out of ten marine scientists work in just three states: California, Maryland, and Virginia.

4. *Diversity*. Again, mastery of more than one specialty gives you an employment edge. Being a plankton connoisseur *and* being able to repair a balky computer while ordering in-port supplies over a radiotelephone in Spanish makes a lasting impression.

Report from a Student

Students in marine science programs graduate, get jobs, and move on. One of the pleasures of being a professor is hearing from them. One of our former students, an employee of the Marine Science Institute at the University of California, Santa Barbara, recently reported his activities as part of a team using the submersible *Alvin* to

Figure 1 Dan Dion attaching hydraulic actuators to water-sampling bottles, in preparation for a dive by *Alvin*. The bottles were part of a sampling program that included measurements of water conductivity, transmissivity, temperature, and iron and manganese ion content near a hydrothermal vent. This information was later merged with data from transponder navigation to obtain a three-dimensional map of plume structure and chemistry.

investigate plumes of warm water issuing from hydrothermal vents along the southern Juan de Fuca Ridge. The nature of his work—and his enthusiasm for it—is clearly evident in this excerpt. Dan Dion writes:

The buoyant plume experiment wasn't going very well. The chemistry dives were pushed back because of technical difficulties and poor weather (rough seas cut two dives). The first two buoyant plume dives ended in failure. The first one because of mechanical/ electrical problems, the second because of a computer crash. Everyone worked around the clock to get things in order for dive 2440. I was scheduled to go down with John Trefrey, from the Florida Institute of Technology. Cindy Van Dover was our pilot. We launched *Alvin* right on schedule at 0800, and descended from the glacier blue water into the bioluminescent snowstorm of the euphotic zone. During the hour and a half descent we listened to the music of Enya in the soft light of the sub as we busily prepared ourselves for the experiment: booting up the computers, loading film in the cameras, tapes in the recorders, etc. We had three laptop computers to deal with in the cramped spaces of the sub. I was in charge of two of them, one that plotted our in-sub navigation (from transponders), and the other that controlled

and recorded data from the [continuous temperature-depth-conductivity probe]. The third laptop was connected to the chemical analyzer and John was in control of that. All of the instruments were operating perfectly. I periodically saved the computer file in the event of another crash. We reached the bottom right on target; Monolith Vent was in sight, 2261 meters below the surface. We did a video survey of the vent, especially a chimney that was rapidly growing back after the geologists had decapitated it just a few days earlier. We ascended to 55 meters-off-bottom and began our drive-throughs. To me, the navigator, it was the ultimate video game. From the computer screen I would guide *Alvin* through a dark abyss, calling out headings that would maneuver us into a "lawn mower" pattern crisscrossing the plume. It was quite visible, and even beautiful; wispy, intricate patterns of "smoke" which seemed to dance like graceful ghosts. We completed passes at 35, 20, 10, and 5 meters above the bottom, then one last one at 45 meters. Eight hours of sub time went by so quickly! Our dive was a huge success; in addition to all the samples we obtained, we generated over 25 megabytes of data. I used everything I learned . . . from computer skills to navigation and marlinspike seamanship (and, of course, chemistry!).

Figure 2 Dan Dion enters the hatch of *Alvin* to visit hydrothermal vents 2,261 meters (7,416 feet) below the surface off the coast of Oregon.

People in Marine Science

The last two decades have seen a significant change in the profile of people involved in oceanographic research in the United States. Some of these changes are summarized here.

1. The number of oceanographers employed in all sectors increased from 1,130 in 1973 to 2,460 in 1989.

2. The percent of oceanographers who consider teaching as their primary activity decreased from 21% in 1973 to 11% in 1989.

3. The percent of oceanographers who consider research as their primary activity remained stable at 40%.

4. In 1989, a majority of Ph.D. oceanographers (60%) were employed at educational institutions.

5. The distribution of oceanographers by discipline (biological, chemical, geological, physical) has remained almost constant.

6. The percentage of women oceanographers has increased from 4% in 1973 to 11% in 1989. In 1992, 30% of students at major oceanographic institutions were women.

7. The two largest oceanographic institutions are the Scripps Institution of Oceanography (affiliated with the University of California, San Diego), and the Woods Hole Oceanographic Institution. In 1970 the faculty at these two institutions constituted 40% of the field. In 1990 they constituted only 25%.

Source: "Workshop Report, National Science Foundation Conference on Undergraduate Programs in Ocean Sciences." Galveston, Texas, 1992.

Oceanography Programs in North America

Here is a list of the largest and most comprehensive marine science teaching and research institutions in North America.

Bedford Institute of Oceanography, Dartmouth, Nova Scotia, Canada
Dalhousie University, Halifax, Nova Scotia, Canada
Louisiana State University, Baton Rouge, LA 70803
Massachusetts Institute of Technology, Cambridge, MA 02139
McGill University, Montreal, Canada
Mississippi State University, Starkville, MS 39759
North Carolina State University, Raleigh, NC 27695
Oregon State University, Corvallis, OR 97331-2134
State University of New York, Stony Brook, NY 11794-5001
Texas A & M University, Galveston, TX 77553-1675
University of Alaska, Fairbanks, AK 99775-5040
University of British Columbia, Vancouver BC, Canada
University of California, San Diego, La Jolla, CA 92093-0232[1]
University of Connecticut, Groton, CT 06340
University of Delaware, Newark, DE 19716
University of Florida, Gainesville, FL 32611-0409
University of Georgia, Athens, GA 30602
University of Hawaii, Honolulu, HI 96822
University of Maine, Orono, ME 04469-5715
University of Maryland, College Park, MD 20742
University of Michigan, Ann Arbor, MI 48109-2099
University of New Hampshire, Durham, NH 03824
University of North Carolina, Chapel Hill, NC 27514
University of Rhode Island, Narragansett, RI 02882-1197
University of South Carolina, Charleston, SC 29401
University of Southern California, Los Angeles, CA 90089-1231
University of Virginia, Charlottesville, VA 22903
University of Washington, Seattle, WA 98105
Woods Hole Oceanographic Institution, Woods Hole, MA 02543

[1]The Scripps Institution of Oceanography is affiliated with the University of California, San Diego.

Marine science is equipment training sessions, long cruises, seminars and lectures, visiting experts, hot sand volleyball games, and chilly labs with classical music. Marine science is a long and demanding road; but it is, quite honestly, great fun. Captain Nemo in his sub never had it this good!

For More Information

1. Consult the catalog of any college or university offering a marine science curriculum.

2. Send for a copy of *Ocean Opportunities, a Guide to What the Oceans Have to Offer*. The pamphlet is available from The Marine Technology Society, 1828 L Street, NW, Washington, DC 20036, (202) 775-5966.

3. See: Wunsch, Carl. 1993. "Marine Science in the Coming Decades." *Science*, January 15, vol. 259, no. 5093:296–297. An expert oceanographer's glimpse into the future.

4. See: Yentsch, C. M. and C. J. Snyderman. 1992. *The Woman Scientist: Meeting the Challenges for a Successful Career*. New York: Plenum. Two marine scientists examine the facets of surviving and succeeding in academia and research science.

5. See: American Meteorological Society, 1992. *Curricula in the Atmospheric, Oceanic, and Related Sciences*. Boston: American Meteorological Society. Source of current information on marine science programs in the United States, Canada, and Puerto Rico.

GLOSSARY

absorption Conversion of sound or light energy into heat.

abyssal hill Small sediment-covered inactive volcano or intrusion of molten rock less than 200 meters (650 feet) high, thought to be associated with seafloor spreading. Abyssal hills punctuate the otherwise flat abyssal plain.

abyssal plain Flat, cold, sediment-covered ocean floor between the continental rise and the oceanic ridge at a depth of 3,700 to 5,500 meters (12,000 to 18,000 feet). Abyssal plains are more extensive in the Atlantic and Indian oceans than in the Pacific.

abyssal zone The ocean between about 4,000 and 5,000 meters (13,000 and 16,500 feet) deep.

accessory pigment One of a class of pigments (such as fucoxanthin, phycobilin, and xanthophyll) present in various photosynthetic plants and which assist in the absorption of light and the transfer of its energy to chlorophyll. Also called masking pigment.

accretion An increase in the mass of a body by accumulation or clumping of smaller particles.

acid A substance that releases a hydrogen ion (H^+) in solution.

acid rain Rain containing acids and acid-forming compounds such as sulfur dioxide and oxides of nitrogen.

acoustical tomography A technique for studying ocean structure that depends on pulses of low-frequency sound to sense differences in water temperature, salinity, and movement beneath the surface.

active margin Continental margin near an area of lithospheric plate convergence. Also called Pacific-type margin.

active sonar A device that generates underwater sound from special transducers and analyzes the returning echoes to gain information of geological, biological, or military importance.

active transport The movement of molecules from a region of low concentration to a region of high concentration through a semi-permeable membrane at the expense of energy.

adaptation An inheritable structural or behavioral modification. A favorable adaptation gives a species an advantage in survival and reproduction. An unfavorable adaptation lessens a species' ability to survive and reproduce.

adhesion Attachment of water molecules to other substances by hydrogen bonds. Wetting.

Agnatha The class of jawless fishes: hagfishes and lampreys.

ahermatypic Describing coral species lacking symbiotic zooxanthellae and incapable of secreting calcium carbonate at a rate suitable for reef production.

air mass A large mass of air with nearly uniform temperature, humidity, and density throughout.

algae Collective term for nonvascular plants possessing chlorophyll and capable of photosynthesis. (Singular, alga.)

algin A mucilaginous commercial product of multicellular marine algae. Widely used as a thickening and emulsifying agent.

alkaline Basic. *See* **base**.

alternation of generations A reproductive cycle in which a plant alternates between sexual and asexual stages.

amphidromic point A "no-tide" point in an ocean caused by basin resonances, friction, and other factors around which tide crests rotate. About a dozen amphidromic points exist in the world ocean. Sometimes called a node.

angiosperm A flowering vascular plant that reproduces by means of a seed-bearing fruit. Examples are sea grasses and mangroves.

angle of incidence In meteorology, the angle of the sun above the horizon.

animal A multicellular organism unable to synthesize its own food and often capable of movement.

Animalia The kingdom to which multicellular heterotrophs belong.

Annelida The phylum of animals to which segmented worms belong.

Antarctic Bottom Water The densest ocean water (1.0279 g/cm^3), formed primarily in Antarctica's Weddell Sea during Southern Hemisphere winters.

Antarctic Circle The imaginary line around the Earth, parallel to the equator at 66°33'S, marking the southernmost limit of sunlight at the June solstice. The Antarctic Circle marks the northern limit of the area within which, for one day or more each year, the sun does not set (around 21 December) or rise (around 21 June).

Antarctic Convergence Convergence zone encircling Antarctica between about 50° and 60°S, marking the boundary between Antarctic Circumpolar Water and Subantarctic Surface Water.

Antarctic Ocean An ocean in the Southern Hemisphere bounded to the north by the Antarctic Convergence and to the south by Antarctica.

aphotic zone The dark ocean below the depth to which light can penetrate.

aquaculture The growing or farming of plants and animals in a water environment under controlled conditions. *Compare* **mariculture**.

Arctic Circle The imaginary line around the Earth, parallel to the equator at 66°33'N, marking the northernmost limit of sunlight at the December solstice. The Arctic Circle marks the southern limit of the area within which, for one day or more each year, the sun does not set (around 21 June) or rise (around 21 December).

Arctic Convergence Convergence zone between Arctic Water and Subarctic Surface Water.

Arctic Ocean An ice-covered ocean north of the continents of North America and Eurasia.

Arthropoda The phylum of animals that includes shrimp, lobsters, krill, barnacles, and insects. The phylum Arthropoda is the world's most successful.

artificial system of classification   A method of classifying an object based on attributes other than its reason for existence, its ancestry, or its origin. *Compare* **natural system of classification**.

Asteroidea   The class of the phylum Echinodermata to which sea stars belong.

asthenosphere   The hot plastic layer of the upper mantle below the lithosphere, extending some 700 kilometers (430 miles) below the surface. Convection currents within the asthenosphere power plate tectonics.

atmospheric circulation cell   Large circuit of air driven by uneven solar heating and the Coriolis effect. Three circulation cells form in each hemisphere. *See also* **Hadley cell; Ferrel cell; polar cell**.

atoll   A ring-shaped island of coral reefs and coral debris enclosing, or almost enclosing, a shallow lagoon from which no land protrudes. Atolls often form over sinking, inactive volcanoes.

atom   The smallest particle of an element that exhibits the characteristics of that element.

ATP   Adenosine triphosphate, the compound that acts as the immediate source of energy for all life on Earth. The energy stored in ATP is provided directly by photosynthesis or by respiration of glucose.

authigenic sediment   Sediment formed directly by precipitation from seawater. Also called *hydrogenous sediment*.

autotroph   An organism that makes its own food by photosynthesis or chemosynthesis.

auxospore   A naked diatom cell without valves. Often a dormant stage in the life cycle following sexual reproduction.

Aves   The class of birds.

backshore   Sand on the shoreward side of the berm crest, sloping away from the ocean.

backwash   Water returning to the ocean from waves washing onto a beach.

baleen   The interleaved, hard, fibrous, hornlike filters within the mouth of baleen whales.

barrier island   A long, narrow, wave-built island lying parallel to the mainland and separated from it by a lagoon or bay. *Compare* **sea island**.

barrier reef   A coral reef surrounding an island or lying parallel to the shore of a continent, separated from land by a deep lagoon. Coral debris islands may form along the reef.

basalt   The relatively heavy crustal rock that forms the seabeds, composed mostly of oxygen, silicon, magnesium, and iron. Its density is about 2.9 g/cm³.

base   A substance that combines with a hydrogen ion (H^+) in solution.

bathyal zone   The ocean between about 200 and 4,000 meters (700 and 13,000 feet) deep.

bathyscaphe   Deep-diving submersible designed like a blimp, which uses gasoline for buoyancy and can reach the bottom of the deepest ocean trenches. From the Greek *batheos* ("depth,") and *skaphidion* ("a small ship").

bay mouth bar   An exposed sand bar attached to a headland adjacent to a bay and extending across the mouth of the bay.

beach   A zone of unconsolidated (loose) particles extending from below water level to the edge of the coastal zone.

beach scarp   Vertical wall of variable height marking the landward limit of the most recent high tides. Corresponds with the berm at extreme high tides.

benthic zone   The zone of the ocean bottom. *See also* **pelagic zone**.

berm   A nearly horizontal accumulation of sediment parallel to shore. Marks the normal limit of sand deposition by wave action.

berm crest   The top of the berm; the highest point on most beaches. Corresponds to the shoreward limit of wave action during most high tides.

big bang   The hypothetical event that started the expansion of the universe from a geometric point. The beginning of time.

bilateral symmetry   Body structure having left and right sides that are approximate mirror images of each other. Examples are crabs and humans. *Compare* **radial symmetry**.

biodegradable   Able to be broken by natural processes into simpler compounds.

biogenous sediment   Sediment of biological origin. Organisms can deposit calcareous (calcium-containing) or siliceous (silicon-containing) residue.

biological amplification   Increase in concentration of certain fat-soluble chemicals such as DDT or heavy-metal compounds, in successively higher trophic levels within a food web.

biological factor   A biologically generated aspect of the environment, such as predation or metabolic waste products, that affects living organisms. Biological factors usually operate in association with purely physical factors such as light and temperature.

biological resource   A living animal or plant collected for human use. Also called a *living resource*.

bioluminescence   Biologically produced light.

biomass   The mass of living material in a given area or volume of habitat.

biosynthesis   The initial formation of life on the Earth.

Bivalvia   The class of the phylum Mollusca that includes clams, oysters, and mussels.

Bjerknes, Vilhelm (1862–1951)   Pioneering Norwegian physicist and discoverer of the nature and formation of extratropical cyclones, which cause most mid-latitude weather.

blade   Algal equivalent of a vascular plant's leaf. Also called a *frond*.

bond   *See* **chemical bond**.

brackish   Describing water intermediate in salinity between seawater and fresh water.

breakwater   An artificial structure of durable material that interrupts the progress of waves to shore. Harbors are often shielded by a breakwater.

buffer   A group of substances that tends to resist change in the pH of a solution by combining with free ions.

buoyancy   The ability of an object to float in a fluid by displacement of a volume of fluid equal in mass to the mass of the floating object.

calcareous ooze   Ooze composed mostly of the hard remains of calcium-carbonate-containing organisms.

calcium carbonate compensation depth   The depth at which the rate of accumulation of calcareous sediments equals the rate of dissolution of those sediments. Below this depth, sediment contains little or no calcium carbonate.

calorie   The amount of heat needed to raise the temperature of 1 gram (0.035 ounce) of pure water by 1°C (1.8°F).

capillary wave   A tiny wave with a wavelength of less than 1.73 centimeters (0.68 inch), whose restoring force is surface tension; the first type of wave to form when the wind blows.

Carnivora   The order of mammals that includes seals, sea lions, walruses, and sea otters.

carrying capacity   The size at which a particular population in a particular environment will stabilize when its supply of resources—including nutrients, energy, and living space—remains constant.

cartilage   A tough, elastic tissue that stiffens or supports.

cartographer   A person who makes maps and charts.

catastrophism The theory that the Earth's surface features are formed by catastrophic forces such as the biblical flood. Catastrophists believe in a young Earth and a literal interpretation of the biblical account of Creation.

celestial navigation The technique of finding one's position on Earth by reference to the apparent positions of stars, planets, the moon, and the sun.

cell The basic organizational unit of life on this planet.

Cephalopoda The class of the phylum Mollusca that includes squid, octopuses, and nautiluses.

Cetacea The order of mammals that includes porpoises, dolphins, and whales.

CFCs *See* **chlorofluorocarbons**.

***Challenger* Expedition** The first wholly scientific oceanographic expedition, 1872–76. Named for the steam corvette used in the voyage.

chart A map that depicts mostly water and the adjoining land areas.

chemical bond An energy relationship that holds two atoms together as a result of changes in their electron distribution.

chemical equilibrium In seawater, the condition in which the proportion and amounts of dissolved salts per unit volume of ocean are nearly constant.

chemosynthesis The synthesis of organic compounds from inorganic compounds using energy stored in inorganic substances such as sulfur, ammonia, and hydrogen. Energy is released when these substances are oxidized by certain organisms.

chitin A complex nitrogen-rich carbohydrate from which parts of arthropod exoskeletons are constructed.

chiton A polyplacophoran mollusk.

chlorinated hydrocarbons The most abundant and dangerous class of halogenated hydrocarbons, synthetic organic chemicals hazardous to the marine environment.

chlorinity A measure of the content of chloride, bromine, and iodide ions in seawater. We may derive salinity from chlorinity by multiplying by 1.80655.

chlorofluorocarbons (CFCs) A class of halogenated hydrocarbons thought to be depleting the Earth's atmospheric ozone. CFCs are used as cleaning agents, refrigerants, fire-extinguishing fluids, spray-can propellants, and insulating foams.

chlorophyll A pigment responsible for trapping sunlight and transferring its energy to electrons, thus initiating photosynthesis.

Chlorophyta Green algae.

Chondrichthyes The class of fishes with cartilaginous skeletons: the sharks, skates, rays, and chimaeras.

Chordata The phylum of animals to which tunicates, *Amphioxus*, fishes, amphibians, reptiles, birds, and mammals belong.

chromatophore A pigmented skin cell that expands or contracts to affect color change.

chronometer A very consistent clock. It doesn't need to tell accurate time, but its rate of gain or loss must be constant and known exactly so that accurate time may be calculated.

clamshell sampler Sampling device used to take shallow samples of the ocean bottom.

classification A way of grouping objects according to some stated criteria.

clay Sediment particle smaller than 0.004 millimeter in diameter; the smallest sediment size category.

climate The long-term average of weather in an area.

climax community A stable, long-established community of self-perpetuating organisms that tends not to change with time.

clockwise Rotation around a point in the direction that clock hands move.

clumped distribution Distribution of organisms within a community in small, patchy aggregations, or clumps; the most common distribution pattern.

Cnidaria The phylum of animals to which corals, jellyfish, and sea anemones belong.

cnidoblast Type of cell found in members of the phylum Cnidaria that contains a stinging capsule. The threads that evert from the capsules assist in capturing prey and repelling aggressors.

coast The zone extending from the ocean inland as far as the environment is immediately affected by marine processes.

coastal cell The natural sector of a coastline in which sand input and sand outflow are balanced.

coastal upwelling Upwelling adjacent to a coast, usually induced by wind.

coccolithophore A very small planktonic alga carrying discs of calcium carbonate, which contributes to biogenous sediments.

cohesion Attachment of water molecules to each other by hydrogen bonds.

colligative properties Those characteristics of a solution that differ from those of pure water because of material held in solution.

Columbus, Christopher (1451–1506) Italian explorer in the service of Spain who discovered islands in the Caribbean in 1492. Although traditionally credited as the discoverer of America, he never actually sighted the North American continent.

commensalism A symbiotic interaction between two species in which only one species benefits and neither is harmed.

commercial extinction Depletion of a resource species to a point where it is no longer profitable to harvest the species.

community The populations of all species that occupy a particular habitat and interact within that habitat.

compass An instrument for showing direction by means of a magnetic needle swinging freely on a pivot and pointing to magnetic north.

compensation depth The depth in the water column at which the production of carbohydrates and oxygen by photosynthesis exactly equals the consumption of carbohydrates and oxygen by respiration. The break-even point for autotrophs. Generally a function of light level.

compound A substance composed of two or more elements in a fixed proportion.

condensation theory Premise that stars and planets accumulate from contracting, accreting clouds of galactic gas, dust, and debris.

conduction The transfer of heat through matter by the collision of one atom with another.

conservative constituent An element that occurs in constant proportion in seawater. For example, chlorine, sodium, and magnesium.

constructive interference The addition of wave energy as waves interact, producing larger waves.

consumer A heterotrophic organism.

continental crust The solid masses of the continents, composed primarily of granite.

continental drift The theory that the continents move slowly across the surface of the Earth.

continental margin The submerged outer edge of a continent, made of granitic crust. Includes the continental shelf and continental slope. *Compare* **ocean basin**.

continental rise The wedge of sediment forming the gentle transition from the outer (lower) edge of the continental slope to the abyssal plain. Usually associated with passive margins.

continental shelf   Gradually sloping submerged extension of a continent, composed of granitic rock overlain by sediments. Has features similar to the edge of the nearby continent.

continental slope   The sloping transition between the granite of the continent and the basalt of the seabed. The true edge of a continent.

contour current   A bottom current made up of dense water that flows around (rather than over) seabed projections.

convection   Movement within a fluid resulting from differential heating and cooling of the fluid. Convection produces mass transport or mixing of the fluid.

convection current   A single closed-flow circuit of rising warm material and falling cool material.

convergence zone   The line along which waters of different density converge. Convergence zones form the boundaries of tropical, subtropical, temperate, and polar areas.

convergent evolution   The evolution of similar characteristics in organisms of different ancestry; the body shape of a porpoise and a shark, for instance.

convergent plate boundary   A region where plates are pushing together and where a mountain range, island arc, and/or trench will eventually form. Often a site of much seismic and volcanic activity.

Cook, James (1728–1779)   Officer in the British Royal Navy who led the first European voyages of scientific discovery.

coral   Any of over 6,000 species of small cnidarians, many of which are capable of generating hard calcareous (aragonite, $CaCO_3$) skeletons.

coral reef   A linear mass of calcium carbonate (aragonite and calcite) assembled from coral organisms, algae, mollusks, worms, and so on. Coral may contribute less than half of the reef material.

core   The innermost layer of the Earth, composed primarily of iron, with nickel and heavy elements. The inner core is thought to be a solid 6,000°C (11,000°F) sphere, the outer core a 5,000°C (9,000°F) liquid mass. The average density of the outer core is about 11.8 g/cm^3, and that of the inner core is about 16 g/cm^3.

Coriolis, Gaspard Gustave de (1792–1843)   The French scientist who in 1835 worked out the mathematics of the motion of bodies on a rotating surface. *See* **Coriolis effect**.

Coriolis effect   The apparent deflection of a moving object from its initial course when its speed and direction are measured in reference to the surface of the rotating Earth. The object is deflected to the right of its anticipated course in the Northern Hemisphere and to the left in the Southern Hemisphere. The deflection occurs for any horizontal movement of objects with mass and has no effect at the equator.

cosmogenous sediment   Sediment of extraterrestrial origin.

counterclockwise   Rotation around a point in the direction opposite to that in which clock hands move. Also called *anticlockwise*.

countercurrent   A surface current flowing in the opposite direction from an adjacent surface current.

covalent bond   A chemical bond formed between two atoms by electron sharing.

crest   *See* **wave crest**.

crust   The outermost solid layer of the Earth, composed mostly of granite and basalt. The top of the lithosphere. The crust has a density of 2.7–2.9 g/cm^3 and accounts for 0.4% of the Earth's mass.

Crustacea   The class of phylum Arthropoda to which lobsters, shrimp, crabs, barnacles, and copepods belong.

cryptic coloration   Camouflage. May be active (under control of the animal) or passive (an unalterable color or shape).

current   Mass flow of water. (The term is usually reserved for horizontal movement.)

cyclone   A weather system with a low-pressure area in the center around which winds blow counterclockwise in the Northern Hemisphere and clockwise in the Southern Hemisphere. Not to be confused with a tornado, a much smaller weather phenomenon associated with severe thunderstorms. *See also* **tropical cyclone; extratropical cyclone**.

deep scattering layer (DSL)   A relatively dense aggregation of fishes, squid, and other mesopelagic organisms capable of reflecting a sonar pulse that resembles a false bottom in the ocean. Its position varies with the time of day.

deep-water wave   A wave in water deeper than one-half its wavelength.

deep zone   The zone of the ocean below the pycnocline, in which there is little additional change of density with increasing depth. Contains about 80% of the world's water.

degree   An arbitrary measure of temperature. One degree Celsius (°C) = 1.8 degrees Fahrenheit (°F).

delta   The deposit of sediments found at a river mouth, sometimes triangular in shape (hence the name after the Greek letter).

density   The mass per unit volume of a substance, usually expressed in grams per cubic centimeter (g/cm^3).

density curve   A graph showing the relationship between a fluid's temperature or salinity and its density.

density stratification   The formation of layers in a material, with each deeper layer being denser (weighing more per unit of volume) than the layer above.

dependency   A feeding relationship in which an organism is limited to feeding on one species or, in extreme cases, on one size phase of one species.

deposition   Accumulation, usually of sediments.

depositional coast   A coast in which processes that deposit sediment exceed erosive processes.

desalination   The process of removing salt from seawater or brackish water.

desiccation   Drying.

destructive interference   The subtraction of wave energy as waves interact, producing smaller waves.

diatom   The Earth's most abundant, successful, and efficient single-celled phytoplankton. Diatoms possess two interlocking valves made primarily of silica. The valves contribute to biogenous sediments.

diffusion   The movement—driven by heat—of molecules from a region of high concentration to a region of low concentration.

dinoflagellate   One of a class of microscopic single-celled flagellates, not all of which are autotrophic. The outer covering is often of stiff cellulose. Planktonic dinoflagellates are responsible for "red tides."

dissolution   The dissolving by water of minerals in rocks.

disthermal zone   The zone of stable temperature below the thermocline.

disturbing force   The energy that causes a wave to form.

diurnal tide   A tidal cycle of one high tide and one low tide per day.

divergent evolution   Evolutionary radiation of different species from a common ancestor.

divergent plate boundary   A region where plates are moving apart and where new ocean or rift valley will eventually form. A spreading center forms the junction.

doldrums   The zone of rising air near the equator known for sultry air and variable breezes. Also known as the *intertropical convergence zone. See also* **intertropical convergence zone** (ITCZ).

downwelling   Circulation pattern in which surface water moves vertically downward.

drag   The resistance to movement of an organism induced by the fluid through which it swims.

drift net   Fine vertically suspended net that may be 7 meters (25 feet) high and 80 kilometers (50 miles) long.

DSL   *See* **deep scattering layer**.

earthquake   A sudden motion of the Earth's crust resulting from waves in the Earth caused by faulting of the rocks or by volcanic activity.

eastern boundary current   Weak, cold, diffuse, slow-moving current at the eastern boundary of an ocean (off the west coast of a continent). Examples include the Canary Current and the Humboldt Current.

ebb current   Water rushing out of an enclosed harbor or bay because of the fall in sea level as a tide trough approaches.

Echinodermata   The phylum of exclusively marine animals to which sea stars, brittle stars, sea urchins, and sea cucumbers belong.

Echinoidea   The class of the phylum Echinodermata to which sea urchins and sand dollars belong.

echolocation   The use of reflected sound to detect environmental objects. Cetaceans use echolocation to detect prey and avoid obstacles.

echo sounder   A device that reflects sound off the ocean bottom to sense water depth. Its accuracy is affected by the variability of the speed of sound through water.

ecology   Study of the interactions of organisms with one another and with their environment.

ectoderm   The outermost layer of cells in a developing embryo.

ectotherm   An organism incapable of generating and maintaining steady internal temperature from metabolic heat and therefore whose internal body temperature is approximately the same as that of the surrounding environment. A cold-blooded organism.

eddy   A circular movement of water usually formed where currents pass obstructions, or between two adjacent currents flowing in opposite directions, or along the edge of a permanent current.

EEZ   *See* **exclusive economic zone**.

Ekman spiral   A theoretical model of the effect on water of wind blowing over the ocean. Because of the Coriolis effect, the surface layer is expected to drift at an angle of 45° to the right of the wind in the Northern Hemisphere and 45° to the left in the Southern Hemisphere. Water at successively lower layers drifts progressively to the right (N), or left (S), though not as swiftly as the surface flow.

Ekman transport   Net water transport, the sum of layer movement due to the Ekman spiral. Theoretical Ekman transport in the Northern Hemisphere is 90° to the right of the wind direction.

electron   A tiny negatively charged particle in an atom responsible for chemical bonding.

element   A substance composed of identical atoms that cannot be broken into simpler substances by chemical means.

El Niño   A southward-flowing nutrient-poor current of warm water off the coast of western South America, caused by a breakdown of trade wind circulation.

endoderm   The innermost layer of cells in a developing embryo.

endotherm   An organism capable of generating and regulating metabolic heat to maintain a steady internal temperature. Birds and mammals are the only animals capable of true endothermy. A warm blooded organism.

energy   The capacity to do work.

ENSO   Acronym for the coupled phenomena of El Niño and the Southern Oscillation. *See also* **El Niño**; **Southern Oscillation**.

entropy   A measure of the disorder in a system.

environmental resistance   All the limiting factors that act together to regulate the maximum allowable size, or carrying capacity, of a population.

epicenter   The point on the Earth's surface directly above the focus of an earthquake.

epipelagic zone   The lighted, or photic, zone in the ocean.

equator   *See* **geographical equator**; **meteorological equator**.

equatorial upwelling   Upwelling in which water moving westward on either side of the geographical equator tends to be deflected slightly poleward and replaced by deep water often rich in nutrients. *See also* **upwelling**.

Eratosthenes of Cyrene (276–192 B.C.) Greek scholar and librarian at Alexandria who first calculated the circumference of the Earth about 230 B.C.

erosion   A process of being gradually worn away.

erosional coast   A coast in which erosive processes exceed depositional ones.

estuary   A body of water partially surrounded by land where fresh water from a river mixes with ocean water, creating an area of remarkable biological productivity.

euphotic zone   The upper layer of the photic zone in which net photosynthetic gain occurs. *Compare* **photic zone**.

euryhaline   Describing an organism able to tolerate a wide range in salinity.

eurythermal   Describing an organism able to tolerate wide variance in temperature.

eurythermal zone   The upper layer of water, where temperature changes with the seasons.

eustatic change   A worldwide change in sea level, as distinct from local changes.

eutrophication   A set of physical, chemical, and biological changes brought about when excessive nutrients are released into water.

evaporite   Deposit formed by the evaporation of ocean water.

evolution   Change. The maintenance of life under constantly changing conditions by continuous adaptation of successive generations of a species to its environment.

excess volatiles   A compound found in the ocean and atmosphere in quantities greater than can be accounted for by the weathering of surface rock. Such compounds probably entered the atmosphere and ocean from deep crustal and upper mantle sources through volcanism.

exclusive economic zone (EEZ)   The offshore zone claimed by signatories to the 1982 United Nations Draft Convention on the Law of the Sea. The EEZ extends 200 nautical miles (370 kilometers) from a contiguous shoreline. *See also* **United States exclusive economic zone**.

exoskeleton   A strong, lightweight, form-fitted external covering and support common to animals of the phylum Arthropoda. The exoskeleton is made partly of chitin and may be strengthened by calcium carbonate.

extratropical cyclone   A low-pressure mid-latitude weather system characterized by converging winds and ascending air rotating counterclockwise in the Northern Hemisphere and clockwise in the Southern Hemisphere. An extratropical cyclone forms at the front between the polar and Ferrel cells.

fault   A fracture in a rock mass along which movement has occurred.

Ferrel, William (1817–1891) The American scientist who discovered the mid-latitude circulation cells of each hemisphere.

Ferrel cell The middle atmospheric circulation cell in each hemisphere. Air in these cells rises at 60° latitude and falls at 30° latitude. *See also* **westerlies**.

fetch The uninterrupted distance over which the wind blows without a significant change in direction, a factor in wind wave development.

Fissipedia The carnivoran suborder that includes sea otters.

fjord A deep, narrow estuary in a valley originally cut by a glacier.

flagellum A whiplike structure used by some small organisms and gametes to move through the environment. (Plural, *flagella*).

float method A method of current study that depends on the movement of a drift bottle or other free-floating object.

flood current Water rushing into an enclosed harbor or bay because of the rise in sea level as a tide crest approaches.

flow method A method of current study that measures the current as it flows past a fixed object.

food General term for organic molecules capable of providing energy to heterotrophs when combined with oxygen during biochemical respiration.

food web A group of organisms associated by a complex set of feeding relationships in which the flow of food energy can be followed from primary producers through consumers.

foraminiferan One of a group of planktonic amoeba-like animals with a calcareous shell, which contributes to biogenous sediments.

Forchhammer's principle *See* **principle of constant proportions**.

foreshore Sand on the seaward side of the berm, sloping toward the ocean, to the low tide mark.

fracture zone Area of irregular, seismically inactive topography marking the position of a once-active transform fault.

freezing point The temperature at which a solid can begin to form as a liquid is cooled.

fringing reef A reef attached to the shore of a continent or island.

front The boundary between two air masses of different density. The density difference can be caused by differences in temperature and/or humidity.

frontal storm Precipitation and wind caused by the meeting of two air masses, associated with an extratropical cyclone. Generally, one air mass will slide over or under the other, and the resulting expansion of air will cause cooling and, consequently, rain or snow.

frustule Siliceous external cell wall of a diatom consisting of two interlocking valves fitted together like the halves of a box.

fucoxanthin A brown or tan accessory pigment found in many species of brown algae and some species of diatoms.

fully developed sea The theoretical maximum height attainable by ocean waves given wind of a specific strength, duration, and fetch. Longer exposure to wind will not increase the size of the waves.

galaxy A large rotating aggregation of stars, dust, gas, and other debris held together by gravity. There are perhaps 50 billion galaxies in the universe and 50 billion stars in each galaxy.

gas bladder In multicellular algae, an air-filled structure that assists in flotation.

gas exchange Simultaneous passage, through a semipermeable membrane, of oxygen into an animal and carbon dioxide out of it.

Gastropoda The class of the phylum Mollusca that includes snails and sea slugs.

geographical equator 0° latitude, an imaginary line equidistant from the geographical poles.

geostrophic Describing a gyre or current in balance between the Coriolis effect and gravity; literally, "turned by the Earth."

glucose A carbohydrate ($C_6H_{12}O_6$) produced by autotrophs during photosynthesis. It can be used for energy or converted into other organic compounds.

granite The relatively light crustal rock—composed mainly of oxygen, silicon, and aluminum—that forms the continents. Its density is about 2.7 g/cm^3.

gravimeter A sensitive device that measures variations in the pull of gravity at different places on the Earth's surface.

gravity wave A wave with wavelength greater than 1.73 centimeters (0.68 inch), whose restoring forces are gravity and momentum.

greenhouse effect Trapping of heat in the atmosphere. Incoming short-wavelength solar radiation penetrates the atmosphere, but the longer-wavelength outgoing radiation is absorbed by greenhouse gases and reradiated to Earth, causing a rise in surface temperature.

greenhouse gases Gases in the Earth's atmosphere that cause the greenhouse effect; include carbon dioxide, methane, and CFCs.

groin A short, artificial projection of durable material placed at a right angle to shore in an attempt to slow longshore transport of sand from a beach. Usually deployed in repeating units.

group velocity Speed of advance of a wave train; for deep-water waves, half the speed of individual waves within the group.

Gulf Stream The strong western boundary current of the North Atlantic, off the east coast of the United States.

guyot A flat-topped, submerged inactive volcano.

gyre Circuit of mid-latitude currents around the periphery of an ocean basin. Most oceanographers recognize five gyres plus the West Wind Drift.

habitat The place where an individual or population of a given species lives. Its "mailing address."

hadal zone The deepest zone of the ocean, below a depth of 5,000 meters (16,500 feet).

Hadley, George (1685–1768) A London lawyer and philosopher who worked out the overall scheme of wind circulation in an effort to explain the trade winds.

Hadley cell The atmospheric circulation cell nearest the equator in each hemisphere. Air in these cells rises near the equator because of strong solar heating there and falls because of cooling at about 30° latitude. *See also* **trade winds**.

halocline The zone of the ocean in which salinity increases rapidly with depth. *See also* **pycnocline**.

Harrison, John (1693—1776) British clockmaker who invented the modern chronometer in 1760.

heat A form of energy produced by the random vibration of atoms or molecules.

heat budget An expression of the total solar energy received on the Earth during some period of time and the total heat lost from the Earth by reflection and radiation into space through the same period.

heat capacity The amount of heat required to raise the temperature of 1 gram of a substance by 1°C (1.8°F).

Henry the Navigator (1394–1460) Prince of Portugal who established a school for the study of geography, seamanship, shipbuilding, and navigation.

hermatypic Describing coral species possessing symbiotic zooxanthellae within their tissues and capable of secreting calcium carbonate at a rate suitable for reef production.

heterotroph   An organism that derives nourishment from other organisms because it is unable to synthesize its own food molecules.

hierarchy   Grouping of objects by degrees of complexity, grade, or class. A hierarchical system of nomenclature is based on distinctions within groups and between groups.

high-energy coast   A coast exposed to large waves.

high seas   That part of the ocean past the exclusive economic zone, which is considered common property to be shared by the citizens of the world. About 60% of the ocean area.

high tide   The high-water position corresponding to a tidal crest.

holdfast   A complex branching structure that anchors many kinds of multicellular algae to the substrate.

holoplankton   Permanent members of the plankton community. Examples are diatoms and copepods. *Compare* **meroplankton**.

Holothuroidea   The class of the phylum Echinodermata to which sea cucumbers belong.

horse latitudes   Zones of erratic horizontal surface air circulation near 30°N and 30°S latitudes. Over land, dry air falling from high altitudes produces deserts at these latitudes (for example, the Sahara).

hot spot   A small stationary heat source in the Earth's mantle. Hot spots are not always located at a plate boundary.

hurricane   A large tropical cyclone in the North Atlantic or eastern Pacific, whose winds exceed 118 kilometers (74 miles) per hour.

hydrogen bond   Relatively weak bond formed between a partially positive hydrogen atom and a partially negative oxygen, fluorine, or nitrogen atom of an adjacent molecule.

hydrogenous sediment   Sediment formed directly by precipitation from seawater. Also called *authigenic sediment*.

hydrostatic pressure   The constant pressure of water around a submerged organism.

hydrothermal vent   Spring of hot, mineral- and gas-rich seawater found on some oceanic ridges in zones of active seafloor spreading.

hypertonic   Referring to a solution having a higher concentration of dissolved substances than the solution that surrounds it.

hypothesis   A speculation about the natural world that may be verified or disproved by observation and experiment.

hypotonic   Referring to a solution having a lower concentration of dissolved substances than the solution that surrounds it.

hypsographic curve   A graph of the area of the Earth's surface above any given elevation or depth above or below sea level.

ice age   One of several periods (lasting several thousand years each) of low temperature during the last million years. Glaciers and polar ice were derived from ocean water, lowering sea level at least 100 meters (328 feet). (See Appendix 2, Geological time.)

iceberg   Large mass of ice floating in the ocean that was formed on or adjacent to land. Tabular icebergs are tablelike or flat; pinnacled icebergs are castellated, or jagged. Southern icebergs are often tabular; northern icebergs are often pinnacled.

ice cap   Permanent cover of ice. Formally limited to ice atop land, but informally applied also to floating ice in the Arctic Ocean.

ice floe   A mass of firm sea ice floating as a unit.

ice pack   A large, floating expanse of broken ice masses pressed and frozen together.

inlet   A passage giving the ocean access to an enclosed lagoon, harbor, or bay.

insolation rate   The amount of solar energy reaching the Earth's surface per unit time.

interference   Addition or subtraction of wave energy as waves interact. Also called *resonance*. *See also* **constructive interference**; **destructive interference**.

intermediate-depth water wave   A wave moving through water deeper than 1/20 but shallower than ½ its wavelength. Also called a *transitional wave*.

internal wave   A progressive wave occurring at the boundary between liquids of different densities.

intertidal zone   The marine zone between the highest high tide point on a shoreline and the lowest low tide point. The intertidal zone is sometimes subdivided into four separate habitats by height above tidal datum, typically numbered 1 to 4, land to sea.

intertropical convergence zone (ITCZ)   The equatorial area at which the trade winds converge. The ITCZ usually lies at or near the meteorological equator. Also called the *doldrums*.

invertebrate   Animal lacking a backbone.

ion   An atom (or small group of atoms) that becomes electrically charged by gaining or losing one or more electrons.

ionic bond   A chemical bond resulting from attraction between oppositely charged ions. These forces are said to be "electrostatic" in nature.

ionizing radiation   Fast-moving particles or high-energy electromagnetic radiation emitted as unstable atomic nuclei disintegrate. The radiation has enough energy to dislodge one or more electrons from atoms it hits to form charged ions, which can react with and damage living tissue.

island arc   Curving chain of volcanic islands and seamounts almost always found paralleling the concave edge of a trench.

isostatic equilibrium   Balanced support of lighter material in a heavier, displaced supporting matrix. Analogous to buoyancy in a liquid.

isotonic   Referring to a solution having the same concentration of dissolved substances as the solution that surrounds it.

ITCZ   *See* **intertropical convergence zone**.

kelp   Informal name for any species of large phaeophyte.

kingdom   The largest category of biological classification. Five kingdoms are presently recognized.

knot   A speed of 1 nautical mile per hour. *See* **nautical mile**.

krill   *Euphausia superba*, a thumb-sized crustacean common in Antarctic waters.

lagoon   A shallow body of seawater generally isolated from the ocean by a barrier island. Also the body of water enclosed within an atoll, or the water within a reverse estuary.

land breeze   Movement of air offshore as marine air heats and rises.

latent heat of evaporation   Heat added to a liquid during evaporation (or released from a gas during condensation) that produces a change in state but not a change in temperature. For pure water, 585 calories per gram at 20°C (68°F).

latent heat of fusion   Heat removed from a liquid during freezing (or added to a solid during thawing) that produces a change in state but not a change in temperature. For pure water, 80 calories per gram at 0°C (32°F).

lateral-line system   A system of sensors and nerves in the head and midbody of fishes and some amphibians that functions to detect low-frequency vibrations in water.

latitude   Regularly spaced imaginary lines on the Earth's surface running parallel to the equator.

law   A large construct explaining events in nature that have been observed to occur with unvarying uniformity under the same conditions.

Library of Alexandria   The greatest collection of writings in the ancient world, founded in the third century B.C. by Alexander the Great. Could be considered the first university.

light   Electromagnetic radiation propagated as small, nearly massless particles that behave like both a wave and a stream of particles.

limiting factor   A physical or biological environmental factor whose absence or presence in an inappropriate amount limits the normal actions of an organism.

Linnaeus, Carolus   Carl von Linné (1707–1778). Swedish "father" of modern taxonomy.

lithification   Conversion of sediment into sedimentary rock by pressure or by the introduction of a mineral cement.

lithosphere   The brittle, relatively cool outer layer of the Earth, consisting of the oceanic and continental crust and the outermost, rigid layer of mantle.

littoral zone   The band of coast alternately covered and uncovered by tidal action; the intertidal zone.

longitude   Regularly spaced imaginary lines on the Earth's surface running north and south and converging at the poles.

longshore bar   A submerged or exposed line of sand lying parallel to shore and accumulated by wave action.

longshore current   A current running parallel to shore in the surf zone, caused by the incomplete refraction of waves approaching the beach at an angle.

longshore drift   Movement of sediments parallel to shore, driven by wave energy.

longshore trough   Submerged excavation parallel to shore adjacent to an exposed sandy beach. Caused by the turbulence of water returning to the ocean after each wave.

low-energy coast   A coast only rarely exposed to large waves.

low tide   The low-water position corresponding to a tidal trough.

low tide terrace   The smooth, hard-packed beach seaward of the beach scarp on which waves expend most of their energy. Site of the most vigorous onshore and offshore movement of sand.

lunar tide   Tide caused by gravitational and inertial interaction of moon and Earth.

macroplankton   Animal plankters larger than 1 to 2 centimeters (½ to 1 inch). An example is the jellyfish.

Magellan, Ferdinand (c. 1480–1521)   Portuguese navigator in the service of Spain who led the first expedition to circumnavigate the Earth, 1519–22. He was killed in the Philippines.

magma   Molten rock capable of fluid flow. Called lava aboveground.

magnetometer   A device that measures the amount and direction of residual magnetism in a rock sample.

Mammalia   The class of mammals.

mangrove   Large flowering shrub or tree that grows in dense thickets or forests along muddy or silty tropical coasts.

mantle   The layer of the Earth between the crust and the core, composed of silicates of iron and magnesium. The mantle has an average density of about 4.5 g/cm³ and accounts for about 68% of the Earth's mass.

map   A representation of the Earth's surface, usually depicting mostly land areas. *See also* **chart**.

mariculture   The farming of marine organisms, usually in estuaries, bays, or nearshore environments or in specially designed structures using circulating seawater. *Compare* **aquaculture**.

marine energy resource   Any resource resulting from the direct extraction of energy from the heat or movement of ocean water.

marine pollution   The introduction by humans of substances or energy into the ocean that change the quality of the water or affect the physical and biological environment.

marine science   The process (or result) of applying the scientific method to the ocean, its surroundings, and the life forms within it. Also called oceanography or oceanology.

masking pigment   *See* **accessory pigment**.

mass   A measure of the quantity of matter.

Maury, Matthew (1806–1873)   "Father" of physical oceanography. Probably the first person to undertake the systematic study of the ocean as a full-time occupation, and probably the first to understand the global interlocking of currents, wind flow, and weather.

maximum sustainable yield   The maximum amount of fish, crustaceans, and mollusks that can be caught without impairing future populations.

mean sea level   The height of the ocean surface averaged over a few years' time.

medusa   Free-swimming body form of many members of the phylum Cnidaria.

membrane   A complex structure of proteins and lipids that forms boundaries around and within the cell. It is usually semipermeable, allowing some kinds of molecules to pass through but not others.

meroplankton   Temporary members of the plankton community. Examples are very young fishes and barnacle larvae. *Compare* **holoplankton**.

mesoderm   The middle layer of cells in a developing embryo.

mesosphere   The rigid inner mantle, similar in chemical composition to the asthenosphere.

metabolic rate   The rate at which energy-releasing reactions proceed within an organism.

metamerism   Segmentation; repeating body parts.

Meteor **Expedition**   German Atlantic expedition begun in 1925; the first to use an echo sounder and other modern optical and electronic instrumentation.

meteorological equator   Also called the thermal equator. The irregular imaginary line of thermal equilibrium between hemispheres. It is situated about 5° north of the geographical equator, and its position changes with the seasons, moving slightly north in northern summer.

meteorological tide   A tide influenced by the weather. Arrival of a storm surge will alter the estimate of a tide's height or arrival time, as will a strong, steady onshore or offshore wind.

Milky Way   The name of our galaxy. Sometimes applied to the field of stars in our home spiral arm, which is correctly called the Orion arm.

mixed layer   *See* **surface zone**.

mixed tide   A complex tidal cycle, usually with two high tides and two low tides of unequal height per day.

mixing time   The time necessary to mix a substance through the ocean, about 1,000 years.

mixture   A close intermingling of different substances that still retain separate identities. The properties of a mixture are heterogeneous; they may vary within the mixture.

molecule   A group of atoms held together by chemical bonds. The smallest unit of a compound that retains the characteristics of the compound.

Mollusca The phylum of animals that includes chitons, snails, clams, and octopuses.

molt To shed an external covering.

monsoon A pattern of wind circulation that changes with the season. Also, the rainy season in areas with monsoon wind patterns.

moon tide *See* **lunar tide**.

motile Able to move about.

multicellular Consisting of more than one cell.

multicellular algae Algae with bodies consisting of more than one cell. Examples are kelp and *Ulva*.

mutation A heritable change in an organisms's genes.

mutualism A symbiotic interaction between two species that is beneficial to both.

Mysticeti The suborder of baleen whales.

nanoplankton Very small members of the plankton community. Examples are coccolithophores and silicoflagellates.

Nansen bottle A water-sampling instrument perfected early in this century by the Norwegian scientist and explorer Fridtjof Nansen.

natural selection A mechanism of evolution that results in the continuation of only those forms of life best adapted to survive and reproduce in their environment.

natural system of classification A method of classifying an organism based on its ancestry or origin.

nautical chart A chart used for marine navigation.

nautical mile The length of 1 minute of latitude, 6,076 feet, 1.15 statute miles, or 1.85 kilometers. (See Appendixes 1 and 4.)

neap tide The time of smallest variation between high and low tides occurring when Earth, moon, and sun align at right angles. Neap tides alternate with spring tides, occurring at two-week intervals.

nebula Diffuse cloud of dust and gas.

Nematoda The phylum of animals to which roundworms belong.

neritic Of the shore or coast. Refers to continental margins and the water covering them, or to nearshore organisms.

neritic zone The zone of open water near shore, over the continental shelf.

niche Description of an organism's functional role in a habitat. Its "job."

node The line or point of no wave action in a standing pattern. *See also* **amphidromic point**.

nodule Solid mass of hydrogenous sediment, most commonly manganese or ferromanganese nodules and phosphorite nodules.

nonconservative constituent An element whose proportion in seawater varies with time and place, depending on biological demand or chemical reactivity. An element with a short residence time. For example, iron, aluminum, silicon, trace nutrients, dissolved oxygen, and carbon dioxide.

nonconservative nutrient A compound or ion needed by autotrophs for primary productivity and which changes in concentration with biological activity.

nonextractive resource Any use of the ocean in place, such as transportation of people and commodities by sea, recreation, or waste disposal.

nonrenewable resource Any resource that is present on Earth in fixed amounts and cannot be replenished.

nonvascular plant Plant having no obvious vessels for the transport of fluid and lacking leaves, stems, and roots. Examples are algae.

nor'easter (northeaster) Any energetic extratropical cyclone that sweeps the eastern seaboard of North America in winter.

notochord Stiffening structure found at some time in the life cycle of all members of the phylum Chordata.

nutrient Any needed substance that an organism obtains from its environment *except* oxygen, carbon dioxide, and water.

ocean (1) The great body of saline water that covers 70.78% of the surface of the Earth. (2) One of its primary subdivisions, bounded by continents, the equator, and other imaginary lines.

ocean basin Deep-ocean floor made of basaltic crust. *Compare* **continental margin**.

oceanic crust The outermost solid surface of the Earth beneath ocean floor sediments, composed primarily of basalt.

oceanic ridge Young seabed at the active spreading center of an ocean, often unmasked by sediment, bulging above the abyssal plain. The boundary between diverging plates. Often called a mid-ocean ridge, though less than 60% of the length exists at mid-ocean.

oceanic zone The zone of open water away from shore, past the continental shelf.

oceanography The science of the ocean. *See also* **marine science**.

oceanus Latin form of *okeanos*, the Greek name for the "ocean river" past Gibraltar.

Odontoceti The suborder of toothed whales.

oolite sand Hydrogenous sediment formed when calcium carbonate precipitates from warmed seawater as pH rises, forming rounded grains around a shell fragment or other particle.

ooze Sediment of at least 30% biological origin.

Ophiuroidea The class of the phylum Echinodermata to which brittle stars belong.

orbit In ocean waves, the circular pattern of water particle movement at the air-sea interface. Orbital motion contrasts with the side-to-side or back-and-forth motion of pure transverse or longitudinal waves.

orbital inclination The 23°27' "tilt" of the Earth's rotational axis relative to the plane of its orbit around the sun.

orbital wave A progressive wave in which particles of the medium move in closed circles.

osmoregulation The ability to adjust internal salt concentration.

osmosis The diffusion of water from a region of high water concentration to a region of lower water concentration through a semipermeable membrane.

Osteichthyes The class of fishes with bony skeletons.

outgassing The volcanic venting of volatile substances.

overfishing Harvesting so many fish that there is not enough breeding stock left to replenish the species.

oxygen minimum zone A zone in which oxygen is depleted by animals and not replaced by phytoplankton.

oxygen revolution The time span, from about 2 billion to 400 million years ago, during which photosynthetic autotrophs changed the composition of the Earth's atmosphere to its current oxygen-rich mixture.

ozone O_3, the triatomic form of oxygen. Ozone in the upper atmosphere protects living things from some of the harmful effects of the sun's ultraviolet radiation.

ozone layer A diffuse layer of ozone mixed with other gases surrounding the world at a height of about 20 to 40 kilometers (12 to 25 miles).

P wave Primary wave. A compressional wave associated with an earthquake and which can move through both liquid and rock.

Pacific Ring of Fire The zone of seismic and volcanic activity that encircles the Pacific Ocean.

paleomagnetism The "fossil," or remanent, magnetic field of a rock.

Pangaea Name given by Alfred Wegener to the original "protocontinent." The breakup of Pangaea gave rise to the Atlantic Ocean and to the continents we see today.

Panthalassa Name given by Alfred Wegener to the ocean surrounding Pangaea.

parasitism A symbiotic relationship in which one species spends part or all of its life cycle on or within another, using the host species (or food within the host) as a source of nutrients. The most common form of symbiosis.

partially mixed estuary An estuary in which an influx of seawater occurs beneath a surface layer of fresh water flowing seaward. Mixing occurs along the junction.

passive margin Continental margin near an area of lithospheric plate divergence. Also called *Atlantic-type margin*.

passive sonar A device that detects the intensity and direction of underwater sounds.

PCBs *See* **polychlorinated biphenyls**.

pelagic Of the open ocean. Refers to the water above the deep-ocean basins, sediments of oceanic origin, or organisms of the open ocean.

pelagic zone The realm of open water. *See also* **benthic zone**.

period *See* **wave period**.

pH A measure of the acidity or alkalinity of a solution. Numerically, the negative logarithm of the concentration of hydrogen ions in an aqueous solution. A pH of 7 is neutral; lower numbers indicate acidity, and higher numbers indicate alkalinity.

Phaeophyta Brown multicellular algae, including kelps.

photic zone The thin film of lighted water at the top of the world ocean. The photic zone rarely extends deeper than 200 meters (660 feet). *Compare* **euphotic zone**.

photon The smallest unit of light energy.

photosynthesis The process by which autotrophs bind light energy into the chemical bonds of food with the aid of chlorophyll and other substances. The process uses carbon dioxide and water as raw materials and yields glucose and oxygen.

phycobilin A reddish accessory pigment found in red algae.

phylum One of the major groups of the animal kingdom whose members share a similar body plan, level of complexity, and evolution-ary history (see Appendix 5). (Plural, *phyla*) (The major groups of the plant kingdom are called divisions.)

physical factor An aspect of the physical environment that affects living organisms, such as light, salinity, or temperature.

physical resource Any resource that has resulted from the deposition, precipitation, or accumulation of a useful nonliving substance in the ocean or seabed. Also called a *nonliving resource*.

phytoplankton Plantlike, usually single-celled members of the plankton community.

Pinnipedia The carnivoran suborder that contains the seals, sea lions, and walruses.

piston corer A seabed-sampling device capable of punching through up to 25 meters (80 feet) of sediment and returning an intact plug of material.

planet A smaller, usually nonluminous body orbiting a star.

plankter Informal name for a member of the plankton community.

plankton Drifting or weakly swimming organisms suspended in water. Their horizontal position is to a large extent dependent on the mass flow of water rather than on their own swimming efforts.

plankton bloom A sudden increase in the number of phytoplankton cells in a volume of water.

plankton net Conical net of fine nylon or dacron fabric used to collect plankton.

Plantae The kingdom to which multicellular autotrophs belong.

plate One of about a dozen rigid segments of the Earth's lithosphere that move independently. The plate consists of continental or oceanic crust and the cool, rigid upper mantle directly below the crust.

plate tectonics The theory that the Earth's lithosphere is fractured into plates, which move relative to each other and are driven by convection currents in the mantle. Most volcanic and seismic activity occurs at plate margins.

Platyhelminthes The phylum of animals to which flatworms belong.

plunging wave Breaking wave in which the upper section topples forward and away from the bottom, forming an air-filled tube.

polar cell The atmospheric circulation cell centered over each pole.

polar front Boundary between the polar cell and the Ferrel cell in each hemisphere.

polar molecule A molecule with unbalanced charge. One end of the molecule has a slight negative charge, and the other end has a slight positive charge.

polar ocean areas Zones to the south of the Antarctic Convergence and to the north of the Arctic Convergence.

polar regions Earth's cold area poleward of either the Arctic or Antarctic Circle.

pollutant A substance that causes damage by interfering directly or indirectly with an organism's biochemical processes.

Polychaeta The largest and most diverse class of phylum Annelida. Nearly all polychaetes are marine.

polychlorinated biphenyls (PCBs) Chlorinated hydrocarbons once widely used to cool and insulate electrical devices and to strengthen wood or concrete. PCBs may be responsible for the changes and declining fertility of some marine mammals.

Polynesia A large group of Pacific islands lying east of Melanesia and Micronesia and extending from the Hawaiian Islands south to New Zealand and east to Easter Island.

polynya A gap in polar pack ice at which liquid water contacts the atmosphere.

polyp One of two body forms of Cnidaria. Polyps are cup-shaped and possess rings of tentacles. Coral animals are polyps.

Polyplacophora The class of the phylum Mollusca that includes the chitons.

poorly sorted sediment A sediment in which particles of many sizes are found.

population A group of individuals of the same species occupying the same area.

population density The number of individuals per unit area.

Porifera The phylum of animals to which sponges belong.

potable water Water suitable for drinking.

precipitate (1) A solid substance formed in an aqueous reaction. (2) The process by which a solute forms in and falls from a solution. The falling of water or ice from the atmosphere.

precipitation Liquid or solid water that falls from the air and reaches the surface as rain, hail, or snowfall.

pressure Force per unit area.

prey An organism consumed by a predator.

primary coast Coasts on which terrestrial influences dominate. *See also* **secondary coast**.

primary consumer   Initial consumer of primary producers. The consumers of autotrophs; the second level in food webs.

primary forces   The forces that induce and maintain water flow in ocean current systems: thermal expansion, wind friction, and density differences.

primary producer   An organism capable of using energy from light or energy-rich chemicals in the environment to produce energy-rich organic compounds. An autotroph.

primary productivity   The synthesis of organic materials from inorganic substances by photosynthesis or chemosynthesis. Expressed in grams of carbon bound into carbohydrate per unit area per unit time (g $C/m^2/yr$).

principle of constant proportions   The proportions of major conservative elements in seawater remain nearly constant, though total salinity may change with location. Also called *Forchhammer's principle*.

progressive wave   A wave of moving energy in which the wave form moves in one direction along the surface (or junction) of the transmission medium (or media).

Protista   The kingdom of single-celled nucleated organisms to which protozoa, diatoms, and dinoflagellates belong. Also called *Protoctista*.

proton   A positively charged particle at the center of an atom.

protostar   Tightly condensed knot of material that has not yet attained fusion temperature.

protozoa   Multiphyletic animal group within the kingdom Protista. Protozoa include amoebas, paramecia, foraminiferans, and radiolarians.

pteropod   Small planktonic mollusk with a calcareous shell, which contributes to biogenous sediments.

pycnocline   The middle zone of the ocean in which density increases rapidly with depth. Temperature falls and salinity rises in this zone.

radial symmetry   Body structure in which the body parts radiate from a central axis like spokes from a wheel. An example is a sea star. *Compare* **bilateral symmetry**.

radioactive decay   The disintegration of unstable forms of elements, which releases subatomic particles and heat.

radiolarian   One of a group of usually planktonic amoeba-like animals with a siliceous shell, which contributes to biogenous sediments.

radiometric dating   A technique using the constant rate at which naturally radioactive elements decay to determine the age of a material containing those elements.

random distribution   Distribution of organisms within a community whereby the position of one organism is in no way influenced by the positions of other organisms or by physical variations within that community. A very rare distribution pattern.

reef   A hazard to navigation. A shoal, a shallow area, or a mass of fish or other marine life.

refraction   Bending of light or sound waves as they move at an angle other than 90° between media of different optical or acoustical densities. *See also* **wave refraction**.

refractive index   The degree of refraction from one medium to another expressed as a ratio. The higher the ratio (refractive index), the greater the bending of waves between media.

refractometer   A compact optical device that determines the salinity of a water sample by comparing the refractive index of the sample to the refractive index of water of known salinity.

renewable resource   Any resource that is naturally replaced on a seasonal basis by the growth of living organisms or by other natural processes.

Reptilia   The class of reptiles, including turtles, crocodiles, iguanas, and snakes.

residence time   The average length of time a dissolved substance spends in the ocean.

respiration   Release of stored energy from chemical bonds in food; carbon dioxide and water are formed as by-products. (Respiration is a biochemical process and is not the same as the mechanical process of breathing.)

restoring force   The dominant force trying to return water to flatness after formation of a wave.

reverse estuary   An estuary along an arid coast in which salinity increases from the ocean to the estuary's upper reaches because of evaporation of seawater and a lack of freshwater input.

Rhodophyta   Red multicellular algae.

Richter scale   A logarithmic measure of earthquake magnitude. A great earthquake measures above 8 on the Richter scale.

rip current   A strong, narrow surface current that flows seaward through the surf zone and is caused by the escape of excess water that has piled up in a longshore trough.

rogue wave   A single wave crest much higher than usual, caused by constructive interference.

S wave   Secondary wave. A transverse wave associated with an earthquake and which cannot move through liquid.

salinity   A measure of the dissolved solids in seawater, usually expressed in grams per kilogram or parts per thousand by weight. Standard seawater has a salinity of 35 ‰ at 0°C (32°F).

salinometer   An electronic device that determines salinity by measuring the electrical conductivity of a seawater sample.

salt gland   Specialized tissue responsible for concentration and excretion of excess salt from blood and other body fluids.

salt wedge estuary   An estuary in which rapid river flow and small tidal range cause an inclined wedge of seawater to form at the mouth.

sand   Sediment particle between 0.062 and 2 millimeters in diameter.

sandbar   A submerged or exposed line of sand accumulated by wave action.

sand spit   An accumulation of sand and gravel deposited downcurrent from a headland. Sand spits often have a curl at the tips.

saturation   State of a solution in which no more of the solute will dissolve in the solvent. The rate at which molecules of the solute are being dissolved equals the rate at which they are being precipitated from the solution.

scattering   The dispersion (or "bounce") of sound or light waves when they strike particles suspended in water or air. The amount of scatter depends on the number, size, and composition of the particles.

schooling   Tendency of small fish of a single species, size, and age to mass in groups. The school moves as a unit, which confuses predators and reduces the effort spent searching for mates.

science   A systematic way of asking questions about the natural world and testing the answers to those questions.

scientific method   The orderly process by which theories explaining the operation of the natural world are verified or rejected.

scientific name   The genus and species name of an organism.

sea   Simultaneous wind waves of many wavelengths forming a chaotic ocean surface. Sea is common in an area of wind wave origin.

sea breeze   Onshore movement of air as inland air heats and rises.

sea cave   A cave near sea level in a sea cliff cut by processes of marine erosion.

sea cliff   Cliff marking the landward limit of marine erosion on an erosional coast.

seafloor spreading   The theory that new ocean crust forms at spreading centers, most of which are on the ocean floor, and pushes the continents aside. Power is thought to be provided by convection currents in the Earth's upper mantle.

sea grass   Any of several marine angiosperms. Examples are *Zostera* (eelgrass) and *Phyllospadix* (surfgrass). Sea grasses are not seaweeds.

sea ice   Ice formed by the freezing of seawater.

sea island   Island whose central core was connected to the mainland when sea level was lower. Rising ocean separates these high points from land, and sedimentary processes surround them with beaches. *Compare* **barrier island**.

sea level   The height of the ocean surface. *See also* **mean sea level**.

seamount   Circular or elliptical projection from the seafloor, more than 1 kilometer (0.6 mile) in height, with a relatively steep slope of 20° to 25°.

Seasat   First satellite dedicated to oceanic research, launched in 1978.

seaweed   Informal term for large marine multicellular algae.

secondary coast   Coasts dominated by marine processes. *See also* **primary coast**.

secondary consumer   Consumer of primary consumers.

secondary forces   The forces that influence the direction and nature of water flow in ocean current systems: the Coriolis effect, gravity, friction, and the shape of ocean basins.

second law of thermodynamics   Disorder (entropy) in a closed system must increase over time. If disorder decreases, it does so at the expense of energy. Since the universe as a whole may be considered a closed system, it follows that an increase in order in one part must result in a decrease in order in another.

sediment   Loose particles of inorganic or organic origin that accumulate on the seabed. Usually sand- or dust-sized, but occasionally as large as gravel, pebbles, or even boulders.

seiche   Pendulum-like rocking of water in an enclosed area; a form of standing wave that can be caused by meteorological or seismic forces, or that may result from normal resonances excited by tides.

seismic   Referring to earthquakes and the shock of earthquakes.

seismic sea wave   Tsunami caused by displacement of the Earth along a fault. (Earthquakes and seismic sea waves are caused by the same phenomenon.)

seismic wave   A low-frequency wave generated by the forces that cause earthquakes. Some kinds of seismic waves can pass through the earth. *See also* **P wave; S wave**.

seismograph   An instrument that detects and records Earth movement associated with earthquakes and other disturbances.

semidiurnal tide   A tidal cycle of two high tides and two low tides each lunar day, with the high tides of nearly equal height.

sensible heat   Heat whose gain or loss is detectable by a thermometer or other sensor.

sessile   Attached. Nonmotile. Unable to move about.

sewage sludge   Semisolid mixture of organic matter, microorganisms, toxic metals, and synthetic organic chemicals removed from wastewater at a sewage treatment plant.

shadow zone   (1) The wide band at the Earth's surface 105°–143° away from an earthquake in which seismic waves are nearly absent. P waves are absent because they are refracted by the Earth's liquid outer core; S waves are absent from this band and the zone immediately opposite the earthquake site because they are absorbed by the outer core. (2) In sonar, the volume of ocean from which sound waves diverge and in which a submarine may hide.

shallow-water wave   A wave in water shallower than 1/20 its wavelength.

shelf break   The abrupt increase in slope at the junction between continental shelf and continental slope.

shore   The place where ocean meets land. On nautical charts, the limit of high tides.

side-scan sonar   A high-resolution sound-imaging system used for geological investigations, archeological studies, and the location of sunken ships and airplanes.

siliceous ooze   Ooze composed mostly of the hard remains of silica-containing organisms.

silicoflagellate   A tiny single-celled phytoplankter with a siliceous skeleton.

silt   Sediment particle between 0.004 and 0.062 millimeter in diameter.

Sirenia   The order of mammals that includes manatees, dugongs, and the extinct sea cows.

sofar   *Sound fixing and ranging.* An experimental U.S. Navy technique for locating survivors on life rafts, based on the fact that sound from explosive charges dropped into the layer of minimum sound velocity can be heard for great distances. *See* **sofar channel**.

sofar channel   Layer of minimum sound velocity in which sound transmission is unusually efficient. Sounds leaving this depth tend to be refracted back into it. The sofar layer usually occurs at mid-latitude depths around 1,200 meters (4,000 feet).

solar nebula   The diffuse cloud of dust and gas from which the solar system originated.

solar system   The sun together with the planets and other bodies that revolve around it.

solar tide   Tide caused by the gravitational and inertial interaction of the sun and Earth.

solstice   One of two times of the year when the overhead position of the sun is farthest from the equator. The time of the solstice is midway between equinoxes.

solute   A substance dissolved in a solvent. *See* **solution**.

solution   A homogeneous substance made of two components, the solvent and the solute.

solvent   A substance able to dissolve other substances. *See* **solution**.

sonar   *Sound navigation and ranging.*

sound   A form of energy transmitted by rapid pressure changes in an elastic medium.

sounding   Measurement of the depth of a body of water.

Southern Oscillation   A reversal of airflow between normally low atmospheric pressure over the western Pacific and normally high pressure over the eastern Pacific. The cause of El Niño. *See* **El Niño**.

speciation   The formation of new species. Charles Darwin suggested that this is accomplished through isolation and natural selection.

species   Any group of actually or potentially interbreeding organisms reproductively isolated from all other groups and capable of producing fertile offspring. (*Note*: The word *species* is both singular and plural.)

species diversity   Number of different species in a given area.

species-specific relationship   An exclusive relationship between two species. Parasites are usually species-specific; that is, they can usually parasitize only one species of host.

spilling wave   A breaking wave whose crest slides down the face of the wave.

spreading center   The junction between diverging plates at which new ocean floor is being made. Also called *spreading zone*.

spring tide   The time of greatest variation between high and low tides occurring when Earth, moon, and sun form a straight line. Spring tides alternate with neap tides throughout the year, occurring at two-week intervals.

standing wave   A wave in which water oscillates without causing progressive wave forward movement. There is no net transmission of energy in a standing wave.

star   A massive sphere of incandescent gases powered by the conversion of hydrogen to helium and other heavier elements.

state   An expression of the internal form of matter. Water exists in three states: solid, liquid, and gas. A solid has a fixed volume and fixed shape; a liquid has a fixed volume but no fixed shape; and a gas has neither fixed volume nor fixed shape.

stenohaline   Describing an organism unable to tolerate a wide range in salinity.

stenothermal   Describing an organism unable to tolerate wide variance in temperature.

stipe   Multicellular algal equivalent of a vascular plant's stem.

storm   Local or regional atmospheric disturbance characterized by strong winds often accompanied by precipitation.

storm surge   An unusual rise in sea level as a result of the low atmospheric pressure and strong winds associated with a tropical cyclone. Onrushing seawater precedes landfall of the tropical cyclone and causes most of the damage to life and property.

stratigraphy   The branch of geology that deals with the definition and description of natural divisions of rocks. Specifically, the analysis of relationships of rock strata.

subduction   The downward movement into the asthenosphere of a lithospheric plate.

subduction zone   An area at which a lithospheric plate is descending into the asthenosphere. The zone is characterized by linear folds (trenches) in the ocean floor and strong deep-focus earthquakes. Also called a *Wadati-Benioff zone*.

sublittoral zone   The ocean floor near shore. The inner sublittoral extends from the littoral (intertidal) zone to the depth at which wind waves have no influence; the outer sublittoral extends to the edge of the continental shelf.

submarine canyon   A deep, V-shaped valley running roughly perpendicular to the shoreline and cutting across the edge of the continental shelf and slope.

subsidence   Sinking, often of tectonic origin.

Subtropical Convergence   Convergence zone marking the boundary between Central Water and either Subarctic or Subantarctic Surface Water. The northern Subtropical Convergence lies at about 45°N in the Pacific and 60°N in the Atlantic; the southern Subtropical Convergence lies at 40°–50°S.

succession   The changes in species composition that lead to a climax community.

sun tide   *See* **solar tide**.

supernova   The explosive collapse of a massive star.

supralittoral zone   The splash zone above the highest high tide; not technically part of the ocean bottom.

surf   The confused mass of agitated water rushing shoreward during and after a wind wave breaks.

surface current   Horizontal flow of water at the ocean's surface.

surface zone   The upper layer of ocean in which temperature and salinity are relatively constant with depth. Depending on local conditions, the surface zone may reach to 1,000 meters (3,300 feet) or be absent entirely. Also called the *mixed layer*.

surface-to-volume ratio   A physical constraint on the size of cells. As a cell's linear dimensions grow, its surface area does not increase at the same rate as its volume. As surface-to-volume ratio falls, each square unit of outer membrane must serve an increasing interior volume.

surf beat   The pattern of constructive and destructive interference that causes successive breaking waves to grow, shrink, and grow again over a few minutes' time.

surf zone   The region between the breaking waves and the shore.

surging wave   A wave that surges ashore without breaking.

suspension feeder   An animal that feeds by straining or otherwise collecting plankton and tiny food particles from the surrounding water.

sverdrup (sv)   A unit of volume transport named in honor of oceanographer Harald U. Sverdrup: 1 million cubic meters of water flowing past a fixed point each second.

swash   Water from waves washing onto a beach.

swell   Mature wind waves of one wavelength that form orderly undulations of the ocean surface.

swim bladder   A gas-filled organ that assists in maintaining neutral buoyancy in some bony fishes.

symbiosis   The co-occurrence of two species in which the life of one is closely interwoven with the life of the other; mutualism, commensalism, or parasitism.

synoptic sampling   Simultaneous sampling at many locations.

taxonomy   In biology, the laws and principles covering the classification of organisms.

tektite   A small, rounded glassy component of cosmogenous sediments, usually less than 1.5 millimeters ($\frac{1}{16}$ inch) in length. Thought to have formed from the impact of an asteroid or meteor on the crust of the Earth or moon.

Teleostei   The osteichthyan order that contains the cod, tuna, halibut, perch, and other species of bony fishes.

temperate zone   The mid-latitude area between the Tropic of Cancer and the Arctic Circle and between the Tropic of Capricorn and the Antarctic Circle.

temperature   The response of a solid, liquid, or gas to the input or removal of heat energy. A measure of the atomic and molecular vibration in a substance, indicated in degrees.

temperature-salinity (T-S) diagram   A graph showing the relationship of temperature and salinity with depth.

terrane   An isolated segment of seafloor, island arc, plateau, continental crust, or sediment transported by seafloor spreading to a position adjacent to a larger continental mass. Usually different in composition from the larger mass.

terrigenous sediment   Sediment derived from the land and transported to the ocean by wind and flowing water.

territorial waters   Waters extending 12 miles from shore and in which a nation has the right to jurisdiction.

thallus   The body of an alga or other simple plant.

theory   A general explanation of a characteristic of nature consistently supported by observation or experiment.

thermal equator   *See* **meteorological equator**.

thermal equilibrium   The condition in which the total heat coming into a system (such as a planet) is balanced by the total heat leaving the system.

thermal inertia Tendency of a substance to resist change in temperature with the gain or loss of heat energy.

thermocline The zone of the ocean in which temperature decreases rapidly with depth. *See also* **pycnocline**.

thermohaline circulation Water circulation produced by differences in temperature and/or salinity (and therefore density).

thermostatic property A property of water that acts to moderate changes in temperature.

tidal bore A high, often breaking wave generated by a tide crest that advances rapidly up an estuary or river.

tidal current Mass flow of water induced by the raising or lowering of sea level owing to passage of tidal crests or troughs. *See also* **ebb current**; **flood current**.

tidal datum The reference level (0.0) from which tidal height is measured.

tidal range The difference in height between consecutive high and low tides.

tidal wave The crest of the wave causing tides. Another name for a tidal bore. *Not* a tsunami or seismic sea wave.

tide Periodic short-term change in the height of the ocean surface at a particular place, generated by long-wavelength progressive waves which are caused by the interaction of gravitational force and inertia. Movement of the Earth beneath tide crests results in the rhythmic rising and falling of sea level.

tombolo Above-water bridge of sand connecting an offshore feature to the mainland.

top consumer An organism at the apex of a trophic pyramid, usually a carnivore.

tornado Localized, narrow, violent funnel of fast-spinning wind, usually generated when two air masses collide. Not to be confused with a cyclone. (The tornado's oceanic equivalent is a waterspout.)

trace element A minor constituent of seawater present in amounts less than 1 part per million.

trade winds Surface winds within the Hadley cells, centered at about 15° latitude, which approach from the northeast in the Northern Hemisphere and from the southeast in the Southern Hemisphere.

transform fault A plane along which rock masses slide horizontally past one another.

transform plate boundary Places where crustal plates shear laterally past one another. Crust is neither produced nor destroyed at this type of junction.

transitional wave *See* **intermediate-depth water wave**.

transverse current East-to-west or west-to-east current linking the eastern and western boundary currents. An example is the North Equatorial Current.

trench An arc-shaped depression in the deep-ocean floor with very steep sides and a flat sediment-filled bottom coinciding with a subduction zone. Most trenches occur in the Pacific.

trophic level A feeding step within a trophic pyramid.

trophic pyramid A model of feeding relationships between organisms. Primary producers form the base of the pyramid; consumers eating one another form the higher levels, with the top consumer at the apex.

tropical cyclone A weather system of low atmospheric pressure around which winds blow counterclockwise in the Northern Hemisphere and clockwise in the Southern Hemisphere. Originates in the tropics within a single air mass, but may move into temperate waters if water temperature is high enough to sustain it. Small tropical cyclones are called tropical depressions, larger ones tropical storms, and great ones hurricanes, typhoons, or willy-willies, depending on location.

tropical ocean area Warm central zone of the ocean equatorward of the subtropical convergences where surface temperature is nearly always above 20°C (69°F).

Tropic of Cancer The imaginary line around the Earth, parallel to the equator at 23°27′N, marking the point where the sun shines directly overhead at the June solstice.

Tropic of Capricorn The imaginary line around the Earth, parallel to the equator at 23°27′S, marking the point where the sun shines directly overhead at the December solstice.

tropics The area between the Tropic of Cancer and the Tropic of Capricorn.

trough *See* **wave trough**.

tsunami Long-wavelength shallow-water wave caused by rapid displacement of water. *See also* **seismic sea wave**.

tunicate A type of suspension-feeding invertebrate chordate.

turbidite A terrigenous sediment deposited by a turbidity current; typically, coarse-grained layers of nearshore origin interleaved with finer sediments.

turbidity current An underwater "avalanche" of abrasive sediments thought responsible for the deep sculpturing of submarine canyons and a means of transport for sediments accumulating on abyssal plains.

turbulence Chaotic fluid flow.

ultraplankton Extremely small plankton, smaller than nanoplankton.

undercurrent A current flowing beneath a surface current, usually in the opposite direction.

unicellular Consisting of a single cell.

unicellular algae Algae with bodies consisting of a single cell. Examples are diatoms and dinoflagellates.

uniform distribution Distribution of organisms within a community characterized by equal space between individuals (the arrangement of trees in an orchard). The rarest natural distribution pattern.

uniformitarianism The theory that all of Earth's geological features and history can be explained by processes occurring today and that these processes must have been at work for a very long time.

United States exclusive economic zone The region extending seaward from the coast of the United States for 200 nautical miles, within which the United States claims sovereign rights and jurisdiction over all marine resources.

United States Exploring Expedition The first U.S. oceanographic research voyage, launched in 1838.

upwelling Circulation pattern in which deep, cold, usually nutrient-laden water moves toward the surface. Upwelling can be caused by winds blowing parallel to shore or offshore.

$V = \sqrt{gd}$ Relation between velocity (V), the acceleration due to gravity (g), and water depth (d) for shallow-water waves.

$V = L/T$ Relation between velocity (V), wavelength (L), and period (T) for deep-water waves; velocity increases as wavelength increases. Typically measured in meters per second.

valve In diatoms, each half of the protective silica-rich outer portion of the cell. The complete outer covering is called the frustule.

vascular plant Plant having vessels for transport of fluid through leaves, stems, and roots. Examples are sea grasses, mangroves, and maple trees.

velocity Speed in a specified direction.

Vertebrata The subphylum of the phylum Chordata that includes animals with segmented backbones.

vertebrate A chordate with a segmented backbone.

Vikings Seafaring Scandinavian raiders who ravaged the coasts of Europe around A.D. 780–1070.

viscosity Resistance to fluid flow. A measure of the internal friction in fluids.

voyaging Traveling (usually by sea) with a specific purpose.

Wadati-Benioff zone *See* **subduction zone.**

water mass A body of water identifiable by its salinity and temperature (and therefore its density) by its gas content or another indicator.

water vapor The gaseous, invisible form of water.

water-vascular system System of water-filled tubes and canals found in some representatives of the phylum Echinodermata and used for movement, defense, and feeding.

wave Disturbance caused by the movement of energy through a medium.

wave crest Highest part of a progressive wave above average water level.

wave-cut platform The smooth, level terrace sometimes found on erosional coasts that marks the submerged limit of rapid marine erosion.

wave diffraction Bending of waves around obstacles.

wave frequency The number of waves passing a fixed point per second.

wave height Vertical distance between a wave crest and the adjacent wave troughs.

wavelength The horizontal distance between two successive wave crests (or troughs) in a progressive wave.

wave period Time for successive wave crests to pass a fixed point.

wave reflection The reflection of progressive waves by a vertical barrier. Reflection occurs with little loss of energy.

wave refraction Slowing and bending of progressive waves in shallow water.

wave shock Physical movement, often sudden, violent, and of great force, caused by the crash of a wave against an organism.

wave steepness Height-to-wavelength ratio of a wave. The theoretical maximum steepness of deep-water waves is 1:7.

wave train A group of waves of similar wavelength and period moving in the same direction across the ocean surface. The group velocity of a wave train is half the velocity of the individual waves.

wave trough The valley between wave crests below the average water level in a progressive wave.

weather The state of the atmosphere at a specific place and time.

Wegener, Alfred (1880–1930) German scientist who proposed the theory of continental drift in 1912.

well-mixed estuary An estuary in which slow river flow and tidal turbulence mix fresh and salt water in a regular pattern through most of its length.

well-sorted sediment A sediment in which particles are of uniform size.

westerlies Surface winds within the Ferrel cells, centered around 45° latitude, which approach from the southwest in the Northern Hemisphere and from the northwest in the Southern Hemisphere.

western boundary current Strong, warm, concentrated, fast-moving current at the western boundary of an ocean (off the east coast of a continent). Examples include the Gulf Stream and the Japan (Kuroshio) Current.

westward intensification The increase in speed of geostrophic currents as they pass along the western boundary of an ocean basin.

West Wind Drift The Antarctic Circumpolar Current, driven by powerful westerly winds north of Antarctica. The largest of all ocean currents, it continues permanently eastward without changing direction.

Wilson, John Tuzo (b. 1908) Canadian geophysicist who proposed the theory of plate tectonics in 1965.

wind The mass movement of air.

wind duration The length of time the wind blows over the ocean surface, a factor in wind wave development.

wind-induced vertical circulation Vertical movement in surface water (upwelling or downwelling) caused by wind.

wind strength Average speed of the wind, a factor in wind wave development.

wind wave Gravity wave formed by transfer of wind energy into water. Wavelengths from 60 to 150 meters (200 to 500 feet) are most common in the open ocean.

world ocean The great body of saline water that covers 70.78% of the Earth's surface.

xanthophyll A yellow or brown accessory pigment that gives some marine autotrophs a yellow or tan appearance.

zone Division or province of the ocean with homogeneous characteristics.

zooplankton Animal members of the plankton community.

zooxanthellae Unicellular dinoflagellates that are symbiotic with coral and that produce the relatively high pH and some of the enzymes essential for rapid calcium carbonate deposition in coral reefs.

INDEX